Lecture Notes in Electrical Engineering

Volume 1023

The book series *Lecture Notes in Electrical Engineering* (LNEE) publishes the latest developments in Electrical Engineering—quickly, informally and in high quality. While original research reported in proceedings and monographs has traditionally formed the core of LNEE, we also encourage authors to submit books devoted to supporting student education and professional training in the various fields and applications areas of electrical engineering. The series cover classical and emerging topics concerning:

- Communication Engineering, Information Theory and Networks
- Electronics Engineering and Microelectronics
- Signal, Image and Speech Processing
- Wireless and Mobile Communication
- Circuits and Systems
- Energy Systems, Power Electronics and Electrical Machines
- Electro-optical Engineering
- Instrumentation Engineering
- Avionics Engineering
- Control Systems
- Internet-of-Things and Cybersecurity
- Biomedical Devices, MEMS and NEMS

For general information about this book series, comments or suggestions, please contact leontina.dicecco@springer.com.

To submit a proposal or request further information, please contact the Publishing Editor in your country:

China

Jasmine Dou, Editor (jasmine.dou@springer.com)

India, Japan, Rest of Asia

Swati Meherishi, Editorial Director (Swati.Meherishi@springer.com)

Southeast Asia, Australia, New Zealand

Ramesh Nath Premnath, Editor (ramesh.premnath@springernature.com)

USA, Canada

Michael Luby, Senior Editor (michael.luby@springer.com)

All other Countries

Leontina Di Cecco, Senior Editor (leontina.dicecco@springer.com)

**** This series is indexed by EI Compendex and Scopus databases. ****

Asha Rani · Bhavnesh Kumar · Vivek Shrivastava ·
Ramesh C. Bansal
Editors

Signals, Machines and Automation

Select Proceedings of SIGMA 2022

Springer

Editors
Asha Rani
Department of Instrumentation and Control Engineering
Netaji Subhas University of Technology
New Delhi, Delhi, India

Bhavnesh Kumar
Department of Instrumentation and Control Engineering
Netaji Subhas University of Technology
New Delhi, Delhi, India

Vivek Shrivastava
Department of Electrical Engineering
National Institute of Technology Delhi
New Delhi, Delhi, India

Ramesh C. Bansal
Department of Electrical Engineering
University of Sharjah
Sharjah, United Arab Emirates

Department of Electrical, Electronic and Computer Engineering
University of Pretoria
Pretoria, South Africa

ISSN 1876-1100 ISSN 1876-1119 (electronic)
Lecture Notes in Electrical Engineering
ISBN 978-981-99-0968-1 ISBN 978-981-99-0969-8 (eBook)
https://doi.org/10.1007/978-981-99-0969-8

This Springer imprint is published by the registered company Springer Nature Singapore Pte Ltd.
The registered company address is: 152 Beach Road, #21-01/04 Gateway East, Singapore 189721, Singapore

Preface

Welcome to the Proceedings of the 2022 International Conference on Signals, Machines and Automation (SIGMA 2022) which was held in virtual mode on August 05–06, 2022. The 2nd International Conference on Signals, Machines and Automation aims to bring together leading academicians and researchers to exchange and share their experiences and research outcomes within the spectrum of conference theme. The conference provided an excellent opportunity for the presentation of interesting new research results and discussion about them, leading to knowledge transfer and the generation of new ideas.

SIGMA 2022 received more than 200 submissions from different countries, and each paper was reviewed by at least three reviewers in a standard peer-review process. Based on the recommendation by impartial and independent reviewers, finally with acceptance rate of 36.6%, 70 papers were accepted for possible inclusion in Springer LNEE.

To ensure the success of SIGMA 2022, a large number of individuals have worked together and put in a lot of effort. First, we would like to express our gratitude to the authors for bringing their research papers to the conference, as well as for the presentations they gave and the conversations they led throughout the event. Our gratitude goes out to the members of the program committee as well as the reviewers, who did the bulk of the onerous job by thoughtfully examining the papers that were submitted. We would like to extend our sincere gratitude to Prof. J. P. Saini, Vice Chancellor of the Netaji Subhas University of Technology in Delhi. We would like to extend our deepest gratitude to the chairmen of the organizing committee for assisting us in developing such a rich technical program.

New Delhi, India	Asha Rani
New Delhi, India	Bhavnesh Kumar
New Delhi, India	Vivek Shrivastava
Sharjah, United Arab Emirates	Ramesh C. Bansal

Contents

Communication and Network Technologies

High-Computing Technologies

About the Editors

Prof. Asha Rani received her B.Tech. in Electrical Engineering from Now National Institute of Technology (NIT) Hamirpur, in 1998. She received her M.E. degree in Electrical Engineering from the Indian Institute of Technology (IIT) Roorkee in 2000 and her Ph.D. in 2013 from the University of Delhi, India. She joined as a lecturer in the Instrumentation and Control Engineering Division at the Netaji Subhas University of Technology, New Delhi in 2001 and currently is a Professor. She has published several papers in journals and conferences of national and international repute. She has guided over 28 M.Tech. and six Ph.D. theses. Her areas of research are renewable energy, intelligent control, adaptive control, soft computing-based adaptive control, biomedical signal processing, and robotic manipulator. She is a senior member of IEEE and IETE.

Dr. Bhavnesh Kumar has over 10 years of diversified experience in teaching and learning, accreditation, research, industrial, and academics. Currently, he is an assistant professor at the Netaji Subhas University of Technology, India. He has published around 50 research papers in journals and conferences of national and international repute. He is a member of various professional bodies such as IEEE, IACSIT, ISCA, and ISTE. Under his guidance, 10 M.Tech. students have completed their dissertations and three research scholars have completed their Ph.D. His research area includes the application of artificial intelligence to the field of renewable energy systems and electric drives.

Prof. Vivek Shrivastava has over 15 years of diversified experience in teaching and learning, accreditation, research, industrial, and academic leadership in India, China, and the USA. Currently, he is the Dean of Research and Consultancy at the National Institute of Technology (NIT) Delhi. He has published over 80 journal articles, presented papers at conferences, and published several chapters in books. He has supervised four Ph.D. and 14 postgraduate students and is currently supervising several Ph.D. students.

Prof. Ramesh C. Bansal has more than 25 years of diversified experience in research, scholarship of teaching and learning, accreditation, industrial, and academic leadership in several countries. Currently, he is a Professor in the Department of Electrical Engineering at the University of Sharjah and an extraordinary professor at the University of Pretoria. Previously he was a Professor and Group Head (Power) in the ECE Department at the University of Pretoria (UP), South Africa. Before his appointment at UP, he was employed by the University of Queensland, Australia; University of the South Pacific, Fiji; BITS Pilani, India; and Civil Construction Wing, All India Radio. Professor Bansal has published over 400 journal articles and presented papers at conferences, books, and chapters in books. He has Google citations of over 16000 and an h-index of 60. He has supervised 25 Ph.D., five Post Docs and is currently supervising five Ph.D. students. His diversified research interests are in the areas of Renewable Energy (wind, PV, microgrid), Power Systems, and Smart Grid. Professor Bansal is an Editor/Associate Editor of several highly regarded journals including *IEEE Systems Journal*, *IET Renewable Power Generation*, and *Technology and Economics of Smart Grids and Sustainable Energy*. He is a Fellow and Chartered Engineer of IET-UK, a Fellow Institution of Engineers (India), a Fellow SAIEE (South Africa), and a Senior Member of IEEE-USA.

Power System Technologies

Investigating the Application of Ethereum Smart Contracts in Energy Exchanges

Nishkar R. Naraindath, Ramesh C. Bansal, and Raj M. Naidoo

Abstract Blockchain technology serves as a sustainable solution to enable decentralised applications. Insufficient research efforts have been directed in the past towards advancing blockchain technology integration in energy systems. This paper demonstrates that the energy exchange transparency and facilitation can be improved by utilising a smart contract deployed on an Ethereum Test network. The capabilities and limitations of the system have been analysed, with suggestions for improvement and potential research areas provided.

Keywords Blockchain · Decentralisation · Demand management · Peer-to-peer · Power market · Token exchange

1 Introduction

Blockchain technology can potentially revolutionise the energy industry by enabling autonomous and decentralised energy ecosystems [1, 2]. However, more significant research efforts and increased awareness are required to promote larger-scale integration of the technology [3, 4]. Smart contracts exhibit desirable properties for energy markets, such as automatic execution, customisation, and tamper-resistance [5, 6]. The most common smart contract applications have been facilitating bidding amounts, determining tariff rates, and bill issuing to market participants [7]. However,

N. R. Naraindath (✉) · R. C. Bansal · R. M. Naidoo
Department of Electrical, Electronic and Computer Engineering, University of Pretoria, Pretoria, South Africa
e-mail: nrnaraindath@gmail.com

R. C. Bansal
e-mail: rcbansal@ieee.org

R. M. Naidoo
e-mail: raj.naidoo@up.ac.za

R. C. Bansal
Department of Electrical Engineering, University of Sharjah, Sharjah, United Arab Emirates

A. Rani et al. (eds.), *Signals, Machines and Automation*, Lecture Notes in Electrical Engineering 1023, https://doi.org/10.1007/978-981-99-0969-8_1

these are modest applications when considering more advanced bidding strategies [8, 9] and energy hub coordination strategies [10, 11].

In this paper, an investigative study on the application of Ethereum smart contracts in energy exchanges is performed to raise awareness of the practical capabilities of blockchain technology. The smart contract is proposed to address requirements of trust, computational logic, and automatic execution of localised energy exchanges by serving as a regulated and neutral exchange platform.

2 Digital Energy Ecosystem

Energy markets are traditionally facilitated through intermediary energy exchange platforms, which are centralised [12]. Smart contracts are attractive replacements for energy exchanges that can serve as a transparent and impartial intermediary between participants [13, 14]. These contracts are defined using computational logic to coordinate energy exchange transactions autonomously and reliably [15]. Before smart contract technology can be integrated, energy markets must first be represented in the digital domain. One approach taken is by representing stored energy units and the local currency as cryptoassets/tokens [16]. A protocol token, referred to as ePower, is being used to represent 100 Wh of energy. On the other hand, the local currency is represented by a stable coin represented by eZAR. The value of eZAR is dictated by the underlying local currency of South Africa (ZAR). For this study, ePower has been selected to have an exchange rate represented by:

$$1 \text{ eZAR} = 10 \text{ ePower} \tag{1}$$

An energy market among three parties (producer, consumer, and prosumer) has been selected for exploration as each party represents a fundamental user in an advanced energy ecosystem. Each party has initial equity balances as described in Table 1. The following transactions are considered:

1. The energy prosumer purchases ePower tokens to the value of 500 eZAR from the energy producer.
2. The energy prosumer sells 1500 ePower tokens to the energy producer.
3. The energy consumer purchases ePower tokens to the value of 200 eZAR from the energy producer.

Table 1 Initialised equity balances in the energy ecosystem

Producer		Prosumer		Consumer	
eZAR	ePower	eZAR	ePower	eZAR	ePower
0	10,000	1000	0	1000	0

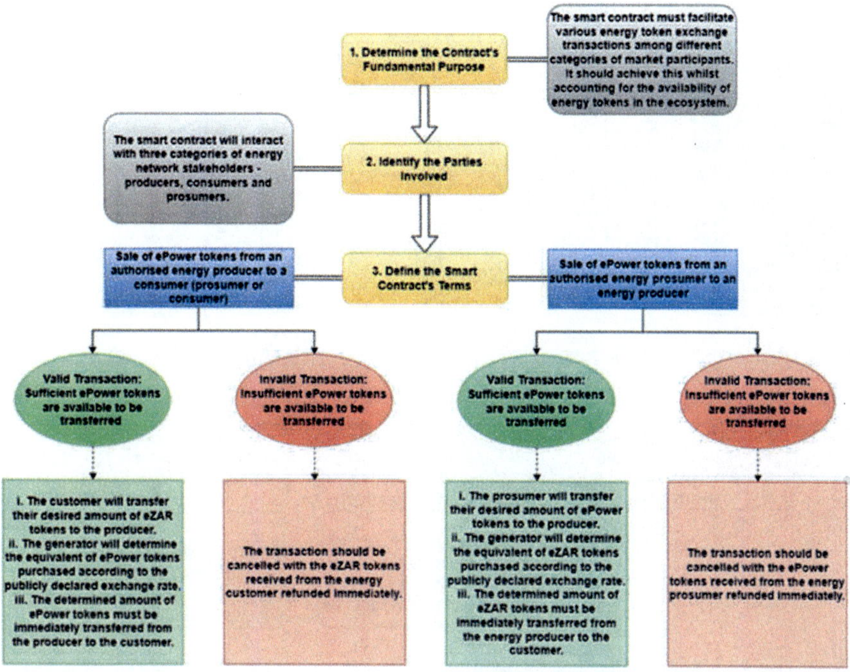

Fig. 1 Energy exchange elements modelled through a proposed framework

3 Smart Contract Creation

3.1 Agreement Modelling

Various energy token exchange scenarios can be modelled as a traditional contract using the framework presented in Fig. 1.

3.2 Smart Contract Definition

The standard contract terms can be integrated into a Solidity program [17] compiled on an open-source online integrated development environment (IDE) such as Remix. Diverse program elements [18] are presented in a single collection of code and data known as the ePower market. An overview of the smart contract's components is presented in Fig. 2. Further details of the smart contract definition process are provided in Fig. 3.

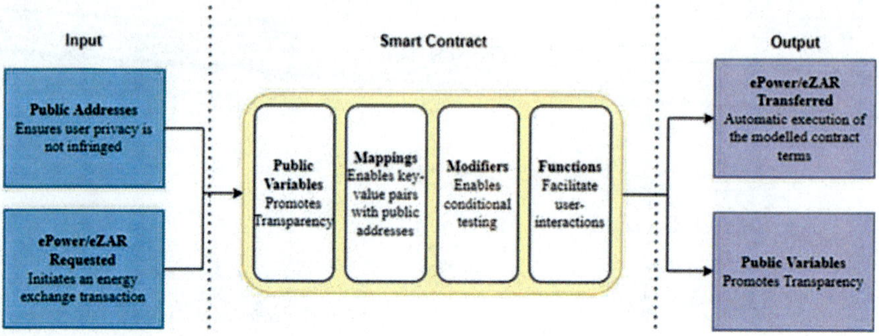

Fig. 2 Representation of the smart contract as a multiple-input multiple-output system

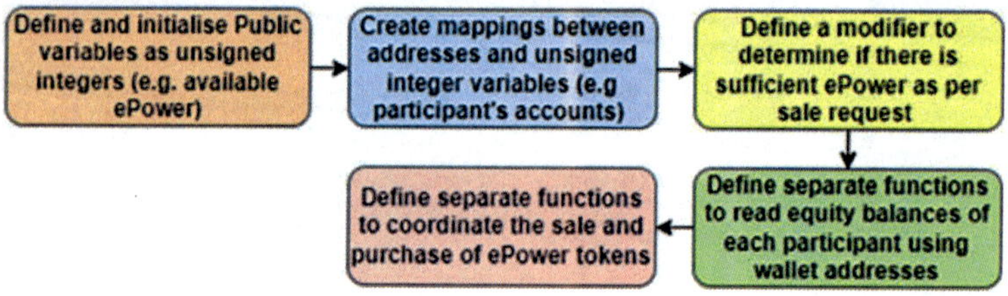

Fig. 3 Overview of the smart contract definition process

4 Smart Contract Deployment

Ganache truffle suite [19] has been employed for blockchain development on an Ethereum Test network. The system provides a user interface capable of interacting with the entire life cycle of a smart contract. MyEtherWallet, an open-source user interface, is then employed to interact directly with the virtual Ethereum network [20]. The interface requires a communication link with the blockchain, which can be obtained by creating a custom node on MyEtherWallet containing the RPC Server and Network ID of the workstation configured in Ganache. Once the node has been configured, the smart contract can be transformed into the byte code equivalent with the aid of Remix IDE. The registered energy producer has been selected to sign the smart contract with the aid of their private key. As a result, this participant incurs an expense, known as a gas fee, to deploy the contract on the Ethereum blockchain.

5 Smart Contract Interaction

The energy ecosystem described in Sect. 2 has been simulated by interacting with the deployed smart contract through read or write commands via MyEtherWallet with an overview of all completed energy exchange transactions depicted in Fig. 4. A

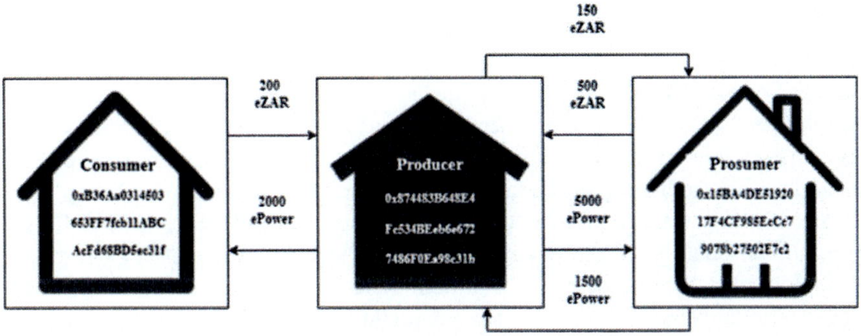

Fig. 4 Energy ecosystem considered in this study

summary of reflected account balances after each transaction is provided in Table 2. The complete blockchain is presented in Fig. 5, highlighting the gas fees associated with each stage of the smart contract's life cycle.

Table 2 Reflected account balances after performing various energy exchange transactions

Transaction number	Producer		Prosumer		Consumer	
	eZAR	ePower	eZAR	ePower	eZAR	ePower
0	0	10,000	1000	0	1000	0
1	500	5000	500	5000	1000	0
2	350	6500	650	3500	1000	0
3	550	4500	650	3500	800	2000

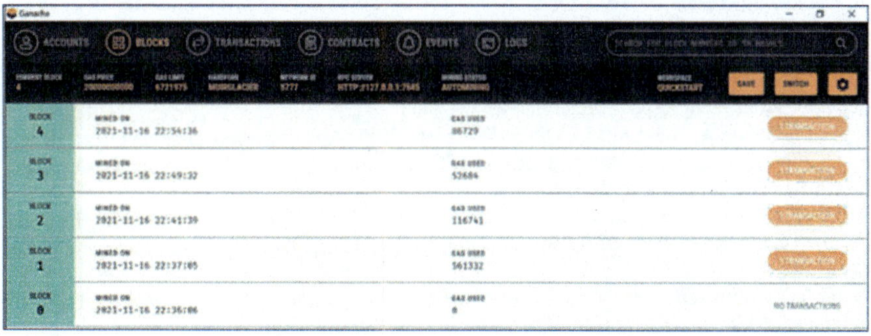

Fig. 5 Blockchain containing the life cycle of the deployed smart contract on an Ethereum Test network

6 Conclusion

A smart contract has been modelled with a generalised framework and deployed on an Ethereum Test network. The energy market has been represented in the digital domain to enable smart contract integration. The smart contract demonstrates precise capabilities in facilitating bidirectional energy exchange transactions among three classes of energy ecosystem users (producers, prosumers, and consumers). Different energy users could initiate exchanges such as the sale and resale of ePower tokens through written commands to the smart contract. The status of eZAR and ePower balances are universally accessible along with the contract, demonstrating the transparency of the energy exchanges in the system. Ethereum caters for modest smart contract applications but is infeasible for the transaction throughput required in real-time energy markets. Furthermore, the network fees were unsustainable, and consequently, developing a specialised blockchain-based system is strongly recommended. Critical limitations to the study are the accuracy of the Ethereum Test network and the centralised configuration of the simulated energy ecosystem. In addition, it suggested that more significant research efforts are focused on the integration of smart contracts in smart power networks to cater for more intricate tasks.

References

1. Rajasekaran AS, Azees M, Al-Turjman F (2022) A comprehensive survey on blockchain technology. Sustain Energy Technol Assess 52:102039
2. Mohanta BK, Panda SS, Jena D (2018) An overview of smart contract and use cases in blockchain technology. In: 9th International conference on computing, communication and networking technologies, pp 1–4
3. Naraindath NR, Bansal RC, Naidoo RM (2022) The uprising of blockchain technology in the energy market industry. In: International conference on recent developments in electrical and electronics engineering (ICRDEEE), Faridabad, India, 15–16 Apr 2022
4. Juszczyk O, Shahzad K (2022) Blockchain technology for renewable energy: principles, applications and prospects. Energies 15(13):4603
5. Wang S, Yuan Y, Wang X, Li J, Qin R, Wang FY (2018) An overview of smart contract: Architecture, applications, and future trends. In: IEEE Intelligent Vehicles Symposium, Proceedings, vol June, no 4, pp 108–113
6. Davelis A, Butt UJ, Pendlebury G, Hussein KE (2022) Emerging technologies: blockchain and smart contracts. In: Blockchain and other emerging technologies for digital business strategies, pp 143–169
7. Kirli D, Couraud B, Robu V, Salgado-Bravo M, Norbu S, Andoni M, Antonopoulos I, Negrete-Pincetic M, Flynn D, Kiprakis A (2022) Smart contracts in energy systems: a systematic review of fundamental approaches and implementations. Renew Sustain Energy Rev 158:112013
8. A. Saxena, R. Kumar, R. C. Bansal, and M. Mahmud, "Bidding strategies of a power producer in power market: measurement indices and evaluation," in Uncertainties in Modern Power Systems. Elsevier, 2021, pp. 635–652.
9. Lupangu C, Justo JJ, Bansal RC (2020) Model predictive for reactive power scheduling control strategy for PV–battery hybrid system in competitive energy market. IEEE Syst J 14(3):4071–4078

10. Heidari A, Bansal RC, Hossain J, Zhu J (2022) Strategic risk aversion of smart energy hubs in the joined energy markets applying a stochastic game approach. J Clean Prod 349(131386):1–15
11. Heidari A, Bansal RC (2021) Probabilistic correlation of renewable energies within energy hubs for cooperative games in integrated energy markets. Electr Power Syst Res 199(May):107397
12. Klement P, Brandt T, Schmeling L, de Bronstein AA, Wehkamp S, Penaherrera Vaca FA, Lanezki M, Schönfeldt P, Hill A, Katic N (2022) Local energy markets in action: Smart integration of national markets, distributed energy resources and incentivisation to promote citizen participation. Energies 15(8):2749
13. Damisa U, Nwulu NI, Siano P (2022) Towards blockchain-based energy trading: A smart contract implementation of energy double auction and spinning reserve trading. Energies 15(11):4084
14. Wu Y, Cimen H, Vasquez JC, Guerrero JM (2022) P2p energy trading: blockchain-enabled p2p energy society with multi-scale flexibility services. Energy Rep 8:3614–3628
15. Górski T (2022) Reconfigurable smart contracts for renewable energy exchange with re-use of verification rules. Appl Sci 12(11):5339
16. Hougan M, Lawant D (2021) Cryptoassets: the guide to bitcoin, blockchain, and cryptocurrency for investment professionals. CFA Institute Research Foundation
17. Dannen C (2017) Introducing ethereum and solidity. Apress, Berkeley, CA
18. Sanchez-Gomez N, Torres-Valderrama J, Risoto MM, Garrido A (2021) Blockchain smart contract meta-modeling. J Web Eng 20(7):2059–2080
19. Anilkumar V, Joji JA, Afzal A, Sheik R (2019) Blockchain simulation and development platforms: survey, issues and challenges. In: IEEE international conference on intelligent computing and control systems (ICCS), pp 935–939
20. Devi Parameswari C, Ilayaraja M (2022) Decentralized public distribution system on the ethereum blockchain interact with myetherwallet. In: Blockchain and deep learning. Springer, pp 107–124

Feasibility Analysis of V2G Operation in India and Cost Minimization for EV Car Parks Using Firefly Algorithm

Prince Bharti and Omveer Singh

Abstract Electric vehicles (EVs) in smart car parks can be employed as energy storage systems and a reserve against unanticipated outrage if charging and discharging are coordinated. A modeling and control framework for EVs in a smart vehicle park has been developed in this work, which encompasses important elements such as charging and discharging expenses, battery degradation costs, driving likelihood, feed in tariff (FIT), and vehicle-to-grid (V2G) rebates are all factors to consider. The charging and discharging activities of each EV are recorded and arranged using an optimization path in order to reduce the cost of parking lot electricity consumption. Comprehensive simulation studies have demonstrated the potential benefits of V2G for automobiles. Multiple EV park systems are susceptible to car and battery characteristics, as well as FIT and regulatory support. Firefly algorithm is implemented for optimization of EV charging costs. Optimization is done using MATLAB/ Simulink software. Feasibility analysis of the system of car park incorporating use of renewable resources is also done. Hence, this work mainly analyzes the grid stability in case an EV charging aggregator is setup and also to reduce the charging cost of EVs using V2G technology.

Keywords Electric vehicle (EV) · State of charge (SoC) · Battery degradation cost · Charging and discharging cost · Vehicle-to-grid (V2G) · Feed in tariff (FIT) · Firefly algorithm · Optimization

1 Introduction

Over the last few years, there has been multiple advancements in the transportation systems all over the world, and the development of electric vehicles is being seen as one of the solutions to the biggest crisis that the environment is facing, i.e., pollution. Developed countries have achieved huge milestones in the research and development of electric transportation systems like high speed railways, smart traction systems,

P. Bharti · O. Singh (✉)
Department of Electrical Engineering, Gautam Buddha University, Gautam Buddha Nagar, Greater Noida, Uttar Pradesh, India
e-mail: omveer.singh@gbu.ac.in

A. Rani et al. (eds.), *Signals, Machines and Automation*, Lecture Notes in Electrical Engineering 1023, https://doi.org/10.1007/978-981-99-0969-8_2

use of renewable energy for electric transportation systems, hybrid vehicles, vehicle charging systems, etc. However, the adoption of electric vehicles in the place of conventional IC engine vehicles is relatively new in Indian context. Smart vehicle charging infrastructure such as V2G does also seem to be required due to sudden increase of electricity demand due to the increasing trend of electric cars among the consumers.

V2G technology incorporates a bidirectional charger for electric cars which enables the flow of electricity from the electricity grid to the vehicle and vice versa. In this paper, an analysis is done for knowing the effects of implementing V2G technology in India and to reduce the charging costs of the EVs using V2G enabled charging station.

2 Literature Review

Electric-drive vehicles can turn into a significant asset for the electric utility framework as well as a solution for air pollution, system reliability, and can have economic benefits. Both consumers and utilities will benefit from efficient V2G. Since there are various advantages of V2G technologies, many researchers have tried to take up the opportunity to work in this field. S.A. Amamra's research to involve bidirectional optimized V2G operation, day ahead scheduling of EV charging/discharging with the help of two case studies had resulted in increased opportunities for frequency and voltage regulation while reducing EV charging costs [1].

Nandini I. Nimalsiri and her fellow researchers have provided proper studies for EV charging coordination in their work which involves a distributed charging control system. They have also proposed algorithm for distributed EV charge scheduling considering the various perspectives of grid operators, aggregators and EV users [2]. A similar work can be seen in the research of Vishu Gupta and his team in which they have worked on a decentralized charging management scheme for EV in a multi-aggregator scenario, and their study was conducted in Jaipur, Rajasthan [3].

Various other researches have been done for coordination management of EV charging facilities, William Infante and Jin Ma have worked in this aspect to analyze the coexistence of multiple and varied charging facilities. Their model has shown the capabilities of demand side management also [4]. EVs with conventional charging strategies can result in undesired impact on distribution network; Yanchog Zheng and his team have a contribution for analyzing CPMM and DDRM. The results show variation in profits for the aggregator with the agreement price to reduce total cost and maximize profit [5]. Yan Cao and his fellow researchers have worked to develop an optimization technique with the help of mixed integer linear programming model to overcome the uncertainty in the market price of electricity for the aggregator and to promote its profits and better charging and discharging scheduling strategies of Evs [6].

In [7], a system is proposed as a coordination model based on a marginal price strategy to coordinate between EV charging system and aggregator. This model also claims to have 78.3% reduction in total cost for EV charging. EV and EV charging system are also coordination based upon user's selection of the route. Governments now place a high premium on energy efficiency due to the rise in global electricity usage. The efficient integration of a significant penetration of electric vehicles (EVs) into the energy markets has so received attention. Solanke et al. [8] offers an evaluation of various EV aggregator strategy options. The smart charging approach suggested in a prior study is taken into consideration in this analysis. The smart charging method takes user preferences into account and uses charging power rate modulation. This study simulates the impact of various tactics using a case study of a distribution system from the Ecuadorian city of Quito. The expenses of EV aggregators and technical circumstances are assessed while various actions are simulated.

The transition from traditional fossil fuel-powered vehicles to zero or extremely low tailpipe emission vehicles is now taking place in the global transportation sector. A suitable charging station (CS) infrastructure, information technology, intelligent distributed energy generating units, and supportive governmental regulations are needed to achieve this shift. The purpose of [9] was to discuss the important considerations that must be made while designing the infrastructure for electric vehicle charging stations. The report also discussed important research and advancements in planning and technology that are being made to improve the architecture and effective administration of charging station infrastructure. It discusses the current state of innovations for electric vehicle charging stations in India. The study specifically offered a critical analysis of the research and improvements in the infrastructure of charging stations, the issues it raises, and the ongoing efforts to standardize it in order to assist researchers in addressing the issues.

3 Development of the Model for Feasibility Analysis of V2G Operation in India

In this section, a basic simulation model is studied for V2G enabled operation using existing conventional and renewable resources and infrastructure. The simulation is being performed on HOMER Pro software. Here, the model is based upon the resources available in Greater Noida, Uttar Pradesh, India.

3.1 Component Information

For the design of the aggregator setup, several resources have been taken into consideration. The list of the resources is as follows:

The grid of 500 KW capacity, a diesel generator of 100 kW capacity, a solar PV system (generic flat plate PV array) of 1 MW, a Li-ion battery for energy storage of

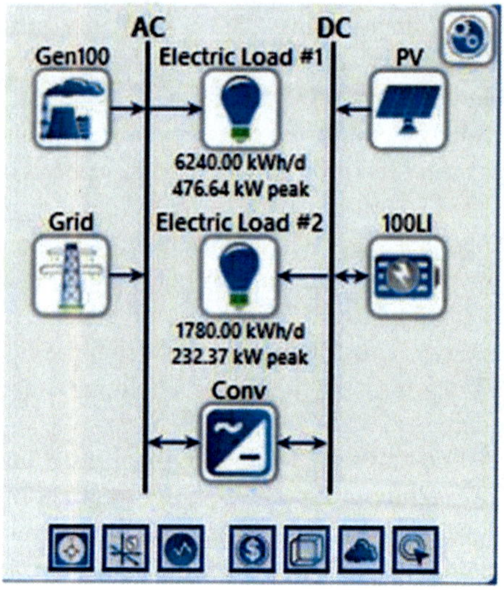

Fig. 1 Simulation model for feasibility analysis of V2G operation

200 KW, a converter for PV system and a battery are connected to a DC bus. A 700 KW (peak) electrical load representing EVs is charged at the aggregator charging station.

3.2 *Simulation Model*

See Fig. 1.

3.3 *Feasibility Analysis Objectives*

The main objectives of this study are:

- To find out if V2G operations using Evs are carried out with the existing infrastructure, what will be the impacts of this.
- To analyze if we need an additional infrastructure in our country for EV penetration.
- To find out whether V2G operations can be a feasible solution for demand side management.

4 Mathematical Model for Cost Minimization of V2G Enabled EV Charging in a Car Park

Since EVs can be used as reserve energy storage, a car park model has been designed for V2G activities. Modeling and control framework has been built-up which includes key factors such as charging, discharging, battery degradation costs, driving probability, feed in tariff, V2G rebates, and vehicle SoC.

Electricity Cost Model

By taking $u(t)$ as the charging status of the electric vehicle at time t such that

$u(t) = 1$ while charging; -1 while discharging; 0 while disconnect.

Total electricity cost (C) has four components:

i.e.,

$$C = C_{\text{charging}} - C_{\text{dischaging}} + C_{\text{loss}} - C_{\text{rebate}} \tag{1}$$

Now define a function S.

$S(x) = 1$when $x > 0$; 0 when $x \leq 0$.

Charging cost:

$$C_{\text{charging}} = \int_{to}^{tf} p * S(u(t)) * P_{\text{ev}} \mathrm{d}t \tag{2}$$

where p = price of electricity and P_{ev} = EV charging power.

Discharging Cost:

$$C_{\text{discharging}} = \int_{to}^{tf} q * S(-u(t)) * P_{\text{ev}} \mathrm{d}t \tag{3}$$

where q = feed in tariff.

Battery Degradation Cost:

$$C_{\text{loss}} = \int_{to}^{tf} D_r * (S(u(t)) + S(-u(t))) * P_{\text{ev}} \mathrm{d}t \tag{4}$$

where D_r = battery degradation cost.

Rebate Cost:

$$C_{\text{rebate}} = \int_{to}^{tf} P_r * P_{ev} * S(-u(t))\, \mathrm{d}t \tag{5}$$

where P_r = rebate price of electricity.

Putting values of (2), (3), (4), and (5) in (1)

$$C = P_{ev} \int_{to}^{tf} [p * S(u(t)) - q * S(-u(t)) + D_r * (S(u(t)) + S(-u(t))) - P_r * S(-u(t))] \, dt$$

Total Charging Cost Considering EV Plugin Probability:

For a fully charged EV if max driving distance is $d_{\max}$ and after driving it over a distance d_d

$$\text{SoC} = 1 - d_d / d_{\max} \tag{6}$$

Considering probability P_j corresponding to driven distance D_j for M number of drives, the decrease in SoC for each travel possibility after each drive will be

$$\Delta\text{SoC} = (P / d_{\max}) \sum_{j=0}^{M} D_j * P_j \tag{7}$$

Here, P will be the probability of EV driving out of the car park so the probability staying inside the car park will be $(1 - P)$.

Now, the total electricity cost of one EV will be:

$$C_{\text{total}} = (1 - P) * P_{ev} \int_{to}^{tf} [p * S(u(t)) - q * S(-u(t)) + D_r * (S(u(t)) + S(-u(t))) - P_r * S(-u(t))] \, dt$$

When multiple EVs are connected in the car park, the total cost will be the sum of costs for each EV, i.e.,

$$C_{\text{total}} = \sum_{i=1}^{N} C_{\text{total}}^{i}$$

Here, N is the number of EVs in the car park.

5 Firefly Algorithm

One of the new metaheuristic algorithms for optimization issues is the firefly method. The flashing behavior of fireflies inspired the algorithm. Randomly produced solutions will be treated as fireflies in the algorithm, and their brightness will be assigned based on their performance on the objective function. A firefly will be drawn to a brighter firefly, and if there is no brighter firefly, it will move randomly, according to one of the algorithms' principles.

Attractiveness, randomization, and absorption are the three parameters that govern the firefly algorithm. The attractiveness parameter is calculated using exponential functions and is based on the light intensity between two fireflies. It employed the idea of how the light of individual fireflies pulled them together, as well as an unpredictability aspect, to encourage participation.

6 Optimization Implementation Procedure of the Proposed System

Considering a car park of typical office area with 50 eV charging slot for this case study. Use of electricity is done for working hours only, i.e., 9 am to 5 pm. Taking TATA Nexon EV as an example whose maximum driving distance at full charge is 312 km. Maximum SoC is taken as 1.00, and min SoC is taken as 0.00. Final SoC at the end of the day should not be less than 0.7. It takes around 10 h for a full charge of its 30.2 kwh battery. Charging power is 3.2 kw with onboard charger. Level 2 chargers are being used which are also known as industrial chargers having a voltage level of 220–240 V, and they are generally used at public charging stations.

Rebate price is selected as Rs 4.53/kwh (15% of usages in UP for FY 2019–20). FIT is chosen at Rs 7.0/kwh. Battery degradation cost is taken as Rs 3.40/kwh. Price of purchasing power from Grid in UP is Rs 7.70/kwh. Number of drivings for each EV during working hours is taken either 0, 1 or 2. Possible driving distances are taken as 1 km, 2 km, 3 km, 4 km, and 5 km. Corresponding values of P_j are 0.47, 0.23, 0.13, 0.12 and 0.05 and P is 0.21.

7 Results

7.1 *Feasibility Analysis of the Proposed System Using HOMER Pro Software*

From the simulation model results, following points can be concluded:

EV penetration in India can be increased while dealing with the energy crisis simultaneously using the V2G technology. The results show that using the existing power system infrastructure V2G operations can be carried out. The power selling capacity of the system is found to be approximately two times of the total power purchased from the grid. 99.5 percent of total energy consumption is supplied by the renewable resources. So, it may also be concluded that V2G will promote clean

Table 1 Tabular representation of HOMER Pro simulation results

Parameters	NPC	Operating cost	Initial capital	Renewable fraction (%)
Case 1	₹22,058	₹1388	₹40,000	99.5
Case 2	₹22,071	₹1388	₹40,019	99.5
Case 3	₹22,566	₹1372	₹40,300	99.5
Case 4	₹22,628	₹1374	₹40,389	99.5
Case 5	₹25,890	₹1478	₹45,000	99.5

energy while dealing with the energy shortage at the same time. Table 1 gives the NPC of different combination of sources, and we can find the best NPC in case 1.

7.2 *Cost Optimization Results Using MATLAB/Simulink Software*

The following graph (Fig. 2) gives the relationship between EV initial SoC and their charging costs when the system is unoptimized.

For an unoptimized system of car parks with V2G technology, we can see the graph tends to depict costs in opposite direction because of the selected SoC of 0.7 above which the V2G service will take place and the effective cost will include charging/discharging costs, FITs, and rebate prices. Since all these factors contribute to the EV user's profit and not only the charging cost, the total cost during the charging phase, i.e., below 0.7 SoC, shows decreasing characteristics and tends to zero when it is close to 0.7 SoC, because the more SoC is close to the threshold SoC, the less

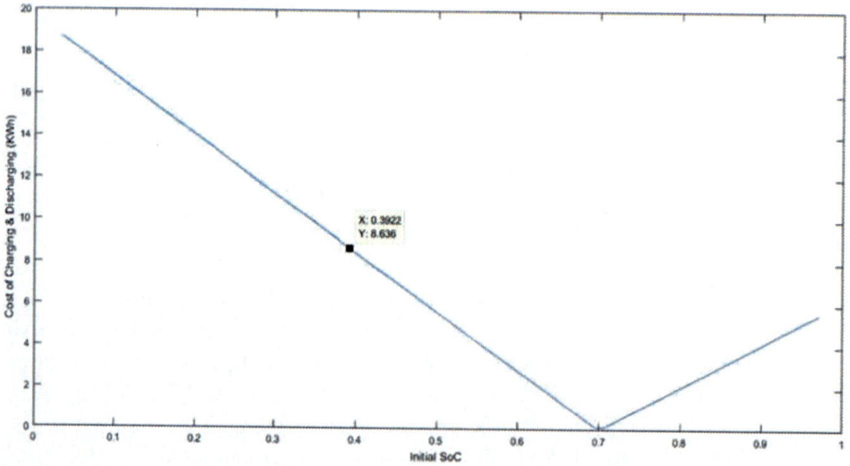

Fig. 2 Comparison of total charging/discharging cost w.r.t. EV SoC for an unoptimized system

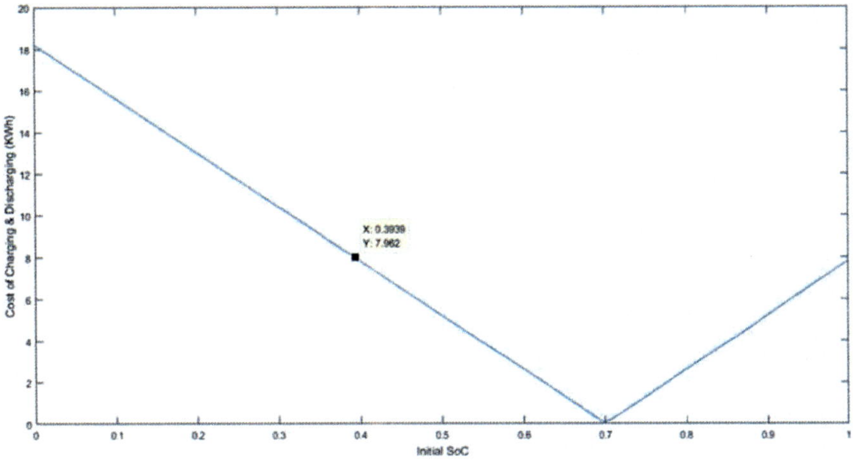

Fig. 3 Results for best cost search for charging/discharging in an optimized system using firefly algorithm

charging is required, while the effective cost during the discharging phase, i.e., above 0.7 SoC, is of the opposite nature (increasing characteristics). A total number of 1000 iterations have been taken into consideration for this study (Fig. 3).

The application of firefly algorithm clearly brings down the charging costs and enhances the discharging costs also. So, we can conclude that the implementation of the firefly algorithm in this case will bring down EV charging costs and maximize EV user benefits.

8 Conclusion

For smart EV car park systems, a control approach for EV charging and discharging is proposed in this paper. The goal is to reduce the cost of car park power by managing charging and discharging activities in which electric vehicles are parked. The possibility of V2G advantages has been suggested by detailed modeling studies. However, various aspects, such as the battery capacities, play a role. The cost of degradation, the rebate price, the FIT, and the electricity price, etc. have also been considered. The primary variables that might encourage V2G are thought to be battery performance. The government's strategy and grid company subsidies, for example, will have an impact on EV consumers' decisions for V2G involvement.

First of all, V2G operation feasibility in India is analyzed using HOMER Pro software based on the power and energy infrastructure of the city of Lucknow, Uttar Pradesh. It has been found that V2G operation will have positive impacts on the grid, and it will also involve the possibility of optimum use of renewable energy sources. Secondly, cost optimization for an EV car park with V2G enabled EV operation is done using MATLAB software by firefly algorithm. Total financial benefits have been seen to increase using firefly algorithm for the EV car park.

References

1. Amamra SA, Marco J (2019) Vehicle-to-grid aggregator to support power grid and reduce electric vehicle charging cost. IEEE Access 7:178528–178538
2. Nimalsiri NI, Mediwaththe CP, Ratnam EL, Shaw M, Smith DB, Halgamuge SK (2020) A survey of algorithms for distributed charging control of electric vehicles in smart grid. IEEE Trans Intell Transp Syst 21(11)
3. Gupta V, Kumar R, Panigrahi BK (2020) User-willingness-based decentralized EV charging management in multiaggregator scheduling. IEEE Trans Ind Appl 56(5):5704–5715
4. Infante W, Ma J (2020) Coordinated management and ratio assessment of electric vehicle charging facilities. IEEE Trans Ind Appl 56(5):5955–5962
5. Zheng Y, Yu H, Shao Z, Jian L (2020) Day-ahead bidding strategy for electric vehicle aggregator enabling multiple agent modes in uncertain electricity markets. Elsevier, Appl Energy 280:15
6. Cao Y, Huang L, Li Y, Jermsittiparsert K, Ahmadi-Nezamabad H, Nojavan S (2020) Optimal scheduling of electric vehicles aggregator under market price uncertainty using robust optimization technique. Int J Electr Power Energy Syst 117:105628
7. Ding Z, Lu Y, Lai K, Yang M, Lee WJ (2020) Optimal coordinated operation scheduling for electric vehicle aggregator and charging stations in an integrated electricity-transportation system. Elsevier, Electr Power Energy Syst
8. Solanke TU, Ramachandaramurthy VK, Yong JY, Pasupuleti J, Kasinathan P, Rajagopalan A (2020) A review of strategic charging–discharging control of grid-connected electric vehicles. Elsevier, J Energy Storage
9. Mishra S, Verma S, Chowdhury S, Gaur A, Mohapatra S, Dwivedi G, Verma P (2021) A comprehensive review on developments in electric vehicle charging station infrastructure and present scenario of India. MDPI, Sustainability

Piezoelectric Sensors for the Conversion of Noise Pollution into Electricity

Arunesh Kumar Singh, Shahida Khatoon, and Kriti

Abstract The use of piezoelectric sensors for noise pollution energy conversion is presented in this paper. Noise pollution is waste energy in the environment that has the potential to get converted into electrical energy using suitable energy conversion techniques. Piezoelectric sensors such as PZT and PVDF thin film have very good power density and stability for the low-frequency weak vibration/pressure signal. Therefore, piezoelectric materials are found suitable for the conversion of noise pollution into electrical energy. Proper amplification and rectification can increase the electrical energy level at the output, and the power management unit can provide storage of the electrical signal. We have also reviewed various energy conversion models and impact of the vehicular noise pollution on human health.

Keywords Noise pollution · Energy conversion · Piezoelectric sensor · PZT · PDVF film · Vehicular noise

1 Introduction

The energy conversion from waste energy to useful energy has been generating much more interest in recent years. The natural dissipated energy such as solar energy, wind energy, fluid flow, temperature gradient, and human motion can be converted into a valuable form of energy that can be collected and supplied to the grid or directly used for running small appliances. Wireless sensor network (WSN) is a wirelessly interconnected network of sensors where the sensors collect data from remote locations and interchange the gathered information. With the development of wireless sensor networks for various applications, such as biomedical devices,

A. K. Singh · S. Khatoon · Kriti (✉)
Department of Electrical Engineering, Jamia Millia Islamia, New Delhi 110025, India
e-mail: tripathikriti@gmail.com

A. K. Singh
e-mail: asingh1@jmi.ac.in

S. Khatoon
e-mail: skhatoon@jmi.ac.in

A. Rani et al. (eds.), *Signals, Machines and Automation*, Lecture Notes in Electrical Engineering 1023, https://doi.org/10.1007/978-981-99-0969-8_3

RF remote sensing, health care monitoring, etc., the challenge is to fulfill the energy requirements of sensor nodes. These sensor nodes are microsensor that only requires a micro-level of power. This novel noise pollution energy converter can fulfill the energy requirement of these remotely located microsensors and help to make them self-sufficient [1].

In the last decade, the field of micro-electromechanical system (MEMS) devices has been developed with rapid growth. MEMS devices are micro-sized structures, incorporating electronic circuitry with the mechanical system. MEMS devices can be utilized in various applications in WSN as sensor nodes. With the advancement in IC technology in recent years, it has been possible to process, amplify, and retain low-level energy signals and fulfill power requirements for energizing micropower sensors. In the present times of the Internet of things (IoT) and artificial intelligence (AI), the self-sufficient micropower generated by MEMS devices can provide a better alternative to rechargeable lithium batteries in implantable devices and a solution to its high cost [2].

1.1 Impact of Noise Pollution on Human Health

Environmental pollution means contamination in the environment. Contamination may be due to manmade activities or natural occurrences. Environmental pollution is mainly categorized as water pollution, soil pollution, air pollution thermal pollution, noise pollution, and radioactive pollution [3].

Noise pollution is one of the most significant types of environmental pollution and is hazardous to human and animal beings. The high-level sound generated by human activities is sometimes detrimental and called anthropogenic noise. Anthropogenic noise may lead to sleeping problems, chronic stress, blood pressure fluctuation, and other issues in humans. Natural habitat reduction, anxiety, masking, and communication breakdown are problems caused by noise pollution among animals and birds. Sources of noise pollution are traffic activities on the roads, ships in the ocean, and air traffic caused by aircraft. Along with this, commercial and industrial activities are also the major source of anthropogenic sound. In the world of technology, modernization and urbanization come along with noise pollution [4, 5].

Noise is an undesirable state of sound. Any acoustic signal can be a noise signal if it is annoying. Loud music can be entertaining for some people but not for all. Noise is based on perception. For example, hummingbirds and rainfall can be smooth and loving to some but not for all. The sound signal is basically a varying wave of air pressure. Human hearing ability discriminates noise from the sound perceived. Sound pressure level (SPL) meter is used to determine the loudness of the sound. The measurement unit of SPL is given in dBA where A refers to the A-weighted frequency filter. Mostly, A-weighting frequency filtering is used in SPL meter rather than C-weighted and Z-weighted because its frequency response is quite similar to the frequency response of the human ear. 60–65 dBA SPL is comfortable for the human ear but any level beyond 70–75dBA may cause noise pollution [6].

1.2 Potential of Noise Pollution

Figure 1 shows the sources contributing to noise pollution. Among the sources of noise pollution, it is found that traffic noise contributes the most. In the industries, the level of noise may be high enough to harm, but usually, safety guards are provided to workers. Also, such industries are located distant from the urban area. In the urban area, the noise level surpasses the noise limit recommended by Central Pollution Control Board (CPCB). This noise level affects people and results in stress, strain, and other disorders. Figure 2 shows the permitted noise levels by the World Health Organization (WHO) and Australia, Japan, the United States, and India, respectively [7].

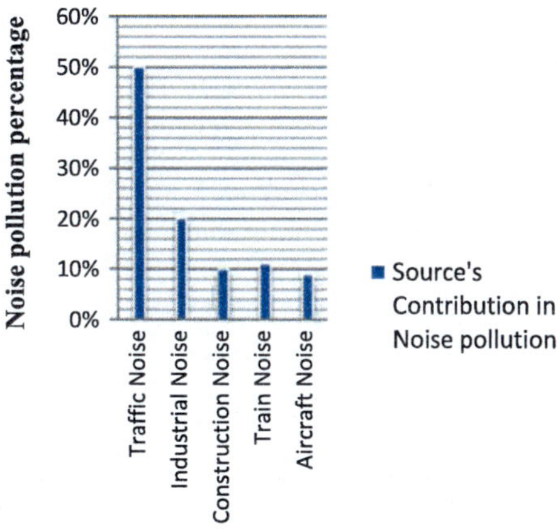

Fig. 1 Sources contribution to noise pollution

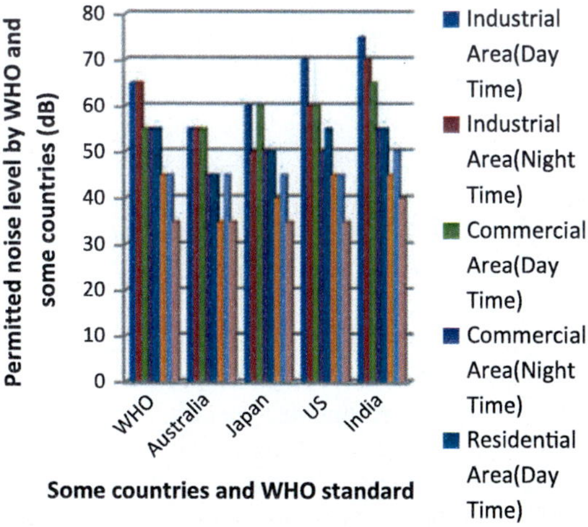

Fig. 2 Permitted noise level by WHO and some countries

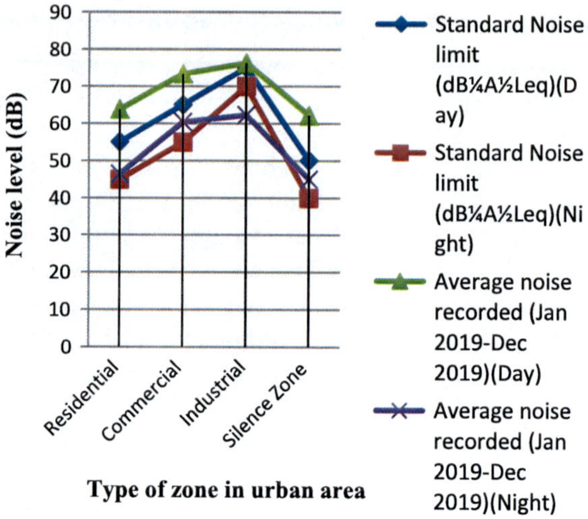

Fig. 3 Permitted noise versus observed in Noida (Jan–Dec, 2019)

It can be observed from Fig. 2 that a comparatively higher noise level is allowed in India; still, noise pollution in India goes beyond the prescribed limit. In the official Website of Uttar Pradesh Pollution Control Board, the noise pollution data is available till March 2020. Consolidating the data for year 2019 for Noida, Uttar Pradesh, it is observed that the noise quality of the city is beyond the prescribed limit. Figure 3 shows the consolidated noise in the year 2019 for Noida city, Uttar Pradesh with respect to the standard permissible noise limit by the government authorities [8].

From the above data, it is observed that the noise quality in the city is not under the prescribed limit except average noise recorded at night in the industrial area. Hence, this quality of noise can be harmful to human beings.

2 Piezoelectric Sensor for Noise Pollution Energy Converter

In the year 1880, the principle of piezoelectricity was invented. An experiment was performed by two French scientists to observe the change in electric potential when piezoelectric substrate deforms. Piezoelectric materials are found to have better response and stability for mechanic-electric power conversion as compared to electrostatic and electro-magnetic systems [9].

Figure 4 shows the basic piezoelectric module to convert mechanical vibrations impacting on the y-axis into a small electric charge developed in the x-axis direction. This effect is called the piezoelectric effect [10]. Figure 5 depicts the commercially available PVDF film with and without attached proof mass.

A very thin layer of PZT material, for example, zirconate titanate or polyvinylidene fluoride (PVDF) thin film is placed between the upper and lower electrodes. Graphene

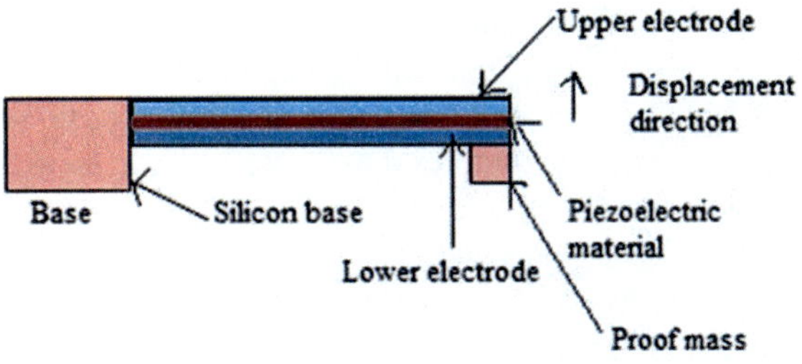

Fig. 4 Basic MEMS piezoelectric converter

Fig. 5 Commercial available PVDF film with and without proof mass

material and metal are found to have outstanding electrical and thermal conductivity. Silicon base is attached as a proof mass for the adjustment of the resonance frequency to match with the frequency of the input acoustic signal. The resonance condition will help to attain maximum output power. Thin-film PVDF device can induce 7.6 $V_{p\text{-}p}$ when exposed to a 105 dB acoustic signal [11]. The following are the set of equations for piezoelectric material according to the definition in IEEE standard on piezoelectric.

$$D_i = e_{ij} S_j + \varepsilon_{ii}^S E_i \tag{1}$$

$$T_j = C_{ij}^E S_j - e_{ij} E_i \tag{2}$$

where T_j is stress given in N/m^2, S_j is mechanical strain, E_i is the electric field, D_i is the electrical displacement, C_{ij}^E is elastic stiffness constant, ε_{ii}^S is permittivity under unit strain S, e is the piezoelectric constant [2].

The piezoelectric transducer works in two modes, such as d_{31} and d_{33}. In d_{31} mode, the electric charge develops in the perpendicular direction of pressure applied. In d_{33} mode, the electric charge develops in the parallel direction as a pressure signal is applied. Figure 4 represents the d_{31} operating mode of the piezo sensor. Using a piezo sensor in d_{33} mode, it can provide high sensitivity and open-circuit voltage. This mode is suitable for PZT material as a piezo sensor. Acoustic energy-based energy harvesting requires the d_{31} mode of the piezoelectric transducer as shown in Fig. 4.

A piezo sensor when placed under stress, a force exerted is given by $F = \delta A$, where δ is the stress, and A is cross sectional area of the PVDF film. Total work done is given by

$$W_D = F\Delta l = 2p^2\delta A/l_0 \tag{3}$$

where p is the pressure signal amplitude and Δl is the change in piezo length.

Electric charge induced by the piezo sensor can be given as

$$Q = d_{31}EA \in \tag{4}$$

where d_{31} is the piezoelectric constant, $\epsilon = \Delta l/l_0$ is the strain.

Energy stored in the capacitor (C) is given by

$$W_{\text{ES}} = Q^2/2C \tag{5}$$

$$C = \varepsilon_d A/t,$$

where ε_d is permittivity of the piezoelectric sensor

Total stored energy W_{ES} is given by

$$W_{\text{ES}} = \left(t\, d_{31}^2 E^2 \varepsilon^2 A\right) / 2\varepsilon_d \tag{6}$$

Efficiency of energy conversion is given by

$$\eta = W_{\text{ES}}/ W_D$$

$$\eta = \left(l_0^2 d_{31}^2 E^2 \varepsilon^2\right)/\left(4\varepsilon_d p^2 \delta\right) \tag{7}$$

3 Role of Resonator as Pre-amplifier

Simply a piezoelectric transducer is not sufficient to generate electrical output alone. Low vibration energy of the noise pollution can be amplified before being processed by a piezo sensor using a resonator module.

Fig. 6 Simple resonator module

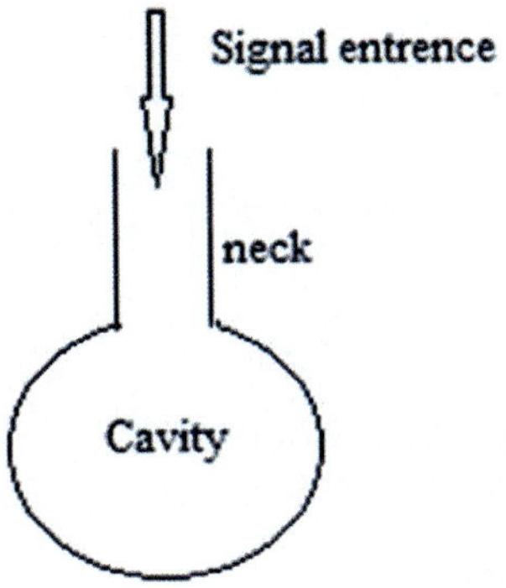

A simple resonator module is shown in Fig. 6. In the resonator module, the piezo sensor is placed inside the cavity. The instance when the incoming vibration frequency matches with the resonance frequency, high output is generated that results in more output power. Using Helmholtz's formula, the resonance frequency is given by

$$f_R = \frac{C}{2\pi}\sqrt{\frac{4r^2}{d^2h(t+kr)}} \tag{8}$$

where f_R is the resonance frequency, c is the sound velocity, r is the radius of cavity entrance, d = diameter, t = thickness of cavity, and k is constant [12, 13].

4 Noise/Sound to Electric Energy Conversion

The block diagram of noise pollution energy converter is shown in Fig. 7. Noise pollution signal gets amplified before reaching the piezo sensor in the resonator cavity utilizing the resonance phenomenon. Piezo sensor converts it into a low electrical signal which further gets rectification and amplification. Super-capacitor or battery can be used for energy storage [14–16]. It is found from prior research that the low acceleration vibrations of train tunnels can harvest up to 395 micro-joule energy. 78 dB SPL signal could generate 26.7 mV RMS voltage and an SPL signal of 100 dB can provide 1.148 mW power with a setup of four PZT plates [17].

5 Practical Model of Noise Pollution Energy Converter

Figure 8 shows the circuit simulation of the noise pollution energy converter. The AC voltage source along with the dependent current source represents the PVDF

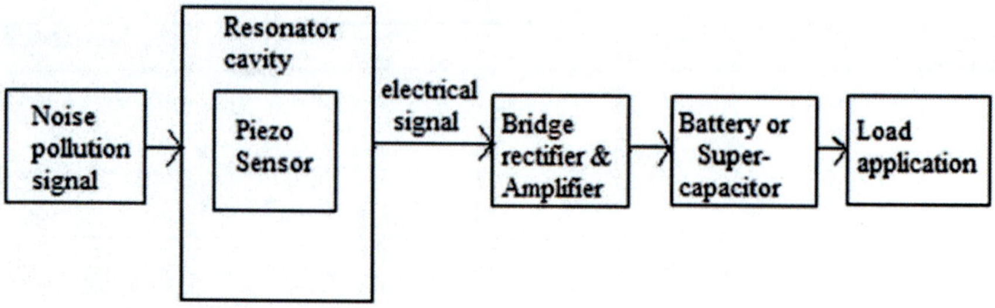

Fig. 7 Block diagram of noise pollution electric energy converter

film. An amplifier circuit based on LM-324 operational amplifier is used here in non-inverting configuration. Arrangement of resistors and capacitor is similar to the breadboard circuit. We have observed that a single PVDF film produces average voltage as depicted in Table 1.

We have simulated the noise pollution energy converter circuit on NI Multisim software as shown in Fig. 8, for various voltages produced by PVDF film as shown in Table 1.

It is observed from the output waveform shown in Fig. 9 that 25.1 V RMS voltage and 20.2 mA RMS current have been converted from an input RMS voltage of

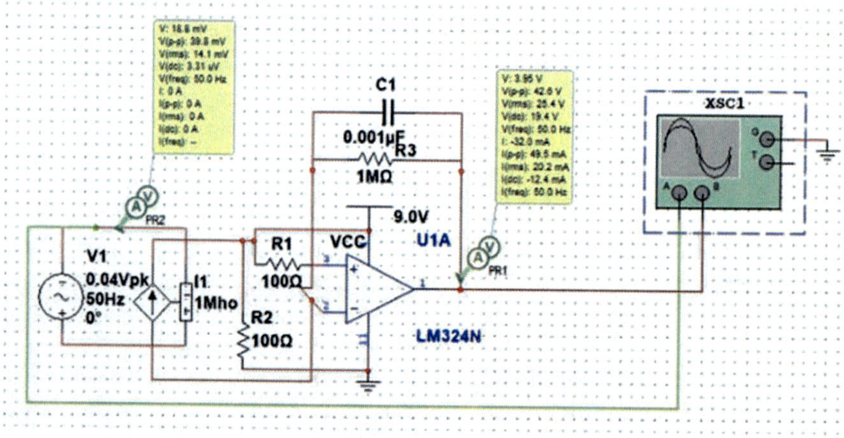

Fig. 8 Circuit simulation of noise pollution energy converter on NI multisim

Table 1 Voltage produced by single PVDF for different noise pollution level

S. No.	Noise SPL level (in dB)	Voltage produced by PVDF (in mV)
1	76	30
2	78	35
3	80	40

14.1 mV and 28.3 mA RMS current at a constant frequency of 50 Hz. The overall output voltage and current data observed for various input noise pollution level are depicted in Table 2. We can conclude that the circuit converts the noise pollution into electrical energy and raise the voltage level to a significant level that is suitable for various low power application. The efficiency graph is observed between output voltage and noise pollution level which is shown in Fig. 10.

For the simulation of piezoelectric sensors, there is some finite element simulation (FEM) software such as ABAQUS, COMSOL, and ANSYS multi-physics available in the market. ABAQUS is mainly used to simulate piezoelectric devices for sensor and actuator applications. The advantage of ABAQUS is its user-friendly interface and effective treatment for nonlinear systems. The limitation of ABAQUS is in the simulation of a coupled electromechanical system which is possible to simulate using ANSYS and COMSOL software [18, 19].

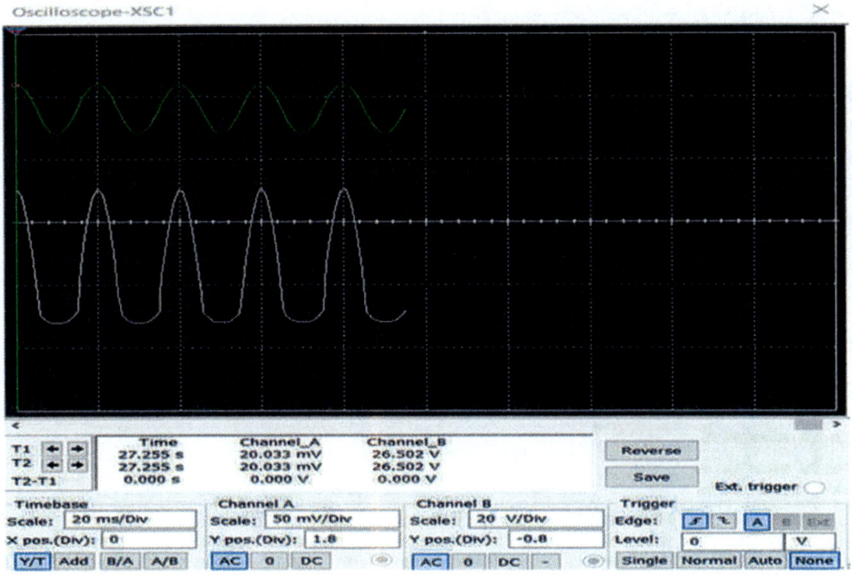

Fig. 9 Output waveform for the circuit simulation shown in Fig. 8

Table 2 Overall circuit response for various level of noise pollution

S. No.	Noise pollution level (in dB)	Input voltage (in mV)	Input current (in mA)	Output voltage (in V)	Output current (in mA)
1	76	10.6	21.2	21.4	15.0
2	78	12.4	24.7	23.4	17.6
3	80	14.1	28.3	25.1	20.2

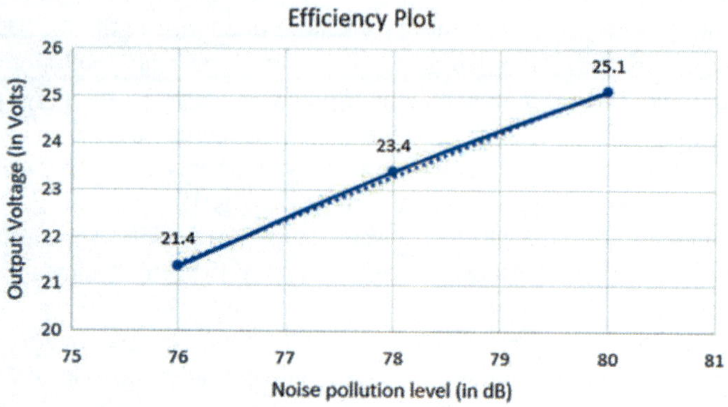

Fig. 10 Efficiency plot between output voltage versus noise pollution level

6 Conclusion

The noise pollution energy converter has the potential to act as a micropower generating system. With the development of an energy harvesting module, it has become possible to convert approximately 80 dB noise signals into microwatt power signals utilizing d_{31} and d_{33} modes of piezoelectric sensors. The level of the electric output signal can be increased further utilizing resonance and power electronics circuits such as instrumentation amplifiers and power amplifiers. The mechanism of noise pollution energy converter utilizes noise pollution which is inevitable to make devices self-powered and sustainable. This technique can be applicable for wireless sensor network in the field of E-health monitoring and biomedical implantable devices, i.e., cardiac pacemakers, insulin pumps, and neuro-stimulators. The noise pollution-based energy converter can be placed roadside near the traffic signal to provide electricity to power the traffic light and surveillance system.

References

1. Choi J, Jung I, Kang CY (2019) A brief review of sound energy harvesting. Nano Energy 56:169–183. https://doi.org/10.1016/j.nanoen.2018.11.036
2. Jong W, Shiun B (2012) Piezoelectric MEMS power generators for vibration energy harvesting. In: Small-scale energy harvesting
3. Goswami S, Swain BK (2017) Environmental noise in India: a review. Curr Pollut Rep 3(3):220–229. https://doi.org/10.1007/s40726-017-0062-8
4. Andersson EM, Ögren M, Molnár P, Segersson D, Rosengren A, Stockfelt L (2020) Road traffic noise, air pollution and cardiovascular events in a Swedish cohort. Environ Res 185(March):109446. https://doi.org/10.1016/j.envres.2020.109446
5. U. Nations and E. Programme (2022) Noise, blazes and mismatches
6. Guarnaccia C, Singh D, Quartieri J, Nigam SP, Kumar M, Mastorakis NE (2018) Honking noise contribution to road traffic noise prediction. In: AIP conference proceedings, vol 1982. https://doi.org/10.1063/1.5045448

7. Chauhan A, Pawar M, Kumar D, Kumar N, Kumar R (2010) Assessment of noise level status in different areas of Moradabad City. Researcher 2(5):88–90
8. Regional Office—U.P. Pollution Control Board, Noida
9. Pillai MA, Deenadayalan E (2014) A review of acoustic energy harvesting. Int J Precis Eng Manuf 15(5):949–965. SpringerOpen. https://doi.org/10.1007/s12541-014-0422-x
10. Jeon YB, Sood R, Jeong JH, Kim SG (2005) MEMS power generator with transverse mode thin film PZT. Sens Act A Phys 122(1):16–22. https://doi.org/10.1016/j.sna.2004.12.032
11. Park S, Kim Y, Jung H, Park JY, Lee N, Seo Y (2017) Energy harvesting efficiency of piezoelectric polymer film with graphene and metal electrodes. Sci Rep 7(1):1–8. https://doi.org/10.1038/s41598-017-17791-3
12. Li Z, Zou H, Li J, Chen H (2021) Research progress of cavity-based acoustic energy harvester. J. Vibroengineering 23(7):1680–1693. https://doi.org/10.21595/JVE.2021.22057
13. Yuan M, Cao Z, Luo J, Zhang J, Chang C (2017) An efficient low-frequency acoustic energy harvester. Sens Act A Phys 264:84–89. https://doi.org/10.1016/j.sna.2017.07.051
14. Fang LH, Hassan SI, Abd Rahim R, Isa M, bin Ismail B (2017) Exploring piezoelectric for sound wave as energy harvester. Energy Procedia 105:459–466. https://doi.org/10.1016/j.egypro.2017.03.341
15. Ma X, Wilson A, Rahn CD, Trolier-McKinstry S (2016) Efficient energy harvesting using piezoelectric compliant mechanisms: theory and experiment. J Vib Acoust Trans ASME 138(2):1–9. https://doi.org/10.1115/1.4032178
16. Fang LH, Rahim RA, Naimah S (2021) Design of artificial piezo-leaf wind energy harvesting system monitoring based on Blynk apps. In: AIP conference proceedings, vol 2339, no May. https://doi.org/10.1063/5.0044292
17. Li B, You JH, Kim YJ (2013) Low frequency acoustic energy harvesting using PZT piezoelectric plates in a straight tube resonator. Smart Mater Struct 22(5):055013. https://doi.org/10.1088/0964-1726/22/5/055013
18. Yuan M, Cao Z, Luo J, Chou X (2019) Recent developments of acoustic energy harvesting: a review. https://doi.org/10.3390/mi10010048
19. Ali B, Ashraf MW, Tayyaba S (2019) Simulation, fuzzy analysis and development of ZnO nanostructure-based piezoelectric MEMS energy harvester. Energies 12(5):807. https://doi.org/10.3390/en12050807

Implementation of Single-Phase Cascaded Seven-Level Inverter with Non-isolated Converter

C. Dinakaran and T. Padmavathi

Abstract This article proposes an inventive cascaded H-bridge single-phase multilevel inverter over a minimal portion based on switches used in favor of solar photovoltaic (PV) utilization. Multilevel inverters (MLI) persist consistently on behalf of the energy innovation system's capability to sustain immense power, including power quality challenging operation. This article proposes to designate a modified cascaded seven-level inverter topology intended as a nearly sinusoidal output voltage, minor numbers of energy switches, reduced complexity, and diminished total harmonic distortion. The design is simple with accessible to develop higher levels. However, its gate driver circuits are still interpreted considering the number related to access switches being depreciated. The pulse width modulation system operates over the multicarrier system to generate the power switches' gating signals. This competence remains accomplished through analyzing the expected topology over the traditional topologies as the primary investigation point, demonstrating the competent certainty of the suggested topology through unique access in propagating voltage levels for a seven-level inverter over simulation, proving MATLAB/SIMULINK, PROTEUS results.

Keywords Cascaded H-bridge MLI · Pulse width modulation (PWM) inverter · Solar PV system · Modified MLI · Total harmonic distortion (THD)

C. Dinakaran (✉)
Department of Electrical and Electronics Engineering, GITAM School of Technology, GITAM (Deemed to be University), Visakhapatnam, Andhra Pradesh, India
e-mail: dina4karan@gmail.com

T. Padmavathi
Department of Electrical, Electronics and Communication Engineering, GITAM School of Technology, GITAM (Deemed to be University), Visakhapatnam, Andhra Pradesh, India
e-mail: ptadi@gitam.edu

A. Rani et al. (eds.), *Signals, Machines and Automation*, Lecture Notes in Electrical Engineering 1023, https://doi.org/10.1007/978-981-99-0969-8_4

1 Introduction

At present, non-conventional energy utilization obtains higher favorable well-adjusted [1] efficiency and environmentally. The solar photovoltaic (SPV) compromise sustains preserved, clean, decisive non-conventional energy resources [2] during the further use of zero fuel imputation [3], no expressive competence, inferior operational regulation, ultimate dependent protection with the expanded investigation [4]. The particular points of consequence frame the exhaustion of SPV in a different situation more in off-grid design [5]. Multilevel inverter terminology is substantiated during universal exploration recognition over the researcher with front-end enterprise in a particular intermediate along with high energy appliance [6] because of their suitability to produce an enormous condition of output waveforms [7], condensation switching stress completed the switches counting decided to switch frequency [8].

The current document promotes a narrative single-phase grid-associated MLI with a PV panel during an enhanced converter, using a substitute for DC resources in MLI [9]. The proposed technology needs fewer switches while interconnected to the critical appearance of MLI [10]. A cascaded multilevel inverter is a variety of contribution voltage sources through output voltage levels be located taking place the implication [11]. The existing inverter grouping is output voltage obtained initiation special input voltage connection with the renewable power utilization [12].

The exploration of cascaded multilevel inverter during switching states of a seven-level inverter is extensive in Sect. 2. The modes of procedure for customized topology are exhaustive in Sect. 3. The outcomes along with analysis are analyzed in Sect. 4. The outcome is correlated in Sect. 5.

2 Cascaded H-Bridge (CHB) MLI

2.1 Cascaded Seven-Level Inverter

CHB inverters are multilevel inverters that comprise a sequence complex of two or more single-phase CHB inverters. Every CHB interacts with three voltage source phase legs, supposing that the line-line voltage endures the converter output. Consequently, a distinct CHB converter is proficient in attaining three different voltage levels. Every segment acquires two reachable switching states to prevent the DC link capacitor from short-circuit. Figure 1 illustrates three H-bridges connected through its series, with an estimated appearance of its potential creature's four-level output voltages. The complete converter output voltage obtains excessively ornamented, generating seven differential voltage levels. In typical conditions, as of n CHB in series, $2n + 1$ different voltage levels continue to be attained through an eventual output voltage of nVdc.

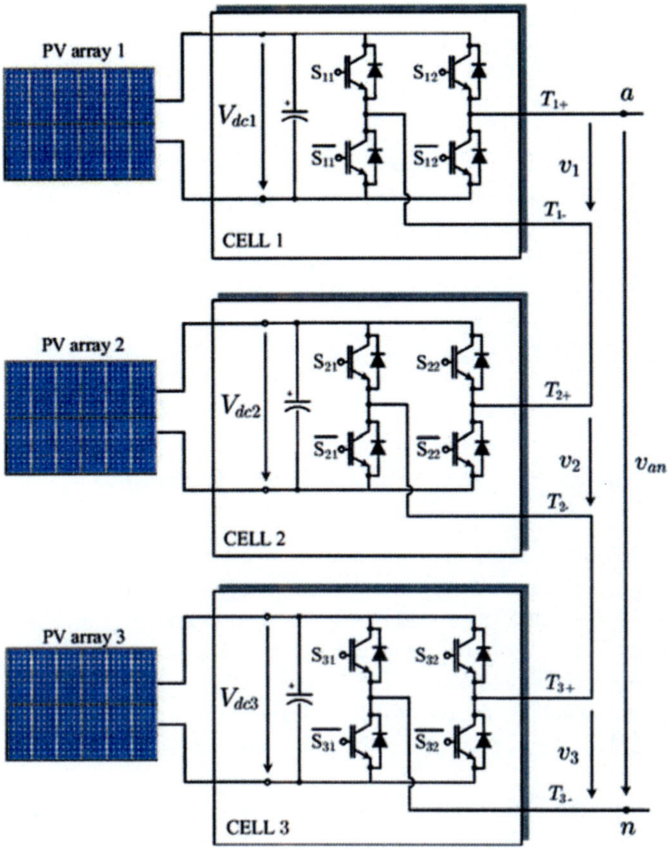

Fig. 1 Power path by solar PV unit

2.2 *Projected Seven-Level Cascaded H-Bridge MLI*

This exertion, projected CHB seven-level inverters together in sequence, requires similar DC input voltages. A cell's suitable assortment of voltage asymmetry can embrace fabrication at various voltage levels measured by diminishing replication. As shown in Fig. 2, an asymmetry when an unconventional voltage ratio of 1:3 destined to facilitate a 2-cell cascaded H-bridge (CHB) has headed for recognition seven levels expert for the period of a 2-cell CHB in Fig. 1.

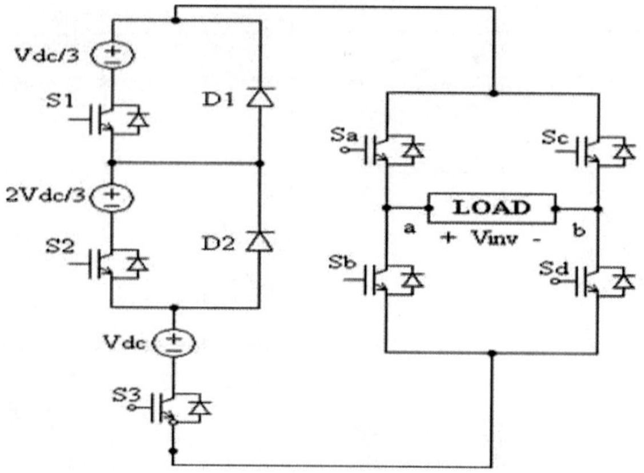

Fig. 2 Designed seven-level cascaded H-bridge MLI

3 Modes of the Process of the Designed Technique

The switching development of a seven-level inverter is defined in Table 1. The production levels continue unidirectional among CHB, altering the production by bidirectional. The output waveforms of the cascaded H-bridge acquire seven levels, as endorsed in Fig. 2. Switches such as S_a and S_d are ON through all during the negative half cycle, S_b and S_c are ON in the CHB MLI.

Appropriate switching relating to the inverter can yield seven output voltage levels ($V_{dc}/3, 2V_{dc}/3, V_{dc}, 0, -V_{dc}/3, -2V_{dc}/3, -V_{dc}$) origination the DC contributes voltage.

Where '0' is OFF and '1' is ON. Table 1 projects the switching progression to extend seven output voltage levels ($V_{dc}/3$, $2V_{dc}/3$, V_{dc}, 0, $-V_{dc}/3$, $-2V_{dc}/3$, $-V_{dc}$).

Table 1 Production voltage present to the switches ON–OFF position

V_0	S_1	S_2	S_3	S_a	S_b	S_c	S_d	D_1	D_2
$V_{dc}/3$	0	0	1	1	0	0	1	1	1
$2V_{dc}/3$	0	1	1	1	0	0	1	1	0
V_{dc}	1	1	1	1	0	0	1	0	0
0	0	0	0	1	0	0	1	1	1
0*	0	0	0	0	1	1	0	1	1
$-V_{dc}/3$	0	0	1	0	1	1	0	1	1
$-2V_{dc}/3$	0	1	1	0	1	1	0	1	0
$-V_{dc}$	1	1	1	0	1	1	0	0	0

4 Results and Discussion

4.1 Proposed Arrangement Simulation Results

The designed arrangement furnished in Fig. 3 over MATLAB/SIMULINK environment terminated the SIMPOWER arrangement toolbox with output waveform as demonstrated in Fig. 4. Moreover, THD is also characterized in Fig. 5, % THD of 7 level CHBMLI is 9.40%.

The percentage reduction of THD using 7 level modified MLI when compared to 5 level modified MLI is indicated in Table 2.

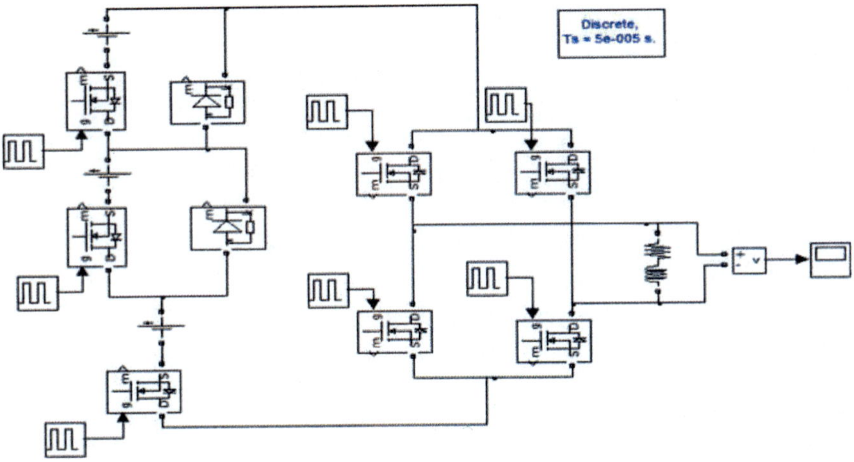

Fig. 3 Simulation circuit intended for designed cascaded H-bridge seven-level inverter

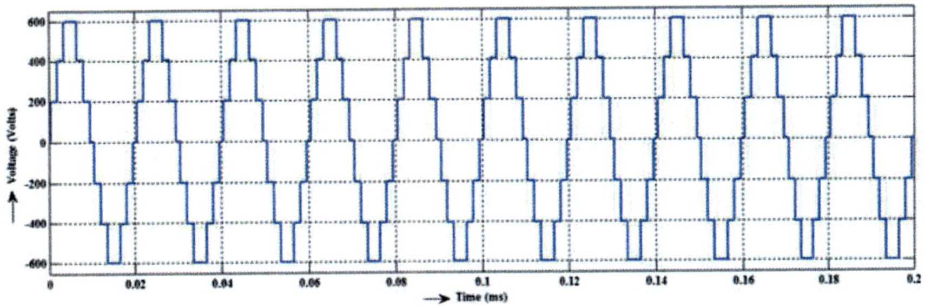

Fig. 4 Productivity voltage intended as designed cascaded H-bridge seven-level inverter

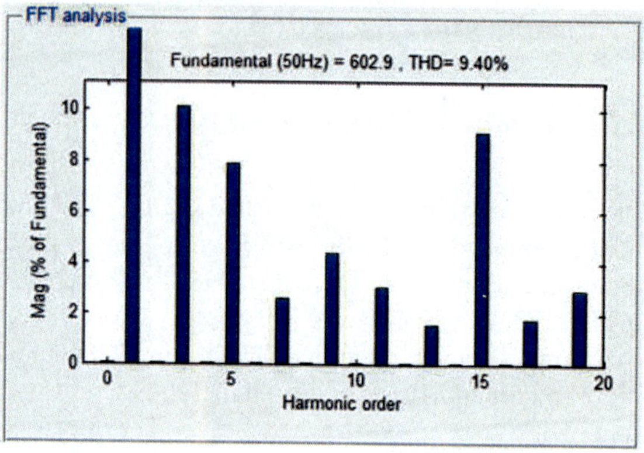

Fig. 5 THD intended in favor of designed cascaded H-bridge seven-level inverter

Table 2 Comparison between modified MLI

S. No.	Voltage level	No. of switches	% THD
1	5	6	11.91
2	7	7	9.40

4.2 Proposed Arrangement PROTEUS Setup

Proteus is one of the great conventional simulators. It conserves used to construct all circuits on stimulating fields. It is symmetrical through promoting the graphical user interface to constitute probably correlated to the substantial prototype panel, as depicted in Fig. 6. Further, it also preserves the waveform to produce printed circuit boards, as demonstrated in Fig. 7.

5 Conclusion

This manuscript has projected the designed cascaded H-bridge MLI over limited power switches in MATLAB/SIMULINK and PROTEUS software demonstration. The single PWM arrangement declines the complexity in the pulse generation periphery, including reducing the THD with a better power factor, this is because of reducing the number of switches in the present CHB 7 level inverter. Suitable for the distinctive presence, the suggested topology is executive to the conventional method. Thus, single PWM is conveniently used for the recommended MLI. Multilevel inverters consist of a developed initiatory, a developing technology against a well-identified, attractive intermediate voltage constancy, and immense power utilization. The device's extended enhancement, including industrial operation, progress firmness contemporary challenges with convenience through stimulating further development of cascaded H-bridge MLI technology. The simulation outcome in engagement

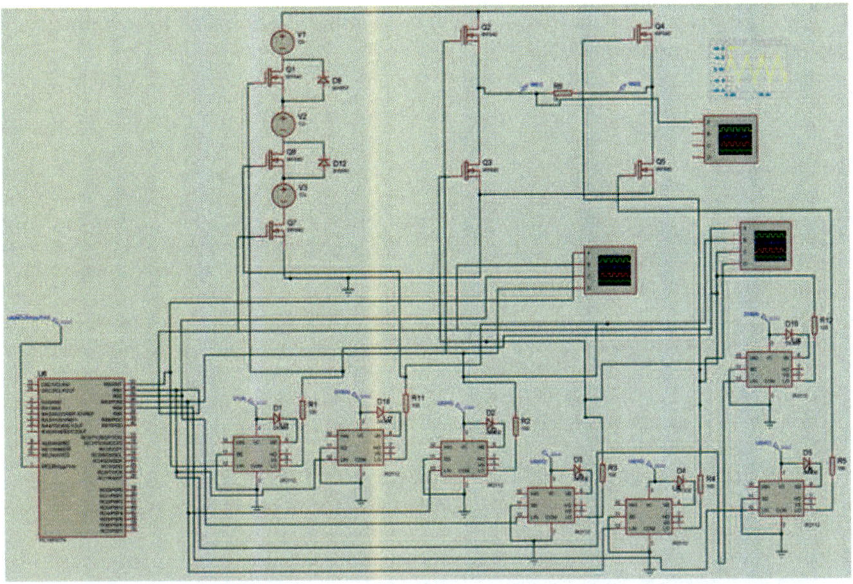

Fig. 6 Expansion of designed cascaded H-bridge seven-level inverter in PROTEUS

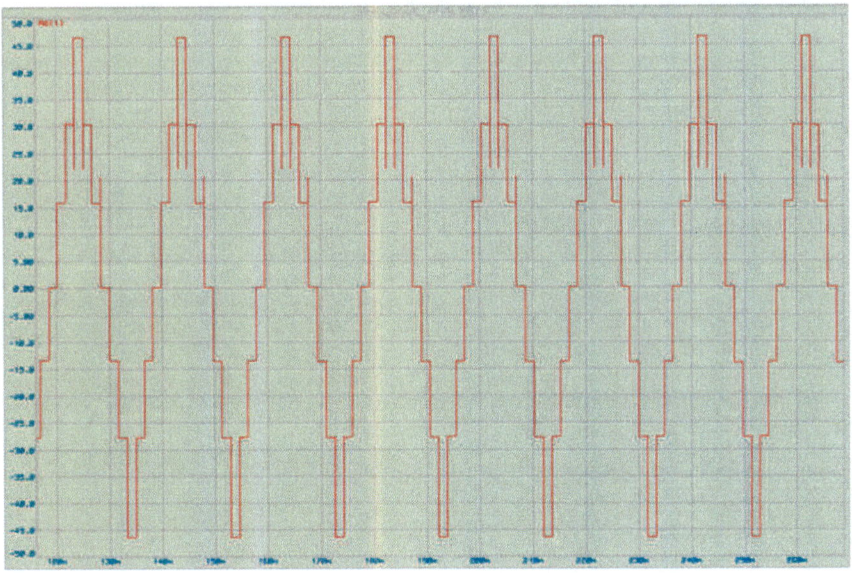

Fig. 7 Output voltage for scheduled cascaded H-bridge seven-level inverter

well at the time subsequent PROTEUS. This layout is anticipated to be immensely convenient in immense power utilization.

References

1. Bharatiraja C, Lakshmikhandan K, Kamalesh MS, Rajasekaran D, Twala B (2021) A non-isolated high-gain DC to DC converter connected multi-level inverter for photo-voltaic energy sources. J Appl Sci Eng 24(3):415–422
2. Huang Q, Huang AQ, Yu R, Liu P, Yu W (2018) High-efficiency and high-density single-phase dual-mode cascaded buck–boost multilevel transformerless PV inverter with GaN AC switches. IEEE Trans Power Electron 34(8):7474–7488
3. Lashab A, Sera D, Hahn F, Camurca L, Terriche Y, Liserre M, Guerrero JM (2020) Cascaded multilevel PV inverter with improved harmonic performance during power imbalance between power cells. IEEE Trans Ind Appl 56(3):2788–2798
4. Henao-Bravo EE, Ramos-Paja CA, Saavedra-Montes AJ, González-Montoya D, Sierra-Perez J (2020) Design method of dual active bridge converters for photovoltaic systems with high voltage gain. Energies 13(7):1711
5. Sheir A, Sood VK (2019)Cascaded modular converter with reduced output voltage ripple. In: 2019 IEEE electrical power and energy conference (EPEC). IEEE, pp 1–7
6. Suroso S, Noguchi T (2014) A single-phase multilevel current-source converter using H-bridge and DC current modules. Int J Power Electron Drive Syst 4(2):165
7. Gorji SA, Sahebi HG, Ektesabi M, Rad AB (2019) Topologies and control schemes of bidirectional DC–DC power converters: an overview. IEEE Access 7:117997–118019
8. Dinakaran C, Padmavathi T (2021) Performance evaluation on several string multilevel inverter considering hybrid renewable energy systems. In: 2021 Innovations in power and advanced computing technologies (i-PACT). IEEE, pp 1–8
9. Pradhan AK, Kar SK, Mohanty MK, Behra N (2021) Design and simulation of cascaded and hybrid multilevel inverter with reduced number of semiconductor switches. Int J Ambient Energy 42(8):950–960
10. Lakshmi TV, George N, Umashankar S, Kothari DP (2013) Cascaded seven level inverter with reduced number of switches using level shifting PWM technique. In: 2013 International conference on power, energy and control (ICPEC). IEEE, pp 676–680
11. Dinakaran C, Padmavathi T (2022) Simulation of single-phase cascaded H-bridge multilevel inverter with non-isolated converter for solar photovoltaic system. In: International conference on electrical and electronics engineering. Springer, Singapore, pp 1–12
12. Siddique MD, Mekhilef S, Shah NM, Sarwar A, Memon MA (2020) A new single-phase cascaded multilevel inverter topology with reduced number of switches and voltage stress. Int Trans Electr Energy Syst 30(2):e12191

An Adaptive and Efficient VSC-MTDC Controller for DC Voltage Regulation and Frequency Stability in AC Network

Ashima Taneja, Radheshyam Saha, and Madhusudan Singh

Abstract In modern power systems, application of VSC-based multi-terminal high voltage direct current (mVSCHVDC) technology is showing a growing research interest with significant power infeed from various renewable energy resources. In effect, mVSCHVDC grid coupled with AC systems enables to enhance controllability, reliability, and voltage-frequency stability of interconnecting networks with an enabling scenario for flexible sharing of energy reserves in power markets among the AC networks. The research reveals that during a frequency excursion in an AC-system, dual-droop control strategies, i.e., power and frequency droops in mVSCHVDC exhibit limited frequency support capability due to negative impact of DC voltage droop. Also, frequency deviations in the healthier networks are held responsible for limited power support from dual-droop controlled mVSCHVDC. In order to have adequate frequency support from mVSCHVDC grid, a new adaptive dual-droop converter control strategy is proposed in this paper that makes a paradigm shift in mVSCHVDC control from the earlier principles reported. Also, it nullifies the impact of DC voltage droop and impact of frequency deviations as well in other networks on the frequency support capability of the converter control. The effectiveness of the proposed control is justified and demonstrated by triggering a significant load increase event in one of the AC areas of a five-terminal mVSCHVDC system simulated on MATLAB/Simulink platform.

Keywords Dual-droop · Droop interaction · Communication-free support

1 Introduction

With several favorable control characteristics presented by VSC-HVDC technology, it has become a preferred choice for configuring multi-terminal high voltage direct current (mVSCHVDC) grid networks [1]. Such an integration of various far-located

A. Taneja (✉) · R. Saha · M. Singh
Delhi Technological University, New Delhi, India
e-mail: tanejaashima.1988@gmail.com

A. Rani et al. (eds.), *Signals, Machines and Automation*, Lecture Notes in Electrical Engineering 1023, https://doi.org/10.1007/978-981-99-0969-8_5

converters into a common HVDC grid increases overall reliability by offering convenience of power flow on different transmission paths. Also, mVSCHVDC systems allow prosumers or consumers-suppliers to have multiple choices for power trading. In addition, such DC grids can be enabled to enhance system security by allowing sharing primary reserves among their integrated AC networks [2].

Existing literature demonstrates the methodology for mitigating frequency excursion via mVSCHVDC systems by utilizing spinning reserves of the interconnected healthier AC networks [2–7]. A dual-droop control technique using DC voltage–power-frequency droop has been proposed in [2] for bidirectional control of converters constituting mVSCHVDC grid so that integrated AC–DC grid can maintain its DC voltage stability as well as participate in mitigating frequency excursion issues of the interconnecting AC networks. However, limited frequency support is provided by the dual-droop control technique due to opposite impact of DC voltage droop on frequency droop. A correction factor for frequency droop coefficient has been proposed in [3] so that the impact of DC voltage droop on frequency droop can be neutralized. However, it is based on the assumption of negligible frequency deviations in supporting networks. In [4], it is proposed to exercise only power–frequency droop of dual-droop controller in VSC interfaced AC areas undergoing frequency disturbance. But such demarcation is based on rate of change of frequency (RoCoF) for differentiating disturbed and supporting networks, which becomes ambiguous if a severe frequency excursion occurs. Application of an inertial droop in respect of RoCoF along with the dual-droop control strategy in converter control of mVSCHVDC system has been proposed in [5], but it does not consider the interactions among the droop. But it involves additional requirement of electrical parameters to be communicated to converter control instead of performing local measurements. An improved dual-droop control technique has been proposed in [6] to reduce the negative impact of DC voltage droop but it has the same complexity of communicating the required parameters same as that of [5]. [7] proposes a non-linear dual-droop control for mVSCHVDC but has limited active power support capability due to mutual interaction among the droops. [8, 9] proposes an interaction-free dual-droop control technique for bipolar-configured VSC-MTDC system which uses P-V_{dc} droop control at one pole and P-f droop control at the other. Although, better frequency regulation can be obtained with respect to traditional dual-droop control [2], but it requires a dedicated slack converter in the grid for power balancing. Also, AC area with this slack VSC is also not allowed to withdraw frequency support from the MTDC grid. Few papers [10, 11] suggest to use electrostatic energy stored in station capacitors for frequency regulation. But, for appreciable support to be withdrawn, it requires use of either extra high DC transmission voltage or super capacitors.

The research work formulated in this paper makes an attempt to improve the efficiency of frequency support achieved from mVSCHVDC grid by proposing an adaptive dual-droop control technique for grid converter terminals with interconnected AC networks participating in maintaining DC voltage and frequency stability. The proposed control aids in nullifying the opposite impact imposed by DC voltage droop on the frequency droop. And, it eliminates the negative impact of frequency deviations (occurring in the supporting networks) upon the frequency support taken by the disturbed network from the DC grid. It is worth noting that the proposed work relies only upon local measurements and is, thus, reliable over the communication-based frequency support techniques proposed in the literature [5, 6]. Factoring the above, a design, simulation, and modeling for a five-terminal, mVSCHVDC grid have been carried out on MATLAB/Simulink platform for bringing forward efficiency of proposed work over traditional techniques.

This paper is outlined as: Sect. 2 presents the inefficiency of frequency support obtained from mVSCHVDC grid controlled via traditional dual-droop control technique. Section 3 presents the proposed control technique for obtaining enhanced frequency support. Section 4 presents the system configuration and provides the results obtained in case of frequency contingency. This paper is concluded in Sect. 5.

2 Dual-Droop Control for Multi-terminal HVDC Grid

The generation of reference direct-axis converter current (i_d^*) for the dual-droop control used with standard vector current control is shown in Fig. 1. ρ_v as DC voltage droop and, ρ_f is the power-frequency droop coefficient, for ith VSC of an n-terminal,

$$P_{\text{vsc}.i} = P^*_{\text{vsc}.i} + \rho_{v.i}\left(V^*_{\text{dc}.i} - V_{\text{dc}.i}\right) - \rho_{f.i}\left(f_i^* - f_i\right) \tag{2.1}$$

Considering a load/generation unbalance in ith AC network,

$$\Delta P_{\text{vsc}.i.pu} = \rho_{f.i}.\Delta f_{i.pu} - \rho_{v.i}.\Delta V_{\text{dc}.i.pu} \tag{2.2}$$

Adding the power deviation of all the n-converters,

$$\sum_{i=1}^{n} \Delta P_{\text{vsc}.i.pu} = \sum_{i=1}^{n} \rho_{f.i}.\Delta f_{i.pu} - \sum_{i=1}^{n} \rho_{v.i}.\Delta V_{\text{dc}.i.pu} \tag{2.3}$$

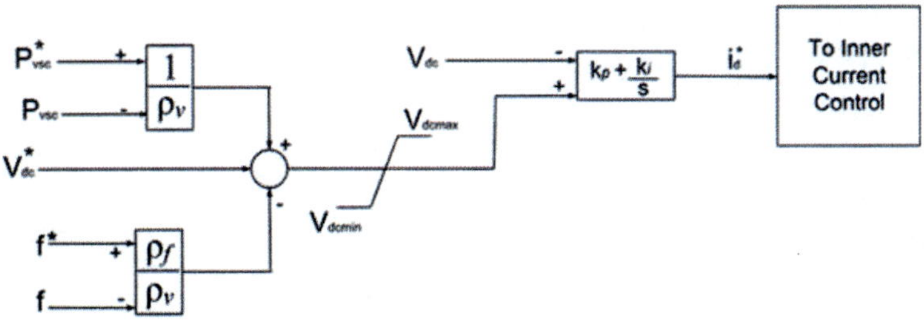

Fig. 1 Generation of i_d^* using the dual-droop control for grid converters

Assuming negligible losses along with equal DC voltage deviation,

$$\Delta V_{dc.pu} = \frac{\sum_{i=1}^{n} \rho_{f.i}.\Delta f_{i.pu}}{\sum_{i=1}^{n} \rho_{v.i}} \tag{2.4}$$

Substituting (2.4) in (2.2) and rearranging,

$$\Delta P_{vsc.i.pu} = \rho_{f.i}.\Delta f_{i.pu}\left(1 - \frac{\rho_{v.i}}{\sum\limits_{i=1}^{n} \rho_{v.i}}\right) - \frac{\rho_{v.i}}{\sum\limits_{i=1}^{n} \rho_{v.i}}.\sum_{\substack{j=1\\ j\neq i}}^{n} \rho_{f.j}.\Delta f_{j.pu} \tag{2.5}$$

Thus, change in power deviation made by the disturbed VSC depends on its local frequency deviation ($\Delta f_{i.pu}$) and the two droop coefficients and frequency deviations occurring in other (supporting) areas ($\Delta f_{j.pu}$). Ideally, change in power flow made by ith VSC:

$$\Delta P_{vsc.i.pu.ideal} = \rho_{f.i}.\Delta f_{i.pu} \tag{2.6}$$

But, actual change in converter power flow for such excursion is given by (2.5), so that,

$$\Delta P_{vsc.i.pu.actual} = \Delta P_{vsc.i.pu.ideal}(1 - k_i) - k_i.\sum_{\substack{j=1\\ j\neq i}}^{n} \rho_{f.j}.\Delta f_{j.pu} \tag{2.7}$$

where $k_i = \frac{\rho_{v.i}}{\sum_{i=1}^{n} \rho_{v.i}}$. It may be noted that subscript 'i' corresponds to disturbed AC network while 'j' is for AC networks supporting the disturbed network, and the ratio 'k_i' is always less than 1.

Also, from (2.6) that $\Delta P_{vsc.i.pu.actual}$ is lesser than $\Delta P_{vsc.i.pu.ideal}$ due to:

a. Degradation of $\Delta P_{vsc.i.pu.ideal}$ by factor '$1 - k$'.
b. Impact of frequency deviation, Δf_j occurring in healthier areas while supporting the disturbed network.

Thus, limited active power gets injected in to the disturbed AC network from mVSCHVDC grid. For a VSC interfaced with the supporting network,

$$\Delta P_{vsc.j.pu} = \rho_{f.j}.\Delta f_{j.pu} - \rho_{v.j}.\Delta V_{dc.j.pu} \tag{2.8}$$

But, for supporting VSCs, change in power flow is a result of local DC voltage deviation,

$$\Delta P_{vsc.j.pu.ideal} = \rho_{v.j}.\Delta V_{dc.j.pu} \tag{2.9}$$

Using (2.9) in (2.8),

$$\Delta P_{\text{vsc}.j.pu.\text{actual}} = -\left(\Delta P_{\text{vsc}.j.pu.\text{ideal}} - \rho_{f.j}.\Delta f_{j.pu}\right) \tag{2.10}$$

Thus, power support offered by any *j*th network to the disturbed (*i*th) network via mVSCHVDC grid gets lessened by the power withdrawn by the *j*th network in itself in response to the frequency deviations, $\Delta f_{j.pu}$ occurring in it.

3 Proposed Dual-Droop Control Strategy

Since power deviation made by the disturbed VSC is opposite to that of the supporting VSC, thus VSC controller can distinguish whether the interconnected AC area is disturbed or supporting area considering no converter outages and also no request for any power flow change, i.e., $\Delta P_{\text{vsc}}^{*} = 0$. It is proposed that whenever converter controller identifies a disturbed area and its frequency falls out of the dead-band range, it should use

$$\rho_{f.\text{modified}} = \frac{\rho_f}{1-k} \tag{3.1}$$

Since $k < 1$, thus, $\rho_{f.modified}$ is more than ρ_f. However, if the controller identifies a supporting area, then it reduces its ρ_f to zero and should exercise only its DC voltage droop. This ensures supporting area is not withdrawing any power support while it is supporting the disturbed area. Table 1 provides the droop coefficients for three possible scenarios: steady state, supporting state, and disturbed state.

Thus, using Table 1, net change in power flow becomes

For disturbed VSC,

$$\Delta P_{\text{vsc}.i.pu.\text{actual}} = \Delta P_{\text{vsc}.i.pu.\text{ideal}} \tag{3.2}$$

For supporting VSC,

$$\Delta P_{\text{vsc}.j.pu.\text{actual}} = -\Delta P_{\text{vsc}.j.pu.\text{ideal}} \tag{3.3}$$

Table 1 Droop coefficients with proposed control for mVSCHVDC system

State of AC network→	Steady state	Disturbed state	Supporting state
ρ_v	Retained as original	Retained as original	Retained as original
ρ_f	Retained as original	Calculated from (3.1)	Zero

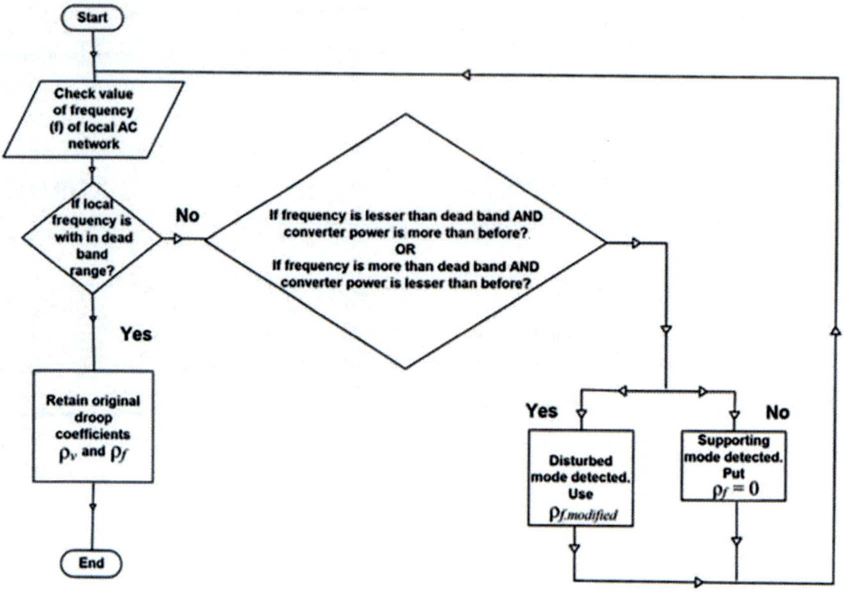

Fig. 2 Flowchart for selecting frequency droop coefficient in the dual-droop controller

Once frequency of the AC areas recovers back within the dead-band range, original frequency droop coefficients are restored. The algorithm for selecting the frequency droop coefficient in the proposed dual-droop controller is represented by flowchart in Fig. 2.

4 Simulations and Result

Figure 3 shows the proposed configuration of five-terminal mVSCHVDC grid system simulated on MATLAB/Simulink platform. All the three onshore converters are dual-droop controlled (DDC). AC–DC systems parameters are listed in Table 2. VSC associated with the wind farms are considered to be in constant power control (CPC) mode.

Three distinct dual-droop control case studies are made:

Case-1: DDC-mVSCHVDC using desired values of ρ_f and ρ_v.

Case-2: Modified DDC-mVSCHVDC using $\rho_{f.\text{modified}}$.

Case-3: Proposed DDC-mVSCHVDC using the droop coefficients given in Table 1.

Simulink representation for selection of ρ_f by proposed controller at VSC-1 is shown in Fig. 4. The frequency support offered by these three control schemes is compared by initiating a load increase of 10% at $t = 45$ s, in ONG-1. Figure 5 shows frequency of three AC grids. With proposed control technique, frequency

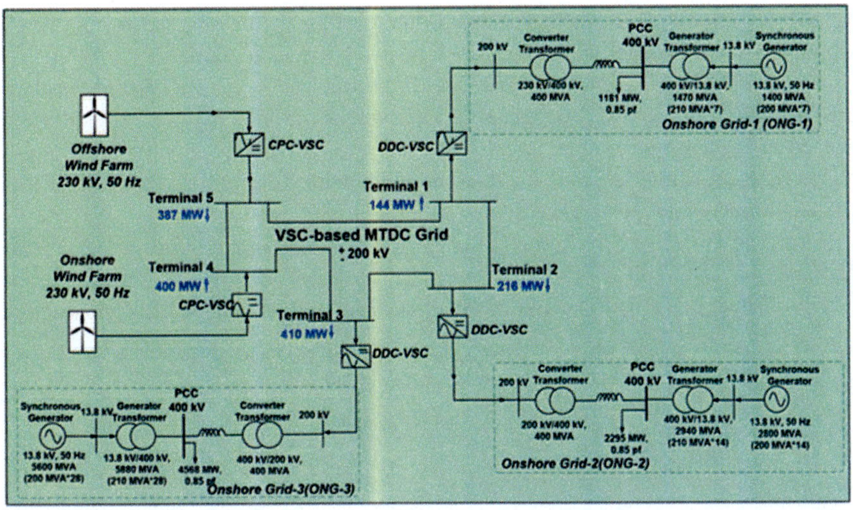

Fig. 3 Five-terminal mVSCHVDC system coupled with AC networks

Table 2 System parameters

System components	Ratings	
Onshore grids (ONG-1, ONG-2, and ONG-3)		
Power generation in ONG-1, ONG-2, and ONG-3	1050 MW, 2100 MW, and 4200 MW	
Governor droop and inertia time constant	0.05 pu, 3.2 s	
Onshore grid converters		
ρ_{dc} for VSC-1,VSC-2, and VSC-3 controllers	0.3, 0.2, 0.1	
Frequency droop coefficient ($\rho_f * \rho_{dc}$)	Desired value	Modified value
VSC-1,VSC-2, and VSC-3	30, 20, 10	36.67, 27.5, 22

nadir is better (49.87 Hz) in comparison with the other techniques (49.72 Hz for Case-1 and 49.59 Hz for Case-2). Although modified frequency droop coefficients of (3.1) are used in Case-2 for negating the impact of DC voltage droop, still its frequency nadir is worse than Case-1 in which no modification is applied. This is because in Case-2, $\rho_{f\text{modified}}$ (higher than desired) is used for all the VSCs (both disturbed as well as supporting). As a result of which, the supporting VSCs could only make limited power deviation as followed by (2.10). Thus, lesser amount of power is inverted by VSC-1 in Case-2 as compared to Case-1. However, for Case-3, since $\rho_{f\text{modified}}$ is used only for VSC-1 terminal along with its value being zero for the supporting VSC terminals (VSC-2 and VSC-3), it enables to withdraw more power from mVSCHVDC grid from supporting networks (ONG-2 and ONG-3). Figure 6 shows the power inverted by the three onshore converters of the mVSCHVDC grid. As explained above, by using the proposed scheme, VSC-1 is able to withdraw more power as shown in Fig. 6a with respect to other two cases. Also, inverted power by

VSC-1 for Case-2 is lesser than Case-1 due to $\rho_{f.\mathrm{modified}}$ being used for all VSCs which are having higher values than their respective desired values.

Figure 7 shows local DC voltage deviation as observed at VSC-1. As higher values are used in Case-2 as $\rho_{f.\mathrm{modified}}$ and also, its frequency drop being more (as shown in Fig. 5) and as asserted mathematically by (2.4), the DC voltage deviation for Case-2 is more than the other cases. Deactivation of frequency droop for supporting VSCs in Case-3 makes its DC voltage deviation least among the three cases.

From Fig. 6b, power deviation made by VSC-2 is maximum in proposed scheme because only DC voltage droop is exercised by it while supporting excursion of ONG-1. Thus, it is able to make power deviation as per its ideal value. Even though

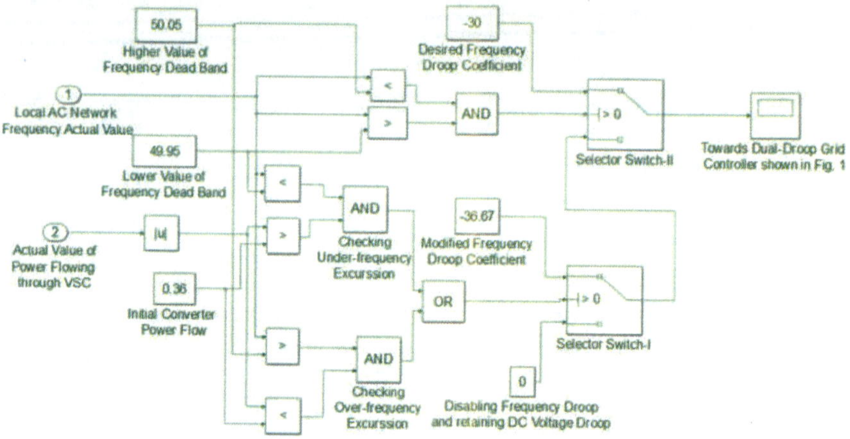

Fig. 4 Simulink model for selection of ρ_f by the proposed controller at terminal-1

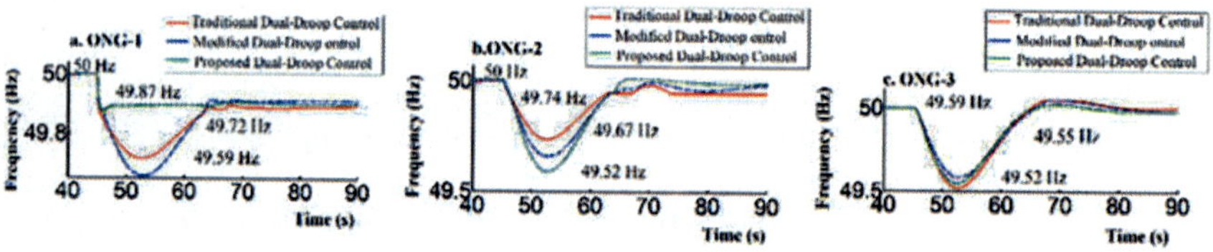

Fig. 5 Frequency of **a** ONG-1, **b** ONG-2, **c** ONG-3 after applying load increase in ONG-1

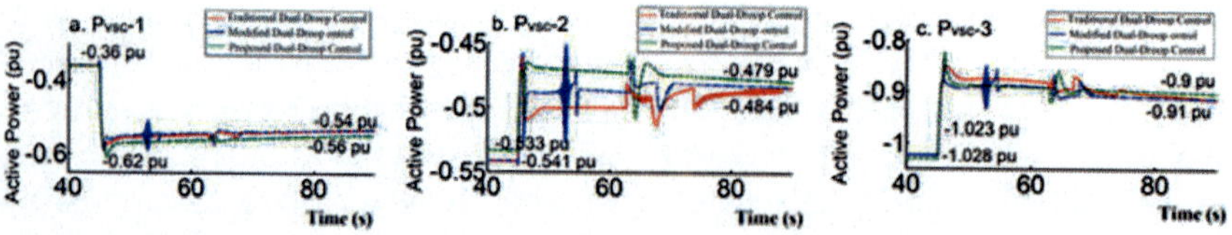

Fig. 6 Power inverted by **a** VSC-1, **b** VSC-2, and **c** VSC-3 after load increase in ONG-1

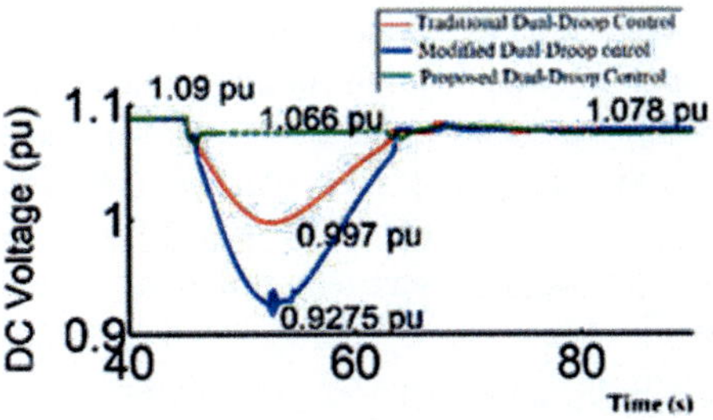

Fig. 7 Local DC bus voltage at VSC-1 after load increase in ONG-1

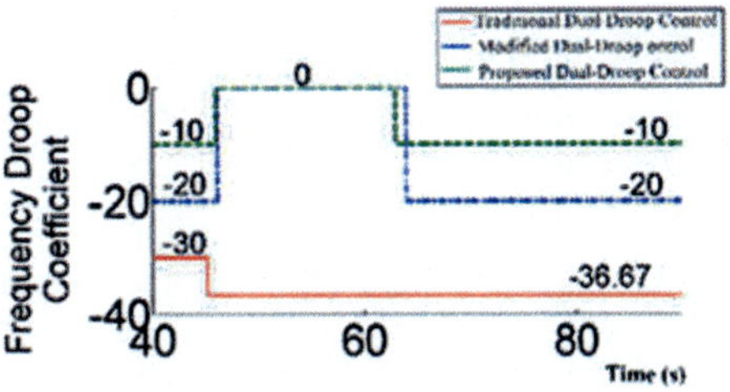

Fig. 8 Frequency droop coefficients exercised by the proposed controller

higher frequency droop coefficients are used in Case-2 than Case-1, still its power deviation is more than Case-1. This is justified by higher DC voltage deviation of Case-2.

From Fig. 6c, power deviation made by VSC-3 is more in Case-2 than in Case-3 even though proposed scheme disables its frequency droop for supporting VSCs. This is justified again due to higher DC voltage deviation of Case-2 than Case-3. At the same time, power deviation of Case-1 is even higher than Case-2. This is because in Case-1, desired values of frequency droop coefficients are employed which are lower than in Case-2. Figure 8 shows ρ_f exercised by proposed controller. It can be seen that modified (higher value = 36.67) is used for the disturbed VSC at terminal-1 only. Also, supporting VSC terminal-2 and VSC terminal-3 exercise only DC voltage droop by making their frequency droop coefficient to 0 during excursion.

5 Conclusion

In this paper, a new adaptive dual-droop control technique for VSC- MTDC grid has been designed for mitigation of frequency excursions in the interconnected AC networks. It has transpired that the frequency support offered by conventional dual-droop control is limited, firstly, on account of mutual interaction between DC voltage droop and frequency droop and secondly, due to the frequency deviations occurring in the supporting networks. The paper has demonstrated the mitigation measure of the unmet effect of DC voltage droop on frequency droop and has shown a new control strategy for nullifying the impact of DC voltage droop by designing appropriate frequency droop coefficient in VSC terminals associated to the disturbed area. In addition, the limited active power supporting capability of conventional dual-droop control in respect of any frequency deviation in the supporting areas has been considerably improved by use of just DC voltage–power droop in VSCs coupled supporting AC areas. The proposed adaptive control strategy has been tested on a typical five-terminal configuration of VSC-based MTDC grid on MATLAB/Simulink platform by creating an event of sudden large load increase in one of the AC networks. The results confirm that the proposed dual-droop control strategy helps in obtaining much-needed frequency support over the earlier control strategies.

References

1. Saha R, Singh M, Taneja A (2021) An adaptive master-slave technique using converter current modulation in VSC-based MTDC system. In: 2021 6th International conference for convergence in technology (I2CT), pp 1–8. https://doi.org/10.1109/I2CT51068.2021.9418191
2. Chaudhuri NR, Majumder R, Chaudhuri B (2013) System frequency support through multi-terminal DC (MVSCHVDC) grids. IEEE Trans Power Syst 28(1):347–356. https://doi.org/10.1109/TPWRS.2012.2196805
3. Akkari S, Dai J, Petit M, Guillaud X (2015) Coupling between the frequency droop and the voltage droop of an AC/DC converter in an MVSCHVDC system. IEEE Eindhoven Power Tech 2015:1–6. https://doi.org/10.1109/PTC.2015.7232285
4. Jose K, Adeuyi O, Liang J, Ugalde-Loo CE (2018) Coordination of fast frequency support from multi-terminal HVDC grids. In: 2018 IEEE international energy conference (ENERGYCON), pp 1–6. https://doi.org/10.1109/ENERGYCON.2018.8398852
5. Vennelaganti SG, Chaudhuri NR (2018) Selective power routing in MVSCHVDC grids for inertial and primary frequency support. IEEE Trans Power Syst 33(6):7020–7030. https://doi.org/10.1109/TPWRS.2018.2854647
6. Zhang Q et al (2021) Primary frequency support through North American continental HVDC interconnections with VSC-MVSCHVDC systems. IEEE Trans Power Syst 36(1):806–817. https://doi.org/10.1109/TPWRS.2020.3013638
7. Shadabi H, Kamwa I (2021) Dual adaptive nonlinear droop control of VSC-MVSCHVDC system for improved transient stability and provision of primary frequency support. IEEE Access 9:76806–76815. https://doi.org/10.1109/ACCESS.2021.3078066
8. Kumar AS, Padhy BP (2020) An interaction less duo control strategy for bi-polar voltage source converter in renewables integrated multi-terminal HVDC (MTDC) grids. In: 2020 IEEE international conference on power electronics, drives and energy systems (PEDES), pp 1–5. https://doi.org/10.1109/PEDES49360.2020.9379680
9. Ancha SK, Padhy BP, An interaction less duo control strategy for bi-polar voltage source converter in renewables integrated multi-terminal HVDC grids. IEEE Trans Ind Appl. https://doi.org/10.1109/TIA.2022.3172065
10. Shen Z et al (2022) Variable-inertia emulation control scheme for VSC-HVDC transmission systems. IEEE Trans Power Syst 37(1):629–639. https://doi.org/10.1109/TPWRS.2021.3088259
11. Zhu J et al (2022) Inertia emulation and fast frequency-droop control strategy of a point-to-point VSC-HVdc transmission system for asynchronous grid interconnection. IEEE Trans Power Electron 37(6):6530–6543. https://doi.org/10.1109/TPEL.2021.3139960

Planning of Load Frequency Control in Two Degrees of Freedom PID Controller Using MFO Technique in 2 Area Power System

Mukesh Pushkarna, Haroon Ashfaq, and Rajveer Singh

Abstract In this paper, the load frequency regulation of a two-area system is investigated in detail. In this research, two regions have been explored in which non-reheated types of turbines are utilized in both areas and in which a new controller known as 2-DOF PID is used in the secondary loop of the secondary system (2-DOF PID). The settings of this controller have been optimized using the newest meta-heuristic approach, commonly known as the moth flame optimization algorithm (MFO), in order to decrease the variance in frequency of both area and tie-line power. The same procedure is followed by PID and integral controllers, whose settings are modified in the same way by the MFO. When the results of these experiments are examined, it is discovered that the 2-DOF PID controller outperforms the other controller in terms of minimizing the variation in frequency of both area and tie-line power.

Keywords Teaching learning-based optimization (TLBO) · Moth flame optimization (MFO)

1 Introduction

The power system is a complicated system that consists primarily of three structure: generation, transmission, and distribution. It consists of variable load, which varies at all times and must be maintained at the set level. Load frequency control (LFC) is a concept that has been introduced here. The frequency variation will be controlled by LFC, which will keep it at its usual value. The frequency will change as the load changes. It will increase if the load lowers and drop if the load increases, resulting in load frequency control. The primary and secondary loops are the two loops in LFC.

The primary loop is unable to restore the frequency to its previous value, necessitating the usage of a secondary loop to accomplish this task. An interconnection line

M. Pushkarna (✉) · H. Ashfaq · R. Singh
Department of Electrical Engineering, Jamia Millia Islamia University, New Delhi, India
e-mail: mukesh.pushkarna@gla.ac.in

M. Pushkarna
Department of Electrical Engineering, GLA University Mathura, Chaumuhan, India

A. Rani et al. (eds.), *Signals, Machines and Automation*, Lecture Notes in Electrical Engineering 1023, https://doi.org/10.1007/978-981-99-0969-8_6

exists in a two-region system, which transports electricity from one area to another in order to satisfy the demands of the consumers. The tie line has two functions: First, it serves to interchange electricity across numerous places, and second, it serves to preserve the frequency. On the LFC, there have been hundreds of studies undertaken throughout the years. In terms of the most dependable controller design for AGC, the Elgerd and Fosa are the first to share their viewpoints. Chang et al. [1] investigates a number of different LFC controllers. The PID controller is well-known in the industry because of its dependability, simplicity, and ability to make on-site adjustments as necessary. An intelligence strategy for the AGC-linked power system, as defined in [1, 2], is being examined, as previously stated. The AGC is controlled and regulate by artificial bee colony (ABC) algorithm [3, 4], as it has the most effective and intelligent local and global hunt capabilities of any algorithm. In a distributed power system, proportional integral derivative (PID) and proportional integral (PI) controllers are used to govern automated generation control (AGC). In a distributed power system, the differential evolution algorithm [5] is used to govern the automatic gain control (AGC). The PSO approach is utilized to handle a wide range of engineering problems, and it may be employed in this situation to handle the optimal control and setting of AGC [6, 7]. The TLBO is used in this paper to implement a technique for AGC. The TLBO is described in detail in [7]. Based on the preceding literature study, [8–10] a two-degree-of-freedom controller will be implemented in a two-area system, with system parameters optimized using the MFO algorithm.

2 System Understudy

Discussions have taken place in this section on the two-area system of LFC [11, 12]. The two-area system is illustrated in Fig. 1. Speed controlling units, turbine units, generator units, and load units are found in both areas of the power system. A ΔP_{ref} (provided to the controller), a PD (load disruption of an area), and a ΔP_{tie} are the three inputs to each area (tie-line power). In each area, the outputs are the frequency deviation F and the area control error (ACE). As an example, consider this equation for ACE:

$$\mathrm{ACE} = \mathrm{B}\Delta\mathrm{F} + \Delta P_{\mathrm{Tie}} \tag{1}$$

The frequency bias parameter is denoted by the letter B in the above equation. A transfer function is used to represent the relationship between all three units in each region.

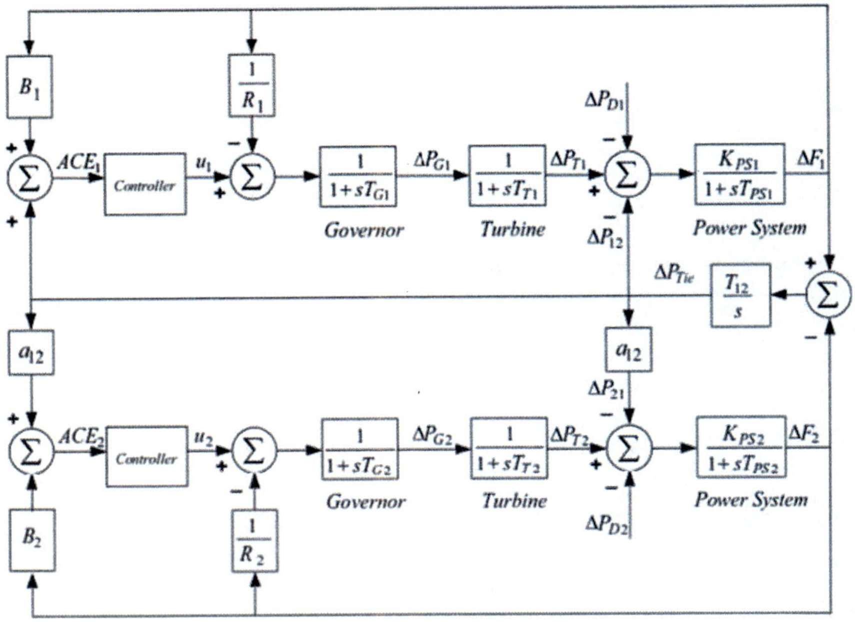

Fig. 1 Two-area system

3 2-DOF and PID Controller

When it comes to control systems, the degree of freedom can be interpreted as the number of different transfer function in closed-loop, which may be implemented independently.

A two-degree-of-freedom PID controller is capable of detecting a quick shock or disturbance without the need for a illustrative boost or maximizing in the pinnacle tracking process.

PID controllers with two degrees of freedom (DOF) are also useful since they may lessen the influence of the reference signal on the control system.

The traditional control architecture depicted in the preceding picture (2) makes use of a 2-DOF PID controller. The output is represented by u, and the two inputs, r and y, are depicted in parallel form. A diagram of the close loop TF of the 2-DOF PID is provided in the following equation (Fig. 2).

$$u = K_p(br - y) + \frac{K_i}{S}(r - y) + \frac{K_d S}{NS + 1}(cr - y) \tag{2}$$

where

K_p—This symbol represents proportionate gain.

K_i—Represent the gain of the integrator.

K_d is used to represent the derivative gain.

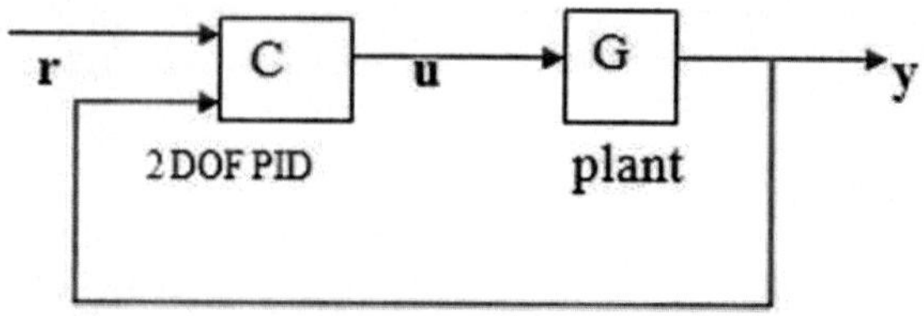

Fig. 2 2-DOF and PID controller

Filter gain is represented by the number N.

The pinnacle weight is represented by the proportional term, while the derivative weight is represented by the secondary term.

4 Moth Flow Algorithm

The optimization is a technical phrase that refers to the ability to identify the most appropriate method for a given problem. The complexity of the questions has increased day by day over the past few years, or a decade, and the necessity for a new optimization tool will become increasingly apparent in future days.

Currently, a new technique known as the MFO algorithm, presented by Mirhalili in 2014–15, is being employed and was energized by natural phenomena. It is intriguing to learn about moths, specifically about their directed navigation mechanism in the dark, which makes use of the moonlight. The majority of moths have a strong preference for flying throughout the night.

The moth flame optimization (MFO) method is the mathematical model that allows for the occurrence of these events to occur. Both the terms moths and flames accurately describe the situation. Specifically, the moth implies the driver looking operator, and flame implies the optimal place for the moth to fulfill that goal. As a result, each moth seeks around a flame and restores it if a more exact answer is available in contrast to the former one.

The proposed MFO algorithm, which estimates the global optimization of the issue, is three-tuple in nature and may be expressed as follows:

$$\text{MFO} = (I,\ P,\ T)$$

where I made a connection between the random number of moths and the corresponding fitness score. According to the following equations, this function may be expressed mathematically.

$$\text{I} : \emptyset \rightarrow \{M, OM\}$$

Specifically, the second word P is the primary function, which is to move with the moth around the circumferences to hunt for available space. This function receives the matrix M as input and uses it to calculate the value of the matrix and return it.

$$P : M \rightarrow M$$

True or false will be returned by the third function F. If the termination requirements are met, the program responds with a true signal, and if the termination criteria are not met, the program replies with a false signal.

$$T : M \rightarrow \{True, False\}$$

These two matrixes, denoted by the letters ub and lb, in which the letters ub and lb denote the top and bottom values of variables, respectively.

$$\mathrm{ub} = \{\mathrm{ub}_1\mathrm{ub}_2\mathrm{ub}_3 \ldots \mathrm{ub}_{n-1}\mathrm{ub}_n\}$$

where ub_i bottom shows the upper bound of the ith variable.

$$\mathrm{lb} = \{\mathrm{lb}_1\mathrm{lb}_2\mathrm{lb}_3 \ldots \mathrm{lb}_{n-1}\mathrm{lb}_n\}$$

where lb_i bottom shows the lower bound of the ith variables.

Using the following equation, each moth will be able to update its location in relation to the flame:

In the above expression, M_i represents the position of the ith moth and jth flame, which is denoted by F_j, and S is the spiral function of the moth and flame.

$$M = S\left(M_i F_j\right) \rightarrow M$$

S depicts a moth fluttering around a flame and how it may be at several locations at the same time. The flames are deemed to be the best answer, and their values are stored in the F matrix. The flowchart for MFO is depicted in Fig. 3.

5 Simulation Result

In two-area systems where the power system that has been adopted is the same, it has been taken into consideration. In two circumstances, the MFO provides benefits to both area controllers. Assume that area one has a known step load deviation, and that area two has a known step load deviation in the first scenario and the second scenario, respectively. A result is produced in both circumstances. The parameters will be the same for both sections because the system is regarded as similar. The system operates at a frequency of 50 Hz (Hz). Table 1 shows two characteristics of the area system that are relevant to this study.

Case 1—When the step disruption is applied to region 1 of the diagram.

This is the situation in which there is a 2% step disruption in area 1. Simulation is carried out using a variety of controllers in its secondary loop, and the resulting

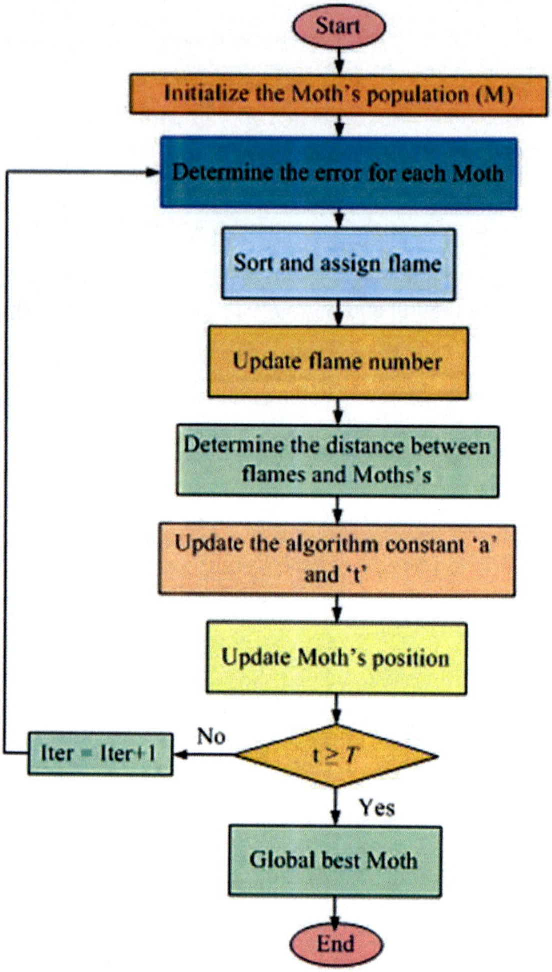

Fig. 3 Flowchart for proposed algorithm

Table 1 Parameter for two area

System parameter	Value
K_{ps}	108
T_{ps}	23 s
T_{sg}	0.4 s
T_T	0.6 s
Speed regulator	3
Bias parameters	0.32
Tie-line power	0.1

Table 2 Disturbance for 2% in first area

Controller	K_p	K_i	K_d	N	B	C	K
2-DOF PID	2.907	5	5	610	11	0.06	0.88
PID	4	4	1.66	–	–	–	–
Integral	–	0.4	–	–	–	–	–

result is determined. As an example, the following are the values of parameters for several controllers:

In Case 2, step disruption is directed to area 2.

As an example, consider the issue of a 2% step disruption in area 2. Simulation is carried out using a variety of controllers in its secondary loop, and the resulting results are determined. Various controllers' parameters have the following values, which are listed in alphabetical order (Table 2, Figs. 4, 5, 6, Table 3, Figs. 7, 8 and 9).

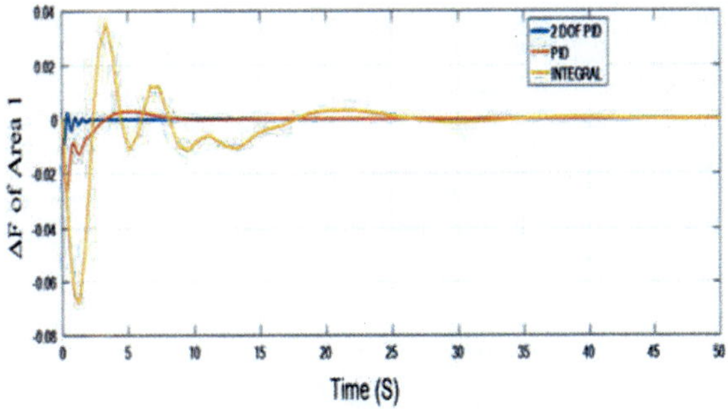

Fig. 4 Deviation of frequency in area 1

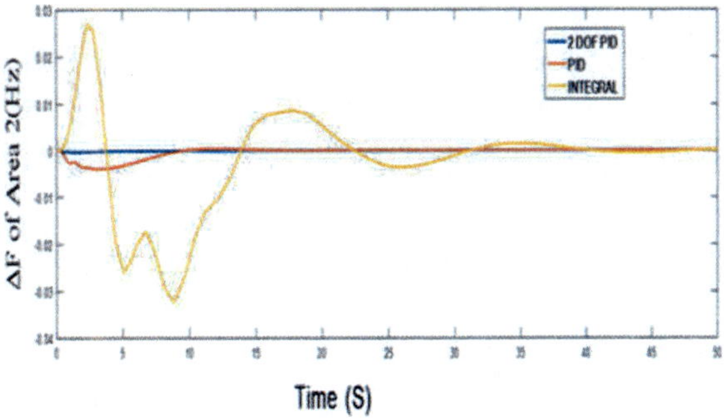

Fig. 5 Deviation of frequency in area 2

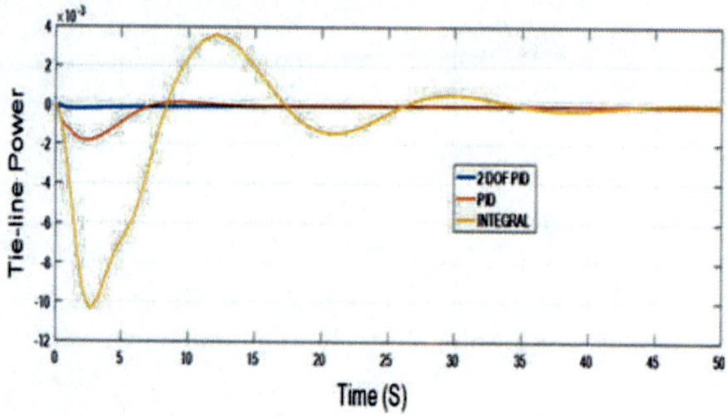

Fig. 6 Tie line power in Area 2

Table 3 Disturbance for 2% in second area

Controller	K_p	K_i	K_d	N	B	C	K
2-DOF PID	2.8722	5	5	610	11	1.06	0.8
PID	3.68	1.87	4.30	–	–	–	–
Integral	–	0.6	–	–	–	–	–

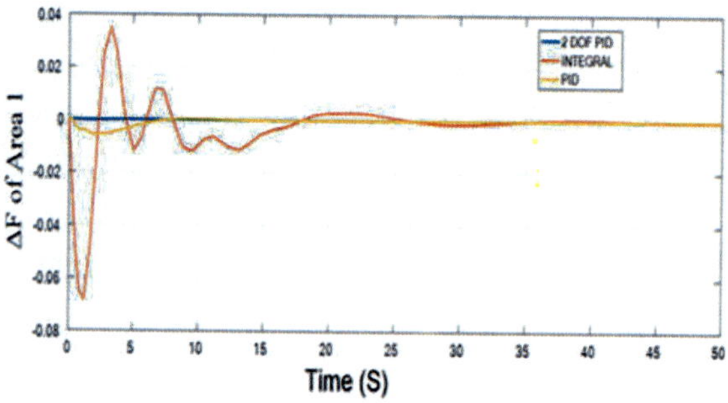

Fig. 7 Drift-frequency of area 1

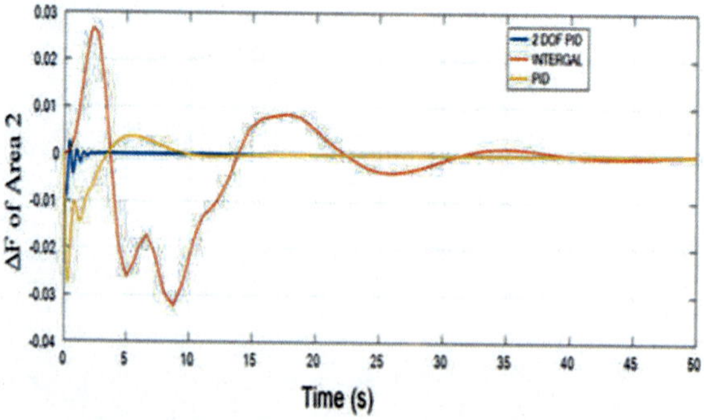

Fig. 8 Drift- frequency of area 2

Fig. 9 Tie line power in Area 1

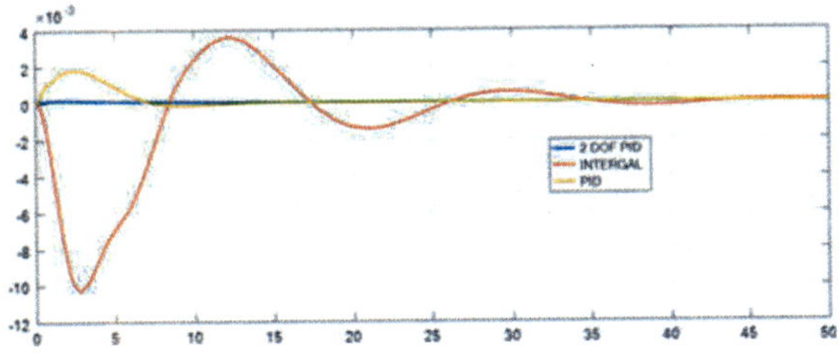

6 Conclusion

This paper presents a section in two scenarios in which a disturbance is caused in two separate locations. The outcomes of the simulations are displayed for both situations. It is discovered from the two scenarios stated above that a 2-DOF PID controller modified with the use of the MFO algorithm outperforms the two other classic controllers known as the PID controller and the integral controller in terms of performance. This novel controller can minimize the settling time, eliminate frequency oscillations, and lower the peak overshoot of the variable by a factor of two. As a result, the architecture with a 2-DOF PID controller obtains increased overall system performance as a result of its use.

References

1. Chang CS, Fu W, Wen F (1998) Load frequency control using genetic-algorithm based fuzzy gain scheduling of PI controllers. Electr Mach Power Syst 26(1):39–52
2. Elgerd OI, Fosha CE (1970) Optimum megawatt-frequency control of multiarea electric energy systems. IEEE Trans Power Apparat Syst (4):556–563
3. Chhabra H, Mohan V, Rani A, Singh V (2019) Multi-objective Cuckoo Search algorithm-based 2-DOF FOPD controller for robotic manipulator. In: Select proceedings of ICSC 2018. https://doi.org/10.1007/978-981-13-2553-3_33
4. Mohan V, Chhabra H, Rani A, Singh V (2019) An expert 2DOF fractional order fuzzy PID controller for nonlinear systems. Neural Comput Appl 31:4253–4270. https://doi.org/10.1007/s00521-017-3330-z
5. Mohanty B, Panda S, Hota PK (2014) Differential evolution algorithm based automatic generation control for interconnected power systems with non-linearity. Alex Eng J 53(3):537–552
6. Ghoshal S (2004) Optimizations of PID gains by particle swarm optimizations in fuzzy based automatic generation control. Electr Power Syst Res 72(3):203–212
7. Sahu RK, Panda S, Rout UK, Sahoo DK (2016) Teaching learning based optimization algorithm for automatic generation control of power system using 2-DOF PID controller. Int J Electr Power Energy Syst 77:287–301
8. Kundur P (2009) Power system stability and control. TMH 8th reprint, New Delhi
9. Elgerd OI (1983) Electric energy systems theory: an introduction. TMH, New Delhi
10. Ahamed TI, Rao PN, Sastry PS (2002) A reinforcement learning approach to automatic generation control. Electr Power Syst Res 63(1):9–26

11. Gozde H, Taplamacioglu MC, Kocaarslan I (2012) Comparative performance analysis of Artificial Bee Colony algorithm in automatic generation control for interconnected reheat thermal power system. Int J Electr Power Energy Syst 42(1):167–178
12. Karnavas YL, Papadopoulos DP (2002) AGC for autonomous power system using combined intelligent techniques. Electric Power Syst Res 62(3):225–239

Optimization of Cost and Sensitivity Analysis of a Standalone Microgrid

Papia Ray and Surender Reddy Salkuti

Abstract This paper discusses the cost optimization and sensitivity analysis of standalone and hybrid microgrids in remote areas. The hybrid microgrid considered in this paper comprises a diesel generator (DG), wind turbine (WT), and solar PV unit. Cost optimization and sensitivity analysis have been performed by changing diesel price, solar irradiation, and wind speed. The proposed approach in this paper can work for the progress of rural places by using a cleaner, eco-friendly environment based on renewable energy sources (RESs) to acquire load requirements. For the simulation studies and economic expediency, a remote village, named Kotapokhari in Khordha District, Odisha, India, is considered. The Hybrid Optimization Model for Electric Renewable (HOMER) Pro software is used for analyzing the economic feasibility of the proposed hybrid microgrid. The best combination of the microgrid for minimal net present cost and cost of energy was found. Further, the sensitivity analysis and emission by pollutants and load sharing of different components of the standalone microgrid were formulated. The suitability and cost-effectiveness of the proposed hybrid power system are validated from simulation results.

Keywords Microgrid · Wind power · Solar PV power · Cost · Sensitivity · Renewable energy

1 Introduction

The software HOMER Pro is used for hybrid optimization model for multiple energy resources (HOMER). It is a standard software for optimizing a microgrid in various sectors. This software is originated at National Renewable Energy Laboratory (NREL) and distributed by HOMER Energy for engineering and economics work

P. Ray
Department of Electrical Engineering, Veer Surendra Sai University of Technology, Burla, India

S. R. Salkuti (✉)
Department of Railroad and Electrical Engineering, Woosong University, Daejeon, Republic of Korea
e-mail: surender@wsu.ac.kr

A. Rani et al. (eds.), *Signals, Machines and Automation*, Lecture Notes in Electrical Engineering 1023, https://doi.org/10.1007/978-981-99-0969-8_7

simultaneously. This software designs and simulates models with various combinations and simulates hybrid microgrid (HM) operation for an entire year [1, 2]. Considering total net present cost (TNPC), a list of configuration results is displayed in a sorted form from lowest to highest in this software. Still, TNPC system configuration has differed with sensitivity variable designer selection. From the HOMER software, optimal results for each selection, including the sensitivity variables for the hybrid renewable energy system (HRES) with different configurations based on the highest to lowest cost, can be tabulated [3].

A simulation study is performed here of a remote village, named Kotapokhari in Khordha District, Odisha (20.1301 °N, 85.4788 °E), India, for its economic expediency. An HM system consisting of a PV panel, wind turbine (WT), converter, battery, and a diesel generator (DG) is planned to maintain that village's required load. As off-grid loads are beneficial for the hybrid system, the analysis is based on this, which will further help the remote end village-like Kotapokhari design the power model [4, 5]. HOMER legacy 2.68 beta version is applied here for the optimization of the system, and sensitivity analyzes are carried out for accurate results. For global system modelling, analysis, and simulation, HOMER Pro was used to acquire net present cost (NPC) and sensitivity analysis (SA). This study aims to give the least total NPC and SA of three cases and find the best combination [6]. The main aim is to develop an accurate cost and sensitivity analysis scheme. The main contributions are as follows:

- Optimization and cost analysis of different components of the standalone HM.
- Quantity of emission production by various pollutants.
- Percentage of power contributed by the different components.
- Exact location, wind speed, and solar radiation data.
- Cost of energy (COE), system NPC, and SA performance.

2 Proposed Scheme

The analysis acknowledges that of a small village in the Khordha district of Odisha state. The objective is to give the least total NPC, COE, and the net quantity of emission produced by various hybrid systems with pollutants for delivering power to acquire the energy requirement of said study place [7]. Odisha lies between 17.780 N and 22.730 N, longitudes 81.37 E and 87.53 E with an area of 155,707 km^2, which is 4.87% of the total area of India and a coastline of 450 km. It comprises 30 districts and 314 blocks—this work is based upon Kotapokhari, Khordha, which has hilly terrain; the plain area is more fertile than the plateau and is primarily a forest area [8, 9]. The study area has 01 unelectrified villages (Kotapokjhari) where the average population is 600 with 250 houses. Our research area's exact geographical location is situated in Khordha district, Odisha in India. Figure 1 depicts the proposed scheme.

Fig. 1 Proposed scheme

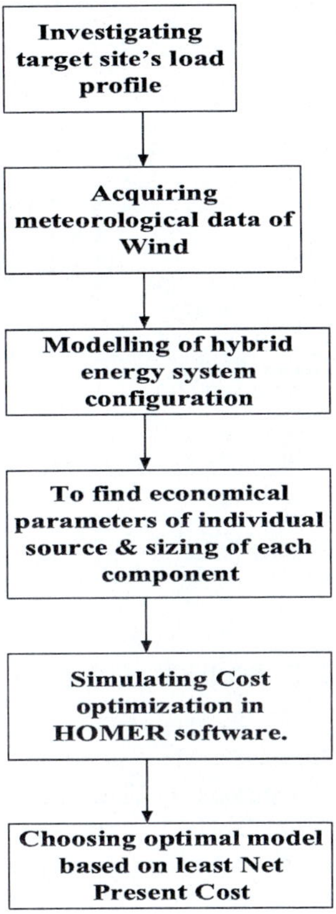

2.1 Determination of Daily Load Profile

A rural village in Khordha, Odisha, has been considered here as the site for analysis purposes whose various loads are fluorescent tubes, fans, televisions, electric irons, refrigerators, washing machines, computers, and grinders. The daily consumption of energy is 345 kWh/d, and the peak load is 64 kW [10]. Figure 2 depicts the hourly load profile in January, whereas the scaled load is depicted in Table 1 with a calculated correction factor.

Table 1 shows that the average kW/day is 345, average kW is 14.38 peak load which is 64.91 kW, and the load factor (LF) is 0.22 [11]. From Fig. 2 (variation of electrical load during the whole day), it can be observed that at 8 pm, the load is maximum; meanwhile, at 9 am, the load is minimal. It can also be observed that the load is comparatively higher than in the daytime during the evening time.

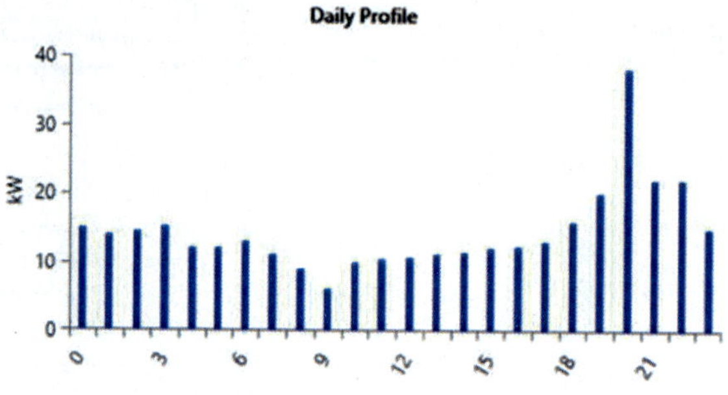

Fig. 2 January's hourly load profile

Table 1 Scaled value of load

Metric	Baseline	Scaled	Metric	Baseline	Scaled
Average (kWh/day)	345	345	Peak (kW)	**64.91**	**64.91**
Average (kW)	14.38	14.3.8	Load factor (LF)	0.22	0.22

2.2 *Metrological Data*

The selected area is located at 20° 11.2′N latitude and 85° 37.1′E longitude. The average value of all day's solar irradiation (SI) and clearness index values is depicted in Fig. 2. The SI value lies between 4.131 kWh/m^2/day to 6.132 kWh/m^2/day and an annual average of 5.03 kWh/m^2/day. Average global horizontal irradiances (GHI) data is around 4.82 kWh/m^2. GHI data can be varied throughout the year. From Fig. 3, it can be observed that the GHI data of the exact project location [12]. It is generally higher during the summer season. Its impact is lower than on a clear day on a cloudy day. Meanwhile, its impact changes from day tonight.

Figure 4 depicts a year's wind speed (WS) at the proposed location. The selected site's average WS profile is 4.32 m/s. Figure 3 shows peak WS which occurs in April

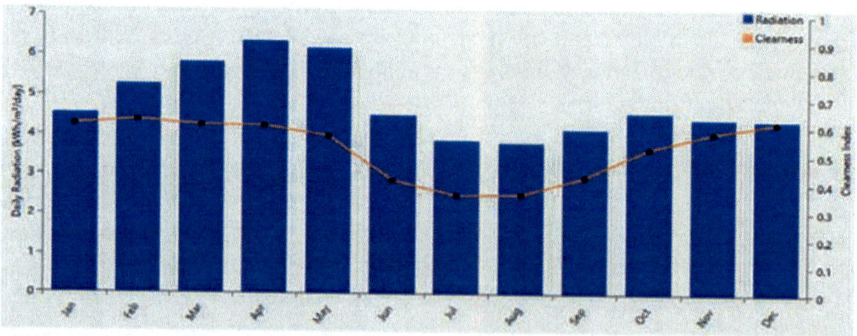

Fig. 3 Monthly average GHI solar value

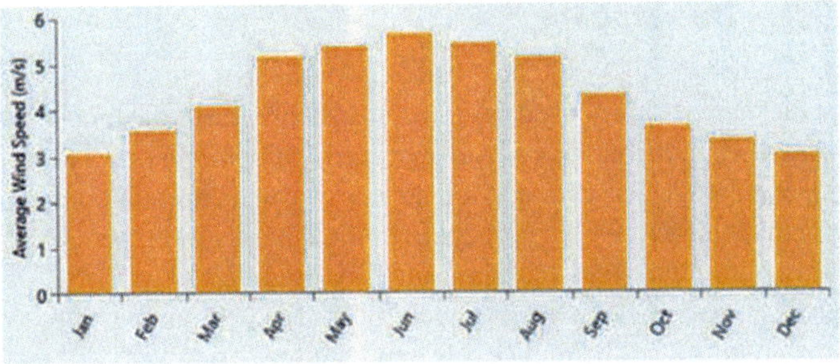

Fig. 4 Monthly average wind speed (WS) data

at 4.680 m/s and lowermost in January at 3.150 m/s. From Fig. 4, it can be observed that the exact project location. The average WS data is around 4.29 m/s. Generally, the WS is usually high during June and low during January.

2.3 Schematic Model Diagram

Figure 5 depicts the schematic model of the microgrid using HOMER software. This schematic has two buses, such as one is AC, and the other is DC. The AC has an electric load of 345 kWh/d, and a 64.91 peak is connected to the microgrid. A generic flat plate PV of 1 kW is connected to the DC bus bar. A 1 kW converter is connected between AC and DC to convert DC to AC. A WT and a DG are connected to an AC line [13]. Figure 5 shows that an AC electric load of 345 kWh/d and 64.91 kW peak can meet the load demand by connecting a 1 kW generator, 1 kW PV, 1 kWh LA battery, and 1 kW turbine.

Fig. 5 HOMER software-based microgrid model

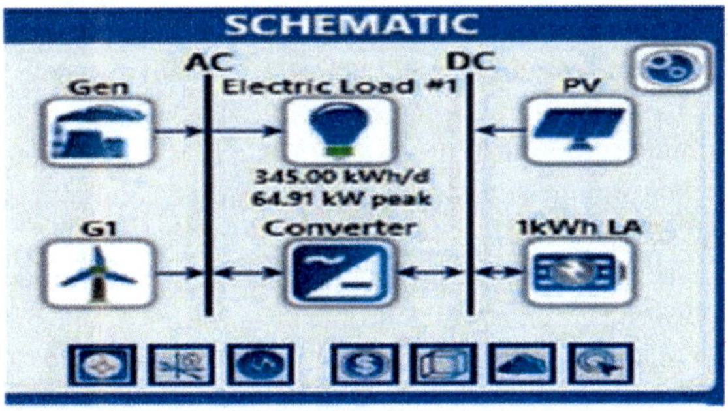

2.4 Problem Description

This section discusses the proposed system components and its detailed analysis, such as cost, rating, and contribution of load sharing. The optimization of cost analysis, sensitivity analysis (SA), percentage of load sharing by a different component of the system, percentage of excess load production, the quantity of emission produced by various pollutants, and system economics (SE) is also discussed [14]. A maximum of 03 AC or DC generators can be designed by HOMER Pro in a power system that consumes different fuels. The fuel cost is expressed by using,

$$F_{\text{fuel}} = C_0 Y_{\text{generator}} + c_1 P_{\text{generator}} \tag{1}$$

where C_0 is the intercept coefficient of the fuel curve, C_1 is the fuel curve slope, $Y_{\text{generator}}$ is generator rated capacity (kW). DC electricity is generated by the HOMER Pro-based PV module when solar irradiance is incident upon it. Solar voltaic is selected when there is less WS and more price of diesel [15, 16]. The PV array's power output is determined using,

$$P_{pv} = C_{\text{PV}} P_{\text{PV}} \frac{I_T}{I_S} \tag{2}$$

where C_{PV} is the PV derating factor, I_S is the standard radiation value (1 kW/m^2), I_T is PV array radiation (kW/m^2), $P_{\text{generator}}$ is generator output power (kW), P_{PV} and PV array rated capacity (kW).

3 Results and Discussion

Optimization results from HOMER Pro by considering sensitive variables such as SI are 4.82 kWh/m^2/d, diesel price is \$1.34/L, and SI is 4.82 kWh/m^2/d. Here, the PV-battery-wind-generator converter system is the most economical model. Figure 6 depicts the optimization results of the best combination of the microgrid, which gives minimum COE and NPC. The hybrid (PV, wind, generator, battery energy sources) model provides the best possible result. The obtained optimum NPC is \$727,306, and COE is \$0.447.

Table 2 shows the percentage of energy produced by different microgrid components. Table 2 shows that the most significant amount of energy is supplied by generic flat plate PV, 66.8%, then WT, which is around 31%. The PV cell produces 194,918 kWh of energy, the generator produces 6575 kWh of energy, and the WT has 90,484 kWh of energy in a calendrical year. The total electricity produced in a year is around 291,977 kWh of energy.

Table 3 presents the percentage of excess energy produced by the system. Table 3 shows that yearly 140,884 kWh of extra energy is produced by our design model,

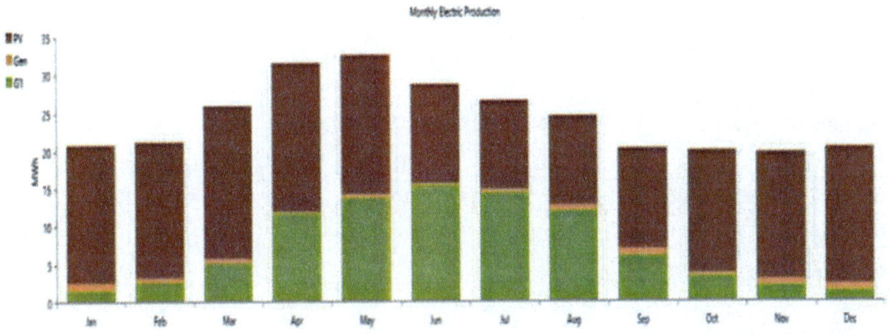

Fig. 6 Electricity production each month

Table 2 Energy production

Production	kWh/yr	Percentage (%)
Generic flat plate PV	**194,918**	**66.8**
Auto size genset (Generator)	6575	2.25
Generic 1 kW (WT)	90,484	31.0
Total	29,1977	100

which is around 48.3% of total energy production. If this system is connected to the grid, it can be sold to the grid, and the extra amount of energy can repay our running cost and reduce our COE. This model has no capacity shortage, and there is no unmet electric load as it is the standalone grid. One can reduce the excess electricity production if it is connected to the grid or the extra load. Table 4 shows how much quantity of emissions is produced by different pollutants when our system is running. Table 4 shows that the affecting environmental materials like carbon dioxide (CO_2) and carbon monoxide (CO) can be noticed. Using HRES, sulphur dioxide (SO_2) becomes minimal.

Table 3 Percentage of excess energy produced

Quantity	kWh/yr	Percentage (%)
Excess electricity	140,884	48.3
Unmet electric load	0	0
Capacity shortage	0	0

Table 4 Emission quantity by various pollutants

Quantity	Value	Units	Quantity	Value	Units
CO_2	5905	Kg/yr	Particulate matter	0.226	Kg/yr
CO	37.2	Kg/yr	SO_2	14.5	Kg/yr
Unburned hydrocarbons	1.62	Kg/yr	Nitrogen oxides	35.0	Kg/yr

Table 5 Analysis of cost of various microgrid components

Component	Capital ($)	Replacement ($)	O&M ($)	Fuel ($)	Salvage ($)	Total ($)
Generator	36,576	0.00	28,187.78	39,080.67	5461.69	98,382.76
WT	94,536	0.00	1749.61	0.00	0.00	96,285.61
Battery	152,751.50	271,314.52	47,353.49	0.00	36,015.20	435,404.31
Generic flat plate PV	87,067.02	0.00	2254.49	0.00	0.00	89,321.50
System converter	**3447.14**	**1098.97**	**3985.01**	**0.00**	**619.34**	**7911.78**
System	374,377.65	272,413.49	83,530.38	39,080.67	42,096.23	72,305.97

Table 6 Results of sensitivity analysis

Diesel fuel price ($/l)	Sensitivity		COE ($)	NPC ($)
	Solar radiation (kWh/m^2)	WS (m/s)		
1.34	4.82	4.29	0.447	727,306
1.0	5.2	5.2	0.382	622,074
0.8	**6**	**5.5**	**0.356**	**579,185**

3.1 *Cost and Sensitivity Analysis*

Table 5 indicates the cost structure of the total system component, their depreciation, and the system's running cost. Table 5 shows that the system's capital, replacement, operation and management (O&M), fuel, and salvage value consist of generic flat PV, grid, and system converter. Out of which, system converter cost is much lower than other system components.

Table 6 presents the SA results by changing the sensitivity parameters such as SI and WS and determining the COE and NPC. Table 6 depicts the variation in diesel fuel price, SI, WS, and their respective COE and net NPC. The optimal result of COE is $0.356, and NPC is $579,185 when diesel price is $0.8, solar radiation is 6 kWh/m^2 and WS is 5.5 m/s.

3.2 *Economic Analysis*

Table 7 presents the SE of proposed HM from where cost analysis that is the present worth return on investment (ROI), annual worth, simple payback (SP), internal rate of return (IRR), and discount payback (DP) can be determined. SE of HM can be depicted in Table 7, where it can be seen that the simple payback period is 1.85 years,

Table 7 System economic analysis

Metric	Value ($)	Metric	Value ($)
Present worth ($)	1,832,556	IRR (%)	52.1
Annual worth ($/year)	141,756	SP (year)	**1.85**
ROI (%)	45.7	DP (year)	2.01

and the return on investment is 45.7% which is quite reasonable for this type of scheme.

3.3 Time Series Analysis

Figure 6 indicates the electricity production of different components throughout the year. Figure 6 depicts the quantity of electricity generated by individual components every month. It can be observed that PV cells and WTs have maximum energy, and the DG is created when there is short of energy. We found that the PV and diesel hybrid model are the best combination for our project. Solar radiance and the WS increased during the summer, so our COE and NPC decreased.

Figure 6 depicts the quantity of electricity generated by individual components every month. It can be observed that PV cells and WTs have maximum energy, and the DG is created when there is short of energy. We found that the PV and diesel hybrid model is the best combination for our project. Solar radiance and the WS increased during the summer, so our COE and NPC decreased.

4 Conclusion

In this paper, an optimized HRES model consists of locally accessible RES, specifically solar, wind and DG, and a battery storage system to distribute power in different Kotapokhari village load sections of Khordha district, Odisha. The proposed HOMER Pro-based optimized model gives minimal overall NPC and COE. The value of COE is 0.447$/kWh, and therefore, the total NPC is $727,306. This work analyzes the cost, SA, NPC, COE, the quantity of emission produced by different pollutants, economic analysis, and load sharing of the various components of a standalone microgrid. We also find out the best combination of the resources for the microgrid. Comparing the cost and SA between standalone grid and the grid-connected microgrid, taking the actual data of any industry, and designing a grid-connected microgrid model to satisfy the required load demand are the scope for the future research work.

References

1. Khare V, Nema S, Baredar P (2013) Status of solar wind renewable energy in India. Renew Sustain Energy Rev 27:1–10
2. Vallati A, Grignaffini S, Romagna M (2015) A new method to energy saving in a micro grid. Sustainability 7:13904–13919
3. Silveira EF, Oliveira TF, Brasil Jr ACP (2015) Hybrid energy scenarios for Fernando de Noronha archipelago. Energy Procedia 75:2833–2838
4. Lal DK, Dash BB, Akella AK (2011) Optimization of PV/wind/micro-hydro/diesel hybrid power system in HOMER. Int J Electr Eng Inform 3(3):307–325
5. Munuswamy S, Nakamura K, Katta A (2011) Comparing the cost of electricity sourced from a fuel cell-based renewable energy system and the national grid to electrify a rural health centre in India: a case study. Renew Energy 36:2978–2983
6. Dash RL, Behera L, Mohanty B, Hota PK (2018) Cost and sensitivity analysis of a microgrid using HOMER-Pro software in both grid-connected and standalone mode. In: International conference on recent innovations in electrical, electronics & communication engineering (ICRIEECE), Bhubaneswar, India, pp 3444–3449
7. Dutta S, Qian A, Adhikari A, Yue W, Yufan Z, Zhaoyu L (2019) Modelling and cost optimization of an islanded microgrid with an existing microhydro using HOMER software. In: 2nd Asia conference on energy and environment engineering, Japan, pp 74–78
8. Vendoti S, Muralidhar M, Kiranmayi R (2018) HOMER based optimization of solar-wind-diesel hybrid system for electrification in a rural village. In: International conference on computer communication and informatics (ICCCI), Coimbatore, India, pp 1–6
9. Mustafa F, Sahito AA, Khatri SA, Kumar L (2022) Hybrid energy system modelling for oil & gas fields: a case study of Pasakhi satellite oil & gas complex. Distrib Gener Alternative Energy J 37(2):281–310
10. Lemence ALG, Tamayao MAM (2021) Energy consumption profile estimation and benefits of hybrid solar energy system adoption for rural health units in the Philippines. Renew Energy 178:651–668
11. Manurung D, Dalimi R (2020) Optimization electricity supply for thousand island Jakarta Province by using hybrid photovoltaic-wind turbine-diesel generator-crude palm oil (CPO) power plant-battery system. J Phys Conf Ser 1500(1)
12. Joshi K, Ashok A, Chandel S (2016) Optimal economic analysis of hybrid renewable energy system. In: International conference on advances in computing, communication, & automation (ICACCA), pp 1–5
13. Sahoo S, Swain SC, Chowdary KVVS, Pradhan A (2022) Cost and feasibility analysis for designing a PV–wind hybrid renewable energy system (A case study for Campus-3, KIIT University, Bhubaneswar). In: Innovation in Electrical Power Engineering, Communication, And Computing Technology, Springer Singapore
14. Hatti M (2018) Artificial intelligence in renewable energetic systems—smart sustainable energy systems. In: Lecture notes in networks and systems, vol 35
15. Eze F, Ogola J, Kivindu R, Egbo M, Obi C (2022) Technical and economic feasibility assessment of hybrid renewable energy system at Kenyan institutional building: a case study. Sustain Energy Technol Assess 51
16. Lemence AL, Tamayao MA (2021) Energy consumption profile estimation and benefits of hybrid solar energy system adoption for rural health units in the Philippines. Renew Energy 178:651–668

Analysis and Mitigation of Hall Sensor Glitch Effects in Brushless DC Motor Based E-Vehicle Controller

Arvind Goswami, Mini Sreejeth, and Madhusudan Singh

Abstract Brushless DC (BLDC) motors operate in both sensored mode and sensorless mode options depending upon the application requirements. Hall effect sensors are the most common prevalent solutions used in sensored mode of operation. This type of arrangement is widely used in e-vehicle applications, particularly in e-rickshaw type of applications where sub 2 kW rating drives are employed. It utilizes the signals from three Hall effect-based sensors to calculate speed, obtain rotor position and timings of inverter switching, and determine switching pattern. The whole operation of the drive is dependent upon the signals obtained from the hall sensors. Sometimes due to BLDC motor manufacturing process or aging effect, there are unwanted glitches at the transition points of the hall sensor signals. This causes unwanted hall signal pattern detection, and consequently, wrong switching sequence can be implemented if this is not accounted in the BLDC drive controller itself. This paper presents the study of such glitch effects on the inverter driving signals by obtaining real-time experimental signals on the e-rickshaw controller and attached motor. Further, to safeguard the drive from wrong switching pattern due to glitch, the software implementation of the methods to handle such cases is presented, and the suitability for commercial applications like e-vehicle controller is explored. The experimental research setup is based on STM32103 series ARM microcontroller based design, and the firmware is developed using STM32CubeMX utility software with IAR Embedded Workbench for arm controller.

Keywords BLDC motor · Hall sensor · Glitch · Hall placement · Switching · Torque ripple · Simulation · Floating phase · STM32

A. Goswami (✉) · M. Sreejeth · M. Singh
Department of Electrical Engineering, Delhi Technological University, Delhi, India
e-mail: arvind_phd2k19@dtu.ac.in

M. Sreejeth
e-mail: minisreejeth@dce.ac.in

M. Singh
e-mail: madhusudan@dce.ac.in

A. Rani et al. (eds.), *Signals, Machines and Automation*, Lecture Notes in Electrical Engineering 1023, https://doi.org/10.1007/978-981-99-0969-8_8

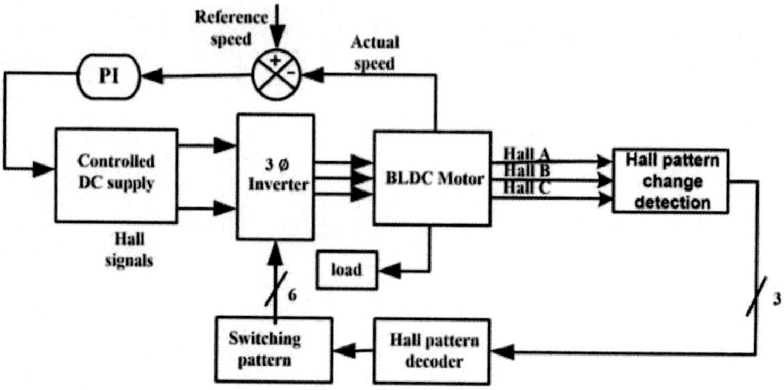

Fig. 1 Block diagram of the control system for BLDC motor with hall sensor feedback

1 Introduction

Brushless DC (BLDC) motors with hall sensors are widely used in e-vehicles, particularly e-rickshaw besides other applications like home appliances and automation industry. The base design of these type of applications requires a controller circuit incorporating speed, current monitoring, and subsequent control functions for PWM duty cycle adjustments [1]. Figure 1 shows the block diagram of the control system for BLDC motor with hall sensor feedback arrangement.

One of the issues that controller design is required to handle is that there are glitches or unwanted pulses along with the normal hall transition signals. Since the complete controller firmware design and operation of the drive are dependent on the hall sensor output signals, any unwanted or noise signal leads to a logic level transition which is detected by the controller firmware, and subsequently, incorrect switching pattern can be applied to the inverter section if the anomaly is not detected by the appropriate methods [2]. This is discussed in this paper with actual e-rickshaw controller and motor set in normal, faulty, and corrected condition. The switching table and associated disturbance in it are discussed in subsequent sections.

2 Background

The functioning and signal integrity of hall sensors are affected by the error in their placement besides aging and mechanical fatigue [1]. The error in placement can be attributed to mounting procedure itself. One of the major effects of these errors is the presence of small duration pulses or glitches at either rising or falling end of hall signal transitions. This results in increase in torque ripples, noise, and current ripples during motor operation. Further, the drive controller design also becomes more complex when the existing design is changed to address these issues. Such

Table 1 Switching pattern with hall signal sequence

S. No.	Hall sensors			Switches					
	H3	H2	H1	AL	AH	BL	BH	CL	CH
1	1	0	1	0	1	1	0	0	0
2	0	0	1	0	1	0	0	1	0
3	0	1	1	0	0	0	1	1	0
4	0	1	0	1	0	0	1	0	0
5	1	1	0	1	0	0	0	0	1
6	1	0	0	0	0	1	0	0	1

type of issues is reported to occur mostly in BLDC motors with higher number of poles and smaller in size [3–5].

The overall functioning of BLDC motor and control is dependent on timings obtained by hall sensor transition signals. These signals are used to detect rotor position and speed as rate of change of rotor position. Thus, hall sensor signal integrity directly affects the calculation of motor speed and control algorithm implementation itself [6]. The timing of inverter switching with switching pattern for motor control is dependent and synchronized with these hall signals [7]. When the hall signals and consequent hall pattern are not accurate, switching signals for inverter section are also not exactly timed according to desired algorithm and calculations. This can cause serious errors in motor control operation [2]. Table 1 shows the switching sequence for three phase inverter with corresponding hall signal values.

Normal hall sensor output signals and signals with glitch effect are shown in Fig. 2. The BLDC motors mass produced and used in applications like e-rickshaw are reported to have inaccurately placed hall sensor issue in many cases [4]. The study of such hall faults and glitch effects along with methods to mitigate their effects on the motor drive operation is one of the major important subjects of research in BLDC motor drive operation and is discussed in detail in this paper.

3 Experimental Setup and Signal Analysis

The experimental setup consists a 1 kW BLDC motor, powered by 48 V, 72 Ah lithium ion battery pack. The hardware setup is shown in Fig. 3 with machine and test set up parameters mentioned in Table 2.

The setup consists of 3-phase BLDC motor controller of e-rickshaw application. The 32 bit STM ARM microcontroller is used in design. The test setup controller is set for 1 kW BLDC motor, peak current capacity of 60 A with 42–56 V input voltage range.

Figures 4, 5 and 6 show the glitch effect on phase voltage profile. The three hall output signals and corresponding voltage profile of one motor phase when hall signal transitions are normal without any glitch are shown in Fig. 4. The three hall signals

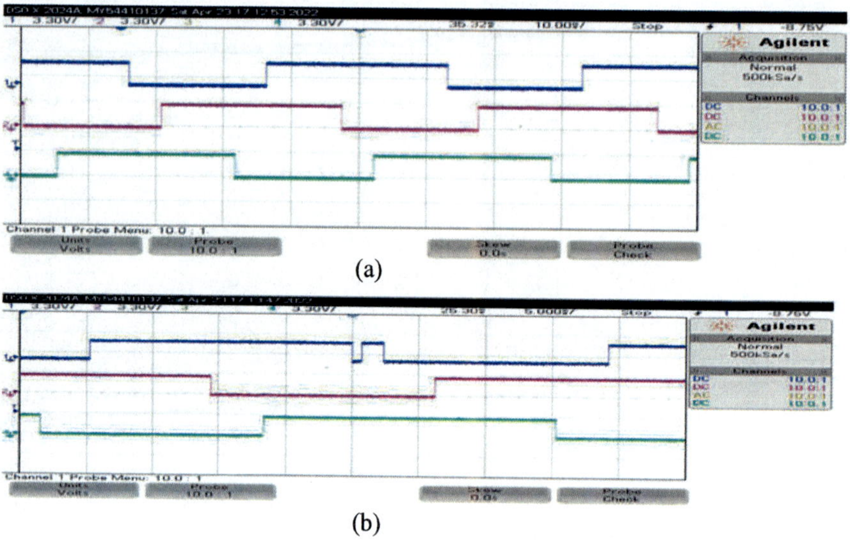

Fig. 2 Normal hall sensor output signals and signals with glitch. **a** Normal hall sensor output signals. **b** Hall sensor output signals with glitch in one hall signal in blue colored signal

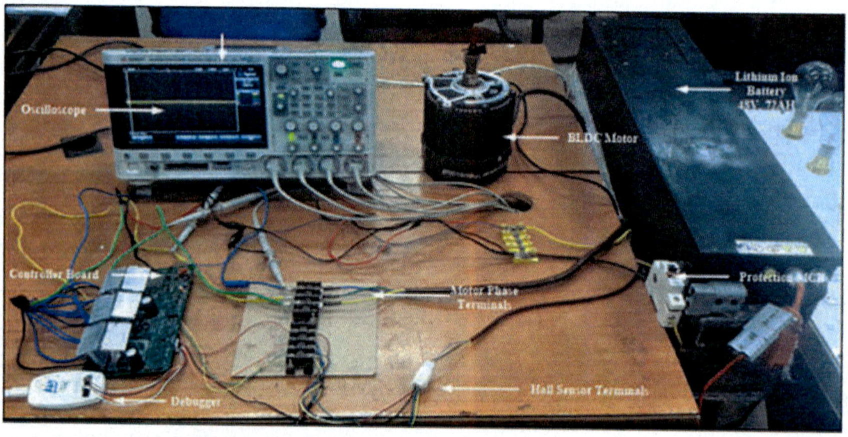

Fig. 3 Experimental setup with controller and motor

with glitch and corresponding effect on voltage profile of one motor phase are shown in Fig. 5.

The zoomed one motor phase voltage profile in Fig. 6 shows the floating phase condition when the first hall signal depicted in yellow colored waveform and second hall signal depicted in green colored waveform have glitch effect in after normal transition. This is due to the fact that glitch effect causes the detection of new hall pattern which is not according to the switching pattern, and if switching off operation

Table 2 Experimental setup details

Machine parameters		Test setup parameters	
Number of pole pairs	4	DC supply	48 V, 72Ah battery
Motor rated power	1 kW	Protection	MCB on DC line
Motor rated speed	3600 RPM	PWM frequency	16.6 kHz
Motor rated voltage	48 V	Commutation	Trapezoidal
Motor winding	3-phase star	Programmer	ST link-V2
Hall sensors	3	PI parameters	$K_p = 0.61$, $K_i = 0.013$

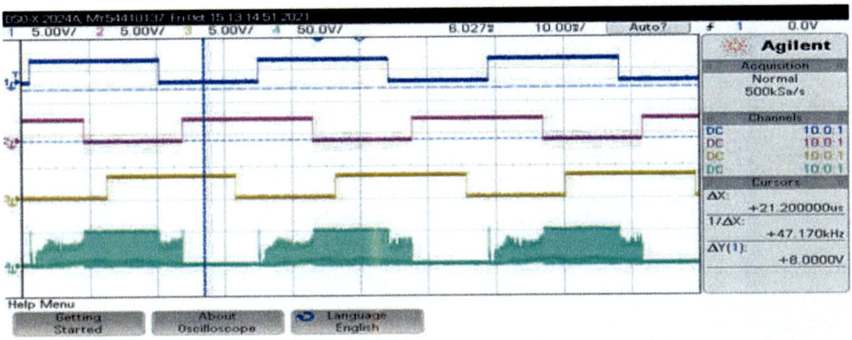

Fig. 4 Normal hall signal transitions without any glitch and corresponding single motor phase voltage waveform

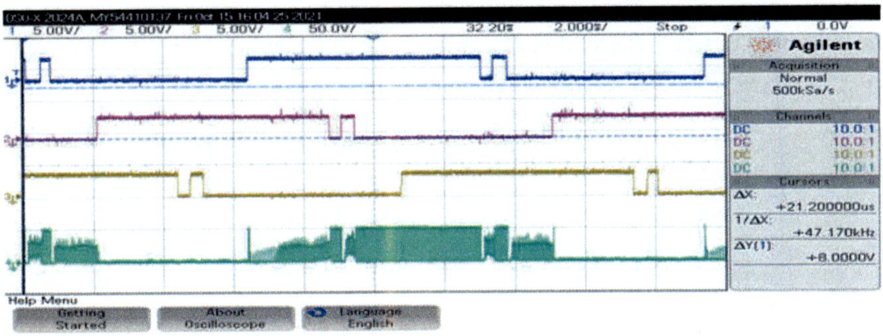

Fig. 5 Hall signal transitions with glitch effect and corresponding single motor phase voltage waveform without corrective mechanism

of one inverter switch is carried out but next switching is avoided because of automatic check in software to avoid short circuit switching in inverter, the motor phase will become floating till new correct switching pattern is executed [8]. Even if a time delay check is implemented for detection of glitch, it will have to be executed by software delay which is interfered and interrupted by other functional interrupts in

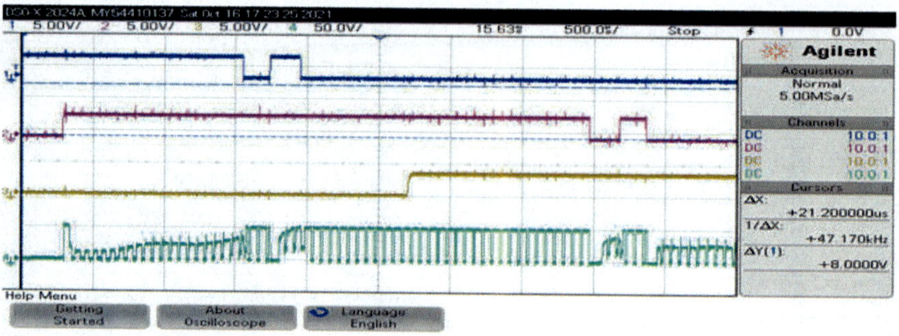

Fig. 6 Hall signal transitions with glitch effect and corresponding single motor phase voltage waveform showing floating phase effect due to wrong switching

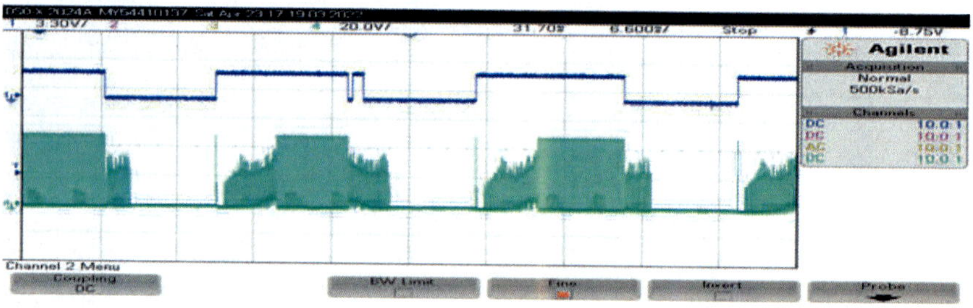

Fig. 7 Voltage waveform of one motor phase without floating phase effect when hall signal glitch fault is mitigated in firmware

the controller design. Hence, it is also not feasible in commercial applications like e-vehicle controllers.

The glitch effect in the presented research work is mitigated by the method of comparing the present hall value pattern with the previous one and going for corrective check only in the condition of this pattern not matching with the expected value as available in Table 1. During the rest of operation time, no controller time is spent on going for corrective check when hall sequence is according to expected value in Table 1. During corrective check, no switching operation is performed even for one leg of inverter phase. Thus, the previous state is maintained and last switching pattern which is correct is extended during the glitch duration also. Thus, floating phase condition is avoided as shown in Figs. 7 and 8 voltage profiles of one motor phase versus hall signal with glitch effect.

4 Comparative Analysis and Discussion

The waveforms for the normal hall sensor output case and the ones with glitch effect are presented in Figs. 4, 5 and 6. The hall signals with transition and switching timings

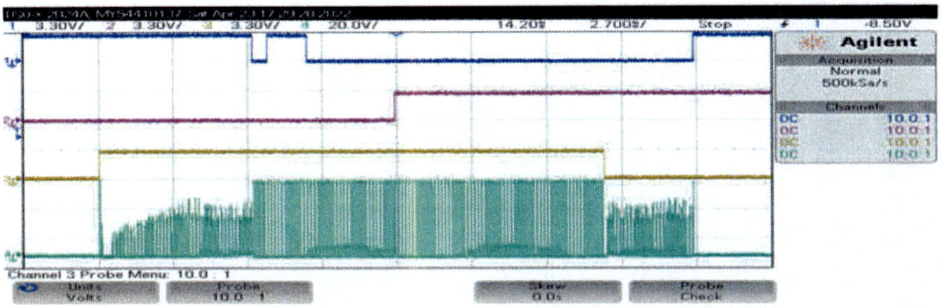

Fig. 8 Voltage waveform with PWM for one motor phase without floating phase effect and hall signal with glitch with corrective mechanism

with one phase voltage profile are presented for analysis. Finally, the waveforms after mitigation of the effect are presented in Figs. 7 and 8. The following set of observations can be inferred due to glitch effect in hall sensor signals in BLDC motor:

I. When the hall signals have glitch effect as shown in Fig. 2b, the hall pattern read by controller will be different for glitch time. In Fig. 2a, the normal binary hall sequence after 101 is 001. In the Fig. 2b, the hall pattern change is from 101 to 001 as expected, but due to glitch effect, next pattern read is again 101. Thus if switching sequence is executed according to the hall values read without any check, it will not be according to expected sequence as in Table 1.
II. Due to this wrong switching pattern, there is a possibility of shorting in the inverter section due to same phase high voltage and low voltage side switches simultaneously getting switched on.
III. Considering the above two points, if a software check is implemented on every hall transition, there will be a delay between switching off of the one switch corresponding to one hall pattern and switching on next switch for next hall pattern. This delay will be over and beyond interrupt latency, pin value settling time, and dead time.
IV. During this time if a, according to software check, no valid switching pattern is executed, the phase will experience a floating phase condition.
V. The floating phase or unwanted switching transitions cause more current ripples, higher current flow, and overheating of controller as well as motor winding.

When the glitch effect is mitigated by suggested method as shown in Figs. 7 and 8, it is observed that:

I. No additional switching delay is created.
II. Correct switching pattern is maintained by controller.
III. Floating phase condition is avoided.
IV. The heating of the MOSFET switches was reduced by around 2 °C when checked at the surface of connected heat sink.
V. The phase current as well as overall current drawn from battery is reduced by around 5%.

5 Conclusion

This paper presented a detailed background and analysis of the glitch effect in hall sensor based sensored BLDC motor drive operation. The glitch effect is shown and compared with the normal case by checking the waveforms using e-rickshaw controller and motor set. The disturbance in phase voltage is clearly visible when wrong switching pattern is avoided using interrupt based method. Consequent floating phase effect, higher current flow, and heating effect are discussed. Further, the corrected waveforms are shown when the glitch effect is avoided by other method which does not cause any additional delay in switching pattern changes corresponding to change in hall value pattern. Thus, floating phase effect is successfully avoided. The whole scheme is implemented on STM32 microcontroller based e-vehicle controller and tested on experimental setup described in paper. The BLDC motor is checked for speed ramping up to the rated value and is able to run at full speed without getting affected by the glitches present in hall sensor output signals.

Acknowledgements The authors acknowledge the support of Ministry of Electronics and Information Technology, Government of India, (MeitY) in the project work.

References

1. Mousmi A, Abbou Y, Houm EL (2020) Binary diagnosis of hall effect sensors in brushless DC motor drives. IEEE Trans Power Electron 35(4):3859–3868
2. Goswami A, Sreejeth M, Upadhyay R, Indu S (2021) Brushless DC motor based E-rickshaw controller design. In: 2021 International conference on recent trends on electronics, information, communication & technology (RTEICT), pp 196–200
3. Goswami A, Sreejeth M, Singh M (2022) Simulation and analysis of hall sensor misalignment in BLDC motor drive. In: 2022 IEEE Delhi section conference (DELCON), pp 1–6
4. Tsotoulidis S, Safacas A (2014) Side-effects of Hall sensors misplacement on BLDC motor drive operation. In: 2014 International conference on electrical machines (ICEM), pp 1825–1830.https://doi.org/10.1109/ICELMACH.2014.6960431
5. Papathanasopoulos DA, Spyropoulos DV, Mitronikas ED (2019) Fault diagnosis of misaligned hall-effect position sensors in brushless DC motor drives using a Goertzel algorithm. In: 2019 IEEE 12th international symposium on diagnostics for electrical machines, power electronics and drives (SDEMPED), pp 167–173
6. Park JS, Choi JH, Lee J (2016) Compensation method of position signal error with misaligned Hall-effect sensors of BLDC motor. J Electr Eng Technol 11(4):889–897. The Korean Institute of Electrical Engineers
7. Nama T, Gogoi AK, Tripathy P (2017) Application of a smart hall effect sensor system for 3-phase BLDC drives. In: 2017 IEEE international symposium on robotics and intelligent sensors (IRIS). Ottawa, ON, Canada, pp 208–212
8. Ebadpour M, Amiri N, Jatskevich J (2021) Fast fault-tolerant control for improved dynamic performance of Hall-sensor-controlled brushless DC motor drives. IEEE Trans Power Electron 36(12):14051–14061

Radial Basis Function Neural Network with Wavelet Transform for Fault Detection in Transmission Line

Debani Prasad Mishra, Prabin Biswal, Smruti Susmita Sahu, Spandan Dash, and Nimay Chandra Giri

Abstract Incipient faults which occur due to the electrical arc occurrence within the facility cables which have insulation problems are hard to detect by normal protective relays, and with passing time can become a permanent fault within the system. Employing Radial Basis Function Neural Network (RBFNN) method, the paper puts forward the method to detect the faults and finding out the efficiency and advantages of RBFNN method over other methods within the facility grid. The result proposed in this paper is based on the differentiation between the wavelet transform method and RBFNN method of the measured voltage and fundamental component of the measured voltage, evaluated by RBFNN computation within the sending end of the cable during the fault using which the incipient fault is detected. This method uses neurons to train the model and increase its accuracy so that with new data set, it can produce super accurate and faster results.

Keywords Radial basis function neural network (RBFNN) · Artificial neural network (ANN) · Discrete wavelet transform (DWT) · High pass filter (HPF) · Low pass filter (LPF)

1 Introduction

The protection of power networks from unpredictable events as well as boosting their reliability in solid performance has always been a critical concern in the end-to-end delivery of energy [1]. In this situation, removing potential mistakes can go a long way toward preventing similar unpleasant incidents and so preserving the power system's reliability [2]. Electrical arcing in sections of electrical wires with closure abnormalities is most likely to blame for the first faults. Re-emerging with an

D. P. Mishra (✉) · P. Biswal · S. S. Sahu · S. Dash
Department of Electronics and Electrical Engineering, International Institute of Information Technology, Bhubaneswar, India
e-mail: debani@iiit-bh.ac.in

N. C. Giri
Department of ECE and CREE, Centurion University of Technology and Management, Rajaseetapuram, Odisha, India

A. Rani et al. (eds.), *Signals, Machines and Automation*, Lecture Notes in Electrical Engineering 1023, https://doi.org/10.1007/978-981-99-0969-8_9

equally elevated aspect of frequency occurring, initial errors over time can damage cable overhaul and can become a permanent flaw in the system. As a result, a precise system is crucial for rapid detection and detection of a recent fault in the system among comparable possible events [3]. Nascent defects are most noticeable in their short duration, as well as low electrical amplitude compared to other system defects, considering which conventional protective transmission can detect these errors [4]. The nascent flaw is recalled by a distorted waveform, similar to square surgery, rather than an error. With the detection of emerging errors, a permanent network error can be predicted [1]. Radial basis work networks are often used by the ANN type for performance measurement problems [5]. RBFNN is differentiated from other neural networks due to its better speculation and faster learning speed. The RBFNN-dependent model [1] uses the output data of the model simulated data as a target and then [6] trains the neural network using it [7]. In Sect. 2, the simulation behind the sensing and categorization in case of [8] wavelet-based transmission line fault is briefly discussed. Further, to overcome the constraints of this method, RBFNN has been proposed in Sect. 3. Its detailed methodology followed by experimental results has been briefly mentioned in Sect. 4. At last, in Sect. 5, the research has been concluded.

2 Wavelet-Based Transmission Line Fault Detection

Electromagnetic transients in energy systems are characterized by the most common short-term factors, such as errors, which are very important. These defects cause serious damage to the electrical system and the problem. Therefore, there is a need to find the error, its type, where it is formed in the line as well as to rectify the error at the earliest so as not to cause the same damage [9]. Wavelet modification is largely created by this strategy, which could be effectively implemented by using two filters HP and LP [10]. HPF was derived after applying the wavelet function while the measured details of the input with the LPF, which brings a smooth translation of the input signal and is taken from the mathematical function associated with the prominent mother wavelet. In this analysis, the results are controlled using db4 as the mother wavelet (prime wavelet) for signal analysis [1]. Basic electrical power and current phasors are calculated using an algorithm based on [11] discrete Fourier transform that minimizes the effects of DC decay significantly. The system is modeled on the MATLAB SimPowerSystems site. The results shown in the proposed system are secure, fast, and very accurate. The frequency has been set to 50 Hz with voltage of 220 V and transmission line of length 120 km. Making amplification of current signals using a threshold power level, the primary power of all current signals by keeping the balance without change to calculate the limit of the change in power, if any power exceeds this level, this means that there is an error condition at that section of the line [12]. The error is generated by selecting combination of different phases of lines from fault generation box (Fig. 1).

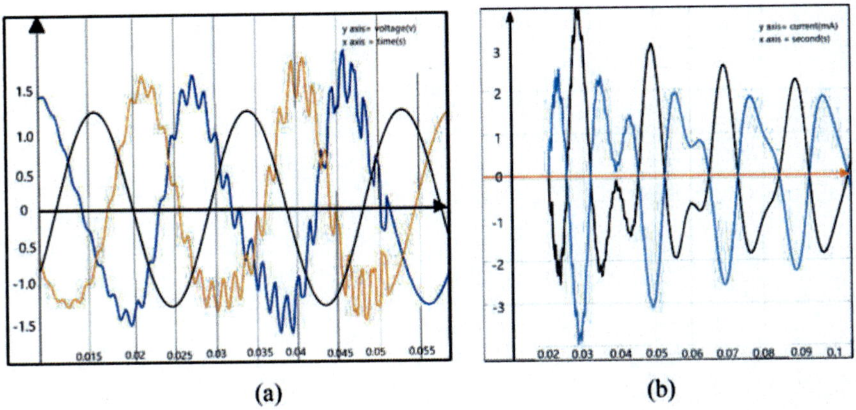

Fig. 1 a Voltage output waveform, b current output waveform

3 Radial Basis Function Neural Network (RBFNN)

RBFNNs are used to compare advanced functions straight from output data to a simple topology [1]. They have a better overall performance compared to most neural feed-forward networks. The most widely used basic function is the Gaussian exp function. In successional learning, the neural network is trained to measure activity while a sequence of training sample dice is randomly drawn, introduced into a network and are read by each network [13]. RBFNN has a special two-layered structure one hidden layer and one output. The input layer does not process the information; it only works to distribute input data between nodes. Each node in the hidden layer is an RBF [1]. The output from *j*th Gaussian node of the x_i input object is calculated, where the vector x_i captures the *i*th input data, and vector c_j represents the central point of the *j*th Gaussian function. $\|x_i - c_j\|$ is the calculated [6] Euclidean distance, and width σ is a criterion that controls the smoothness of the function. The output is calculated by a combination of linear radial functions and bias w_0 [7].

4 Experiment and Result

Accuracy of artificial neural network depends upon the input and output data. Larger the data, higher will be the accuracy and vice-versa. The simulation is run for 138 times. The maximum coefficient of all simulations is stored in form of a table. In order to classify all types of faults, we need output signal for each phase current as well as for ground current. There are 4 variables which represents phase A, B, C, and ground. In normal condition, the output is encountered as "0" and during fault it is "1" (Table 1).

Table 1 Maximum coefficient of different phases in different types of faults

Type of fault	Max. coefficient of phase A current	Max. coefficient of phase B current	Max. coefficient of phase C current	Max. coefficient of ground current
ABC-G	1.6097e+07	4.0725e+07	1.6097e+07	7.1824e+05
ABC	1.6097e+07	4.0725e+07	1.6097e+07	0.0081
AB-G	1.0796e+07	2.1332e+07	103.9772	7.7574e+05
AC-G	1.9807e+07	103.9772	8.6730e+06	1.9393e+06
BC-G	103.9784	4.0725e+07	8.1478e+06	9.7619e+05

We have obtained 138 cases by applying all 12 fault conditions before the transmission line as well as after the transmission line position and by changing the fault ground resistance in the form of a table as shown here (Table 2).

Normal output condition is represented by assigning “0” and that of fault condition is represented by “1”. At this stage, we have defined the input and output data for the RBFNN. We have a total of 138 cases. The normal proportion for division of data into training and testing is 70% and 30%, respectively. However, as a rough estimate, we have chosen 30 cases for testing purpose and remaining 108 cases for training purpose. After successful training, the performance of the RBFNN is tested for these unseen 30 cases. If neural network classifies correctly, its efficiency will be 100% (Table 3).

Table 2 Output values of different phases in different types of faults

Type of fault	Output value for phase A	Output value for phase B	Output value for phase C	Output value for ground current
ABC-G fault	1	1	1	1
ABC fault	1	1	1	0
AB-G fault	1	1	0	1
AC-G fault	1	0	1	1
BC-G fault	0	1	1	1

Table 3 Comparison of RBFNN output with actual output

Type of fault	Actual output value for phase A	Actual output value for phase B	Actual output value for phase C	RBFNN output for phase A	RBFNN output for phase B	RBFNN output for phase C
ABC-G	1	1	1	1.0000	1.0000	1.0000
ABC	1	1	1	1.0000	1.0000	1.0000
AB-G	1	1	0	1.0000	1.0000	−0.0000
AC-G	1	0	1	1.0000	−0.0000	1.0000
BC-G	0	1	1	−0.0000	1.0000	1.0000

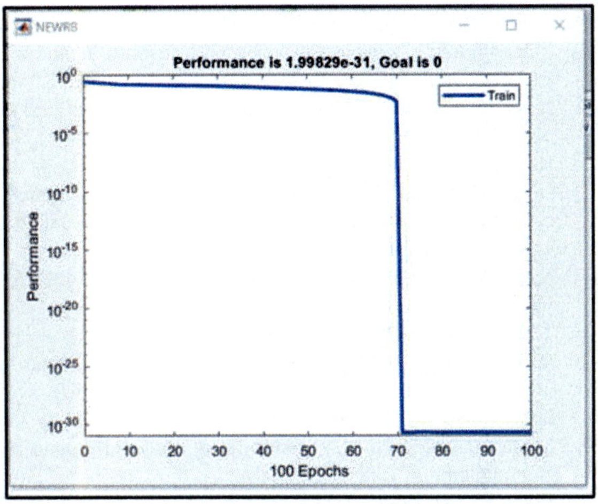

Fig. 2 Performance of RBFNN training data

RBFNN has provided the output of all 30 cases, which was compared with the actual output of all the 30 cases (Fig. 2).

Radial basis function neural network is very fast and has completed its training within short period of time. With 100 neurons, it has provided highest successful training with error equal to 1.99829e−31.

5 Conclusion

RBFNN has successfully and accurately detected the output of all the 30 cases. This shows that it has accuracy of 99.16%. RBFNN is much faster from the traditional wavelet method. Wavelet transform depends on the threshold value to classify the different types of faults in power system, which is not generalized and cannot be used for all types of power system, leading to lack of directional selectivity. Therefore, to overcome this limitation, RBFNN method has been used. We can see this from the graph where 108 cases have been trained in very short period of time.

References

1. Ray P, Mishra DP, Budumuru GK (2016) Hybrid technique for underground cable to locate the fault. In: 2016 International conference on information technology (ICIT), pp 235–240. https://doi.org/10.1109/ICIT.2016.055
2. Sudhir KS (2013) Wavelet based fault detection & ANN based fault classification in transmission line protection scheme. Int J Eng Res Technol 2(5):1492–1500
3. Kirubadevi S, Sutha S (2017) Wavelet based transmission line fault identification and classification. In: 2017 International conference on computation of power, energy information and communication (ICCPEIC). IEEE

4. Rao YS, Kumar GR, Rao GK (2017) A new approach for classification of fault in transmission line with combination of wavelet multi resolution analysis and neural networks. Int J Power Electron Drive Syst 8(1):505
5. Jin N, Liu D (2008) Wavelet basis function neural networks for sequential learning. IEEE Trans Neural Netw 19(3):523–528
6. Ray P, Mishra DP, Dey K, Mishra P (2017) Fault detection and classification of a transmission line using discrete wavelet transform & artificial neural network. In: 2017 International conference on information technology (ICIT), Bhubaneswar, pp 178–183. https://doi.org/10.1109/ICIT.2017.24
7. Ray P, Mishra DP, Panda DD (2015) Hybrid technique for fault location of a distribution line. In: 2015 Annual IEEE India conference (INDICON), New Delhi, pp 1–6. https://doi.org/10.1109/INDICON.2015.7443134
8. Ren S, Gao L (2009) Linking direct orthogonal signal correction and wavelet transform with radial basis function neural network to analyze overlapping spectra. In: 2009 Second international workshop on knowledge discovery and data mining. IEEE
9. Mishra DP, Ray P (2018) Fault detection, location and classification of a transmission line. Neural Comput Appl 30:1377–1424
10. Sidhu TS, Xu Z (2010) Detection of incipient faults in distribution underground cables. IEEE Trans Power Deliv 25(3):1363–1371
11. Borghetti A, et al (2008) Continuous-wavelet transform for fault location in distribution power networks: definition of mother wavelets inferred from fault originated transients. IEEE Trans Power Syst 23(2):380–388
12. Baqui I et al (2011) High impedance fault detection methodology using wavelet transform and artificial neural networks. Electr Power Syst Res 81(7):1325–1333
13. Lin Y, Wang FY (2005) Modular structure of fuzzy system modeling using wavelet networks. In: Proceedings. 2005 IEEE networking, sensing and control. IEEE

Study of DVR Controller for Mitigation of Three-Phase System Short Duration Fault

Lipsa Pani, Abhimanyu Mohapatra, Prakash K. Ray, and Asit Mohanty

Abstract In the recent years due to the increase of sensitive elements in the power system grid, power quality faces common distortions such as harmonics, voltage sag, voltage swell, high inrush current, and voltage flickering. The most occurring phenomenon is voltage sag in transmission and distribution line. Owing to these problems, power quality grabs major attention of researchers to provide enhanced quality of power that can give the most business answer for relieve voltage list by infusing voltage as well as power into the framework. This paper portrays the viability of utilizing dynamic voltage restorers (DVRs) to relieve voltage drops in power dissemination frameworks at basic burdens by using PI and purposed ANN controller in *three-phase* system. The DVR is one of the desired elements to compensate voltage sag or swell occurred in the system. It is a power electronic-based gadget based on VSI that gives three-stage controllable voltage source, whose voltage vector (magnitude and phase) adds to the source voltage during drop (sag) occasion, to reestablish the heap voltage to pre-list conditions. The DVR can reestablish the heap voltage within couple of milliseconds.

Keywords ANN controller · DVR · PI · Power quality · Voltage sag · VSC

1 Introduction

Electrical power system works efficiently and undisturbed power flows through it.

But in present era due to high demand of power, the complexity of power system grid increases to meet the need of end users and use of sensitive electronics devices in transmission line enhances the chance of power disturbances in the system which degrades the life span of the equipment. These disturbances like voltage sag, swell, harmonics, flickering, and frequency deviation costs power quality issues affecting end users and industries.

L. Pani (✉) · A. Mohapatra · P. K. Ray · A. Mohanty
Department of Electrical Engineering, Odisha University of Technology and Research, Bhubaneswar 751029, India
e-mail: priyalipani@gmail.com

A. Rani et al. (eds.), *Signals, Machines and Automation*, Lecture Notes in Electrical Engineering 1023, https://doi.org/10.1007/978-981-99-0969-8_10

Power electronics devices like D-STATCOM and DVR are most effective to eliminate voltage sag/swell. Voltage sag and swell can cause sensitive equipment (such as in semiconductor) to fail, or shutdown, as well as create a large current unbalance that could blow fuses or trip breakers. Voltage swell is less occurring in transmission system while voltage sag is most frequently occurring problem [1].

As the size is small and price is cost effective for DVR in comparison with the other costume devices with more energy capacity. It injects only the required voltage to upgrade the voltage sag to the normal voltage level. It requires less maintenance.

It has the capability to generate the required power by using the converter from DC to AC.

Voltage sag is a short duration drop in voltage magnitude caused for some milli seconds due to sudden induced fault and high inrush current. Single line to ground fault is most occurring nearly 80%, whereas in two-phase and three-phase, fault occurs nearly about 20%. It reduces the efficiency of the system. When a fault or short circuit occurs in a feeder high impedance current flow through it. By the conservation law, the faulty current affects the voltage causing voltage drop in the other feeder. This voltage drop is known as voltage sag where the voltage rms is dropped by 10–90% from the nominal voltage [1].

2 DVR System Modelling

DVR is a series compensating device connected in series between the common coupling point and load voltage. It is used to overcome voltage sag and swell that occurs in transmission and distribution system. It is a solid-state DC–AC converter that can generate and absorb real and reactive power for generating desired load voltage at load side by injecting required synchronous three-phase AC voltage [2]. DVR is effective for critical loads and requires less maintenance with easy implementation into the system [3, 4] (Fig. 1).

Fig. 1 DVR installment in power system in series

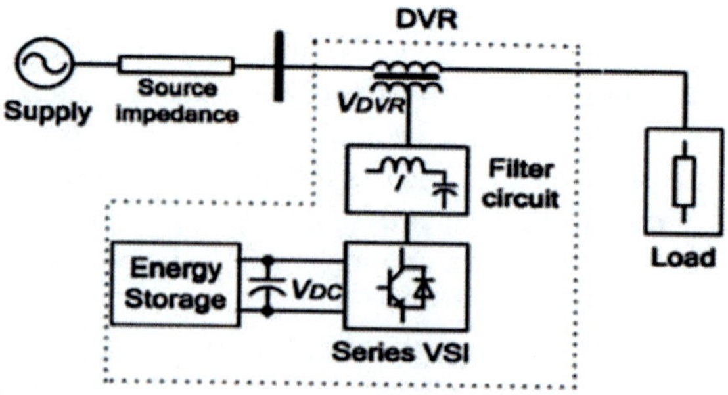

2.1 DVR Components

DVR consists of voltage source inverter, energy storage devices, and filter circuit [4].

(a) **The Converter**: Voltage source inverter is used to convert DC voltage to AC voltage. VSI consists of semiconductor devices (e.g., IGBT, MOSFET, rectifier, and thyristor) which are also source of harmonics. Thus, to eliminate these, filter circuit has been placed.
(b) **Filter Circuit**: The output of inverter is distorted and contains harmonics. So, to filter high order switching harmonics generated by the pulse width modulation (PWM), filter circuit is used. So as to improve power quality.
(c) **Energy storage**: It is a DC power storage device (e.g., battery and ultra-capacitor) that presents as DC-link to synthesize required injected AC voltage. The depth and duration of sag decide the capacity of energy storage device.

2.2 DVR Operation

(a) **By-pass mode**: It isolates the system and gives alternative path to the fault to flow. Thus, system is protected from fault trap.
(b) **Stand-by mode**: Injection voltage is zero as full load current passes through the primary side of transformer. It is ready compensation when the source voltage is at rated level.
(c) **Injection/Boosting mode**: When voltage drops from nominal voltage, DVR starts to work to boost voltage rms through the booster transformer. Injection voltage is the difference between pre-sag voltage rms and sag voltage rms. It supplies real power to improve the voltage profile [5].

2.3 DVR Performance Calculation

See Fig. 2.

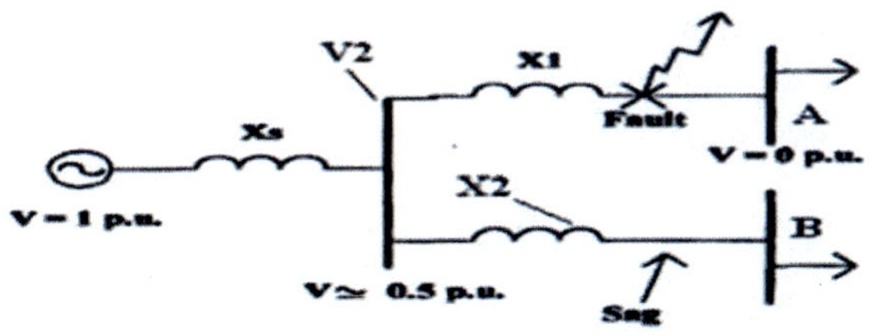

Fig. 2 Voltage sag calculation

(a) **Sagging voltage calculation**

Let us consider.
Supply voltage = V_S, Load voltage = V_L, Injection voltage = V_{inj}.
For feeder 1:
Load A = Z_{loadA}, Reactance = x_1, Current = I_1.
For feeder 2:
Load B = Z_{loadB}, Reactance = x_2, Current = I_2, $I = I_1 + I_2$.
At no fault condition;

$$I = \frac{V_2}{x_1 + Z_{\text{loadA}}} + \frac{V_2}{x_2 + Z_{\text{loadB}}} \cdot (V_2 = V_1) \tag{1}$$

At fault condition;

$$I = \frac{V_2}{x_2 + Z_{\text{loadB}}} + \frac{V_2}{x_1} \tag{2}$$

Thus,

$$\text{sagging voltage} = V_2 = V_s - I x_s \tag{3}$$

(b) **Now for injection voltage**

$$Z_{\text{th}} = R_{\text{th}} + jX_{\text{th}}$$

$$V_{\text{DVR}} + V_{\text{th}} = V_L + Z_{\text{th}} I_L$$

DVR voltage for injection is

$$V_{\text{DVR}} = V_L - V_{\text{th}} + Z_{\text{th}} I_L \tag{4}$$

For injection, V_L is considered as reference voltage.

$$V_{\text{inj}}\angle\alpha = V_L\angle 0^\circ + Z_{\text{th}} I_L \angle(\beta - \theta) - V_{\text{th}}\angle\delta \tag{5}$$

where α, β, and δ are angles of V_{DVR}, Z_{th}, and V_{th}
Load power factor angle is

$$\theta = \tan^{-1}\left(\frac{\text{reactive power}}{\text{real power}}\right) \tag{6}$$

Table 1 System parameters

Elements	Ratings
Source voltage	11 kV
Load voltage	440
Transformer rating	11 kV/440 V
Fault timing	(0.05–0.08) μs
Filter circuit (RL)	1 Ω, 10e−3H
Filter circuit (RC)	1 Ω, 10e−6F

Thus, power injection of DVR:

$$S_{\text{DVR}} = V_{\text{DVR}} \cdot I_L^* \tag{7}$$

2.4 *System Modeling Parameters*

See Table 1.

3 Artificial Neural Network-Based Controller

3.1 *Controlling Scheme*

1. ANN controller

ANN is used to find the optimal injection voltage; this controller is used with a unit delay block to generate error signal and a reference signal that is generated from abc to dq0. It is passed through dq0 to abc transformation block to generate three-phase signal which provided as an input to PWM for pulse generation so as to meet the desired voltage injection and frequency. ANN does not need any mathematical modeling which provides optimized output.

2. PI controller

Here, two discrete PI controllers have been used in the controlling scheme to regulate the error between the actual and reference signal voltage [6]. Proportional response is adjusted by multiplying the error by constant proportional gain. Integral term [7, 8] is used to integrate magnitude and duration of error. Three-phase PLL is used for phase shifting, and output of PLL is used to give signal to abc to dq0 transformation (direct-quadrature-zero) which converts load voltage from abc frame to dq0 frame (Figs. 3 and 4).

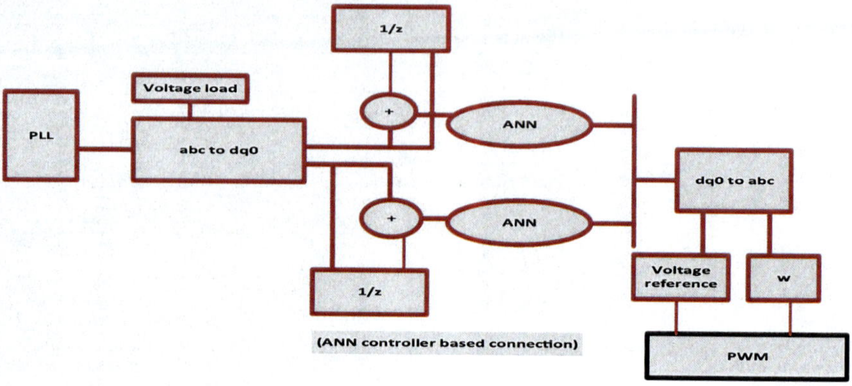

Fig. 3 ANN controller-based circuit

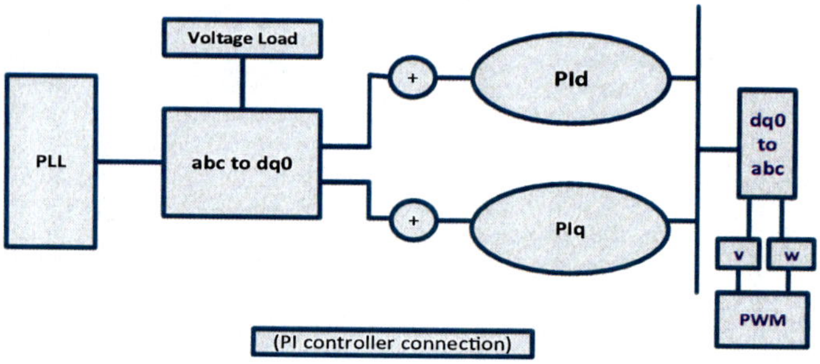

Fig. 4 PI controller-based circuit

3.2 *Mathematical Control Technique of DVR*

The prior role of DVR is to detect voltage sag and uplift the sagging voltage to its nominal voltage; two-level converter PWM is used for pulse triggering during the sag period [6]. The starting and ending time period of voltage sag can be calculated by dq0 method where d is depth and q is phase shift.

$$V_0 = \frac{1}{3}(V_a + V_b + V_c) = 0. \tag{8}$$

$$V_d = \frac{2}{3}\left[\sin wt + V_b \sin\left(wt - \frac{2\pi}{3}\right) + V_c \sin\left(wt + \frac{2\pi}{3}\right)\right] \tag{9}$$

$$V_q = \frac{2}{3}\left[\cos wt + V_b \cos\left(wt - \frac{2\pi}{3}\right) + V_c \cos\left(wt + \frac{2\pi}{3}\right)\right] \tag{10}$$

Then, the three-phase voltage is converted into two quadrature constant voltages V_d and V_q.

4 Simulation and Result

At normal condition, the voltage magnitude remains constant but during the fault condition period, i.e., 0.05–0.08 μs, the voltage magnitude decreases, i.e., sag ranges to near about 64.8% of the nominal voltage (Fig. 5).

As we are using electronics semiconductor devices and controllers in the system, it also generates harmonics by itself. The below graphs (Figs. 7 and 8) show harmonics induced by the two controllers. ANN controller shows in phase harmonics throughout the waveform, whereas the amplitude of the harmonics produced by the PI controller during the sagging period is higher (Fig. 6).

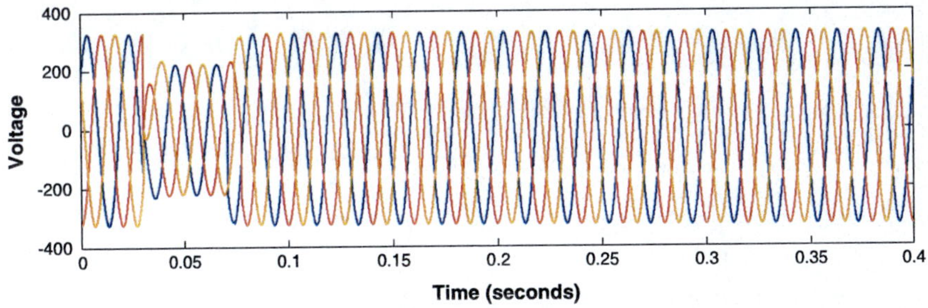

Fig. 5 Sagging voltage

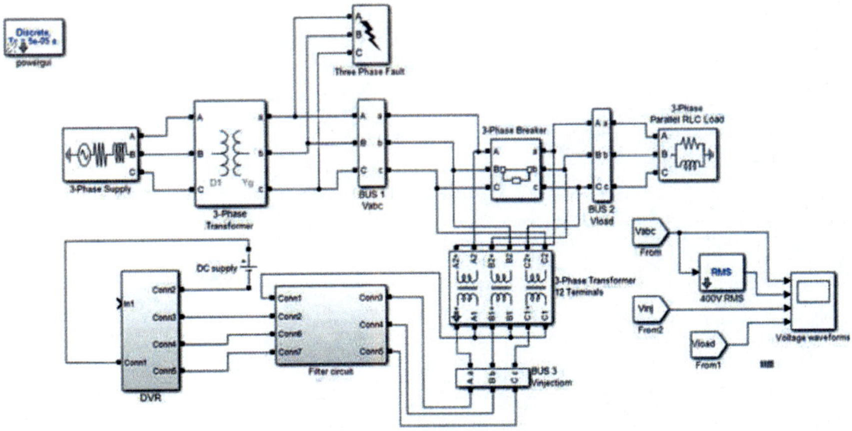

Fig. 6 DVR connected transmission line with fault

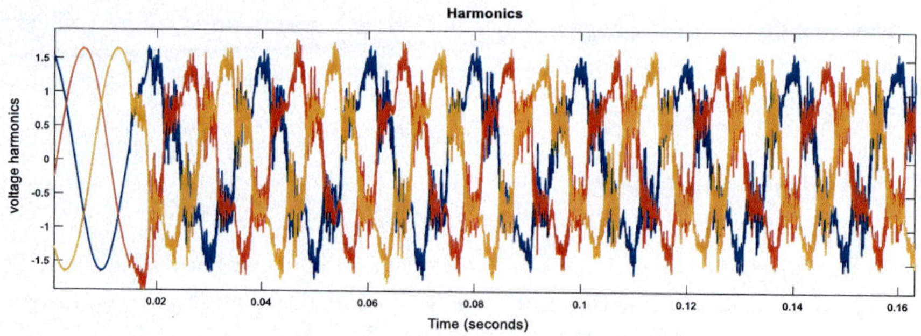

Fig. 7 Harmonics by ANN control

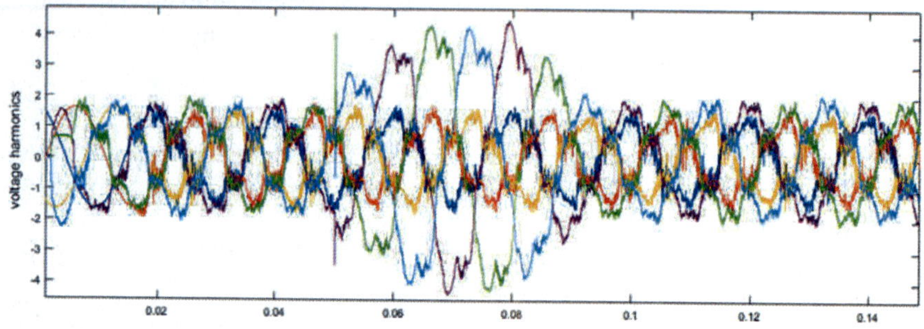

Fig. 8 Harmonics by PI control

Injection voltage depends upon the voltage sag. The difference of the pre-sag voltage magnitude and sag voltage magnitude is injected through the series transformer to boost the sagging voltage toward its nominal voltage magnitude.

The below figures clearly show the difference in injection voltage [5] (Figs. 9 and 10).

Post-sag, the ANN controller, the waveforms contain less distortion with less harmonics than PI controller waveform (Figs. 11 and 12).

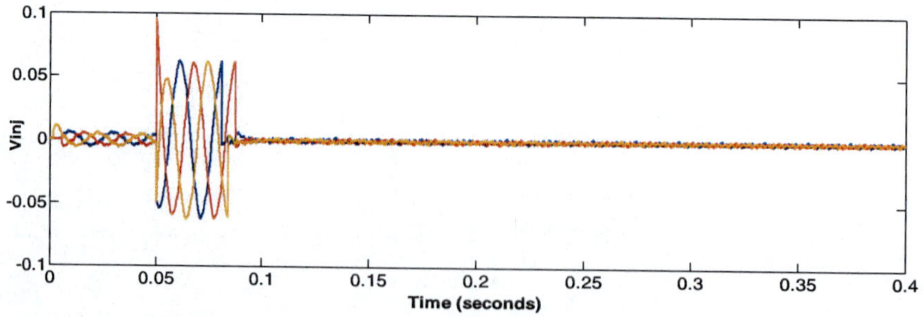

Fig. 9 Voltage injection by PI control

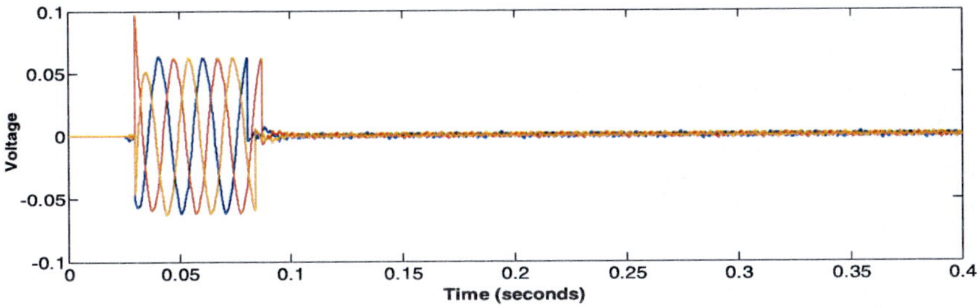

Fig. 10 Voltage injection by ANN control

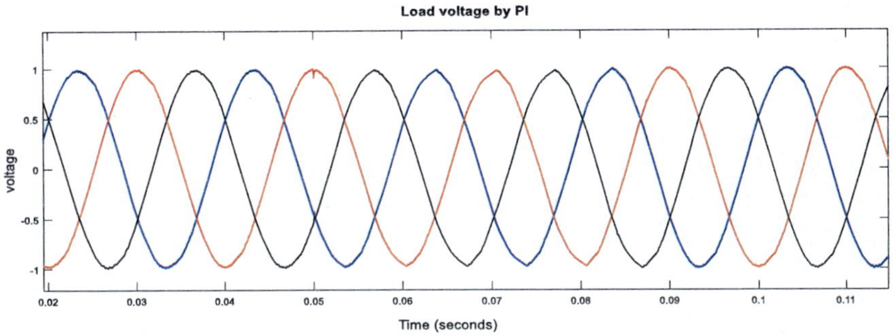

Fig. 11 Post sagging load voltage by PI controller

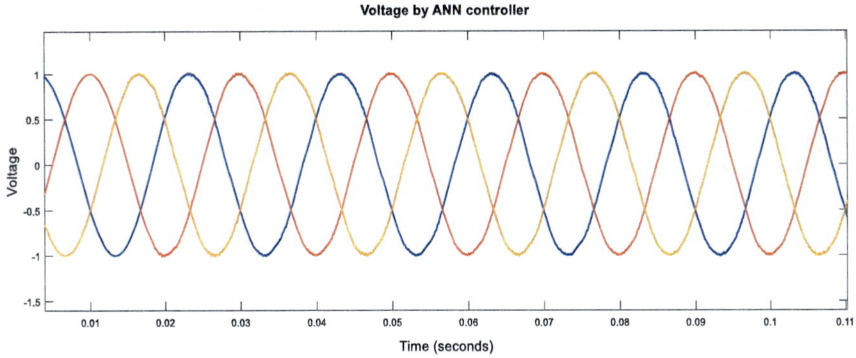

Fig. 12 Post sagging load voltage by ANN controller

4.1 Analysis and Discussion

Voltage sag occurred for duration 0.3–0.08 s. The DVR is injected with a different voltage, i.e., more than nominal voltage to have a continuous supply in load side in synchronism with line voltage. PI controller is one of the existing controllers, whereas

ANN is purposed controller used for comparison. Total harmonics distortion (THD) with frequency 50 Hz and 5-cycles by FFT analysis is done to compare the sag compensation of DVR with PI is 2.57% and ANN controller is 1.69%.

4.1.: Voltage magnitude comparison in per unit system

Sl. No.	Situation	V_{rms} (pu)
1	System without fault	1.00
2	System connected to DVR with PI controller	0.944
3	System connected to DVR with ANN controller	0.996

5 Conclusion

The operation of DVR is validated by creating a MATLAB/Simulink model. VSI is connected in feedback configuration in series with the coupling transformer to compensate the sagging voltage by injecting active power to the series transformer to boost the voltage profile. For both the controllers (PI and ANN), results have been discussed with proper comparisons. While comparing the PI controller to the ANN controller, ANN controller gives more accurate and desired output voltage.

References

1. Abas N, Dilshad S, Khalid A, Saleem MS, Khan N (2020) Power quality improvement using dynamic voltage restorer. IEEE Access 8:164325–164339
2. Soomro AH, Larik AS, Mahar MA, Sahito AA, Soomro AM, Kaloi GS (2021) Dynamic voltage restorer—A comprehensive review. Energy Rep 7:6786–6805
3. Prakash Y, Sankar S (2014) IEEE 2014 Power and energy systems conference: towards sustainable energy (PESTSE)—Bangalore, India (2014.03.13–2014.03.15). Power quality improvement using DVR in power system, pp 1–6
4. Wahab SW, Yusof AM (2006) Voltage sag and mitigation using dynamic voltage restorer (DVR) system. Elektrika J Electr Eng 8(2):32–37
5. Vaidya HD et al (2020) Single-phase series compensator circuit for mitigating voltage sag or swell in the power system networks—methodology and modelling. In: 2020 IEEE international conference on environment and electrical engineering and 2020 IEEE industrial and commercial power systems Europe (EEEIC/I&CPS Europe), pp 1–5
6. Joga SRK, Avinash R, Ramu M (2014) Analysis of DVR performance for voltage flickers mitigations in distribution system using fuzzy logic as controller. Int J Sci Eng Res 5(4)
7. Ramasamy AK, Krishnan Iyer R, Ramachandaramuthy VK, Mukerjee RN (2005) Dynamic voltage restorer for voltage sag compensation. In: 2005 International conference on power electronics and drives systems, pp 1289–1294
8. Ray PK, Das SR, Mohanty A, Eddy FY, Krishnan A, Gooi HB (2018) Improvement in power quality using hybrid power filters based on robust extended Kalman filter. In: 2018 19th International Carpathian control conference (ICCC), 28May 2018. IEEE, pp 585–590

Annual Cost and Power Loss Minimization in a Radial Distribution Network by Capacitor Allocation

Ankush Tandon and Sarfaraz Nawaz

Abstract In this paper, an imperialistic competitive algorithm (ICA) is presented to identify the best possible size and site of capacitors. The primary objective of this work is to minimize the cost of energy losses while maintaining voltage profile within limits. In order to illustrate the ICA's effectiveness and feasibility, IEEE 33-bus system is used. The analysis has been carried out at three different load levels namely nominal load (100% of load), light load (50% of load), and heavy load level (150% of load). Furthermore, the results of adopted method are compared to other recent approaches for highlighting the benefits of the algorithm in order to minimize the cost of energy losses as well as maximizing the net saving.

Keywords Distribution system · Cost of energy loss · Capacitor placement · Voltage profile · Power loss

1 Introduction

The classic power system comprises of three significant components, power generation, power transmission lines, and distribution system (DS). The increase in load demand leads to more losses and voltage profile reduction. The reduction in voltage at nodes is increases as moves away from the substation end to the consumer end in the DS. The insufficient reactive power is mainly responsible for the voltage drop. It may even cause voltage collapse in certain areas of critical loading.

To deal with the increasing demand and improving the system efficiency in distribution system, capacitor placement has becomes one of the most favorable alternatives [1]. To improve voltage and reduce the cost of energy loss in distribution system, capacitor placement has become one of the most recommended options. Capacitor placement is a suitable method in reducing the cost incurred in terms of energy loss as it provides reactive power compensation to the DS. The amount of compensation of reactive power in DS is very much linked to the insertion of capacitors with

A. Tandon (✉) · S. Nawaz
Department of Electrical Engineering, Swami Keshvanand Institute of Technology, Management and Gramothan, Jaipur, India
e-mail: ankush@skit.ac.in

A. Rani et al. (eds.), *Signals, Machines and Automation*, Lecture Notes in Electrical Engineering 1023, https://doi.org/10.1007/978-981-99-0969-8_11

optimal location and size [2]. Mustafa et al. [3] incorporated genetic algorithm (GA) to trace the optimal site of the capacitor with suitable size to improve the voltage profile and reduce power losses in the DS. In [4], RND Costa presented a new approach quantum behaved PSOwhich is inspired by the natural behavior of swarm and quantum mechanics theories. Bhadoria et al. [5] solved the capacitor placement problem with the help of iterative search technique in the radial DS to enhance the voltage of the buses and reduce the power losses. To diminish the cost of capacitor installation and losses, a new approach based on mimics of flower pollination is introduced by Tamilselvan et al. in [6]. EA Al Ammar et al. [7] suggested a hybrid technique which is blend of genetic algorithm and fuzzy expert system approach for allocation of capacitors in the distorted RDNto reduce harmonic distortion, increase in the voltage profile, and reduce power loss. In [8], author incorporated sine cosine algorithm (SCA) for the allocation of capacitor in DS to enhance the voltage stability and mitigate the cost incurred due to real power losses. Sultana et al. [9] formulate the problems of capacitor placement and solved them by implementing oppositional cuckoo optimization algorithm (OCOA) to minimize the total cost. In [10], reduction in cost of energy losses, voltage deviation, power losses, and improving the voltage stability index is achieved using water cycle algorithm (WCA).

In this paper, a popular technique known as imperialist competitive algorithm (ICA) is introduced. ICA is employed for the determination of optimal site of the capacitor with optimal size for maximizing the net saving as well as improving voltage profile. The effectiveness and feasibility of the suggested ICA technique have been examined on 33-bus at different loading level, i.e., nominal, light, and heavy load level with the help of MATLAB software. It can be observed from results that proposed approach has superior performance as compared to other techniques mentioned in this paper.

2 Problem Formulation

The main objective of this paper is to reduce the cost of energy loss in the DS after capacitor allocation. The cost function consists of two parts as shown in Eq. 1. The first one represents the cost of real power loss, and the second is the cost of reactive power compensation by capacitor [11].

Total cost can be computed by given equation:

$$\text{Minimum Cost} = K_p\, P_{\text{loss}} + \sum_{k=1}^{n} K_{fc}\, Q_{fc} \tag{1}$$

where, K_p is the cost of real power loss which is 168 \$/kW (11,928 rupees/kW) [7]. P_{loss} is the power loss in kW. K_{fc} is the purchase cost of the capacitor is 3 \$/kVAr (213 rupees/kVAr) [12]. The size of the capacitor is denoted by Q_{fc} in KVAr. n is the total number of capacitors installed in the distribution system.

2.1 Constraints

a. *Apparent power flow constraint*

$$S_i \leq S_{(i,\max)}$$

b. *Bus voltage constraint*

$$V_{(m,\min)} \leq V_m \leq V_{(m,\max)}$$

c. *Power factor*

$$\mathrm{PF}_{\min} \leq \mathrm{PF}_{\mathrm{real}} \leq \mathrm{PF}_{\max}$$

d. *Capacitor size*

$$Q_c^{\max} \leq 0.7 * Q_{load}$$

3 Imperialist Competitive Algorithm

Lucas and Gargari were the first to introduce ICA, which is based on the imperialist competitions, a social-political process [13]. Figure 1 shows the ICA's flowchart.

The proposed ICA algorithm considers an initial population, named as countries similar to the other evolutionary algorithm. Imperialists are chosen from among the best countries. The remaining populations become the colonies. Colonies from the generated population are distributed among the imperialists according to imperialist power. In this problem, candidate bus for capacitor allocation is represented by imperialist and other location designated as colonies. Candidate bus and remaining bus together form an empire.

Now, all the colonies begin to travel on the way to suitable imperialists. The sum of power imperialist and the power of its relevant colonies are used to measure an empire's overall power. Then, the competition of imperialists starts between all the empires. If any of the empire is not capable to increase their power or cannot compete the competition will be terminated from the competition. Imperialist competition leads to a decline in the power of a weaker empire and a progressive increase in the strength of a powerful empire.

The empire which losing its power continuously will collapse. Because of this imperialistic competition, all countries will consolidate into a situation where there is only one empire which possesses all colonies. In this newly updated empire, imperialist and colonies have the same power.

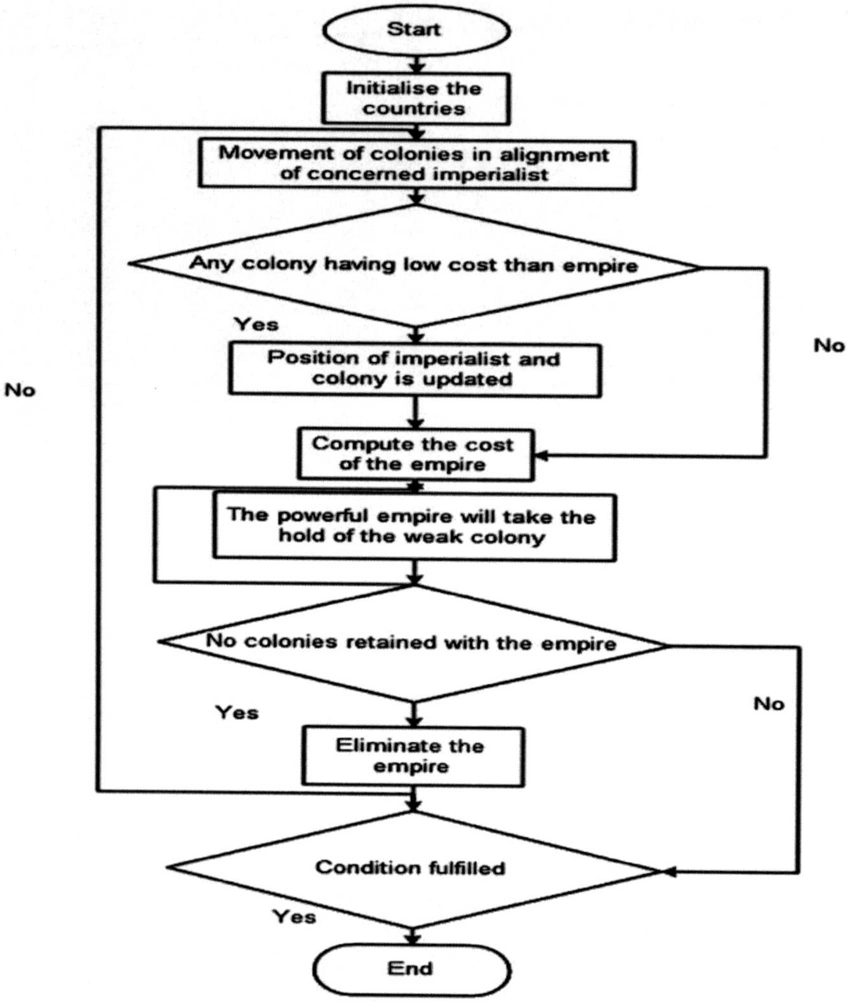

Fig. 1 Flowchart of ICA

4 Results and Discussion

The suggested ICA technique has been applied on standard 33-bus to identify the suitable site of capacitor units with optimal size to minimize the cost of energy loss would be minimum. Three different loading levels, i.e., nominal, light, and heavy load levels, are also considered to check the effectiveness and feasibility of the ICA technique. The results are compared with other existing algorithms. The MATLAB software is used for coding of ICA algorithm.

4.1 IEEE 33-Bus System

The 33-bus radial DS consists of 32 branches and 33 buses. The reactive and real power loads are 2300 kVAr and 3715 kW, respectively, with 12.66 kV and 100 mVA as the base values of the 33-bus test system. The system has reactive and real power losses 136.60 kVAr and 202.4 kW, respectively.

The numerical results for capacitor placement are shown in Table 1 at nominal, light, and heavy load level. The percentage loss reduction and net saving are 32.53 and 20.74, 30.99 and 5.7, and 34.69 and 27.41 for nominal, light, and heavy load level, respectively, with improving voltage profile as compared to base case. Bus no 13 and 30 are identified as suitable site for placement of capacitor for different levels of load.

Table 2 shows the comparison of numerical results of the proposed ICA algorithm with other existing algorithms such as Chu and Beasley genetic algorithm (CBGA) [12], flower pollination algorithm (FPA) [6], and sine cosine algorithm (SCA) [8] to reveal the feasibility and effectiveness of proposed technique at nominal

Table 1 Result of 33-bus system

			Load level					
			Light load level		Nominal load level		Heavy load level	
(Base case) without capacitor	Loss	kW	47.07		202.49		496.33	
		kVAr	31.35		135.01		331.39	
	V (min) pu		0.9583		0.9131		0.8634	
	Cost of energy loss (lakhs)		5.61		24.15		59.20	
	Total cost (lakhs)		5.61		24.15		59.20	
With capacitor	Location and size (kVAr)		13	202.4	13	407	13	616
			30	465	30	930	301	410
	Loss	kW	32.48		136.60		324.13	
		kVAr	21.62		90.96		216	
	V (min)		0.9693		0.9366		0.9019	
	Loss reduction in %		30.99		32.53		34.69	
	Cost of energy loss (lakhs)		3.87		16.29		38.66	
	Capacitor cost (lakhs)		1.42		2.84		4.31	
	Total cost (lakhs)		5.29		19.14		42.97	
	Net saving (lakhs)		0.32		5.01		16.23	
	Net saving in %		5.7		20.74		27.41	

Table 2 Comparative result analysis for 33-bus system at nominal load

	Base case (without capacitor)	With capacitor								
		CBGA [12] (2021)		FPA [6] (2018)			SCA [8] (2018)		Proposed ICA	
Optimal location and size (MVAr) of capacitors	–	12	0.45	13	0.45	24	14	0.35	**13**	**0.407**
		24	0.45	0.45	30	0.9	30	1	**30**	**0.930**
		30	1.05							
Total kVAr		1950		1800			1350		**1337**	
Power loss (kW)	202.49	138.41		139.07			142.551		**136.60**	
Loss reduction in %		31.64		31.32			29.60		**32.53**	
V_{min} (pu)		–		0.9327			0.9300		**0.9366**	
Cost of energy loss (lakhs)	24.15	16.50		16.58			17.00		**16.29**	
Capacitor costs (lakhs)	–	4.15		3.83			2.87		**2.84**	
Total costs (lakhs)	24.15	20.65		20.41			19.87		**19.13**	
Net saving (lakhs)	–	3.50		3.74			4.28		**5.01**	
Net saving in %	–	14.49		15.48			17.72		**20.74**	

The bold is signifying the proposed results of ICA

load level. It can be observed from Table 2 that the suggested ICA has a higher reduction in cost of energy loss in comparison with other. The voltage profile is depicted in (Fig. 2).

5 Conclusion

This paper exhibits the development and implementation of a unique approach for optimal capacitor allocation to reduce the cost of energy losses in distribution system. A new technique known as imperialist competitive algorithm (ICA) is introduced which is inspired by the competition of imperialist. ICA is implemented for the determination of suitable site of the capacitor. The effectiveness and feasibility of the proposed ICA technique have been examined on 33-bus test systems at different loading level, i.e., light, nominal, and heavy load level with the help of MATLAB

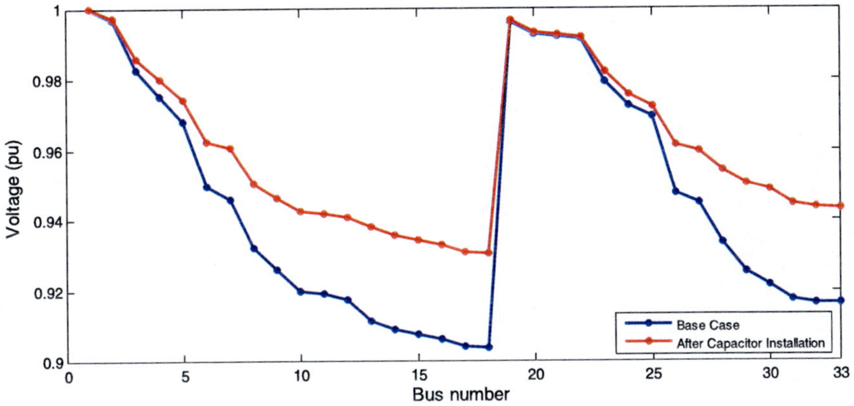

Fig. 2 Variation of voltage profile at nominal loading (IEEE 33-bus system)

software. The obtained numerical results compared with other existing methods in literatures, and result proves the superiority of the ICA technique in all scenarios.

References

1. Mostafa SMG, Shingh JG, Haque HME (2020) An extensive literature review and new proposal on optimal capacitor placement in distribution systems. J Eng Adv 1(04):150–169
2. Selim A, Kamel S, Jurado F (2020) Capacitors allocation in distribution systems using a hybrid formulation based on analytical and two metaheuristic optimization techniques. Comput Electr Eng 85:106675
3. Mustafa S, Arief A, Nappu MB (2018) Optimal capacitor placement and economic analysis for reactive power compensation to improve system's efficiency at Bosowa Cement Industry, Maros. In: 2018 international conference on information and communications technology (ICOIACT). IEEE, pp 778–783
4. Costa Filho RND (2018) Optimal capacitor placement in radial distribution system using QPSO. In: 2018 SimposioBrasileiro de SistemasEletricos (SBSE). IEEE, pp 1–6
5. Bhadoria VS, Pal NS, Shrivastava V, Jaiswal SP (2018) Optimal sitting and sizing of capacitor using iterative search method for enhancement of reliability of distribution system. In: Advances in energy and power systems. Springer, Singapore, pp 123–129
6. Tamilselvan V, Jayabarathi T, Raghunathan T, Yang X-S (2018) Optimal capacitor placement in radial distribution systems using flower pollination algorithm. Alexandria Eng J 57(4):2775–2786
7. Al-Ammar EA, Ghazi GA, Ko W (2018) Impact of ambient temperature on shunt capacitor placement in a distorted radial distribution system. Energies 11(6):1585
8. Biswal SR, Shankar G (2018) Optimal sizing and allocation of capacitors in radial distribution system using sine cosine algorithm. In: 2018 IEEE international conference on power electronics, drives and energy systems (PEDES). IEEE, pp 1–4
9. Sultana S, Roy PK (2018) Capacitor placement in radial distribution system using oppositional cuckoo optimization algorithm. Int J Swarm Intell Res (IJSIR) 9(3):64–95
10. Abou El-Ela AA, El-Sehiemy RA, Abbas AS (2018) Optimal placement and sizing of distributed generation and capacitor banks in distribution systems using water cycle algorithm. IEEE Syst J 12(4):3629–3636

11. George T, Youssef A-R, Ebeed M, Kamel S (2018) Ant lion optimization technique for optimal capacitor placement based on total cost and power loss minimization. In: 2018 international conference on innovative trends in computer engineering (ITCE). IEEE, pp 350–356
12. Riaño FE, Cruz JF, Montoya OD, Chamorro HR, Alvarado-Barrios L (2021) Reduction of losses and operating costs in distribution networks using a genetic algorithm and mathematical optimization. Electronics 10(4):419
13. Atashpaz-Gargari E, Lucas C (2007) Imperialist competitive algorithm: an algorithm for optimization inspired by imperialistic competition. In: 2007 IEEE congress on evolutionary computation. IEEE, pp 4661–4667

Automated Signal Monitoring of LT8228 Buck–Boost Converter

Rahul Raj, Bhanu Pratap, and Chandana HB

Abstract The use of DC systems has increased with the increase in renewable energy systems. DC power converters are widely used for transferring electrical energy with minimal losses. The power converters used in industries have different types of monitoring pins whose readings need to be measured automatically to keep track of power flow. This paper will discuss the automation of monitoring pins of the LT8228 bidirectional buck–boost controller using Arduino Uno to estimate the actual current flow through the device. For this purpose, DC2351A is used, which is a demo board for LT8228. Evaluating the current flowing through the circuit is necessary to find the actual power consumed by the load, which can be stored in a database for billing energy consumed.

Keywords Bidirectional DC–DC converter · LT8228 · DC2351A · Arduino Uno

1 Introduction

Energy can be harnessed directly from the sun, even in cloudy weather. Solar energy is used worldwide and is increasingly popular for generating electricity or heating and desalinating water as it is affordable, reliable, and sustainable. According to Sustainable Development Goal—SDG7, the world must ensure access to affordable and clean energy by 2030 [1]. Solar energy can play a significant role in achieving this goal and can be accessed easily in any part of the world without much initial cost. In 2020, the total installed capacity of solar energy was 773,200 MW [2].

R. Raj (✉) · B. Pratap
Department of Electrical Engineering, National Institute of Technology Kurukshetra, Kurukshetra, India
e-mail: raj.rahul2012@gmail.com

B. Pratap
e-mail: bhanu@nitkkr.ac.in

C. HB
Department of Electrical Engineering, Manipal Institute of Technology, Manipal, India

A. Rani et al. (eds.), *Signals, Machines and Automation*, Lecture Notes in Electrical Engineering 1023, https://doi.org/10.1007/978-981-99-0969-8_12

To transmit solar energy, a DC–DC converter is widely used. The DC–DC converter is an electromagnetic device used to convert one voltage level to another voltage level [3]. Its power level ranges from a low level of small batteries to a high level, like high voltage transmission. In transmission, high voltage is preferred over low voltage to decrease transmission losses; for this purpose, DC–DC converter is used. DC transmission is preferred over AC transmission as losses in DC transmission are negligible compared to AC transmission [4].

There are various types of DC–DC converters, such as buck converter, boost converter, buck–boost converter. For smooth transfer of power in both the direction for buck and boost mode, bidirectional DC–DC converters are used. Bidirectional converters can be used to step up the voltage as well as step down the voltage.

Voltages step up when the power is injected into the transmission line and step down when the power has to be taken from the nano-grid. One of the types of bidirectional controllers is LT8228 [5] from analog devices. This controller features buck and boost mode up to 100 V, and current and voltage can be controlled using various pins in the controller.

UNO is probably the most popular Arduino board in the world which is equipped with an Atmega328P processor. Arduino Uno boards operate at 16 MHz. Its 32 KB of program memory, 1 KB of EEPROM, and 2 kB of RAM [6]. It quickly became the industry standard for development boards. This pin header makes the Arduino Uno compatible with most development board shields.

When the power is to be transferred, it is necessary to record the amount of power being transferred for billing of energy used. For this purpose, the controller has two monitoring pins that estimate the output and input current as measuring high current from a microcontroller requires extra setup, which will increase the system price. The signal from these pins need to be observed automatically and record the power being transferred. The 10-bit ADC of Arduino UNO is one of the limitations in measuring the low voltages in these monitoring pins [7].

This paper will discuss the automation of monitoring of signals using Arduino Uno to measure the very low voltage which will estimate the actual current flowing through the circuit. Section 2 presents a brief idea about the bidirectional DC-DC converter and its structure and operation. In Sect. 3, the LT8228 controller and DC2351A demo board connection to the Arduino board have been discussed. The software requirements and calculations for comparing results have been explained in Sect. 4. Experimental results are presented in Sect. 5. Finally, the conclusion is given in Sect. 6, followed up by references.

2 Working of Bidirectional DC–DC Converter

A common DC–DC conventional converter can only transfer power in one direction, which is not helpful when a battery needs to be interfaced, or two sources are to be connected with each other. A bidirectional DC–DC converter is used to transfer power in both directions in a situation like an energy source connected to a motor,

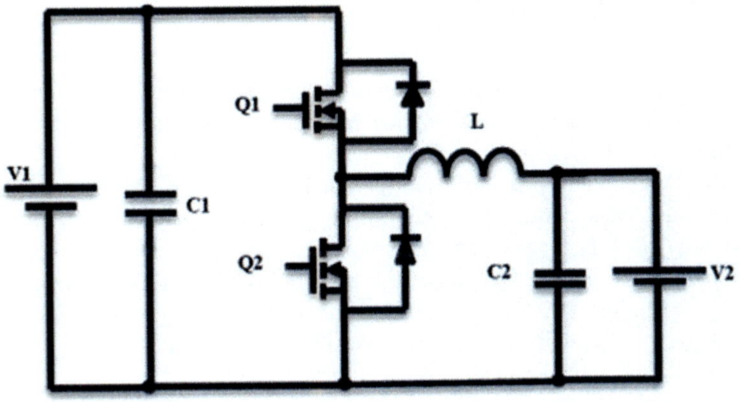

Fig.1 DC–DC bidirectional buck–boost converter [8]

usually in electric vehicles. This bidirectional converter controls the voltage level, and it is used for bucking and boosting the voltage level by controlling the switching circuit. A bidirectional DC–DC converter is also known as a half-bridge DC–DC converter. Many topologies are available, like Cuk, SEPIC, and flyback. The basic buck and boost circuit are connected anti-parallel in this type of converter. It can work in both directions [8].

Mode 1: Buck Mode

In this mode, the voltage at the output side will be lesser than on the input side, but the power on both sides remains the same in ideal conditions. This can be achieved by making the switch in series with the parallel combination of diode and inductor.

In Fig. 1, input supply is given at V1 and output at V2. The switch Q1 and Diode D2 begin conduction, and the circuit operates as the buck converter. Switch Q2 and Diode D1 remain off all the time.

Mode 2: Boost Mode

In this mode, the voltage at the output side will be more than the voltage on the input side, but the power remains the same in ideal conditions. This can be achieved by making the inductor in series with the parallel combination of the switch and the diode.

In Fig. 1, input supply is given at V2 and output at V1. Switch Q2 and Diode D1 begin conduction, and the circuit operates as the boost converter. Switch Q1 and Diode D2 remain off all the time.

3 Hardware Setup Description

Demo board DC2351A is used to measure the monitoring pin voltages [9], as shown in Fig. 2. This board features an LT8228 controller for controlling the output current and voltage.

The DC2351A board is a 500W bidirectional DC-DC converter that can be used as a boost and buck converter by controlling the ground pin. It consists of four protection MOSFETs that offer reverse voltage, reverse current, and short circuit protection on both sides. It also consists of 8 operating MOSFETs that support buck and boost operation. DC2351A board has all the monitoring pins at the bottom, which can be connected to analog inputs of Arduino Uno for calculating the current in the circuit. The pin diagram of LT8228 is shown in Fig. 3. LT8228 consists of gate control and control to the operation of DC2351A, including enabling and switching frequency. It offers features like under-voltage protection and overvoltage protection. It is convenient as the output can be controlled by controlling the input to the feedback pins of the controller. The monitoring outputs are sourced by this controller, which senses input current to the board and gives scaled output.

The simulation of DC2351A has been shown in Fig. 4. The principal aim of this setup is to keep continuous track and measure the IMON1 pin and IMON2 pin. IMON1 pin will give the equivalent voltage for current flowing from the V1 side, and the IMON2 pin will provide the equivalent voltage for current flowing from the V2 side.

Monitoring pins of DC2351A are connected to the 10-bit ADC of Arduino, and the reference voltage of Arduino is set to 1.1 V internally. Since the 10-bit ADC cannot measure low voltage accurately so, using oversampling library, the analog

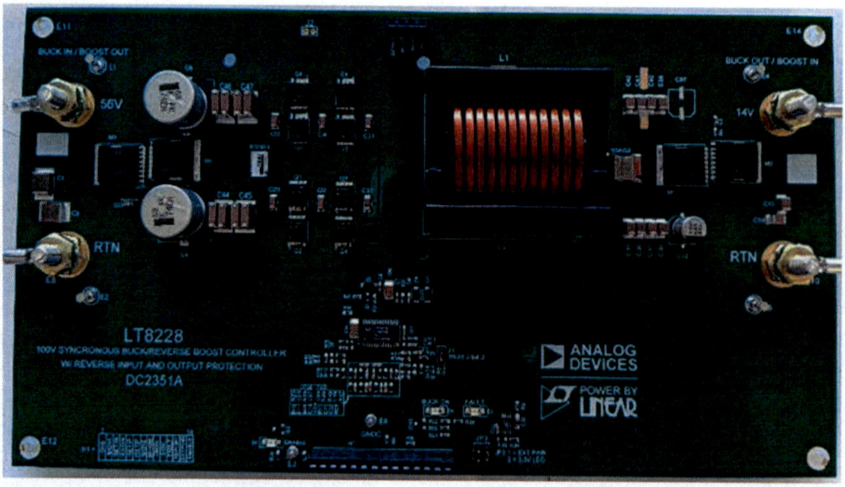

Fig. 2 DC2351A demo board [9]

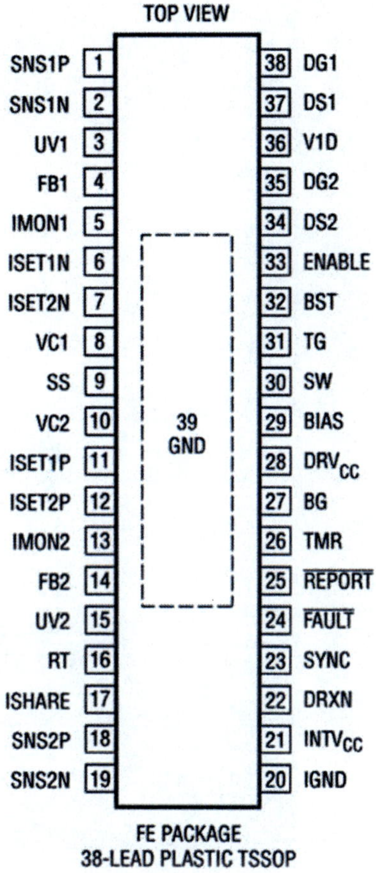

Fig. 3 LT8228 pin configuration [5]

readings are oversampled to 16-bit, which can measure the low voltage accurately. In Fig. 5, the whole setup for measuring monitoring pins with Arduino Uno is shown.

4 Software Requirements

Arduino IDE is an open-source software where electronics can be built easily and efficiently with lower costs in a microcontroller platform and is used for writing the code and for uploading it to the microcontroller. Arduino UNO includes features like ROM, RAM, and flash memory. It also consists of a 5 and 3.3 V supplies along with a few GND points [10].

The method used to get the monitoring pin output is oversampling. Library for oversampling was included named Oversample.h [11]. The internal reference was set

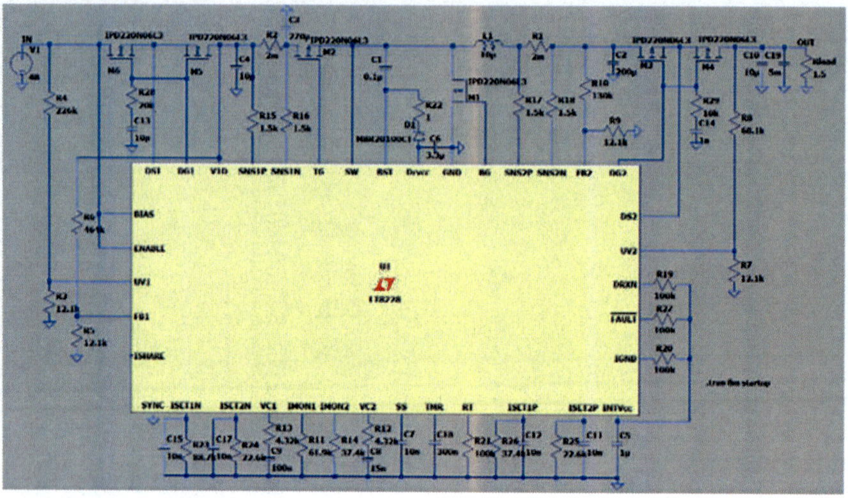

Fig. 4 Circuit diagram of DC2351A

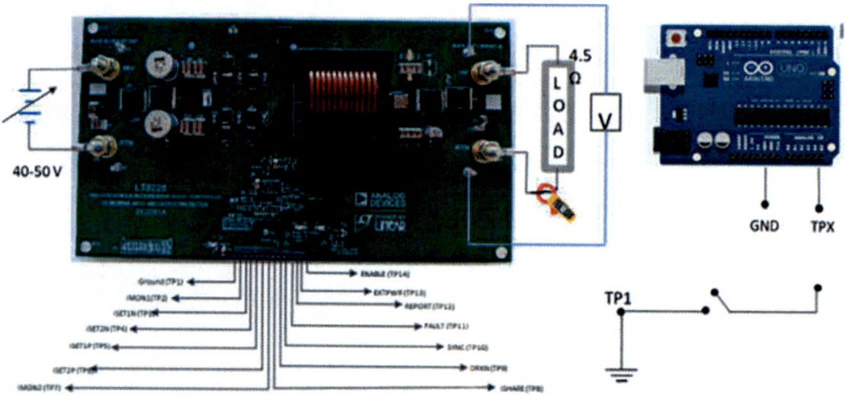

Fig. 5 Hardware setup (proposed work)

for analog reference voltage, and the measured voltage was automatically recorded in an Excel sheet. Part of the Arduino IDE code is shown in Fig. 6.

Following formula has been used to convert the analog reading from Arduino to the actual voltage across pins, which is used when oversampler is set to 16-bit.

$$\text{Pin Voltage(mV)} = \frac{\text{Analog Reading}}{65{,}536} \times 1.1 \times 1000$$

Following formula is used to estimate the actual current flowing through the circuit from the voltage measured from IMON1 and IMON2 pins.

```
void loop()
{
  long scaled0 = sampler0->readDecimated();
  long scaled1 = sampler1->readDecimated();
  long scaled2 = sampler2->readDecimated();
  long scaled3 = sampler3->readDecimated();
  long scaled4 = sampler4->readDecimated();
  long scaled5 = sampler5->readDecimated();
 IMON1 = (((scaled0/65536.00000)*1.1)*1000);
 ISET1N = (((scaled1/65536.00000)*1.1)*1000);
 ISET2N = (((scaled2/65536.00000)*1.1)*1000);
 ISET1P = (((scaled3/65536.00000)*1.1)*1000);
 ISET2P = (((scaled4/65536.00000)*1.1)*1000);
 IMON2 = (((scaled5/65536.00000)*1.1)*1000);
```

Fig. 6 Arduino code on Arduino IDE

$$\text{Computed Current(mA)} = \frac{\text{Pin Voltage}}{40{,}200} \times 750{,}000$$

The milli-volt is measured across the monitoring pins using Arduino ADC using the above method. This voltage is converted to the current flowing through the resistance of monitoring pins. The current then can be scaled to the actual current flowing through the circuit using the scaling factor as per the datasheet of DC2351A.

The current is estimated from the above formula. As the output voltage is already set by choosing the appropriate resistance in the feedback pin while designing the converter, the total power transferred can be calculated using the estimated current and known output voltage.

This calculated power can be transferred to the database and can be used for the billing of power consumed by the user.

5 Experimental Result

The graph for comparing the output current measured using an oscilloscope and using monitoring pins was plotted while keeping the input voltage to DC2351A board constant and varying the load resistance. The current plot for buck mode is shown in Fig. 7, and for boost, mode is shown in Fig. 8.

Similarly, in Figs. 9 and 10, the input current versus input voltage is plotted while keeping the load resistance of DC2351A as constant.

It is evident from the above figures that the computed current converging to the actual current in both buck mode and boost mode. Thus, the difference

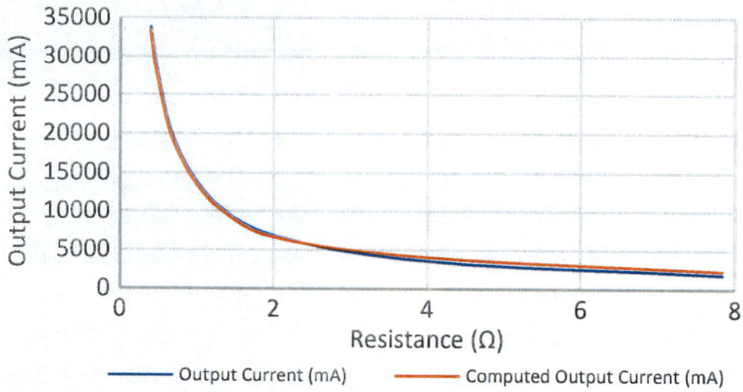

Fig. 7 Comparison between computed and actual current versus load resistance in buck mode

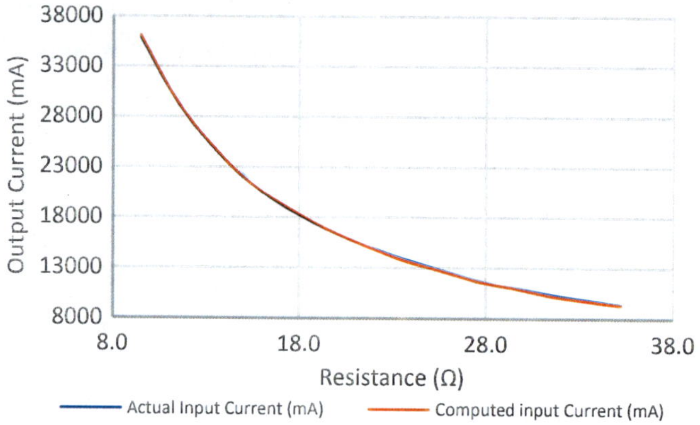

Fig. 8 Comparison between computed and actual current versus load resistance in boost mode

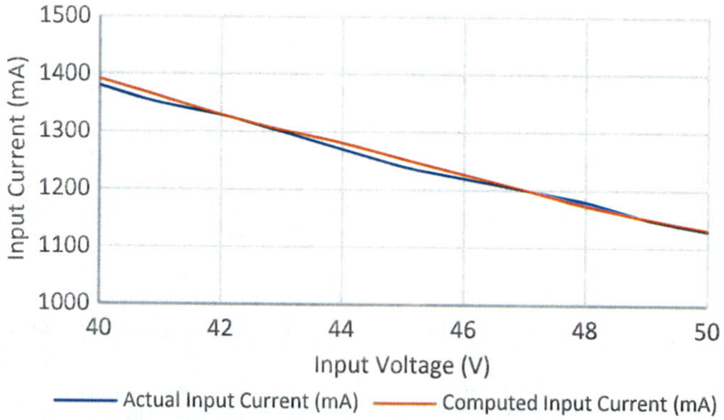

Fig. 9 Comparison between computed and actual current versus input voltage in buck mode

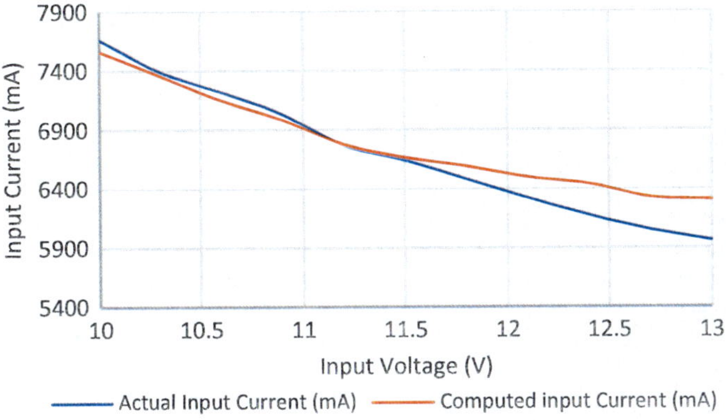

Fig. 10 Comparison between computed and actual current versus input voltage in boost mode

between computed current using Arduino ADC readings and current measured using oscilloscope, i.e., actual current is very less.

6 Conclusion

Measurement of power is a necessary factor when transferring the power. For measuring high current, additional equipment is required conventionally. By using monitoring pins of LT8228, the output current can easily be calculated with good accuracy and very low cost. Arduino Uno was able to calculate the output current, and it has been observed that by increasing the resolution of ADC by oversampling the readings, the overall calculated current is almost equal to the actual current flowing in the circuit measured using the oscilloscope's current probe. By oversampling the ADC pin of Arduino, the power transferred using a DC–DC converter can be easily calculated using signal monitoring on monitoring pins of DC2351A without using expensive current sensors separately.

References

1. Villanueva Ulfgard R (2017) Mexico from the millennium development goals to the sustainable development goals: congruence and dissonance in development compromises. In: Villanueva Ulfgard R (ed) Palgrave Macmillan US, New York, pp 17–51. https://doi.org/10.1057/978-1-137-58582-02
2. Jaganmohan M (2021) Global cumulative installed solar PV capacity 2000–2020
3. Sorouri H, Sedighizadeh M, Oshnoei A, Khezri R (2022) An intelligent adaptive control of dc–dc power buck converters. Int J Electr Power Energy Syst 141:108099
4. Venter C, Raji A, Adonis M (2015) The dc house for low power households dc-dc converter analysis. In: 2015 international conference on the domestic use of energy (DUE). IEEE, pp 143–146

5. Analog Devices: Bidirectional Synchronous 100 V Buck/Boost Controller with Reverse Supply, Reverse Current and Fault Protection (2019) Analog devices. Rev. 0
6. Arduino: Arduino Uno R3 (2022) Arduino
7. Kondaveeti HK, Kumaravelu NK, Vanambathina SD, Mathe SE, Vappangi S (2021) A systematic literature review on prototyping with arduino: applications, challenges, advantages, and limitations. Comput Sci Rev 40:100364
8. Somavanshi C. What is a bidirectional DC–DC converter, circuit diagram, working. https://www.cselectricalandelectronics.com/what-is-a-bidirectional-dc-dc-converter-circuit-diagram-working/
9. Analog Devices: LT8228 High Voltage, High Efficiency Bidirectional Synchronous Buck or Boost Converter with Input and Output Circuit Breakers (2019) Analog devices. Rev. 0
10. Richa AKKMS, Das A (2021) An iot based health monitoring system using arduino uno. Int J Eng Res Tchnol 10(03)
11. Chris: Arduino ADC Oversampling (2017). https://github.com/stylesuxx/Oversample

Energy Management System of Standalone DC Microgrid

Supriya Sharma and Pankaj Gupta

Abstract Solar photovoltaic (PV) energy usage that is stochastic and intermittent necessitates energy storage systems for continuous power delivery to loads. This research offers an energy management system (EMS) for a hybrid PV-battery system supplying local load in an independent DC microgrid (MG). The developed EMS algorithm ensures uninterrupted power supply to both critical and non-critical loads via storage unit during nighttime or cloudy days following load shedding in order to maximize power continuity to critical loads based on battery state of charge (SoC) levels when PV generation is low or unavailable. During peak solar hours, the load is supplied by PV, and the battery enters charging mode. If the battery is fully charged, the dump load is attached to absorb the extra energy. To test various scenarios, the Simulink model designed with MATLAB is tested under changing irradiance levels and battery SOC. The results demonstrate a seamless transition between PV and battery charging/discharging modes while maintaining a consistent DC connection voltage.

Keywords Isolated DC microgrid · PV · Renewable energy · Battery energy storage systems · Bidirectional buck–boost converter

1 Introduction

Microgrids (MGs) are yet another development made possible by the rising usage of renewable energy sources. According to the IEEE standard 2030.7, a MG is defined as follows: A collection of loads and distributed energy sources (DESs) coupled to form a single, manageable unit with respect to the grid, with clearly defined electrical boundaries and the ability to connect and detach from the grid to enable operation in both grid-connected and islanded modes [1, 2]. DC MG is preferred not only due to the increasing popularity of DC loads like as electric vehicles, LEDs, and battery storages, but also because of lower transmission losses, greater coordination among

S. Sharma (✉) · P. Gupta
Department of Electronics and Communication Engineering, Indira Gandhi Delhi Technical University for Women, Kashmere Gate, Delhi 110006, India
e-mail: ssharma0026@gmail.com

A. Rani et al. (eds.), *Signals, Machines and Automation*, Lecture Notes in Electrical Engineering 1023, https://doi.org/10.1007/978-981-99-0969-8_13

DESs, enhanced dependability, simpler connections, and the absence of frequency control difficulties [3]. EMS is utilized by MG to effectively manage the load and the available resources. Using a collection of computer-aided tools known as an EMS [4, 5], a generating and transmission system can be monitored, managed, and optimized. This is achieved while adhering to specific restrictions. Numerous EMS methods for DC MG with PV and battery storage systems (BSSs) for both AC and DC loads have been proposed in the past. Some of the work on EMS of islanded DC MG are discussed here. The author of [6] focused on off-grid DC MG systems comprised mostly of hydrogen, PV, and batteries. In order to achieve decentralized operation of the EMS, a two-level control with droop control is implemented. The goal of developing a PV-based MG with a rule-based prioritizing system is to give a consistent and continuous supply to critical loads [7]. The usage of a fuzzy controller for energy management in a MG with PV in order to extend the battery life is discussed in [8]. In the same context, [9] proposes an optimal control strategy for a freestanding PV-battery-supercapacitor system based on hybrid approaches using fuzzy logic control (FLC) and particle swarm optimization (PSO), with the aim of extending the battery's life. The EMS presented in [10] is capable of maintaining a constant DC bus voltage across the whole system. Another FLC-based EMS was designed and implemented in islanded mode [11] in order to govern the switching of the DC–DC converters and maintain power balance as well as a constant DC bus voltage. On the other hand, the primary focus of this effort will be the design of a rule-based EMS for an isolated mode-operating PV-battery hybrid DC MG. Another rule-based approach for DC standalone MG with PV, wind, battery, and supercapacitor feeding dump and local load is provided in [12].

The primary purpose of energy management in this developed isolated DC MGs is to accomplish the following objectives: (a) Maintaining continuity of power supply to the load for the maximum duration; (b) Maximizing power from PV system with Perturb and Observe (P&O) MPPT algorithm; (c) Keeping the DC bus voltage fixated at desired voltage levels, i.e., 300 V; and (d) Managing the battery operation mode (charging/discharging) based on the battery's SoC.

In the process of creating EMS, one of the main regulating factors that is frequently employed as a decision criterion is the battery's SoC. This research takes into account lithium-ion batteries since they are the most technologically sophisticated and display superior qualities in terms of reliability, power management, energy density, charging/discharging cycles, and other similar aspects. This paper aims to design a rule-based EMS for an isolated PV-battery DC MG with the objective to provide continuous supply to critical loads for the longest duration and maintain the DC bus voltage stable under various conditions. This has been accomplished by optimizing the PV generation using the P&O method as well as balancing the power by operating the battery in charging/discharging mode. Ultimately, the goal of this strategy is to provide continuous supply to critical loads for the longest duration possible. The following is the structure of the paper: The design of the MG is discussed in Sect. 2. In the following Sect. 3, we will talk about the proposed EMS strategy. In the following section, the results of the simulation will be critically examined for a variety of various operating situations. The concluding section of the paper is Sect. 5.

2 DC MG Structure

The PV system is connected into the DC MG via a DC–DC boost converter in order to supply the DC load. There are three loads, each consuming 500 W, and they are all connected, with the first load being the most important and the third load being the least important. This analysis takes into consideration an additional resistive load, which is thus treated as a dump load. In order to keep the power balance stable, a battery that is coupled by a bidirectional DC–DC buck/boost converter is used. The controlled switches that are linked to the load are used to achieve load shedding in the event that the power deficiency mode is active. This is done in order to preserve power continuity to the vital load and to keep the DC bus voltage stable. The planned structure of the MG is seen in Fig. 1, along with the control signals.

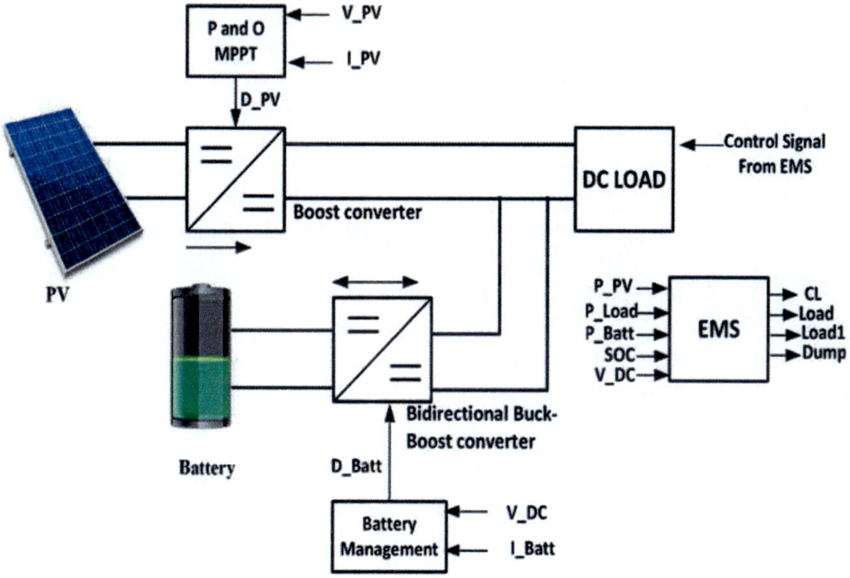

Fig. 1 Proposed DC MG layout

2.1 PV System

In the conduct of this study, a photovoltaic (PV) system is constructed utilizing sixteen modules, two of which are connected in series while the remaining eight are connected in parallel. The total power output is roughly five kilowatts. This PV system is integrated with the help of a DC–DC boost converter to the DC bus, and the Perturb and Observe Maximum Power Point Tracking (MPPT) technique is utilized in order to operate this module at its maximum capacity. This approach is often utilized due to its simplicity of implementation [13].

2.2 Battery System

Due to intermittent nature of solar, it is not possible to rely on this generation completely. The fluctuations in the solar generation are compensated with the help of battery. During day, when the irradiance levels are high and the PV generation is supplying the load, the battery is charged depending on its SoC. Otherwise, during night or cloudy days, the battery is discharged to maintain the power supply continuity to load. The battery is connected with the bidirectional buck–boost DC–DC converter to DC bus to maintain the flow of energy in either direction during charging/discharging mode. An inbuilt battery module available in MATLAB of 150 nominal voltage, and 78 AH capacity is considered in this study. The bidirectional buck–boost converter is used to connect this battery with 300 V DC bus. The main goal of this converter is to maintain a constant DC bus voltage by charging or discharging battery according to the PV generation. Dual-loop PI controller is preferred for the converter as it provides stability to the system. The outer loop tracks the DC bus reference voltage and supplies the inner loop with reference current. The parameters of the controller have been calculated as discussed in [14]. Figure 2 depicts the cascaded PI controller.

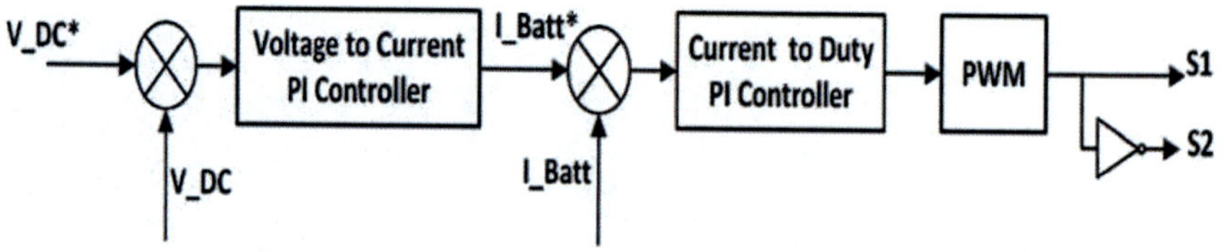

Fig. 2 Dual-loop PI controller for bidirectional buck–boost converter

3 Proposed Energy Management Strategy

The EMS that was designed and is depicted in Fig. 3 operates in two distinct modes: the sufficient PV power mode and the deficient PV power mode.

Sufficient photovoltaic energy: In this mode, for enough PV supply during daytime hours, all of the load, including critical and non-critical, is supplied by this generation when the battery SoC falls between the lowest and highest ranges. While for SoC above the maximum limit, which in this case is 90%, a dump load is linked into the system with a controlled switch so that the surplus PV power may be harvested. If the percentage of charge on the SoC falls below the minimal mark of 10%; however, it will begin charging. The load that is considered to be of the least vital importance is taken off, and the remaining load is fed by the PV.

Deficient PV generation: EMS will operate in this mode whenever the irradiance is either very low or completely absent. By operating in the phase known as discharging, the battery fulfills its role as a source. For a SoC more than 70%, the entire load is kept linked. However, load shedding is chosen by the EMS because the SoC is decreasing, and the generation from PV is either very low or nonexistent. When the SoC drops below 70% again, the load that is considered to be of the least critical importance is turned off so that the vital loads receive a continuous supply for as long as is

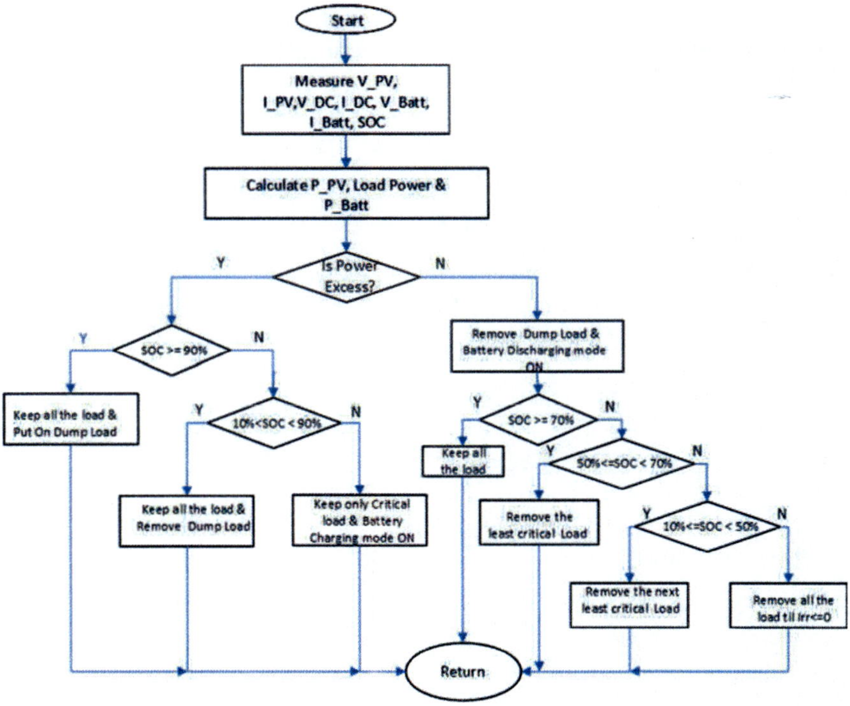

Fig. 3 Proposed EMS flowchart

humanly practicable. When the SoC falls further below 50%, the next least critical load will be taken offline. In the event that the PV system continues to produce an insufficient amount of power and the SOC falls below the threshold level, all of the load is disconnected, and the system remains in this mode until PV generation is restarted.

4 Simulation Results

The proposed algorithm is simulated on MATLAB/Simulink, and the effectiveness has been tested for different irradiance levels and battery's SoC in PV surplus as well as deficient mode. The results are illustrated in Fig. 4. The simulation results for PV surplus mode are obtained by varying the irradiance levels as mentioned in Fig. 4a. and keeping the battery's SoC > 90%. The dump load is removed only when the PV system is generating at its 50% of full capacity as otherwise, the system is operating in excess PV generation mode. When irradiance is reduced to nil in deficient mode (Fig. 4b), battery starts discharging to maintain the supply to load. In both the modes, DC link voltage is maintained at 300 V. The load shedding also improved the duration of continuous supply to critical loads. The results in Fig. 4c, d depict the load shedding and restoring conditions at SoC near 50 and 70%. As Soc is below 70%, the dump load and least critical load is already in OFF mode. Now, when SoC further falls below 50%, the next non-critical load turns off but turns back ON when SoC rises above 50% as depicted in Fig. 4c.

5 Conclusion

The proposed EMS achieved constant DC link voltage for variable irradiation levels and various SoC. Using a twin-loop PI controller, charging and discharging have been conducted smoothly, as evidenced by the results. Moreover, the continuity of supply is maintained to critical loads for as long as possible during periods of low irradiance by following load shedding of the least critical load in steps, and when PV generation is high, dump load is used to capture the excess generation if the battery is fully charged to minimize energy loss. This work can be remodeled in the future for grid-tied system, and instead of a rule-based approach, intelligent EMS based on optimization approaches such as FLC can be implemented while exploring more distributed generations.

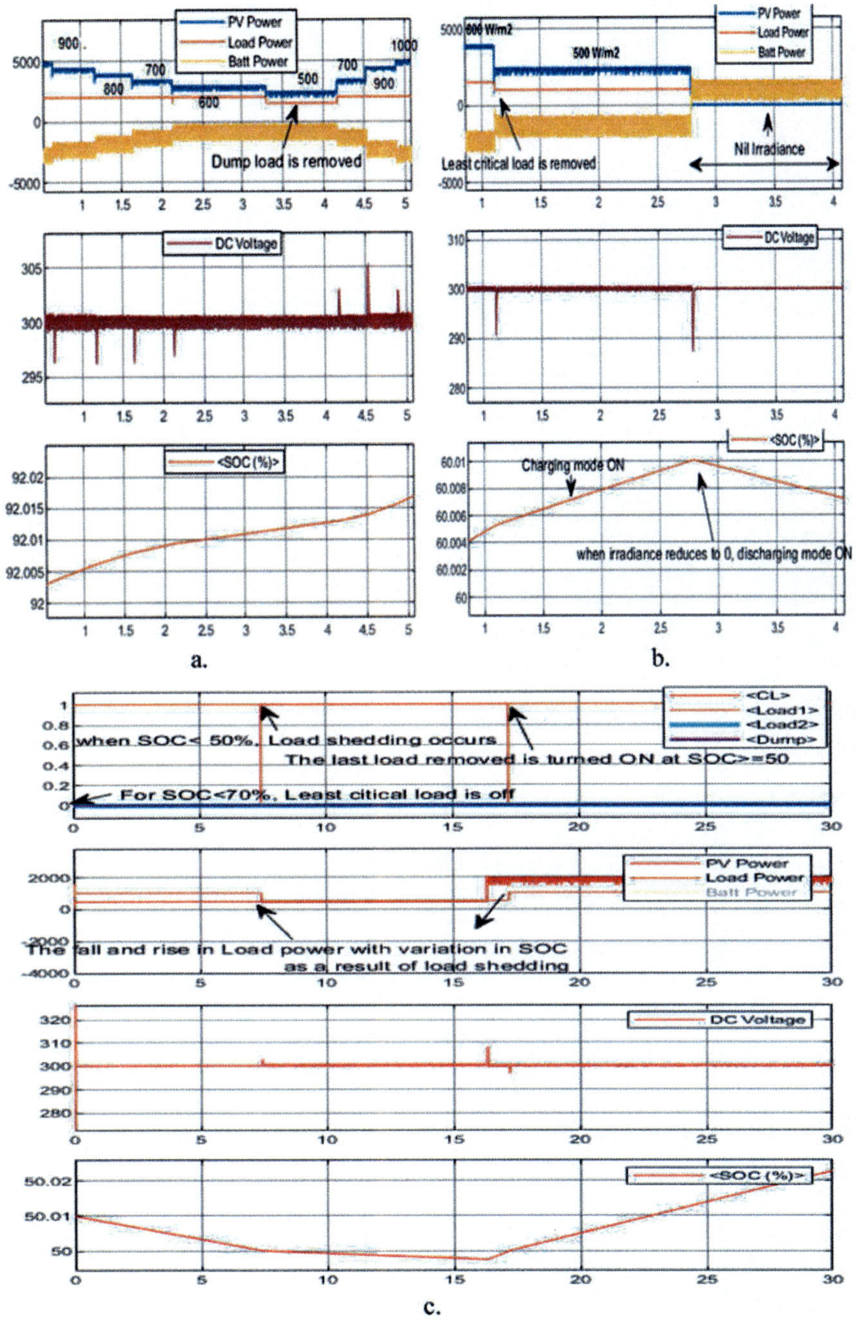

Fig. 4 Simulation results for **a** Surplus mode. **b** Deficient mode. **c** SoC near 50%. **d** SoC near 70%, showing PV power, load power, battery power, DC bus voltage, and SoC

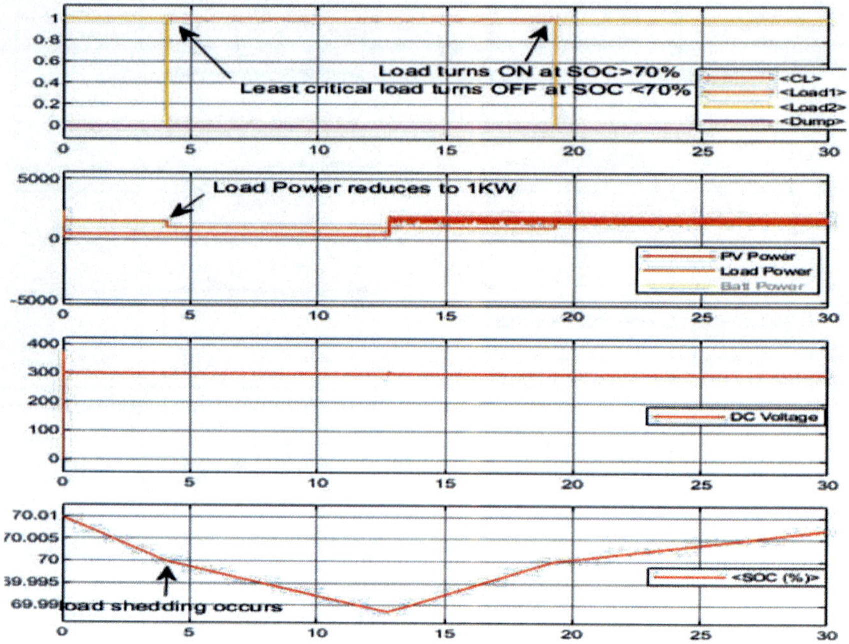

d.

Fig. 4 (continued)

References

1. Lasseter RH (2002) Micro grids. In: IEEE power engineering society winter meeting. conference proceedings (Cat. No.02CH37309), 305-8. https://doi.org/10.1109/pesw.2002.985003
2. IEEE PES (2018) IEEE standard for the specification of microgrid controllers
3. Shah T, Ansari ZA (2018) An overview of intelligent energy management system for DC microgrid: system and communication architecture and application in power distribtion system. In: 13th international conference on industrial information system ICIIS 2018, pp 339–42. https://doi.org/10.1109/ICIINFS.2018.8721384
4. Kueck JD, Staunton RH, Labinov SD, Kirby BJ (2003) Microgrid energy management system. CERTS
5. Razeghi G, Gu F, Neal R, Samuelsen S (2018) A generic microgrid controller: concept, testing, and insights. Appl Energy 2018(229):660–671. https://doi.org/10.1016/j.apenergy.2018.08.014
6. Han Y, Yang H, Li Q, Chen W, Zare F, Guerrero JM (2020) Mode-triggered droop method for the decentralized energy management of an islanded hybrid PV/hydrogen/battery DC microgrid. Energy 2020:199. https://doi.org/10.1016/j.energy.2020.117441
7. Ayodele TR, Ogunjuyigbe ASO, Akpeji KO, Akinola OO (2017) Prioritized rule-based load management technique for residential building powered by PV/battery system. Eng Sci Technol Int J 20(3):859–873. https://doi.org/10.1016/j.jestch.2017.04.003

8. Athira GR, Pandi VR (2017) Energy management in islanded DC microgrid using fuzzy controller to improve battery performance. In: 2017 international conference on technological advancements in power and energy (TAP energy), pp 1–6. https://doi.org/10.1109/TAPENERGY.2017.8397369
9. Chong LW, Wong YW, Rajkumar RK, Isa D (2016) An optimal control strategy for standalone PV system with battery-supercapacitor hybrid energy storage system. J Power Sour 331:553–565. https://doi.org/10.1016/j.jpowsour.2016.09.061
10. Rajasekar N, Bilakanti N, Miyatake M (2017) Energy management technique for home micro grid system. In: TENCON 2017, IEEE region 10 conference, pp 1433–1438
11. De Bosio F, Luna AC, Ribeiro LADS, Graells M, Saavedra OR, Guerrero JM (2016) Analysis and improvement of the energy management of an isolated microgrid in Lencois island based on a linear optimization approach. In: ECCE 2016—IEEE energy convers congress exposition proceeding 2016, 0–6. https://doi.org/10.1109/ECCE.2016.7854871
12. Korada DMR, Mishra MK, Yallamilli RS (2020) Dynamic energy management in DC microgrid using composite energy storage system. In: 2020 IEEE international conference on power electron smart grid renew energy, PESGRE 2020, pp 1–6 (2020). https://doi.org/10.1109/PESGRE45664.2020.9070693
13. Gupta AK, Saxena R (2016) Review on widely-used MPPT techniques for PV applications. In: 2016 international conference on innovation and challenges in cyber security (ICICCS-INBUSH), pp 270–273. https://doi.org/10.1109/ICICCS.2016.7542321
14. Özdemir A, Erdem Z (2017) Double-loop PI controller design of the DC–DC boost converter with a proposed approach for calculation of the controller parameters. Proc Inst Mech Eng Part I J Syst Control Eng 232(2):137–148

State Estimation in Active Power Distribution Systems Using Phasor Measurement Units

V. Aarthi, Ram Krishan, and Himanshu Grover

Abstract Real-time monitoring of the states of power systems is crucial and becoming a very challenging task with increase in the penetration of renewable generation. Now days, the use of phasor estimation units (PMUs) is becoming very crucial in power systems monitoring and control. However, it is not feasible to place PMUs in each and every location as it is very expensive. The use of pseudo and virtual measurements techniques to estimate the power system states with better accuracy is very effective. In this paper, first, optimal location of the PMUs in a simple active power distribution system (APDS) is determined using a simple method. Further, the APDS is built in PSCAD, whose signals are fed into the PMUs, to obtain the magnitude and phase angle of the signals, and these values are used for performing the state estimation.

Keywords State estimation · PMUs · Optimal PMU placement · Active power distribution systems

1 Introduction

Operation of modern power systems is very complex because it consists of various dynamic loads and generation sources like conventional and renewable sources, spinning and non-spinning reserves [1]. A power system needs to be operated with high reliability at minimum cost. For a reliable operation, knowledge of the system states is very important. The part of power system which can be operated even in islanded mode of operation with the help of available distributed generations (DGs) including renewable energy sources is termed as active power distribution systems (APDSs).

V. Aarthi (✉) · R. Krishan
Department of Electrical Engineering, National Institute of Technology Warangal, Warangal, India
e-mail: avee21328@student.nitw.ac.in

R. Krishan
e-mail: rkrishan@nitw.ac.in

H. Grover
Department of Energy Science and Engineering, Indian Institute of Technology Delhi, Delhi, India

A. Rani et al. (eds.), *Signals, Machines and Automation*, Lecture Notes in Electrical Engineering 1023, https://doi.org/10.1007/978-981-99-0969-8_14

It is very tedious process to know the states of APDS at a particular instant of time. To solve this problem, state estimation is very useful where the states of the power system are estimated using the available flow and injection measurements [2].

Power system state estimation refers to the process of collection of measurements across the system and to extract the best estimation of system operational condition [3]. Measurements, which are nonlinear functions of the system state, are collected, and a load flow-like calculations are performed iteratively to determine the most probable system state from the known information.

Earlier, remote terminal units (RTUs) are used to sense data though the measuring transformers (CT's and PT's), and these sensed analog signals/values are provided to the supervisory control and data acquisition (SCADA) system for the monitoring and control. If there is any disturbance in the system, the energy management system (EMS) present in the control center gives the effective control action [4]. The output data rate of SCADA is once in 4–6 s, but updates of much higher rates are required to capture power system dynamics, which is essential for analyzing disturbances and post-disturbance scenarios and taking fast control actions. Data at sub-second rate is required to monitor the dynamics of power system more closely. Also, time synchronized wide area system data is required to have an accurate view of the entire power system. Figure 1 shows a general flow of information and communication and how the control happens in power system.

The output of SCADA is the RMS value of voltage and current, but bus voltage angle data is required as a local measurement. The angle separation is a key indicator of system stress and needs to be monitored at important transmission corridors in real time. In order to overcome all these disadvantages, we are shifting toward phasor measurement units (PMUs) to get good accuracy and significant reporting rate in view of the power system dynamics. In the last decade, the large use of PMUs for the monitoring of the transmission networks has influenced the operational practices of these systems concerning the following aspects: (i) system stability; (ii) state estimation (SE); (iii) system reliability; and (iv) protections schemes [5]. The use of low-cost hardware platforms is contributing to the massive use of PMUs in distribution networks [6]. Dedicated installations in real distribution networks have already demonstrated the feasibility of this solution [7, 8]. As an example, the active distribution network (ADN) in [7] is equipped with PMUs in every bus measuring nodal voltages and injected current synchro phasors. By leveraging the distribution

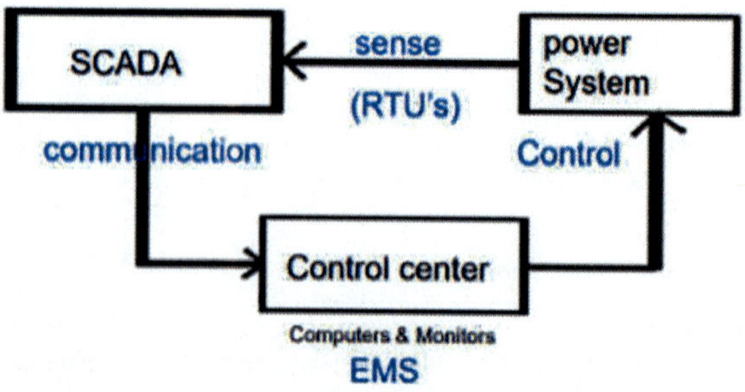

Fig. 1 Power system operational paradigm

network operator real time monitoring such as fault detection and location, might be developed using this same metering infrastructure [8–10].

In this paper, a simple APDS model is considered in PSCAD, whose output voltage and current are sent to PMU to compute the states of some of the buses in which PMU is present. The change in frequency and rate of change of frequency waveforms obtained from PMU are compared with the waveforms of frequency and differentiation of the frequency obtained using PLL. The states of other buses are calculated by forming the measurement matrix, Jacobian matrix, and the effective Jacobian and measurement matrixes and by computing the diagonal covariance matrix. The results obtained from normal state estimation are compared with the results obtained from hybrid state estimation. Also, a fault is created in the system, and the states of the system are estimated during the instant of fault using the PMU measurement. Thus, the states of the system can be obtained on real-time basis without necessarily using PMU in all buses.

2 PMU Modeling

A synchro phasor is a precision time-tagged positive sequence phasor measured at different locations. A PMU is a high-speed device which is used to provide the synchronized measurement of phasors. It converts three-phase analog signal of voltage or current into synchro phasors. It produces synchronized phasor, frequency, rate of change of frequency (ROCOF), and real and reactive powers, etc., estimates from voltage and/or current signals and a time synchronizing signal [11]. PMUs directly measure the state variables, i.e., magnitude and phase angle of bus voltages, with a very high accuracy. They can consequently increase the robustness and precision of the estimation process [12]. The recent literature has also discussed the use of phasor measurement units (PMUs) to develop low-latency and high-refresh rate real-time state estimators (SEs) for distribution networks [13–15].

Figure 2 shows the single-phase block diagram of PMU phasor signal. The input signal is discretized using sample and hold circuit, and the output of this signal is multiplied with the time synchronized sine and cosine signals which are passed through a low-pass filter to obtain the real and imaginary parts of the signal [16]. The parameters of the low-pass filter are designed using the following equations. The filter length and filter cut-off frequency is calculated using (1) and (2).

$$\text{Filter length } (L) = \left\{\frac{f_0}{F_s}\right\} \times \left\{5.8943 + 1.467 * \ln\left\{\frac{f_0}{F_s}\right\}\right\} \text{ cycles} \tag{1}$$

$$F_{\text{cuttoff}} = F_s(0.1981 - 0.0005F_s) \approx F_s/5 \tag{2}$$

Here, f_0 is the fundamental frequency, and F_s is the sampling frequency.

The change in frequency can be mathematically obtained as:

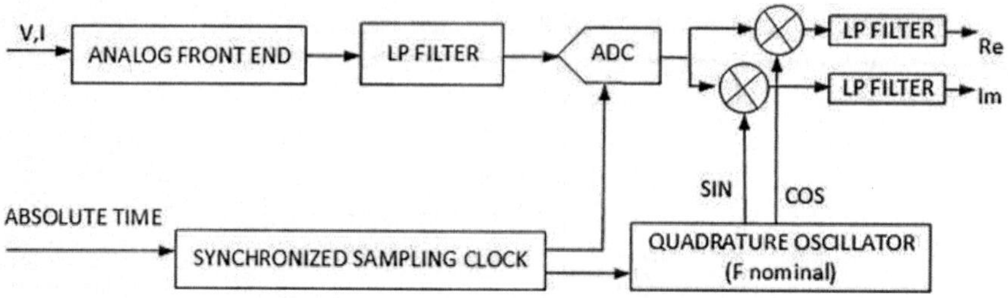

Fig. 2 Single-phase section of the PMU phasor signal processing model

$$\Delta F(i) = \{6[\theta(i) - \theta(i-1)] + 3[\theta(i-1) - \theta(i-2)] \\ +[\theta(i-2) - \theta(i-3)]\}/[20\pi t] \tag{3}$$

And the rate of change of frequency (ROCOF) can be framed as:

$$\mathrm{DF}(i) = [F(i) - F(i-1)]/[t] \tag{4}$$

where Δt is the sampling time ($1/F_s$).

3 State Estimation Model

The system model to estimate the true values of the states in terms of measurement and error can be expressed as:

$$Z = Hx + e \tag{5}$$

where Z is the measurement matrix, Hx is the true values obtained from system model, x is the state vector, and e is the error matrix.

Measurements can be of a variety of types. Most commonly used measurements are the line power flows, bus power injections, bus voltage magnitudes, and line current flow magnitudes. For a system containing N buses, the state vector will have $2N - 1$ elements, N bus voltage magnitudes, and $(N - 1)$ phase angles. Hence, the state vector x will have the following form assuming bus 1 as the reference

$$x^T = \theta_2, \theta_3, \ldots\ldots.\theta_n\ V_1, V_2, V_3, \ldots\ldots V_n \tag{6}$$

The expressions for the injection power P_i and Q_i are given below, assuming the general two-port π-model for the network branches.

$$P_i = V_i \sum_{j \in Ni} V_j\left(G_{ij}\cos\left(\theta_{ij}\right) + B_{ij}\sin\left(\theta_{ij}\right)\right) \tag{7}$$

$$Q_i = V_i \sum_{j \in Ni} V_j \left(G_{ij} \cos(\theta_{ij}) - B_{ij} \sin(\theta_{ij}) \right) \tag{8}$$

The real power P_{ij} and reactive power Q_{ij} flow from bus i to bus j:

$$P_{ij} = V_i^2 (g_{si} + g_{ii}) - V_i V_j \left(g_{ii} \cos\theta_{ij} + b_{ij} \sin\theta_{ij} \right) \tag{9}$$

$$Q_{ij} = -V_i^2 (b_{si} + b_{ii}) - V_i V_j \left(g_{ii} \sin\theta_{ij} - b_{ij} \cos\theta_{ij} \right) \tag{10}$$

The measurement Jacobian matrix, H can be formed as:

$$H = \begin{bmatrix} \frac{dP_{inj}}{d\theta} & \frac{dP_{inj}}{dV} \\ \frac{dP_{\text{flow}}}{d\theta} & \frac{dP_{\text{flow}}}{dV} \\ \frac{dQ_{inj}}{d\theta} & \frac{dQ_{inj}}{dV} \\ \frac{dQ_{\text{flow}}}{d\theta} & \frac{dQ_{\text{flow}}}{dV} \end{bmatrix} \tag{11}$$

Figure 3 shows the flowchart for state estimation. The PMUs give the magnitude and angle of voltage and current to estimate the states which are unknown in the system. State estimation also helps in detection of bad data and in which bus or line the bad data has occurred. State estimation results are processed to identify errors in the meters and parameters given that there are enough redundant measurements. In addition, failure or loss of a meter can be detected. Topology of a network changes when a line is lost due to overloading or equipment failure. In such a case, the security of the system might be in danger since some other lines might get overloaded as well. State estimation results, when analyzed carefully, can give early warnings of such cases.

For performing state estimation using PMU measurements, direct angle measurements are included along with the classical measurement set. The proposed model is based on enhancing the output of the WLS estimator by modeling a linear measurement set composed of PMU measurements and WLS state estimate.

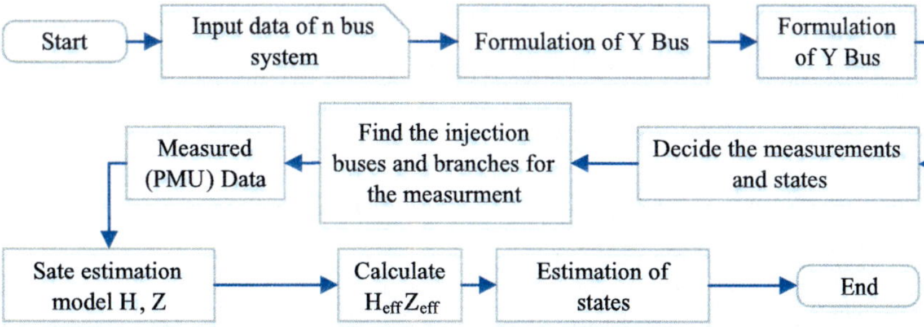

Fig. 3 Flowchart for state estimation

The proposed measurement model for estimating the states using PMU measurements is as follows:

$$M = H^c x_{\text{pest}} + e \tag{12}$$

Here, M is the measurement vector which comprises of the direct state output from WLS (V_{se}, θ_{se}) and PMU voltage measurements ($V_{\text{PMU}}, \theta_{\text{PMU}}$).

$$M = \begin{bmatrix} \begin{bmatrix} V_{se} \\ \theta_{se} \end{bmatrix} \\ \begin{bmatrix} V_{\text{PMU}} \\ \theta_{\text{PMU}} \end{bmatrix} \end{bmatrix} \tag{13}$$

$$H^c = \begin{bmatrix} H_{11} & H_{12} \\ H_{21} & H_{22} \\ H_{31} & H_{32} \\ H_{41} & H_{42} \end{bmatrix} \tag{14}$$

Here, the coefficients of H^c matrix can be framed as:

$$\mathbf{H_{11} = I; H_{12} = 0; H_{21} = 0; H_{22} = I;}$$

where $\mathbf{I}$ is an $N \times N$ identity matrix and N is the number of buses in the system. $\mathbf{0}$ is the $N \times N$ null matrix.

$\mathbf{H_{31}}$ is a $P \times N$ matrix, P being the number of PMUs. Each row k corresponds to pmu k and has all zeros except at the jth column corresponding to the index of Vkpmu in the state vector.

$\mathbf{H_{32} = 0}$ $(P \times N)$; $\mathbf{H_{41} = 0}$ $(P \times N$; $\mathbf{H_{42}}$ is a $P \times N$ matrix, P being the number of PMUs). Each row k corresponds to pmu k and has all zeros except at the jth column corresponding to the index of Vkpmu in the state vector.

The x_{pest} in (12) is computed as:

$$x_{\text{pest}} = (H^{cT} R^{-1} H^c)(H^{cT} R^{-1} M) \tag{15}$$

Here, R is the diagonal covariance matrix and is defined by:

$$R = \text{diag}\{[\sigma_{Vse}^2, \sigma_{\theta se}^2, \sigma_{V\text{PMU}}^2, \sigma_{\theta \text{PMU}}^2]\} \tag{16}$$

Here, $\sigma_{Vse}^2, \sigma_{\theta se}^2, \sigma_{V\text{PMU}}^2, \sigma_{\theta\text{PMU}}^2$ are the standard deviation of the errors of V_{se}, θ_{se}, $V_{\text{PMU}}, \theta_{\text{PMU}}$, respectively.

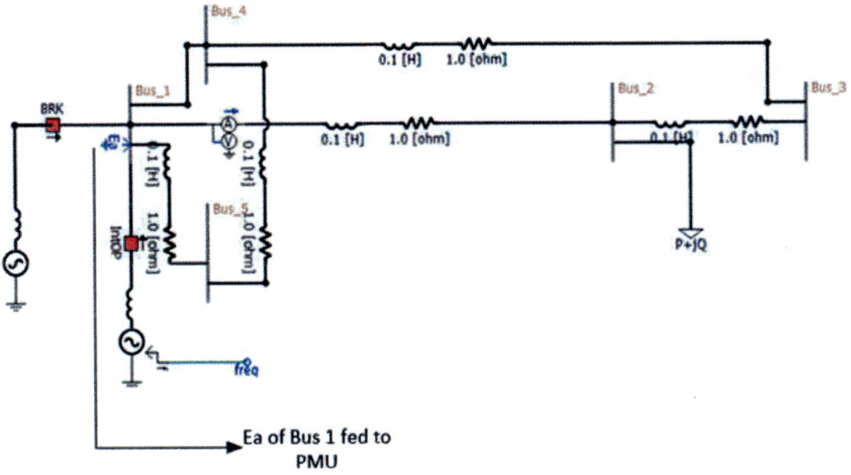

Fig. 4 Simulation model considered

4 Simulation Model

An IEEE 5-bus system is considered which has a normal conventional generation source of 132 kV 100 MVA, a DG source of 132 kV connected to bus 1, a circuit breaker to isolate the conventional source and another breaker to switch on the DG source when needed. The load considered is of 100 MW and 25 MVAr, and the circuit breaker trips from 1.0 to 1.3 s, and the simulation is run for 2 s. Only one PMU is considered in the system, and this PMU is placed in bus 1. As direct PMU model is not available in PSCAD, a virtual PMU is implemented in MATLAB, and the voltage data obtained from bus 1 in this simulation is fed into the virtual PMU implemented in MATLAB for obtaining the states of bus 1. The block diagram of the considered system is shown in Fig. 4, and the block diagram of PMU implemented in MATLAB is shown in Fig. 5.

The input V_1, V_2, V_3 in Fig. 5 correspond to the V_a, V_b, V_c of E_a (bus 1 Voltage) shown in Fig. 4. Figure 2 is implemented in subsystem, subsystem1 and subsystem2, and (1), (2), and (3) are implemented in change in frequency block, and (4) is implemented in ROCOF block of Fig. 5.

5 Simulation Results and Discussions

The simulation model shown in Fig. 4 is simulated in PSCAD, and the three-phase voltage and current output of bus 1 obtained for a fault from 1 to 1.3 s are shown in Figs. 6 and 7.

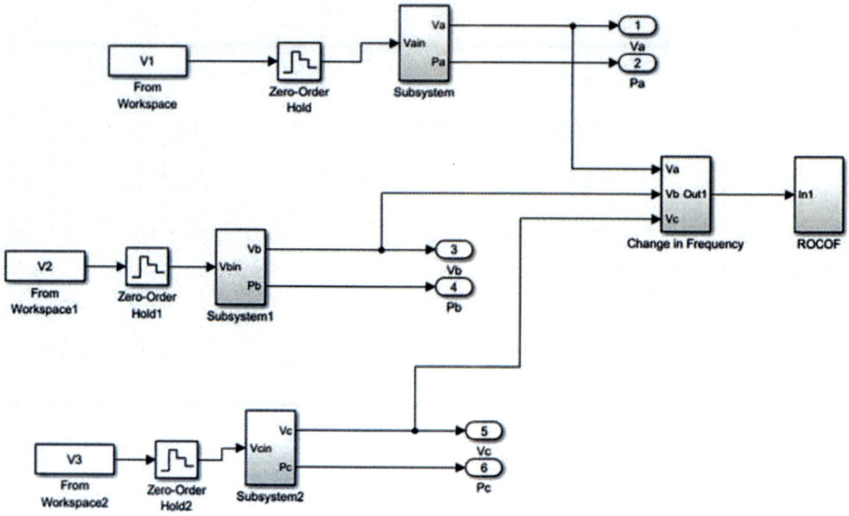

Fig. 5 Block diagram of PMU implementation in Simulink

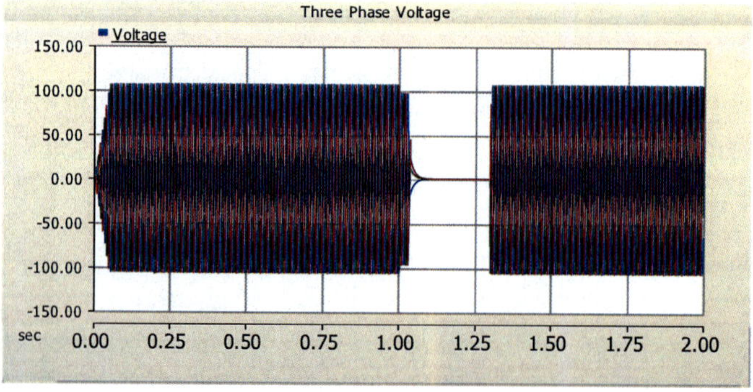

Fig. 6 Three-phase voltage output of bus 1

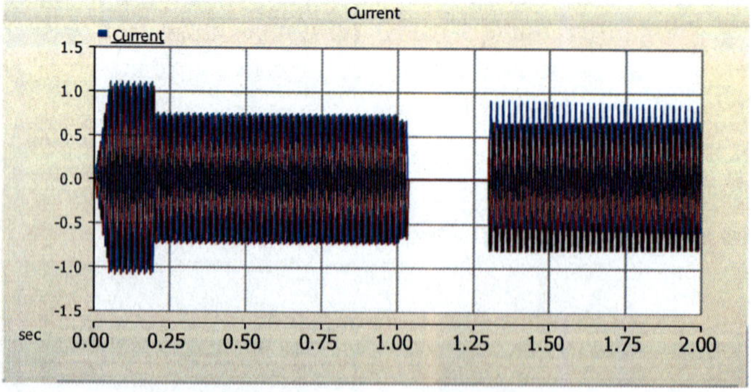

Fig. 7 Current flowing through the line

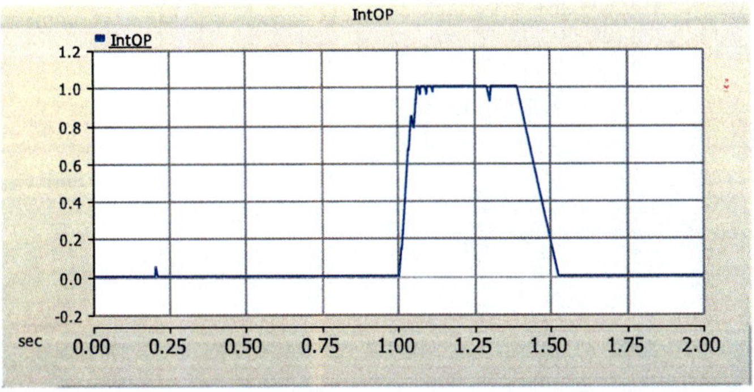

Fig. 8 Trip signal issued to the circuit breaker

It can be observed from Fig. 8 that the trip signal has been given to the circuit breaker from 1.0 to 1.3 s, but there is a transient observed till 1.05 s.

It can be seen from Fig. 9 that there is a transient initially till 0.25 s, and after that the power is maintained at 100 MW till 1 s, and it abruptly drops to zero when the circuit breaker trips, and in 1.3 s, the power output is back to 100 MW from getting the power from the DG source.

The change in frequency and rate of change of frequency (ROCOF) obtained by comparing and differentiating the frequency signal obtained from PLL (Fig. 10) are shown in Figs. 11 and 12.

The output data of bus 1 for the model shown in Fig. 4 is obtained from PSCAD are fed to the PMU model in Simulink shown in Fig. 5, and the transition results are observed. The magnitude of three-phase voltage output for bus 1 (V_{abc}) is shown in Fig. 13.

It can be observed from the waveforms that similar transitions are observed by the PMU and the simulation in PSCAD. Figures 6 and 13 show a similar behavior from 1 to 1.3 s. Figure 14 shows the positive sequence component of V_{abc}. The magnitude

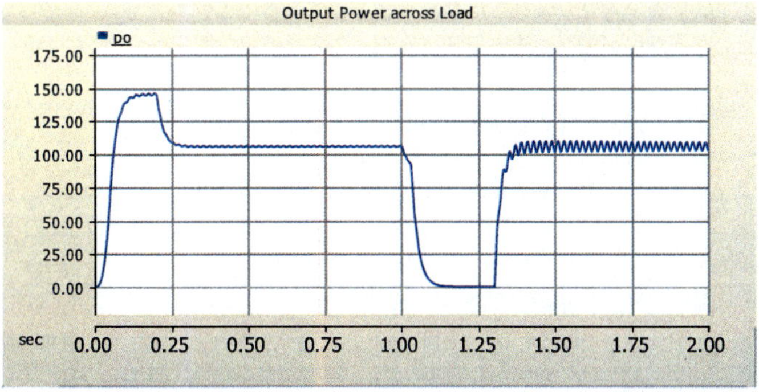

Fig. 9 Change in power observed during the transition

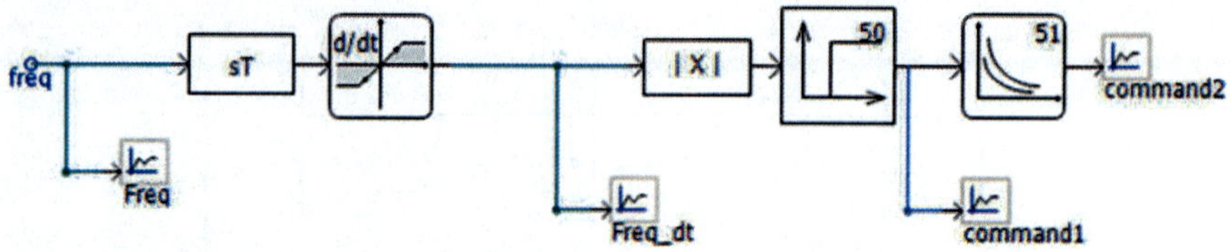

Fig. 10 Differentiation of frequency signal

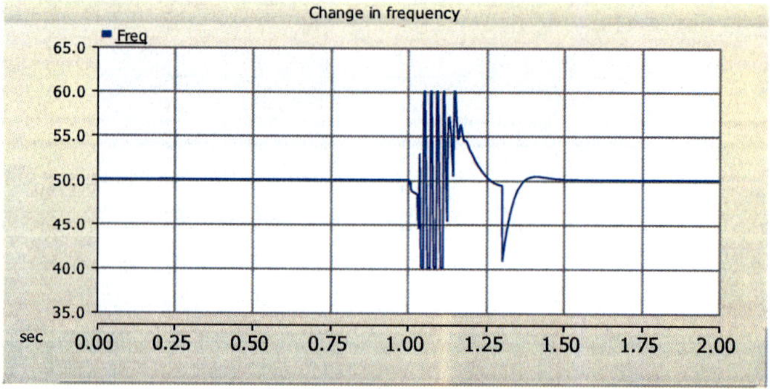

Fig. 11 Change in frequency obtained due to transition

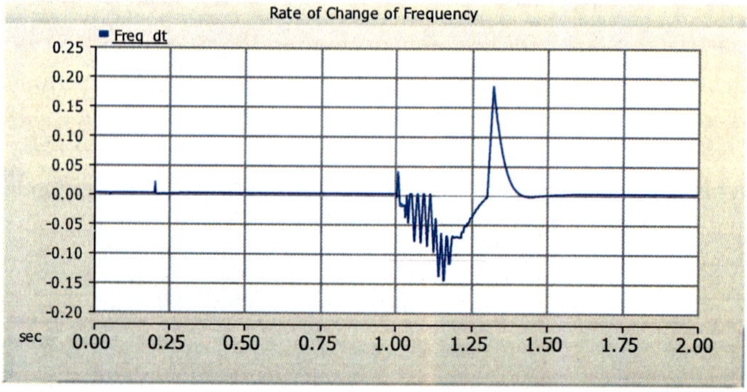

Fig. 12 Rate of change of frequency (ROCOF) observed

of V_{abc} signal is shown in Fig. 13, and the phase angle of V_{abc} is shown in Fig. 15. We can see that the there is a large variation in the phase angle during the disturbance in all 3 phases. Figure 11 shows the change in frequency which comes to around 1.2 times (60 Hz) the nominal frequency (50 Hz). The similar change is observed in Fig. 16 from 1 to 1.3 s. In case of Fig. 11, it can be seen that there are small peaks before a high shoot up in the waveform. But, in case of PMU (Fig. 16), the shoot up is very high indicating a large disturbance in the system. Thus, the system can act more faster with PMU. The ROCOF, shown in Fig. 12, shoots up to around 0.2 Hz/s,

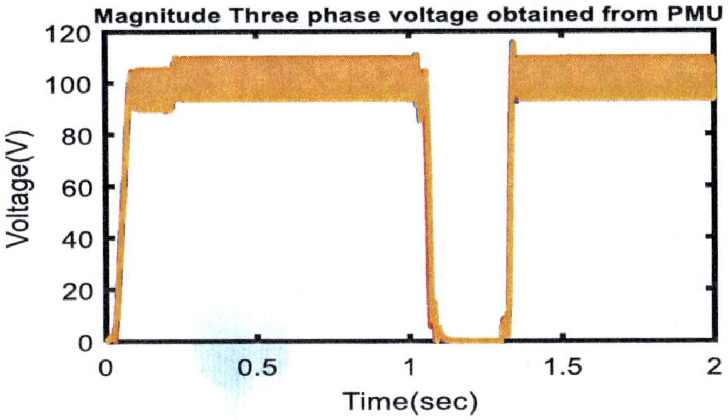

Fig. 13 Magnitude of three-phase voltage output obtained from PMU

and this similar change is observed in Fig. 17 where there is a shoot up to around 60 Hz/s, two times, for the period of 1–1.3 s.

From Figs. 16 and 17, it can be observed that the changes seen in the power system are detected by the PMUs as in PSCAD. Hence, the data obtained from PMU can be used to estimate the unknown states of the system.

Using the data obtained from the PMU, we can now perform state estimation. The Z matrix and H matrix are formulated as shown in (5) and (11), a random error W is introduced into the system, and the H_{eff} and Z_{eff} are computed as follows:

$$H_{\text{eff}} = H^T W H \tag{17}$$

$$Z_{\text{eff}} = H^T W Z \tag{18}$$

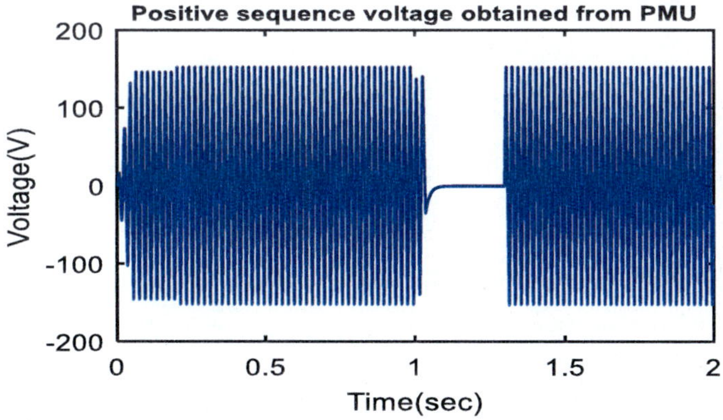

Fig. 14 Positive sequence voltage output obtained from PMU

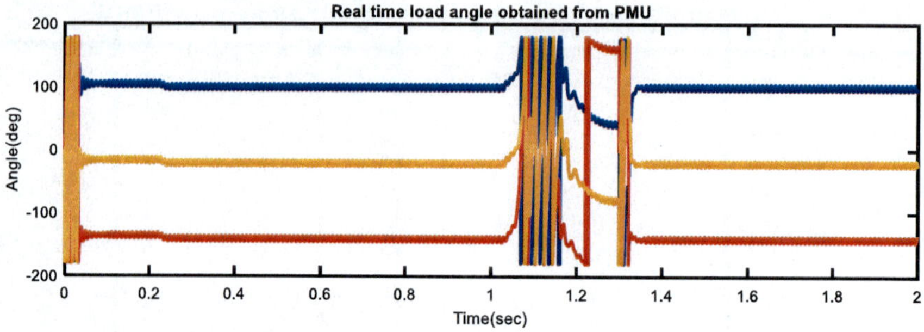

Fig. 15 Phase angle of V_{abc} obtained from PMU

Fig. 16 Change in frequency obtained from PMU

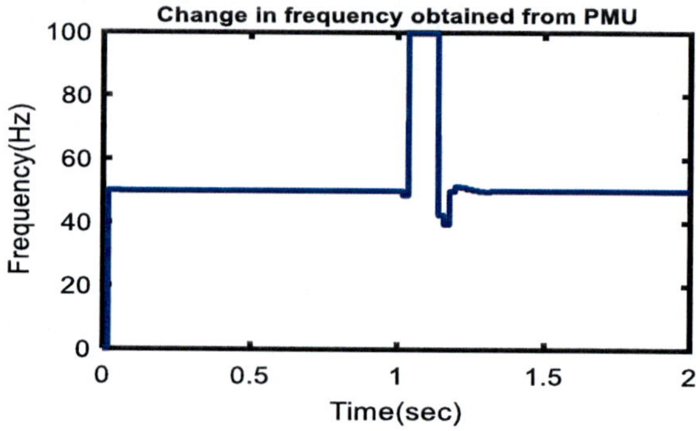

Fig. 17 ROCOF obtained from PMU

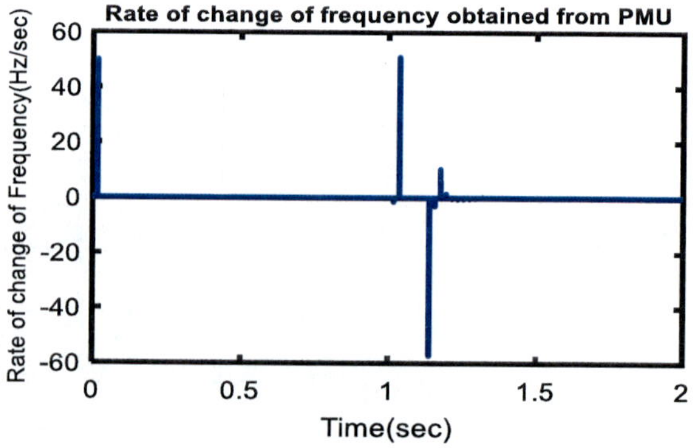

From the above formulations, the states can be estimated as:

$$X = H_{\text{eff}}^{-1} Z_{\text{eff}} \tag{19}$$

The system shown in Fig. 4 is considered in MATLAB for estimating the states of the system. The Z_{meas} and states corresponding to bus 1 are taken from the PMU data along with the DG source, and the states for other buses are estimated using the hybrid state estimation technique.

Here, the number of measurements is considered as 16, and number of states of the system are 9 states considered: $[\theta_2|\theta_3|\theta_4|\theta_5|V_1|V_2|V_3|V_4|V_5]$.

The 5-bus system has 4 generators at bus at 2, 3, 4, and 5, and there are 6 branches present. Here, bus 1 is considered as the slack bus.

The measurement matrix (Z) considered for this system is shown in (20).

$$Z = [P_1 P_3 P_4 P_5 Q_1 Q_3 Q_4 Q_5 P_{12} P_{23} P_{14} P_{15} Q_{12} Q_{23} Q_{14} Q_{15}]^T \tag{20}$$

Here, $P_1 P_3 P_4 P_5 Q_1 Q_3 Q_4 Q_5$ are the injection real and reactive power flow measurements, and $P_{12} P_{23} P_{14} P_{15} Q_{12} Q_{23} Q_{14} Q_{15}$ are the flow real and reactive power measurements.

The Jacobian matrix is calculated using (11) as:

$$H = \begin{bmatrix} 222.5069 & -35.2348 & 0 & -154.703 & 22.2507 & -3.5235 & 0 & -3.2569 & -15.4703 \\ 0 & -91.6758 & 125.0125 & 0 & 0 & -9.1676 & 12.5013 & -3.3337 & 0 \\ -32.569 & 0 & -33.3367 & -33.3367 & -3.2569 & 0 & -3.3337 & 9.9242 & -3.3337 \\ -154.703 & 0 & 0 & 188.0396 & -15.4703 & 0 & 0 & -3.3337 & 18.804 \\ -22.2507 & 3.5235 & 0 & 15.4703 & 222.4619 & -35.2348 & 0 & -32.569 & -154.703 \\ 0 & 9.1676 & -12.5013 & 0 & 0 & -91.6758 & 124.9872 & 99.2223 & 0 \\ 3.2569 & 0 & 3.3337 & 3.3337 & -32.569 & 0 & -33.3367 & 99.2223 & -33.3367 \\ 15.4703 & 0 & 0 & -18.804 & -154.703 & 0 & 0 & -33.3367 & 188.0016 \\ 35.2348 & -35.2348 & 0 & 0 & 35.2348 & -35.2348 & 0 & 0 & 0 \\ 0.0000 & 91.6758 & -91.6758 & 0.0000 & 0.0000 & 91.6758 & -91.6758 & 0.0000 & 0.000 \\ 32.569 & 0 & 0 & 0 & 32.569 & 0 & 0 & 0 & 0 \\ 154.703 & 0 & 0 & -154.703 & 154.703 & 0 & 0 & 0 & -154.703 \\ -3.5235 & 3.5235 & 0 & 0 & 35.2348 & -35.2348 & 0 & 0 & 0 \\ 0 & -9.1676 & 9.1676 & 0 & 0 & 91.6758 & -91.6758 & 0 & 0 \\ -3.2569 & 0 & 0 & 0 & 32.569 & 0 & 0 & -32.569 & 0 \\ -15.4703 & 0 & 0 & 15.4703 & 154.703 & 0 & 0 & 0 & -154.703 \end{bmatrix} \tag{21}$$

The measurement matrix Z can be obtained using (20) as:

$$z = \begin{bmatrix} 0.4 \; 0.2349 \; -3.9497 \; 4.6651 \quad 0.0585 \; 0.9604 \quad 0.5265 \; -0.3821 \; \ldots \\ \ldots \; 0.216 \quad -0.9397 \; -0.1361 \; 0.2274 \; -2.4801 \; 0.5212 \; -1.8544 \; 2.266 \end{bmatrix}^T \tag{22}$$

Table 1 Actual and estimated values of states without using PMU data

Values	Actual values		Estimated values		Error	
States	Voltage (pu)	Angle (°)	Voltage (pu)	Angle (°)	Error in V	Error in θ
1	1.00	0	1.0585	0	– 0.0585	0
2	0.959	3.273	0.9785	3.2190	– 0.0195	0.054
3	1.00	– 0.759	1.0101	– 0.7515	– 0.0101	– 0.0075
4	1.00	– 0.492	1.0021	– 0.4900	– 0.0021	– 0.002
5	1.00	4.112	1.0051	4.0446	– 0.0051	0.0674

Table 2 Actual and estimated values of states using PMU data

Values	Actual values		Estimated values		Error	
States	Voltage (pu)	Angle (°)	Voltage (pu)	Angle (°)	Error in V	Error in θ
1	1.00	0	1.01	0	– 0.01	0
2	0.959	3.273	0.961	3.269	– 0.002	0.004
3	1.00	– 0.759	1.002	– 0.7578	– 0.002	– 0.0012
4	1.00	– 0.492	1.001	– 0.4919	– 0.001	– 1E – 04
5	1.00	4.112	1.0	4.11	0	0.002

The H_{eff} matrix using (17) is obtained as:

$$H_{\text{eff}} = \begin{bmatrix} 0.3793 & -0.0367 & 0.0017 & -0.3115 & -0.0132 & -0.0040 & 0.0000 & -0.0015 & 0.0204 \\ -0.0367 & 0.0376 & -0.0350 & 0.0238 & -0.0040 & -0.0439 & 0.0636 & -0.0169 & 0.0012 \\ 0.0017 & -0.0350 & 0.0440 & 0.0017 & 0.0000 & 0.0636 & -0.0869 & 0.0233 & 0.0000 \\ -0.3115 & 0.0238 & 0.0017 & 0.2758 & 0.0204 & 0.0012 & 0.0000 & 0.0011 & -0.0228 \\ -0.0132 & -0.0040 & 0.0000 & 0.0204 & 1.7063 & -0.0219 & 0.0015 & -0.0160 & -1.6681 \\ -0.0040 & -0.0439 & 0.0636 & 0.0012 & -0.0219 & 0.6331 & -0.8003 & 0.1775 & 0.0117 \\ 0.0000 & 0.0636 & -0.0869 & 0.0000 & 0.0015 & -0.8003 & 1.0403 & -0.2432 & 0.0015 \\ -0.0015 & -0.0169 & 0.0233 & 0.0011 & -0.0160 & 0.1775 & -0.2432 & 0.0825 & -0.0008 \\ 0.0204 & 0.0012 & 0.0000 & -0.0228 & -1.6681 & 0.0117 & 0.0015 & -0.0008 & 1.6558 \end{bmatrix} * 10^{14} \quad (23)$$

The Z_{eff} matrix obtained using (18) is given as:

$$Z_{\text{eff}} = \left[-1.37\text{E}+12\ -9.72\text{E}+10\ 1.19\text{E}+11\ 1.77\text{E}+12\ -4.24\text{E}+10\ -1.09\text{E}+10\ 1.50\text{E}+11\ -4.18\text{E}+10\ -5.58\text{E}+10\right]^T \quad (24)$$

The actual values obtained from load flow and the estimated values obtained from state estimation during normal and with PMU measurements are shown in Tables 1 and 2.

To perform the state estimation using PMU, the H^c matrix is obtained using (14) as:

$$H^c = \begin{bmatrix} 1\,0\,0\,0\,0\,0\,0\,0\,0\,0 \\ 0\,1\,0\,0\,0\,0\,0\,0\,0\,0 \\ 0\,0\,1\,0\,0\,0\,0\,0\,0\,0 \\ 0\,0\,0\,1\,0\,0\,0\,0\,0\,0 \\ 0\,0\,0\,0\,1\,0\,0\,0\,0\,0 \\ 0\,0\,0\,0\,0\,1\,0\,0\,0\,0 \\ 0\,0\,0\,0\,0\,0\,1\,0\,0\,0 \\ 0\,0\,0\,0\,0\,0\,0\,1\,0\,0 \\ 0\,0\,0\,0\,0\,0\,0\,0\,1\,0 \\ 0\,0\,0\,0\,0\,0\,0\,0\,0\,1 \\ 1\,0\,0\,0\,0\,0\,0\,0\,0\,0 \\ 0\,0\,0\,0\,0\,1\,0\,0\,0\,0 \end{bmatrix} \tag{25}$$

It can be observed from Tables 1 and 2 that the error obtained in the states is very less when the states are estimated using PMU data. The comparison of errors in voltage and magnitude is shown in Figs. 18 and 19. Here, as the number of synchronized measurements is increased, high accuracy is achieved.

Fig. 18 Voltage error comparison with and without

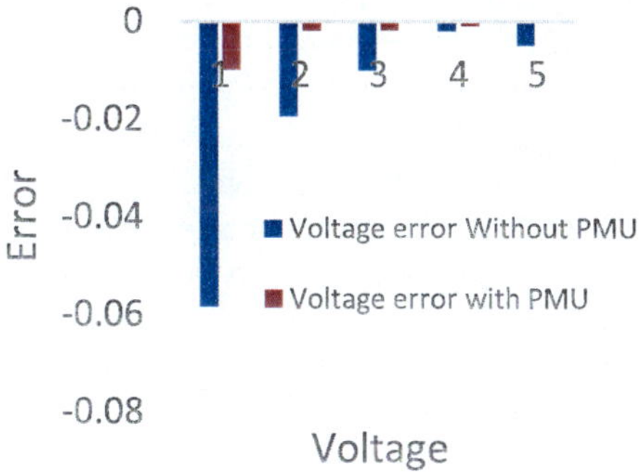

Fig. 19 Phase angle error with and without PMU

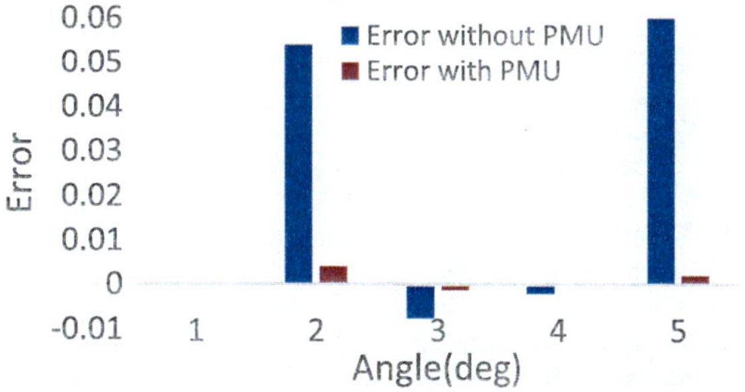

Table 3 Estimated states during fault duration

Values	Estimated values during fault	
States	Voltage (pu)	Angle (°)
1	0.231	0.998
2	0.453	2.346
3	0.864	− 0.4578
4	0.748	− 0.7836
5	0.697	2.146

Table 3 shows the estimated states during fault condition obtained using (13), (14), (15).

6 Conclusion

In this paper, the states of all the buses are obtained on real-time basis. In an APDS system, the states of the buses in which PMUs are placed are directly obtained along with the measurement data. But, for buses where PMUs are not present, hybrid state estimation can be used for this purpose, which is shown in this paper. The usage of PMU helps in faster detection of disturbances and faults that has occurred in the system as show in Sect. 5 due to faster detection of large overshoots. The necessary matrices needed are obtained from the PMU measurement, and from the measurements already available, the initial states are considered as flat start, and the states of other buses are obtained based on the equations. It can be seen that the estimated state values obtained using PMU data are more accurate compared to the states obtained without using PMU data. It is observed that more the number of synchronized measurements considered, higher the accuracy obtained in the states estimated. It is observed that the estimated states are similar to the actual states after introducing some error into the system; hence, we can say that our state estimation algorithm can adapt to sudden disturbances. Hence, these measurements are used on real-time basis for observability analysis of power system. The losses were neglected in this paper. Future scope can be to consider losses while computing the measurement data.

References

1. Milis GM, Asprou M, Kyriakides E, Panayiotou CG, Polycarpou MM (2015) Semantically-enhanced configurability in state estimation structures of power systems. In: IEEE symposium series on computational intelligence, pp 679–686
2. Thomas MS, McDonald JD (2015) Power system scada and smart grids. CRC Press
3. James J, Bindu S (2015) Hybrid state estimation including PMU measurements. In: International conference on control communication & computing India (ICCC), pp 309–313

4. Chandra A, Pradhan AK, Sinha AK (2016) PMU based real time power system state estimation using ePHASORsim. In: National power systems conference (NPSC), pp 1–6
5. Sarri S, Paolone M, Cherkaoui R, Borghetti A, Napolitano F, Nucci CA (2012) State estimation of active distribution networks: comparison between WLS and iterated Kalman-filter algorithm integrating PMUS. In: 3rd IEEE PES innovative smart grid technologies Europe (ISGT Europe), pp 1–8
6. Romano P, Paolone M (2014) Enhanced interpolated-DFT for synchrophasor estimation in FPGAs: theory, implementation, and validation of a PMU prototype. IEEE Trans Instrum Meas 63(12):2824–2836
7. Pignati M et al (2015) Real-time state estimation of the EPFL-campus medium-voltage grid by using PMUS. In: IEEE power & energy society innovative smart grid technologies conference (ISGT), pp 1–5
8. von Meier A, Culler D, McEachern A, Arghandeh R (2014) Micro-synchrophasors for distribution systems. ISGT, pp 1–5
9. Shiroei M, Daniar S, Akhbari M (2009) A new algorithm for fault location on transmission lines. In: IEEE power & energy society general meeting, pp 1–5
10. Janssen P, Maun JC (2013) Monitoring, protection and fault location in power distribution networks using system-wide measurements
11. Gholami M, Tehrani Fard AA, Moeini-Aghtaie M (2018) Linear voltage-based state estimator for active distribution system including phasor measurement unit (PMU). In: Electrical power distribution conference (EPDC), pp 1–6
12. Aminifar F, Shahidehpour M, Fotuhi-Firuzabad M, Kamalinia S (2014) Power system dynamic state estimation with synchronized phasor measurements. IEEE Trans Instrum Meas 63(2):352–363
13. Haughton DA, Heydt GT (2013) A Linear state estimation formulation for smart distribution systems. IEEE Trans Power Syst 28(2):1187–1195
14. Liu J, Tang J, Ponci F, Monti A, Muscas C, Pegoraro PA (2012) trade-offs in PMU deployment for state estimation in active distribution grids. IEEE Trans Smart Grid 3(2):915–924
15. Sarri S, Zanni L, Popovic M, Le Boudec J-Y, Paolone M (2016) Performance assessment of linear state estimators using synchrophasor measurements. IEEE Trans Instrum Meas 65(3):535–548
16. Roscoe AJ, Abdulhadi IF, Burt GM (2013) P and M class phasor measurement unit algorithms using adaptive cascaded filters. IEEE Trans Power Delivery 28(3):1447–1459

Online Estimation of Voltage Stability Margin Using PMU Data

Lohitha Avula, Ram Krishan, and Krishan Kumar

Abstract For secure and reliable operation, online monitoring of voltage stability margin is important in smart grid. It is critical to determine how close a power system is to voltage breakdown online in order to operate it safely. Based on the basic definition of voltage stability margin, a simple method is proposed to determine the voltage stability index (VSI) using PMU data instead of local measurement. This VSI is used for anticipating the voltage collapse in power systems. The index is calculated using the bus's maximum load capacity and the Thevenin equivalent approach for the system's aggregated representation. The index for line voltage measurement can be computed by computing the Thevenin based on fast load flow or phasor measurement units (PMUs) data. Proposed technique is demonstrated with a standard IEEE 14-bus test system.

Keywords Voltage collapse · Local measurements · Online voltage stability · Load voltage capacity · Thevenin equivalent · Phasor measurement units

1 Introduction

In order to meet the growing demand, utilities must take an advantage of using the existing transmission facilities in the power system rather than building the new ones. The factors contributing to power systems operating near their limits include increased complexity, multiple sources, and heavy loads. Consequently, the voltage profile of the stressed grid is deteriorated. This has resulted in the problem of voltage stability and sometimes leads to the voltage collapse. Voltage instability is one of the major problems that could result from such a stressed grid [1].

L. Avula (✉) · R. Krishan
Departmentof Electrical Engineering, National Institute of Technology Warangal, Warangal, India
e-mail: laee21304@student.nitw.ac.in

R. Krishan
e-mail: rkrishan@nitw.ac.in

K. Kumar
DEHN India, Gurugram, India

A. Rani et al. (eds.), *Signals, Machines and Automation*, Lecture Notes in Electrical Engineering 1023, https://doi.org/10.1007/978-981-99-0969-8_15

The power system is a complex network which consists of various kinds of loads and sources. Due to the integration of renewable energy sources (RESs), the complexity of the power system network further increased. Also, due to different kind of loads, like leading and lagging loads, the power system network always is prone to major problems like voltage stability and voltage collapse [2]. The voltage stability problem can be defined as the inability of the power system network to maintain the voltage levels steady and acceptable at all buses in the network during normal operating conditions. This definition can be also being applied to the power system after getting subjected to disturbance. Voltage stability problem occurs with a local disturbance leading to decreasing in the voltage of all the buses. If this continues progressively over a period of time, it may lead to blackout situation.

Voltage collapse refers to the rapid and uncontrollable drop in voltage of a bus or group of buses associated with an increase in load, generally caused by insufficient reactive support in high-load areas. A sudden surge in voltage could also cause voltage collapse, change in the system, such as a line outage [3]. If the load at a particular bus is varied while the rest of the system conditions remain unchanged, the voltage at that bus will also change, changing the node voltage at other buses. Therefore, the voltage at a load bus is partially dependent on the power delivered to that node. The active and reactive powers can be separated into segments such as $\Delta V/\delta P$ and $\Delta V/\delta Q$ where P is active power, Q is reactive power, and V is the voltage at that node.

Voltage collapse and voltage instability are often mistaken for the same phenomenon. Here, it is necessary to clarify the thin line between these terms. Several buses may see a gradual decrease in voltage profile as a result of disturbances in a power system. This is called voltage instability, while voltage collapse is an unfeasible value of voltage whose magnitude is rapidly decreasing. Power system instability can result in voltage collapse. During a disturbance in the power system, there is a point in time where the voltage becomes uncontrollable. In the case of actual voltage collapse, this occurs later than voltage instabilities [4, 5]. Voltage collapse can be categorized into two categories: static and dynamic. The probability and speed of voltage collapse can be affected by many different events such as equipment outages or faults caused by equipment outages, load disturbances, etc. Depending on the load disturbance, it may be a sudden or gradual outage of a large block of load. As the voltage fluctuates in small discrete steps of steady states with slow load fluctuation, so can be considered a static phenomenon. However, the fast load fluctuation as well as equipment outages or faults due to it comprise the dynamic phenomenon. Transient instability is caused by disturbances that require dynamic analysis, but voltage instability can only occur if the voltage values after the disturbance are low, the transient voltage dips last a long time, or the voltage equilibrium achieved after the disturbance is unstable, and adding any reactive power support to that bus will lower the voltage there [6].

Even though the exact calculation of instability onset requires knowledge of the complete power system state including generator over excitation conditions [2, 7, 8], an approximate estimation based exclusively on local measurements is still possible. In order to set up such a problem of approximate decentralized, measurement-based voltage instability monitoring, it is useful to define an upstream, as well as a downstream power network. Several research efforts attempt to calculate an equivalent upstream and downstream power system [9, 10] relying on PMU measurements, based on which to estimate voltage instability onset. A good estimation of voltage instability onset for a load area can be provided in real time, if the boundary buses are monitored through PMU measurements [11], where the sensitivity of transferred active power to apparent conductance is used as an index. Voltage stability improvement can also be done by optimal reactive power measures using a sensitivity index called power transfer stability index (PTSI) [12]. A stability margin estimation technique called the Kantorovich-based voltage stability margin (KVSM) is proposed in [13]. To provide a fast and accurate static voltage stability assessment, many data-driven-based methods have been proposed in [14]. Fast voltage stability index (FVSI) is a mathematical method which in turn speeds up the technique of voltage stability analysis in the prevalence of contingency evaluation due to line outage in a power system [15].

In this paper, the online voltage stability estimation is done using the PMU data by calculating the voltage stability index (VSI). VSI calculation is based on the Thevenin equivalent theory which is used to predict the proximity to voltage collapse using the PMU data. The buses with lowest VSI are the most vulnerable buses and prone to voltage collapse.

2 Voltage Collapse Indicators

In static simulators, things like reactive support requirements and system loading capabilities are determined during planning and operation. Simulations in the time domain are also used to analyze voltage stability [16]. At higher loads and near to the point of voltage collapse, there is no real steady-state solution to load flow; therefore, it was very difficult to arrive at a solution.

2.1 AC State Estimation Model

Load flow analysis or power flow analysis is the procedure required to determine the steady-state operating characteristics of the power system network from the given bus data and line data [17].

Power system AC state estimation is the process of collection of measurements from the system and computing a state vector of the voltage (magnitude and angle) at each observed bus. Measurements which are nonlinear functions of the system state are collected, and load flow-like calculations are performed iteratively to determine the most probable system state from the known information [18].

State estimation results are processed to identify errors in the meters and parameters given that there are enough redundant measurements. In addition, failure or loss of a meter can be detected [19]. In this work, the weighted least square state estimation method is used for the estimation of the voltage stability margin.

Considering the measurement vector as 'z' which contains the 'm' number of measurements and the state vector 'X' denotes the 'n' number of the state variables. Conventional state estimation techniques employ measurement sets which are nonlinear functions of the system state vector and can be represented using $h(X)$. However, all of these measurements have their own unknown error associated with them denoted by 'e.' The measurement errors are assumed to be independent of one another and have an expected value of zero.

$$z = [h(X)] + [e] \tag{1}$$

3 Voltage Stability Index

Voltage stability index is an index which predicts the closeness to voltage instability by assessing the voltage stability of power system. The magnitude changes as the system parameter changes and can be used online or offline to help operators determine closeness to voltage collapse. With the advancement in the technology of smart grid and the phasor measurement unit (PMU) technology, the voltage index can be calculated directly from the PMU measurement data [20]. The development of synchrophasor measurement technology along with other advances in computational facilities, networking infrastructure and communications has given us the new perspectives for wide-area monitoring and control. Phasor measurement-based voltage instability monitoring can be classified in two broad categories:

i. Local measurements-based methods

The local measurements-based methods use the Thevenin equivalent method in assessment of the voltage instability. The Thevenin voltage and the impedance remain constant throughout the measurement.

ii. Observability based methods

The observability-based methods require time-synchronized measurements; it is obtained from the PMU data.

3.1 *Proposed Methodology*

In this work, the voltage stability index is calculated based on the maximum loading capacity and Thevenin equivalent circuit. In order to use this proposed methodology to the transmission system, the whole network should be reduced into a system with two nodes (Figs. 1 and 2).

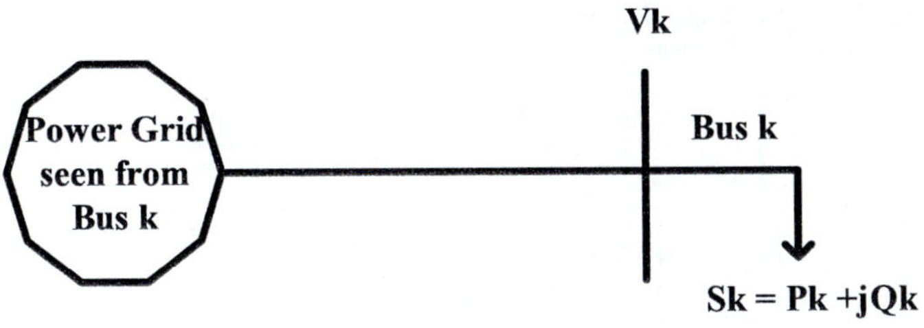

Fig. 1 Thevenin equivalent of the power system

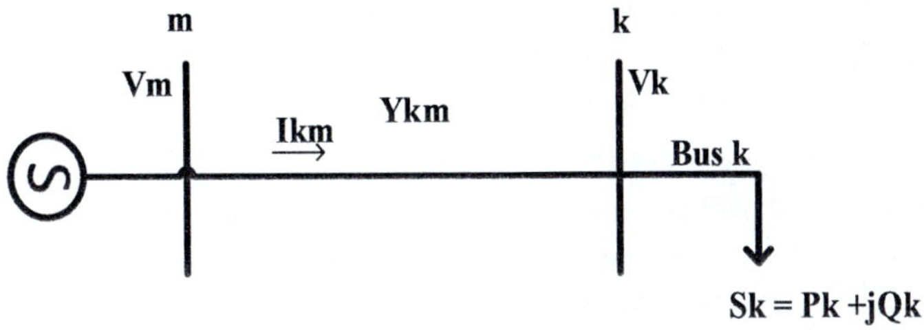

Fig. 2 Thevenin equivalent represented as two node system

From the above diagram, the load impedance at bus k can be determined as

$$Z_k^L = \frac{|V_k|^2}{P_k - jQ_k} \tag{2}$$

$$Z_{kk} = Z_k^L // Z_{\text{th}} = \frac{Z_k^L Z_{\text{th}}}{Z_k^L - Z_{\text{th}}} \tag{3}$$

Figure 3 is the schematic diagram of generator model in power system with constant terminal voltage. The Thevenin impedance and voltage is derived as following with the help of Fig. 3 and Eq. 3

$$Z_{\text{th}} = \left(\frac{1}{Z_{kk}} - \frac{1}{Z_k^L}\right)^{-1} V_{\text{th}} = \left(1 - \frac{Z_{\text{th}}}{Z_k^L}\right) V_k \tag{4}$$

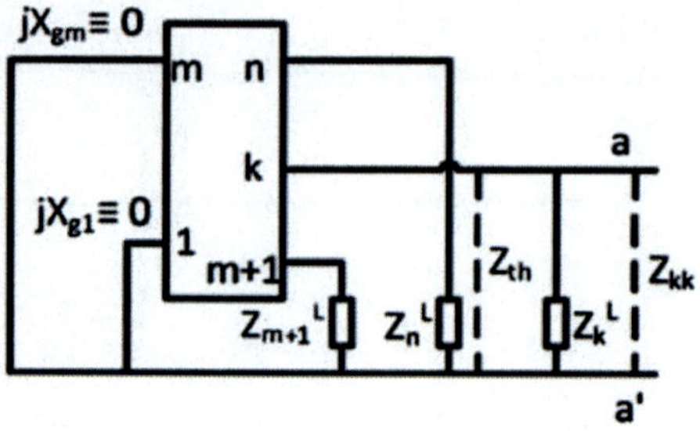

Fig. 3 Generator model and the Thevenin impedance of load bus

3.2 Flowchart

Flowchart of the proposed algorithm is given in Fig. 4.

The active power and reactive power of system at bus k is as follows and are considered for the calculation of VSI

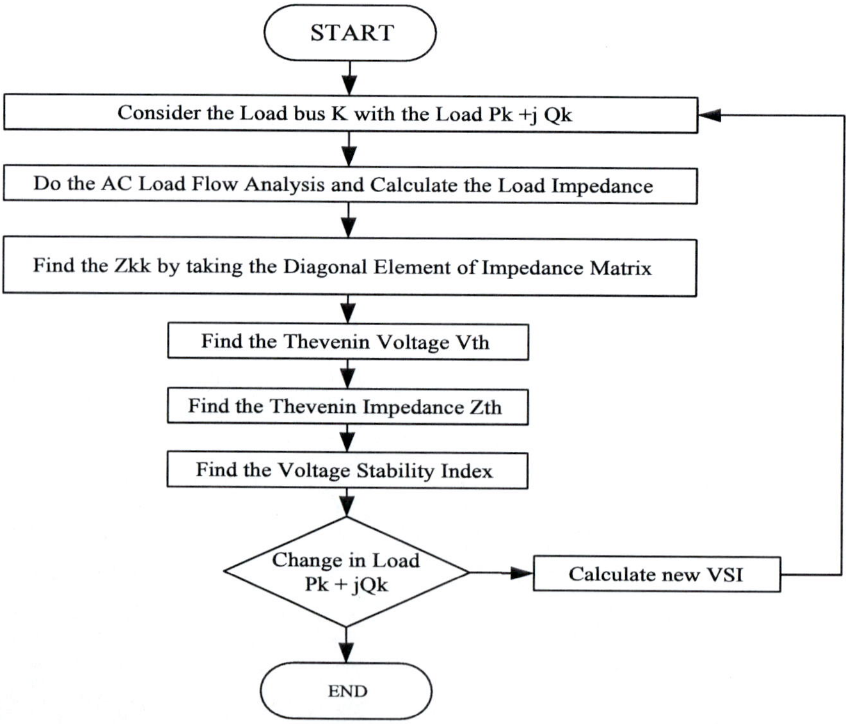

Fig. 4 Flowchart for VSI calculation

$$P_k = \sum_{m=1}^{n} |V_k||V_m||Y_{km}| \cos(\delta_m - \delta_k + \theta_{km}) \tag{5}$$

$$Q_k = -\sum_{m=1}^{n} |V_k||V_m||Y_{km}| \sin(\delta_m - \delta_k + \theta_{km}) \tag{6}$$

The equations can be modified according to the above figure, and squaring Eqs. (5) and (6), we get

$$\begin{aligned} &P_k^2 + V_k^4 Y_{kk}^2 \cos^2\theta_{kk} - 2V_k^2 Y_{kk} \cos\theta_{kk} P_k \\ &= (V_k V_m Y_{km})^2 \cos^2(\delta_m - \delta_k + \theta_{km}) \end{aligned} \tag{7}$$

$$\begin{aligned} &Q_k^2 + V_k^4 Y_{kk}^2 \sin^2\theta_{kk} + 2V_k^2 Y_{kk} \sin\theta_{kk} Q_k \\ &= (V_k V_m Y_{km})^2 \sin^2(\delta_m - \delta_k + \theta_{km}) \end{aligned} \tag{8}$$

Adding the above Eqs. (7) and (8) and rearranging, we get

$$V_k^4 Y_{kk}^2 - (V_k)^2\left\{(V_m Y_{km})^2 + 2Y_{kk}(P_k \cos\theta_{kk} - Q_k \sin\theta_{kk})\right\} + P_k^2 + Q_k^2 = 0 \tag{9}$$

Equation (9) can be formed into a quadratic equation of variable V_k^2, and it is solved to get the solution of V_k^2 as in (10)

$$V_k^2 = \frac{-b \pm \sqrt{b^2 - 4ac}}{2a} \tag{10}$$

where

$$a = Y_{kk}^2;\ b = -\left\{(V_m Y_{km})^2 + 2Y_{kk}(P_k \cos\theta_{kk} - Q_k \sin\theta_{kk})\right\};\ c = P_k^2 + Q_k^2 \tag{11}$$

If $\left(b^2 - 4ac\right)$ is there is no real solution, therefore $b^2 - 4ac \geq 0$. Putting a, b, c values in the equation,

$$\begin{aligned} &\left\{(V_m Y_{km})^2 + 2Y_{kk}(P_k \cos\theta_{kk} - Q_k \sin\theta_{kk})^2\right\} \\ &\quad - 4Y_{kk}^2\left(P_k^2 + Q_k^2\right) \geq 0 \end{aligned} \tag{12}$$

Maximum loading point of the system is reached when the $V_k^2 = 0$, and if it is seen from the voltage stability point of view, we can say that maximum power transfer represents more proximity to the voltage instability.

Therefore, a VSI which is a constant load factor multiplied to $P_k + jQ_k$ to make it VSI $* (P_k + jQ_k)$.

VSI represents the closeness to voltage instability or collapse.

- If VSI = 1, the line has reached to its maximum power transfer capability.

- If VSI < 1, the line violates maximum power transfer capacity, and the voltage becomes unstable.

$$\left\{(V_m Y_{km})^2 + 2Y_{kk}(\text{VSI}P_k \cos\theta_{kk} - \text{VSI}Q_k \sin\theta_{kk})^2\right\} - 4Y_{kk}^2\text{VSI}^2\left(P_k^2 + Q_k^2\right) \geq 0 \tag{13}$$

Rearranging the terms in (13) and finding the value of VSI, we get

$$\text{VSI} = \frac{-b \pm \sqrt{b^2 - 4ac}}{2a} \tag{14}$$

where

$$a = 4Y_{kk}^2\,[(P_k \cos\theta_{kk} - Q_k \sin\theta_{kk}))^2 - P_k^2 + Q_k^2$$
$$b = \left\{4(V_m Y_{km})^2 Y_{kk}(P_k \cos\theta_{kk} - Q_k \sin\theta_{kk})\right\}$$
$$c = (V_m Y_{km})^4$$

According to Thevenin equivalent circuit, V_m is replaced with V_{th}. Y_{km} is replaced with Z_{th} and Y_{kk} by Z_{kk}.

4 Result and Discussion

4.1 Two-Bus System

The two-bus system, which comprises of a generator that feeds local load bus k over the branch Z_{line}, is considered the test system. $V_m = (1.0 + j0.0)$ pu, $Z_{\text{line}} = (0.028 + j0.096)$ pu, and $S_k = (150 + j50)$ MVA are the voltages for generator terminal m.

Thevenin impedance is equivalent to line impedance in a two-bus system and has a constant value of 0.1 pu; here, the VSI is calculated by using (14) and is 0.763. It is high and not close to zero implies that it is not a voltage collapse point. Here, load impedance is substantially greater than the Thevenin impedance under normal loading. As in the case of maximum power transfer, load impedance equals Thevenin impedance at maximum loading; the VSI is very high. At maximum power transfer point, if the voltage is decreased to 0.55 pu and VSI is 0.002, which is the collapse point, in the voltage stability analysis, merely considering the collapse point indicated by very small VSI values close to zero will be an optimistic approach. The voltages in both circumstances are roughly 10% less than the normal voltage when VSI = 1.

4.2 IEEE 14-Bus System

Obtain Y-bus, V_m, and δ by performing load flow analysis with the line data and adjusted bus data. Get a rectangular version of S_{load} (KVA). Except for bus K, represent the load at every load bus by an admittance $Y_{\text{load}} = S/V_m^2$. Modify the Y_{ii}, it should be changed to $Y_{ii} = Y_{ii}, + Y_{\text{load}(i)}$. Delete all axes associated with generator buses, including the slack bus. To get a Z-bus with load, invert it. Bus K's Z-Thevenin is Z_{kk}. The above method should be repeated for each bus except for buses whose load power is zero. Voltage stability index is calculated according to above modeling equations where $Y_{km} = 1/Z$-Thevenin and $Y_{kk} = 1/Z_{kk}$. V_k is the bus K's voltage obtained from the state estimation. The IEEE 14-bus system is converted into the Thevenin equivalent circuit, and VSI is calculated. Except for the buses 1, 7, and 8, all other buses can be considered as load bus i.e., Kth bus as loads at these buses is zero. When buses 3 and 4 are considered as the load bus, the VSI obtained is very close to zero. This implies that there is a chance for voltage collapse. In case of bus 3, the V_m (V_{Thevenin}) is very less, and load impedance and Thevenin impedance are nearly equal, that is the reason for lesser VSI. But, in case of bus 4, the Thevenin voltage is not too less; the VSI is lower because the load impedance is not much greater than the Thevenin impedance.

Table 1 represents the voltages and phase angles at each bus of the IEEE–14-bus system obtained from load flow analysis (before state estimation) and state estimation. The voltages and phase angles obtained from load flow analysis are used in state estimation for finding the error-free voltages and phase angles of the buses. State estimation results are used for calculating the voltage stability index.

Table 2 represents the values of load bus voltage (V_k) which is obtained from the state estimation. Thevenin voltage (V_m) is calculated from Eq. (4) with respect to load bus. Load bus impedance (Z_k) is calculated from Eq. (2), and the Thevenin impedance is calculated from Eqs. (3) and (4). The voltage stability index (VSI) is calculated using Eq. (14).

Figure 5 shows the variation of VSI with respect to the load voltage. Figure 6 shows the variation of VSI with Thevenin voltage. The effect of V_m (Thevenin voltage) is more on the calculation of VSI than the load bus voltage. The value of VSI also depends on the Thevenin impedance and load impedance. The load impedance at buses 3 and 4 is less than the Thevenin impedance, making the buses 3 and 4 as vulnerable buses. The Thevenin voltage (V_m) at buses 3 and 4 is also less compared to other buses Thevenin voltage. Therefore, the VSI of bus 3 and bus 4 is as low as 0.0071 and 0.0049, respectively. The closeness of VSI to zero implies voltage collapse. Higher the VSI, higher is the voltage stability. Thus, the VSI is measure of voltage stability margin, i.e., how close the system is to instability.

Table 1 Voltage magnitude and angle before and after the state estimation

	Without state estimation		With state estimation	
Bus number	Magnitude	Angle (°)	Magnitude	Angle (°)
1	1.060	0.000*	1.0958	0.000*
2	1.045	– 4.983	1.0786	– 2.6228
3	1.010	– 12.725	1.0736	– 3.9666
4	1.018	– 10.313	1.0608	– 3.9069
5	1.020	– 8.774	1.0623	– 3.4489
6	1.070	– 14.221	1.0294	– 5.5366
7	1.062	– 13.360	1.0480	– 5.9744
8	1.090	– 13.360	1.0699	– 5.1270
9	1.056	– 14.939	1.0334	– 6.7800
10	1.051	– 15.097	1.0284	– 6.8106
11	1.057	– 14.791	1.0284	– 5.7644
12	1.055	– 15.076	1.0236	– 5.2591
13	1.050	– 15.156	1.0150	– 5.7493
14	1.036	– 16.034	1.0155	– 7.4784

* represents Slack Bus

Table 2 VSI calculation at load buses

Load bus	V_k	V_m (V_{th})	Z_k	Z_{th}	VSI
2	1.0786	0.7712	4.6268	1.3184	1.0606
3	1.0736	0.1578	1.1995	1.3758	0.0071
4	1.0608	0.4324	1.2464	1.3899	0.0049
5	1.0623	0.9593	14.5294	1.4079	0.2857
6	1.0294	0.8447	7.8608	1.4102	1.2020
9	1.0334	0.5507	3.1549	1.4737	0.5089
10	1.0284	0.8756	9.8773	1.4673	1.3603
11	1.0284	0.9724	26.8724	1.4623	1.4088
12	1.0236	0.9411	16.6128	1.339	0.4423
13	1.0150	0.8185	7.0119	1.3578	0.6713
14	1.0155	0.7954	6.562	1.4226	0.5352

5 Conclusion

Online estimation of voltage stability margin is proposed using the PMU data. The voltage stability index presented in this paper is based on maximum loading capability and can be used in online voltage analysis. Data on load flow or measurement can be used to determine the index. The index can be employed in networks using a quick

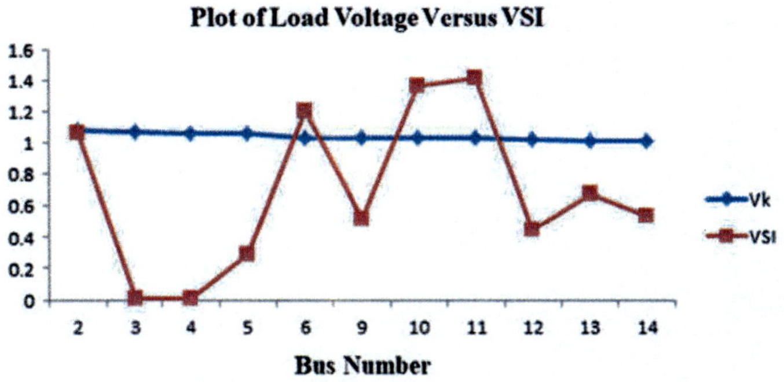

Fig. 5 VSI and load voltage of IEEE 14-bus system

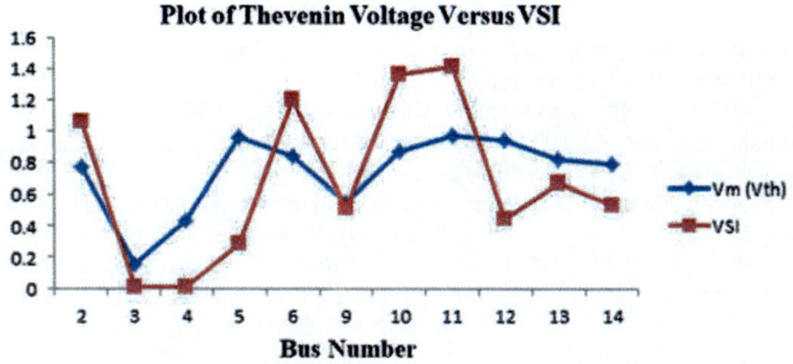

Fig. 6 VSI and Thevenin voltage of IEEE 14-bus system

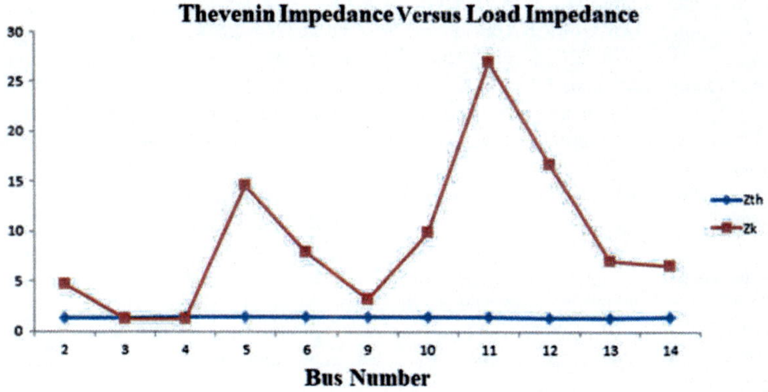

Fig. 7 Thevenin impedance and load impedance of IEEE 14-bus system

technique of Thevenin equivalent calculation, as demonstrated by implementing the algorithm with IEEE 14-bus and two-bus test systems. Here, the VSI is calculated based on the Thevenin voltage of the system, load voltage, Thevenin impedance, and load impedance. The closeness of VSI toward zero indicates the voltage collapse. In IEEE 14-bus system, the most vulnerable buses are bus 3 and 4 as the VSI values are 0.0071 and 0.0049, respectively. The future scope of this approach is to find the real power as well as reactive power loading margin of the system which can be obtained by continuation power flow method.

References

1. Taylor CW (1994) Power system voltage stability. McGraw-Hill Education, New York
2. Van Cutsem T, Vournas C (2008) Voltage stability of electric power systems. Springer
3. Kundur P, Taylor CW (2007) IEEE task force on blackout experience, mitigation, and role of new technologies. In: IEEE/PES task force, power system dynamic performance committee, PES
4. Maharjan R, Kamalasadan S (2015) Voltage stability index for online voltage stability assessment. North Am Power Symp (NAPS) 2015:1–6
5. Kundur P (1994) Power system stability and control. McGraw-Hill
6. Mandoulidis P, Vournas C (2020) A PMU-based real-time estimation of voltage stability and margin. Electr Power Syst Res 178:1–12
7. Glavic M, Van Cutsem T (2009) Wide-area detection of voltage instability from synchronized phasor measurements. Part I: principle. IEEE Trans Power Syst 24:1408–1416
8. Glavic M, Cutsem TV (2009) Wide-area detection of voltage instability from synchronized phasor measurements. Part II: simulation results. IEEE Trans Power Syst 24:1417–1425
9. Hu F, Sun K, Rosso AD, Farantatos E, Bhatt N (2016) Measurement-based real-time voltage stability monitoring for load areas. IEEE Trans Power Syst 31:2787–2798
10. Ramirez L, Dobson I (2014) Monitoring voltage collapse margin by measuring the area voltage across several transmission lines with synchrophasors. In: IEEE PES general meeting, National Harbor MD USA, July 2014
11. Vournas CD, Lambrou C, Mandoulidis P (2017) Voltage stability monitoring from a transmission bus PMU. IEEE Trans Power Syst 32(4):3266–3274
12. Ziegler C, Wolter M (2021) Voltage stability enhancement by reactive power changes based on voltage stability index PTSI. ETG—Kongress, May 2021
13. Weng Y, Yu S, Dvijotham K, Nguyen HD (2022) Fixed-point theorem-based voltage stability margin estimation techniques for distribution systems with renewables. IEEE Trans Industr Inform 18(6)
14. Su H-Y, Hong H-H (2021) An intelligent data-driven learning approach to enhance online probabilistic voltage stability margin prediction. IEEE Trans Power Syst 36(4)
15. Gupta A, Lakra P (2022) A combined voltage and frequency stability enhancement using artificial neural network and fast voltage stability index based load shedding. In: IEEE international conference on communication systems and network technologies, Apr 2022
16. Shair J, Li H, Hu J, Xie X (2021) Power system stability issues, classifications and research prospects in the context of high-penetration of renewables and power electronics. Renew Sustain Energy Rev 145:1–16
17. Sahu P, Verma MK. Online monitoring of voltage stability margin using PMU measurements. Int J Electr Comput Eng (IJECE) 10:1156–1168
18. Leonardi B, Ajjarapu V, Djukanovic (2011) A practical two-stage online voltage stability margin estimation method for utility-scale systems. IEEE Xplore, Oct 2011
19. Vu K, Begovic MM, Novosel D, Saha MM (1999) Use of local measurements to estimate voltage-stability margin. Power Syst IEEE Trans 14(3):1029–1035
20. Zambroni de Souza A, Stacchini de Souza JC, Leite da Silva AM (2000) On-line voltage stability monitoring. IEEE Trans Power Syst 15

Design and Implementation the Active Clamp Forward Converter for Electric Vehicles Application

Arya Singh and Madhusudan Singh

Abstract This paper presents a detailed examination of the circuit functioning and design analysis of an active clamp forward converter (DC–DC) with asynchronous rectification. Forward converter with active clamp (ACFC), to assist reset the flux in the transformer core, one clamp switch and clamp capacitor are used to absorb leakage energy of inductor, and decrease voltage and current spikes on the primary side of transformer and further lowering the voltage stress on the main switch (S_M). By using the switch's output capacitance and transformer leakage inductance, the clamped circuit helps the main switch in turning on during zero-voltage switching (ZVS). Switching the power switch to zero-voltage switching (ZVS) decreases switching losses and increases the converter's overall efficiency. To verify the zero-voltage switching (ZVS) and operation of active clamp forward converter (ACFC), the design procedure and simulation results for ACFC with DC supply voltage of 80–120 V and have a DC output voltage (V_0) of 13.41 V which is operating at rated load current (I_0) of 32.82 A and a switching frequency (F_s) of 250 kHz is presented.

Keywords Zero voltage switching (ZVS) · Active clamp forward converter (ACFC) · Zero derivative switching (ZDS) · Switching frequency (F_s) · Pulse width modulation (PWM)

1 Introduction

A buck-derived architecture with transformer isolation is used in the ACFC. For low-power applications, the ACFC is a common choice for single and/or multiple output power supply. Active clamp forward converters are smaller and more efficient than passive clamp forward converters [1]. In forward converter, transformers are utilised to accomplish circuit isolation and energy transformation from primary side

A. Singh (✉) · M. Singh
Department of Electrical Engineering, Delhi Technological University, Delhi, India
e-mail: aryasingh607@gmail.com

M. Singh
e-mail: madhusudan@dce.ac.in

A. Rani et al. (eds.), *Signals, Machines and Automation*, Lecture Notes in Electrical Engineering 1023, https://doi.org/10.1007/978-981-99-0969-8_16

to secondary side, as well as to reset the transformer's magnetising current using the active clamp approach. While there are a variety of approaches for accomplishing transformer reset, the active clamp approach is both simple and effective. Zero-voltage switching (ZVS), decreased switch voltage stress, wider duty cycle range, and reduced electro-magnetic interference (EMI) are just a few of the benefits of active switching. One of the constraints of the active clamp is the requirement for a precise duty clamp [2]. Increased duty cycle, if not limited to a maximum amount, might cause transformer saturation or additional voltage stress on the main switch, both of which will be dangerous. Another constraint is the requirement for sophisticated control to synchronise delay timing between the active clamp and primary switch gate drive [3]. When the switch (S_M), i.e. main switch is ON, its voltage emerges across the primary winding of transformer, causing the magnetising current (I_M) to increase linearly and reaches to its maximum positive peak value. For periodic voltage, current functioning of transformer circuit, this established magnetising current must be reset when the switch (S_M) is OFF. At the end of every switching phase, the magnetising current/flux in core should revert to its initial value. If the magnetising current/flux will not revert to its initial state at the conclusion of each switching phase, the magnetising current (I_M) will continue to rise, leading to transformer saturation [4–6]. By using an active clamp converter, the overall efficiency of the circuit is considerably increased. A variety of ways have been developed to reduce switching losses and increase system efficiency. With the help of clamp switch (S_{CL}) and clamp capacitor (C_{CL}), the active clamp approach has been developed to collect the excess stored energy inside leakage inductance (L_L) and decrease current, voltage stress of switch (S_M). The active clamp approach uses a normal pulse width modulation (PWM) system to transmit power, and the switch is switched on at zero-voltage switching (ZVS) utilising transformers leakage inductance (L_L) and parasitic capacitance and/or switch output capacitance (C_{DS}) [7, 8].

To show ZVS and enhanced efficiency, an active clamp zero-voltage switching (ZVS) forward converter is built and implemented in MATLAB/Simulink in this study. Sections II and III present a complete analysis of system and design approach and the creation of a 440.1162 W ACFC zero-voltage switching (ZVS) converter. Using clamp switch (S_{CL}), clamping capacitor (C_{CL}) and resonant circuit, the ACFC circuit will reset the surge energy held in leaky inductance, and the main switch's voltage stress is also decreased. To analyse the performance characteristics of the ACFC zero-voltage switching (ZVS) converter, the circuit operation for closed loop system is analysed, and the MATLAB/Simulation results based upon a 440.1162 W prototype circuit with DC link voltage of 80–120 V and output voltage (V_0) of 13.41 V at operating frequency (F_s) of 250 kHz are illustrated. The active clamp forward converter's uses include auxiliary power supply for electric cars.

2 Circuit Description and Analysis

The ACFC circuit is illustrated in Fig. 1. The clamp switch (S_{CL}) and clamping capacitor (C_{CL}) in the active clamped circuit are used to reset the magnetising flux of transformer and absorb the surplus energy created by the leakage inductance (L_L) to reduce voltage stress on the main switch (S_M). To accomplish ZVS functioning for the main switch, the drain-source capacitance (C_{DS}) and leakage inductance (L_L) are in resonant (S_M). The inductance of magnetization is stated as (L_m). The drain-source capacitance (C_{DS}) is equal to the parasitic capacitance of the switch (S_M), i.e. main switch and the switch (S_{CL}), i.e. clamp switch combined in parallel. The converter makes the following assumptions for circuit analysis.

- The drain-source capacitance (C_{DS}) should be smaller compared to the clamp capacitance (C_{CL}).
- The capacitance C_0 is big enough to maintain an approximately constant steady output voltage.
- The inductance (L_0) of the secondary side filter is large enough so that approximately constant current flow through it.
- The magnetising inductance (L_m) must be greater than leakage inductance (L_L).
- Turn ratio of transformer is $n = \frac{N_P}{N_S}$
- Zero-voltage switching (ZVS) is accomplished when the inductive energy is larger enough than the energy stored in capacitor of the drain-source capacitance (C_{DS}).

Mode I: Power is transferred to the secondary side of transformer in this mode of operation, as illustrated in Fig. 2. The magnetising current (I_m) of transformer is added to reflected secondary current (I_s') to generate the input current (I_{in}) and flows via the channel resistance of the switch (S_M). On the secondary side, forward bias causes diode ($D1$) to be ON, while reverse bias causes diode ($D2$) to be OFF. The

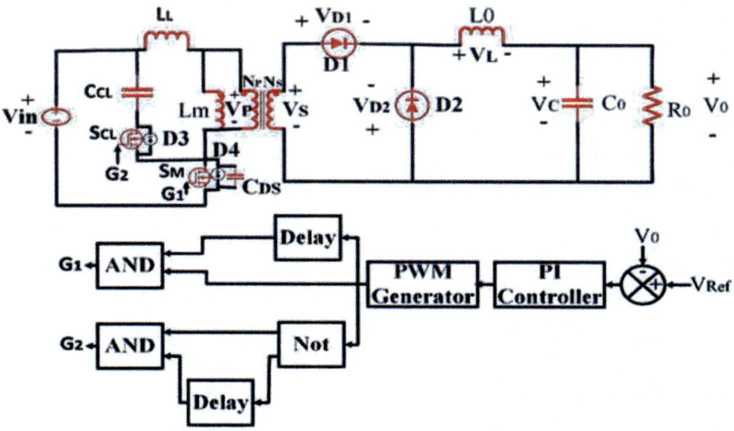

Fig. 1 ACFC circuit diagram

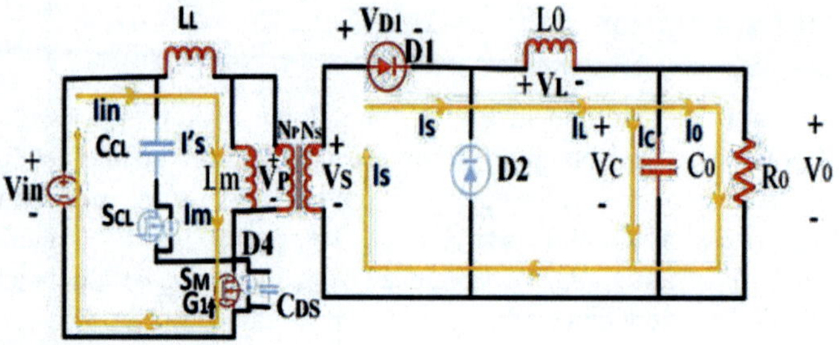

Fig. 2 Circuit diagram of ACFC, switch (S_M) ON, and switch (S_{SC}) OFF

supply voltage is the voltage that appears across the magnetising inductance (V_{in}).

$$I_{\text{in}} = I_m + I_s' \tag{1}$$

Voltage across magnetising inductance L_m is

$$V_{Lm} = V_{\text{in}} = L_m \frac{\mathrm{d}I_m}{\mathrm{d}t} \tag{2}$$

$$\Delta I_m = \frac{\mathrm{D}V_{\text{in}}}{F_s L_m} \tag{3}$$

Voltage across the output inductor L_0 is

$$V_{L0} = V_s - V_0 = V_{\text{in}} \frac{N_S}{N_P} - V_0 \tag{4}$$

$$\Delta I_{L0} = \frac{\left(V_{\text{in}} \frac{N_S}{N_P} - V_0\right) D}{F_s L_0} \tag{5}$$

where V_{Lm} is the voltage of magnetising inductance voltage, V_{L0} will be the voltage across output inductor, D will be the duty cycle, F_s is switching frequency.

Equations (2)–(5) show that current through magnetising inductance (L_m) and output inductance (L_0) increases linearly when the switch (S_M) is ON.

Mode II: In this mode of operation, switch (S_M) is switched off; the primary current (I_P) continues to charge up the main switch's drain-to-source capacitor (C_{DS}), before being diverted through the body diode ($D3$) of clamp switch and charging clamp capacitor (C_{CL}), as illustrated in Fig. 3. Because the secondary side load current (I_0) is freewheeling, there will be no reflected primary current; hence, only transformer magnetising current (I_m) is flowing through $D3$. As a result, the clamp switch ($D3$) body-diode's conduction loss is minimal, and the circumstances are established for clamp switch (S_{CL}) to turn on under ZVS. The resonant period is the duration

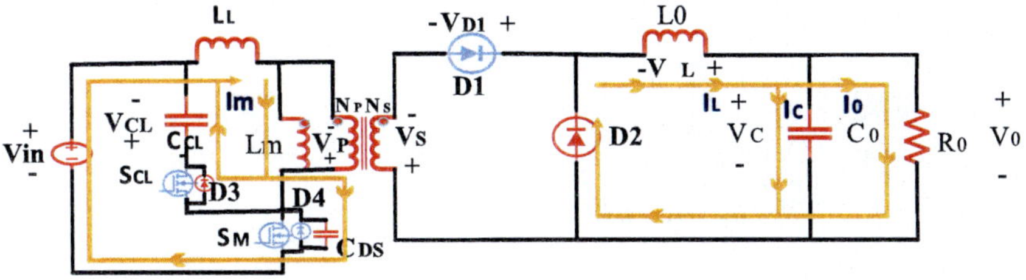

Fig. 3 ACFC circuit when switch (S_M) OFF, switch (S_{SC}) OFF

between turning off the switch (S_M) and turning on the switch ($S_{\rm CL}$). Diode ($D1$) is off on the secondary side because the transformer secondary applies reverse voltage across it, causing diode ($D2$) to conduct and the load current to freewheel through it. Conduction losses of $D1$ and $D2$ can be a significant component to overall power loss in high current applications and are frequently one of the limiting considerations for higher frequency operation.

Voltage across the output inductor L_0 is

$$V_{L0} = -V_0 \tag{6}$$

$$\Delta I_{L0} = \frac{-V_0(1-D)}{F_s L_0} \tag{7}$$

Equation (7) shows that, current through the output inductor decreases linearly when switch S_M, diode $D1$ is off and $D2$ is ON.

Mode III: This is the condition of active clamping, in which the primary side winding of the transformer is reset. Regardless of the fact that the primary current seems to reverse almost instantly in Fig. 4, the change from positive to negative current flow is smooth. The switch ($S_{\rm CL}$) is now completely ON; the voltage appear across transformer primary side is the difference of clamp capacitor voltage ($V_{\rm CL}$) and supply voltage ($V_{\rm in}$). The clamp capacitor ($C_{\rm CL}$) and drain voltage ($V_{\rm DS}$) started to fall, while the magnetising current (I_m) started to rise in the opposite direction. On the other hand, the diode ($D2$) on the secondary side of the transformer, which is carrying entire load current, does have a large conduction loss.

Mode IV: In this mode of operation, under ZVS, the clamp switch ($S_{\rm CL}$) is switched OFF. The main switch drain-to-source capacitance ($C_{\rm DS}$) and transformer magnetising current (I_m) continue to discharge their energy to the supply voltage ($V_{\rm in}$), as illustrated in Fig. 5. The magnetising current goes via the main switch's body diode once the $C_{\rm DS}$ is discharged. After then, the switch (S_M) is then switched ON.

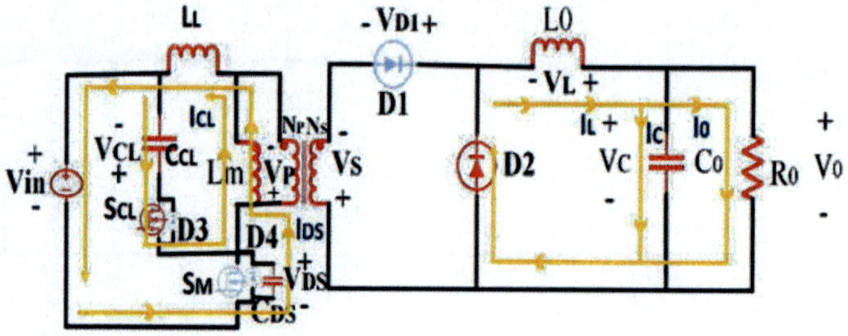

Fig. 4 ACFC circuit when switch (S_M) OFF and switch (S_{SC}) ON

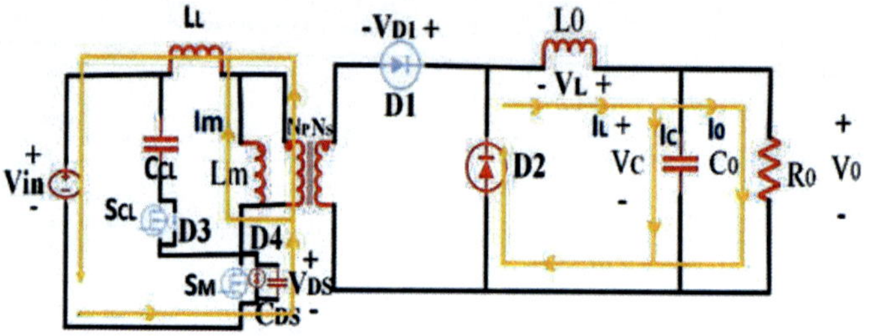

Fig. 5 Magnetising current and drain-source capacitance is discharging to the supply

3 Procedure for Designing Active Clamp Forward Converter

*D*max is the maximum duty cycle, occurs when voltage is maximum. *D*min is the minimum duty cycle, occurs when voltage is minimum. Then, the value of filter inductor

$$L_0 = \frac{V_0(1 - D_{\min})}{\Delta I_{L0} I_0 F_S} \tag{8}$$

where Δ I_L is the ripple in the inductor current, and it is expected to be 5% of the present load I_0.

$$Q = \frac{0.5 \Delta I_{LO}}{2 F_S} \tag{9}$$

$$C_{0\,\min} = \frac{Q}{\Delta V_0} \tag{10}$$

where Δ V_0 is ripple in output voltage (V_0), which is expected to 1% of output voltage.

$$R_0 = \frac{V_0^2}{P_0} \tag{11}$$

$$C_{CL} \geq \frac{10(1 - D_{min})^2}{L_{mag}(2\pi F_s)^2} \tag{12}$$

Here, F_s is the switching frequency; L_0 is output inductance; Q is charge; C_{0min} is the minimum value of output capacitance; R_0 is the output resistance; L_{mag} is the magnetising inductance of transformer, and C_{CL} is clamp capacitance.

4 Simulation Results and Discussion

In this study, active clamp forward converter (ACFC) for transfer of about 488.6 W power at a supply voltage of $V_{in} = 80$ V and output voltage $V_0 = 13.41$ V is designed and simulated in MATLAB/Simulink. The design specifications of ACFC are illustrated in Tables 1 and 2. A MATLAB model of the proposed scheme is developed for simulation.

A. Parameters and Component Values

An active clamp forward converter's circuit (ACFC) modelling has been introduced here. The input power supply for ACFC prototype is a 80 V, 6.53 A DC power source as illustrated in Fig. 6a, b. Figure 6c illustrates that, when the switch (S_M), i.e. main switch is ON, switch (S_{CL}), i.e. clamp switch is OFF, the voltage across the S_{CL}, and winding (N_P) is the supply voltage (V_{in}), the current increases linearly

Table 1 Design specifications of active clamp forward converter

Parameters	Value
Supply voltage (V_{in})	80 V
Supply current (I_{in})	6.536 A
Switching frequency (F_s)	250 kHz
Output voltage (V_0)	13.41 V
Output current (I_0)	32.82 A
Output power (P_0)	440.1162 W

Table 2 Component values and setting of controller in active clamp forward converter

Components	Values
C_{CL}	133.7 nF
K_p	0.0001
K_i	5
C_0	622.4 μF
L_0	14 μH
R_0	0.4 Ω

through transformer primary winding. Figure 6d illustrates, zero-voltage switching across the switch (S_M). Figure 6e illustrates current and voltage from diode ($D1$). Figure 6f, g is the DC output current and voltage, can be used to power up various parts of electric vehicles, and Fig. 6h illustrates the efficiency of the converter to be 85.89%.

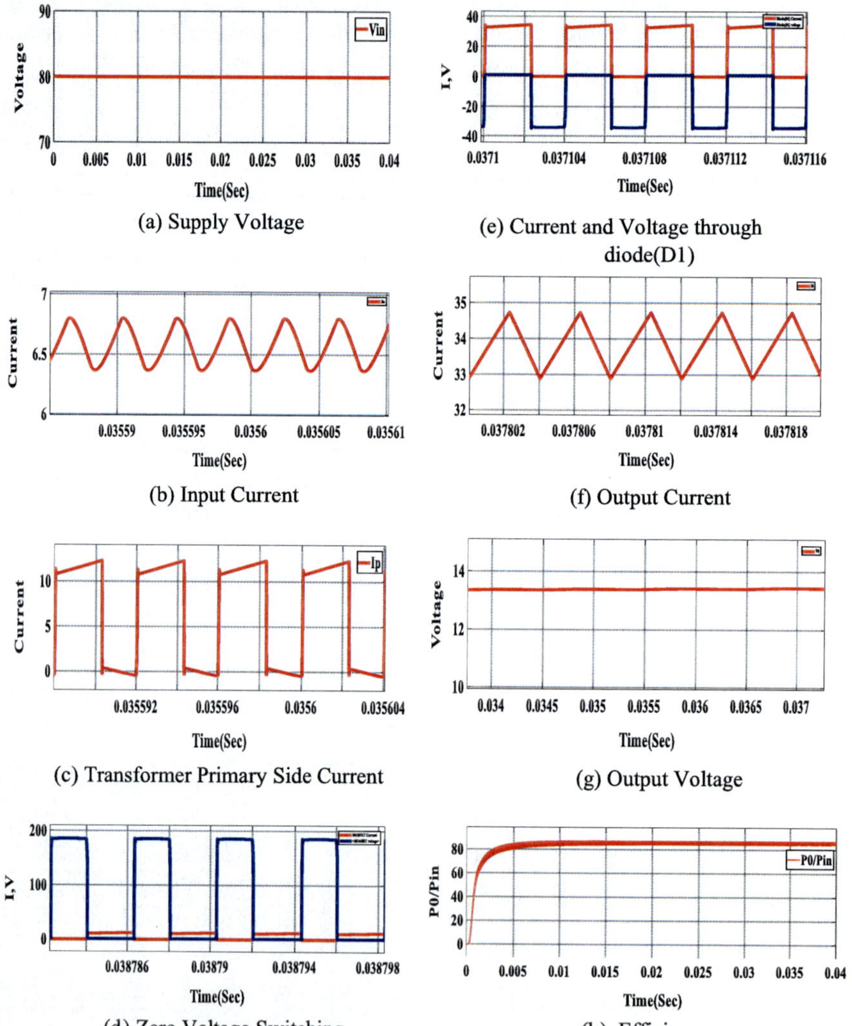

(a) Supply Voltage

(e) Current and Voltage through diode(D1)

(b) Input Current

(f) Output Current

(c) Transformer Primary Side Current

(g) Output Voltage

(d) Zero Voltage Switching

(h) Efficiency

Fig. 6 **a** Supply voltage. **b** Input current. **c** Transformer primary side current. **d** Zero-voltage switching. **e** Current and voltage through diode ($D1$). **f** Output current. **g** Output voltage. **h** Efficiency

5 Conclusion

The comprehensive circuit report of active clamp forward converter (ACFC) is described in this study. To manage output voltage (V_0), a PI-controller might be used and decrease damping in the system, and the converter functioning in closed loop is proven for a fixed output voltage of 13.6 V. The proportional-gain (K_p) is set at 0.0001 for this setup, while integral-gain (K_i) is set to 5. To accomplish ZVS functioning of the main switch, to reset the stored energy held in magnetising and leaking inductances, the clamp circuit is used. An active clamp forward converter (ACFC) is being implemented using a voltage supply of 80–120 V and a supply current of 6.536 A, an output voltage (V_0) of 13.41 V, an output current (I_0) of 32.82 A, and an output power of 440.1162 W is proven through a MATLAB/Simulink simulation study.

The basic difference between forward converter and active clamp forward converter is that in forward converter when switch is off, the residual energy stored in primary of the transformer has no path to go anywhere and within some on–off cycle of transformer the core get saturated and might destroy the switch due to high stress on it; to avoid this problem, we can use one extra winding, i.e. tertiary winding with an anti-parallel diode, but this will increase the size of forward converter. That is why active clamp forward converter is a best choice for EV application.

References

1. Shoyama M, Li G, Ninomiya T (2003) Application of common- source active-clamp circuit to various DC–DC converter topologies. In: IEEE-PESC conference, vol 3, pp 1321–1326
2. Erickson RW. Fundamentals of power electronics, 2nd edn
3. Watson R, Lee FC, Hua GC (1996) Utilization of an active- clamp circuit to achieve soft switching in flyback converters. IEEE Trans Power Electron 11(1):162–169
4. Li QM, Lee FC (2003) Design consideration of the active- clamp forward converter with current mode control during large-signal transient. IEEE Trans Power Electron 18(4):958–965
5. Torrico-Bascop R, Barbi N (2001) A double ZVS-PWM active-clamping forward converter: analysis, design, and experimentation. IEEE Trans Power Electron 16(6):745–751
6. Lim BS, Lee KW, Woo SH, Hee JK (2002) A new self- driven active clamp forward converter using the auxiliary winding of transformer. In: IEEE INTELEC conference, pp 164–168
7. Jitaru D, Birca-Galateanu S (1998) Small-signal characterization of the forward-flyback converters with active clamp. In: IEEE-APEC conference, vol 2, pp 626–632
8. Choi CT, Li CK, Kok SK (1999) Modeling of an active clamp discontinuous conduction mode flyback converter under variation of operating conditions. In: IEEE-PEDS conference, vol 2, pp 730–733

Renewable Energy Technologies

A Comparative Study on Metaheuristic-Based Reconfiguration Strategies for Non-uniformly Shaded PV Array

Reeva Dhariwal and Bhavnesh Kumar

Abstract In photovoltaic power plants, operating environmental conditions have a profound effect on the behavior and the quality of the photovoltaic (PV) energy produced. The non-homogenous insolation reduces the power producing capacity, introduces multiple peaks in the PV curve, and produces hot spots. Metaheuristic algorithms have emerged as an effective tool in reconfiguring the panels to disperse the shading uniformly. The actual locations of the modules remain unchanged during reconfiguration while the electrical connection is changed. Therefore, this paper presents a comprehensive study on two different metaheuristic optimization algorithms, namely particle swarm optimization (PSO) and grasshopper optimization algorithm (GOA). These optimization tools provide a connecting matrix for the new electrical connection that gives high output power to the same PV array. In addition, a comparison of these reconfiguration methods is done to assess their suitability of these methods. Based on the result, PSO gives better output as compared to GOA both in terms of power produced as well as the PV curve.

Keywords Array reconfiguration · Particle swarm optimization · And grasshopper optimization algorithm

1 Introduction

Due to the increase in environmental issues, renewable energy is gaining importance. Solar energy is a popular source of renewable energy and is widely used because of its abundance and accessibility throughout the year. In photovoltaic panels, the output depends on the amount of solar radiation received by the panel. A single PV module is not capable to meet the load requirements, and hence, multiple modules are used.

R. Dhariwal (✉)
Department of Electrical Engineering, Netaji Subhas University of Technology, Delhi, India
e-mail: reevad.ee20@nsut.ac.in

B. Kumar
Department of Instrumentation and Control Engineering, Netaji Subhas University of Technology, Delhi, India
e-mail: kumar_bhavesh@nsut.ac.in

A. Rani et al. (eds.), *Signals, Machines and Automation*, Lecture Notes in Electrical Engineering 1023, https://doi.org/10.1007/978-981-99-0969-8_17

Fixed interconnection schemes such as (i) Series–Parallel (SP), (ii) Bridged-Link (BL), (iii) Honey-Comb (HC), and (iv) Total Cross Tied (TCT) had been presented to counter the non-uniform shading [1]. TCT arrangement gives better solution under partial shading among these interconnection schemes. Despite providing a better performance compared to conventional configuration schemes, the TCT configuration method fails to disperse the shade evenly. Some reconfiguration approaches based on physical relocation are Su Do Ku [2], odd even [3], and zig-zag [4]. However, the drawbacks of these approaches are that they require skilled and labor-intensive people, longer cables, thus increasing losses in the system. Contrary to this are the electrical array reconfiguration (EAR) schemes, fast switching matrix (FSM) [5], and dynamic electrical scheme (DES) [6].

Finding the right switching combination is a complex and difficult task in the electrical switching arrangement. Therefore, metaheuristic optimization algorithms are used as an alternative to solve multi-modal problems. Particle swarm optimization (PSO) proposed in [7, 8] shows that this technique can be used with large solution space which makes this technique better for the EAR problem. Ref. [9] utilizes PSO and genetic algorithm to solve the issues related to PV array. Grasshopper optimization algorithm (GOA) is discussed in [10], which uses the social interaction between the grasshoppers in catching the prey to solve the problem of partial shading. In this paper, a detailed comparative study is done on the above two metaheuristic optimization algorithms namely PSO and GOA.

2 System Description

2.1 *Modeling of PV Cell*

A single diode model of a PV cell is shown in Fig. 1. Applying KCL at node '*P*,'

$$I = I_{\text{ph}} - I_D - \frac{V + IR_S}{R_P} \tag{1}$$

where I is photon current, I_{ph} is PV cell current, I_D is diode current, R_S is series resistance, R_P is parallel resistance.

The ideal diode current equation is written as,

$$I_D = I_o\left(e^{\frac{V_D}{\alpha V_T}-1}\right) \tag{2}$$

where I_0 is the reverse saturation current, 'α' is the diode ideality factor, V_D is the applied voltage across the terminals of the diode, and V_T is the thermal voltage. The thermal voltage subjected to any temperature is given by,

$$V_T = \frac{N_S KT}{q} \tag{3}$$

The PV array current is calculated using the following equation:

$$I = N_{PP}\left(I_{ph} - I_o\left[\exp\left(\frac{V + IR_S}{V_T N_{SS}}\right) - 1\right]\right) - \frac{V + IR_S}{R_P} \tag{4}$$

where N_{SS} and N_{pp} are the number of modules connected in series and parallel. For experimentation, PV module BP Solar BP3170N is taken.

2.2 TCT Configured PV Array

The TCT configuration is formed by connecting cross ties in a series–parallel (SP) configuration to each row. Figure 2 shows a 6 × 4 TCT configured PV array.

The current output of each module depends on the amount of illumination received by the surface as follows:

$$I_{ij} = K_{ij} I_m \tag{5}$$

where I_m is the current generated by the module at standard irradiance and $k_{ij} = G_{ij}/G_o$. By applying KCL, current at each node is represented as:

$$I_a = \sum_{j=1}^{4} (I_{ij} - I_{(i+1)j}) = 0, i = 1, 2, \ldots 6 \tag{6}$$

The array voltage can be calculated by applying KVL as follows:

$$V_a = \sum_{i=1}^{6} V_{mi} \tag{7}$$

where V_{mi} is the voltage of the panels at the ith row.

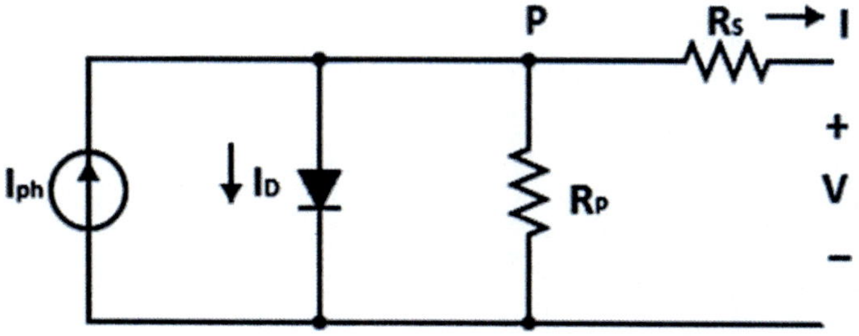

Fig. 1 Equivalent circuit of PV cell

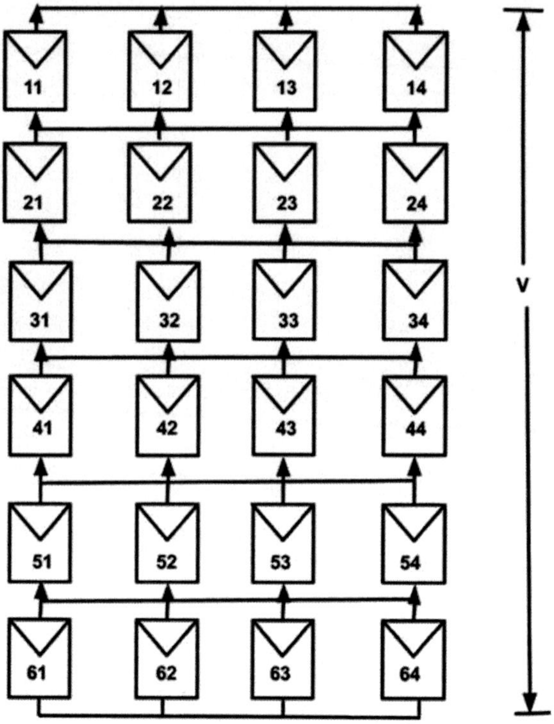

Fig. 2 PV array interconnected in TCT configuration

3 Metaheuristic Optimization Algorithms

There are two basic components of meta-heuristic strategies, namely, exploitation and exploration. Exploitation offers a variety of solutions by exploring the search area, and exploitation searches the local area knowing that a better solution is available in the area. In this paper, GOA and PSO techniques are implemented, and the procedural steps for implementing these techniques are described in the next section (Table 1).

Table 1 Parameters for GOA, PSO

GOA		PSO	
N	20	w	0.9
c_{max}, c_{min}	0.00004, 1	c_1	1.2
t_{max}	10	c_2	1.8

3.1 Grasshopper Optimization Algorithm (GOA)

The grasshoppers have two forces, i.e., attraction and repulsion between them, the repulsive force allows them to explore the search area while the attraction force allows them to remain in the promising regions. Hence, a comfort zone is formed where the two forces are equal. The position having the best fitness function is nearest to the target, and the rest is trying to get to that point through network interaction; the comfort zone is constantly declining until a better solution is found. This behavior can be mathematically modeled as follows:

$$X_i(t+1) = c \sum_{j=1, j\neq i}^{N} c\frac{\mathrm{ub}_d - \mathrm{lb}_d}{c} s\left(\left|X_j(t) - X_i(t)\right|\right)\frac{x_j(t) - x_i(t)}{d_{ij}} + \hat{T}_d \quad (8)$$

$$s(r) = f\exp\left(\frac{-r}{l_s}\right) - \exp(-r) \quad (9)$$

where $X_i(t+1)$ is the position of grasshopper at (t + 1) th iteration, s is a function representing social force, f is attraction force, r is distance, l_s is attractive length, ub_d and lb_d are the upper and lower limits, d_{ij} is distance between ith and jth grasshopper, T_d is a target of d-dimensional position, and c is the diminishing feature representing a descending comfort zone. The equation for c is given by:

$$c(t) = c_{\max} - t\frac{c_{\max} - c_{\min}}{t_{\max}} \quad (10)$$

3.2 Particle Swarm Optimization

The conditions considered while using PSO are initial particle velocity, updating the velocity, social constant, cognitive constant, and weights. The steps to implement PSO for reconfiguration of PV array under partial shading condition are as shown below:

Step 1: Value estimation: Initializing various parameters involved in PSO such as social constant (C_c), cognitive constant (C_g),, and weight (W). The initial velocity of the particle is given by the equation

$$\mathrm{vel}(j) = 1 + \mathrm{round}(\mathrm{rand}() * 8) \quad (11)$$

where j is particle number.

Step 2: Irradiance calculation: The irradiance can be measured using a sensor arrangement. This can also be calculated using the following formula:

$$G_{ij} = \alpha\left[I_{ij} + I_o\left(e^{\frac{V_{ij}}{\alpha V_T}-1}\right)\right] \tag{12}$$

where I_{ij} and V_{ij} are voltage and current corresponding to ith row and jth column.

Step 3: Fitness function: Based on the output power, the fitness of each particle is assessed. The fitness function used in this case is given by the following equation:

$$\text{Max}(f(i)) = \text{sum}(P) + \left(\frac{W_e}{E_e}\right) + \left(W_p * P_a\right). \tag{13}$$

where $f(i)$ is the fitness of the ith element, sum(P) is the total power, E_e is the sum of the error difference between maximum row current, i.e., $E_e = \sum_{K=1}^{6} |I_m - I_k|$, I_m is the maximum value of current when bypassing is considered, P_a is the output power without bypassing, W_e and W_p are the weights.

Step 4: Velocity updation: The velocity for the particles is updated using the following equation:

$$V_i^{t+1} = w * V_i^t + \text{rand}^* C_1 * \left(P_{\text{best}} - X_i^t\right) + \text{rand}^* C_2 * \left(G_{\text{best}} - X_i^t\right) \tag{14}$$

where 'i' corresponds to optimization vector variable, 't' is iteration number, X_i^t and V_i^t are position and velocity of 't'th iteration, C_1 is socail coefficient, C_2 is cognitive coefficient.

Step 5: Finishing stage: The termination criteria is getting an optimal switching pattern, and there is no insolation change. In case of change in insolation, the PSO method must be reinitialized.

4 Results and Discussion

The effectiveness of GOA and PSO for shade dispersion are examined with the help of a 6 × 4 PV array connected in TCT configuration. The theoretical analysis is validated using simulation in the MATLAB/Simulink. Comparison of shade dispersion capability is done on the basis of shading scenario shown in Fig. 3a. In the test case, the PV module is divided into three distinct groups receiving solar radiation of 900, 700, and 300 W/m^2. Results of shade dispersion obtained with considered metaheuristic techniques are shown in Fig. 3b, c. In addition, to support the validity, the IV curve and the PV curve are shown in Fig. 4.

The performance of various techniques can be evaluated using several parameters such as mismatch losses, fill factor, and power loss percentage. The mathematical formulas of these parameters can be written as follows:

$$\text{Mismatch power loss}(P_{\text{MLL}}) = P_{\text{Max}_{\text{STC}}} - P_{\text{GP}_{\text{PSC}}} \tag{15}$$

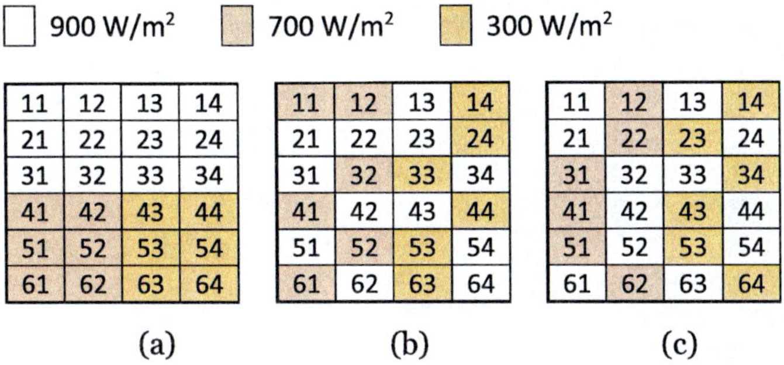

Fig. 3 Shading scenario: **a** before shade dispersion, **b** after shade dispersion using GOA, and **c** after shade dispersion using PSO

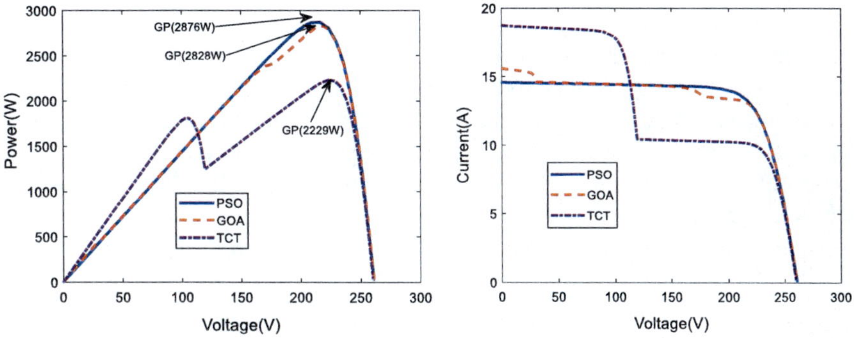

Fig. 4 PV and IV curves with TCT (fixed connection), GOA, and PSO

$$\text{Fill Factor(FF)} = \frac{(V_{\text{mp}} I_{\text{mp}})_{\text{PSC}}}{V_{\text{OC}} I_{\text{SC}}} \tag{16}$$

$$\text{Power loss}(P_L)\% = \frac{\text{GP}_{\text{STC}} - \text{GP}_{\text{PSC}}}{\text{GP}_{\text{STC}}} \tag{17}$$

where $P_{\text{Max}_{\text{STC}}}$ and $P_{\text{GP}_{\text{PSC}}}$ are the maximum power at STC and PSC, respectively, V_{mp} and I_{mp} are the voltage and current at maximum power at PSC, V_{OC} is the open circuit voltage, I_{SC} is the short circuit current, GP_{STC} and GP_{PSC} are the global power produced at STC and PSC, respectively. The values of these parameters are calculated and mentioned in Table 2.

Table 2 Performance parameters of PV array for different techniques

Technique	V_{OC} (V)	I_{SC} (A)	Global peak parameters			Performance parameters		
			GP (W)	V_{mp} (V)	I_{mp} (A)	P_{MML} (W)	FF	P_L (%)
TCT	260.2	18.75	2229	223.3	9.98	1872	0.457	45.647
GOA	261.3	15.59	2828	217	13.03	1273	0.694	31.041
PSO	261.1	14.59	2876	212.6	13.53	1225	0.755	29.870

5 Conclusion

In this paper, a comparative study of two metaheuristic optimization algorithms for shade dispersion is presented. Particle swarm optimization (PSO) and grasshopper optimization algorithm (GOA) for optimal shade dispersion to mitigate the effect of partial shading conditions on PV array are developed and examined. Considered techniques only change the electrical connections, and the physical position in the PV array remains the same. The system performance is analyzed for these two techniques, and the mismatch power loss, fill factor, and power loss percentage are achieved. Hence, it is evident that the PSO method is superior and produces better results compared to the GOA, and TCT configuration with the power loss % of 29.8 and the value of fill factor is 0.755.

References

1. La Manna D, Vigni VL, Sanseverino ER, Di Dio V, Romano P (2014) Reconfigurable electrical interconnection strategies for photovoltaic arrays: a review. Renew Sustain Energy Rev33:412–426
2. Rani BI, Ilango GS, Nagamani C (2013) Enhanced power generation from PV array under partial shading conditions by shade dispersion using Su Do Ku configuration. IEEE Trans Sustain Energy 4(3):594–601
3. Yadav K, Kumar B, Swaroop D (2020) Mitigation of mismatch power losses of PV array under partial shading condition using novel odd even configuration. Energy Rep 6:427–437
4. Vijayalekshmy S, Bindu GR, Iyer SR (2016) A novel zig-zag scheme for power enhancement of partially shaded solar arrays. Sol Energy 135:92–102
5. Alahmad M, Chaaban MA, Kit Lau S, Shi J, Neal J (2012) An adaptive utility interactive photovoltaic system based on a flexible switch matrix to optimize performance in real-time. Sol Energy 86:951–963
6. Romano P, Candela R, Cardinale M, Li Vigni V, Musso D, Riva Sanseverino E (2013) Optimization of photovoltaic energy production through an efficient switching matrix. J Sustain Dev Energy Water Environ Syst 1(3):227–236
7. Bansal JC (2019) Particle swarm optimization. In: Evolutionary and swarm intelligence algorithms. Springer, Cham, pp 11–23
8. Babu TS, Ram JP, Dragičević T, Miyatake M, Blaabjerg F, Rajasekar N (2017) Particle swarm optimization based solar PV array reconfiguration of the maximum power extraction under partial shading conditions. IEEE Trans Sustain Energy 9:74–85

9. Nomani MF, Swaroop D, Kumar B (2020) Photovoltaic cells rearrangement using meta-heuristic optimization techniques for shade dispersion. In: IEEE students conference on engineering & systems (SCES), pp 1–6
10. Fathy A (2018) Recent meta-heuristic grasshopper optimization algorithm for optimal reconfiguration of partially shaded PV array. Sol Energy 171:638–651

Modeling and Analysis of Wind-Driven PMSG for Healthy and Unhealthy Conditions

Nirmal Kumar Agarwal, Neeta Singh, and Abhinav Saxena

Abstract This paper shows the analysis and modeling of wind-driven permanent magnet synchronous generator. The maximum power is extracted from wind turbine by controlling pitch angle and tip to speed ratio. The modeling of permanent magnet synchronous machine is assessed. Thereafter, healthy and unhealthy analysis of PMSG is assessed. The unhealthy condition is being specified in terms of different faults like LLL, LG. Subsequently, power quality issues like THD and MSE are being analyzed for both the conditions.

Keywords THD · Wind · PMSG · Healthy · MSE

1 Introduction

Over the years, the wind generator sector has become increasingly popular. Conventional power generators did not reach the megawatts system. As a result, the majority of the early models used permanent magnet synchronous generators (PMSGs) or a common asynchronous generator. A gearbox is usually connecting an asynchronous generator to a turbine. If the generator has a large number of poles, a permanent compatible generator (PMSG) can be connected to a turbine with a gearbox or directly outside the gearbox [1, 2]. Because of the increase in power per megawatt system, which is now up to 10 MW, PMSG changes necessitate an increase in converter size and weight. Permanent magnetic generators with synchronous generators having the advantages of being more durable, smaller in size, requiring no additional power supply to stimulate the magnetic field, and requiring less adjustment than conventional generators. Furthermore, when compared to the constant-speed technique, variable-speed wind power has advantages such as magnitude, the ability to track

N. K. Agarwal · N. Singh
SOES, GD Goenka University-Sohna, Gurugram, Haryana 122103, India
e-mail: neeta.singh@gdgu.org

A. Saxena (✉)
Department of Electrical Engineering, JSS Academy of Technical Education, Noida, India
e-mail: abhinaviitroorkee@gmail.com

A. Rani et al. (eds.), *Signals, Machines and Automation*, Lecture Notes in Electrical Engineering 1023, https://doi.org/10.1007/978-981-99-0969-8_18

spots, and reduced acoustic noise at low wind speeds [3, 4]. The modeling and control approaches used in two permanent magnet synchronous generator farms for wind applications are described in this paper. A completely integrated rear-turning converter, consisting of two three-phase capacitors, a central DC bus, and an inverter, is used to link generators to the power grid [5, 6]. The entire system is phase connected to the electrical grid. Maximum power point tracking (MPPT) for PMSG speed control, active power control, and DC bus power management are among the proposed control solutions. Some simulation results are shown and examined using the MATLAB/Simulink programming to demonstrate the effectiveness of control schemes [7, 8].

2 Modeling of Wind Turbine

An aerodynamic model of the wind turbines is a basic part of the dynamic models of the electricity producing wind turbines. Theoretical power generated by the turbine is given by,

$$P_m = \frac{1}{2}\rho A V^3 C_p(\lambda, \beta) \tag{1}$$

where

P_m mechanical power developed in turbine.
ρ air density 1.223 kg/m^3.
A area swept by rotor blades.
C_p coefficient of power.
λ ratio between blade tip speed and wind speed at hub height.
β pitch angle,
C_p (λ, β) can be determined as

$$C_p = \frac{1}{2}\left[\frac{116}{\lambda} - 0.4\beta - 5\right]e^{\left(\frac{-21}{\lambda_i}\right)} \tag{2}$$

where λi is defined as,

$$\frac{1}{\lambda_i} = \frac{1}{\lambda} + 0.08\beta - \frac{0.035}{(1+\beta^3)} \tag{3}$$

3 Modeling of PMSG

In this research paper, permanent magnet synchronous generator (PMSG) is used as the wind turbine generator due to its property of self-excitation (by permanent magnet) which eliminates the excitation loss, i.e., excitation losses are not increases as number of poles doubled. Two-phase synchronous reference rotating frame (*d-q* frame) is used to derive the dynamic model of the *d*-axis with PMSG in which the *q*-axis is 90° ahead with respect to the direction of rotation [9, 10]. The electrical model of permanent magnet synchronous generator in synchronous reference rotating frame is represented by the differential equations,

$$\frac{\mathrm{d}}{\mathrm{d}t}(i_d) = \frac{-R_a}{L_d} i_d + \omega_e \frac{L_q}{L_d} i_q + \frac{1}{L_d} U_d \tag{4}$$

$$\frac{\mathrm{d}}{\mathrm{d}t}\left(i_q\right) = \frac{-R_a}{L_q} i_q - \omega_e\left(\frac{L_d}{L_q} i_d + \frac{1}{L_d}\lambda_o\right) + \frac{1}{L_q} U_q$$

$$\omega_e = P\omega_g, e_q = \omega_e\lambda_o, T_e = 1.5P\left[\left(L_d - L_q\right)i_d i_q + i_q\lambda_o\right] \tag{5}$$

where R_a is resistance of stator winding, ωe and ωg are electrical and mechanical rotating speed, λ∘ is flux produced by the permanent magnets, *P* is number of pole pairs, U_d and U_q are *d* and *q*-axis voltages, L_d and L_q are *d* and *q*-axis inductances, Te is electromagnetic torque, and e_q is *q*-axis counter electrical potential [11–15].

4 Performance Assessment with Balanced and Unbalanced Condition

The various performance characteristics like voltage and current at stator terminal and grid have been plotted for healthy and unbalanced condition are shown from Figs. 1, 2, 3, 4, 5, 6, 7, 8, 9, 10, 11 and 12. It is observed that THD or harmonics are found to be less with unbalanced condition in comparison to healthy condition. Such comparative analysis is also shown in Table 1.

Similar kinds of results are also obtained for MSE which is shown in Table 2.

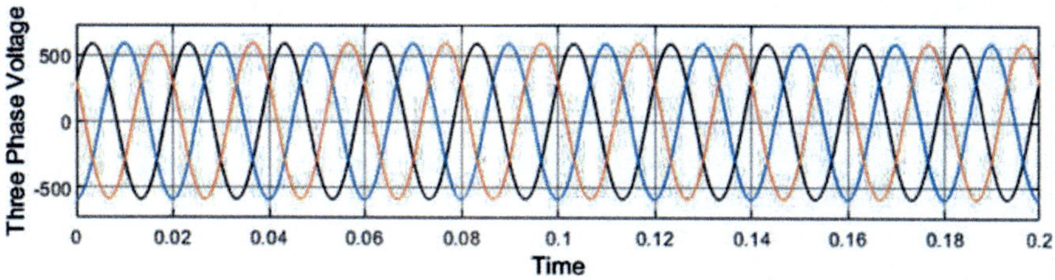

Fig. 1 Balance three-phase voltage at stator terminal

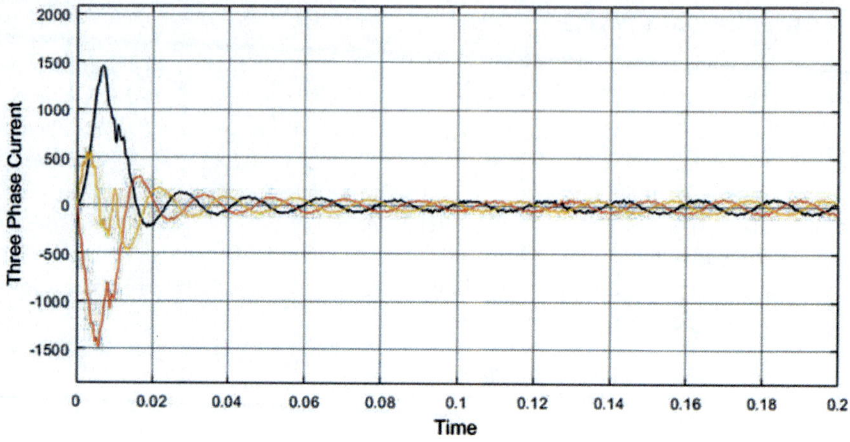

Fig. 2 Balance three-phase current at stator terminal

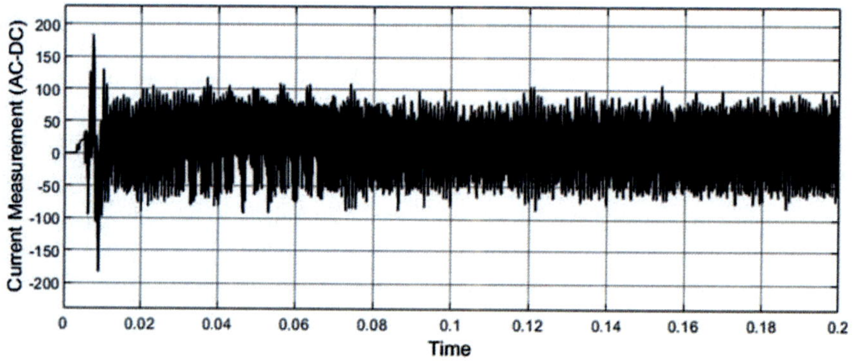

Fig. 3 Current measurement at stator terminal

5 Conclusions

The analysis and modeling of a wind-driven permanent magnet synchronous generator are presented in this study. Controlling the pitch angle and tip to speed ratio of a wind turbine allows it to produce the most power. A permanent magnet synchronous machine's modeling is evaluated. After that, PMSG is analyzed to see if it is healthy or not. Variable loading conditions are used to define the unhealthy situation. Following that, power quality issues such as THD and MSE are investigated for both circumstances. It is observed that MSE and THD are found to be less with healthy condition in comparison with unhealthy condition.

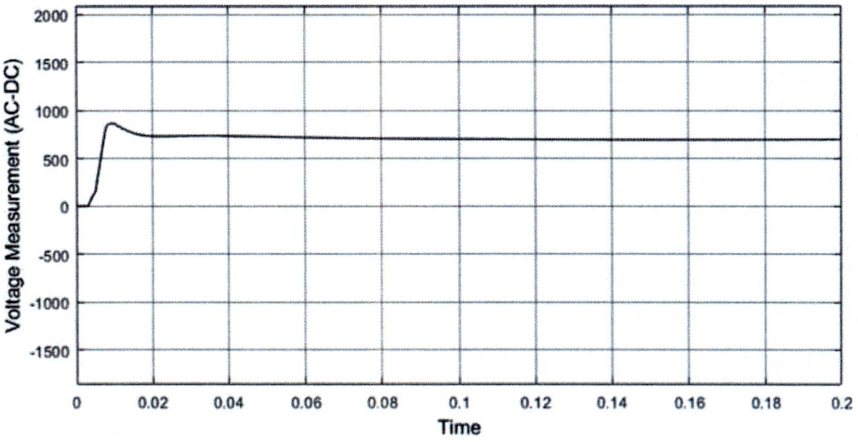

Fig. 4 Voltage measurement at DC link

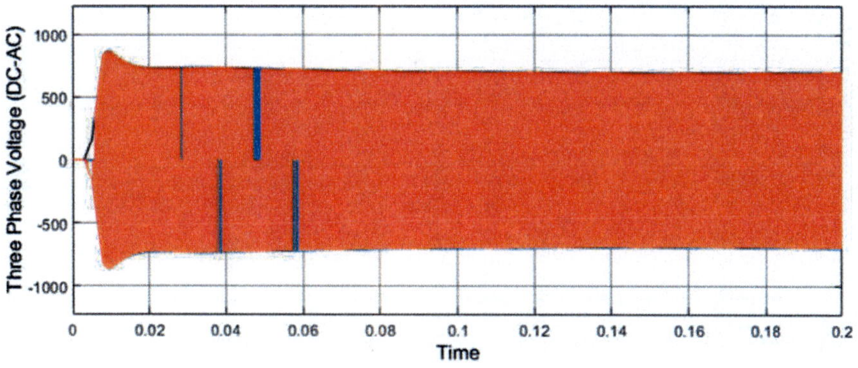

Fig. 5 Three-phase voltage at grid

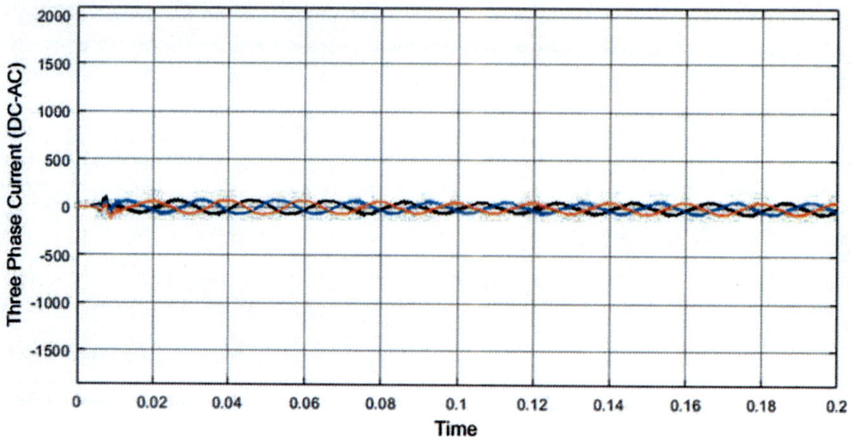

Fig. 6 Three-phase current at grid

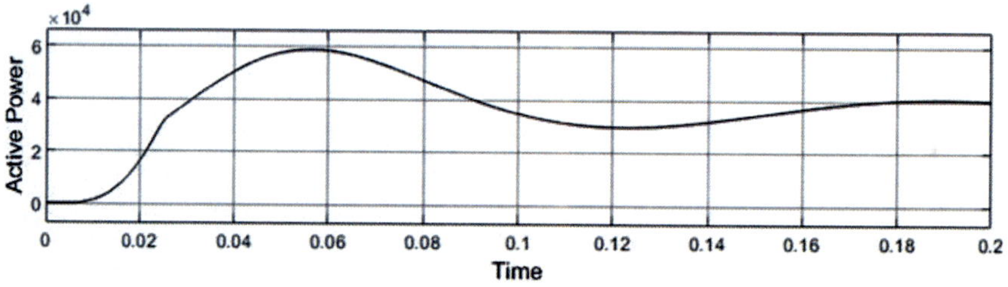

Fig. 7 Active power at stator terminal

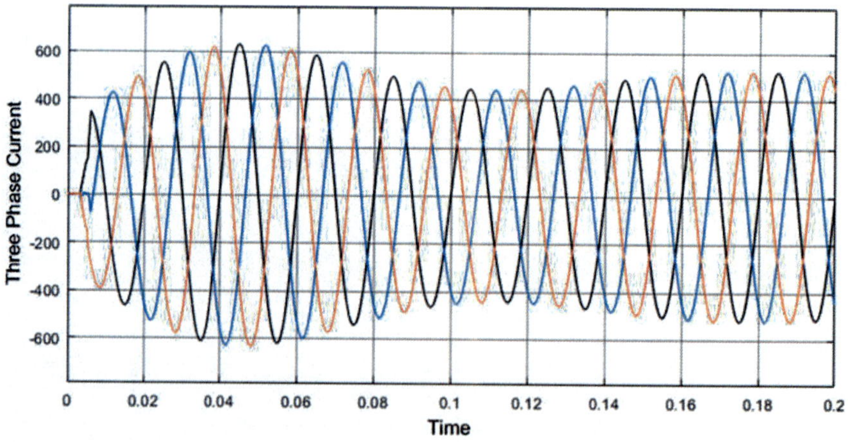

Fig. 8 Unbalanced three-phase current (LLL Fault) at stator terminal

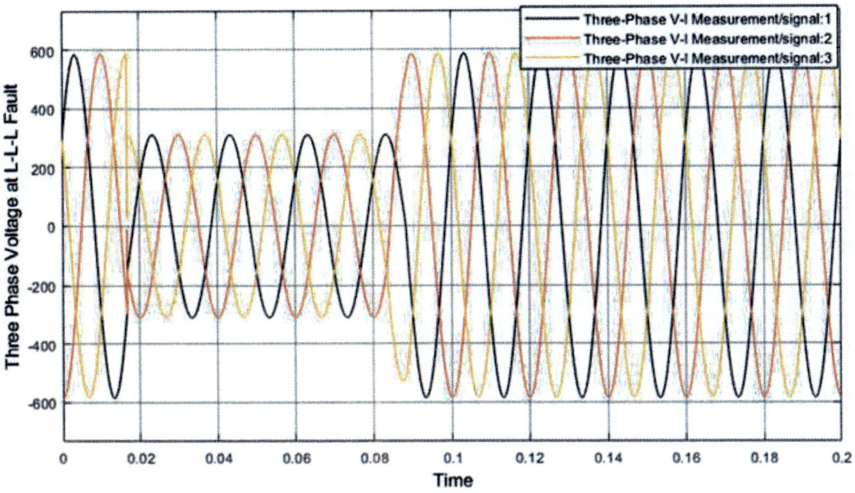

Fig. 9 Unbalanced three-phase voltage (LLL Fault) at stator terminal

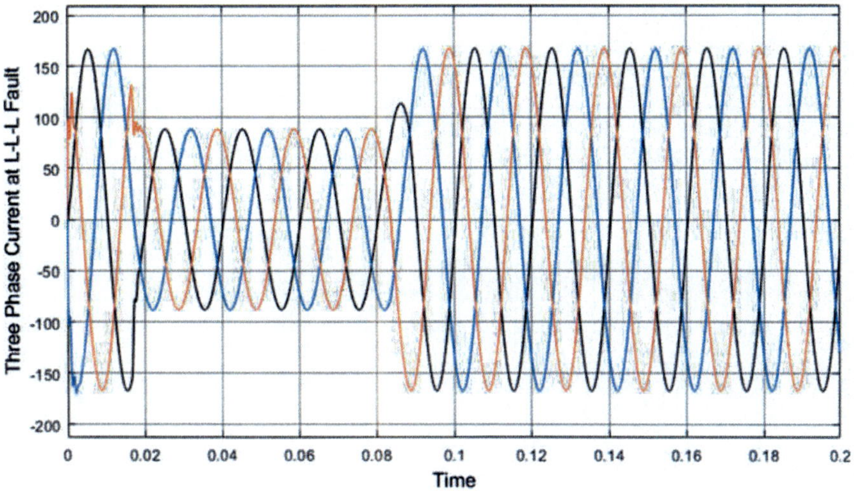

Fig. 10 Unbalanced three-phase current (LLL Fault) at grid

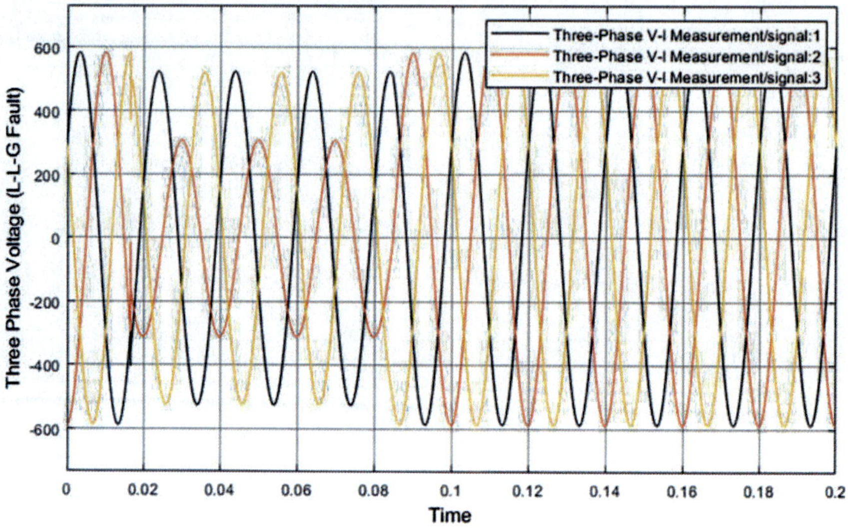

Fig. 11 Unbalanced three-phase voltage (LG Fault) at stator terminal

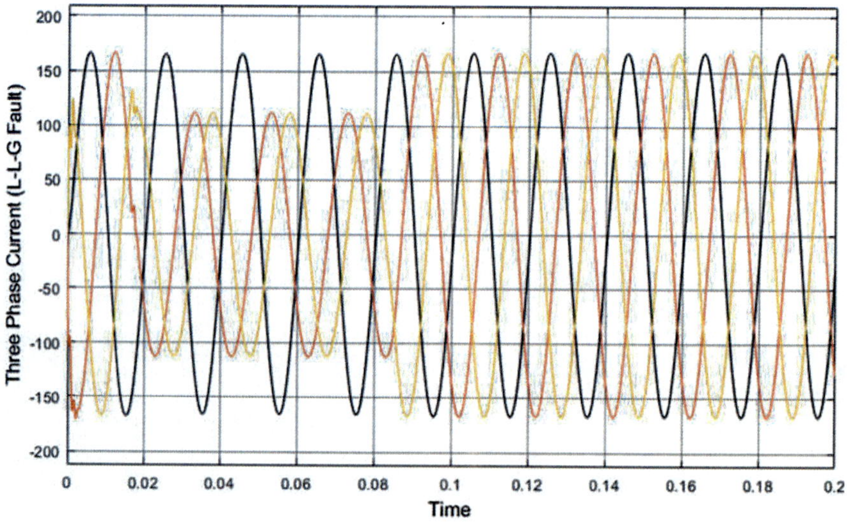

Fig. 12 Unbalanced three-phase current (LG Fault) at stator terminal

Table 1 THD (%) comparison with balanced and unbalanced condition

THD (%)	Healthy condition	Unhealthy condition
Voltage at stator terminal (LLL)	4	8.9
Current at stator terminal (LLL)	5.5	9.1
Voltage at grid (LG)	4.3	9.6
Current at grid (LG)	5.1	9.5
Current at stator terminal (LLG)	4.9	10.2

Table 2 MSE comparison with balanced and unbalanced condition

MSE	Healthy condition	Unhealthy condition
Voltage at stator terminal (LLL)	6	9.8
Current at stator terminal (LLL)	7.2	10.6
Voltage at grid (LG)	5.6	9.8
Current at grid (LG)	4.8	10.8
Current at stator terminal (LLG)	6.1	11.6

References

1. Gupta RA, Singh B, Jain BB (2015) Wind energy conversion system using PMSG. In: 2015 international conference on recent developments in control, automation and power engineering (RDCAPE), pp 199–203. https://doi.org/10.1109/RDCAPE.2015.7281395
2. Tiwari R, Babu NR, Padmanaban S, Martirano L, Siano P (2017) Coordinated DTC and VOC control for PMSG based grid connected wind energy conversion system. In: 2017 IEEE international conference on environment and electrical engineering and 2017 IEEE industrial and commercial power systems Europe (EEEIC/I&CPS Europe), pp 1–6. https://doi.org/10.1109/EEEIC.2017.7977792
3. Alizadeh O, Yazdani A (2013) A strategy for real power control in a direct-drive PMSG-based wind energy conversion system. IEEE Trans Power Delivery 28(3):1297–1305. https://doi.org/10.1109/TPWRD.2013.2258177
4. Vadi S, Gürbüz FB, Bayindir R, Hossain E (2020) Design and simulation of a grid connected wind turbine with permanent magnet synchronous generator. In: 2020 8th international conference on smart grid (icSmartGrid), pp 169–175. https://doi.org/10.1109/icSmartGrid49881.2020.9144762
5. Ndirangu JG, Nderu JN, Muhia AM, Maina CM (2018) Power quality challenges and mitigation measures in grid integration of wind energy conversion systems. IEEE Int Energy Conf (ENERGYCON) 2018:1–6. https://doi.org/10.1109/ENERGYCON.2018.8398823

6. Mora A et al (2019) Model-predictive-control-based capacitor voltage balancing strategies for modular multilevel converters. IEEE Trans Industr Electron 66(3):2432–2443. https://doi.org/10.1109/TIE.2018.2844842
7. Bhimte R, Bhole K, Shah P (2018) Fractional order fuzzy PID controller for a rotary servo system. In: 2018 2nd international conference on trends in electronics and informatics (ICOEI), pp 538–542. https://doi.org/10.1109/ICOEI.2018.8553867
8. Nouman K, Asim Z, Qasim K (2018) Comprehensive study on performance of PID controller and its applications. In: 2018 2nd IEEE advanced information management, communicates, electronic and automation control conference (IMCEC), pp 1574–1579.https://doi.org/10.1109/IMCEC.2018.8469267
9. Zhang Z, Fang H, Gao F, Rodríguez J, Kennel R (2017) Multiple-vector model predictive power control for grid-tied wind turbine system with enhanced steady-state control performance. IEEE Trans Industr Electron 64(8):6287–6298. https://doi.org/10.1109/TIE.2017.2682000
10. Zhang Z, Wang F, Si G, Kennel R (2016) Predictive encoderless control of back-to-back converter PMSG wind turbine systems with extended Kalman filter. In: 2016 IEEE 2nd annual southern power electronics conference (SPEC), pp 1–6. https://doi.org/10.1109/SPEC.2016.7846202
11. Zhang Z, Kennel R (2015) Direct model predictive control of three-level NPC back-to-back power converter PMSG wind turbine systems under unbalanced grid. IEEE Int Symp Predictive Control Electr Drives Power Electron (PRECEDE) 2015:97–102. https://doi.org/10.1109/PRECEDE.2015.7395590
12. Sayritupac J, Albánez E, Rengifo J, Aller JM, Restrepo J (2015) Predictive control strategy for DFIG wind turbines with maximum power point tracking using multilevel converters. IEEE Works Power Electron Power Quality Appl (PEPQA) 2015:1–6. https://doi.org/10.1109/PEPQA.2015.7168207
13. Maaoui-Ben Hassine I, Naouar MW, Mrabet-Bellaaj N (2015) Model based predictive control strategies for wind turbine system based on PMSG. In: IREC2015 the sixth international renewable energy congress, pp 1–6. https://doi.org/10.1109/IREC.2015.7110914
14. Rahmani S, Hamadi A, Ndtoungou A, Al-Haddad K, Kanaan HY (2012) Performance evaluation of a PMSG-based variable speed wind generation system using maximum power point tracking. IEEE Electr Power Energy Conf 2012:223–228. https://doi.org/10.1109/EPEC.2012.6474955
15. Nasr El-Khoury C, Kanaan HY, Mougharbel I (2014) A review of modulation and control strategies for matrix converters applied to PMSG based wind energy conversion systems. In: 2014 IEEE 23rd international symposium on industrial electronics (ISIE), pp 2138–2142. https://doi.org/10.1109/ISIE.2014.6864948

Performance Analysis of Fuzzy Logic-Based MPPT Controller

Vivek Bhardwaj, Rajneesh Sharma, and Tushar Shikhola

Abstract In a smart grid, photovoltaic system integration is becoming crucially influential. However, the system faces nonlinear characteristics, which causes difficulty in obtaining maximum power. To obtain maximum power from the PV system, numerous methods are employed to track the maximum power point (MPP), and these methods are called maximum power point tracking methods. The proposed work contains a comparative analysis between two conventional algorithms, namely, perturb and observe (P&O) and incremental conductance (INC), and an intelligent algorithm fuzzy logic control (FLC), under dynamic environmental conditions. For performance analysis, simulation is employed on MATLAB/SIMULINK software, where FLC showed a superior result in tracking MPP when compared with both conventional techniques INC and P&O.

Keywords Photovoltaic (PV) system · Maximum power point tracking (MPPT) · Fuzzy logic control (FLC) · Incremental conductance (INC) · Perturb and observe (P&O) · DC-DC boost converter · Partial shading condition (PSC)

1 Introduction

To deal with the increase in global warming, exhaustion of fossil fuels and technological advancements have allowed renewable sources to come into the light and become a reliable source of energy. Renewable electricity contribution has increased by about 7% by 2020, and both solar and wind energy stand for over 60% of this rise. Renewables accounted for about 29% of global electricity generation in 2020 [1]. PV systems' characteristic curves are nonlinear due to dynamic environmental conditions and have a low conversion efficiency. Thus, for strengthening the overall performance

V. Bhardwaj (✉) · R. Sharma
Department of Instrumentation and Control, Engineering, Netaji Subhas University of Technology, Delhi, India
e-mail: vivekbhardwajacd@gmail.com

T. Shikhola
Delhi Metro Rail Corporation Limited, Delhi, India

A. Rani et al. (eds.), *Signals, Machines and Automation*, Lecture Notes in Electrical Engineering 1023, https://doi.org/10.1007/978-981-99-0969-8_19

of the PV system and maximum power extraction under all environmental conditions, an MPPT controller is requisite. There are numerous methods proposed for MPPT, namely classified into conventional, intelligent, and hybrid methods which combine both conventional and intelligent methods [2, 3]. Methods tracking MPP under partial shading conditions (PSCs) is also studied in the literature. Conventional methods like P&O [4], and INC [5] are easy to implement, simple, economical, and give satisfactory results under standard testing condition (STC) [5]. But, in a dynamic environment and PSC, the PV system generates multiple local peaks and a global-maxima which cause hindrance for conventional methods to compute and track global peak, that too with low conversion efficiency. These drawbacks could be compensated by the implementation of intelligent methods like artificial neural network (ANN) [6], fuzzy logic control (FLC) [7], particle swarm optimization (PSO), and genetic algorithm (GA), etc. Compare to conventional methods, intelligent methods perform superior in dynamic environmental conditions. In papers [2, 3], all distinct types of proposed methods are studied, comprehensively. The literature gives an in-depth study on different MPPT methods and the design of the MPPT tracker. The P&O, INC, FLC, PSO, and ANN are standard and widely used methods for MPP tracking. All these techniques perform differently in terms of, oscillation around MPP, energy convergence efficiency and speed, cost, and stability at MPP, etc.

2 PV System Modeling

Figure 1 illustrates the PV system modeling basic model, which comprises a PV generator (module), DC converter, MPPT controller, and load. PWM technique helps to achieve the MPP, by varying the duty cycle, of the DC converter.

2.1 PV Cell

To achieve the desired output, in terms of voltage and current from a PV panel, PV cells can be coupled in two combinations, one is, in series, and the other is, in parallel.

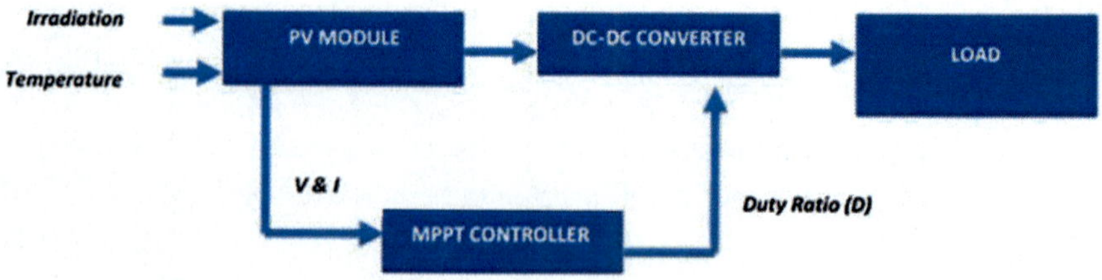

Fig. 1 Basic PV system model

PV cell is a semiconductor device that comprises a diode, in parallel with the current source, a series resistance (R_{se}), and shunt resistance (R_p), as a single diode model. In the form of current, the output Eqs. (1, 2) are given below [8].

$$I_{pv} = I_d + I_{sh} + I \tag{1}$$

$$I = I_{pv} - I_0\left(e^{\left(\frac{V+IR_{se}}{nV_t}\right)} - 1\right) - \frac{V + IR_{se}}{R_p} \tag{2}$$

Advanced renewable energy AREi-230W-M6-G, PV module, which has 230.4 W, 30.72 V, and 7.5 A of maximum power, voltage, and current, respectively, are used.

2.2 DC–DC Converter

The PV model comprises a DC–DC converter, coupled between the solar module and load, which works as a power interface, to make an operating point around or at the MPP. This work utilizes the boost converter impedance matching application, which coordinates between load and the input impedance of the PV module. So, whatever would be the load, the maximum point can be tracked with the given relation between boost converter input impedance (R_t) and load (R_o) in Eq. 3 [9].

$$R_o = \frac{R_t}{(1 - D)^2} \tag{3}$$

For converter designing, the parameters are calculated through the following Eqs. 4 and 5.

$$\text{Duty cycle, } D = 1 - \frac{V_o}{V_s} \quad \text{where } 0 < D < 1 \tag{4}$$

$$\text{Inductor, } L = \frac{D(1 - D)^2 R_o}{2 \cdot F_s}, \text{ Capacitor } C = \frac{D}{R_o \cdot F_s \cdot 0.01} \tag{5}$$

Boost converter parameters are inductor (L) = 400 μH, capacitor (C_{in}) = 1000 μF, capacitor (C) = 50 μF, switching frequency (f) = 5 kHz, and load (R) = 10 Ω.

3 MPPT Algorithm

3.1 *Perturb and Observe (P&O)*

In tracking the maximum power of the PV system, P&O is the most used and the simplest algorithm, which can be described in Fig. 2; if at the PV curve, power at any instant n, $P_{pv}(n)$ is larger than the earlier received power, $P_{pv}(n-1)$ ($\Delta P_{pv} > 0$), the perturbation continues in the same direction. After reaching MPP, the next perturbation may decrease PV power at this instant ($\Delta P_{pv} < 0$), and perturbation starts in opposite direction.

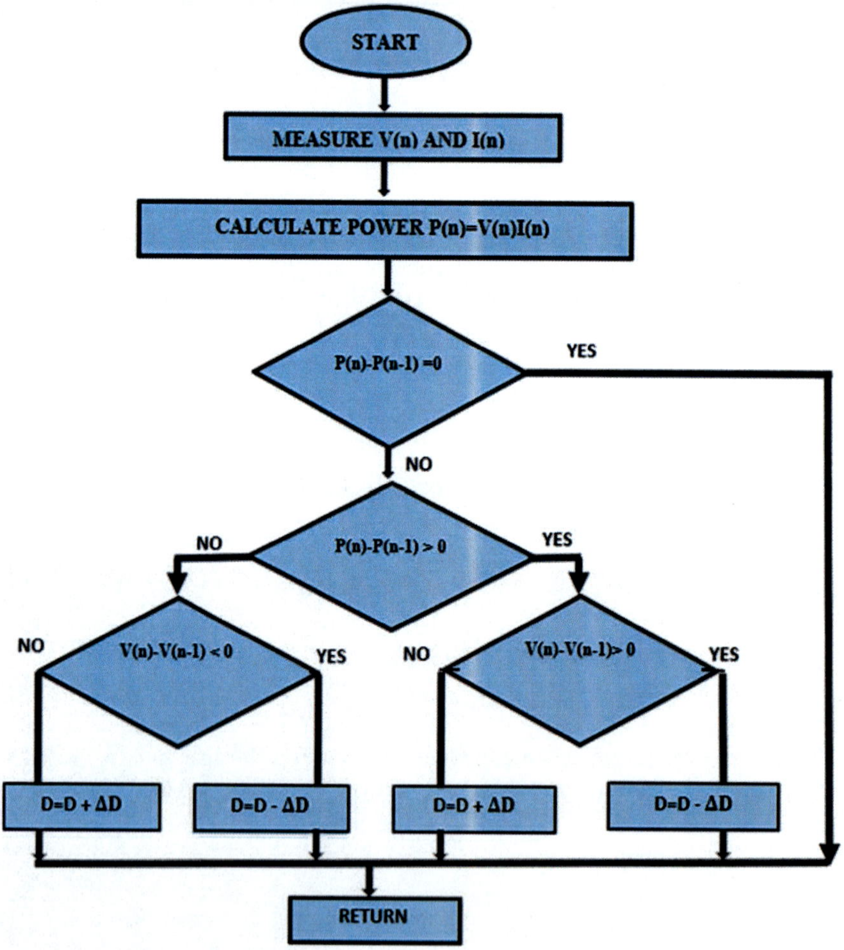

Fig. 2 Flow graph for P&O algorithm

3.2 Incremental Conductance (INC)

The INC approach can be understood through the slop of PV curve characteristics which explains, at MPP slop of the curve is zero ($dP_{pv}/dV_{pv} = 0$), negative ($dP_{pv}/dV_{pv} < 0$) on right side and positive ($dP_{pv}/dV_{pv} > 0$) on left side. The following equation below represents the mathematical relations and flowchart in Fig. 3.

$$P_{PV} = V_{PV} * I_{PV}, \frac{dP_{PV}}{dV_{PV}} = V_{PV}\left(\frac{dI_{PV}}{dV_{PV}}\right) + I_{PV}$$

$$\frac{dP_{PV}}{dV_{PV}} = 0, \frac{dP_{PV}}{dV_{PV}} = -\frac{I_{PV}}{V_{PV}} \quad \text{Operating point at MPP}$$

$$\frac{dP_{PV}}{dV_{PV}} < 0, \quad \frac{dP_{PV}}{dV_{PV}} < \frac{I_{PV}}{V_{PV}} \quad \text{when Operating point is right side on PV curve}$$

$$\frac{dP_{PV}}{dV_{PV}} > 0, \quad \frac{dP_{PV}}{dV_{PV}} > \frac{I_{PV}}{V_{PV}} \quad \text{when Operating point is left side on PV curve}$$

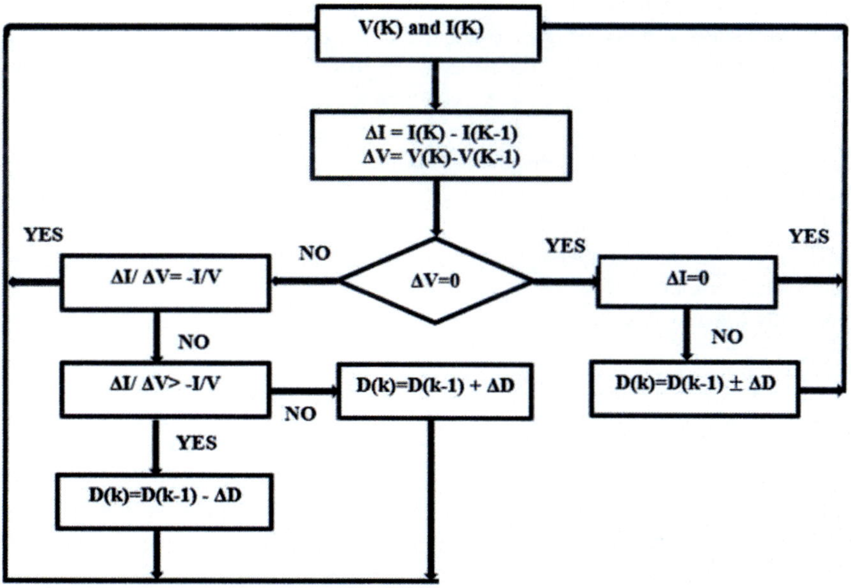

Fig. 3 Flow graph for INC algorithm

3.3 Fuzzy Logic Control (FLC)

FLC process comprises three steps, which are, fuzzification, fuzzy rules (interference engine), and defuzzification. Using different established membership functions, the input variables are translated into linguistic variables and assigned a fuzzy value to them in the first stage of fuzzification. Considering the rules and membership functions of linguistic variables, the inference engine translates data, and the variables are then changed depending on the "if–then" rules. Finally, getting data from the inference engine, the defuzzification block transforms fuzzy data into crisp (non-fuzzy) data in the numerical variables are created from these variables, that may be used to regulate the process. At sampling time n, the following Eqs. (6–9) describe, two inputs and an output Er (error), ΔEr (change in error), and D, respectively (Fig. 4).

$$\Delta P_{\text{PV}}(n) = P_{\text{PV}}(n) - P_{\text{PV}}(n-1) \tag{6}$$

$$\Delta V_{\text{PV}}(n) = V_{\text{PV}}(n) - V_{\text{PV}}(n-1) \tag{7}$$

$$E(n) = \frac{\Delta P_{\text{PV}}(n)}{\Delta V_{\text{PV}}(n)} \tag{8}$$

$$\Delta E(n) = E(n) - E(n-1) \tag{9}$$

Triangular membership functions were used in the fuzzification procedure. Regarding the linguistic variables described in the following, 5 membership functions, which are, negative mini ($N\alpha$), negative large ($N\beta$), zero (Zr), positive mini ($P\alpha$), and positive large ($P\beta$), describe two inputs E and ΔE and output D (duty cycle) and are shown in Fig. 5. Rules for FL controller are described in Table 1.

Fuzzification Method: Generally, the fuzzy control employs one of the following techniques: Max–Min, Max Prod (Somme-Prod). In this instance, we employed Mamdani's max–min fuzzy combinations inference approach.

Defuzzification Methods: Center of area (COA) and the max criterion method (MCM) are two algorithms that are often used for defuzzification. In the evaluation of the final combined fuzzy sets, COA is the defuzzification technique that is most frequently utilized (Fig. 6).

4 Simulations and Results

The designed simulation model to analyze the MPPT algorithms is illustrated below in Fig. 7 with MPPT controller designs for P&O, INC, and FLC, which comprises three parts, MPPT controller block, PV generator (module), and boost converter connected to the resistive load to track MPP in standard testing condition (STC) and

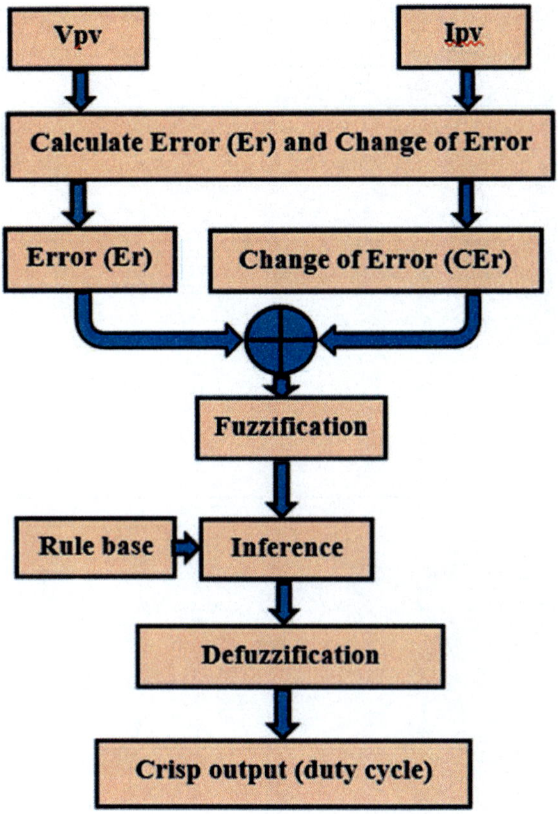

Fig. 4 Flow graph for FL algorithm

dynamic irradiation and temperature. The output characteristics of the PV module illustrated in Fig. 8a, b, respectively. Comparison results are shown in Fig. 9.

Temperature variation—(0–0.25 s) – 40 °C, (0.25–0.5 s) – 35 °C and (0.5–1 s) – 25 °C

Irradiation variation—(0–0.25 s) – 1000 W/m^2, (0.25–0.5 s) – 600 W/m^2, (0.5–0.75 s) – 800 W/m^2, (0.75–1 s) – 1000 W/m^2.

5 Conclusion

In this work, two conventional algorithms, namely P&O and INC, are compared with an intelligent algorithm, i.e., FLC, for tracking MPP. A comprehensive study has been made on the implementation and analysis of these algorithms. A comparison of obtained results shows that, the FLC can track MPP more efficiently than the conventional methods. The study shows that the fuzzy logic controller reaches the maximum power value by compensating for the oscillation around MPP and also

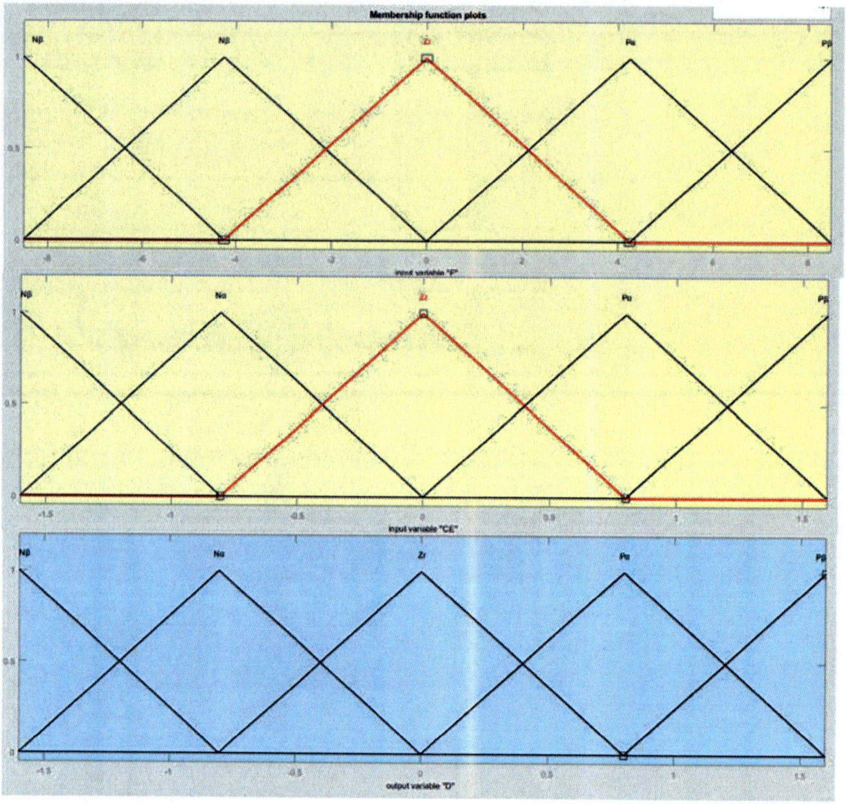

Fig. 5 Membership functions *E*, CE, and *D*, respectively

Table 1 Fuzzy logic rules

Er/ΔEr	$N\beta$	$N\alpha$	Zr	$P\alpha$	$P\beta$
$N\beta$	$P\beta$	$P\alpha$	$N\beta$	$N\alpha$	$N\alpha$
$N\alpha$	$P\alpha$	$P\alpha$	$N\beta$	$N\alpha$	$N\alpha$
Zr	$N\alpha$	$N\alpha$	Zr	$P\beta$	$P\beta$
$P\alpha$	$N\alpha$	$P\beta$	$P\alpha$	$N\beta$	$P\beta$
$P\beta$	$N\beta$	$N\beta$	$P\beta$	$P\beta$	$P\beta$

offers faster tracking and good characteristics. Other advantages of the fuzzy logic controller are that it does not require model training as other intelligent techniques like ANN, GA, PSO, ML, etc. FLC can be used easily to develop hybrid techniques for MPPT controllers.

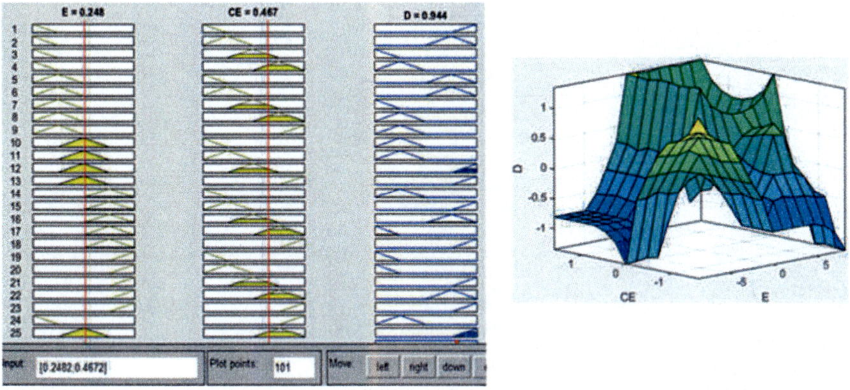

Fig. 6 Rules viewer of FLC and surface viewer of FLC

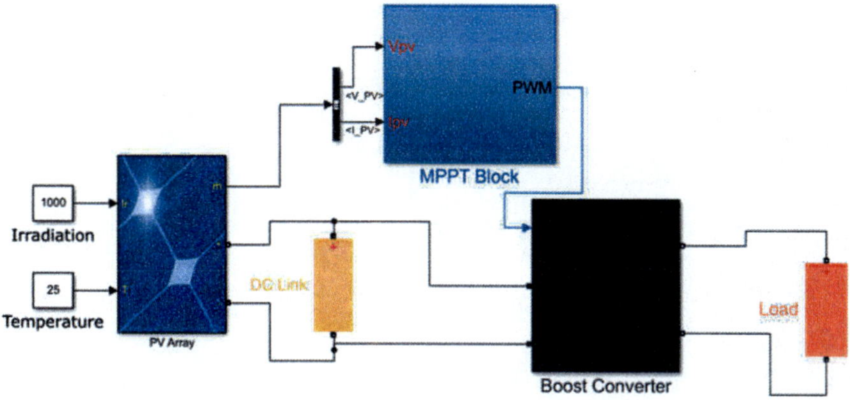

Fig. 7 Simulation block diagram of PV system

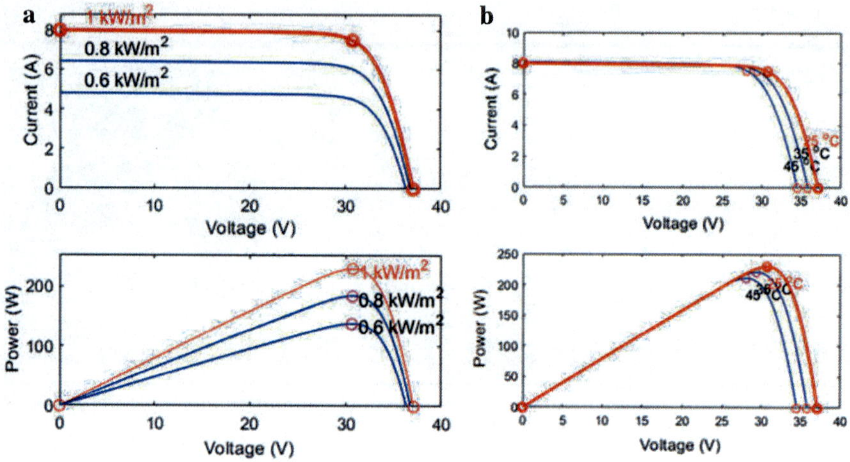

Fig. 8 **a** PV system characteristics at 25 °C, **b** characteristics at 1000 W/m^2

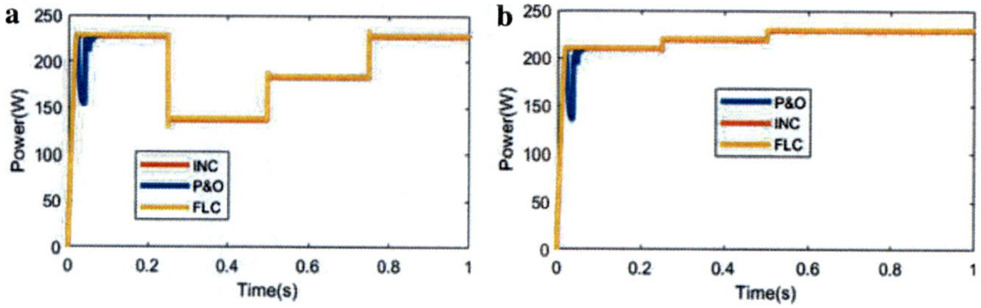

Fig. 9 **a** Power versus time at distinct irradiation, **b** power versus time at distinct temperature

References

1. International Energy Agency IEA (2021) Renewable power. IEA, Paris. https://www.iea.org/reports/renewable
2. Mao M, Cui L, Zhang Q, Guo K, Zhou L, Huang H (2020) Classification and summarization of solar photovoltaic MPPT techniques: a review based on traditional and intelligent control strategies. Energy Rep 6:1312–1327
3. Yap KY, Sarimuthu CR, Lim JM-Y (2020) Artificial intelligence-based MPPT techniques for solar power system: a review. J Mod Power Syst Clean Energy 8:1043–1059
4. Alik R, Jusoh A, Sutikno T (2015) A review on perturb and observe maximum power point tracking in photovoltaic system. Telkomnika 13:745
5. Harrag A, Messalti S (2019) IC-based variable step size neuro-fuzzy MPPT improving PV system performances. Energy Procedia 157:362–374
6. Roy RB, Rokonuzzaman M, Amin N, Mishu MK, Alahakoon S, Rahman S, Mithulananthan N, Rahman KS (2020) A comparative performance analysis of ANN algorithms for MPPT energy harvesting in solar PV system. IEEE Access 9:102137–102152

7. Ge X, Ahmed FW, Rezvani A, Aljojo N, Samad S, Foong LK (2020) Implementation of a novel hybrid BAT-fuzzy controller based MPPT for grid-connected PV-battery system. Control Eng Practice 98:104380
8. Dubey K, Shah MT (2016) Design and simulation of solar PV system. In: 2016 international conference on automatic control and dynamic optimization techniques (ICACDOT), Pune, pp 568–573
9. Umanand L (2009) Power electronics: essentials and applications. Wiley India Pvt. Limited

Optimum Power Tracking of DFIG-Based Wind Energy Conversion System

Shivanshu Nayak, Anil Kumar Yadav, and Pawan Kumar Pathak

Abstract This paper focuses on the design, analysis and control of grid interfaced doubly fed induction generator (DFIG)-based wind energy conversion system (WECS) with maximum power point tracking (MPPT) capability. The modified control of grid side converter (GSC) is adapted to supply regulated power to the grid by maintaining DC link voltage. The control of rotor side converter (RSC) is modified for achieving MPPT and unity power factor (UPF) operation at the stator terminals. The dynamic modeling of a wind turbine driven DFIG is simulated in this study employing MATLAB/SIMULINK, as well as the outcomes are studied for given wind speeds.

Keywords Wind turbine · DFIG · RSC · GSC · MPPT · Vector control

1 Introduction

Renewable energy systems have piqued public attention in the recent years due to the scarcity of conventional energy sources and a slew of issues related with their use, such as pollution, high grid demands, and so on. Alternative energy sources such as wind, solar, and modest hydroelectric power are being pushed by governments all over the world [1]. Among the natural energy collection options outlined above, wind energy is a potential option. In the recent years, wind energy has been extensively studied in many countries, leading to various configurations such as a squirrel cage induction generators (SCIG) system operating at a fixed speed, a permanent magnet synchronous generator (PMSG) system operating at a variable speed, and a DFIG-based variable speed system to improve the efficiency, power rating, and cost benefits among other things [2]. Interconnection of other renewable with wind is planned in all developing countries, and implementation is accelerating. The concerns that are

S. Nayak · A. K. Yadav (✉)
Department of Electrical Engineering, NIT Hamirpur, Hamirpur, HP 177005, India
e-mail: anilkyadav@nith.ac.in

P. K. Pathak
School of Automation, Banasthali Vidyapith, Banasthali, Rajasthan 304022, India

A. Rani et al. (eds.), *Signals, Machines and Automation*, Lecture Notes in Electrical Engineering 1023, https://doi.org/10.1007/978-981-99-0969-8_20

driving this trend are undoubtedly climate change and energy security. Solar and wind energy are the most sought-after renewable energy sources [3, 4]. On a utility scale, despite solar is a more viable choice for household usage, it is not as popular as wind. India had 150 GW of renewable capacity as of November 2021, with solar and wind contributing 48.55 GW and 40.03 GW, respectively. India has pledged to build 450 GW of renewable energy capacity by 2030. Variable speed control WECS based on DFIG has several advantages over fixed speed SCIG, including improved performance, lower rating of converter, costing, and variable losses, and simple power factor correction integration can operate at varying speed and has capability to control active and reactive power at all four quadrant operations. Because of the variable speed operation in DFIG-based WECS, total energy output is much higher, resulting in a higher capacity utilization factor and lower cost per unit energy [5, 6].

A simple DFIG-based WECS device is shown in Fig. 1. The DFIG's stator is instantly linked to the network, while the rotor is linked to the grid via a back-to-back PWM voltage source converter (VSC). The application of DFIG in a wind energy conversion system has numerous advantages. The primary benefit is the generator's ability to deliver electricity at both trailing and leading power factors. The induction machine can also maintain grid synchronization as the wind turbine speed changes since the rotor voltages and currents are regulated. The ability of DFIG to supply power at a constant voltage and frequency when speed of rotor changes is the primary reason for their popularity. As a result, DFIG has become increasingly popular in wind energy applications. When the rotor converters are linked at the rotor side, they deal with substantially less power on comparison if they are linked at the stator side, controlling the DFIG from the rotor side saves money. The fuzzy logic-based wind power generation model is described in [7, 8].

This study employs a DFIG control strategy. The sub-synchronous generating phase of the DFIG is investigated, and a space vector control method is used to control the active and reactive power of the stator via the rotor side converter (RSC). MATLAB/SIMULINK is used to create a model of a turbine driven. An MPPT methodology is used on the wind turbine to record adequate power throughout sub-synchronous mode. Also, there is a resemblance with and without MPPT, and also an examination of the effect of stator active power.

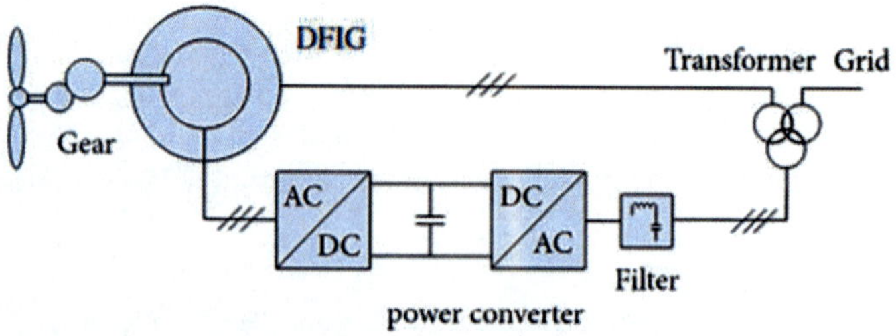

Fig. 1 DFIG-based WECS [4]

2 Wind Turbine Modeling

According to the Betz restriction [4, 5], the amount of wind power that a wind turbine can capture is essentially limited to 55% of the total power content of the wind. The capacity factor of a wind turbine is usually below 0.45. The wind turbine's generalized mechanical formula is as follows:

$$J_s \frac{d\omega_r}{dt} + B_s \omega_r = T_m - T_e \tag{1}$$

For which J_s is the shaft's total inertia, B_s is the friction coefficient, T_m is the torque with wind origin, and T_e is the generator's electromagnetic torque. The torque created is defined by

$$T_m = \frac{P_m}{\omega_r} \tag{2}$$

The mechanical power produced by the wind turbine is measured in Pm.

$$P_m = 0.5\rho A C_p(\lambda, \beta) V^3 \tag{3}$$

where ρ is defined as density of air, A signifies area swept by wind turbine blades, C_p is the power coefficient, and λ implies tip speed ratio.

$$\lambda = \frac{\omega_m R}{V^3} \tag{4}$$

There are a variety of criterion available for coefficient of performance (C_p) [9].

$$C_p = 0.5176 \left[\frac{116}{\lambda_i} - 0.4\beta - 5 \right] e^{\frac{21}{\lambda_i}} + 0.006795\lambda \tag{5}$$

Constant wind has a C_p of 0.44. The machine runs below synchronous speed from cut in speed to rated speed. Various MPPT approaches are used during this time to track optimum power.

3 Maximum Power Point Tracking Techniques

Every wind turbine maximum power variation with rotation speed is predetermined [10–12]. As a result, the control method for MPPT should follow the wind turbine's curve (TCC). Every wind turbine has a TCC that is identical to the one depicted in the illustration. Because the wind is forever shifting, it is critical to include a controller that can monitor peak value regardless of wind speed [6]. Many

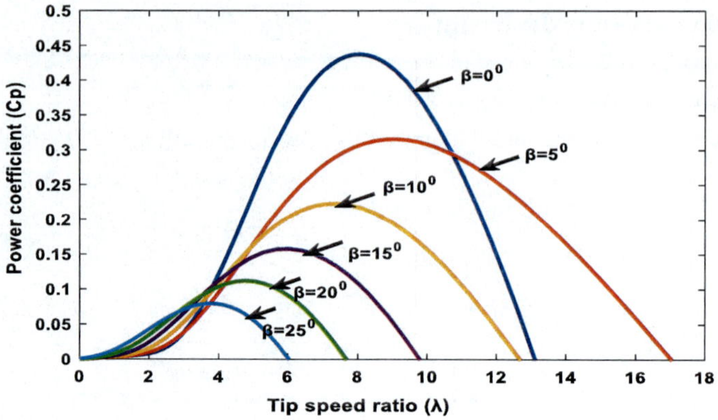

Fig. 2 C_p versus λ at different β [6]

MPPT systems have been proposed [6, 13], and their iterative tracking and self-tuning abilities make them the ideal solution. Here is a comparison of some of the existing MPPT algorithms for wind energy systems. The literatures investigate the principles underlying existing MPPT algorithms for wind energy systems and compare three MPPT strategies. Concluding that the Indirect MPPT technique method is simple, quick, and efficient. Figure 2 represents the relation between C_p and λ.

3.1 Indirect MPPT Technique

Managing wind turbine operating at the optimal level guarantees that the maximum amount of additional wind energy is transformed into mechanical energy [13].

$$T_{\text{mopt}} = \frac{1}{2}\rho\pi R^5 \frac{C_{p\max}}{\lambda_{\text{opt}}^3}\omega_m^2 \tag{6}$$

$$T = K_{\text{opt}}\omega_m^2 \tag{7}$$

The pitch angle (β) is maintained at its optimum value β_{opt}, while TSR (λ) is tuned to the optimal value λ_{opt} over different wind speeds in order to extract maximum power from the wind as shown in Fig. 3. Speed control block is used to generate an optimum torque at $\beta = 0$ and variable TSR. This optimum torque is given to asynchronous motor used as DFIG in simulation.

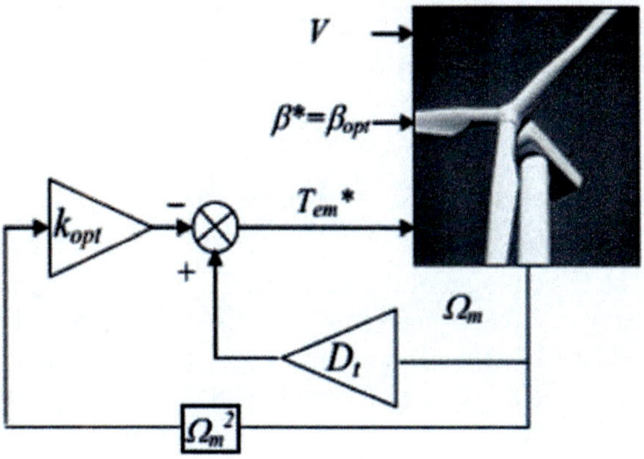

Fig. 3 Indirect MPPT control [13]

3.2 *Optimal Torque Control Method*

Optimizing wind turbine operation guarantees that as much additional wind energy as possible is transformed into mechanical energy. As discussed earlier, once optimum TSR (λ_{opt}) is obtained, it ensures that the maximum amount of available energy from the wind can be used to run the turbine, hence obtaining maximum torque [9, 14]. The basic principle used in OT control method of MPPT is to adjust the DFIG torque according to maximum T_{ref} of the turbine at a given wind speed. Wind speed in term of 'λ' can be obtained as:

$$V_m = \frac{\omega_m R}{\lambda} \tag{8}$$

Using (3) and (8), yield (9).

$$P_m = \frac{1}{2}\rho\pi R^5 \frac{\omega_m^3}{\lambda^3} C_p \tag{9}$$

By using optimum value of 'λ' wind turbine will also run at $C_{p\,max}$.Thus, using $\lambda = \lambda_{opt}$ and $C_p = C_{p\,max}$. From Eq. (9) yields:

$$P_{m\text{ - }opt} = \frac{1}{2}\rho\pi R^5 \frac{C_{p\,max}}{\lambda_{opt}^3}\omega_m^3 = K_{P\text{ - }opt}\omega_m^3 \tag{10}$$

Considering $P_m = \omega_m T_m$;

$$T_{m\text{ - }opt} = \frac{1}{2}\rho\pi R^5 \frac{C_{p\,max}}{\lambda_{opt}^3}\omega_m^2 = K_{opt}\omega_m^2 \tag{11}$$

Since it is a torque control-based method, the optimum torque obtained from Eq. (11) is given as a reference torque to the controller connected to wind turbine. In this torque control approach, the ideal tension is produced as the source torque for the controller connected to the wind turbine.

4 Control of DFIG Based WECS

The basic design of the wind energy conversion system used here includes a wind turbine, a DFIG generator, two power converters at the rotor and grid sides, and control schemes. The rotor voltage obtained by space vector control method in a ‘dq’ reference frame as a function of currents and stator flux is given as:

$$V_{dr} = R_r i_{dr} + \sigma L_r \frac{\mathrm{d}}{\mathrm{d}t} i_{dr} - \omega_r \sigma L_r i_{qr} + \frac{L_m}{L_r}\frac{\mathrm{d}}{\mathrm{d}t}\psi_s \tag{12}$$

$$V_{qr} = R_r i_{qr} + \sigma L_r \frac{\mathrm{d}}{\mathrm{d}t} i_{dr} + \omega_r \sigma L_r i_{dr} + \omega_r \frac{L_m}{L_r}\frac{\mathrm{d}}{\mathrm{d}t}\psi_s \tag{13}$$

Here, stator and rotor inductance is given as: $L_r = L_s = L_{si} + L_m = (0.087 + 2.5)\,\text{mH} = 2.587\,\text{mH}$, and leakage coefficient: $\sigma = 1 - \frac{L_m^2}{L_s * L_r} = 1 - \frac{2.5^2}{2.587^2} = 0.066$

The torque expression can be simplified as:

$$T_{em} = \frac{3}{2}\rho \frac{L_m}{L_s}\psi_s i_{qr} = K_T i_{qr} \tag{14}$$

$$Q_s = \frac{3}{2}\left(V_{qs} i_{ds} - V_{ds} i_{qs}\right) \tag{15}$$

Thus, both rotor current can be controlled independently for active and reactive power control. Similarly, in grid side aligning d-axis with grid voltage:

$$V_{dg} = V_g;\ V_{qg} = 0 \tag{16}$$

Two power equations will become:

$$P_g = \frac{3}{2} V_{dg} i_{dg} \tag{17}$$

$$Q_g = -\frac{3}{2} V_{dg} i_{qg} \tag{18}$$

Thus, d-component of current is used to maintain DC link voltage. The ratings of the various components used in this article are given in Table 1.

Table 1 DFIG parameters

Parameter	Ratings
Air density	1.225 kg/m^3
Pitch angle	0
Max. power coefficient (C_p)	0.44
Optimum TSR	7.2
Switching frequency	4000 Hz
Stator L&R	0.087 mH and 2.6 $m\Omega$
Rotor L&R	0.087mH and 2.9 mΩ
Rated stator power and voltage	2 MW and 690 V
DC link voltage reference	1150 V
Magnetic inductance (L_m)	2.5 mH
Stator to rotor turns ratio	1/3
Pole numbers	4
Maximum slip	1/3

4.1 Design of Battery Storage System

BESS is installed in the DC link of two back-to-back converters. When power produced is more than required, then it gets stored in battery. Because the size of the battery increases the investment, proper battery design is required for satisfactory operation. The battery voltage is selected by considering voltages at both GSC and RSC. The BESS is designed in such a way that DC output voltage of RSC can be regulated by using a BUCK-BOOST converter employed with PI controller with constants value as $k_p = 0.5$ and $k_i = 100$ with output limits of 10^6. Regulated voltage is stored using a capacitor of 80mF.

4.2 Design and Control of RSC

Stator voltage-oriented control (SVOC) scheme is used to control and design RSC in which we convert voltages from *a-b-c* reference frame to decoupled *d-q* reference frame and control individually. It is achieved by aligning d-axis with stator voltage V_s such that $V_{ds} = V_s$ and $V_{qs} = 0$. Here, stator reactive power (Q) is controlled by I_d and active power (P) by I_q. In active power control loop, the ω_e(rotor) is compared with ω_e(ref) obtained by MPPT and error obtained is fed to conventional PI speed regulator to generate the I_q(ref). Similarly, in reactive power control loop, I_d(ref) is generated by comparing stator reactive power to $Q_{\mathrm{ref}} = 0$ by passing error to another PI controller (VAR regulator). Now, the measured value of I_d and I_q at the rotor side is compared with these reference values, and error generated is given to PI current regulator to obtain V_{dr} and V_{qr} with the help of ω_m (mechanical speed of rotor).

Now, these decoupled *d-q* axis voltages is converted to *a-b-c* reference frame using park-Clarke transformation with the help of rotor angle (θ_r) which can be obtained by enhanced phase locked loop (EPLL) technique which requires converting of *a-b-c* reference of stator voltages to $\alpha\beta$ stationary reference frame known as Clarke transformation $\theta_v = \tan^{-1}\left(\frac{V_\beta}{V_\alpha}\right)$; $\theta_s = \theta_v - 90 \therefore \theta_r = \theta_s - \theta_m$. Now with the help of rotor *d-q* reference rotor, voltage is converted to three-phase voltages, and when these reference three-phase voltages are compared with actual three-phase rotor voltages to generate error which is input to hysteresis current controller and output of controller gives PWM pulses for RSC converter by passing through a 3rd harmonic injector block to reduce THD by using min–max technique of minimizing short PWM pulses. In min–max method, the magnitude of a common mode signal (V_{cm}) is subtracted from all the three-phase voltages. The subtraction of V_{cm} introduces of third harmonic component in phase voltages. It has advantages of reducing number of short pulses for a given modulation index so that line-to-line voltages along grid would be same as fundamental line-to-line supply voltages.

$$V_{\text{cm}} = \frac{\max(Va, Vb, Vc) + \min(Va, Vb, Vc)}{2} \tag{19}$$

4.3 Design and Control of GSC

GSC primary duty is to maintain a constant DC voltage across the link capacitor independent of rotor power direction while maintaining UPF. The reactive power is usually set to zero so that all of the energy is exchanged through the stator. To control GSC, we adopt stator voltage-oriented (SVO) control scheme, where DC bus voltage is controlled by I_d and $I_q = 0$ for UPF. Here, I_d(ref) is obtained from PI controller output whose input is the error which is produced by comparing actual DC bus voltage to reference DC voltage. Then I_d(ref) is compared to I_d and error is fed to PI controller to generate control signal of V_{dg} for GSC. Similarly, control signal V_{qg} for GSC is obtained by comparing I_q(ref) and I_q.Now, both V_{qg} and V_{dg} are the input to controller to generate PWM pulses after passing through 3rd harmonic injection block. The control method of is same min–max method as it has advantages of reducing number of short pulses for a given modulation index. The whole control scheme is simulated in MATLAB/SIMULINK at 10 m/s.

Table 2 PI controller parameters

RSC	Speed regulator	$K_p = 0.25,\ K_i = 25$
	VAR regulator	$K_p = 1.3,\ K_i = 100$
GSC	DC voltage regulator	$K_p = 0.25,\ K_i = 50$
	Current regulator	$K_p = 0.25,\ K_i = 120$

4.4 PI Controller Design

PI controller output signal used in RSC and GSC can be designed as:

$$U = K_p * e + K_I \int e * \mathrm{d}t \tag{20}$$

Here, e = error signal, K_p = proportional gain, $K_I = i$ gain.

Choosing the correct parameters for RSC and GSC control is very significant for nice performance. By choosing proper parameters, we can maintain the system stability. Once the parameters are selected, these can be tuned according to required output performance. In this paper, tuning of controller is done by Butterworth method which has an advantage over the protection of DFIG from over-currents. The transfer function can be written as: $P^2 + \sqrt{2} * \omega * p + \omega^2$; where, $P = \frac{\mathrm{d}}{\mathrm{d}t}$; ω = bandwidth of controller. The parameter of PI controller is given in Table 2.

5 Results and Discussion

Here, the DFIG-based wind turbine is simulated in MATLAB/SIMULINK at a constant wind speed of 10 m/s by using a proper control strategy in which a proper decoupling of active and reactive power is achieved in order to maintain stable DC link voltage and to obtain a well-regulated stator power. Figure 4 represents the variation in TSR and torque at a given speed.

We can observe that if the wind speed is kept below the rated wind speed of 13 m/s, the pitch angle of the turbine is maintained at 0 degrees, and the tip speed ratio (TSR) and power coefficient (C_p) also maintain their optimum value. As DFIG is modeled here at a wind speed of 10 m/s which is well below the rated speed, we can see that TSR is maintaining its optimum value of 7.2 with a settling time of approximately 2 s, and similarly, it is maintained at 0.44 with a settling time of 1.6 s. Torque is observed as 10,000 N-m. Figure 5 shows the variation of DC link voltage with time. A proper control strategy at GSC is achieved in order to maintain a constant DC link voltage in order to ensure a constant flow of active power from generator to grid. Initially, transients are quite significant, but with a settling time of approximately 1.5 s, transients die out and a steady DC link voltage is maintained at a ref. value of 1150 V.

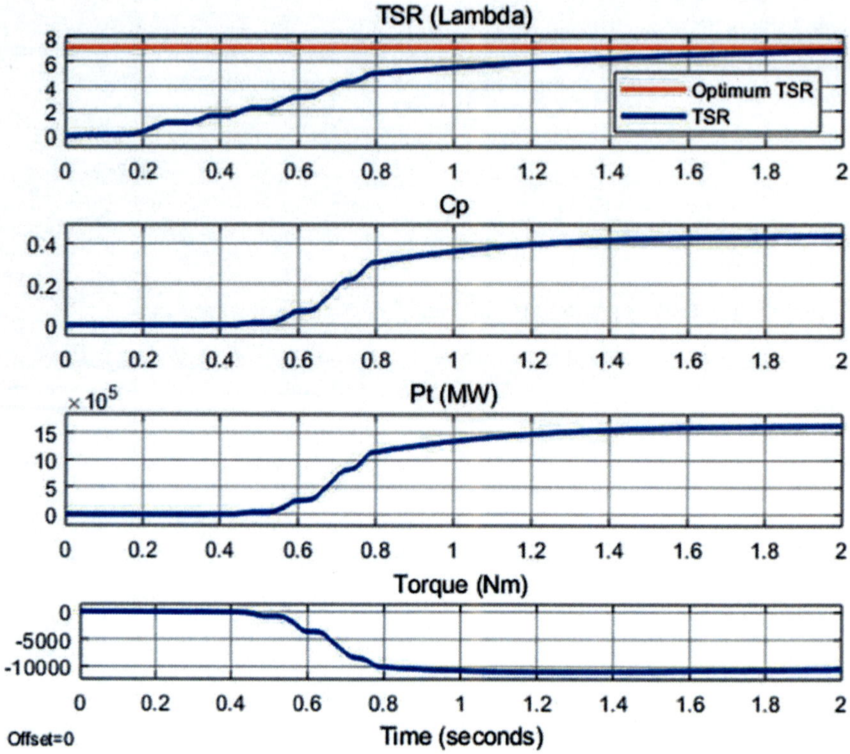

Fig. 4 Wind turbine responses at wind speed of 10 m/s

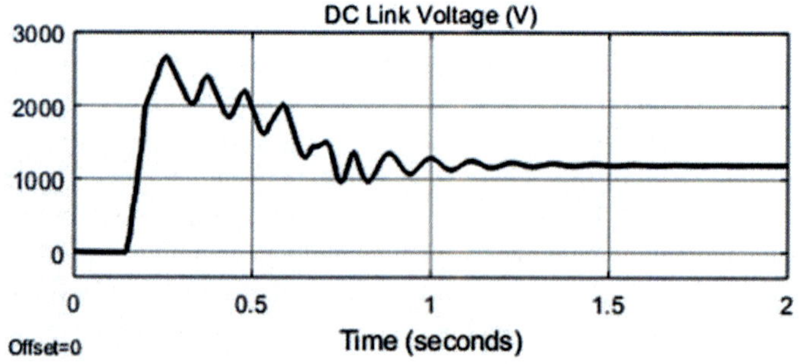

Fig. 5 DC link voltage (V)

Figure 6 depicts the variation of active and reactive power carried by the stator and rotor. Initially, approximately 6MVAr reactive power is required for the magnetization of stator flux in DFIG such that 2 MW of active power can be transmitted from generator to grid with a settling time of approximately 1.6 s. Figure 7 shows the electric power output of DFIG at wind speed of 10 m/s.

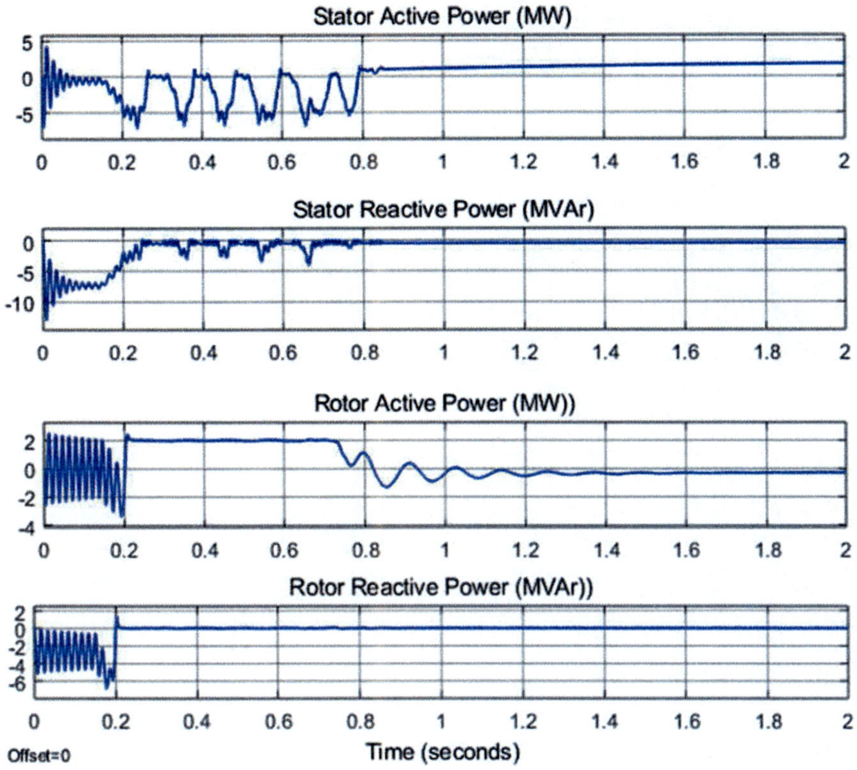

Fig. 6 Stator and rotor power

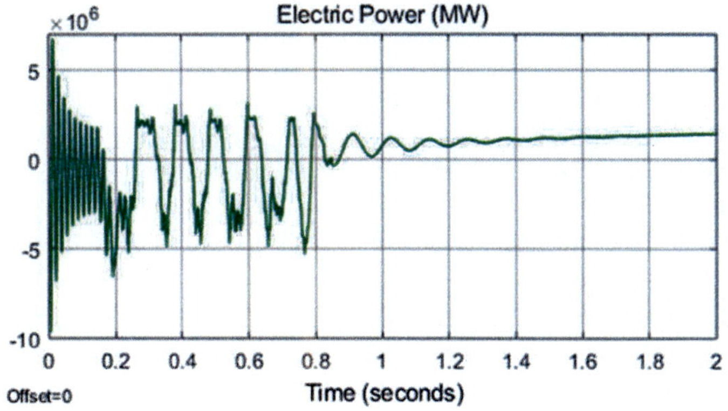

Fig. 7 Electric power output

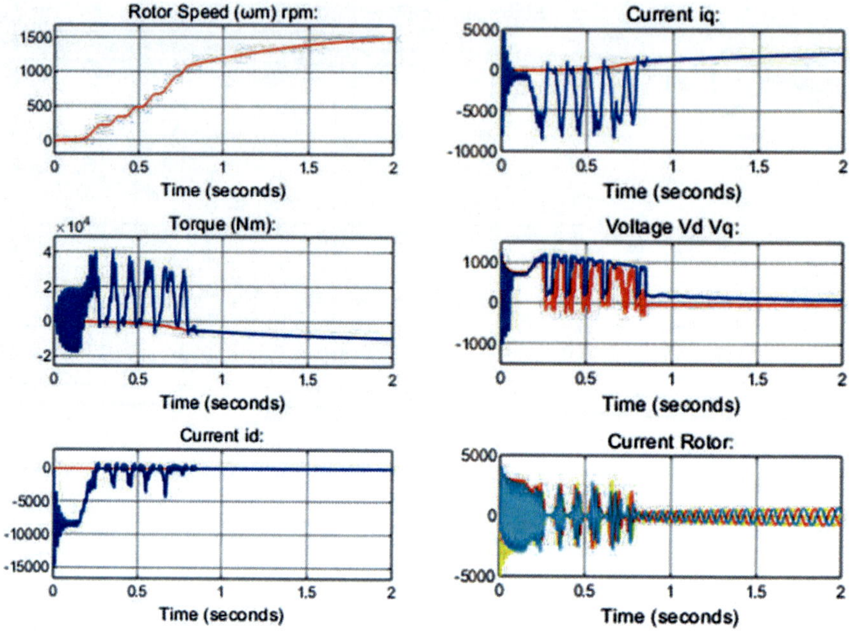

Fig. 8 RSC response

In Fig. 8, RSC response is presented as the time history of the q-axis of grid current, which drives reactive power. The magnitude of q-axis grid current is low since reactive power is adjusted to zero to ensure unity power factor in the rotor circuit.

PI regulator, the magnitude of this current is nearly.

Similarly, because we set the d-axis current in the rotor side to zero and provided it as an input to the PI regulator, the magnitude of this current is nearly nil. The time plot of d-q frame voltages obtained after transformation is shown. These decoupled voltages are used to generate PWM pulses for controlling RSC switches and hence able to regulate the stator power by reducing harmonics. By injecting 3rd harmonic, rotor current is made nearly sinusoid. Figure 7 shows torque v/s time plot which is optimized by extracting maximum power from the wind turbine. Figure 9 represents the three-phase current waveform of stator and rotor which is obtained as output of voltage source inverter (VSC) after going through a filter, and we can observe that they are not perfectly sinusoidal due to some harmonics still present.

As reference current I_d(ref) and I_q(ref) is kept as zero and given to PI regulator, magnitude of this current is seen as zero. The variation of bus voltage with time is shown and it is observed that bus DC voltage achieved its reference optimum value after going through transients. These transients can be reduced by reducing rise time which can be achieved by enhancing the bandwidth of filter used in PI controller. After going through a transient for 1 s, it achieves its steady state. The transient in the power can be reduced by selecting adequate parameters of PI controller or by selecting better

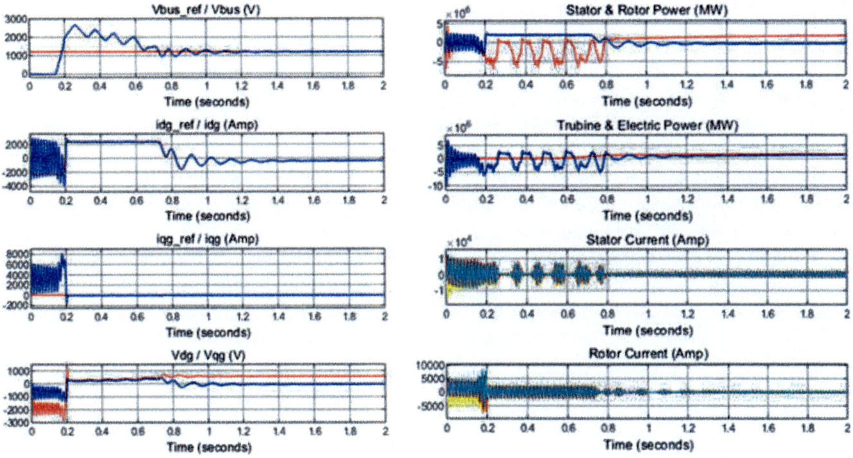

Fig. 9 GSC response

performing or tuned PID controller. The figures also show the variation of stator and rotor three-phase currents with time which is made nearly sinusoid by introducing PWM width controlling of pulses at each half wave of alternating voltage.

6 Conclusions

A dynamic model of a DFIG-based WECS is simulated using MATLAB/SIMULINK in this study, and the simulation results are investigated for various wind speeds. To manage the stator's active and reactive power, the rotor side converter employs a vector technique. The wind turbine also employs the indirect control MPPT technique during sub-synchronous mode to transmit the most of the power from the stator to the grid. When the results are compared, the proposed method improves the active power flow from stator to grid in terms of response time, rise time, and settling time. Utilizing innovative optimization techniques on the DFIG wind energy conversion system will be a significant task in the future.

References

1. Niassati N, Mohseni M, Amiri H, Seyedtabaei K, Hajihosseinlu Niassati A (2012) A new maximum power point tracking technique for wind power conversion systems. In: 15th IEEE power electronics and motion control conference (EPE/PEMC)
2. Liserre M, Cardenas R, Molinas M, Rodriguez J (2011) Overview of multi-MW wind turbines and wind parks. IEEE Trans Ind Electron 58(4):1081–1095

3. Muller S, Deicke M, De Doncker RW (2002) Doubly fed induction generator systems for wind turbines. IEEE Ind Appl Mag 8(3):26–33
4. Mwanik J, Lin H, Dai Z (2017) A concise presentation of doubly fed induction generator wind energy conversion systems challenges and solutions. J Eng 2017:401510
5. Tapia A, Tapia G, Ostolaza J (2003) Modeling and control of a wind turbine driven doubly fed induction generator. IEEE Trans on Energy Conv 18(2):194–204
6. Ramesh M, Yadav AK, Kumar R, Pathak PK (2022) Energy conversion: methods, technology and future directions: an extensive study of frequency supported-wind energy conversion system. Nova Science Publisher's
7. Ramesh M, Yadav AK, Pathak PK (2021) Intelligent adaptive LFC via power flow management of integrated standalone micro-grid system. ISA Trans 112:234–250
8. Pathak PK, Yadav AK (2021) Control of standalone microgrid: SM- and FL-based MRALFC schemes for solar-wind based micro-grid. Academic Press, Elsevier, pp 217–242
9. Pathak PK, Yadav AK, Shastri A, Alvi PA (2022) BWOA assisted PIDF-(1 + I) controller for intelligent load frequency management of standalone micro-grid. ISA Trans 132:387–401
10. Yao J, Li H, Liao Y, Chen Z (2008) An improved control strategy of limiting the DC-link voltage fluctuation for a doubly fed induction wind generator. IEEE Trans Power Electron 23(3):1205–1213
11. Abad G, Iwanski G, López J, Marroyo L, Rodríguez MA (2011) Doubly fed induction machine: modeling and control for wind energy generation applications. Wiley, Hoboken, NJ
12. Raza Kazmi SM, Goto H, Hai-Jiao G, Ichinokura O (2010) Review and critical analysis of the research papers published till date on maximum power point tracking in wind energy conversion system. In: IEEE energy conversion congress and exposition (ECCE), pp 4075–4082
13. Abdullah MA et al (2012) A review of maximum power point tracking algorithms for wind energy systems. Renew Sustain Energy Rev 16(5):3220–3227
14. Kazmi SMR, Goto H, Hai-Jiao G, Ichinokura O (2011) A novel algorithm for fast and efficient speed-sensorless maximum power point tracking in wind energy conversion systems. IEEE Trans Industr Electron 58:29–36

To Design an Optimal PV/Diesel/Battery Hybrid Energy System for Havelock Island in India

Mohammad Shariz Ansari, Aishwary Gautam, Bhavishya Tomar, Madhuram Gautam, and Mohd. Faisal Jalil

Abstract In the case of hybrid model design, the Hybrid Optimization Model for Electric Renewables (HOMER) software is employed for optimization. This work aims to reduce overpriced net present cost (NPC) and cost of energy (COE) at Havelock Island in India and lower the air pollution index by utilizing non-conventional energy sources. Havelock Island's average daily radiation is 5.30 kWh/m^2/day. The HOMER simulates, optimizes, and does sensitivity analysis on the proposed model for thousands of combinations to find the best answer. We are employing diesel generators, solar panels, and battery storage to develop the optimal hybrid system possible for modeling reasons. This model yields the lowest NPC and COE. In this analysis, the sensitivity variables for simulation are diesel generators' run time and fuel cost. The diesel price of Rs. 77/L and the generator run periods of 720 min were used as sensitivity variables for the optimized model. For these values of sensitivity variables, the NPC and COE come out to be Rs. 1.82 B and Rs. 19.87/kWh, respectively. Using an optimal hybrid system instead of an existing diesel-only system at Havelock Island, the pollution will also be reduced by 83.21%.

Keywords Hybrid model · Net present cost (NPC) · Cost of energy (COE) · Renewable energy · HOMER

1 Introduction

The non-renewable fuel source accounts for a large utilization ratio in the world. The development of fossil energy as a fuel is of grave concern for the world in achieving sustainable growth and development, and also, the use of fossil fuel as an

M. S. Ansari (✉) · A. Gautam · B. Tomar · M. Gautam
Department of Electrical and Electronics Engineering, KIET Group of Institutions, Delhi-NCR, Ghaziabad, India
e-mail: mohdshariz@gmail.com

Mohd. F. Jalil
Department of Electrical Engineering, FET, ZHCET, AMU, Aligarh, India

A. Rani et al. (eds.), *Signals, Machines and Automation*, Lecture Notes in Electrical Engineering 1023, https://doi.org/10.1007/978-981-99-0969-8_21

energy source is a worthy cause of worldwide water contamination, air contamination, desertification, and the worldwide crisis of climate [1]. However, around the previous decagons, a reduction in hydrocarbon deposits is detected worldwide [2]. If the depletion rate of hydrocarbon deposits remains at the same pace, it is being predicted by proper estimation methods that the amount of non-renewable sources will get empty by 2070. There is a strong requirement to bring the world under a transition phase from a non-renewable energy source to a renewable energy source. This process includes proper modeling, analysis, calculating, and simulating the various factors so that the dependence on the hydrocarbon deposits gets reduced, and we can pace toward the upcoming green and clean energy sources [3]. Statistical data shows that there are still 1.8 billion people under the outreach to power. Those occupants need to be provided electricity either through laying transmission lines of a grid network or a remote power generation system in remote areas where laying a grid is difficult due to hilly terrains or in island territories [4]. In the Havelock Island of Andaman & Nicobar, the major providers of power sources are the diesel-based generator. There are two negative impacts of using diesel appliances. The first one is the high generation cost of power due to inflation in diesel prices. The second reason is hurting the ecological system and causing air contamination, noise disturbance, and water contamination. However, this remote region has high substantiality rational source of power, so using those available energy resources can be viable to decrease the power generation cost [5]. Due to the high inflation in the price of distillate fuel oil, there is a solid requirement to increase the supply of energy through renewable and sustainable sources of power. Using this methodology, more and more energy can be harnessed without disturbing the ecological disturbance. The balance of nature can be ensured, and there is a way forward to discover more potential in standalone or hybrid systems [6]. The standalone system includes a single system like PV based, whereas a hybrid amalgamates two or more standalone designs like PV-diesel and many other types. The hybrid system includes two or more combinations of renewable sources that may be standalone or maybe in the form of a grid connection. Renewable sources like solar, biomass, hydro, and wind are examples of these sources [7]. We can use the above sources as standalone or hybrid systems depending upon the geological factors determining the actual combination of these systems. There are some places where we can use a variety of two or maybe more than two. At some other sites, we also use renewable hybrid with non-renewable sources like a hybrid diesel-solar hybrid for Havelock Island [8]. Hybrid Optimization Multiple Energy Resources (HOMER) software is a system that programs the economic-based approach so that a feasible programming methodology can be adopted by inserting the appropriate parameter so that a better scheme of implementation can be achieved [9]. This HOMER model has given appropriate analysis in case of rural electrification in the remote areas of Bangladesh and also in the isolated areas of Algeria and North Africa [10].

2 Overview of Havelock Island

2.1 Havelock Profile

Havelock Island is the pioneered Island of Andaman & Nicobar. Havelock belongs to the south administrative district part of the union territory. The Havelock Island, now renamed Swaraj Dweep, is 41 km northeast of the capital city, Port Blair. The area of Havelock Island is 92.2 km^2. It is the major tourist destination in the union territory having beach hotels and is famous for its coral reefs and scuba diving experience. The maximum length of the island is 18 km, and its breadth is 8 km. Island is a sub-group of the Ritchie Archipelago [11].

2.2 Consumption Estimation of Demand Projection for Domestic Consumers

The Havelock Island analytical report suggests that it has an installed potential of 1.6 MW (the fiscal year 2015), but the maximum requirement of the Island is more than 2.5 MW. Being the central tourist spot in the region of Andaman & Nicobar, there is an essential requirement of the continuous supply of electricity for the better economic prospect of the area. Electric energy's present potential is insufficient to fulfill the current demand, especially during the maximum load hours. It is evident from the various reports published that power cut is very much regular during the maximal demand for electricity. There is a requirement for an additional reliable system to avoid this disruption to the electricity, which can lead to a better tourism spot and has a high chance of attracting a large number of tourists so that Havelock Island's economy can be boosted significant factor.

3 Exquisite Hybrid Representation of Havelock Island

3.1 About HOMER

The National Renewable Energy Laboratory in the United States created the HOMER software. HOMER, a computer model, is used in various technologies and applications, including manufacturing, storage, load analysis, and microgrid or off-grid micropower systems [12]. The link can be designed throughout the project cycle and includes several cost-related items, such as the cost of investment, replacement, and operation. It helps us assess and compare different options, technical and economic advantages, and disadvantages of other parts of the projects with HOMER; the design process is simple. Simulation, optimization, and execution were performed in three stages and sensitivity analysis. The first level, which is the most basic, is a simulation, technical analysis, and cost–benefit analysis of the micropower system has been completed. The optimization phase begins following the simulation phase [13].

3.2 *Details of Havelock Island*

NASA dataset is used here to obtain the insolation data for Havelock Island on a mean of per month basis. The per month mean value of solar insolation is obtained from NASA. Table 1, gives the monthly mean value for the global irradiance value of Havelock Island. Table 1 also depicts the clearness index. It is a dimensionless number between 0 and 1 and is defined as the ratio of surface radiation to terrestrial radiation.

The maximal value of the clearness index is 0.690, and the minimal value of the clearness index is 0.420. The maximal and minimal value of the clearness index is reported in April and December, respectively. The maximal and minimal value of per day irradiation is 4.530 kWh/m^2/day and 0.210 kWh/m^2/day, respectively. The month of minimal and maximal value of solar insolation occurs in December and June, respectively. The value of the mean annual irradiation is 2.32 kWh/m^2/day.

Table 1 Clearness index and per day irradiation index for Havelock Island

Month	Clearness index	Daily irradiation (kWh/m^2/day)
January	0.655	5.65
February	0.69	6.47
March	0.674	6.83
April	0.631	6.65
May	0.482	5.08
June	0.42	4.38
July	0.426	4.45
August	0.425	4.45
September	0.454	4.64
October	0.522	4.99
November	0.555	4.86
December	0.621	5.19
Mean value	0.546	5.3

3.3 *Load Profile of Havelock Island*

The shape of the load is an essential parameter when performing optimization by the HOMER because the different power sources depending on the load shape. The selected island also works as a significant dependent parameter for the model design by the HOMER, and we get the best optimization of the proposed model. The monthly solar global irradiance (SGI) and load profile of Havelock Island are shown in Figs. 1 and 2 respectively. On Havelock Island, the primary energy use is to meet the residents and businesses. As a result, only these two loads will be required to handle in HOMER. HOMER uses these data for simulation and optimization. The island will be supplied 24 * 7 h per week, and the demand is expected to rise. The island is also a significant tourist destination in Andaman & Nicobar Island, requiring reliable and continuous power.

4 Architectonics Structure Representation of Havelock Island

On Havelock Island, the power generation is mainly from the diesel generator. Still, it has a vast potential for transitioning from diesel generators to solar PV panels. Since diesel is a very costly, non-renewable, polluting source, this is a time to focus on a safe and reliable form of solar energy. The proposed model consists of solar PV, diesel generator, converter, and battery, as shown in Fig. 3. The HOMER tells us the best possible combination of the mentioned energy-producing sources. It gives us

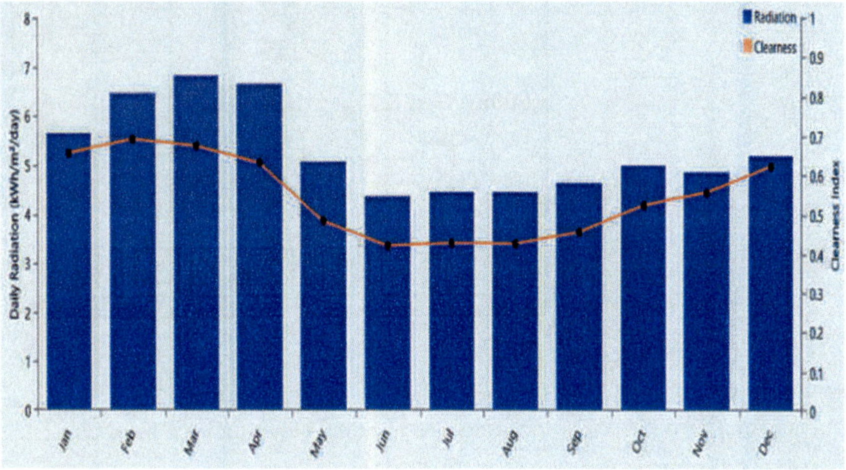

Fig. 1 Monthly average SGI for Havelock Island

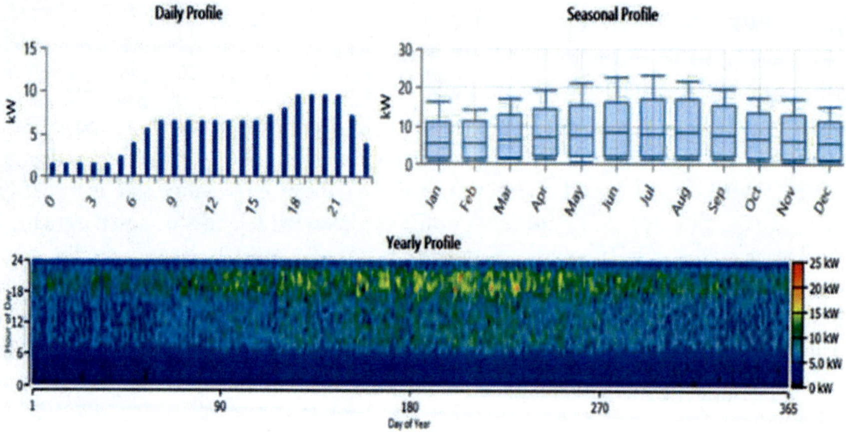

Fig. 2 Load profile of Havelock Island

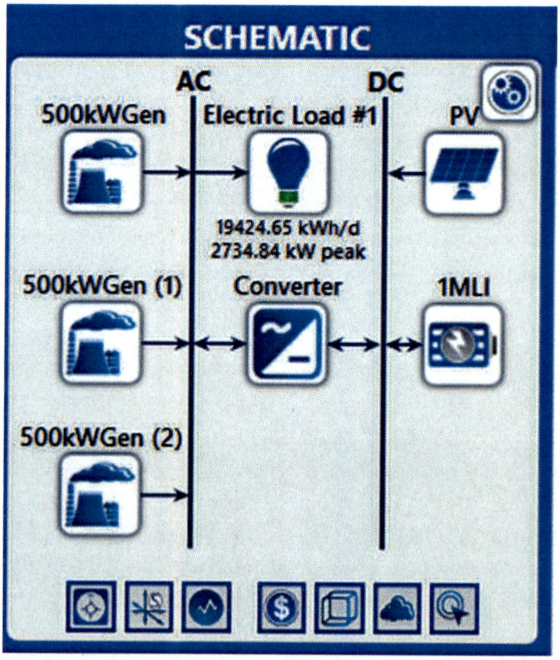

Fig. 3 Schematic diagram of suggested model

the best result by undergoing proper optimization techniques. It requires adequate data insertion for a particular location to retrieve the best-simulated result.

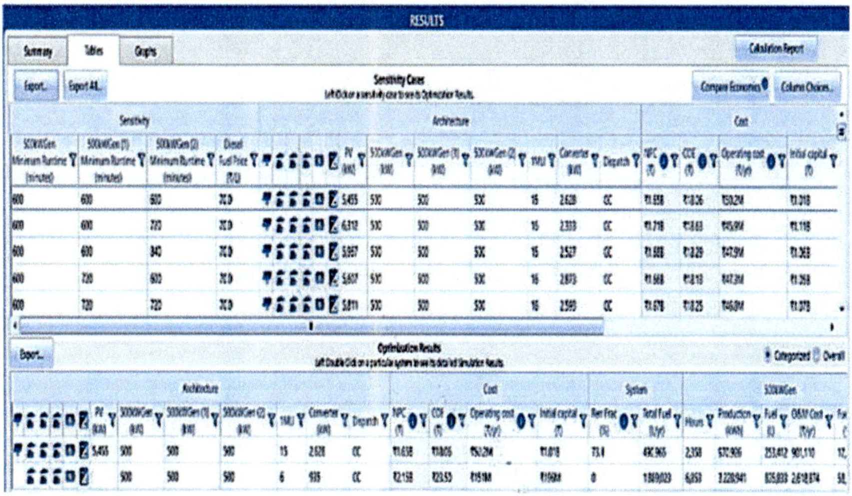

Fig. 4 Optimization and sensitivity results

5 Results Obtained for the Suggested Model at Havelock Island

5.1 *Stimulated and Computed Outcome*

We have obtained stimulated and computed outcomes from HOMER as presented in Fig. 4.

The outcome of the different stimulated parameters is obtained from the upper part of the figure and the calculated result for a stimulated variable from the lower part. The sensitivity factors for diesel pricing are Rs. 70/L, Rs. 77/L, and Rs. 84/L. The sensitivity variable for minimum runtime of diesel generator has been taken as 600 min, 720 min, and 840 min.

5.2 *Net Present Cost (NPC)*

Net present cost is depicted in a bar graph and tabular forms, as shown in Table 2. Net present cost includes the cost of operation, cost of capital, cost of replacement, cost of salvage, and cost of resource that comes out to be for the proposed model. The cost of operation, cost of capital, cost of replacement, cost of salvage, and cost of the resource is for the proposed model Rs. 18,23,97,815.31, Rs. 1,29,34,75,781.06, Rs. 3,83,31,784.37, Rs. 86,78,701.81, and Rs. 31,48,37,708.05, respectively.

Table 2 Net present cost of the proposed system at Havelock Island

Equipment	Capital cost (Rs.)	Operating cost (Rs.)	Replacement cost (Rs.)	Salvage value (Rs.)	Resource cost (Rs.)	Total cost (Rs.)
Generic 500 kW fixed capacity genset-1	99,30,095.76	71,53,482.65	68,90,722.91	−11,27,200.58	15,45,24,699.54	17,73,71,800.29
Generic 500 kW fixed capacity genset-2	99,30,095.76	42,63,436.14	29,69,853.37	−10,79,166.46	9,38,42,376.21	10,99,26,595.02
Generic 500 kW fixed capacity genset-3	99,30,095.76	29,24,030.77	19,30,507.75	−19,14,960.08	6,64,70,632.29	7,93,90,306.50
Generic 100 kWh Li-ion	43,26,12,260.34	16,79,69,183.36	53,76,632.09	−5,74,083.45	0	60,53,83,992.34
Generic flat plate PV	77,60,63,649.15	37,682.40	0	0	0	77,61,01,331.55
System converter	5,50,09,584.29	0	2,11,64,068.25	−39,83,291.24	0	7,21,90,361.30
System	1,29,34,75,781.06	18,23,97,815.31	3,83,31,784.37	−86,78,701.81	31,48,37,708.05	1,82,03,64,386.99

Total NPC includes Rs. 1.29 B as cost of capital, Rs 182.3 M as cost of operation, Rs 38 M as cost of replacement, Rs 8.6 M as cost of salvage, and Rs. 314 M as cost of the resource. Out of these expenses, the system capital cost is maximum, around Rs. 1.29 B. This system's capital cost includes the capital cost of generator sets, which comes out to be Rs. 29.8 M; PV cost of capital is Rs. 776 M, and the capital cost in case of the converter is Rs. 55 M, and capital cost of the battery is Rs. 432 M. The cost of operation is Rs. 182 M; the total cost of replacement is Rs. 38 M; the total cost of salvage is Rs. −8.67 M, and the total cost of the resource is Rs. 314 M.

6 Emission Due to Distinct Contaminating Particles at Havelock Island

The emission due to various kinds of contaminating particles because of only the diesel and hybrid system is given in Table 3. This table presents the emission of carbon dioxide, carbon monoxide, and hydrocarbon without complete combustion. From Table 3, it is also observed that a hybrid system can reduce a considerable amount of pollution. The CO_2 emission in the diesel-only system is 4940254 kg/year, while the CO_2 emission in the hybrid system is 829375 kg/year. If the hybrid system is adopted, a large amount of cut in CO_2, i.e., 4,110,879 kg per year, can be seen. Cut in the emission of other contaminating particles such as CO, SO2, and NO_2 is also presented in Table 3.

Table 3 Comparison of emission of the hybrid system and diesel-only system

Contamination causing particles	Diesel generator emission (kg/year)	Emission by diesel, PV, battery for the diesel price of Rs77/L (kg/year)	Decrease in emission value (kg/year)
Carbon dioxide	4,940,254	829,375	4,110,879
Carbon monoxide	25,558	4291	21,267
Unburned hydrocarbons	1356	228	1128
Particulate matter	219	36.7	182.3
Sulfur dioxide	12,076	2027	10,049
Nitrogen oxides	4898	822	4076

7 Comparative Analysis of Diesel-Only System and Optimal Hybrid System

Comparing the economic estimation of the two given systems is depicted in Table 4. The NPC of Rs. 1.65 B and COE of Rs. 18.06 /kWh are minimum for a hybrid system, including all the possible sources at a diesel cost of Rs. 70/L, and the minimum runtime of all three generators is 720 min. In today's scenario, the actual cost of diesel is Rs. 77/L on Havelock Island. Because of this, the diesel price of Rs. 77/L and minimum run time of all three generators are taken as 720 min as standard, which has resulted in the NPC and COE of Rs. 1.82 B and Rs 19.87 /kWh, respectively, for the system consisting of diesel-PV-battery. Table 4 here depicts that the mere-diesel generator system for the standard value of diesel price Rs. 77/L, and minimum runtime of 720 min for all three generators has the NPC and COE for the stimulated varying values Rs. 2.29 B and Rs. 24.97/kWh, respectively. If we observe both the system values for hybrid and the mere-diesel generator set system at diesel price, i.e., Rs. 77/L, we can clarify that NPC has decreased from Rs. 2.29 B to 1.82 B, and COE has decreased from 24.97/kWh to 19.87/kWh. It says that if a hybrid system is used instead of a diesel-only system, there will be a significant reduction in NPC and COE. If we use a hybrid approach in place of only a diesel system, CO_2 emission will be reduced from 4,940,254 kg per year to 829,375 kg per year, which is a significant reduction.

8 Conclusions

Optimization and sensitivity results are depicted in Fig. 4 for a given set of parameters. The optimal system includes diesel generators, PV, battery storage, and converters. The optimized value of NPC and COE is Rs. 1.82 B and Rs. 19.87/kWh, respectively. The NPC and COE have been calculated for diesel rates of Rs. 70/L, Rs. 77/L, Rs. 84/L, and the minimum runtime of generators as 600 min, 720 min, 840 min, respectively. For optimal solution, the actual rate of diesel, i.e., Rs. 77/L, and 720 min of the minimum runtime of each diesel generator are considered. If we use diesel-only system, the CO_2 emission is 4940254 kg/year; if we use a hybrid approach, the CO_2 emission is 829375 kg/year. From the above, it is seen that if we use the optimal hybrid system, there will be a significant reduction in CO_2 emissions.

Table 4 Comparative analysis of diesel-only system and optimal hybrid system

Sensitivity variables				Diesel-only system (base system)				Optimal hybrid system			
Diesel price (Rs./L)	Running time of DG sets (min)										
	500 kW	500 kW	500 kW	NPC (Rs.)	COE (Rs.)	Ren fraction %	CO_2 emission (kg/year)	NPC (Rs)	COE (Rs)	Ren fraction %	CO_2 emission (kg/year)
70	600	600	600	2.15B	23.5	0	12,87,422	1.65B	18.06	73.4	12,87,422
	720	720	720	2.12B	23.14	0	49,48,076	1.80B	19.63	74.3	12,47,850
	840	840	840	2.12B	23.11	0	49,35,782	1.71B	18.75	72.8	13,21,712
77	600	600	600	2.29B	24.98	0	49,46,031	1.70B	18.61	72.9	1,311,06
	720	720	720	2.29B	24.97	0	49,40,254	1.82B	19.87	82.9	8,29,375
	840	840	840	2.29B	24.98	0	49,34,101	1.75B	19.13	74.4	12,41,848
84	600	600	600	2.46B	26.82	0	49,41,961	1.75B	19.05	78	10,64,265
	720	720	720	2.46B	26.83	0	49,39,450	1.85B	20.2	83	8,25,890
	840	840	840	2.46B	26.79	0	4,928,67	1.80B	19.6	74.4	1,241,84

References

1. Legros G, Havet I, Bruce N, Bonjour S (2009) The energy access situation in developing countries, USA. https://doi.org/10.1503/cmaj.160656
2. BP Energy Outlook (2019) The Energy Outlook explores the forces shaping the global energy transition out to 2040 and the key uncertainties surrounding that
3. Ansari MS, Manaullah, Jalil MF (2016) Investigation of renewable energy potential in union territory of Lakshadweep islands. In: 2016 second international innovative applications of computational intelligence on power, energy and controls with their impact on humanity (CIPECH). IEEE, Ghaziabad, India, pp 209–213. https://doi.org/10.1109/CIPECH.2016.7918768
4. Raghuwanshi SS, Arya R (2019) Economic and reliability evaluation of hybrid photovoltaic energy systems for rural electrification. Int J Renew Energy Res 9:516–524
5. Nasiruddin I, Khatoon S, Jalil MF, Bansal RC (2019) Shade diffusion of partial shaded PV array by using odd-even structure. Sol Energy 181:519–529. https://doi.org/10.1016/j.solener.2019.01.076
6. Manaullah, Ansari MS (2019) Techno-economic feasibility of PV-wind-diesel battery hybrid energy system for Lakshadweep Island in India. Indian J Power River Val Dev 69:141–148
7. Purohit P, Michaelowa A (2007) Potential of wind power projects under the Clean Development Mechanism in India. Carbon Balance Manag 2:8. https://doi.org/10.1186/1750-0680-2-8
8. Manaullah MS (2019) Ansari: Feasibility analysis of renewable energy options for the union territory of Lakshadweep Islands. Int J Glob Energy Issues 42:63. https://doi.org/10.1504/IJGEI.2019.100690
9. Vaisi B (2022) A review of optimization models and applications in robotic manufacturing systems: Industry 4.0 and beyond. Decis Anal J 2:100031. https://doi.org/10.1016/j.dajour.2022.100031
10. Singh R, Bansal RC, Singh AR (2018) Optimization of an isolated photo-voltaic generating unit with battery energy storage system using electric system cascade analysis. Electr Power Syst Res 164:188–200. https://doi.org/10.1016/j.epsr.2018.08.005
11. Stackhouse P, Whitlock C, Zhang T (2004) Solar renewable energy data sets from NASA satellites and research. In: World renewable energy congress VIII (WREC 2004), pp 279–283
12. NREL (2015) Hybrid optimization model for electrical renewable (HOMER)
13. Ansari MS (2022) Feasibility analysis of standalone hybrid renewable energy system for Kiltan Island in India. In: Kumar A, Srivastava SC, Singh SN (ed) Renewable energy towards smart grid. Springer, Singapore, Singapore, pp 79–93. https://doi.org/10.1007/978-981-16-7472-3_7

To Design an Optimal Hybrid Energy System for Agatti Island in India

Mohammad Shariz Ansari, Aditya Srivastava, Aditya Singh, Adesh Gupta, Ameer Faisal, and Mohd. Faisal Jalil

Abstract Since diesel generators are used to generate electricity mostly on smaller, isolated islands, the net present cost (NPC) and cost of energy (COE) are very high. The use of diesel generators also causes increased pollution. As a result, by using renewable energy sources such as solar, this initiative intends to reduce NPC, COE, and pollution. In India's union territory of Lakshadweep, Agatti Island would be the research site (UTL). The daily radiation on Agatti Island is 5.68 kWh/m^2/day. Using HOMER software, modeling, simulation, and optimization of an ideal hybrid model were all done. The optimum solution is to combine diesel, PV, and battery power. In this approach, the NPC and COE are the least. The investigator has included the fuel rate (Rs) as a sensitivity factor. The best system uses the Rs. 72/L diesel price as a sensitivity variable. Furthermore, this optimal hybrid system reduces the pollutants by 90%.

Keywords Hybrid model · Net present cost (NPC) · Cost of energy (COE) · Renewable energy · HOMER

1 Introduction

Energy, as we know, can neither be created nor destroyed is one of the essential needs for every action, the reaction of all things being equal. On the other hand, energy has been and continues to be the greatest threat to people, as most of the energy used on the planet comes from conventional resources [1]. According to current research and consumption rates, some will be exhausted soon. In addition, approximately 1.5 billion people globally still don't have access to electricity. Diesel generators regularly provide energy to these isolated islands and villages [2]. In any event, they

M. S. Ansari (✉) · A. Srivastava · A. Singh · A. Gupta · A. Faisal
Department of Electrical and Electronics Engineering, KIET Group of Institutions, Delhi-NCR, Ghaziabad, India
e-mail: mohdshariz@gmail.com

Mohd. F. Jalil
Department of Electrical Engineering, FET, ZHCET, AMU, Aligarh, India

A. Rani et al. (eds.), *Signals, Machines and Automation*, Lecture Notes in Electrical Engineering 1023, https://doi.org/10.1007/978-981-99-0969-8_22

were becoming increasingly anxious because they were frequently confronted with high fuel prices due to the massive increase in the cost of petroleum products and freight charges [3].

Furthermore, the detrimental natural impacts wrought by diesel harm the neighboring environmental system, contaminating air, water, and soil. They suffer from the negative repercussions of an energy deficit or power outage [4]. With the rising cost of fuel and the rapidly falling cost of sustainable power resources, renewable energy is becoming more competitive with non-renewable energy, thus wide spreading use of renewable power sources for independent systems, such as PV battery or hybrid systems [5]. Looking into RES has primarily been done in system modeling, element sizing, simulation, economic evaluation, and most notably, system optimization. Simulation systems are typically necessary for this type of research. HOMER software is one of the most widely used simulation devices for independent renewable sources [6]. Using software for renewable energy techno-economic, simulation, and modeling analysis has become the subject of many previous studies. The likelihood of achieving energy sovereignty on an island using PV, diesel generators, and batteries was analyzed and conducted [7].

2 Overview of Agatti Island

2.1 Agatti Profile

The Island on which we perform our project is in the union territory Lakshadweep. The Island is about 7.6 km long. From Kochi, the distance of the Island is about 459 km to the west. The population of this Island is about 9000. The people living there mostly speak English, Tamil, and Malayalam languages. The Island has a tourist spot with only two hotels and resorts [8]. The weather on the Island in summer nearly touches 350 °C on average, and during the rainy season, the average rainfall is approx. 62 inches. Every year the climate of Agatti is moderate, neither too hot nor too cold. Therefore, it is one reason that attracts tourists to this Island.

2.2 Consumption Estimation of Demand Projection for Domestic Consumers

According to the reports, the energy demand of Agatti Island in 2015 was 5.51 MU. The power supply is being given 24 h, and in the future, consumption will increase [9]. Daily consumption has risen from 2.82 kWh in 2011 to 3.97kWh in 2015 at a CAGR of 8.97%. Although, taking into consideration that consumption increment on an average basis is less than 1% in the year 2019. And the availability increased in recent years along with other statistical data of that area containing population, the daily consumption of different household has been pushed by 3% only. Therefore, the geographical information of the Agatti plays a significant role in predicting the power demand of the present and the future[10].

2.3 About HOMER

The National Renewable Energy Laboratory in the United States created the HOMER software. HOMER, a computer model, is used in various technologies and applications, including manufacturing, storage, and analysis load. Microgrid or off-grid micropower systems. It is possible to design a link. The project cycle includes several cost-related items, such as the cost of investment, replacement, and operation. It helps you assess and compare different options, technical and economic advantages, and disadvantages of other parts of the projects with HOMER, and the design process is straightforward [11]. Simulation, optimization, and execution were performed in three stages and sensitivity analysis. The first level, which is the most basic, is a simulation, technical analysis, and cost–benefit analysis of the micropower system has been completed—following that, the optimization phase begins following the simulation phase [12]. Clearness index and solar radiation in kWh/m^2/day at Agatti Island is shown in Table 1 and the monthly average solar global irradiation (SGI) is shown in Fig. 1.

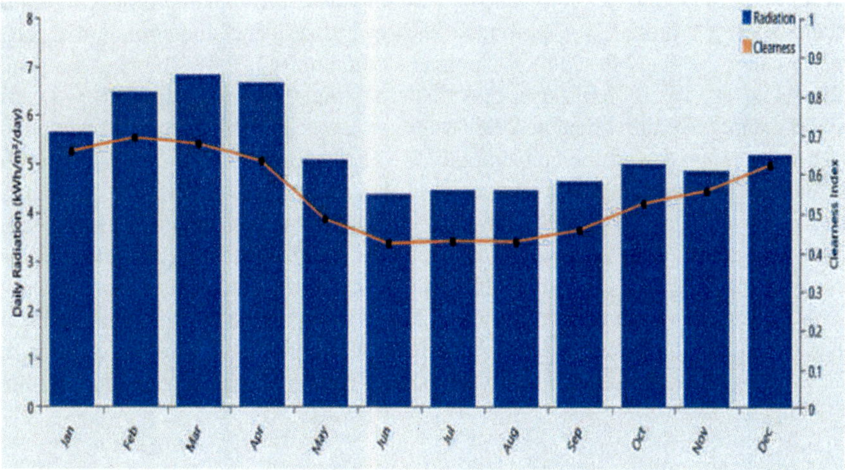

Fig. 1 Monthly average solar global irradiation (SGI) for Agatti Island

Table 1 Clearness index and solar radiation in kWh/m²/day at Agatti Island

Month	Clearness index	Daily radiation (kWh/m^2/day)
Jan.	0.596	5.965
Feb.	0.617	5.847
Mar.	0.593	6.039
Apr.	0.546	5.752
May	0.463	4.848
Jun.	0.379	3.923
Jul.	0.388	4.021
Aug.	0.406	4.24
Sep.	0.438	4.488
Oct.	0.473	4.566
Nov.	0.558	4.069
Dec.	0.574	4.888
Avg.	0.503	4.887

2.4 *Load Profile of Agatti Island*

Since most energy sources depend on the load shape for optimal sizing, the depicted form of a load of that selected Island is essential input data for optimizing the proposed model in HOMER software. Agatti Island's load shape is illustrated in Fig. 2. Figure 2 shows a community load's daily load and monthly and annual load profile. On Agatti Island, energy is primarily used to meet the need of residents and commercial. The yearly energy demand based on this Island's community is 7.88 MU, and the peak demand is 1.4 MW. HOMER uses these data for simulation and optimization. Agatti Island's energy requirement in 2015 was 5.51 MU. The Island will be supplied 24 * 7 h per week, and the consumption is expected to rise. The consumption on the daily basis of the number of electric households has increased [13]. Accordingly, the figure was calculated by dividing the actual sales by the number of people with access to electricity consumers in 2015. Last year, daily household consumption was less than 1%, despite higher availability in previous years and other factors. Only a 3% increase in consumption has been made. However, the Island's geographical characteristics (i.e., location, accessibility, and weather) should also be considered. The existing tariff levels significantly impact determining the current and future state of affairs demands.

3 Architectonics Structure Representation of Agatti Island

Power is generated primarily by diesel generators on Agatti Island. However, there is also considerable potential for solar PV. Because fuel is expensive and exhausts quickly, the goal is to reduce diesel consumption and increase the use of renewable

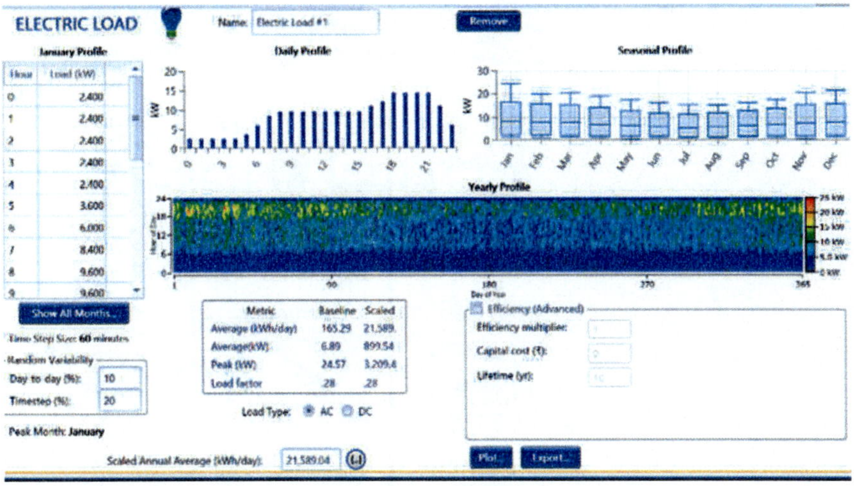

Fig. 2 Load profile of Agatti Island

energy. Since all the non-renewable energy resources heavily pollute the environment when used on a large scale. The author proposes a model that includes solar PV, emergency diesel generators, batteries, converters, and loads. With this hybrid system's use, the pollution emission will reduce. The HOMER software simulates the proposed model to determine the optimal hybrid model. The schematic diagram of the proposed model is shown in Fig. 3.

4 Results Obtained for the Suggested MODEL at Agatti Island

4.1 Stimulated and Computed Outcome

The results from HOMER for optimization and sensitivity are displayed in Fig. 4. The results of various sensitivity variables are shown in the upper half of the picture. In contrast, the optimization result for a specific sensitivity variable is shown in the lower section. Rs. 68/L, Rs. 72/L, and Rs. 76/L were chosen as sensitivity factors for diesel pricing. Other sensitivity variables are also used to calculate the runtime of different rating generators. The software has used all the possible combinations of the given sensitivity variables for simulation and optimization. A total of 22,254 solutions were simulated. Out of them, 3244 were left, 1788 were not taken for lacking a converter, and 855 were rejected for having an unnecessary converter. HOMER software uses only diesel generators, solar PV, and storage systems like batteries. The optimum solution for this combination comes out to be a total net present cost of ₹988 million and a charge of energy levelized as ₹9.71/kWh.

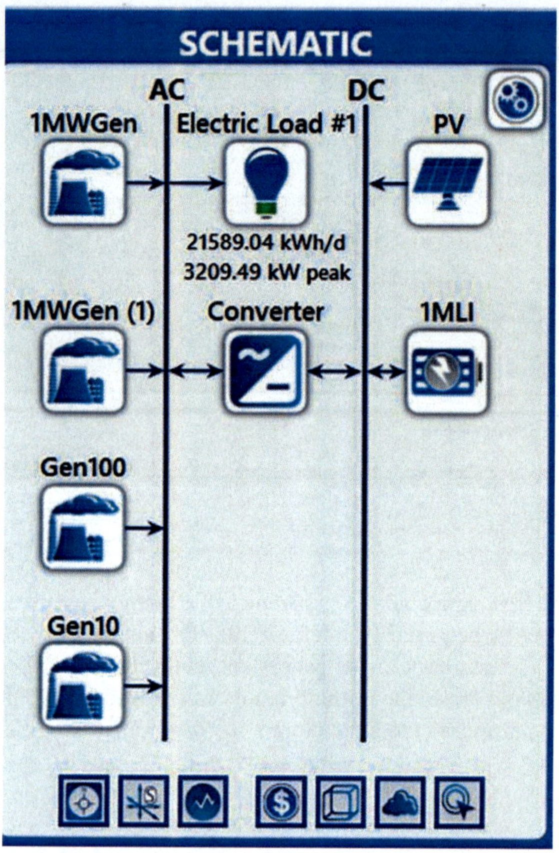

Fig. 3 Schematic diagram of the proposed model

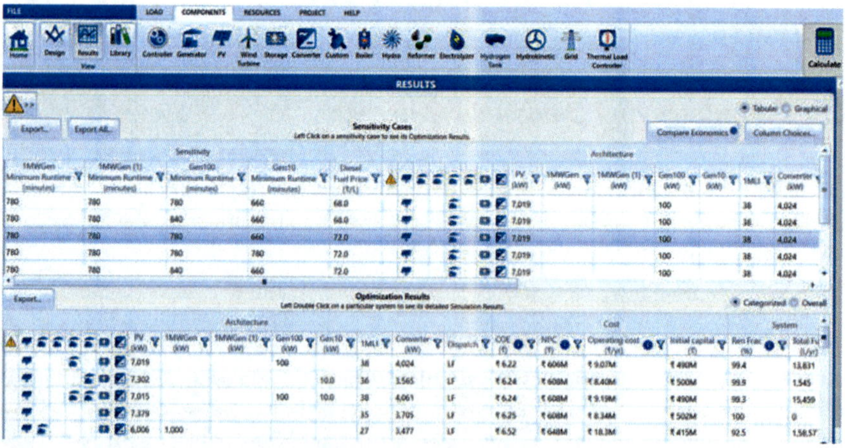

Fig. 4 Sensitivity and optimization results for Agatti Island

4.2 Net Present Cost (NPC)

The net present cost based on different categories has been tabulated in the given Table 2. The total net present cost was found to be ₹988 million, consisting of salvage value, operating cost, capital cost, replacement cost, and resource cost for the proposed model.

Total NPC includes ₹988 M as capital cost, ₹23.9 M as operating cost, ₹128.4 M as replacement cost, −₹25,246,262.16 as salvage value. The capital cost is maximum which is ₹988 M, including all the generator sets of capital cost (₹212 million), Li-ion batteries capital cost (₹40 million), PV capital cost (₹33 million), and converters capital cost (₹4 million). The total operating cost is ₹2 million, the total replacement cost is ₹12 million, and the total salvage value is −₹2 million.

5 Emission of the Diesel-Only System and Optimal Hybrid System at Agatti Island

Table 3, shown below, gives the emission of various types of pollutant materials by different systems. The pollutants contain CO_2, CO, unburned hydrocarbon, particulate matter, sulfur dioxide, and nitrogen oxide. It can be observed from the table that pollution will be decreased if a hybrid system is used. The non-hybrid system emits CO_2 of 5,619,560 kg/year, and in the hybrid system, the emission of CO_2 is 542658 kg/year.

The reduction in CO2 emission is 5076902 kg/year, which is a considerable reduction in value. The decrease in value of other polluting materials like sulfur dioxide, carbon monoxide, and nitrogen dioxide is mentioned in Table 3.

6 Comparative Analysis of Diesel-Only System and Optimal Hybrid System

Comparing economic estimation of the two given systems is depicted in Table 4 for the distinct parameter of stimulated varying values, that is, the fluctuating cost of diesel and run time of diesel generators. The NPC of Rs. 955 M and COE of Rs 9.77 ₹/kWh is minimum for a hybrid system, including all the possible sources at a diesel cost of Rs. 68/L, and the minimum runtime of 1 MW generators is 900 min, 100 Kw generator is 840 min, and for 10 kW generator is 780 min. In today's scenario, the actual cost of diesel is even more than Rs 76/L on Agatti Island. Therefore, diesel price of Rs. 76/L and minimum run time of generators as 840 min, 900 min, and 780 min for 1 MW, 1 MW, 100 kW, and 10 kW, respectively, are considered for Agatti Island. For these combination of sensitivity variables, the NPC and COE of diesel-only system are Rs. 2.28 B and Rs. 22.4/kWh and for hybrid system are Rs.

Table 2 Net present cost of proposed system at Agatti Island

Components	Capital cost	Replacement cost	O&M cost	Fuel cost	Salvage value	Total cost
	(₹)	(₹)	(₹)	(₹)	(₹)	(₹)
100 kW Genset	5,50,000.00	0.00	12,10,015.55	51,52,065.81	−84,446.88	68,27,634.49
10 kW Genset	2,00,000.00	0.00	78,211.48	2,66,319.56	−31,993.92	5,12,537.12
1 MW Genset	45,00,000.00	0.00	1,34,76,936.00	12,14,61,849.44	−4,40,511.15	13,89,98,274.29
1 MW Genset	45,00,000.00	0.00	66,99,685.45	5,50,53,818.36	-5,20,158.67	6,57,33,345.14
1MWh Li-Ion	30,00,00,000.00	12,72,82,148.89	25,85,503.31	0.00	−2,39,55,785.01	40,59,11,867.19
PV	32,94,94,559.53	0.00	79,866.49	0.00	0.00	32,95,74,426.03
Converter	4,00,80,064.60	11,33,661.50	0.00	0.00	−2,13,366.54	4,10,00,359.57
System	67,93,24,624.13	12,84,15,810.39	2,41,30,218.28	18,19,34,053.17	−2,52,46,262.16	98,85,58,443.81

Table 3 Comparison of emission of the hybrid system and diesel-only system

Pollutants	Emission due to diesel-only generator (kg/year)	Emission due to diesel–PV–battery (kg/year)	Reduction in emission (kg/year)
Carbon dioxide	5,619,560	542,658	5,076,902
Carbon monoxide	29,412	2834	26,578
Unburned hydrocarbons	1543	149	1394
Particulate matter	254	24	230
Sulfur dioxide	13,738	1327	12,411
Nitrogen oxides	5640	537	5103

1.01 B and Rs. 9.94/kWh. It says that if a hybrid system is used instead of a diesel-only system, there will be a significant reduction in NPC and COE. If we use a hybrid approach in place of only a diesel system, CO_2 emission will be reduced from 5,622,833 kg per year to 442,194 kg per year, which is a significant reduction.

7 Conclusions

The sensitivity and optimization results are shown in Fig. 4. The software simulates the proposed model for all possible combinations to find the best result. Diesel–PV–battery is the best combination for optimal results. This proposed combination gives the lowest NPC and COE, ₹989 million and 9.71 ₹/kWh, respectively. The fuel rate was included as a sensitivity variable in the sensitivity analysis. Taking a diesel price of ₹72/L, the above is the best solution, i.e., the lowest NPC of ₹989 million and the least COE of 9.71 ₹/kWh were determined. The PV–diesel–battery system is economical, as shown in Fig. 4. Table 2 gives the NPC and cost summary, including operating cost, capital cost, salvage value, replacement cost, and resource cost. The highest cost is capital cost, which is ₹988 million and includes the price of Gen sets (₹212 million), Li-ion batteries (₹40 million), PV capital cost (₹32 million), and converter capital cost (₹4 million). The entire running cost is ₹988million, the total replacement cost is ₹12 million, and the total salvage value is ₹–2 million.

Table 4 Comparative analysis of diesel-only system and optimal hybrid system

Sensitivity variable					DG only system (base system)				Hybrid system			
Diesel Price ₹/L	Running time of DG set (minutes)											
	1 MW	1 MW	100 kW	10 kW	NPC	COE (₹/kWh)	Ren fraction (%)	Emission CO_2 (kg/year)	NPC (₹)	COE (₹/kWh)	Ren fraction (%)	Emission CO_2 (kg/year)
68	900	900	840	780	2.06B	20.2	0	5,623,790	995 M	9.77	90.4	568,685
72	840	900	780	780	2.16B	20.2	0	5,619,629	1B	9.86	90.4	566,957
76	840	900	780	660	2.28B	22.4	0	5,622,833	1.01B	9.94	91	442,194

References

1. BP Energy Outlook (2019) The energy outlook explores the forces shaping the global energy transition out to 2040 and the key uncertainties surrounding that
2. Legros G, Havet I, Bruce N, Bonjour S (2009) The energy access situation in developing countries. USA. https://doi.org/10.1503/cmaj.160656
3. Manaullah, Ansari MS (2014) Solar photo voltaic power generation in union territory of Lakshadweep Island: projected level dissemination using technology diffusion models. In: 2014 innovative applications of computational intelligence on power, energy and controls with their impact on humanity (CIPECH). IEEE, Ghaziabad, India, pp 395–399. https://doi.org/10.1109/CIPECH.2014.7019036
4. Abd Rahman NA, Kamaruzzaman SN, Akashah FW (2022) A review of optimization techniques application for building performance analysis. Civil Eng J 8:823–842. https://doi.org/10.28991/CEJ-2022-08-04-014
5. Nasiruddin I, Khatoon S, Jalil MF, Bansal RC (2019) Shade diffusion of partial shaded PV array by using odd-even structure. Sol Energy 181:519–529. https://doi.org/10.1016/j.solener.2019.01.076
6. Ansari MS, Manaullah, Jalil MF (2016) Investigation of renewable energy potential in union territory of Lakshadweep islands. In: 2016 second international innovative applications of computational intelligence on power, energy and controls with their impact on humanity (CIPECH). IEEE, Ghaziabad, India, pp 209–213. https://doi.org/10.1109/CIPECH.2016.7918768
7. Elsen C, Darses F, Leclercq P (2011) An Anthropo-based standpoint on mediating objects: evolution and extension of industrial design practices. Paris, France
8. Manaullah, Ansari MS (2019) Feasibility analysis of renewable energy options for the union territory of Lakshadweep Islands. Int J Glob Energy Issues 42:63. https://doi.org/10.1504/IJGEI.2019.100690
9. Feldman D, Barbose G, James T, Weaver S, Fu R, Davidson C (2015) Photovoltaic system pricing trends (2015). NREL/PR-6A20-64898
10. Manaullah, Ansari MS (2019) Techno-economic feasibility of PV-wind-diesel battery hybrid energy system for Lakshadweep Island in India. Indian J Power River Val Dev 69:141–148 (2019)
11. NREL (2015) Hybrid optimization model for electrical renewable (HOMER)
12. Roy P, He J, Zhao T, Singh YV (2022) Recent advances of wind-solar hybrid renewable energy systems for power generation: a review. IEEE Open J Ind Electron Soc 3:81–104. https://doi.org/10.1109/OJIES.2022.3144093
13. Ansari MS (2022) Feasibility analysis of standalone hybrid renewable energy system for Kiltan Island in India. In: Kumar A, Srivastava SC, Singh SN (ed) Renewable energy towards smart grid. Springer, Singapore, Singapore, pp 79–93. https://doi.org/10.1007/978-981-16-7472-3_7

Access to Solar Energy for Livelihood Security in Odisha, India

Nimay Chandra Giri, Sima Das, Divya Pant, Vikas Singh Bhadoria, Debani Prasad Mishra, Gyanranjan Mahalik, and Rachid Mrabet

Abstract Fossil fuel-based energy generation contributes to increased greenhouse gas emissions, leading to climate change in both developed and developing countries such as the USA, China, India, Indonesia, and Bangladesh. In India, around 200 million people are deprived of a clean power supply by 2021, highly in hilly and rural locations of the country. The above issues can be mitigated by clean and affordable electricity access for the development of livelihood in the state of Odisha, India. In this present research, a few major off-grid systems such as photovoltaic lighting systems and water pumps have been designed and implemented in urban and rural locations of Odisha to enhance the livelihood security of the people. In a mini solar street light, a 20 Wp polycrystalline solar panel has been used to charge a 12 V 10 Ah

N. C. Giri
Department of ECE & CREE, Centurion University of Technology and Management, Bhubaneswar, Odisha 752050, India

S. Das
Department of CSE, Camellia Institute of Technology and Management, Bainchi, West Bengal 712134, India
e-mail: simadas@ieee.org

D. Pant
Department of Plant Science, Pennsylvania State University, University Park, PA 16802, USA

V. S. Bhadoria (✉)
Shri Vishwakarma Skill University, Palwal, Haryana 121102, India
e-mail: vikasbhadoria@gmail.com

D. P. Mishra
Department of EE, IIIT Bhubaneswar, Odisha 751003, India
e-mail: debani@iiit-bh.ac.in

G. Mahalik
Department of Botany, Centurion University of Technology and Management, Bhubaneswar, Odisha 752050, India
e-mail: gyanranjan.mahalik@cutm.ac.in

R. Mrabet
Department of A&AM, National Institute of Agricultural Research "INRA", 60000 Rabat, Morocco
e-mail: rachid.mrabet@inra.ma

A. Rani et al. (eds.), *Signals, Machines and Automation*, Lecture Notes in Electrical Engineering 1023, https://doi.org/10.1007/978-981-99-0969-8_23

Li-ion battery in 6–7 sunny hours and can run a 12 V 9 W LED light up to 10 h per a day. Further, the efficient design and access of different off-grid systems will fulfill the renewable energy targets in the country. This prospect will help both academia and industry experts in the research and development of eco-friendly solar off-grid systems worldwide.

Keywords Solar energy · Photovoltaic (PV) · Off-grid system · Livelihood

1 Introduction

The energy demand is accelerating exponentially to fulfill human needs. Currently, 78% of the energy demand is reached by fossil fuels and leads to carbon dioxide emissions in the world [1, 2]. This issue can be avoided by the help of solar energy-based power generation. Solar energy is a renewable, affordable, and abundant form of energy in the earth. The solar energy that we see is light, and the energy we feel is heat. The light energy is used by solar photovoltaic panel(s) to convert electrical energy, whereas heat energy used by solar thermal system to produce heat or electricity for further applications. In the recent years, the world is speedily moving toward generating electricity from solar photovoltaic (SPV) system to bridge the gaps between energy demand, supply, and production of the country [3, 4]. Solar energy-based systems has ability to fill-up Sustainable Development Goal 7 (SDG7) by 2030 [5, 6].

Around 5000 trillion kilowatt-hours (kWh) of energy is incident on India's land area in a year. In most parts of the country, clear sunny weather is experienced 280 to 310 days/year [4]. Till March 2021, the solar power installed capacity was 40 GW in India against the target of 100 GW (including 40 GW rooftop) by 2022 [7]. Currently in the state Odisha, around 550 MW of on-grid and 1.5 MW off-grids solar photovoltaic systems have been installed [4]. The SPV system includes solar panel(s), charge controller (CCR) or inverter, battery, and electrical appliances. The off-grid systems are not fed to the normal supply grid [8, 9]. The power generated can be stored in batteries for further applications. The solar emergency and street lighting system consists of suitable solar panel, solar charge controller, battery, LED light, mechanical, and civil structure [10]. The designer or stakeholder or service provider needs to select suitable devices for the making of efficient systems [11, 12]. In solar power plants, maximum power point tracking (MPPT) is also used [13].

In this research, both solar lighting and pumping systems have been designed and installed in urban and rural locations of Odisha to enhance the livelihood security of the people. Basically, solar emergency and street lights are available in 5–100 W to provide low–high intensity of light for different applications such as studying, lighting room, and walking in road. The Odisha Renewable Energy Development Agency (OREDA) has invited bids to develop 700 MW of ground-mounted solar pump projects under the Pradhan Mantri Kisan Urja Suraksha evam Utthaan Mahabhiyan (PM KUSUM) scheme [4]. Under this scheme, farmers can get subsidy of

30–60% for the installation of solar pumping system in the country. The affordable electricity access future will require new policy, awareness, and new energy systems.

2 Materials and Methods

A SPV system is an electrical power system used to convert solar energy to usable electrical energy. Basically, solar PV systems are two types, namely on-grid and off-grid systems. On-grid systems are SPV systems that only generate power when the power grid is accessible. Off-grid systems allow storing the solar power in batteries for use when the power grid goes down [4, 6].

2.1 Solar PV System Components

The SPV system composed of solar module, charge controller (CCR) or inverter (DC-AC), solar battery, and load (LED light) as illustrated in Fig. 1.

From the above said block diagram, a SPV system consists of following components;

Solar PV Cell and Module

The majority solar cell is made up by silicon material to produce electricity [1, 6]. A solar panel is composed of more than one solar cell, aluminum frame, back sheet, connectors, junction box, and connecting wire, as illustrated in Fig. 2. To achieve an essential voltage and current, a group of PV modules are underwired into big array that called PV array [3].

Mathematically,

$$I = I_{\mathrm{L}} - I_{\mathrm{D}} \quad (1)$$

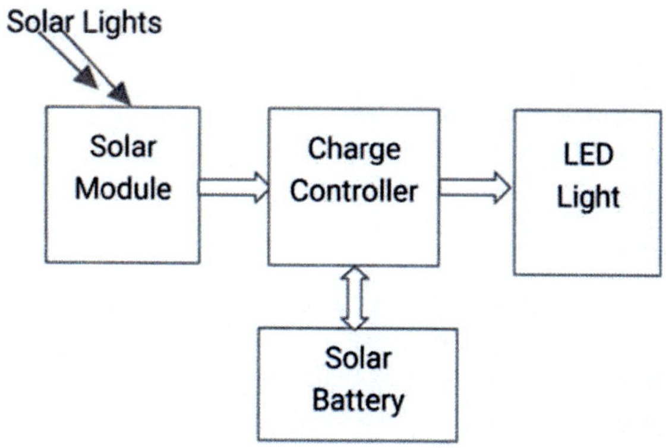

Fig. 1 Basic block diagram of a SPV system

Fig. 2 Solar PV module (5 W)

$$I = I_{\mathrm{L}} - I_0\left[\exp\left(\frac{eV}{KT}\right) - 1\right] \quad (2)$$

where

I	electric current,
I_{L}	solar light generated current,
I_{D}	diode current,
I_0	saturation current,
e	electron charge,
V	voltage across the junction,
K	Boltzmann's constant,
and T	absolute temperature.

Solar Charge Controller

A solar charge controller controls the flow of current within the SPV system. It is a DC device with three pair of terminals for the connection of solar module, battery, and load as given in Fig. 3. For domestic application, pulse width modulation (PWM) type charge controller is used, whereas maximum power point tracker (MPPT) is used in industrial applications [8].

Solar Battery

A solar battery is an electrical device that stored electrical energy for later use and which is charged by a solar module [9]. It is rated in volt ampere-hour (V Ah). Basically, three types of batteries such as lead acid, maintenance free, and li-ion battery are widely used in the market as illustrated in Fig. 4.

Fig. 3 Solar charge controller or regulator (PWM type)

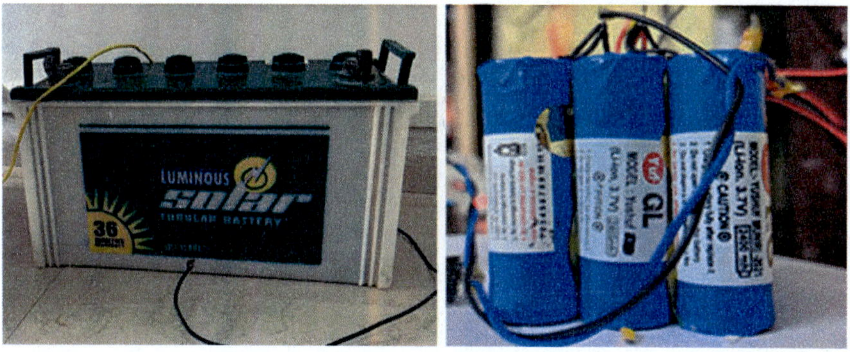

Fig. 4 Solar tubular (12 V 45 Ah) and Li-ion battery (12 V 10 Ah)

3 Results and Discussion

Access to clean energy is a very vital for the wellbeing of the people as well as economic development and poverty alleviation in the country. The design, connection practice, and applications with affordable energy access system (lighting system) are designed and installed in Centurion University of Technology and Management, Odisha, as given in Fig. 5.

The performance of a solar off-grid system directly depends upon the solar irradiation incident on the panels. From the field research, it is found that the average irradiation falling on the panels is highest (6.89 kWh/m^2/day) and lowest (3.90 kWh/m^2/day) in May and December, respectively. It is observed that a 20 Wp polycrystalline solar panel can charge a 12 V 10 Ah Li-ion battery in 6–7 sunny hours and the 12 V 9 W LED light will run up to 10 h per a day.

The government of Odisha also promoting SPV-based irrigation system under scheme of Soura Jalanidhi and Kusum Yojana [4, 7]. These scheme's objective is to

Fig. 5 Solar off-grid systems (lighting system) at CUTM, Odisha

use of green energy, reduce carbon emission and ensure that farmers have adequate electricity for farming activities. Under the scheme, government distributed 5000 solar pumps to the farmers in the state. The solar water pumping systems capacity of 0.5 HP and 1 HP are displayed in Fig. 6. It consists of solar panel, DC motor, lead acid battery (12 V 45 Ah), and water tank and plastic pipes.

The cost of solar pumps depends upon its design, components, and installations. The low cost pumps as per local market price in Odisha are 0.5 HP direct current pump cost of USD 550 to 590 including installation without subsidy, 1 HP direct current pump cost of USD 1000 to 1160. Therefore, the energy crisis can be overcome by efficient use of SPV system in the country.

Fig. 6 0.5 HP and 1 HP solar water pumping system at CUTM, Odisha

4 Conclusion

Solar energy has been rising as an efficient renewable and alternative energy source for power production in any location of the world. The SPV systems are used to convert solar energy to usable electricity. In this study, the design and implementation practice of solar lighting, drinking, and irrigation systems has been analyzed in CUTM, Odisha, India. Based on the empirical study and observation, we found that the solar off-grid livelihood systems are easy to design, implement, and operate in any place of Odisha. The solar mini street lights' battery takes 6–7 h of sunlight to fully charge the battery and can deliver power up to 10 h per day. A 0.5 HP solar drinking system is operated by a 75 Wp solar panel with a 12 V 10 A MPPT charge controller. Similarly, a 1 HP solar water pump for irrigation is operated up to 7 h/sunny day in the renewable energy laboratory of CUTM, Odisha. The above design, analysis data, and government scheme will help to fulfill the government targets and enhance the livelihoods of peoples.

References

1. Akter S, Bagchi K (2021) Is off-grid residential solar power inclusive? Solar power adoption, energy poverty, and social inequality in India. Energy Res Soc Sci 82:102314
2. Giri NC, Shah KK, Ray SS, Vennila H, Das S, Sandhya YB, Pradhan I (2022) Photovoltaic pumping system vs livelihoods and sustainability. Ambient Sci 9(1):27–30
3. Giri NC, Mishra SP, Mohanty RC (2020) Performance parameters, optimization, and recommendation in large scale on-grid SPV power plant, Odisha, India. Int J Mod Agric 9(4):159–167
4. Government of India (GoI) (2021) Physical progress report 2021–22. Ministry of New and Renewable Energy (MNRE). Retrieved from: https://mnre.gov.in/the-ministry/physical-progress
5. Government of India and the United Nations (UN): Sustainable development framework. 2018–2022. Available at: https://in.one.un.org/wp-content/uploads/2019/05/UNSDF_Print-Revised-To-be-Approved-low-res.pdf
6. IRENA (2019) Global energy transformation: a roadmap to 2050. Available at: https://www.irena.org/
7. Government of India (GOI) (2018) National Action Plan on Climate Change Government of India (2018). Available at: http://pmindia.nic.in/climatechange.htm
8. Pani BB, Giri NC, Nayak SR, Mishra SP (2021) Fault detection and troubleshooting in a PV grid-tied inverter. Indian J Sci Technol 14(x):1–10
9. Renewables: "global status report" by renewable energy policy network for the 21st century (REN21) (2019). Available at: https://www.ren21.net/gsr-2019/chapters/chapter_01/chapter_01/
10. Shrimali G, Sen V (2020) Scaling reliable electricity access in India: a public-private partnership model. Energy Sustain Dev 55:69–81
11. Yadav D, Singh N, Bhadoria VS (2021) Comparison of MPPT algorithms in stand-alone photovoltaic (PV) system on resistive load. In: Machine intelligence and smart systems. Springer, Singapore, pp 385–400. https://doi.org/10.1007/978-981-33-4893-6_34

12. Yadav RK, Bhadoria VS, Hrisheekesha PN (2019) Technical and financial assessment of a grid connected solar PV net metering system for residential community. In: 2019 2nd international conference on power energy, environment and intelligent control (PEEIC). IEEE, pp 299–303. https://doi.org/10.1109/PEEIC47157.2019.8976657
13. Govindharaj A, Mariappan A, Ambikapathy A, Bhadoria VS, Alhelou HH (2021) Real-time implementation of adaptive neuro backstepping controller for maximum power point tracking in photo voltaic systems. IEEE Access 9:105859–105875. https://doi.org/10.1109/ACCESS.2021.3099158

Grid-Connected Fuel Cell with Upgraded Voltage Profile

Purvaa Saxena and S. K. Jha

Abstract In this work, an electrical network connected to proton exchange membrane fuel cell (PEMFC) is considered. The voltage profile of the system is enhanced by connecting the fuel cell with the boost converter and a seven-level cascaded multilevel inverter. The traditional PID controller does not responds to the nonlinearities of the system. Hence, PID controller is dynamically tuned by the Whale Optimization Algorithm (WOA). The whole architecture is developed in SIMULINK. For testing the effectiveness of the proposed model, the transients are given to the system for short duration and the results are then compared. It is observed that the proposed model has improved voltage response and reduced harmonics. The analysis for the power quality enhancement is confirmed through Total Harmonic Distortion (THD).

Keywords Whale Optimization Algorithm (WOA) · Proton exchange membrane fuel cell (PEMFC) · Boost converter (BC) · Multilevel inverter (MLI) · Cascaded H-Bridge (CHB) · Total Harmonic Distortion (THD)

1 Introduction

Fuel cell is a device responsible for the conversion of chemical energy of hydrogen to electricity [1]. There are various types of FC that can be used in electrical grid. However, PEMFC has been used as energy source in this paper due its high-power density, light weight and less starting time [2].

Conventional PID controller is a linear device which is used to minimize the error of the system [3]. The main drawback of the conventional PID controller is that it does not respond to the nonlinearities of the system. So, to encounter this, the parameters of the PID controller are dynamically tuned by various algorithms such as PSO [4], Genetic Algorithm [5], Bat Search [6] and evolutionary algorithms

P. Saxena (✉) · S. K. Jha
Department of Instrumentation and Control Engineering, Netaji Subhas University of Technology, New Delhi, India
e-mail: purvaas.ie20@nsut.ac.in

A. Rani et al. (eds.), *Signals, Machines and Automation*, Lecture Notes in Electrical Engineering 1023, https://doi.org/10.1007/978-981-99-0969-8_24

[7]. These algorithms have some shortcomings such as slow convergence speed and precision. So, to overcome these issues, WOA has been implemented in this paper.

Furthermore, the low-level output voltage of the fuel cell is stepped up by the boost converter. The DC voltage of the boost converter is converted to the AC voltage using a three-phase seven-level cascaded multilevel inverter. An MLI gives better system stability along with decreased harmonics. Also, the THD values is reduced by this model. The gate signal to the inverter is given through the PID controller. The parameters of the PID controller are dynamically tuned using WOA algorithm. It has been observed that this controller gives better voltage characteristics and improved efficiency.

This paper shows PEMFC-based grid connected model. The model has been developed in MATLAB/SIMULINK. The proposed model promises to improve the voltage profile during the transient stage. It has been observed that this controller gives better voltage characteristics and improved efficiency when the system is subjected to transients.

2 Mathematical Modelling

2.1 *Modelling of PEMFC*

This section deals with the electrochemical model of the PEMFC. PEMFC has more advantages over other FC such as it has less starting time, design is compact and low cost. So, PEMFC has been considered in this paper.

A mathematical model is used to predict the dynamic behaviour of the PEMFC. The V_{FC} of the fuel cell is given by:

$$V_{FC} = E_{Nernst} - V_{act} - V_{ohm} - V_{con} \tag{1}$$

where E_{Nernst} is thermodynamic potential and represents reversible voltage, V_{act} is activation voltage, V_{ohmic} is ohmic voltage drop and V_{con} represents concentration losses [9].

$$E_{Nernst} = 1.229 - 0.85 \times 10^{-3}(T - 298.15) + \frac{R \cdot T}{2 \cdot F} \ln\left(P_{O_2}^{0.5} P_{H_2}\right) \tag{2}$$

$$V_{act} = -[\xi_1 + \xi_2 \cdot T_3 \cdot T \cdot \ln(CO_2) + \xi_4 \cdot T \cdot \ln(I_{FC})] \tag{3}$$

$$V_{ohm} = I_{fc}(R_M + R_C) \tag{4}$$

$$V_{conc} = -B \ln\left(1 - \left(\frac{J}{J_{max}}\right)\right) \tag{5}$$

where P_{O_2} and P_{H_2} are the partial pressures (atm) of oxygen and hydrogen, T is the absolute cell temperature (K), I_{fc} is the cell operating current (A), ξ_i $(i = 1\ldots0.4)$ represents the parametric coefficients for each cell model, R_M is equivalent membrane resistance to proton conduction, R_C is the equivalent contact resistance to electron conduction, J_{max} is then maximum current density and J is actual current density.

2.2 *Modelling of Boost Converter*

Boost converter is a DC-to-DC power electronic converter. It is used to enhance the input voltage. It is also known as step up chopper. The switching of the switch is controlled by the PWM signal. The model consists of input voltage source (V_s), diode (D), inductor (L), switch (S) and capacitor (C). When the switch is in ON state, it offers low resistance path to the current. So, the current flows through switch and back to the source. Meanwhile, the inductor stores the energy. When the switch is in OFF state, the inductor reverses its polarity and the stored energy of inductor is dissipated through the load [8].

3 Control Structure

3.1 *PID Controller*

Conventional PID controller is a linear device which is used to minimize the error of the system. The most commonly used controller is the PID controller. PID has three basic control behaviours which are proportional, integrator and derivative. The controller has three parameters, namely K_p, K_i and K_d. The parameters of the controller are then dynamically tuned by WOA.

3.2 *Seven-Level Cascaded Multilevel Inverter*

MLI are the widely used inverters. One of the most common MLI is CHB inverter as it gives better signal quality with reduced THD. The seven-level CHB MLI gives seven steps in the output voltage waveform. If the input voltage is V_{DC} then output voltage will be 3 V_{DC}, 2 V_{DC}, V_{DC}, 0, $-V_{DC}$, -2 V_{DC} and -3 V_{DC} [9].

3.3 Whale Optimization Algorithm

WOA is based on the foraging behaviour of the humpback whales. Foraging behaviour refers to the technique of bubble making. Since whales have slow speed so they cannot catch the fishes. So, they came up with a technique, wherein they form bubbles for catching the fishes. The bubble is basically created in a spiral format around the fishes. This forces the fishes to move towards the surface and the radius of the spiral keeps decreasing [10]. The mathematical model is given as-

(A) Random Prey: It forces the search agent to move away from the present location. It moves randomly in the space for searching. The formula is

$$\vec{D} = \left|\vec{C} \times \overrightarrow{P_{\text{rand}}} - \overrightarrow{P_l^t}\right| \tag{6}$$

$$\overrightarrow{P_l^t} = \overrightarrow{P_{\text{rand}}} - \vec{A} \times \vec{D} \tag{7}$$

where $\vec{P}$ is the various position vectors.

(B) Encircling prey: When $p < 0.5$ and $\left|\vec{A}\right| \leq 1$, it finds the prey and surrounds it. It is given by

$$\vec{D} = \left|\vec{C} \times \overrightarrow{P_*^t} - \overrightarrow{P_l^t}\right| \tag{8}$$

$$\overrightarrow{P_l^{t+1}} = \overrightarrow{P_l^t} - \vec{A} \times \vec{D} \tag{9}$$

where $\overrightarrow{P_*^t}$ is the position vector of tth generation.

(C) Bubble-net attacking: When $p \geq 0.5$, it calculates the distance between search agent and optimal search agent. It then calculates the spiral model to update the position of the search agent.

$$\overrightarrow{P_l^{t+1}} = \overrightarrow{D^l} \cdot e^{bl} \cdot \cos(2\pi l) + \overrightarrow{P_*^t} \tag{10}$$

The pseudocode for WOA is given as

Initialize the whale's population
Set algorithm parameters
Set performance index
Calculate fitness of all search agent
While (termination is not satisfied)
Encircle prey
Search the prey

(continued)

(continued)
Compute the fitness by index
End While
Return the best result
End

4 MATLAB/Simulink Model

In this work, PEMFC is used. PEMFC is coupled with boost converter which is used for increasing the FC output. The boost converter is further connected to a three-phase seven-level CHB MLI which inverts the DC voltage into seven-level AC voltage. The gate signal of the inverter is given through the PID controller. In order to evaluate the performance of the power network, the system is subjected to oscillatory transient fault ($t = 0.3$–0.4 s). The obtained results show that proposed technique gives better results than the conventional PID controller (Fig. 1 and Table 1).

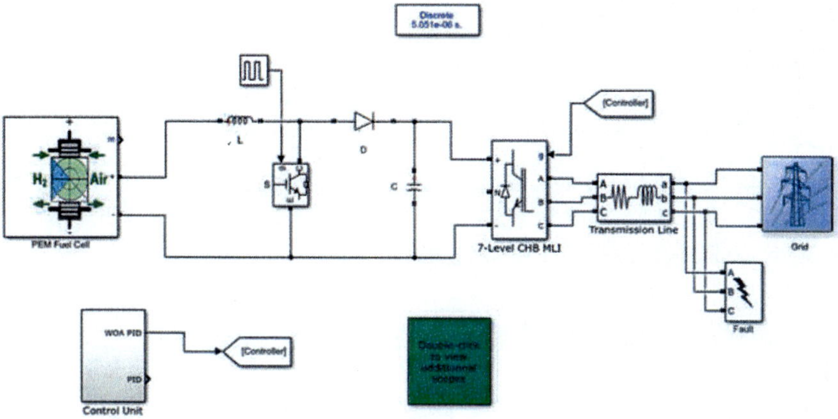

Fig. 1 Simulink model of the proposed system

Table 1 Values of different parameters

Parameters	Values
SOFC	$V_o = 60$ V, $R_1 + R_2 = 1.4\ \Omega$, $R_3 = 1.06\ \Omega$, $C = 1.25$ F
Boost	L = 0.8 mH, c = 100 µF, $F_s = 30$ KHz, $V_o = 100$ V, $V_{in} = 40$ V
Grid	$v = 600$, $x/r = 7, f = 50$ Hz

5 Results

Figures 2 and 3 depict the voltage waveforms of the power network during faults for PID and WOA optimized PID, respectively. It is seen from the figures that the harmonic content has been reduced tremendously when the parameters of the PID controller are tuned with WOA algorithm. Moreover, the system tends to become more stable.

Figure 4 show the THD values for PID and WOA-PID. The THD values obtained for PID and the proposed model are 9.08% and 1.15%, respectively. It is seen that the proposed model is more effective in reducing the harmonic contents and gives transient stability.

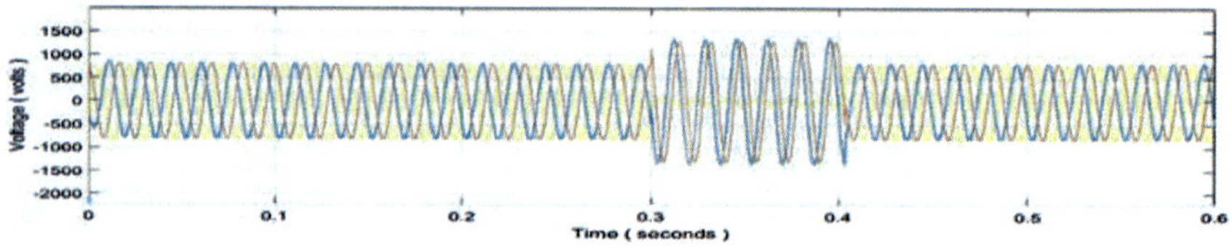

Fig. 2 Voltage profile during fault for PID

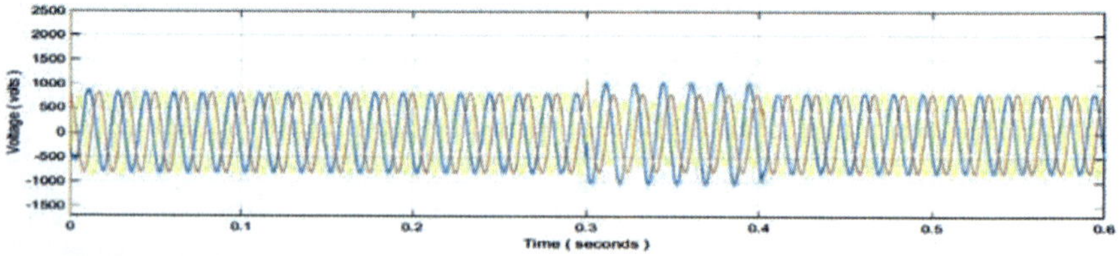

Fig. 3 Voltage profile during fault for WOA optimized PID

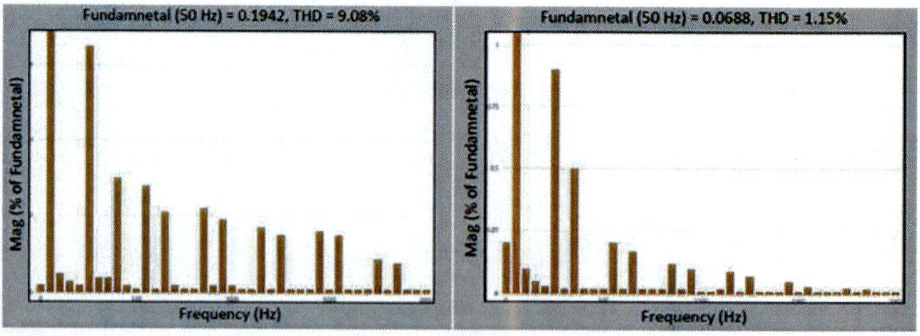

Fig. 4 THD of the voltage for PID and WOA optimized PID, respectively

6 Conclusion

In this paper, a fuel cell connected grid network is proposed, wherein the parameters of the PID controller are dynamically tuned with WOA algorithm. Due to dynamic tuning of the parameters, the PID controller can robustly respond to the nonlinearities of the system. The proposed technique compares the conventional PID controller with the WOA-PID controller. The system is given transients for short duration. By comparing the results, it is observed that during the transients fault, the proposed model is more effective in reducing the harmonic contents and gives transient stability. Moreover, the WOA optimization technique gives better power quality and voltage profile with reduced peak time and rise time thereby, improving the dynamic profile of the system. So, the fuel cell connected grid stands to be a potential candidate for generating electricity in near future.

References

1. Akinyele D, Olabode E, Amole A (2020) Review of fuel cell technologies and applications for sustainable microgrid systems. In: Inventions, pp 5–42
2. Kavya VR, Padmavathy KS, Shaneeth M (2013) Steady state analysis and control of PEM fuel cell power plant. In: International conference on control communication and computing (ICCC), pp 233–237
3. Chopra V, Singla SK, Dewan L (2014) Comparative analysis of tuning a PID controller using intelligent methods. ACTA Polytech Hung, 235–249
4. Ruan J, Zhang B, Liu B, Wang S (2021) The multi-objective optimization of cost, energy consumption and battery degradation for fuel cell-battery hybrid electric vehicle. In: 2021 11th international conference on power, energy and electrical engineering (CPEEE), pp 50–55
5. Lianfeng L, Ting-Cheng C (2019) Parameter optimization of hybrid fuel cell system based on genetic algorithm. In: 2019 IEEE Eurasia conference on biomedical engineering, healthcare and sustainability (ECBIOS), pp 102–104
6. Yang XS (2010) A new metaheuristic bat -inspired algorithm. In: Gonzalez JR et al (eds) Nature-inspired cooperative strategies for optimization (NICSO2010). Springer, SCI 284, pp 65–74
7. Buchaniec S, Sciazko A, Mozdzierz M, Brus G (2019) A novel approach to the optimization of a solid oxide fuel cell anode using evolutionary algorithms. IEEE Access 7:34361–34372
8. Hu X, Ma P, Wang J, Tan G (2020) A hybrid cascaded DC–DC boost converter with ripple reduction and large conversion ratio. IEEE J Emerg Sel Top Power Electron 2020:761–770
9. Abhilash T, Kirubakaran A, Somasekhar VT (2017) A novel three-phase seven-level inverter. In: Innovations in power and advanced computing technologies (i-PACT), pp 1–5
10. Zhai Q, Xia X, Feng S, Huang M (2020) Optimization design of LQR controller based on improved whale optimization algorithm: In: 2020 3rd international conference on information and computer technologies (ICICT), pp 380–384

Performance Analysis and Comparison of a Solar Tree with Stand-Alone System

Saurabh Chaudhary, Vinaya Rana, and Nidhi Singh

Abstract In this paper, the design and performance analysis of a 50 W solar tree with a cost-effective installation is analyzed at Basti, Uttar Pradesh, India. The energy demand is increasing day by day and the fossil fuels are decreasing. So, an alternative option is required to fulfill the energy demand. The renewable energy-based source is best option to fulfill the energy scarcity. Solar tree is designed in a way that there is no effect of shading on the panel, also panels are rearranged according to azimuth angle. The technical and economical assessment of PV solar tree results a future adoptive technology for highly populated area. Solar PV systems are highly cost-effective and helps in the reduction of greenhouse gases. Five PV panels of 10 W are used to form the solar tree and compered with a single 50 W stand-alone PV panel. The maximum power and efficiency of the solar tree is more as compared to stand-alone PV panel with all same conditions, i.e., irradiation level temperature.

Keywords Solar PV system · Solar tree · Renewable energy · Stand-alone system

1 Introduction

Energy consumption is increasing day by day. The rising demand puts pressure on natural resources that leading to global warming. The reduction in electricity generation has forced the adoption of alternative and sustainable electricity generation all over the world. Renewable energy (RE) technologies are one of the best and sustainable sources of energy. The total renewable energy generation capacity in the country is estimated at 105,854 MW as of February 2022, which includes solar power, wind power, small hydro, biomass power and power from oil mills and waste. Energy consumption is currently highest in urban areas [1].

Nowadays, the RE trend is increasing with the initiatives and policies offered by the Indian government. The solar PV system can be designed as either a stand-alone grid-connected system or both [3–6]. In stand-alone PV system, a battery storage

S. Chaudhary · V. Rana (✉) · N. Singh
EED, Gautam Buddha University, Greater Noida, UP 201312, India
e-mail: vinayrana80@gmail.com

A. Rani et al. (eds.), *Signals, Machines and Automation*, Lecture Notes in Electrical Engineering 1023, https://doi.org/10.1007/978-981-99-0969-8_25

system is required to fulfill the energy demand day and night. The average lifespan of solar PV panel is around 20–25 years and the payback period are estimated to be 9 years. In the grid-connected solar PV system, the system directly connected to the grid so that the extra energy can be transferred into gird through net metering [2–5]. India a highly populated country that ranks 2nd in world population. Since population density is high in metro cites, which consist of high-rise buildings, land area required for a traditional stand-alone system is scarce, hence, installation of solar trees is the appropriate option as it requires significantly lesser area than the stand-alone system.

2 Design of Solar Tree

In this paper, analysis of 50 W solar tree with stand-alone system has been done. A solar tree is constructed using following components:

(a) Solar panel
(b) Long pole
(c) LEDs (for load)
(d) Small pole (for branches of tree).

A five feet long solar tree is design and installed on the rooftop with 5 branches. Each branch consists of 10 W panel and all panels are connected in parallel to fulfill the load demand. The performance of the solar tree is analyzed on the basis irradiation level change in real time.

2.1 Solar Tree Stem and Branches

The support structure for the solar panel in the solar PV tree can be similar to the stem design of a natural tree. The leaves of solar tree are arranged in a "phyllotaxy" pattern. The solar tree stem is fixed at one place then branches of the tree are welded in it properly. After that the solar panel is installed on each branch of the tree at the same azimuth angle 23° so that the maximum sun intensity is achieved on the solar tree [6, 7]. The solar tree consists of 5 branches, four of which are project outwards in 4 directions and one is in the center of the pole. The load is connected with 8 LED bulbs with the rating of 7 W and 12 V DC each. The component required in solar tree is given in Table 1.

Table 1 Components rating for solar tree

S. No.	Component name	Quantity	Rating
1	Solar PV panel	5	10 W, 12 V each
2	LEDs bulb	8	7 W, 12 V each
3	Inferred thermometer	1	−50 to 550 °C
4	Multimeter	2	0–10 A DC, 0–1000 V DC
5	Lux meter	1	0–200,000 lx

Table 2 Rating of solar panel of solar tree and stand-alone system

S. No.	Parameter	Solar tree	Stand-alone
1	Model no.	12 V/10 W	12 V/50 W
2	Power max (P_m)	10 Wp	50 W
3	Open-circuit voltage (V_{oc})	21.65 V	21.56 V
4	Voltage at P_m (V_m)	17.5 V	17.6 V
5	Short-circuit current (I_{sc})	0.61 A	3.11 A
6	Current at P_m (I_m)	0.57 A	2.89 A
7	Fuse rating	2 A	6 A
8	Panel cell area	812.5 cm^2	3898.68 cm^2
9	Quantity of panel	5 panel	1 panel
10	Total cell area	0.406 m^2	0.3898 m^2

2.2 *Stand-Alone System*

A 50 W solar panel is used for this analysis. The azimuth angle is 23°. The solar module is installed on the rooftop adjacent to the solar tree with same irradiation level and same temperature (Table 2).

3 Modeling of Solar PV Panel

The various PV cells connected in series and parallel to produce specified output power in a solar PV panel. The equivalent diagram of a cell is shown in Fig. 1 as:

The output current and voltage of the PV cell are expressed [7, 8] in (1), (2) and (3), respectively.

$$I_{pv} = I_{ph} - I_d - I_{sh} \tag{1}$$

Fig. 1 Equivalent circuit of a solar cell

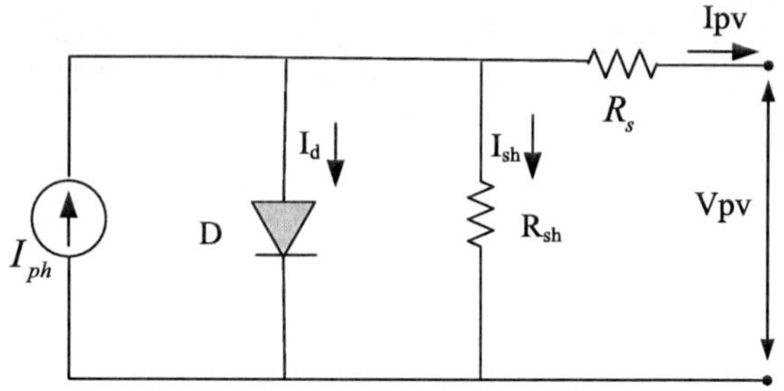

$$I_{pv} = I_{ph} - I_0\left[\exp\left(\frac{q(V_c + R_s I_{pv})}{AkT_c}\right) - 1\right] - \left(\frac{V_{pv} + R_s I_{pv}}{R_{sh}}\right) \tag{2}$$

$$V_{pv} = \frac{AkT_c}{q}\left(\frac{I_{ph} + I_0 - I_{pv}}{I_0}\right) - R_s I_c \tag{3}$$

The short-circuit current and open-circuit voltages of a PV cell are in (4) and (5) as,

$$I_{sc} = I_{ph} - I_0\left[\exp\left(\frac{qR_s I_{sc}}{AkT_c}\right) - 1\right] - \left(\frac{R_s I_{sc}}{R_{sh}}\right) \tag{4}$$

$$V_{oc} = \frac{AkT_c}{q}\ln\left(\frac{I_{ph} + I_0}{R_{sh}}\right) \tag{5}$$

where T_c—temperature (K), q—electron charge (1.602×10^{-19} C), I_d—diode current (A), I_o—reverse saturation current of diode (A), I_{sc}—short-circuit current, I_{ph}—photo current (A), k-Boltzmann constant (1.38×10^{-23} J/K) and A is ideality factor (1.2).

$$\text{Fill factor} = \frac{V_m \times I_m}{V_{oc} \times I_{sc}} \tag{6}$$

$$\text{Input Power } (P_{in}) = \text{Irr.} \times \text{solar cell area} \tag{7}$$

$$\text{Efficiency } \eta\ (\%) = \frac{P_m}{P_{in}} \tag{8}$$

4 Results and Discussion

The power output of the solar tree is fed to the LED load for analyzing the overall performance of the solar tree and compared it with a stand-alone (50 W) solar panel. The complete designed of the proposed solar tree and stand-alone PV panel is shown in Fig. 2. The various parameters (V_{oc}, I_{sc} and P_m) are calculated for analysis and comparison purpose.

In Fig. 3a, the current in the I-V curve increases and reaches a peak at 2.3 A along with increase in voltage then it starts decreasing continuously and reaches down to 0.38 A as voltage is increases. When the current reaches to its maximum value, then it is known as the maximum current (I_m) along the maximum voltage (V_m). In Fig. 3b, the power in P–V curve of the solar tree increases and voltage increases and reaches to maximum power, i.e., 32 W and then decreases as the voltage increases. When the power reaches to its maximum value, then the point is known as maximum power point (P_m).

Fig. 2 Real time diagram of solar tree and a solar stand-alone PV module

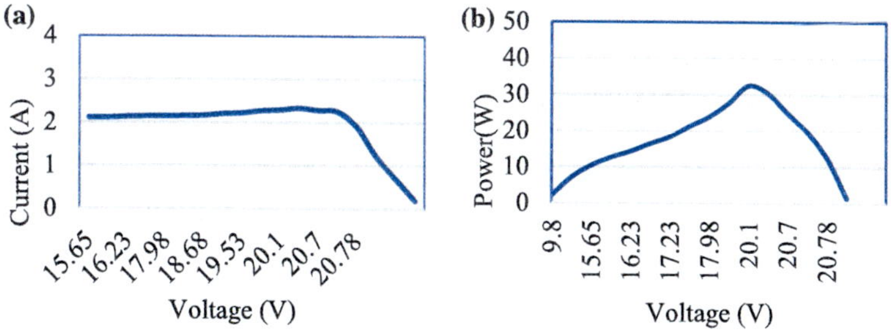

Fig. 3 **a** I-V graph of the solar tree. **b** P–V graph of the solar tree

The LED bulbs are parallel connected to the load, and initially, 2 bulbs are connected and measured the voltage and current of the solar tree. Then increasing the LEDs bulb up to 8 bulbs, and measured the voltage and current of the solar tree. Now, the P_{max} efficiency and fill factor (FF) of the solar tree are calculated with the help of V_{oc}, I_{sc}, V_m and I_m. The proposed system is then compared with a 50W stand-alone PV panel. The FF and η of the solar tree is calculated on every time interval. Initially, the FF and η of the solar tree is 0.67 and 8.10%, respectively. As the irradiation level changes, the output power of the solar tree changes accordingly as shown in Table 3.

The analysis of a stand-alone module has been done from different time scale, i.e., 10:00 A.M. to 5:00 P.M. with one-hour time interval. The FF and η of the panel is calculated on every time interval. Initially, the FF and η of the panel is 0.60 and 7.72%, respectively. As the irradiation level increases or decreases, the output power of the solar stand-alone system increases or decreases, respectively, as given in Table 4.

Figure 4 shows the graph between power and time. It is observed that P_m of solar tree (P_m-ST) is 32 W and P_m of stand-alone solar module (P_{m}_Sa) is 29.7 W at noon time. Figure 5 shows the graph between efficiency and time. Maximum efficiency of

Table 3 Performance analysis of solar tree (50 W)

Time	Irr. (W/m^2)	Temp (°C)	P (W)	P_m (W)	FF	η (%)
10:00	724	39.50	35.60	23.84	0.67	8.11
11:00	841	40.00	42.36	27.64	0.65	8.10
12:00	993	43.50	47.83	32.42	0.68	8.04
13:00	952	46.00	47.64	30.44	0.63	7.88
14:00	950	41.30	43.47	25.04	0.58	6.49
15:00	651	35.60	29.05	19.76	0.68	7.47
16:00	436	24.90	18.70	12.64	0.68	7.13
17:00	188	19.40	7.9	4.68	0.59	6.13

Table 4 Analysis of stand-alone solar module (50 W)

Time	Irr. (W/m^2)	Temp (°C)	P (W)	P_m (W)	FF	η (%)
10:00	724	38.50	36.23	21.78	0.60	7.72
11:00	841	39.80	41.77	25.97	0.62	7.93
12:00	993	43.90	45.24	29.79	0.66	7.70
13:00	952	46.20	45.16	27.81	0.61	7.50
14:00	950	41.50	34.34	22.25	0.65	6.01
15:00	651	35.90	24.36	14.85	0.61	5.85
16:00	436	25.10	16.26	11.52	0.71	6.77
17:00	188	20.30	6.92	4.34	0.63	5.92

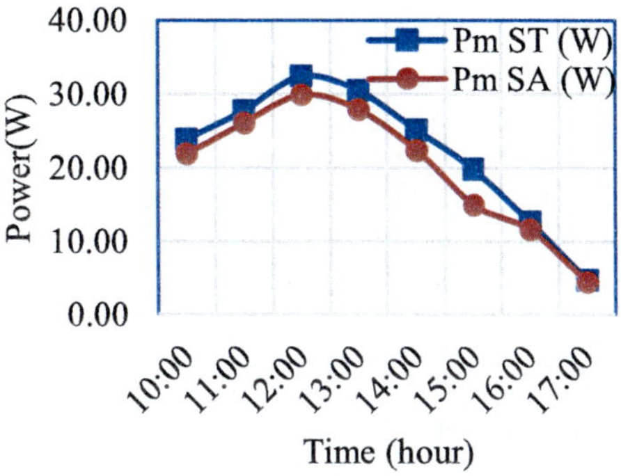

Fig. 4 Output power variation in solar tree and stand-alone solar module

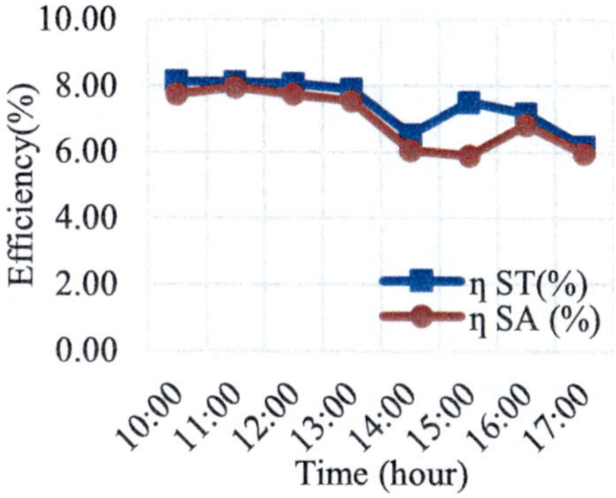

Fig. 5 Efficiency variation in solar tree and stand-alone solar module

solar tree (η_St) is 8% approximately, while efficiency of stand-alone solar module (η_Sa) is approx.7.9%.

From Table 5, it is observed that power shows an increase from 10 A.M. to 12 P.M. and starts decreasing and is significantly lower at 5 P.M. in both stand-alone and solar tree systems, however, overall value of power is higher in solar tree as compared to stand-alone system. Efficiency does not show a constantly decreasing pattern; however, there is an overall decrease from 10 A.M. to 5 P.M.

5 Conclusion

In this paper, a solar tree of 50 W has been designed and compared with a stand-alone solar module of 50 W. The performance parameter is calculated on the basis of temperature and irradiation on different time scale. The V_{oc}, I_{sc} and P_m are 17.98 V, 1.98 A and 23.84 W, when the irradiance is 724 W/m^2 and temperature of panel is 39.5 °C. As temperature and irradiation level increase/decrease, the output power of solar tree increases/decreases, respectively. At the same time, the P_m, V_{oc} and I_{sc} of stand-alone PV module are 17.85 V, 2.03 A and 21.78 W. The output power and

Table 5 Time versus solar tree (P_{max} and η) and stand-alone system (P_{max} and η)

Time	Solar tree			Stand-alone		
	P_m (W)	FF	η (%)	P_m (W)	FF	η (%)
10:00	23.84	0.67	8.11	21.78	0.60	7.72
11:00	27.64	0.65	8.10	25.97	0.62	7.93
12:00	32.42	0.68	8.04	29.79	0.66	7.70
13:00	30.44	0.63	7.88	27.81	0.61	7.50
14:00	25.04	0.58	6.49	22.25	0.65	6.01
15:00	19.76	0.68	7.47	14.85	0.61	5.85
16:00	12.64	0.68	7.13	11.52	0.71	6.77
17:00	4.68	0.59	6.13	4.34	0.63	5.92

efficiency of solar tree are more as compared to stand-alone PV panel, i.e., 23.84 W, 8.1% of solar tree and 21.78 W, 7.72% stand-alone. So, the solar tree power and efficiency is more and the installation area of solar tree is less as compared to stand-alone PV system.

References

1. Ministry of power, India, https://powermin.gov.in/en/content/power-sector-glance-all-india
2. Mohanty P, Muneer T (2014) Smart design of stand-alone solar PV system for off grid electrification projects. In: Bhattacharyya S, Palit D (eds) Mini-grids for rural electrification of developing countries. Green energy and technology. Springer, Cham, pp 63–93
3. Chaurey A, Kandpal TC (2010) A techno-economic comparison of rural electrification based on solar home systems and PV microgrids. Energy Policy 38:3118–3129
4. Chandel M, Agrawal GD, Mathur S, Mathur A (2014) Techno-economic analysis of solar photovoltaic power plant for garment zone of Jaipur city. Case Stud Therm Eng 2:1–7
5. Singh R, Rawat N, Srivastava R (2019) Performance evaluation of a solar tree design and a fixed solar panel for effective solar power harnessing. Int J Appl Eng Res 14(11):2616–2621. ISSN 0973-4562
6. Michael PR, Johnston DE, Moreno W (2020) A conversion guide: solar irradiance and lux illuminance, Copyright © 2020. This is an open access article. https://doi.org/10.21595/jme.2020.21667
7. Rana V, Chauhan YK, Ansari, MA (2016) A multilevel inverter fed induction motor driven water pumping system based on solar photovoltaic. In: India international conference on power electronics, IICPE
8. Kumar K, Ansari MA, Varshney SK, Rana V, Tyagi A (2018) An efficient technique for power management in hybrid solar PV and fuel cell system. Smart Sci, 234–244

Comparative Analysis of Series–Parallel and Bridge Link Configurations Under Various Partial Shading Conditions

Ashwani Srivastav, Vinod Kumar Yadav, and M. Rizwan

Abstract Photovoltaic system is most popular renewable source of energy due to its widespread availability and comparatively easier conversion into electrical energy. Solar energy is also environmentally friendly in comparison with traditional energy sources. Solar energy is also expected to become viable source of energy in coming future. However, there are still some serious problems with photovoltaic systems that must be investigated and solutions sought. Partial shading conditions (PSCs) are one of the most serious issues with photovoltaic systems, as they can cause serious damage to the PV array by causing hotspots and reducing the PV array's ability to generate maximum power. In this paper, series–parallel and bridge link configuration of 3 $\times$ 3 PV array are modeled and simulated under different partial shading conditions, the resulting current voltage and power voltage characteristics are analyzed, and photovoltaic system's performance is compared. For the simulation, MATLAB/Simulink software is used.

Keywords Photovoltaic array configurations · Series–parallel · Bridge link · Bypass diode · Partial shading conditions (PSCs)

1 Introduction

Combination of multiple PV cells serially and parallelly makes one PV array which are then connected in different configuration to maximize the output generated. To enhance the terminal voltage photovoltaic cells are connected serially, similarly to enhance the output current photovoltaic cells are connected parallelly. The solar

A. Srivastav (✉) · V. K. Yadav · M. Rizwan
Department of Electrical Engineering, Delhi Technological University, New Delhi, India
e-mail: ashwanisrivastav_2k20pes06@dtu.ac.in

V. K. Yadav
e-mail: vinodkumar@dtu.ac.in

M. Rizwan
e-mail: rizwan@dce.ac.in

A. Rani et al. (eds.), *Signals, Machines and Automation*, Lecture Notes in Electrical Engineering 1023, https://doi.org/10.1007/978-981-99-0969-8_26

energy is the most popular form of renewable energy as it is available most abundantly, lower cost of generation, produce clean and green energy but along with all the advantages it possess, it also has certain serious issues associated with it like deposition of aerosol, partial shading condition (PSC) and generation of hotspot. Researches and study are being done to resolve these issues in order to produce a PV system with high efficiency, low cost and high reliability [1].

Reliability of the photovoltaic system is affected by the non-uniform illumination over the PV panel. Partial shading condition could be caused by various factors like snow, clouds, tree or building shadow, etc. [2]. Along with reliability, partial shading also effects maximum power generation, causes hotspot generation and electrical power mismatch [3]. Under partial shading condition, the shaded cell experiences the reverse biased voltage which causes generation of hotspot and lead to sudden rise in temperature which in turn can permanently damage the PV cell, can cause fire, joints may become fragile or shattering of protective glass [4, 5].

So the study of partial shading condition is necessary. To decrease the effects of the partial shading condition, bypass diode is use which is connected in antiparallel manner across a panel. Under full illumination, the bypass diode is reversed biased and the entire output current is passed through panel itself but under the condition of partial shading the bypass diode get forward biased due to the reversal of voltage across the shaded cell and bypass the entire PV panel across which it is connected [6–9].

The effects of the partial shading can also be reduced effectively with the proper selection of different configuration in which different panels are to be connected to form one array [10]. The most basic one is series configuration, but as per different study, it is already stated that the effect of partial shading condition is worse in it also the generated output is very low. So different configurations are suggested by the scholars, out of which the series–parallel and bridge link configuration are studied and analyzed in this work.

2 Mathematical Modeling

A PV panel contains number of photovoltaic cell which are interconnected serially or parallelly as per the output voltage and output current requirements. Figure 1 shows the circuit diagram of a photovoltaic cell. In Fig. 1, I_{p} is photocurrent, I_{d} is photodiode current, R_{sh} is shunt resistance depicting leakage current and R_{se} is series resistance depicting internal resistance of the cell, I is cell current and V is voltage produce by one cell [11] (Fig. 2).

Current and voltage equations of a photovoltaic cell are as follows [10]:

$$I = I_{\mathrm{p}} - I_{\mathrm{d}} - I_{\mathrm{sh}} \tag{1}$$

$$I_d = I_o\left[e^{\frac{V+IR_{se}}{V\alpha}} - 1\right] \tag{2}$$

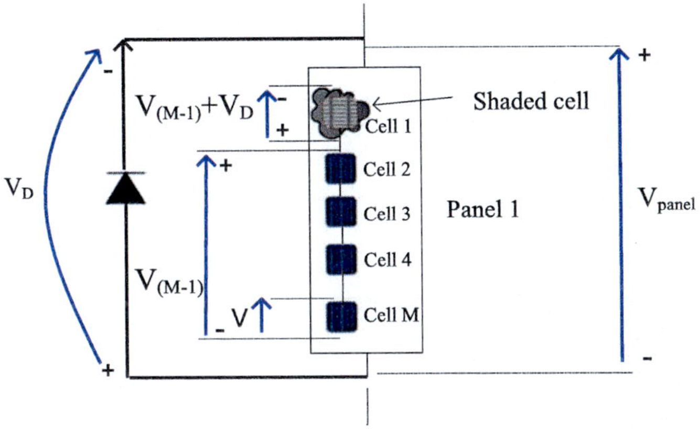

Fig. 1 Distribution of voltage in closed loop circuit of a sub-panel with M number of cell and a bypass diode

Fig. 2 Circuit diagram of PV cell

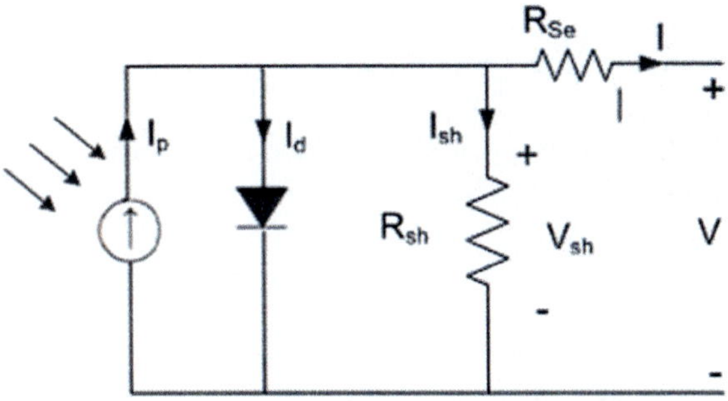

$$I = \frac{V + IR_{se}}{R_{sh}} \tag{3}$$

From Eqs. 1, 2 and 3, we get:

$$I = I_p - I_o\left[e^{\frac{V+IR_{se}}{V_t}} - 1\right] + \frac{V + IR_{se}}{R_{sh}} \tag{4}$$

where I_o is saturation current of the photodiode, $V_t = \frac{AkT}{q}$ in which A is ideality factor, k is Boltzmann constant, T is temperature of the cell (Kelvin) and q is elementary charge [12, 13]. From above equations, it can be inferred that photocurrent is dependent upon the temperature of the cell. Generated photocurrent rise slightly with rise in temperature but the saturation current reduces exponentially with rise in temperature hence decreasing cell voltage hence the efficiency and reliability of PV system decrease [14].

If N_S numbers of cells are connected serially and N_P numbers of cells are connected parallelly in the photovoltaic array than array current I_A is:

$$I_A = N_p I_p - N_p I_O \left[e^{\frac{V + I R_{se}}{V_t}} - 1 \right] - \frac{N_P V + I N_S R_{se}}{N_S R_{sh}} \quad (5)$$

3 Simulation Modeling and Results

Simulation of a 3 × 3 PV system is modeled for series–parallel and bridge link. As the name suggests in series–parallel configuration, some panels are connected serially to form one string and then different strings are reconnected parallelly. Bridge link configuration modifies version of series–parallel configuration in which alternate panels are connected together like a bridge. Connections of both configurations are shown in Fig. 3.

For the study, following different cases of partial shading are taken [15, 16].

Non-shaded Condition (NS): In this case, all the panels are fully and uniformly illuminated.

Row Shading Condition (RS): In this case, first panel of each string which are connected parallelly are shaded, i.e., one row is shaded.

Column Shading Condition (CS): In this case, one entire string, i.e., serially connected panels are shaded.

Diagonal Shading Condition (DS): In this case, diagonal panels of PV array matrix are shaded.

Random Shading Condition (RAS): In this case, panels are shaded randomly, i.e., they do not form any specific pattern.

Non-Uniform shading (NUS): In this case, a PV panel of the array receives irregular irradiance. This case may arise due to passing clouds etc.

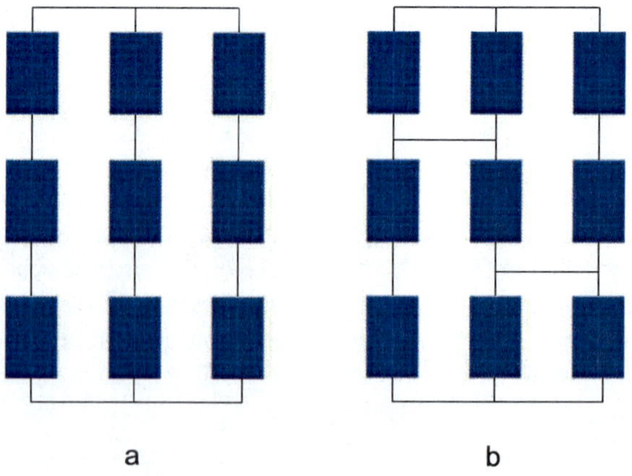

Fig. 3 **a** Series–parallel configuration and **b** bridge link configuration of a 3 × 3 PV array

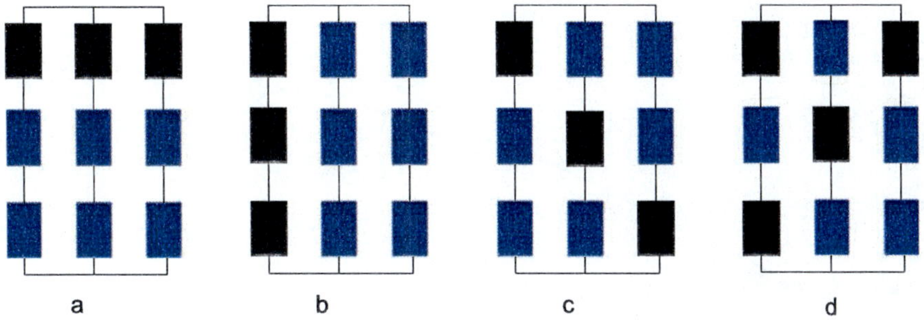

Fig. 4 Different shading patterns considered: **a** row shading, **b** column shading, **c** diagonal shading and **d** random shading

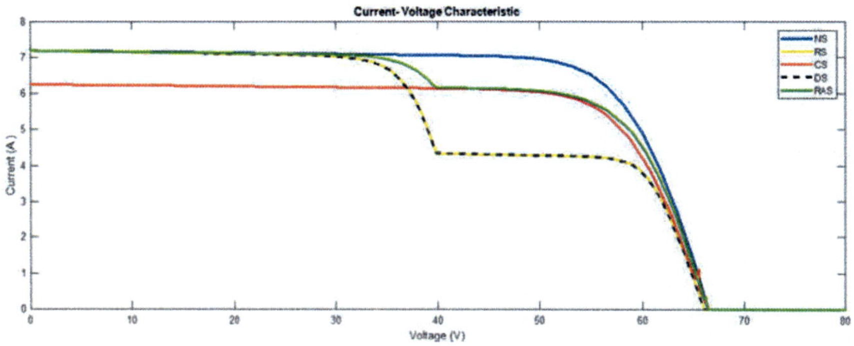

Fig. 5 I-V plot for series–parallel configuration under different PSC for 1000 W/m^2

All the cases of shading are taken for the analysis of PV system. For fully illuminated condition, irradiance is taken 1000 W/m^2. For the shading condition, the 600 W/m^2 irradiance is taken. For all the test condition, 25 °C temperature is taken (Fig. 4).

Simulation results for series–parallel and bridge link configuration for different partial shading condition stated above are shown in Figs. 5, 6, 7 and 8. Results include current voltage and power voltage characteristics of both configuration.

From Table 1, it can be observed that there are different power points, also we get global maximum and local maximum power points which cause problem for conventional power tracing algorithms.

4 Conclusion

The power generated depends upon the partial shading condition, as no. of modules under shading increases power generated decreases. Also it is observed from the

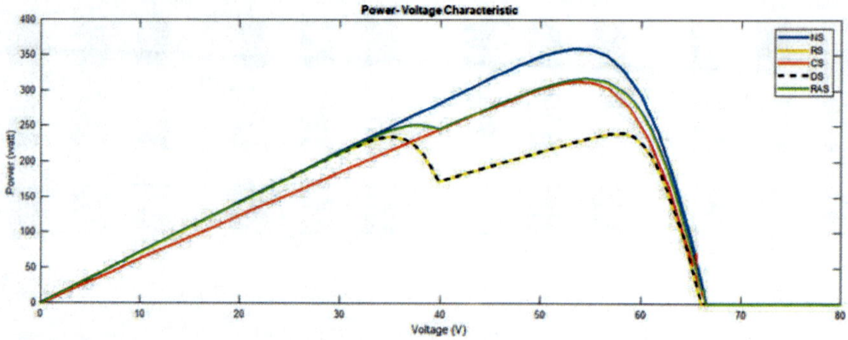

Fig.6 P–V plot for series–parallel configuration under different PSC for 1000 W/m^2

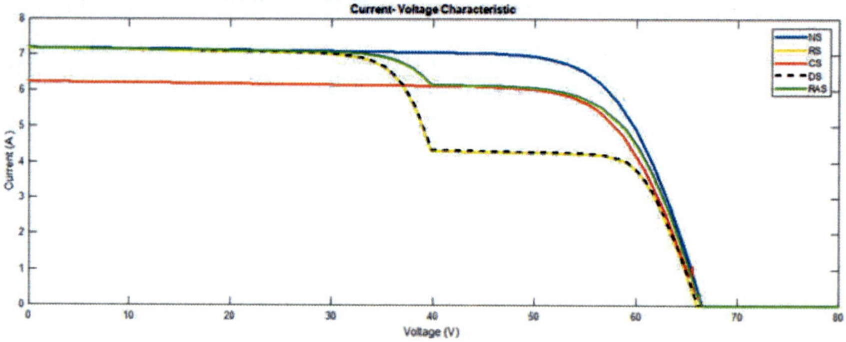

Fig.7 I-V plot for bridge link configuration under different PSC for 1000 W/m^2

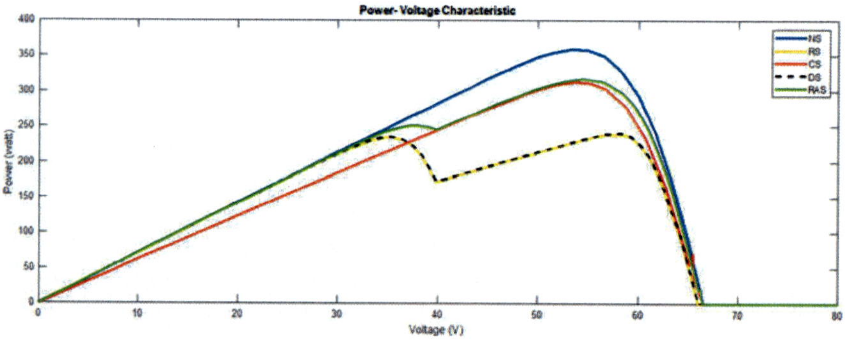

Fig. 8 P–V plot for bridge link configuration under different PSC for 1000 W/m^2

Table 1 GMPP and LMPP during various partial shading conditions

Topology	PSCs	GMPP			LMPP		
		Voltage (V)	Current (A)	Power (W)	Voltage (V)	Current (A)	Power (W)
Series–parallel	NS	53.44	6.732	359.7	–	–	–
	RS	58	4.152	235.8	34.86	6.707	231.3
	CS	54.17	5.770	305.5	–	–	–
	DS	54.17	5.770	305.5	–	–	–
	RAS	54.17	5.997	325.9	34.86	6.707	231.4
Bridge link	NS	53.42	6.727	359.7	–	–	–
	RS	58	4.159	242.8	34.86	6.707	236.3
	CS	53.79	5.808	312.5	–	–	–
	DS	53.79	5.808	312.5	–	–	–
	RAS	56	5.854	332.1	34.86	6.707	236.4

characteristics that for diagonal and row shading, we get similar results. From the above results, it can be analyzed that generated power and output voltage is more for bridge link configuration, as in bridge link comparatively more panels are interconnected which provides an extra path for the current to flow preventing decrease in current. So it can be concluded that bridge link is better than series–parallel in terms of power generation, impact of partial shading and hence efficiency.

References

1. Joshi AS, Dincer I, Reddy BV (2009) Performance analysis of photovoltaic systems: a review. Renew Sustain Energy Rev 13(8):1884–1897
2. Ghosh S, Yadav VK, Mukherjee V (2019) Impact of environmental factors on photovoltaic performance and their mitigation strategies—a holistic review. Renew Energy Focus, 153–172
3. Hamdi S, Saigaa D, Drif M (2014) Modeling and simulation of photovoltaic array with different interconnection configurations under partial shading conditions for fill factor evaluation. In: Proceedings of the 2014 international renewable and sustainable energy conference (IRSEC 2014), pp 25–31
4. Barreiro C, Jansson PM, Thompson A, Schmalzel JL (2011) PV by-pass diode performance in landscape and portrait modalities. In: 37th IEEE photovoltaic specialists conference, Seattle, WA, USA, pp 003097–003102, June 2011
5. Hu Y, Zhang J, Li P, Yu D, Jiang L (2017) Non-uniform aged modules reconfiguration for large-scale PV array. IEEE Trans Device Mater Reliab 17(3):560–569
6. Guerriero P, Daliento S (2019) Toward a hot spot free PV module. IEEE J Photovoltaics 9(3):796–802
7. Guerriero P, Di Napoli F, Coppola M, Daliento S (2016) A new bypass circuit for hot spot mitigation. In: 2016 international symposium on power electronics, electrical drives, automation and motion (SPEEDAM), pp 1067–1072
8. Hui L, Yunmei C, Xiangwei L (2019) Study of bypass diodes configuration on PV modules with partial shaded. In: Chinese control and decision conference (CCDC), pp 511–515

9. Ghosh S, Yadav VK, Mukherjee V (2020) A novel hot spot mitigation circuit for improved reliability of PV module. IEEE Trans Device Mater Reliab, 191–198
10. Ghosh S, Yadav VK, Mukherjee V (2019) Improvement of partial shading resilience of PV array though modified bypass arrangement. Renew Energy 143:079–1093
11. Bingöl O, Özkaya B (2018) Analysis and comparison of different PV array configurations under partial shading conditions. Sol Energy 160:336–343
12. Zhao Z, Cheng R, Yan B, Zhang J, Zhang Z, Zhang M, Lai LL (2020) A dynamic particles MPPT method for photovoltaic systems under partial shading conditions. Energy Convers Manag 220:113070. https://doi.org/10.1016/j.enconman.2020.113070
13. Walker G (2001) Evaluating MPPT converter topologies using a matlab PV model. J Electr Electron Eng Aust 21:49–55
14. Touati FA, Al-Hitmi MA, Bouchech HJ (2013) Study of the effects of dust, relative humidity, and temperature on solar PV performance in Doha: comparison between monocrystalline and amorphous. PVS, Int J Green Energy 10:680–689
15. Palawat KP, Yadav VK, Meena RL, Ghosh S (2021) Comparative performance of different PV array topologies under partial shading condition. In: Innovations in electrical and electronic engineering. Springer, Singapore, pp 367–380
16. Palawat KP, Yadav VK, Meena RL, Ghosh S (2021) Experimental investigation of performance of PV array topologies under simulated PSCs. In: Applied soft computing and embedded system applications in solar energy. CRC Press, pp 47–64

Advanced Control Technologies

Necessary and Sufficient Delay-Independent Stability Conditions for Commensurate Time Delay Systems

Pooja Sharma and Satyanarayana Neeli

Abstract This paper presents an algebraic delay-independent stability (DIS) test for the commensurate multiple time delay systems (CMTDSs). We provide necessary and sufficient conditions for stability test. The stability is analysed using stability tests of two univariate polynomials and a generalized eigenvalue problem. The proposed stability approach is advantageous compared to the existing methodologies because it includes less computational complexity. Numerical examples are given to demonstrate the applicability and effectiveness of the proposed approach.

Keywords Delay-independent stability · Algebraic · Linear systems · Commensurate · Multiple time delays

1 Introduction

The motivation to investigate the stability of time delay systems (TDSs) in the context of practical and commercial applications comes from the fact that instability and unexpected changes in system performance [1]. The examples of TDSs are such as gyroscopic system [2] and load frequency control system [3]. Consequently, the stability analysis of TDS has become the most important for qualitative study of TDS [1–7]. The stability of multiple time delay systems (MTDSs) has also gained ample attention amidst TDSs (see [4–7], and references therein). The class of MTDS generally grouped into commensurate [4] and noncommensurate [6] multiple time delay systems where commensurate multiple time delay systems (CMTDS) [4], delay terms rationally depend on a single delay otherwise, it is known as noncommensurate MTDS [6]. The investigation on commensurate multiple time delay system (CMTDS) has become increasingly relevant research because systems having com-

P. Sharma (✉) · S. Neeli
Department of Electrical Engineering, Malaviya National Institute of Technology, Jaipur, India
e-mail: 2017ree9082@mnit.ac.in

S. Neeli
e-mail: nsnarayana.ee@mnit.ac.in

A. Rani et al. (eds.), *Signals, Machines and Automation*, Lecture Notes in Electrical Engineering 1023, https://doi.org/10.1007/978-981-99-0969-8_27

mensurate type delays can be found in various engineering applications, for instance, N-machine-M-Bus power system [9] and unmanned underwater vehicle [10]. The stability of CMTDSs has been widely investigated over the years [9–14]. Roughly speaking, the stability criteria of TDS can be divided into two broad categories: delay-dependent stability (DDS) [9] and delay-independent stability (DIS) [4], whether or not, the information of delay term is required or not, respectively. In most of the practical systems, either it is not feasible to compute delay terms or delay parameters cannot determine exactly or both; hence, the DIS criterion is preferred [9].

Some of the DIS tests for CMTDS available in the literature are: frequency sweeping tests [4], a polynomial (auxiliary characteristic equation) method [6], a graphical approach using bivariate polynomial [12], zero criteria-based bivariable polynomial approach [13, 15], frequency-dependent one-dimensional (1D) Lyapunov equation [16], and 2D Lyapunov equation [17]. In [14], the stability condition of TDS is given in terms of LMI. The necessary and sufficient stability conditions for DIS of CMTDS were proposed in [8, 13], but later it was established in [15] that these conditions are sufficient. Some sufficient conditions for CMTDS are derived in terms of frequency-dependent 1D Lyapunov equation [16]. In [17], an algebraic procedure for a 2D continuous system was developed using the Lyapunov equation and Kronecker product. These conditions were established to provide Hurwitz stability of polynomial merely. A 2D algebraic technique to check the Routh–Schur stability of multiple TDS was established as sufficient conditions in [18]. The Hurwitz–Schur stability test was derived algebraically in [19] for TDS, but this procedure provided merely sufficient conditions for stability. Thus, Lyapunov approaches, 2D Hurwitz–Schur stability, and LMI techniques result in distinct conservatisms and sufficient conditions. Therefore, the above-discussed stability tests provide merely sufficient conditions. Different from these approaches, an algebraic method for necessary and sufficient conditions for TDSs is required. In [20], an algebraic approach for 2D filters (discrete domain) was established as stability tests of three 1D conditions and one generalized eigenvalue problem (GEVP). Some necessary and sufficient stability conditions to test the stability of 2D linear systems (continuous, discrete, and hybrid) are developed using two polynomials and one GEVP [21].

Motivated by this observation and the stability tests of 2D linear systems in [21], this stability approach can be applied to CMTDSs. In this paper, we investigate the necessary and sufficient conditions for DIS of CMTDS in algebraic manner using this novel stability test. This method has a lower computing complexity and a limited number of 1D tests when compared to previous results [16–19]. The proposed DIS test for CMTDS is based on the bivariate polynomial that is derived using the system's characteristic polynomial. Only two univariate polynomial stability tests and a GEVP are required. To demonstrate the procedure and validity of the proposed approach, numerical examples are given.

Notations. The real and complex number sets are used to represent $\mathbb{R}$ and $\mathbb{C}$ in this study. $\mathbb{R}_+$ states the set of positive real number set. The determinant and transpose of matrix Z are represented by $\det(Z)$ and $(Z)^T$, respectively. $\mathrm{Re}(y)$ stands for real part of y, and conjugate of y is represented by $\bar{y}$. The open left half complex plane and the closed right half complex plane are represented by $\mathbb{C}_-$ and $\mathbb{C}_+$, respectively. The closed unit disc is represented by $\mathbb{D}$, and its complementary set is denoted by $\overline{\mathbb{D}}$. $\delta\mathbb{D}$ states the boundary of unit disc. $*$ denotes the symmetric matrix blocks and the notation iff stands for the if and only if. The $n \times n$ identity matrix is represented by I_n, and 0_n is used for $n \times n$ zero matrix. l is used to represent $\sqrt{-1}$ in this paper.

2 Problem Formulation

In this section, the DIS problem of CMTDS using an algebraic approach is discussed. To illustrate this goal, consider general form of CMTDS represented by delay differential equation as

$$\dot{x}(t) = A_0 x(t) + \sum_{k=1}^{m} B_k x(t - \tau_k), \tag{1}$$

where $x(t) \in \mathbb{R}^n$ is the state, $A_0, B_1, \ldots, B_m \in \mathbb{R}^{n \times n}$ are known matrices, $\tau_k = k\tau$ ($\tau \in \mathbb{R}_+\{\tau \geq 0\}$) are the commensurate delays, and $k = 1, 2, \ldots, m$. The CMTDS (1) is said to be asymptotically stable if and only if the roots of the characteristic polynomial given by

$$\tilde{P}(s, e^{-\tau s}) \triangleq \det\left(sI - A_0 - \sum_{k=1}^{m} B_k e^{-k\tau s}\right), \tag{2}$$

are located in the left half of the complex plane. Where s is the Laplace operator. By considering $z = e^{-\tau s}$, the bivariate polynomial for characteristic polynomial (2) can be rewritten as follows:

$$\tilde{P}(s, z) \triangleq \det\left(sI - A_0 - \sum_{k=1}^{m} B_k z^k\right). \tag{3}$$

Our objective in this paper is to propose algebraic and simple necessary and sufficient conditions for checking the DIS of CMTDS (1) with its corresponding bivariate polynomial (3). In the next section, we recall some useful definitions for the stability analysis of CMTDS using characteristic polynomial in the next section.

3 Preliminaries

In this paper, we are interested in the DIS analysis of CMTDS (1) using an algebraic procedure. To this end, we interpret here some definitions of stability which are indispensable to derive the main results of this paper. The polynomial (3) of CMTDS (1) can be rewritten as irreducible bivariate polynomial as

$$P(s, z) \triangleq \sum_{i=0}^{n_2} \sum_{j=0}^{n_1} c_{ij} s^i z^j, \tag{4}$$

where n_1 and n_2 are the orders of s and z variables, respectively. For the abovementioned bivariate polynomial (4) of CMTDS (1), we introduce the following definitions.

Definition 1 *A univariate polynomial $d(\lambda)$ is Hurwitz when stability region is $\mathbb{C}_-$ and Schur when stability region is $\mathbb{D}$ with respect to λ iff*

$$d(\lambda) \neq 0, \forall \lambda \in \mathbb{C}_+ \, or \, \lambda \in \overline{\mathbb{D}}. \tag{5}$$

This paper deals with DIS of the system (1) with bivariate polynomial $\tilde{P}(s, z)$ (4).

Definition 2 *The CMTDS given in (1) is DIS if*

$$\tilde{P}(s, z) \neq 0, \forall s \in \mathbb{C}_+ \, and \, z \in \mathbb{D}. \tag{6}$$

Remark 1 In Definition (2), the discrete TDSs or noncommensurate TDSs consist of $D_1 = D_2 = \mathbb{D}$ (closed unit disc) and stability region for z_1, z_2 variables will be in $D_1 \times D_2$ region.

The stability test for CMTDS (1) is derived using the characteristic bivariate polynomial (4), which will be presented in the next section.

4 Algebraic Necessary and Sufficient Delay-Independent Stability Conditions

In this section, we use the results from Sect. 3 to determine the necessary and sufficient DIS conditions for CMTDS (1).

Lemma 1 *[22] The CMTDS in (1) with its associated characteristic bivariate polynomial (4) is stable if, it holds the following equivalent conditions*

$$P(\hat{s}, z) \text{ is stable in terms of } z, \text{ for a } \hat{s} \in \mathbb{C}_-, \tag{7}$$

$$P(s, \hat{z}) \text{ is stable in terms of } s, \ \forall \hat{z} \in \mathbb{D}. \tag{8}$$

The conditions (7) and (8) are equal to each other. Further, we merely discuss here condition (8). The stability test of univariate polynomial in z can be done using (7). The main problem in Lemma 1 is with condition (8), where stability is checked for bivariate polynomial (4). This problem can be solved by transforming some univariate polynomial's stability tests and then solving a generalized eigenvalue problem (GEVP). The bivariate polynomial (4) can be rewritten as follows

$$P(s,z) = P_{a(z)}(s) = \sum_{i=0}^{n_2} c_i(\mathbf{Z})\, s^i = \sum_{i=0}^{n_2} C_i^T \mathbf{Z}\, s^i, \tag{9}$$

where $C_i \triangleq \left[c_{i0},\ c_{i1}\ ,c_{i2},\ \ldots,\ c_{in_1}\right]^T$ and $\mathbf{Z} = \left[1,\ z,\ z^2,\ \ldots\ z^{n_1}\right]^T$. Consider that $\overline{P}_{a(z)}(s)$ is equal to its conjugate $P_{a(z)}(s)$ (due to real coefficients) (9) and $\overline{P}_{a(z)}(s)$ can be represented as

$$\overline{P}_{a(z)}(s) = \sum_{i=0}^{n_2} \overline{C}_i(z) s^i = \sum_{i=0}^{n_2} \overline{C}_i^T \bar{Z} s^i. \tag{10}$$

By multiplying (9) and (10), we get new polynomial that is defined as

$$F_{a(z)}(s) \triangleq P_{a(z)}(s)\overline{P}_{a(z)}(s) = \sum_{i'=0}^{n_2}\sum_{i=0}^{n_2} C_i^T \Psi(z)\overline{C}_i s^{i+i'}, \tag{11}$$

where $\Psi(z) \triangleq \frac{1}{2}(\overline{Z}Z^T + Z\overline{Z}^T)$. For stability tests of the (7) and (8) conditions, $F_{a(z)}(s)$ and $\Psi(z)$ are needed to be computed over $z \in \overline{D}$. The matrix $\Psi(z)$ for $z \in \overline{\mathbb{D}}$ can be represented as follows:

$$\Psi(z) = \frac{1}{2} \times \begin{bmatrix} 2 & & & \\ z+\overline{z} & 2(z\overline{z}) & & * \\ z^2+\overline{z}^2 & z\overline{z}(z+\overline{z}) & 2(z\overline{z})^2 & \\ \vdots & \vdots & & \ddots \\ z^{n_1}+\overline{z}^{n_1} & z\overline{z}(z^{n_1-1}+\overline{z}^{n_1-1}) & \cdots & 2(z\overline{z})^{n_1} \end{bmatrix}. \tag{12}$$

Lemma 2 *For CMTDS in (1) $z = \overline{\mathbb{D}} = (e^{i\theta} : 0 \le \theta < 2\pi)$ then $\Psi(z) = \gamma(\tilde{z})$, where $\tilde{z} = Re(z)$, $\tilde{z}$ should not belong to $T \triangleq [-1, 1]$, and $\gamma_{i,j}(\tilde{z}) = S_{|i-j|}(\tilde{z})$, $S_0(\tilde{z}) = 1$,*

$$\begin{aligned} &S_1(\tilde{z}) = \tilde{z},\ S_2(\tilde{z}) = 2\tilde{z}^2 - 1, \ldots, S_{n_1+1}(\tilde{z}) = 2\tilde{z}S_{n_1}(\tilde{z}) - S_{n_1-1}(\tilde{z}), \\ &\qquad j = 1, 2, \ldots, n_1 - 1. \end{aligned} \tag{13}$$

Proof From $z = \mathbb{D} = (e^{l\theta} : 0 \le \theta < 2\pi)$, $l = \sqrt{-1}$ it follows that $\tilde{z} \notin T$, and $S_j(x)$ is the first type Tchebyshev polynomial [22], we get,

$$\Psi(z)|_{z\in\overline{\mathbb{D}}} = \gamma(\tilde{z}) = \begin{bmatrix} S_0(\tilde{z}) & & & & \\ S_1(\tilde{z}) & S_0(\tilde{z}) & & * & \\ S_2(\tilde{z}) & S_1(\tilde{z}) & S_0(\tilde{z}) & & \\ \vdots & \vdots & & \ddots & \\ S_{n_1}(\tilde{z}) & S_{n_1-1}(\tilde{z}) & \cdots & S_1(\tilde{z}) & S_0(\tilde{z}) \end{bmatrix}. \tag{14}$$

By using (11) and Lemma 2, we get

$$F_{a(z)}(s)|_{\tilde{z}\in\overline{\mathbb{D}}} = F_{a(\tilde{z})}(s) \triangleq \sum_{i=0}^{n_2}\sum_{i'=0}^{n_2} C_i^T \gamma(\tilde{z})\overline{C}_i s^{i+i'}. \tag{15}$$

Remark 2 From (11), it is noted that $F_{a(z)}(s)$ is of order $2n_2$ and its characteristics will be as same as $P(s,z)$ for stability test. Hence, (8) can be tested by using $F_{a(z)}(s)$, rather than $\tilde{P}(s,z)$. Further, using (12) it is noted that $\Psi(z)$ is real and symmetric matrix. Hence, we observe using (11) and (12), $F_{a(z)}(s)$ is a real polynomial. In our case, conditions (7) and (8) are equal to the stability test of $F_{a(\tilde{z})}(s)$ (15) with respect to $s\forall\tilde{z} \in T$ where $T \triangleq [-1, 1]$. For computation of $F_{a(\tilde{z})}(s)$ (15) and condition (11), we follow the next lemmas and definitions.

Definition 3 *[22] The Jury matrix of dimensions* $(2r-1)\times(2r-1)$ *denoted by* $\Delta[V(\sigma)]$ *for the polynomial* $V(\sigma) = \sum_{h=0}^{2r} V_h\sigma^h$ *with the specified stability region* $\mathbb{C}_- \times \overline{\mathbb{D}}$*, where* $V_h \in \mathbb{R}$ *is as follows:*

$$z = \overline{\mathbb{D}}, \text{ then } \Delta[V(\sigma)] \triangleq K - L, \tag{16}$$

$$\text{where } K = \begin{bmatrix} V_{2r} & V_{2r-1} & V_{2r-2} & \cdots & V_2 \\ & V_{2r} & V_{2r-1} & \cdots & V_3 \\ & & V_{2r} & \cdots & V_4 \\ & 0 & & \ddots & \vdots \\ & & & \cdots & V_{2r} \end{bmatrix}, \text{ and } L = \begin{bmatrix} & & & & V_0 \\ 0 & & & \cdot^{\cdot^{\cdot}} & \vdots \\ & & V_0 & \cdots & V_{2r-4} \\ & V_0 & V_1 & \cdots & V_{2r-3} \\ V_0 & V_1 & V_2 & \cdots & V_{2r-2} \end{bmatrix}.$$

Lemma 3 *Consider* $F_{a(z)}(s)$ *in (15) is stable in respect to* s, $\forall z\in\overline{\mathbb{D}}$ *iff det*$(\Delta[F_{a(\tilde{z})}(s)])$ *and all elements of* $R(\tilde{z})$ *are positive* $\forall\tilde{z} \in T$*, where*

$$R(\tilde{z}) \triangleq \left\{F_{a(\tilde{z})}(0) \text{ for } s = \mathbb{C}_-\right\}. \tag{17}$$

The condition (8) is checked by finding the roots of univariate polynomials and with the solution of generalized eigenvalue problem (GEVP) in next lemma and theorem.

Lemma 4 *Consider* $T \triangleq [-1,\ 1]$ *and condition (7) and following statements are equal,*

(a) $P(s,\ \tilde{z})$ *is stable in respect to* s *for a* $\tilde{z} \in \delta\mathbb{D}$,
(b) All the elements of $R(\tilde{z})$ *(17) are not equal to zero* $\forall\ \tilde{z} \in T$,
(c) and the GEVP,

$$\begin{bmatrix} 0 & I & & & \\ & & I & 0 & \\ & 0 & & I & \\ & & & \ddots & \\ \alpha_0 & \alpha_1 & & \cdots & \alpha_{n_1-1} \end{bmatrix} \begin{bmatrix} w_1 \\ w_2 \\ \vdots \\ w_{n_1-1} \\ w_{n_1} \end{bmatrix} = \tilde{z} \begin{bmatrix} I & & & & \\ & I & 0 & & \\ & & \ddots & & \\ & & & I & 0 \\ 0 & 0 & & \cdots & -\alpha_{n_1-1} \end{bmatrix} \begin{bmatrix} w_1 \\ w_2 \\ \vdots \\ w_{n_1-1} \\ w_{n_1} \end{bmatrix} \tag{18}$$

of dimensions $n_1(2n_2 - 1) \times n_1(2n_2 - 1)$ does not have any roots of GEVP in T, where

$$\alpha_j \triangleq \Delta[C_j(s)], \quad j = 0, 1, \ldots, n_1. \tag{19}$$

When $F_{a(\tilde{z})}(s),\ i = 0, 1, 2, \ldots, n_2$ (15) can be rewritten as follows

$$F_{a(\tilde{z})}(s) = F_{a(s)}(\tilde{z}) \triangleq \sum_{j=0}^{n_1} C_j(s)\tilde{z}^j. \tag{20}$$

Proof The Remark (2) states that stability region for condition (7) in respect to $s, \forall\tilde{z} \in \delta\mathbb{D}$, is equal to the stability test of $F_{a(\tilde{z})}(s)$ in (15) in respect to s can be violated for a $\tilde{z} \in T$ so that one of the elements of $R(\tilde{z})$ is zero. Therefore, $F_{a(\tilde{z})}(s)$ is stable with respect to $s, \forall\tilde{z} \in T$ iff *(a)*. $F_{a(\tilde{z})}(s)$ is stable with respect to s for $\tilde{z} \in T$, (*b*). All elements of $R(\tilde{z})$ are not equal to zero $\forall\tilde{z} \in T$, and (*c*). $\det(\Delta[F_{a(\tilde{z})}(s)]) = 0$ does not have any root in T. Using (16) and (20), the Jury matrix for $F_{a(\tilde{z})}(s)$ can be rewritten as

$$\det(\Delta[F_{a(\tilde{z})}(s)]) = \det\left(\sum_{j=0}^{n_1} \Delta[C_j(s)]\tilde{z}^j\right) = 0. \tag{21}$$

We consider the expansion of polynomial (19) as GEVP as follows:

$$(\alpha_0 + \alpha_1\tilde{z} + \alpha_2\tilde{z}^2 + \cdots + \alpha_{n_1}\tilde{z}^{n_1})w = 0, \tag{22}$$

where $w \in \mathbb{R}^{2n_1-1}$ and we define α_j are considered here $j = 0, 1, \ldots, n_1$, where $w_1 \triangleq w,\ w_2 \triangleq \tilde{z}w_1, \ldots, w_{n_1} = \tilde{z}w_{n_1-1}$. Therefore, the DIS condition of CMTDS (1) can derive using the following theorem.

Theorem 1 *Assume that CMTDS (1) with bivariate polynomial (4), and* $T \triangleq [-1, 1]$, *for* $s \in \mathbb{C}_-$ *and* $z \in \overline{\mathbb{D}}$, *respectively. By considering Definitions 1 and 2, CMTDS (1) is delay-independent stable iff, it holds the following statements,*

(a) $P(\hat{s}, z)$ *is stable in respect to* z *for a* $\hat{s} \in \mathbb{C}_{-}$.
(b) $P(s, \hat{z})$ *is stable in respect to* s, *for a* $\hat{z} \in \delta\mathbb{D}$.
(c) All the members of $R(\tilde{z})$ *in (17) do not have any real roots in* T.
(d) The GEVP in (18) contains no eigenvalue in $T \in [-1, 1]$.

Proof *Using the aforementioned lemmas and definitions, we prove theorem.*

Using above-mentioned Theorem 1 and Lemmas 1–4, the DIS of CMTDS (1) can be algebraically tested using the following algorithm.

Algorithm 1 An algorithm for delay-independent stability for CMTDS (1)

1: Obtain the bivariate polynomial of CMTDS (1) as (4).
2: State the stability region for both variables ($\mathbb{C}_{-} \times \overline{\mathbb{D}}$), and check DIS condition (a) of theorem 1 and follow the next step. If it is not satisfied, then CMTDS is unstable.
3: Rewrite the bivariate polynomial (4) as (9) and compute C_i.
4: Using Lemma 2 compute $\gamma(\tilde{z})$, $F_{a(\tilde{z})(s)}$, and $R(\tilde{z})$ using (17).
5: Test condition (c) of Theorem 1, If condition is not satisfied, then CMTDS (1) is unstable otherwise follow the next step.
6: Compute $F_{a(\tilde{z})}(s)$ and $C_j(s)$ (20) for $j = 0, 1, \ldots, n_1$.
7: Using Lemma 2 and Definition 3 compute Jury matrix.
8: Obtain the GEVP (18), check the condition (d) of theorem 1, if condition is satisfied then CMTDS (1) is stable, otherwise unstable.

Furthermore, the applicability of the Algorithm **1** can be seen in the next section using numerical example.

5 Numerical Example and Discussion

In this section, numerical example demonstrates the applicability of the proposed algebraic method, and the approach is also compared to existing methods.

Example 1 Consider the following second-order CMTDS (1) (motor-driven pendulum with multiple delays in feedback [25]) with following data,

$$A_0 = \begin{bmatrix} 0 & 1 \\ -59.9568 & -1.4584 \end{bmatrix}, \quad B_1 = \begin{bmatrix} 0 & 0 \\ -13.5602 & 0 \end{bmatrix}, \quad B_2 = \begin{bmatrix} 0 & 0 \\ 0 & 1.2339 \end{bmatrix}, \tag{23}$$

and $\tau_2 = 2\tau_1$.The CMTDS with given data is delay-free stable, and roots are $-0.1122 \pm \iota 8.5735$. We investigate the necessary and sufficient conditions of DIS of CMTDS using algorithm **1** in the following steps as

(i) Step 1, the characteristic bivariate polynomial for CMTDS is obtained as

$$P(s,\ z) = s^2 - 1.2339sz^2 + 1.4584s + 13.5602z + 59.9568 \tag{24}$$

(ii) Step 2, the stability region $\mathbb{C}_- \times \overline{\mathbb{D}}$ and univariate polynomial $P(0,\ z)$ is stable and its roots are $-0.7292 \pm l7.7088$.
(iii) Step 3, for $P(s,\ 1)$ polynomial the roots are $-0.1122 + l8.5735$ and $-0.1122 - 8.5735l$. Therefore, $P(s,\ 1)$ is stable.
(iv) Step 4, $F_{a(\tilde{z})}(s)$ is $-91.9528s^2z - 1.7997s^2 + 27.1205sz^2 + 19.7769sz + 983.3501s + 10141.1348z^2 + 564.7471z + 4280.6336$ and roots of $F_{a(\tilde{z})}(0)$ are $-0.0278 \pm l0.649$. It shows that the $R(\tilde{z})$ does not have any real roots in $T \in [-1,\ 1]$.
(v) Step 5, by solving GEVP in (18) we find the roots of GEVP are 525.13, $-232.2 \pm l222.3$, $228.3 \pm l236.6$ and 6.20, respectively. Hence, there is no real eigenvalue lie in $T \in [-1,\ 1]$.

CMTDS (23) is delay-independent stable. For the comparison, the DIS of CMTDS is also investigated using delay Lyapunov matrix in [24], and the results discussed in [24] conclude the sufficient conditions of stability only with more computational time. The algebraic DIS approach developed into single formalism with complex computations and required the computation of α free parameter [23].

6 Concluding Remarks

In this paper, we have proposed necessary and sufficient conditions for testing the delay-independent stability of time delay systems of commensurate type algebraically. The approach we followed here requires merely two univariate polynomials stability tests and a generalized eigenvalue problem. Using this approach, we reduce the mathematical computational burden over existing methods, even if the multiplicity of delay increases. The effectiveness and applicability of the proposed approach to CMTDS are demonstrated using numerical example. The future work will extend the proposed approach to linear systems with noncommensurate, distributed, and time-varying delays.

References

1. Sipahi R, Niculescu SI, Abdallah CT, Michels W, Gu K (2011) Stability and stabilization of systems with time delay-limitations and opportunities. IEEE Control Syst Mag 31(1):38–65
2. Freitas P (2000) Delay-induced instabilities in gyroscopic systems. SIAM J Control Optim 39(1):196–207
3. Sonmez S, Ayasun S, Nwankpa CO (2016) An exact method for computing delay margin for stability of load frequency control systems with constant communication delays. IEEE Trans Power Syst 31(1):370–377

4. Chen J, Latchman HA (1995) Frequency sweeping tests for stability independent of delay. IEEE Trans Autom Control 40(9):1640–1645
5. Bouzidi Y, Poteaux A, Quadrat A (2016) Computer algebra methods for the stability analysis of differential systems with commensurate time-delays. IFAC Papersonline 49(10):194–199
6. Alikoc B, Ergenc AF (2017) A polynomial method for stability analysis of LTI systems independent of delays. SIAM J Control Optim 55(4):2661–2683
7. Darabsah IA, Janaideh MA, Campbell SA (2021) Stability of connected autonomous vehicle networks with commensurate time delays. In: Proceeding of 2021 American control conference, pp 3308–3315
8. Kamen EW (1982) Linear systems with commensurate time delays: stability and stabilization independent of delay. IEEE Trans Autom Control 27(2):367–375
9. Li J, Chen Z, Cai D, Zhen W, Huang Q (2016) Delay-dependent stability control for power system with multiple time-delays. IEEE Trans Power Syst 31(3):2316–2326
10. Yan Z, Liu X, Zhou J, Wu D (2018) Coordinated target tracking strategy for multiple unmanned underwater vehicles with time delays. IEEE Access 6:10348–10357
11. Hu X, Cong Y, Hu GD (2020) Delay-dependent stability of Runge-Kutta methods for linear delay differential-algebraic equations. Journal Comput Appl Math 363:300–311
12. Delice II, Sipahi R (2011) Delay-independent stability test for systems with multiple time-delays. IEEE Trans Autom Control 57(4):963–972
13. Kamen EW (1980) On the relationship between zero criteria for two- variable polynomials and asymptotic stability of delay differential equations. IEEE Trans Autom Control 25(5):983–984
14. Chesi G, Middleton RH (2014) Necessary and sufficient LMI conditions for stability and performance analysis of 2-D mixed continuous-discrete-time systems. IEEE Trans Autom Control 59(4):996–1007
15. Kamen EW (1983) Correction to Linear systems with commensurate time delays stability and stabilization independent of delay. IEEE Trans Autom Control 28(2):248–249
16. Agathoklis P, Foda SG (1989) Stability and the matrix Lyapunov equation for delay differential systems. Int J Control 49(2):417–432
17. Agathoklis P, Jury EI, Mansour M (1991) Algebraic necessary and sufficient conditions for the very strict Hurwitz property of a 2D polynomial. Multidimensional Syst Signal Process 2(1):45–53
18. Yang X (2001) 2D stability test for time-delay systems. In: Proceeding of 2001 international conference on Info-Tech and Info-Net, vol 4, pp 203–208
19. Gao Q, Zhu Y, Xiao Y (2013) Two dimensional Hurwitz-Schur stability test of linear systems with interval delays. J Syst Eng Elect 24(4):666–673
20. Kraus FJ, Agathoklis P (2020) Stability testing of 2D filters based on Tschebyscheff polynomials and generalized eigenvalues. In: Proceeding of 2020 IEEE international symposium on circuits and systems, pp 1–5
21. Mohsenipour R, Agathoklis P (2021) Algebraic necessary and sufficient conditions for testing stability of 2-D linear systems. IEEE Trans Autom Control 66(4):1825–1831
22. Jury EI (1974) Inners and stability of dynamic systems. Wiley, New York, USA
23. Wei P, Guan Q, Yu W, Wang L (2006) Easily testable necessary and sufficient algebraic criteria for delay-independent stability of a class of neutral differential systems. Syst Control Lett 57(2):165–174
24. Campos ER, Mondie S, Loreto MD (2018) Necessary stability conditions for linear difference equations in continuous time. IEEE Trans Autom Control 63(12):4405–4412
25. Olgac N, Ergenc AF, Sipahi R (2005) Delay scheduling: a new concept for stabilization in multiple delay systems. J Vib Control 11(5):1159–1172

Adaptive Neuro Fuzzy Control of Triple Inverted Pendulum System

Ashwani Kharola, Rahul, and Varun Pokhriyal

Abstract This study presents proportional-integral-derivative (PID) and adaptive neuro fuzzy inference system (ANFIS) control strategy to stabilize of highly nonlinear triple inverted pendulum system. A nonlinear dynamic representation of the system has been proposed and simulated in MATLAB/Simulink platform. The result indicates better performance of ANFIS controller compared to PID controller. Both ANFIS and PID controllers were able to stabilize complete system with desirable overshoot and steady state response. In order to minimize if–then fuzzy rules, the ANFIS controller has been designed using only three membership functions of triangular shape. The proposed ANFIS controller aided in solving the problem of fuzzy rule explosion commonly associated with fuzzy controllers.

Keywords Triple inverted pendulum · Nonlinear systems · PID · ANFIS · MATLAB · Simulink · Simulation

1 Introduction

Triple inverted pendulum is an exemplar of highly nonlinear underactuated mechanical system having eight different equilibrium positions out of which only lower position is stable [1]. It is a famous testing bed mechanism for new control algorithms. These systems have been keen source of interest for researchers since last four decades [2]. Researchers have been adopting various control techniques for stabilization of these nonlinear systems. For an instance, Chen and Theodomile [3]

A. Kharola (✉)
Department of Mechanical Engineering, Graphic Era (Deemed to Be University), Dehradun, India
e-mail: ashwanidaa@gmail.com

Rahul
Department of Electronics and Communication Engineering, Delhi Technological University, New Delhi, India
e-mail: rahulverma@dtu.ac.in

V. Pokhriyal
Scabula Labs Private Limited, Dehradun, India

A. Rani et al. (eds.), *Signals, Machines and Automation*, Lecture Notes in Electrical Engineering 1023, https://doi.org/10.1007/978-981-99-0969-8_28

considered fuzzy reasoning-based approach for stability control of triple inverted pendulum. The study considered an optimal linear quadratic regulator (LQR) for determining parameters of fuzzy controller. Simulations demonstrated the soundness of anticipated approach indicating a good dynamic performance with simpler parameter selection. Further, Huang et al. [4] proposed an optimized method based on motion vision for control of triple inverted pendulum. Real-time images of the system during swing-up motion were collected and optimized using Harris algorithm. The results indicate that stabilization of proposed system can be successfully realized through proposed technique. Arkhipova [5] realized stabilization of triple inverted pendulum in upper unstable orientation via parametric excitation of the support. The authors adopted multiple scale method and the Floquet theory for obtaining the desired results. The results showed that stabilization is possible at different excitation frequencies.

In a study by Jahn et al. [6], an inversion-based control approach has been adopted for stabilization of triple inverted pendulum. The authors analyzed the problem as a two-point boundary value problem and proposed a linear-quadratic-Gaussian (LQG) controller which compensated noises and external disturbances associated with the system. Masrom et al. [7] integrated interval type-2 fuzzy technique with particle swarm optimization (PSO) and spiral dynamic algorithm for control of triple inverted pendulum on wheels. PSO and spiral dynamic algorithm were used for enhancing performance of proposed fuzzy controller. Several tests were conducted which demonstrated the robustness of proposed approach. In this study, initially a proportional-integral-derivative (PID) controller has been designed for control of triple inverted pendulum system. A Simulink model of proposed system has been created and simulated in MATLAB/Simulink. The outputs of PID have been used for learning of an adaptive neuro fuzzy inference system (ANFIS) controller. A novel ANFIS controller has been proposed which can remove the problem of rule explosion associated with conventional fuzzy controller [8]. The ANFIS controller has been designed using only three membership functions of triangular shape (constant type) giving only nine if–then fuzzy rules.

2 Nonlinear Dynamical Equations of Triple Inverted Pendulum

The proposed system constitutes three rigid pendulums each of mass ($m = 0.2$ kg) and length ($l = 0.1$ m) mounted on a mutable cart of mass ($M = 1.0$ kg) as shown in Fig. 1 [9]. A control force (F) is desired to move the cart in linear path against frictional force ($b = 0.1$ Ns/m^2) and gravity ($g = 9.81$ m/s^2). Initially, the bottom, middle and top pendulums are inclined at an angle θ_1, θ_2 and θ_3 with the vertical direction, respectively [10].

The expressions for linear acceleration of the cart ($\ddot{x}$) and angular accelerations of pendulums ($\ddot{\theta}$) were evaluated through Newton's second law considering forces

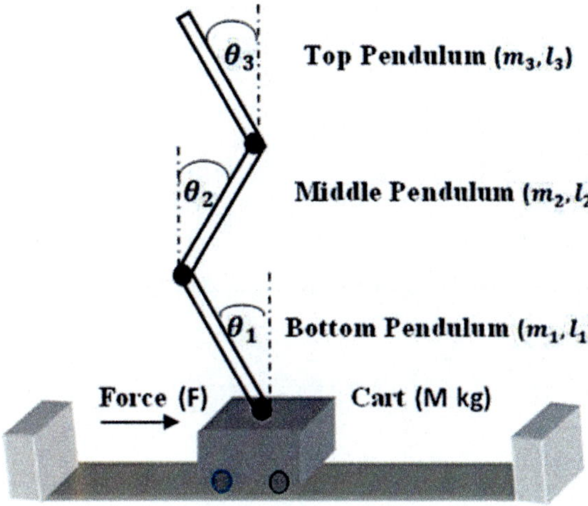

Fig. 1 Triple inverted pendulum on cart

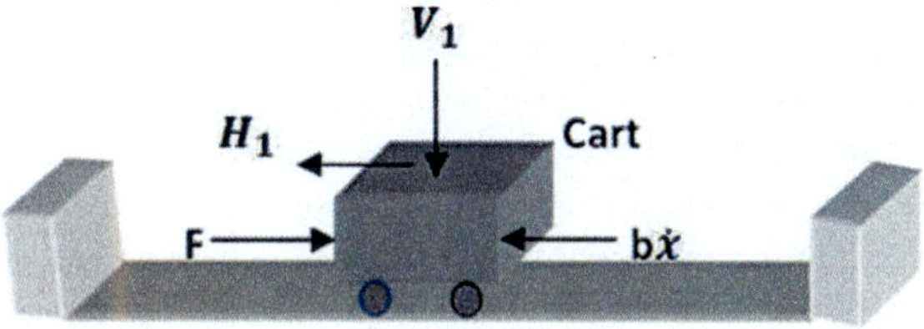

Fig. 2 Forces acting on carriage during motion

acting on each sub-system separately [11]. The resultant forces on cart are shown in Fig. 2, where H and V indicate the interface forces between bottom pendulum and cart in x and y directions, respectively. Further, F is the control force, $b\,\dot{x}$ is the frictional force and $\dot{x}$ denotes linear velocity of the cart.

Balancing forces in horizontal direction give following expression for linear acceleration ($\ddot{x}$) of cart:

$$\ddot{x} = \frac{1}{M}(F - H_1 - b\dot{x}) \tag{1}$$

The forces acting on bottom, middle and top pendulum are shown in Fig. 3. Again H and V indicate the contact forces between pendulums in x and y directions, and 'mg' corresponds to force of gravity on pendulums.

After balancing forces acting on bottom, middle and top pendulum following expressions for angular acceleration of bottom ($\ddot{\theta}_1$), middle ($\ddot{\theta}_2$) and top ($\ddot{\theta}_3$) pendulums were obtained,

$$\ddot{\theta}_1 = \frac{1}{I_1}\left(H_1 l_1 \cos\theta_1 + V_1 l_1 \sin\theta_1 - b_1\dot{\theta}_1 + H_2 l_1 \cos\theta_1 + V_2 l_1 \sin\theta_1\right) \tag{2}$$

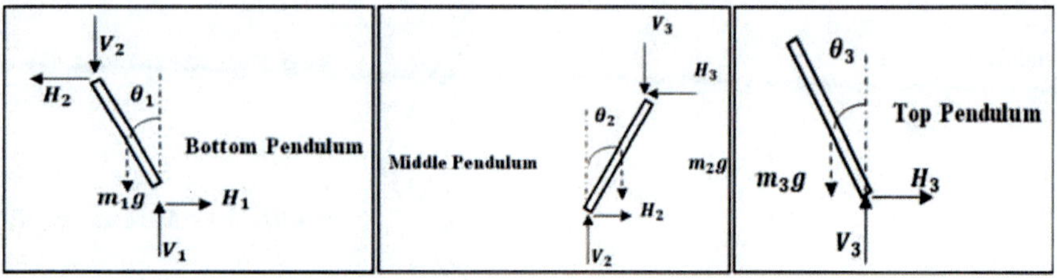

Fig. 3 Forces acting on bottom, middle and top pendulum sub-system

$$\ddot{\theta}_2 = \frac{1}{I_2}\left(H_2 l_2 \cos\theta_2 + V_2 l_2 \sin\theta_2 - b_2\dot{\theta}_2 + H_3 l_2 \cos\theta_2 + V_3 l_2 \sin\theta_2\right) \quad (3)$$

$$\ddot{\theta}_3 = \frac{1}{I_3}\left(H_3 l_3 \cos\theta_3 + V_3 l_3 \sin\theta_3 - b_3\dot{\theta}_3\right) \quad (4)$$

where $\dot{\theta}$ and I represent angular velocity and moment of inertia of pendulums. Based on above equations, a Simulink of the proposed system has been developed as shown in Fig. 4.

In Fig. 4, eight different outputs have been considered for monitoring the performance of complete system. Each output has been represented with the help of a separate output block. The outputs considered for analysis were as follows: cart position (block 1), cart velocity (block 2), bottom pendulum angle (block 3), bottom pendulum angular velocity (block 4), middle pendulum angle (block 5), middle pendulum angular velocity (block 6), top pendulum angle (block 7) and top pendulum angular velocity (block 8).

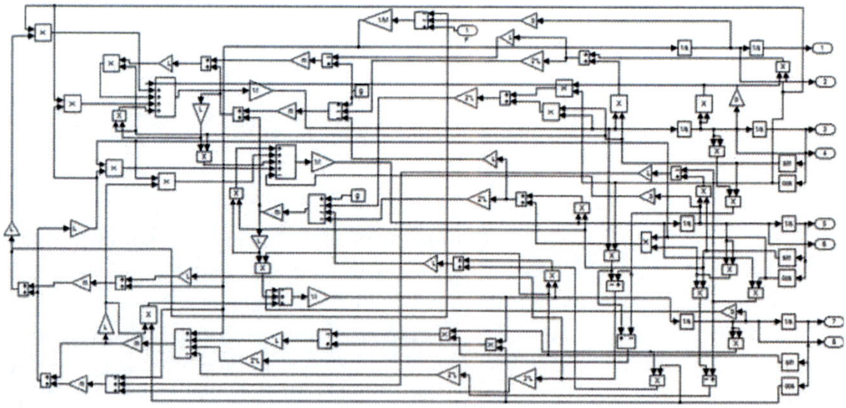

Fig. 4 Simulink model of triple inverted pendulum

3 PID-Based ANFIS Control of Triple Inverted Pendulum

PID is a closed loop feedback controller which computes an error amid desired value and output value [12]. PID controller applies a correction to its control function depending on proportional (K_p), integral (K_i) and derivative (K_d) gains [13]. The PID gain has been auto tuned using PID tuner in MATLAB/Simulink. The gains attained for each sub-system are given in Table 1. The output of PID has been considered as inputs for learning of ANFIS controller [14]. Four separate ANFIS controllers have been designed using only 3 triangular shape membership functions of constant type which helped in reducing the number of if–then fuzzy rules to 9. The training error obtained for ANFIS controller of different sub-systems is also highlighted in Table 1 [15].

The triangular shape membership functions and 3D rule surface attained after tuning for cart position are shown in Fig. 5. The three different grades of membership which were considered for designing of membership functions are low, medium and high. In Fig. 5, horizontal axis of membership functions indicates cart position, whereas vertical axis indicates degree of membership which ranges from 0 to 1.

Table 1 Training error obtained for different sub-systems

Sub-system	Training error	PID gains		
		K_p	K_i	K_d
Bottom pendulum	4.3e−007	1	0	−1
Middle pendulum	6.2e−007	1	0	−1
Top pendulum	1.3e−006	100	1	−10
Cart	2.4e−006	10	0	−10

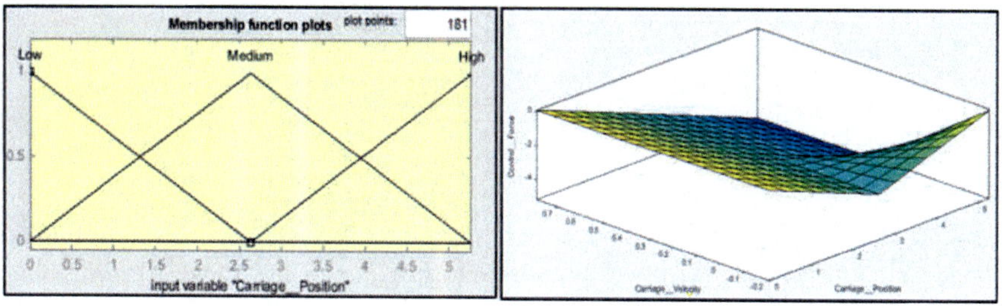

Fig. 5 Tuned membership functions and 3D rule surface obtained for output 'cart position'

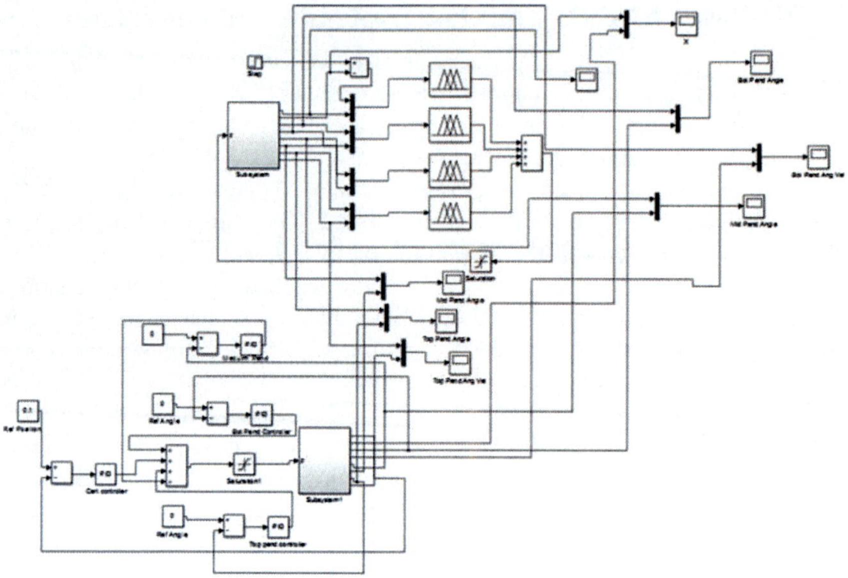

Fig. 6 Simulink sub-system model for comparison of controllers

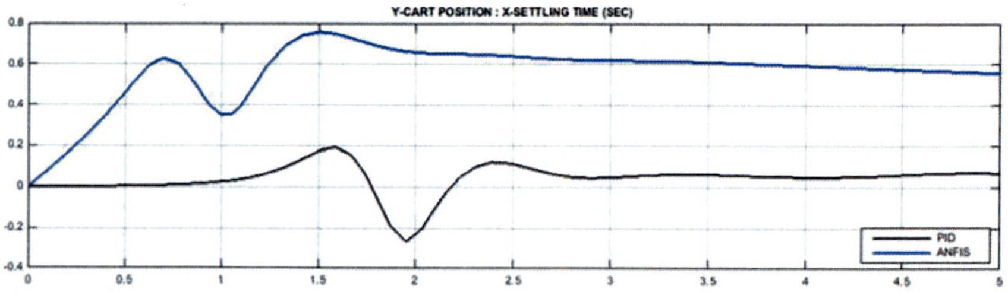

Fig. 7 Simulation graph for cart position

4 Simulation Results

The PID and ANFIS controllers were masked into a sub-system as shown in Fig. 6. The simulation results are shown with the help of Figs 7, 8, 9 and 10. Finally, a comparison of between the two control strategies is given in Table 2.

5 Conclusion

The study successfully presents two robust control strategies, i.e., PID and ANFIS for stabilization of highly nonlinear triple inverted pendulum system. The result of PID has been successfully used for training and optimization of ANFIS controller using

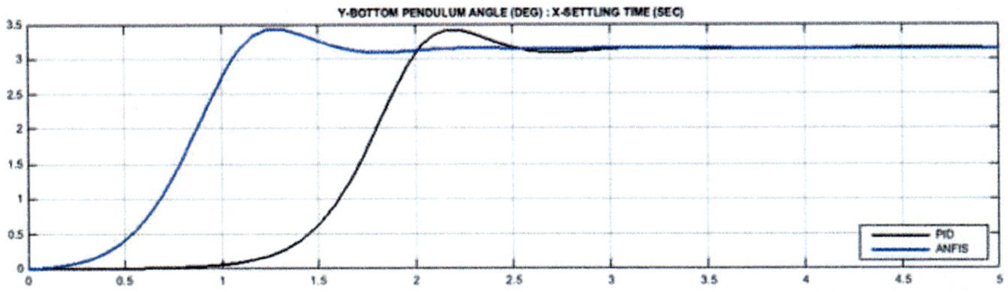

Fig. 8 Simulation graph for bottom pendulum angle

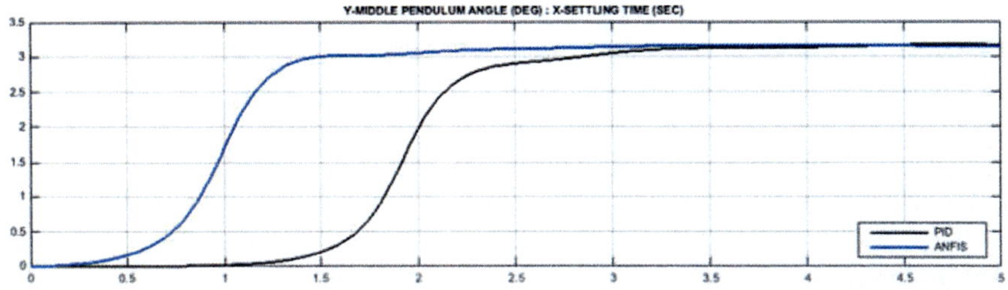

Fig. 9 Simulation graph for middle pendulum angle

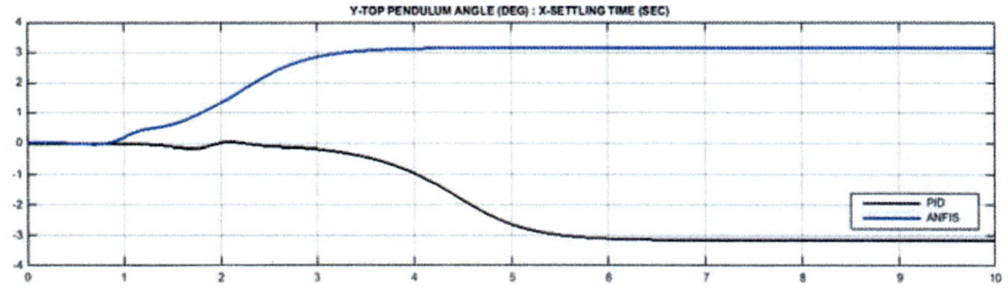

Fig. 10 Simulation graph for top pendulum angle

Levenberg–Marquardt learning algorithm. The objective of designing an ANFIS controller which can aid in eradicating the problem of fuzzy rule explosion has been successfully achieved. ANFIS controller has been designed using only three triangular shape membership functions. The simulation result indicates better performance of ANFIS compared to PID controller. ANFIS controller stabilizes the system within 3.1 s which is 2.4 s less compared to PID controller. Further, the PID controller provides better overshoot response preferably for cart position and bottom pendulum angle. Lastly, both the controllers provided excellent steady state error response. As an extension for future work, attempts can be made for real-time control of proposed system. Further efforts can be made to reduce the steady state errors and consider machine learning approaches like artificial neural networks for control of other variants of nonlinear systems.

Table 2 Simulation results and comparison

Controller	Settling time (s)	Overshoot ranges	Steady state error
Cart position			
ANFIS	2.0	0.75	0.58
PID	2.8	0.2 to −0.25	0.08
Bottom pendulum angle			
ANFIS	1.7	3.4°	3.1°
PID	2.5	3.35°	3.1°
Middle pendulum angle			
ANFIS	1.5	3.2°	3.2°
PID	3.2	3.2°	3.2°
Top pendulum angle			
ANFIS	3.1	3.1°	3.1°
PID	5.5	0.1° to −3.1°	−3.1°

References

1. Ananevskii IM (2018) The control of a three-link inverted pendulum near the equilibrium point. Mech Solids 53:16–21. https://doi.org/10.3103/S0025654418030020
2. Furut K, Ochiai T, Ono N (1984) Attitude control of a triple inverted pendulum. Int J Control 39(6):1351–1365. https://doi.org/10.1080/00207178408933251
3. Chen W, Theodomile N (2016) Simulation of a Triple inverted pendulum based on fuzzy controller. World J Eng Technol 4(2):267–272. https://doi.org/10.4236/wjet.2016.42026
4. Huang X, Wen F, Wei Z (2018) Optimization of triple inverted pendulum control process based on motion vision. EURASIP J Image Video Process 73:1–8. https://doi.org/10.1186/s13640-018-0294-6
5. Arkhipova IM (2019) On stabilization of a triple inverted pendulum via vibration of a support with an arbitrary frequency. Vestnik St. Petersburg University, Mathematics 52:194–198. https://doi.org/10.1134/S1063454119020031
6. Jahn B, Watermann L, Reger J (2021) On the design of stable periodic orbits of a triple pendulum on a cart with experimental validation. Automatica 125:1–7. https://doi.org/10.1016/j.automatica.2020.109403
7. Masrom MF, Ghani NMA, Tokhi MO (2021) Particle swarm optimization and spiral dynamic algorithm based interval type-2 fuzzy logic control of triple link inverted pendulum system: a comparative assessment. J Low Freq Noise Vib Control 40(1):367–382. https://doi.org/10.1177/1461348419873780
8. Hussain K, Salleh M (2015) Analysis of techniques for ANFIS rule-base minimization and accuracy maximization. ARPN J Eng Appl Sci 10(20):9739–9746
9. Su H, Woodham CA (2003) On the uncontrollable damped triple inverted pendulum. J Comput Appl Math 151(2):425–443. https://doi.org/10.1016/S0377-0427(02)00663-5
10. Medrano-Cerda GA (1997) Robust stabilization of a triple inverted pendulum cart. Int J Control 68(4):849–866. https://doi.org/10.1080/002071797223361
11. Pant S, Kumar A, Ram M (2019) Solution of nonlinear systems of equations via metaheuristics. Int J Math Eng Manage Sci 4(5):1108–1126
12. Kim MH, Lee SU (2021) PID with a switching action controller for nonlinear systems of second order controller canonical form. Int J Control Autom Syst 19:2343–2356. https://doi.org/10.1007/s12555-020-0346-4

13. Kouba NEY, Menaa M, Hasni M, Boudour M (2019) A new robust fuzzy-PID controller design using gravitational search algorithm. Int J Comput Aided Eng Technol 11(3):331–360. https://doi.org/10.1504/IJCAET.2019.099327
14. Tsai CC, Tai FC, Chang YL, Tsai CT (2017) Adaptive predictive PID control using fuzzy wavelet neural networks for nonlinear discrete-time time-delay systems. Int J Fuzzy Syst 19:1718–1730. https://doi.org/10.1007/s40815-017-0405-z
15. Malik P, Nautiyal L, Ram M (2019) A method for considering error propagation in reliability estimation of component-based software systems. Int J Math Eng Manage Sci 4(3):635–653

Adaptive Control of DC Microgrid Using EMS and Meterological Data

Birendra Krishna Ghosh, Nasim Ali Khan, Rajat Kumar Mandal, Ankon Das, Aniruddhwa Ghosh, Chiranjit Bhunia, and Saakshyadip Bhowmick

Abstract DC Micro-grids (DCMGs) are becoming more common as the need and demand of renewable energy is exponentially increasing. Many works going for the same, some of them including optimization of Battery Energy Storage Systems (BESSs), Artificial Intelligence (AI)-based algorithms, Droop Control Management, etc. In this paper, based on estimation of power generation ability by means of Energy Management System (EMS) via meteorological statistics results of load scheduling has been done. Authors are using Python for coding and Weather Map API to get the data, to develop a program. Using this program, authors are able to calculate our power generation capacity (Solar and Wind power in this case) for a future of 24 h. This pre-determination of generation capacity will help us in scheduling the power requirement with the AC Grid more effectively and efficiently. The proposal can reduce subsequent time regulator approach in view of remote locations as of grid instabilities. The perception permits the use of discrete and flexible power supply and intake configurations for apiece constituent in the micro-grid that may be fixed to decrease the set up at effective price. To develop relative study on network-based analysis for economical execution of micro-grid EMS is used. The present micro-grid escalated to energy management conception for micro-grids bus parameters are utilized as a networking window with the micro-grid devices.

Keywords DCMG · ACG · EMS · Meteorological data · Weather Map API · Solar and wind power generation capacity

B. K. Ghosh (✉) · R. K. Mandal · A. Das · A. Ghosh · C. Bhunia · S. Bhowmick
Techno International New Town, 1/1, DG Block, Action Area I, Newtown, Kolkata 700156, India
e-mail: birendra.krishna.ghosh@tict.edu.in

N. A. Khan
Aliah University, Action Area II, Plot No. IIA/27, Newtown, Kolkata 700160, India

A. Rani et al. (eds.), *Signals, Machines and Automation*, Lecture Notes in Electrical Engineering 1023, https://doi.org/10.1007/978-981-99-0969-8_29

1 Introduction

DC Micro-grids (DCMGs) offers the uncomplicated, consistent, low cost, accessible and extremely efficient resolution to afford electricity for publics who are alive without utilizing the facility of electricity. Different renewable energy sources linked in series to produce energy at 230 V DC, which can be transferred terminated about 1–2 km distance same as 230 V AC power supply system with double core copper cable with specific number of RCC poles. Communities are connected to the energy distribution scheme all the way through connection boxes mounted on RCC poles. Charge regulator mounted in micro-grid each household end, once 230 V DC stage down 14 V DC for battery storage and fuel-saving DC supply to domestic loads. Some remarkable benefits of DCMG like higher efficiency, less pollution which makes it eco-friendly. DCMG systems are more stable, cost efficient due to which it is widely used in power sector. Also some drawbacks like undefined environment of renewable distributed energy resources like wind and solar generations, market rates, suitable load distribution. For the specific benefits of DC supply over AC supply like suited with renewable distributed energy resources, battery bank and distributed loads, DC micro-grid significant study are as since limited years. Managing the energy and power in the DC micro-grid scheme has been a contest for the scholars. Micro-grid arrangement and mechanism were the incorporated portion of power and EMS. Over the course of years, micro-grid's EMS has been thoroughly studies to tackle these problems and many techniques such as uncertainty in modeling practices, targeted function and limits, methods to get optimize result have been devised. It was stating that assuming numerous targeted function including various methodology and financial restraints has a huge impact for achieved EMS outcomes. Renewable energy resources scheduled a huge measure to encounter the necessities of improved demand power, moderate the eco-friendly contaminants, and attain social and financial advantages for maintainable improvement [1, 2]. Main objectives of EMS are to enhance the task, energy optimum forecast and scheme consistency in together islanding mode and grid related mode micro-grids for supportable growth. Various technologies were applied for economical enactment of EMS in micro-grid [2, 3].

For improving the financial side and the robust task the opportunities accepting DC controller of DCMG has discussed and a hierarchical control method was proposed. Case study on micro-grid at Illinois Institute of Technology, Chicago, it was functioned as DCMG and outcomes for respectively event were associated with AC model [1]. Newton Raphson method for DC grid analysis was used for optimization calculations [4]. Several EMS with different level of control strategy was implemented for a DCMG process. The major objective behind this kind of operation was to confirm the consistency, output speed, accurateness control and economic operation of the system [5]. The concept of multi-micro-grid using renewable distributed energy resources and battery bank was introduced to make a linkage with energy supervision and functional development by double way energy flow network [4, 6]. In the grid, there are several micro-grids to control of energy dissemination largely contributes in effective performance of each micro-grid [7, 8]. The major objectives of this proposed system were the controlling of the solar power, controlling of step-up chopper and the implementation of the battery charging & discharging converters. The simulation output turns out to be as expected and thus can be termed almost precise [9, 10]. The proposed method was built on the basis of distributed model predictive control. The advantages of the scheme was controlled action in a distributed manner and successful power balancing to maintain under variable wind and load condition [3]. In order to feed power to the distant localities in an observational and measureable way, the control of power and managing approach was proposed for the renewable integrated DCMG [11, 12]. Intelligent energy management system (IEMS) [13, 14] were to ensure the load sharing between different resources, to decrease significant loss in a particular system to improve reliability and quality of power. IEMS permits uninterrupted and perfect observation with smart control of distribution scheme performance and it offers enhanced functional conditions as mean of load sharing, consistency improvement and loss reduction of DCMG. In anticipated IEMS for DCMG has been considered with the aim of survey the battery bank performance in respect of discharging and charging [15].

An advanced Micro-grid Supervisory Controller (MGSC) & an EMS was suggested for a micro-grid control. The advantages of these proposed controls were based on architectures of the system as de-centralized architecture provides flexibility to the system while centralized architecture ensures secure and reliable comprehensive analysis of the system. Artificial neural networks (ANN)-based control is introduced, which is an interconnected group of nodes [11, 13]. For diverse unsteady situations ANN supervisor in DCMG and implements fit to path voltage situations promptly and endure load distribution. ANN in DCMG displays that offered manager has capability to sustain potential difference stability of unconnected DCMG and achieve load sharing between the paralleled coupled dispersed generation components [2, 4]. To reduce the price was expressed to smart scheduling-based power generations aimed at micro-grids. Improvement of chance restraint approximation and vigorous optimization methods toward primarily convert besides explain major difficulty [6, 11]. The outcome of probabilistic producing features of dispersed generation with the interrelated MG scheme shall be restricted through cutting-edge artificial intelligence by using optimization techniques [14, 15]. Novelty of this paper is to reduce grid instabilities in remote location using API via time regulator approach as well as by using EMS–discrete and flexible power in DCMG with constituent and operational price will be improved in compare with conventional system.

2 Scope of Work

Figure 1 shows the approach to solution. It is to be noted that, only show Solar and Wind Power sources as renewable energy sources because considering only these two sources while developing program. Authors are also considering that previously developed technologies like MPPT, BESS, IEMS, etc. are already present in. In this platform comparative data have been investigated with present and earlier

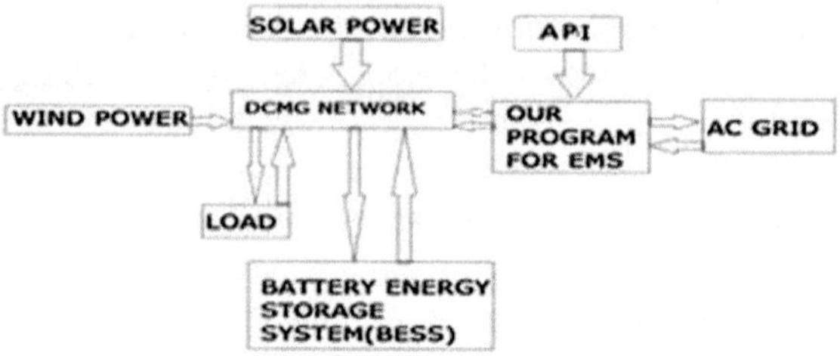

Fig. 1 Proposed block diagram of adapted model

DCMG steady state conditions and effectiveness in respect of demand and finest forecast. Managing different scattered renewable energy sources can be optimized by EMS. Protected, consistent and quality of power is preferred as well as modern technique by machine learning in remote locations. The present micro-grid intensified to energy management concept using machine learning for DC micro-grids, bus parameters are utilized as a networking window with the micro-grid apparatuses. Develop too intelligently ANN-based control power sharing using machine learning in the multiple-source configuration of DCMG, appliance can be substantiated.

3 Coding for Optimal Scheduling

3.1 Logic

(a) Get the location of renewable energy sources. (b) Obtain the required meteorological data. (c) Calculate energy production capacity d) Estimate load demand. (e) Calculate energy required to be Import/Export. (f) Interact with AC Grid for scheduling.

3.2 Flow Chart of Proposed Work

See Fig. 2.

3.3 Explanation

Program can be described in 4 parts.

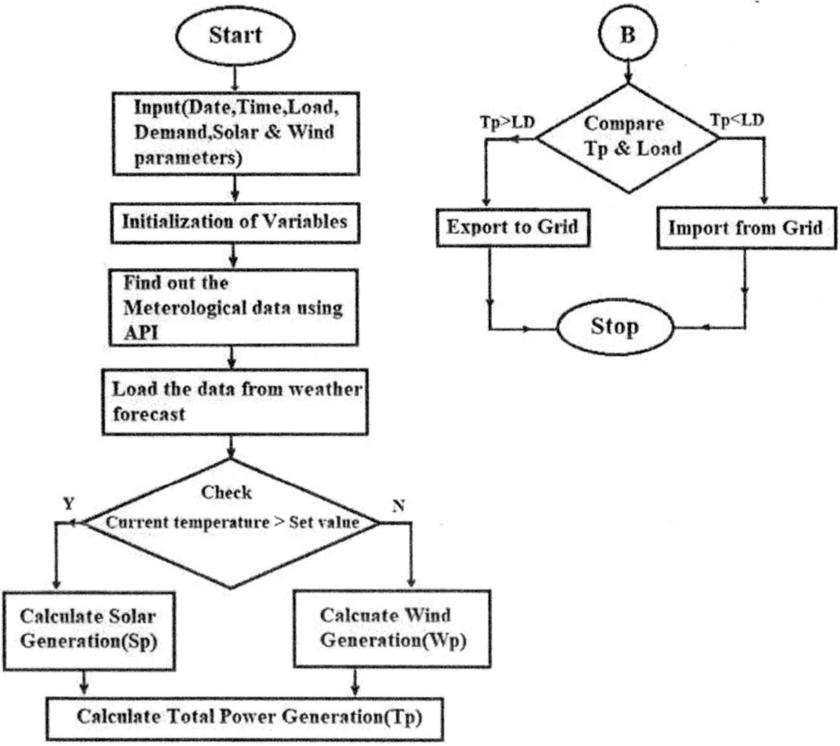

Fig. 2 Flow chart representation of proposed work

3.3.1 User Input

In this part the user need to (one-time) enter the variables and constraints of the Generating source. For example: Solar panel efficiency, Area covered by solar panels, Wind Turbine Generator efficiency, Rotor swept area, etc. In the program, also taking the load demand, Air Density and Average wind blowing time from user because these data can be extracted only when the system is connected to real system and authors have premium access to open weather map (The API using in the map).

3.3.2 Get Location and Meteorological Data

Using Weather Map API to get meteorological data. Using the free version consequently get data using city name. Also access data using the Geo-Location which will make data more accurate but it needs premium version of the API.

3.3.3 Calculation of Power Generation Capacity

Solar Power

As per knowledge that solar power (P) available when cloud percentage is F%

$$P = 990\left(1 - 0.75F^3\right)\ \mathrm{W/m^2} \tag{1}$$

Now if the efficiency of solar panel is η, then

$$P = 990\left(1 - 0.75F^3\right) \times \eta\ \mathrm{W/m^2} \tag{2}$$

Now if the area of solar panel installation is b m^2, then Solar Power Generation

$$S_P = 990\left(1 - 0.75F^3\right) \times \eta \times b\ \mathrm{W} \tag{3}$$

Assuming that sun shines for t hrs on that day, then

$$S_P = 990\left(1 - 0.75F^3\right) \times \eta \times b \times t\ \mathrm{W\,h} \tag{4}$$

Wind Power

The equation for wind power

$$W_P = 0.5 \times \rho_{\mathrm{A}} \times A_{\mathrm{s}} \times C_{\mathrm{cp}} \times V_{\mathrm{w}} \times \eta_{\mathrm{g}} \times \eta_{\mathrm{bb}}\ \mathrm{W} \tag{5}$$

wherever $\rho_{\mathrm{A}} =$ Density of air (kg/m^3) $A_{\mathrm{s}} =$ Swept area of rotor (m^2)
$C_{\mathrm{cp}} =$ Performance coeff., $V_{\mathrm{w}} =$ Linear velocity of wind (m/s),
$\eta_{\mathrm{g}} =$ Efficacy of generator, $\eta_{\mathrm{bb}} =$ gear box bearing efficacy.
Now, if average wind blowing time is t h,

$$W_P = 0.5 \times \rho_{\mathrm{A}} \times A_{\mathrm{s}} \times C_{\mathrm{cp}} \times V_{\mathrm{w}} \times \eta_{\mathrm{g}} \times \eta_{\mathrm{bb}} \times t\ \mathrm{W\,h} \tag{6}$$

Now using Eqs. (4) and (6)
Total power generation of the system,

$$T_P = S_P + W_P \tag{7}$$

3.3.4 Interaction with Grid

Using Eq. (7) total production capacity (T_p) and using load power (L_p) demand consumption capacity. Difference of T_p and L_p ($T_p - L_p$) will give total power needed to be Import/Export and thus scheduling with AC Grid can be done.

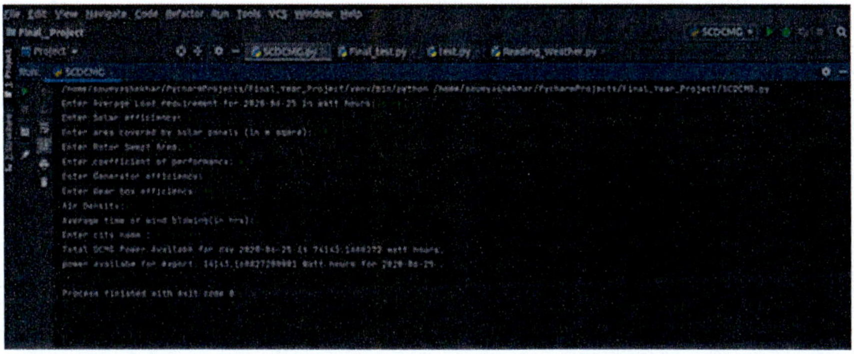

Fig. 3 Program showing output

Table 1 Internal calculation in the program

Type	Power demand (in W)
Load consumption	60,000
Solar	60,237.03
Wind	13,906.13
Total	74,143.16
Total for export	14,143.14

4 Output and Results

The output of the prototype program when run it by providing initial constraints is shown below.

Figure 3 shows the following result: For June 2021, assuming the given constraints of the equipment and a total load of 60 kW for a given area in Kolkata, the amount of power available for export to ACG will be 14.16 kW because of weather condition of that day in Kolkata (Table 1).

Graph in Fig. 4 arrangements by earlier IEMS scheme shows that an uncontrolled utilization and charge in. During night and morning of the daylight, solar is unavailable in probable IEMS scheme. This diagram predictable IEMS shows utmost actual purpose solar power for during daylight. In aforementioned work, comparative data have been analyzed in 24 h duration with present and earlier DCMG steadiness conditions and usefulness in respect of demand response and optimum scheduling.

5 Conclusion

Adequate knowledge to develop a program which can help us to predict the generation capacity of renewable energy sources connected to the DCMG is successfully accomplished. The program is just a prototype program, so, naturally there is a lot of

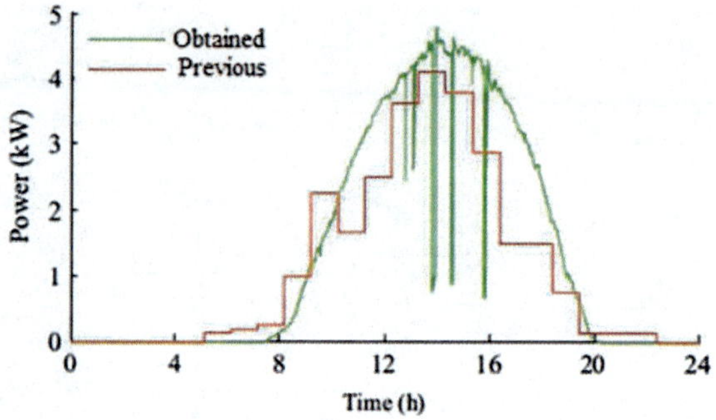

Fig. 4 Analysis of intelligent EMS system

scope for development of the same. In this platform considers only wind and solar power sources. It may be modified through adding more renewable energy sources like Wave Energy, Tidal Energy and Geothermal Energy. Using Weather Map API therefore limited access like Location by city only, daily and weekly predictions only. Therefore, the program can be modified to have Geo-Location Data, Hourly or Minutely data, etc. by using either the premium version or use some another API. Introduction of Artificial Neural Network in the program will make it more efficient and effective. Using the program, the scheduling between DCMG and AC Grid can be more effective as authors have access to already forecasted data. Novelty of this work is detection of actual weather from metrological data, the load demand (60 kW) is distributed in optimized way by solar (60.23 kW) and wind (13.9 kW). It is also seen almost 14.41 kW energy can be exported to reduce the outside demand congestion. To optimize the solar power and wind power generation EMS system is used. Observing the real time data of solar radiation intensity and present air thrust EMS decide, the optimum power generation to meet the actual demand. There are lots of scope for further development in conjunction with accumulating distributed energy sources by optimized demand response.

References

1. Che L, Shahidehpour M (2014) DC microgrids: economic operation and enhancement of resilience by hierarchical control. IEEE Trans Smart Grid 5(5):2517–2526
2. Sanjeev P, Padhy NP, Agarwal P (2018) Autonomous power control and management between standalone DC microgrids. IEEE Trans Industr Inf 14(7):2941–2950
3. Munoz MA, Morales JM, Pineda S (2020) Feature-driven improvement of renewable energy forecasting and trading. IEEE Trans Power Syst 35(5):3753–3763
4. Int J Emerg Electr Power Syst 17(3):339–350. ISSN 1553-779X, ISSN 2194-5756
5. Ullah MH, Paul S, Park J (2018) Real-time electricity price forecasting for energy management in grid-tied MTDC microgrids. In: 2018 IEEE energy conversion congress and exposition (ECCE), Portland, OR, pp 73–80
6. Kou P, Liang D, Gao L (2017) Distributed coordination of multiple PMSGs in an islanded DC microgrid for load sharing. IEEE Trans Energy Convers 32(2):471–485

7. Xiao J, Wang P, Setyawan L, Xu Q (2016) Multi-level energy management system for real-time scheduling of DC microgrids with multiple slack terminals. IEEE Trans Energy Convers 31(1):392–400
8. Wang M, Tan S, Lee C, Hui SY (2018) A configuration of storage system for DC microgrids. IEEE Trans Power Electron 33(5):3722–3733
9. Chettibi N, Mellit A, Sulligoi G, Massi Pavan A (2018) Adaptive neural network-based control of a hybrid AC/DC microgrid. IEEE Trans Smart Grid 9(3):1667–1679
10. Dong W, Li S, Fu X (2018) Artificial neural network control of a standalone DC microgrid. In: 2018 Clemson University power systems conference (PSC), Charleston, SC, USA, pp 1–5
11. Guo F, Xu Q, Wen C, Wang L, Wang P (2018) Distributed secondary control for power allocation and voltage restoration in islanded DC microgrids. IEEE Trans Sustain Energy 9(4):1857–1869
12. Sah SK, Kumar R, Biswas S, Ghosh S, Mandal RK, Ghosh BK, Biswas M, Maji D (2018) Reactive power control of modified IEEE 14 bus system using STATCOM. In: 2018 second international conference on intelligent computing and control systems (ICICCS). IEEE, pp 335–339
13. Ghosh BK, Sen S, Chanda S (2020) Intelligent energy management systems for optimal techno-commercial benefit in DC micro-grids: a review. In: 2020 IEEE VLSI device circuit and system (VLSI DCS). IEEE, pp 1–6
14. Roy S, Goswami S, Pal A, Kumar A, Singh HK, Biswas M, Ghosh BK, Mandal RK (2018) Application of modified particle swarm optimization technique for economic scheduling of a complex micro grid with renewable energy sources. In: 2018 2nd international conference on trends in electronics and informatics (ICOEI). IEEE, pp 77–83
15. Brahmia I, Wang J, Xu H, Wang H, Turci LDO (2021) Robust data predictive control framework for smart multi-microgrid energy dispatch considering electricity market uncertainty. IEEE Access 9:32390–32404

Model Order Reduction of Linear Interval System by Using Firefly Algorithm and Extension of Differentiation Method

Raj Anand and Amarnath Jha

Abstract This article represents the model order reduction of a large-scale linear interval system. The desired order of numerator coefficients is measured using the Firefly algorithm based on integral square error (ISE) reduction as an objective function related to the unit step as input. The reduced-order system's denominator coefficients are calculated using the extension of the differentiation method. Stable reduced-order models are always generated using the suggested strategy. The numerical examples are presented to simplify the method's accuracy, and the computational simplicity and system stability are confirmed using Kharitonov's theorem.

Keywords Linear interval systems · ISE minimization approach · Modified Routh approximation · Firefly algorithm · The extension of differentiation method · Kharitonov's theorem

1 Introduction

The design and implementation of a controller for a large–system are always an important issue because the presence of uncertain variables makes it tedious and costly. Generally, mathematical approaches are employed to realize the models of higher-order interval systems. Minimizing large-scale intervals to lower order is a crucial analysis, synthesis, and simulation for a practical system. The primary benefit of the simplified model is that it permits a better comprehension of the higher-order system, more accessible design, and faster solutions. Some widespread model reduction approaches for continuous linear systems [1–5] and linear interval systems are Routh-Pade approximation [6], Gamma-delta Routh approximation [7–11]. The two limitations of the above interval approximation methods that Hwang and Yang

R. Anand (✉)
Department of Electrical Engineering, Netaji Subhas University of Technology, New Delhi, India
e-mail: raja.ee20@nsut.ac.in

A. Jha
Department of Instrumentation and Control Engineering, Netaji Subhas University of Technology, New Delhi, India
e-mail: anjha@nsut.ac.in

A. Rani et al. (eds.), *Signals, Machines and Automation*, Lecture Notes in Electrical Engineering 1023, https://doi.org/10.1007/978-981-99-0969-8_30

[8] claimed were (1) it does not ensure success in producing a complete interval Routh array and (2) even if the initial interval is stable, certain Routh approximation intervals might not be stable. To reduce the complexities, some other methods got proposed, modified Routh array [9], comment on the computation of interval system [8], linear interval system by minimization of ISE using Genetic algorithm and PSO. Most of the methods mentioned above are that they do not generate a stable simplified model even if the novel model is stable. The suggested approaches ensure a lower-order system's stability in which an extension of the differentiation approach is employed to minimize the denominator of the original system. Based on ISE reduction as an objective function with a unit step as input, the Firefly algorithm [12] is used to reduce the numerator coefficients [12–19]. A numerical example does validation of the technique is solved and compared with reduced models of different methods. The paper is structured as follows: Sect. 2 defines the problem methodology of the proposed method. In Sect. 3, numerical testing done by the proposed method and compared with other approaches and conclusions are reported in Sect. 4.

2 Problem Methodologies

Consider the case of a stable higher-order linear interval system of order n given by

$$F_n(s) = \frac{[X_0^-, X_0^+] + [X_1^-, X_1^+]s + \cdots + [X_{n-1}^-, X_{n-1}^+]s^{n-1}}{[Y_0^-, Y_0^+] + [Y_1^-, Y_1^+]s + \cdots + [Y_n^-, Y_n^+]s^n} = \frac{x(s)}{y(s)} \tag{1}$$

where $x_i^- \leq x_i \leq x_i^+$ for $i = 0, 1, 2 \ldots n - 1$ and $y_j^- \leq y_j \leq y_j^+$ for $j = 0, 1, 2 \ldots n$ represents lower as well as upper bounds of $x(s)$ and $y(s)$ of the original model. After writing Kharitonov's polynomials of the original model, $F_n(s)$ hat transfer function will be selected for the order reduction whose transient time is less

$$G_n(s) = \frac{x_1 s^{n-1} + x_2 s^{n-2} + \cdots + x_n}{s^n + y_1 s^{n-1} + y_2 s^{n-2} + \cdots + y_n} \tag{2}$$

$G_n(s)$ is the selected transfer function model for order reduction. If k represents the reduced order, then the reduced model of (2) is represented as

$$G_k(s) = \frac{N_1(s)}{D_1(s)} = \frac{f_1 s^{k-1} + f_2 s^{k-2} + \cdots + f_k}{s^k + h_1 s^{k-1} + h_2 s^{k-2} + \cdots + h_k} \tag{3}$$

The unknown coefficients of $N_1(s)$, i.e., $f_1, f_2, f_3 \ldots f_n$ are to be assessed with the Firefly method applied to reductions of "Integral Square Error" (ISE) define in Eq. (4). Error represents the variation of step response with the original system and the simplified representation in Fig. 1.

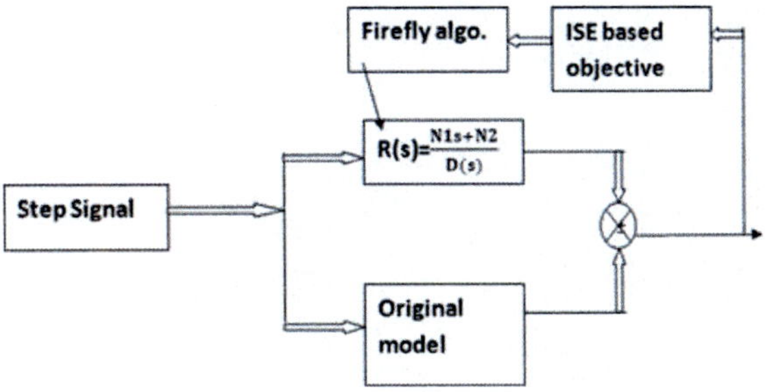

Fig. 1 The complete layout of steps to be followed

$$I = \int_{t=0}^{t=\infty} |e(t)|^2 \tag{4}$$

2.1 *The Method for Obtaining Reduced-Order Denominator Polynomials $D_1(s)$ from the Denominator of the Original "Interval Transfer Function" Model Y(s)*

(1) The first row from Table 1 [20] is the denominator coefficients of $F_n(s)$.
(2) Differentiation of the first row results in row 2.
(3) The 3rd row is derived using a modified version of the Routh approximation, which provides an order n-1 reduced-order denominator using the following relation.$[y_{ij}^-, y_{ij}^+] = [y_{(i-2,j+1)}^-, y_{(i-2,j+1)}^+] - \frac{y_{(i-2,1)}}{y_{(i-1,1)}}[y_{(i-1,j+1)}^-, y_{(i-1,j+1)}^+]$
For, $i = 3, 5, 7, 9 \ldots, j = 1, 2, 3, 4, 5 \ldots n - 1$.

$$y_{i-2,1} = \frac{[y_{i-2,1}^-, y_{i+2,1}^+]}{2};\ y_{i-1,1} = \frac{[y_{i-1,1}^-, y_{i+1,1}^+]}{2},$$

are midpoints of coefficients to overcome dependency property.
(4) The 4th row is derived by differentiating row 3 and applying the modified Routh approximation used in rows 3 & 4 to obtain a minimized order $n-2$ denominator.
(5) Reduced-order denominator $D_1(s)$ is determined by

$$[y_{n-1}^-, y_{n-1}^+]s^{n-1} + [y_{n-2}^-, y_{n-2}^+]s^{n-2} + \cdots + [y_0^-, y_0^+]$$

Table 1 Parameters of step responses example 1

Parameters	original model	proposed method	$\gamma - \delta$ method	Impulse-energy	Mihalov-Cauer method	PSO
Overshoot	2.495	0.2553	1.1241	1.7426	1.2193	9.4305
Settling time	2.249	1.6248	1.7439	0.6892	2.4058	2.8186
Rise time	0.832	1.0008	1.1345	0.9488	1.6284	0.4906
Peak time	1.836	2.7659	2.9024	2.4551	3.7269	1.2555
Steady-state	0.697	0.6977	0.7422	0.6775	0.6527	0.7266
ISE	–	$0.00013549 \times e^{-5}$	$0.0068115942 \times e^{-5}$	$0.0047199905 \times e^{-5}$	$0.0050823168 \times e^{-5}$	$0.00085458 \times e^{-5}$

2.2 Procedure to Achieve Reduced-Order Numerator Polynomial $N_1(s)$ Using Firefly Optimization Algorithm

The Firefly algorithm [12] is a bio-inspired meta-heuristics method. This algorithm is influenced by the rhythmic flashing pattern and behavior of Fireflies at night. The whole algorithm can be concluded in the following steps for numerator reduction.

(1) Generate random solutions set *X*, i.e., Fireflies, Compute Intensity (*I*), i.e., *I* proportional to fitness function $f(x)$.

(2) Update each Firefly by the position using $x_i^{t+1} = x_i^t + \beta_0 e^{-\gamma r_{ij}^2}(x_j^t - x_i^t) + \alpha_t \in_i^t$
Here, β_0 attractiveness constant is taken as 1, γ is absorption coefficient equal to 0.01, α represents randomness strength value taken as 1, ε represents random number derived from Gaussian distribution, t is the number of iterations taken as 20, and population size are 100 for the entire problem.

(3) Perform greedy selection or update of old position X_i with X_i^{t+1} and if the above step improves the solution then terminate otherwise step 2 must be followed again.

3 Results and Discussion

Example1.Suppose a 3rdorder interval system *G*(*s*) [7] minimized in the 2nd order by the proposed approach.

$$G(s) = \frac{[2, 3]s^2 + [17.8, 18.5]s + [15, 16]}{[2, 3]s^3 + [17, 18]s^2 + [35, 36]s + [20.5, 21.5]}$$

Step1.After writing Kharitonov's polynomial of *G*(*s*), the model with less transient time is represented as

$$G_o(s) = \frac{3s^2 + 18.5s + 15}{3s^3 + 17s^2 + 35s + 21.5},$$

The 2nd order selected model is represented as

$$R_2(s) = \frac{N_2(s)}{D_2(s)} = \frac{x_1 s + x_2}{s^2 + 3.44s + 3.223}$$

where $D_2(s)$ is obtained by the extension of the differentiation method and the free coefficients x_1 and x_2 is determined by using the Firefly algorithm for which ranges of coefficients are considered between $1.1 \le x_1 \le 2.24$ and $2.2 \ge x_1 \ge 2.3$. An estimation of ranges based on the integral square error. Therefore the lower & upper bounds are described in the Firefly algorithm according to the ranges in vector format. x_1 and x_2 are obtained as 1.37 and 2.24. So, the obtained second-order model by the proposed method and reduced models by some additional approaches given as

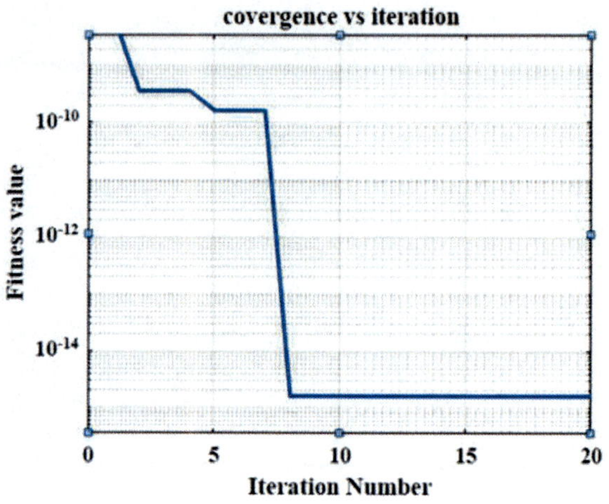

Fig. 2 The performance of Firefly algorithm to obtain free parameters x_1 and x_2

$$R_2(s)|_{\substack{\text{Proposed}\\\text{method}}} = \frac{N_2(s)}{D_2(s)} = \frac{1.37s + 2.24}{s^2 + 3.448s + 3.223},$$

$$R_2(s)|_{\substack{\gamma-\delta\\\text{method}}} = \frac{N_2(s)}{D_2(s)} = \frac{1.255s + 1.117}{s^2 + 2.4435s + 1.505}$$

$$R_2(s)|_{\substack{\text{Impulse}\\\text{Energy}\\\text{method}}} = \frac{N_2(s)}{D_2(s)} = \frac{1.368s + 1.027}{s^2 + 2.623s + 1.516},$$

$$R_2(s)|_{\substack{\text{Mihalov}\\\text{Cauer}\\\text{method}}} = \frac{N_2(s)}{D_2(s)} = \frac{11.19s + 14.17}{17s^2 + 33.61s + 21.71}$$

$$R_2(s)|_{\substack{\text{By}\\\text{Partical}\\\text{Swarn}\\\text{Optimization}}} = \frac{N_2(s)}{D_2(s)} = \frac{2.2615s + 2.3421}{s^2 + 3.448s + 3.223}$$

Figure 2 depicts the Firefly algorithm's performance for the fitness function value during coefficient calculation for 20 iterations. Figure 3 represents the evaluation of the original model step response with the reduced model obtained by using the suggested approach and the reduced models obtained by some different conventional and soft computing methods, i.e., gamma-delta, Impulse-Energy, Mihalov-Cauer, and Particle Swarm optimization algorithm for the same problem. The comparison of various performance factors such as ISE, settling time, rise time, and overshoot are presented in Table 1.

4 Conclusion

Based on the techniques mentioned above, this paper presents how to reduce higher-order linear interval systems to optimize the system's performance. The numerator

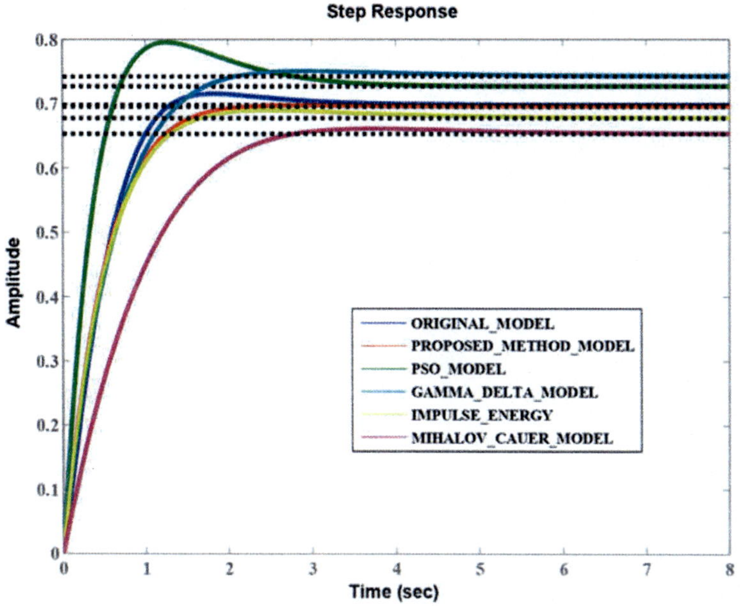

Fig. 3 The comparison of the proposed model's step with the reduced model acquired by using different methods and the original model

of the higher-order system is reduced by applying the ISE-based Firefly algorithm relating to a unit step input is used to get the best optimal numerator-free coefficient, and the denominator is reduced by using an extension of the differentiation approach. Evaluating the minimized model's step responses with the original model provides the most accurate approximation as testing done in example 1, after comparisons of the proposed approach with other approaches (i.e., Gamma-Delta approximation, impulse energy method, Mihalov-Cauer method, and particle swarm optimization). So, the suggested way is highly efficient in providing minimum ISE, undistorted steady-state response, and assures the overall model's stability.

References

1. Shieh L, Goldman M (1974) Continued fraction expansion and inversion of the Cauer third form. IEEE Trans Circ Syst 21(3):341–345
2. Hutton M (1975) Routh approximations for reducing order of linear, time-invariant systems. IEEE Trans Autom Control 20(3):329–337
3. Shamash Y (1975) Model reduction using the Routh stability criterion and the Padé approximation technique. Int J Control 21(3):475–484
4. Krishnamurthy V, Seshadri V (1978) Model reduction using the Routh stability criterion. IEEE Trans Autom Control 23(4):729–731

5. Lucas TN (1983) Factor division: a useful algorithm in model reduction. IEEE Proc D-Control Theory Appl 130(6):362–364
6. Ismail O, Bandyopadhyay B (1994) Routh-Pade approximation for interval systems. IEEE Trans Autom Control 39(12):2454–2456
7. Upadhye A, Ismail O, Bandyopadhyay (1997) SPL gamma/-/spl delta/Routh approximation for interval systems. IEEE Trans Autom Control 42(8):1127–1130
8. Hwang C, Yang S-F (1999) Comments on the computation of interval Routh approximants. IEEE Trans Autom Control 44(9):1728–1787
9. Dolgin Y, Zeheb E (2003) On Routh-Pade model reduction of interval system. IEEE Trans Autom Control 48(1):10–12
10. Yang S-F (2005) Comments on Routh-Pade model reduction of interval systems. IEEE Trans Autom Control 50(2):273–274
11. Nirmalkumar, Gurusamy (2008) Order reduction by error minimization technique. In: International conference on computing, communication and networking, pp 1–6
12. Yang XS (2009) Firefly algorithms for multimodal optimization. In: International symposium on stochastic algorithms, Berlin, Heidelberg, pp 169–178
13. Saini DK, Prasad R (2010) Order reduction of linear interval systems using genetic algorithm. Int J Eng Technol 2(5):316–319
14. Janardhanan S (2013) Model order reduction and controller design techniques. In: Systems and control engineering. IIT Bombay, Bombay
15. Kumar D, Nagar SK (2014) Model reduction by extended minimal optimal Hakal norm approximation. Appl Math Model 38(11):2922–2933
16. Jha AN, Gupta JRP, Hote YV (2014) Reduced-order Modeling for some class of interval system. Int J Model Simul 34(2):63–69
17. Sharma O, Sambariya (2016) Model order reduction using Routh approximation and cuckoo search algorithm. J Autom Control 4(1):1–9
18. Kumar SM, Vijaya Anand N, Srinivasa Rao R (2016) Impulse energy approximation of higher-order interval systems using Kharitonov's polynomials. Trans Inst Measur Control 38(10):1225–1235
19. Sambariya DK, Manohar (2016) Preservation of stability for reduced-order model large scale systems using differentiation method. Br J Math Sci 13:1–17
20. Kumar K, Nagar SK (2020) Model order reduction of interval systems using an arithmetic operation. Int J Syst Sci 51(5):886–902

Optimum LQR Controller for Inverted Pendulum Using Whale Optimization Algorithm

Bharti Panjwani, Vipul Kumar, Jyoti Yadav, and Vijay Mohan

Abstract In this work, a Linear-Quadratic Regulator (LQR)-based control scheme is designed for a highly nonlinear and unstable inverted pendulum system. The system is linearised about its vertical position based on certain assumptions. Initially, weight matrices of the LQR controller are selected based on a trial and error method. These matrices are then optimised using a multi-objective genetic algorithm (GA) and whale optimization algorithm (WOA). The robustness of designed controllers is tested by reference tracking and parametric uncertainty analysis. The results reveal that optimisation of LQR by WOA provides superior performance compared to GA.

Keywords LQR · GA · WOA · Inverted pendulum · Robustness analysis

1 Introduction

The inverted pendulum is a platform for testing several control algorithms because of its nonlinear and unstable behaviour [1]. An inverted pendulum has its centre of mass above its pivot point, due to which it falls when released from a slight angle about its vertical position. It is a classical control theory problem for verifying different control techniques [2, 3]. The aim is to move the cart to prevent the pendulum from falling [4]. It is done with the help of a DC motor and a control technique [5–7]. Modern and advanced control techniques are available to control such systems [8–11].

LQR designed by trial and error method gives a satisfactory response close to the desired response but not optimal. The optimum values of weight matrices are

B. Panjwani (✉) · V. Kumar
Division of ICE, Netaji Subhas Institute of Technology, University of Delhi, Delhi, India
e-mail: bhartinitj.07@gmail.com

J. Yadav
Department of ICE, Netaji Subhas University of Technology, Delhi, India

V. Mohan
Department of Mechatronics, Manipal Institute of Technology, Manipal Academy of Higher Education, Manipal, Karnataka 576104, India
e-mail: vijay.mohan@manipal.edu

A. Rani et al. (eds.), *Signals, Machines and Automation*, Lecture Notes in Electrical Engineering 1023, https://doi.org/10.1007/978-981-99-0969-8_31

required for an efficient response, which is achieved using GA and PSO [4, 12]. A genetic algorithm works on the reactions from its environment and can solve multiple dimensions optimisation problems [13, 14].

In this paper, the whale optimisation algorithm [15] is also utilised to obtain the optimal feedback gain matrix '*K*' of LQR controller. WOA copies the social behaviour of Humpback whales and follows their hunting method known as a bubble net method. Simulation results for inverted pendulum show that the optimal solution obtained from WOA leads to superior performance compared to GA and trial and error method in overshoot, settling time and Integral Absolute Error (IAE). Furthermore, uncertainty analysis is performed to show the robustness of the WOA optimised LQR controller over others.

The paper is organised as follows. Section 2 describes the mathematical modelling of the inverted pendulum. Controller design and its optimisation are presented in Sects. 3 and 4, respectively. Section 5 summarises the simulation results. Lastly, Sect. 6 concludes the research work.

2 Mathematical Modelling of Inverted Pendulum

The structure of the inverted pendulum is shown in Fig. 1. It consists of a pendulum connected to a movable cart that can move left and right on a rail to prevent the pendulum from falling [16]. The system's parameters, their nominal values, and SI units are given in Table 1.

The dynamic behaviour of angle and position of the system varies proportionally to the control force '*F*' [2]. Differential equations relate the kinetic and potential energy of the system with control force [5]. The state space model of the system can be derived from 'LaGrange mechanics' as,

$$\frac{d}{dt}\left[\frac{\partial}{\partial q^{\circ}}L\right]-\frac{\partial}{\partial q}L=\tau \tag{1}$$

where Lagrange Function $L = K - V$. K is the kinetic energy and V is potential energy of the system given as

Fig. 1 Schematic diagram of inverted pendulum

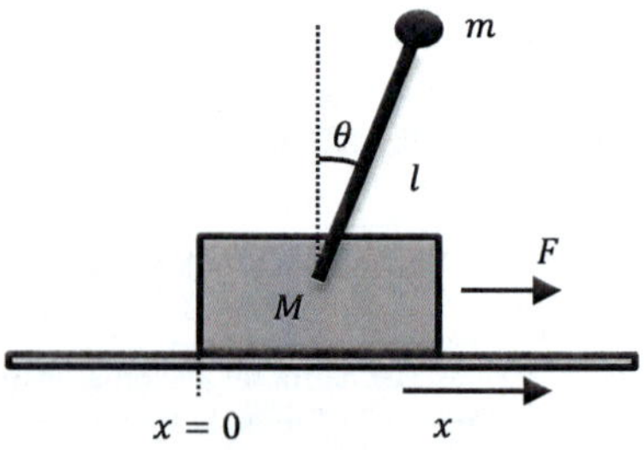

Table 1 Parameters of inverted pendulum [17]

Symbol	Quantity	Value/SI unit
M	Mass of the cart	2.4 kg
m	Mass of the pendulum	0.23 kg
l	Length of pendulum	0.36 m
f	Coefficient of friction	0.1 N/m/s
g	Acceleration due to gravity	9.8 m/s^2
F	Control force applied to the cart	Newton
θ	Angle between pole and vertical direction	Radian
x	Position of the cart	Metre

$$K = \frac{1}{2}M\dot{x}^2 + \frac{1}{2}m[(\dot{x} + l\dot{\theta}\cos\theta)^2 + \left(l\dot{\theta}\sin\theta\right)^2\big] \tag{2}$$

$$V = mgl\cos\dot{\theta} \tag{3}$$

Solving Eq. (1) by substituting Eqs. (2) and (3) yields

$$(M+m)\ddot{x} + ml\ddot{\theta}\cos\theta - ml\dot{\theta}^2\sin\theta = F \tag{4}$$

$$l\ddot{\theta} + \ddot{x}\cos\theta - g\sin\theta = -f\dot{\theta} \tag{5}$$

where $f\dot{\theta}$ denotes the friction in the rotational link of system. The state obtained are $X(t) = [x\dot{x}\theta\dot{\theta}]^{\mathrm{T}}$. After eliminating algebraic loops, the state model is

$$\begin{cases} \dot{x}_1 = \dot{x} = x_2 \\ \dot{x}_2 = \ddot{x} = \frac{-mg\sin x_3\cos x_3 + mlx_4^2\sin x_3 + fmx_4\cos x_3 + F}{M + \left(1 - \cos^2 x_3\right)m} \\ \dot{x}_3 = \dot{\theta} = x_4 \\ \dot{x}_4 = \frac{(M+m)(g\sin x_3 - fx_4) - \left(lmx_4^2\sin x_3 + F\right)\cos x_3}{l\left(M + \left(1 - \cos^2 x_3\right)m\right)} \end{cases} \tag{6}$$

The state space obtained above is nonlinear model that needs to be linearised in order to introduce a modern control scheme. Thus, system is linearised around its equilibrium point $\left[x\dot{x}\theta\dot{\theta}\right] = \left[0000\right]$. The linear model can be obtained by approximation of certain terms. These approximations are $\sin x_3 = x_3$, $\cos x_3 = 1$, $x_4^2 = 0$, $x_3x_4 = 0$.

$$\begin{cases} \begin{bmatrix} \dot{x} \\ \ddot{x} \\ \dot{\theta} \\ \ddot{\theta} \end{bmatrix} = \begin{bmatrix} 0\,0 & 1 & 0 \\ 0\,0 & -0.9392 & 0.0096 \\ 0\,0 & 0 & 1 \\ 0\,0 & 311.2802 & -0.3044 \end{bmatrix} \begin{bmatrix} x \\ \dot{x} \\ \theta \\ \dot{\theta} \end{bmatrix} + \begin{bmatrix} 0 \\ 0.4167 \\ 0 \\ -1.1574 \end{bmatrix} u(t) \\ y(t) = \begin{bmatrix} 1\,0\,0\,0 \\ 0\,0\,1\,0 \end{bmatrix} \begin{bmatrix} x \\ \dot{x} \\ \theta \\ \dot{\theta} \end{bmatrix} + \begin{bmatrix} 0 \\ 0 \end{bmatrix} u(t) \end{cases} \tag{7}$$

3 Controller Design

The control objective is to move the cart so that it keeps the pendulum in an upright position. This means that at $\theta = \dot{\theta} = 0$ [4]. An optimal LQR controller is designed to stabilise the system state feedback. The control input $u(t)$ is obtained by reducing the cost function J.

$$J = \int_0^\infty \left(x^T Q x + u^T R u\right) \mathrm{d}t \tag{8}$$

Here Q is a positive semi-definite matrix known as state variable weighting matrix, and R is positive definite matrix as input variable weighting matrix [12]. Hence, input vector u is designed which reduces the cost function J [1]. Therefore, control signal $u(t)$ is called optimal control [2]:

$$u = -K{*}x = -R^{-1}B^{\mathrm{T}}P{*}x \tag{9}$$

where P is obtained by solving Riccati equation and K is the feedback gain matrix. Now, solving Riccati equation:

$$A^{\mathrm{T}}P + PA - PBR^{-1}B^{\mathrm{T}}P + Q = 0 \tag{10}$$

where [13] $Q = \mathrm{diag}\left(\left[\,100\ 1\ 200\ 1\,\right]\right)$ and $R = 1$. Then,

$$\begin{aligned} K &= -R^{-1}B^T P = \left[K1\ K2\ K3\ K4\right] \\ &= \left[\,-10.0\ -8.10\ -633.86\ -35.64\right] \end{aligned} \tag{11}$$

This feedback gain matrix is then used in the system model of the inverted pendulum to get the results. The system model of inverted pendulum stabilisation using LQR controller is given in Fig. 2.

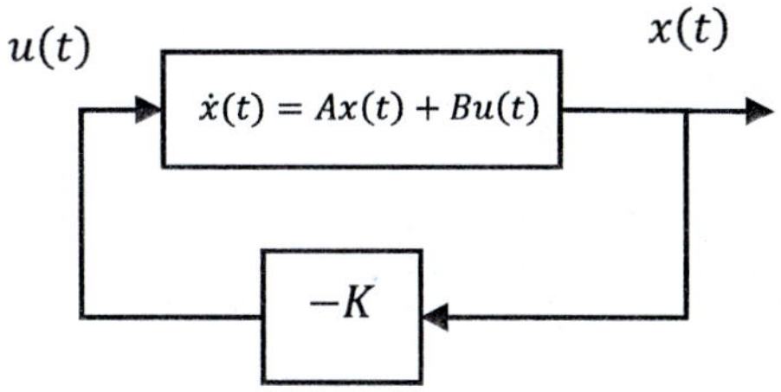

Fig. 2 LQR control structure

4 Optimisation of Controller

The feedback gain matrix is obtained using two weighting matrices Q and R as defined above. The elements of Q and R matrices emphasis on the two main state variables–angle and position. These values are varied to get desired response. To get an optimal feedback gain matrix, we have used two optimisation algorithms: GA and WOA. Aim is to obtain the optimal values of matrices Q and R, which minimises the settling time and overshoot of the response.

4.1 *Genetic Algorithm*

The genetic algorithm (GA) is one of the evolutionary algorithms that work on the principle of laws of natural selection and evolution. This algorithm creates a virtual environment where better responses are emphasised, whereas other responses are disregarded. The basis of the algorithm is the three operations of evolution: reproduction, crossover, and mutation [1]. The design steps for GA are given in literature [1]. The two objectives of minimisation in the fitness function are: settling time and overshoot. This algorithm is carried out by using optimisation toolbox in MATLAB. Convergence plot is shown in Fig. 3 and the optimum value of K obtained is:

$$K = [-12.0458 - 14.0369 - 707.8372 - 39.8890]$$

4.2 *Whale Optimization Algorithm*

This algorithm is a nature inspired algorithm which follows a biological phenomenon of impersonating the behaviour of Humpback whales. It is found that they are intelligent animals and have emotions as well. However, they are predators and preys on small fishes near the surface of the ocean. Their method of hunting is known as bubble net feeding method. Humpback whale first goes deep inside the water and

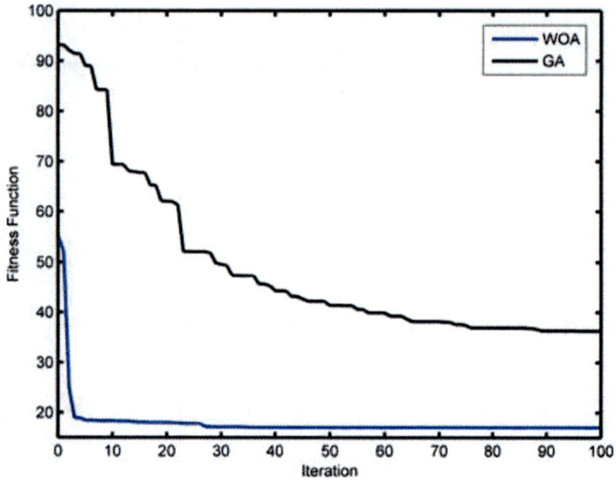

Fig. 3 Convergence curve for GA and WOA

form bubbles in the shape of spirals. These spirals look like small circles encircling the group of small fishes present near the surface of water. Whale then swims up to the surface to kill its prey. The key mathematical formulation and design procedure of the WOA algorithm is presented in literature [15]. The convergence curve obtained is shown in Fig. 3 and the value of K obtained using WOA is:

$$K = [-1.2015 - 19.3238 - 839.1297 - 58.2497]$$

5 Simulation and Results

The system model of the inverted pendulum is designed in MATLAB, and the results are analysed for reference tracking and parametric uncertainties. First, LQR is developed and then optimal feedback gain matrix is then obtained using GA and WOA. Comparative analysis of LQR, GA-LQR, and WOA-LQR controller is shown in Table 2. Figure 4 shows the open loop response of the pendulum on the cart (Fig. 5).

From Table 2, it is clear that the best response is achieved by WOA-LQR, as it reduces the settling time by 99% and 53.9%, respectively. The use of WOA-LQR

Table 2 Time-domain specifications for LQR, GA-LQR, and WOA-LQR

Time-domain specifications	Angle of the pendulum			Position of the cart		
	LQR	GA	WOA	LQR	GA	WOA
Settling time (s)	17.2	9.54	0.16	57.18	33.88	26.32
Overshoot	0.012	0.0114	0.0025	1.52	1.39	1.02
IAE	0.149	0.149	0.104	8.42	5.42	4.39

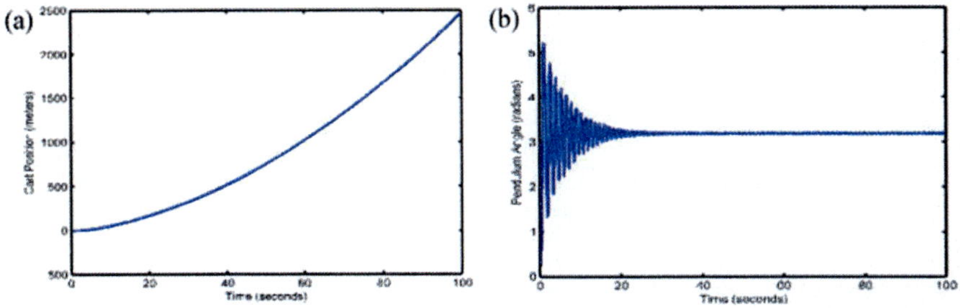

Fig. 4 Open loop response **a** cart position **b** pendulum angle

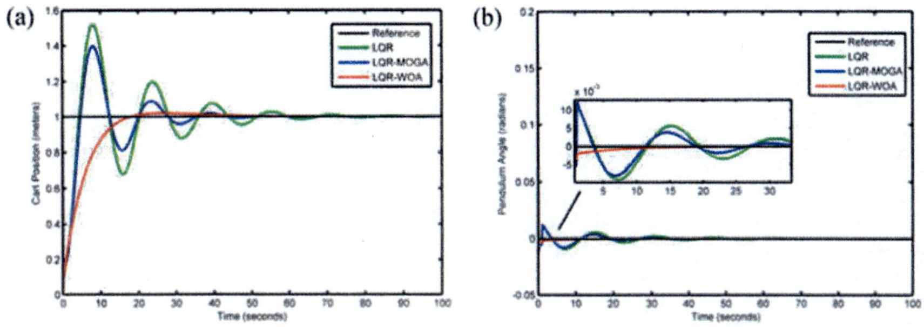

Fig. 5 Reference tracking graph **a** cart position **b** pendulum angle

significantly improves the overshoot of the cart and pendulum as compared to GA-LQR and LQR. Also, Integral Absolute Error (IAE) is minimum for WOA-LQR optimised controller. Results are also compared with the literature [13].

5.1 Uncertainty Analysis

Uncertainty analysis is carried out for the mass of cart alone, mass of the pendulum alone, both the masses, length of pendulum alone, and both masses and length together. IAE of different controllers are obtained for parametric uncertainty. Controller performance under parameter uncertainties is shown in Table 3 and Fig. 6.

6 Conclusion

In this paper, the dynamics of a nonlinear inverted pendulum is derived using the Euler–Lagrange formulation and linearised about its operating point. Further, an LQR controller is designed using GA and WOA optimisation. The performance of

Table 3 IAE for ±50% parametric uncertainty

Uncertainty in parameter	IAE of pendulum					
	LQR		GA-LQR		WOA-LQR	
	+50%	−50%	+50%	−50%	+50%	−50%
M	0.152	0.146	0.107	0.101	0.030	0.029
m	0.148	0.149	0.104	0.104	0.029	0.028
L	0.145	0.154	0.102	0.107	0.030	0.029
M & *m*	0.151	0.146	0.106	0.102	0.030	0.028
M, *m* & *l*	0.149	0.153	0.107	0.106	0.033	0.028

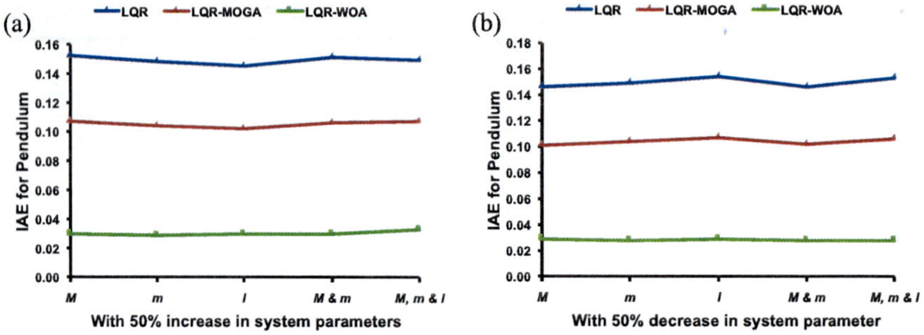

Fig. 6 IAE variations for model parameter uncertainties **a** 50% increase **b** 50% decrease

WOA is superior to GA in reference tracking. Furthermore, uncertainty analysis is successfully carried out, showing that WOA-LQR is more robust than GA-LQR controller.

References

1. Mohan V, Rani A, Singh V (2017) Robust adaptive fuzzy controller applied to double inverted pendulum. J Intell Fuzzy Syst 32(5):3669–3687
2. Mohan V, Singh N (2013) Performance comparison of LQR and ANFIS controller for stabilising double inverted pendulum system. In: 2013 IEEE international conference on signal processing, computing and control (ISPCC), 2013. IEEE, pp 1–6
3. Lakmesari SH, Mahmoodabadi M, Ibrahim MY (2021) Fuzzy logic and gradient descent-based optimal adaptive robust controller with inverted pendulum verification. Chaos, Solitons Fractals 151:111257
4. Omatu S, Deris S (1996) Stabilization of inverted pendulum by the genetic algorithm. In: Proceedings of IEEE international conference on evolutionary computation, 1996. IEEE, pp 700–705
5. Chen X, Zhou H, Ma R, Zuo F, Zhai G, Gong M Linear motor driven inverted pendulum and lqr controller design. In: 2007 IEEE International Conference on Automation and Logistics, 2007. IEEE, pp 1750–1754

6. Chhabra H, Mohan V, Rani A, Singh V (2020) Robust nonlinear fractional order fuzzy PD plus fuzzy I controller applied to robotic manipulator. Neural Comput Appl 32(7):2055–2079
7. Nguyen NP, Oh H, Kim Y, Moon J (2021) A nonlinear hybrid controller for swinging-up and stabilising the rotary inverted pendulum. Nonlinear Dyn 104(2):1117–1137
8. Mohan V, Chhabra H, Rani A, Singh V (2019) An expert 2DOF fractional order fuzzy PID controller for nonlinear systems. Neural Comput Appl 31(8):4253–4270
9. Mohan V, Chhabra H, Rani A, Singh V (2018) Robust self-tuning fractional order PID controller dedicated to nonlinear dynamic system. Journal of Intelligent & Fuzzy Systems 34(3):1467–1478
10. Mohan V, Pachauri N, Panjwani B, Kamath DV (2022) A novel cascaded fractional fuzzy approach for control of fermentation process. Bioresour Technol 127–377
11. Dao PN, Liu Y-C (2021) Adaptive reinforcement learning strategy with sliding mode control for unknown and disturbed wheeled inverted pendulum. Int J Control Autom Syst 19(2):1139–1150
12. Ghoreishi SA, Nekoui MA (2012) Optimal weighting matrices design for LQR controller based on genetic algorithm and PSO. Advanced Materials Research. Trans Tech Publ, pp 7546–7553
13. Panjwani B, Singh V, Rani A, Mohan V (2021) Optimum multi-drug regime for compartment model of tumour: cell-cycle-specific dynamics in the presence of resistance. J Pharmacokinet Pharmacodyn 48(4):543–562
14. Aly AA (2011) PID parameters optimisation using genetic algorithm technique for electrohydraulic servo control system. Intell Control Autom 2(02):69
15. Mirjalili S, Lewis A (2016) The whale optimisation algorithm. Adv Eng Softw 95:51–67
16. Li Q-R, Tao W-H, Sun N, Zhang C-Y, Yao L-H (2008) Stabilization control of double inverted pendulum system. In: 2008 3rd international conference on innovative computing information and control, 2008. IEEE, pp 417–417
17. Control of an inverted pendulum (2012) https://www.control.isy.liu.se/student/tsrt03/files/invpendpmenglish.pdf

Indirect Adaptive Control of Nonlinear System Using Recurrent Neural Network

Prashasti Srivastava and Rajesh Kumar

Abstract In this article, a locally recurrent neural network with input feed through (LRNNIFT) is presented for control of nonlinear dynamical systems. The rationale of using LRNNIFT is due to its modest structure and mathematical model, which gives it an edge over the existing Elman neural networks (ENN) and feed forward neural networks (FFNN). Results from simulation showed that LRNNIFT-based controller is able to achieve adaptive control in a nonlinear system. It is also tested and observed to counterbalance the effects of disturbances. A comparative analysis is presented with the help of simulation, and it is deduced that overall performance of LRNNIFT controller is better than that of FFNN and ENN controllers.

Keywords Neural-based control · Back-propagation · Nonlinear system · Robustness

1 Introduction

The world of science has grown drastically in the past few decades with an increased focus on solving complex real world problems. The growth of such a mindset led the researchers to realise that conventional modelling or control techniques are no longer applicable to complex dynamical systems. This paved way for the rise of the era of soft computing techniques, algorithms based on biological phenomena. One of the earliest significant control techniques in conventional control was the Ziegler Nichols method [1] which was further followed by several modifications [2, 3] and advanced forced oscillation techniques [4, 5] to achieve control on linear plants. Since most processes that require control nowadays are nonlinear, the use of conventional proportional- integral-derivative controller has seen decrease in relevance as majority of the plants are nonlinear in nature or whose dynamics are not fully known. Artificial

P. Srivastava (✉) · R. Kumar
Electrical Engineering Department, Delhi Technological University, Delhi, India
e-mail: prashastisrivastava_2k20ci06@dtu.ac.in

R. Kumar
e-mail: rajeshkumar@dtu.ac.in

A. Rani et al. (eds.), *Signals, Machines and Automation*, Lecture Notes in Electrical Engineering 1023, https://doi.org/10.1007/978-981-99-0969-8_32

neural networks (ANNs) are a major breakthrough in this area as these structures help estimate as well as control such systems. Various structures of ANNs have been developed over the time based on the loops, activation function. Simplest ones are feed forward neural networks [6] having low convergence rates which were followed by multi-layered feed forward neural networks [7] which again had issues related to slow learning rates. Recurrent neural networks (RNNs) [8] have gained traction in the recent years for the control [9, 10] of nonlinear systems due to the presence of recurrent loops leading to faster learning. RNNs are further implemented with modifications for adaptive control as well [11].

1.1 Contributions and Novelties of the Paper

1. This paper presents a detailed simulation of the LRNNIFT for control of a nonlinear time-delayed plant. Initial and final responses of a system with LRNNIFT controller are shown.
2. The proposed controller is also compared with controllers based some of the existing neural network structures such as feed forward neural network (FFNN) and Elman neural network (ENN).
3. The popular and robust back-propagation algorithm is used to tune all the controller models.
4. Stringent analysis and comparison of robustness of LRNNIFT, ENN and FFNN controllers is performed by considering disturbance signal effects which indicates the adaptive nature of the controller.

2 Overview of LRNNIFT

LRNNIFT is essentially a locally recurrent neural network in which each input node is connected directly to the output neuron via weights called as feed through weights. The objective is to compare the performance of the novel controller based on LRNNIFT with FFNN and ENN controllers while considering uniform parameter values for all of them. This helps us gauge the actual behaviour of the controller while having the above two models as a basis for comparison for the control of nonlinear systems. The structure of the proposed controller is shown in Fig. 1. In the figure, the bright red arrows represent the local recurrent weights generated as output of hidden neuron or node and propagated through a lag of a unit instant as connected back to the same neuron. This leads to the formation of a locally recurrent structure. The maroon arrows represent feed through weights connecting the input layer neurons to the output layer node through weights denoted as $N = N_1, N_2, \ldots, N_q$. The recurrent weights are defined as $W_L = w_1, w_2, \ldots, w_p$ while W_a represents the input weight vector. All weights can be updated. From the figure, it can be seen that if W^L and N are removed or made equal to, the structure is reduced to a FFNN. Furthermore,

FFNN and ENN structures are considered to compare the LRNNIFT controller structure. The reason behind these structures to be chosen for comparison was primarily for their proven performance in the fields of both estimation and control. FFNN is a simple structure with no recurrent weights whereas ENN has a rather complex structure with every recurrent weight is fed back to each hidden node. The output weight vector is given as, $W_b = w_b^1, w_b^2, \ldots, w_b^P$.

The input vector is defined as $X = x_1(k), x_2(k), \ldots, x_q(k)$. Subsequently, the output of any pth recurrent node at any kth instant can be calculated as:

$$O_p(k) = f\left[U_p(k)\right] \tag{1}$$

The function f is the hyperbolic tangent function. The induced field (IF) of any pth recurrent node can be given as:

$$U_p(k) = W_p^L(k) O_p(k-1) + \sum_q W_{pq}^b(k) x_q(k) \tag{2}$$

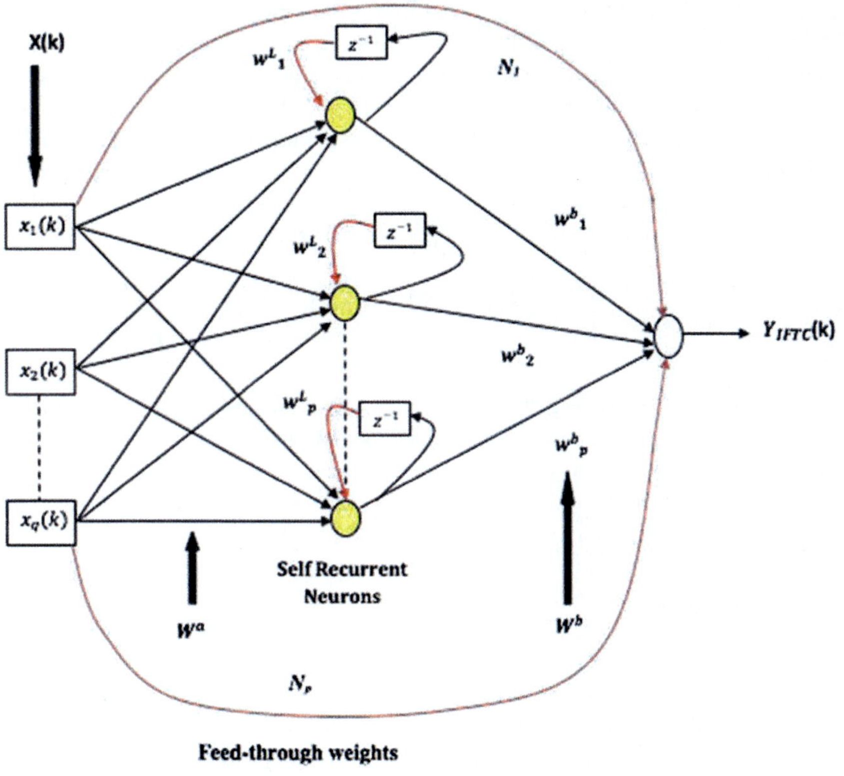

Fig. 1 Structure of the proposed controller

The IF of the output node is described below:

$$S(k) = \sum_{p} W_b^p(k) O_p(k) \tag{3}$$

The output value from the output neuron will be equal to the sum of its own IF and the feed through factor (from input) as a linear function has been considered as the activation function. Hence, we write it as:

$$S_{\mathrm{IFTC}}(k) = S(k) = \sum_{p} W_0^p(k) U_p(k) + \sum_{q} N_{pq}(k) x_q(k) \tag{4}$$

2.1 *Indirect Adaptive Control of a Time-Delayed Nonlinear System*

As discussed previously, designing a controller for a system which is dynamic and nonlinear in nature is a complicated task as linear control techniques fail on such plants. Artificial neural networks have majorly solved this problem due to their flexible nature as there is a vast majority of structures from which an appropriate structure can be chosen whose parameters can be tuned based on system requirements. In this brief, a modified recurrent neural network is used as a controller for a dynamic plant which is to be controlled along a reference model. The general mathematical formulation of a nonlinear time-delayed plant can be given as:

$$Y_{\mathrm{LRN}}(k) = F\begin{bmatrix} Y_{\mathrm{LRN}}^q(k-1), Y_{\mathrm{LRN}}^q(k-2) \ldots Y_{\mathrm{LRN}}^q \\ (k-O), u_c(k-1), u_c(k-2) \ldots u_c(k-D) \end{bmatrix} \tag{5}$$

In the above equation, $Y_{\mathrm{LRN}}{}^q\ (k-1)$ represents a previous value of the system delayed by an instant. In a similar fashion, all the outputs of y_{LRN} are mentioned till qth instant. Similarly, input past values are written $u_c(k-1)$ to $u_c(k-D)$. Here, u_c is essentially the controller output which will act as an input to the plant. The actual input which will be fed to the controller is $r(k)$.The aim here is to control the above plant, that is, to align its response with the reference model (desired response of the plant). This simply translates to $F \approx Fm$, where Fm is the reference model. The potency of any nonlinear control method is established only if it reduces the dependency on plant parameters and structural complexity along with providing faster control response. Therefore, in case of LRNNIFT controller, only three inputs are taken from the vast array of system variables–the present value of input to plant, $u_c(k)$, one previous value of output, $Y_{\mathrm{LRN}}^q(k-1)$, and a previous value of external input, $r(k-1)$. The control scheme is estimated relying on these three inputs only and is calculated as Y_{LRN}(k). Motivation behind selecting few inputs (here, three) is

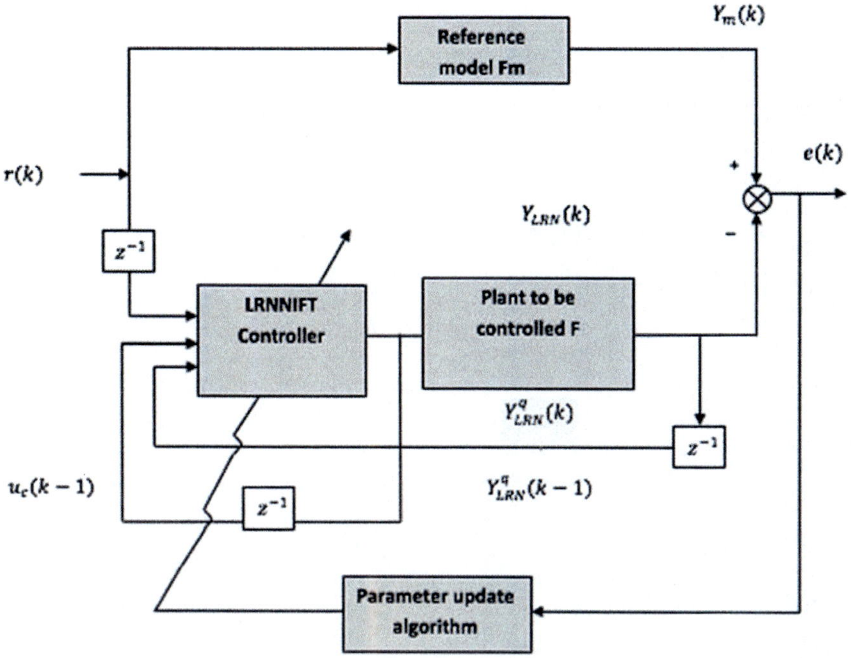

Fig. 2 Adaptive control scheme using LRNNIFT model (Proposed)

that this minimises plant parameter dependency of the controller along with reducing the computation load by lowering the number of weights to be adjusted. Figure 2 represents the block diagram of proposed LRNNIFT controller.

Further, the versatile back-propagation method is applied for adjusting the weights.

3 Simulation Study

In order to evaluate the efficacy of the proposed LRNNIFT-based control strategy, the scheme is implemented on a complex dynamic system. Furthermore, the results obtained from the proposed controller are compared with the FFNN and ENN controllers. Structurally, a single input, hidden and output layer, 4 hidden neurons, uniform learning rate and instantaneous training is applicable to all the three controllers. The reason why we considered uniformity among structure parameters for our analysis is to better judge the performance of LRNNIFT controller. For simulation, the following nonlinear dynamical plant has been considered:

$$y_o(k) = \frac{y_o(k-1)y_o(k-2)[y_o(k-1)+2.5]}{1+y_o^2(k-1)+y_o^2(k-2)} + u_c(k-1) \quad (6)$$

where $u_c(k)$ denotes the input to the plant. The reference model is given as,

$$y_m(k) = 0.6y_m(k-1) + 0.3y_m(k-2) + r(k) \quad (7)$$

where $r(k)$ is the BIBO stable external input to the system given as,

$$r(k) = \sin\frac{2\pi k}{25} \quad (8)$$

The control objective here is to bring the difference between reference model and plant's response $e_c(k) = ym(k)y_o(k)$ approximately equal to zero by introducing an optimal control signal $u_c(k)$ at every instant, to the plant via LRNNIFT as a rectified input to it. $u_c(k)$ can be computed from the knowledge of $y'_{(k)}$ and its past values as

$$u_c(k) = F[y_o(k), y_o(k-1)] + 0.6y_o(k) + 0.3y_o(k-1) + r(k) \quad (9)$$

Figure 3 represents the plant output response (in dotted pink) along with reference model response (in solid green) without control scheme implementation. From the plot, it can be clearly observed that the two responses do not coincide (as desired). Therefore, we use the adaptive control configuration shown in Fig. 2 and apply it to the plant. The value of learning rate is taken as 0.028. The total number of hidden neurons are 4.

Figure 4 shows the response of LRNNIFT controller compared with FFNN and ENN based controllers and plant during the early stages of training. The instantaneous training was done for 60,000 time steps after which it was terminated. Post training, the controllers started tracking the reference plant's output.

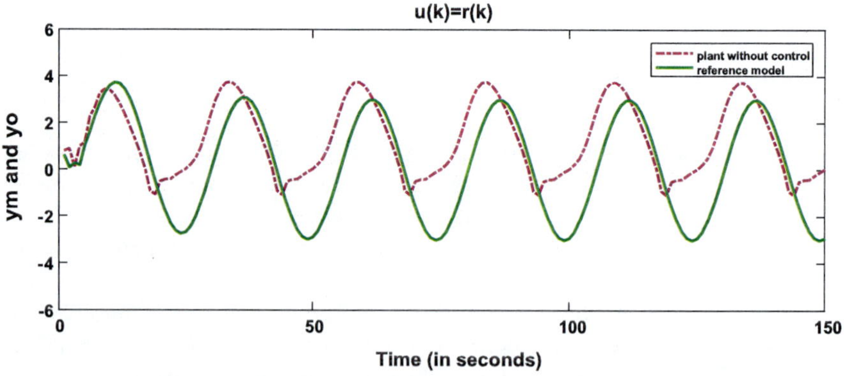

Fig. 3 Plant response without control scheme

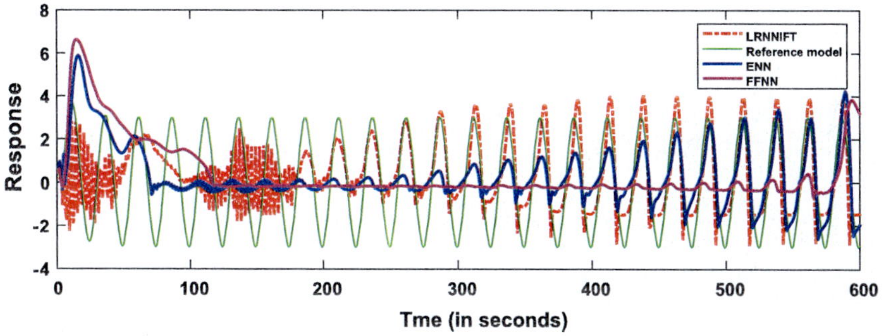

Fig. 4 Response of controllers during initial phases of training

which can be seen in Fig. 5. From Fig. 4, we can clearly observe that LRNNIFT controller has the fastest response among all three. Additionally, it is able to force the plant to track the reference model from the very first instant. Time of response of being a critical aspect in controller design makes the proposed controller better than ENN and FFNN-based control. Table 1. shows the Average Mean Square Error (AMSE) and Total Mean Average Error values for all the three controllers which is also the least for the proposed controller. The proposed controller is also checked for robustness against disturbance signals in the system. This is one of the key aspects of closed loop control. A step signal of amplitude 5 is added as disturbance to the plant at $k = 55{,}000$th instant.

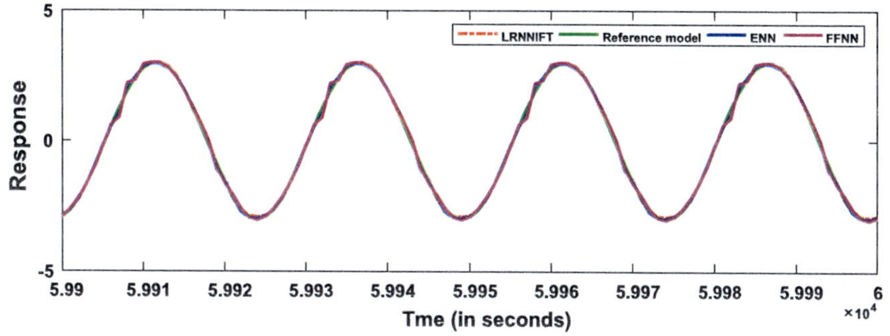

Fig. 5 Response of controllers after successful training

Table 1 Output error comparison of ENN, FFNN and LRNNIFT controllers

Error	Controllers		
	ENN	FFNN	LRNNIFT
Average MSE	0.0469	0.0907	0.0453
Total MAE	0.1329	0.1956	0.1261

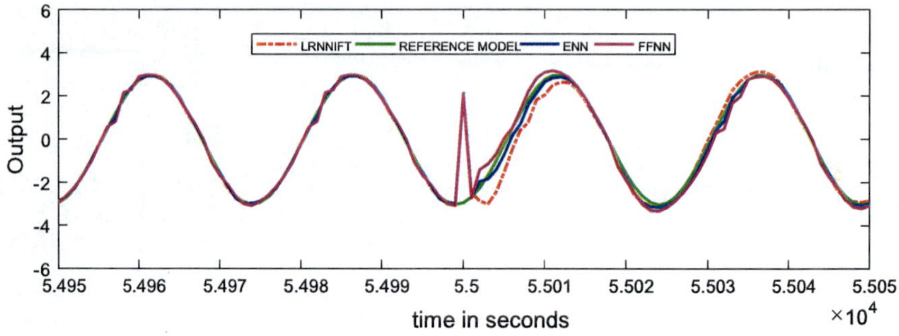

Fig. 6 Disturbance analysis for robustness

The disturbance leads to spike in controller response and the instantaneous mean square errors and mean average errors also experience the same. In Fig. 6, we can see the noise signal causing disturbance at $k = 55{,}000$ instant but as the training went on it rapidly recovered and went back on the track within few instants. This proves the robust or adaptive nature of the controller. On comparison, we can observe that the FFNN controller has under-performed whereas the ENN controller has performed only slightly better than LRNNIFT controller. This is because the ENN is an extremely complex structure, that is for equal number of inputs and hidden neurons, the number of update parameters (or weights) for ENN is 32 whereas for LRNNIFT it is only 20. FFNN controller has 16 weights only but it is slow and less accurate as it cannot track the plant's past values.

4 Conclusion

In this paper, an adaptive controller for nonlinear plants is proposed based on a locally recurrent network that has input fed through weights to the output (LRNNIFT). Parameter tuning by minimising the error function is done via back-propagation method. The controller is implemented on a nonlinear complex system and its results are compared with FFNN and ENN controllers. The simulation results clearly depict that the proposed controller performs better than the other two controllers both in terms of error mitigation and speed of tracking. The controllers are also tested for robustness by introducing a disturbance signal in the plant equation. It is observed that the proposed controller successfully adapts by moving back to the original track. Although the results of ENN are slightly better than LRNNIFT in b terms of robustness but the drawback here would be the high complexity of ENN network which again leads to the proposed controller to be a better choice. After extensive mathematical analysis and simulation results, we can conclude that the proposed controller provides better control over plants along with having a simpler structure.

References

1. Hägglund T, Åström KJ (2004) Revisiting the Ziegler-Nicholstuning rules for PI control—Part II the frequency response method. Asian J Contr 6.4:469–482
2. Gude JJ, Kahoraho E (2010) Modified Ziegler-Nichols method for fractional PIcontrollers. In: 2010 IEEE 15th conference on emerging technologies factory automation (ETFA 2010), 2010, pp 1–5. https://doi.org/10.1109/ETFA.2010.5641074
3. Wang QG, Lee TH, Zhang Y (1998) Multiloop version of the modified Ziegler Nichols method for two input two output processes. Ind Eng Chem Res 37(12):4725–4733
4. Lorenzini C, Alves Pereira LF, Bazanella AS (2016) A generalized forced oscillation method for tuning proportional-resonant controllers. IEEE Transactions on Control Systems Technology 28.3 (2019): 1108–1115.
5. Bazanella AS, Alves Pereira LF, Parraga A (2016) A new method for PID tuning including plants without ultimate frequency. IEEE Trans Contr Syst Technol 25(2):637–644
6. Sanger TD (1989) Optimal unsupervised learning in a single-layer linear feedforward neural network. Neural Netw 2(6):459–473
7. Svozil D, Kvasnicka V, Pospichal J (1997) Introduction to multi-layerfeed-forward neural networks. Chemom Intell Lab Syst 39(1):43–62
8. Medsker LR, Jain LC (2001) Recurrent neural networks. Design Appl 5:64–67
9. Hunt KJ et al (1992) Neural networks for control systems—a survey. Automatica 28(6):1083–1112
10. Isidori A, Sontag ED, Thoma M (1995) Nonlinear control systems, Vol 3. Springer, London
11. Perrusquía A, Yu W (2021) Identification and optimal control of nonlinear systems using recurrent neural networks and reinforcement learning: an overview. Neurocomputing 438:145–154

Solar Fed Speed Control of Water Pumping System Using Constant Voltage Controlled MPPT Technique and PI Controller

Aditya Nath Jha, Bhavnesh Kumar, and Arjun Tyagi

Abstract In this paper, regulation of solar fed water pumping system by Maximum Power Point Technique (MPPT) and Proportional–Integral (PI) controller is modeled. This system contains a Photovoltaic (PV) array, Buck converter controls by constant voltage controlled technique for extracting maximum power from PV array. Chopper A-type controlled by PI controller used for speed regulation of DC separately excited motor. In this system, constant voltage maximum power point tracking is employed because of its easy implementation on digital as well as on analog circuits. PI controller is used because of its simple structure and produces small steady-state error, also able to maintain reference speed. This system is developed and simulated in MATLAB/SIMULINK to analyze the performance. The outcomes obtained for the system from simulation have demonstrated satisfactory performance for different solar irradiance. It can maintain reference speed for different solar irradiance.

Keywords Buck converter · Maximum power point tracking (MPPT) · Photovoltaic (PV) · Proportional–integral (PI) controller

1 Introduction

In the current situation where the price of conventional energy is growing gradually, along with the requirement for electrical energy. To fulfill the demand for electrical energy one needs to search for a cheaper and cleaner source of energy. Renewable energies like solar, wind, tidal, geothermal, etc. can meet these requirements. In countries like India where sunlight is present in abundance, the utilization of solar energy should be prioritized [1]. An increase in the price of fuels greatly affects the agriculture sector where the diesel-based generator is used for water pumping,

A. N. Jha (✉) · B. Kumar
Department of Instrument and Control Engineering, Netaji Subhas University of Technology, New Delhi, India
e-mail: adityan.ie20@nsut.ac.in

A. Tyagi
Department of Electrical Engineering, Netaji Subhas University of Technology, New Delhi, India

A. Rani et al. (eds.), *Signals, Machines and Automation*, Lecture Notes in Electrical Engineering 1023, https://doi.org/10.1007/978-981-99-0969-8_33

PV standalone system can able replace diesel-based generator which not only saves money for farmers but also helps to reduce carbon footprints.

In industries, more than 40% of electrical power is consumed by electrical machines as compared to other electrical equipment present in the industry [2]. DC motor is one of the mostly used electrical machine in industries due to advantages like cheaper in cost, efficient design, full torque available at zero speed, free from harmonics, great speed regulations [3, 4]. Due to these advantages, it is mostly used in paper mills, water pumping, electrical cars and home applications. DC motor are used where the requirements are low speed variations and accuracy is high.

To extract maximum available power from solar panel, one need to employ the maximum power point technique like the constant voltage controlled MPPT algorithm [5]. It is easy to devise in analog as well as in digital circuits. Without MPPT algorithm, when DC motor is directly connected to PV array it can lose a substantial part of accessible solar energy from the PV array due to a mismatch of the meeting point of the current–voltage (I–V) curve of PV array and DC motor load line [6].

Speed control of DC motor is done by using a Proportional and Integral (PI) controller [7, 8] which triggers the gate pulse of the chopper circuit. Several researchers have proposed different techniques to control the speed of DC motor but the advantage of PI controller such as robust in terms of tracking of speed, easy to design and implementation according to the system, makes PI controller still popular in the industries.

In this paper, speed control of solar fed separately excited DC motor is done by PI controller and constant voltage controlled MPPT algorithm. The system modeling is presented in Sect. 2. Section 3 describe the results and analysis, followed by the conclusion in Sect. 4.

2 System Modeling

The suitable configuration of separately excited DC motor, PV array, Buck converter and PI controller for Solar fed system is shown in Fig. 1. The selection of building block in the system is such that it should not get affected by any external disturbances [9].

2.1 Design of PV Array

PV modules are configured together to form PV array to serve the electrical demand of separately Excited DC motor, Buck converter and losses of overall system by changing sunlight into electrical power [10]. PV cells are coupled in series and parallel arrangement to match the voltage, current and power demand of the overall system.

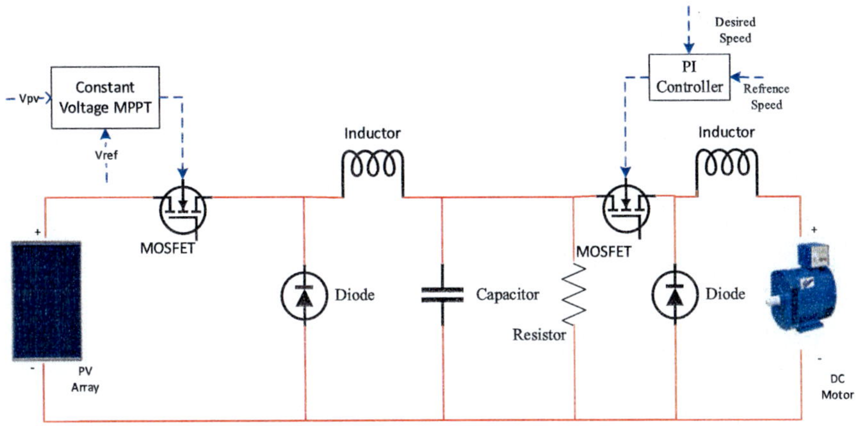

Fig. 1 Block diagram of Solar fed separately excited motor

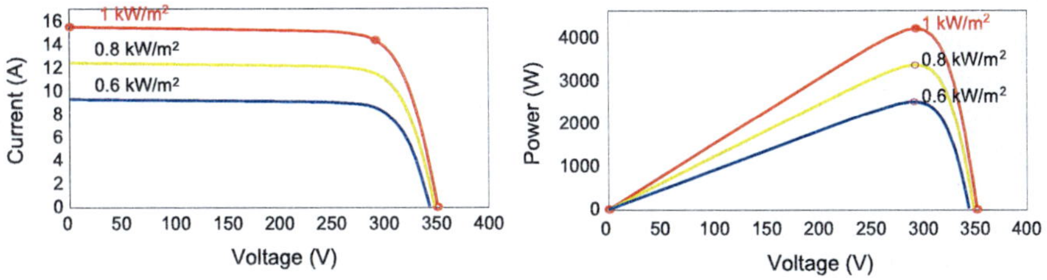

Fig. 2 V–I and V–W characteristic of 8 × 3 A10 green technology A10J-S72-175 for 1000, 800 and 600 W/m^2

The PV array model is chosen for this study is A10 Green Technology A10J-S72-175. It consist of eight series and three parallel strings to match the demand. The characteristics of voltage (V) versus current (I) and voltage (V) versus power (W) of this PV array model for the irritation of 1000, 800, 600 W/m^2 at 25 °C are shown in Fig. 2. The energy obtained from selected PV array module at 1000 W/m^2 is 4.2 kW by using constant controlled voltage algorithm. Gate pulse of buck converter is triggered to acquire maximum available power. The data of selected module A10 Green technology A10J-S72-175 is shown in Table 1. The 4.2 kW PV array is designed for 3.7 kW separately excited DC motor, the excess power is need for compensating losses of motor and buck converter.

2.2 Constant Voltage Controlled MPPT Algorithm

In constant voltage controlled MPPT technique assume a reference voltage for MPP voltage according to voltage calculated at standard test condition (STC) given by

Table 1 Details parameters of module A10 green technology A10J-S72-175

S. No.	Parameters	Value
1	Maximum power, P_{max}	175 W
2	Cells per module, n	72
3	Open circuit voltage, V_{oc}	44 V
4	Short circuit current, I_{sc}	5.17 Amp
5	Voltage at maximum power point, V_{PM}	37 V
6	Current at maximum power point, I_{PM}	4.78 Amp

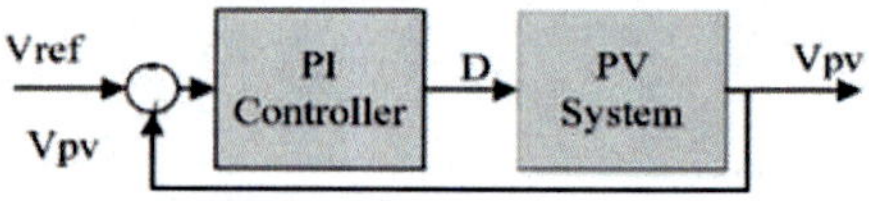

Fig. 3 Block diagram of constant excited motor voltage control

manufacturer. The reference voltage is used for feedback control loop that employs PI controller to trigger the gate pulse of buck converter to extract maximum power available on PV array as shown in Fig. 3. In constant voltage controlled MPPT algorithm, only required measurement is voltage of PV array that makes this algorithm very easy to implement in both digital and analog circuit.

2.3 *Separately Excited DC Motor*

The circuit diagram of separately excited DC motor is shown in Fig. 4. Where V is armature voltage, R is resistance of armature, L is armature inductance, E is back emf of separately excited DC motor and i_a is armature current flown in DC motor.

In separately excited DC motor field circuit is supplied from separate voltage source due to this field flux is always constant [11]. The parameter of separately excited DC motor selected in this study are as follows: Armature voltage = 240 V, Field voltage = 300 V, Speed = 1740 RPM, Power = 3740 W.

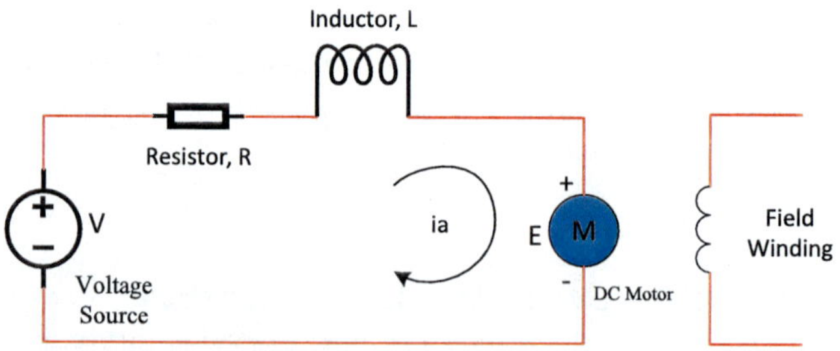

Fig. 4 Circuit diagram of separately excited DC motor

The mathematical model of separately excited DC motor is given by Eqs. (1)–(5).

$$V - E = Ri_a + L\frac{di_a}{dt} \tag{1}$$

$$E = \alpha w \tag{2}$$

$$I_a = \frac{V}{L} - \frac{a}{L}W - \frac{R}{L}i_a \tag{3}$$

$$T = T_1 + M\frac{dw}{dt} + Bw \tag{4}$$

$$T = \beta i_a \tag{5}$$

where *w* represents speed of DC motor in rad/sec. *T* represent torque of separately excited DC motor in N–m. T_1 represent load torque in N–m applied to separately excited DC motor. B represents friction of the DC motor. *M* represent inertia of separately excited DC motor in kg-m^2/s^2. β represent the coefficient of torque in N-m/A and α represent the coefficient of back emf in Vs/rad.

3 Simulation Result

The speed regulation of PV array fed separately excited DC motor is done by using MATLAB/SIMULINK package. A10 Green technology A10J-S72-175 PV array system is used for this study, for this eight modules in series and three strings in parallel are connected. The solar irradiance varies from 600, 800 to 1000 W/m^2 as shown in Fig. 5. To study the behavior of PV array and separately excited DC motor different solar irradiance is applied.

Fig. 5 Solar irradiance profile

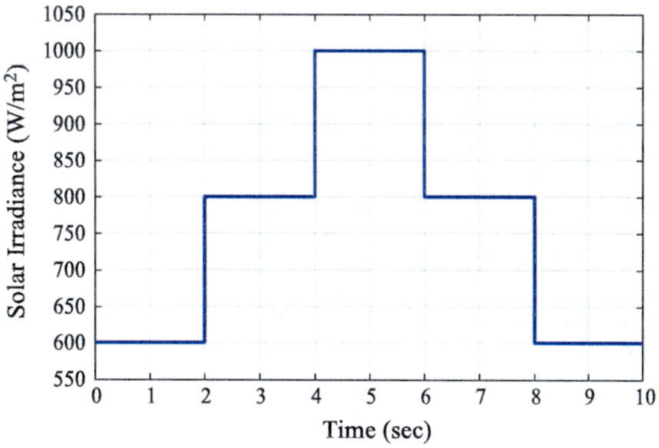

To simulate water pumping system the load torque applied to DC motor is square function of speed of motor. The ratings of separately excited DC motor of permanent magnet or wound type is 240 V armature voltage, speed of rated DC motor is 1740 RPM, 300 V field voltage. Output of chopper is powering the armature of separately excited DC motor which also controls the voltage of motor, in turn control the speed of DC motor. The field circuit of separately excited DC motor is supplied by 300 V DC.

The voltage and current response of PV array is shown in Fig. 6 for varying solar irradiance and constant cell temperature at 25 °C. The PV array voltage is constant at 293 V for different irradiance and PV array current is maximum at 1000 W/m^2 which is 14.4 A. To achieve MPPT, constant voltage controlled MPPT algorithm is applied, because it only requires voltage measurement. Using mentioned MPPT technique and PI controller tuned the duty ratio of buck converter to extract maximum power available in PV array [12]. The gate circuit receive the tuned duty ratio and triggers the buck converter. The step-down voltage response of converter and input voltage applied to the converter is shown in Fig. 7.

To extract the maximum power from PV array using constant voltage controlled MPPT algorithm, it needs standard test condition (STC) reference voltage in this case it is 293 V at 1000 W/m^2. With constant voltage technique PV array reaches STC reference value in less than 0.1 s.

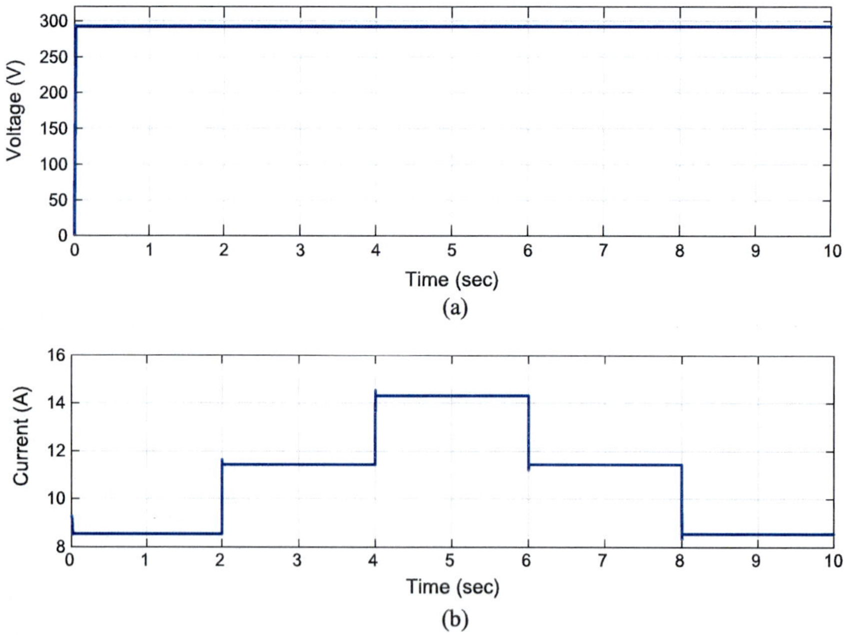

Fig. 6 **a** PV voltage current **b** PV array current response for varying solar irradiation

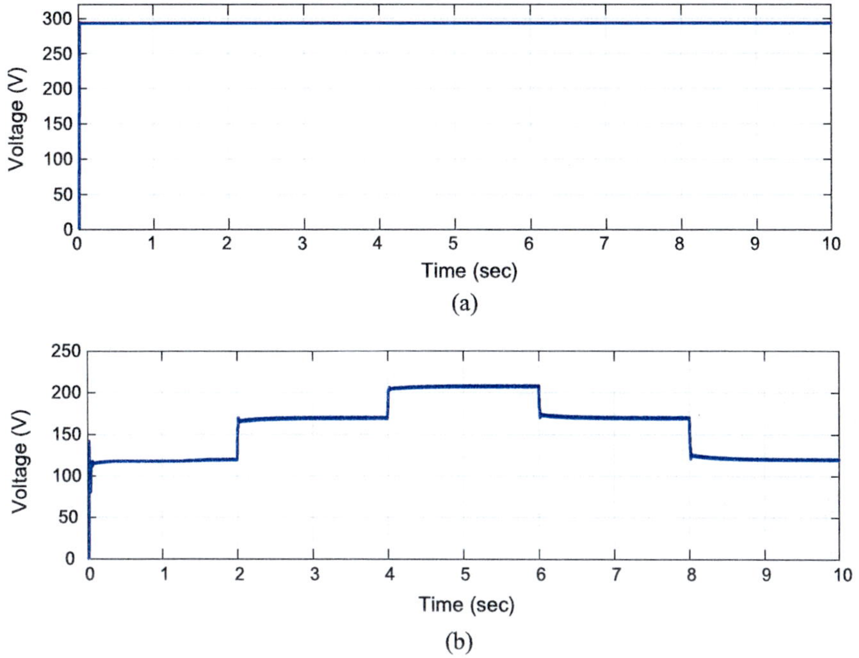

Fig. 7 **a** Input voltage response to buck converter and **b** output voltage response of buck converter for varying solar irradiations

The step-down voltage generated from buck converter becomes the input voltage for chopper circuit. Triggering of chopper circuit is done by conventional Proportional and Integral (PI) controller which control the voltage across the armature of separately excited DC motor, in turn to control the speed of motor. For triggering of gate pulse of chopper, the reference speed is set at 1400 RPM and with the help of PI controller and Pulse Width Modulation (PWM) generator it compare with actual speed of separately excited DC motor. At starting, DC motor draws high current from DC link capacitor and settle downs when voltage in capacitor starting to reduce in turn actual speeding is increasing but at time $t = 1.5$ s, DC motor speed reaches to selected desired speed without any fault. Insolation of PV panel varies from 600 to 800 to 1000 W/m^2 due to study the behavior of DC motor at different irradiation. In Fig. 8 speed, torque and current response of separately excited DC motor is shown.

4 Conclusion

The PV fed speed control of separately excited DC motor using constant voltage controlled MPPT and PI controller is implemented in this work. The solar irradiance of PV array varies in step size from 600 to 800 to 1000 W/m^2, it increases as well

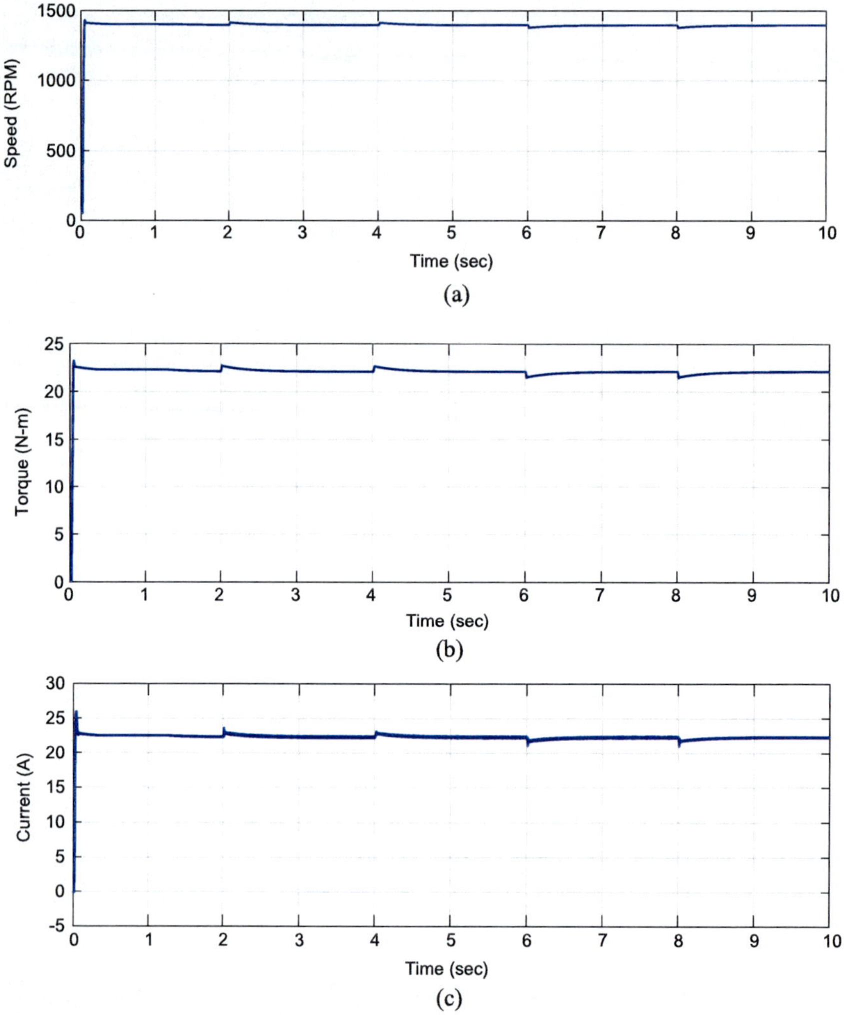

Fig. 8 **a** Speed response of DC motor **b** load torque response of DC motor and **c** armature current response of DC motor for varying solar irradiations

as decrease in given time. In both cases, using MPPT algorithm maximum power attained by attaining standard test condition voltage 293 V and using conventional PI controller actual speed of DC motor is able to reach reference speed 1400 RPM.

References

1. Shukla S, Singh B (2018) Reduced current sensor based solar PV fed motion sensorless induction motor drive for water pumping. IEEE Trans Industr Inf 15(7):3973–3986
2. Chauhan U, Singh V, Kumar B, Rani A (2020) An improved MVO assisted global MPPT algorithm for partially shaded PV system. J Intell Fuzzy Syst 38(5):6715–6726
3. Munje RK, Roda MR, Kushare BE (2010) Speed control of DC motor using PI and SMC. In: Conference proceedings IPEC, pp 945–950
4. Shobha Rani D, Mahankali M (2017) Boost converter fed high performance BLDC drive for solar PV array powered air cooling system. Adv Electr Electron Eng 15(2):154–168
5. Mohammed AE, Zahawi B, John Atkinson D (2010) Comparison of directly connected and constant voltage controlled photovoltaic pumping systems. IEEE Trans Sustain Energy 1(3):184–192
6. Rajan K, Singh B (2017) Solar PV powered BLDC motor drive for water pumping using Cuk converter. IET Electr Power Appl 11:222–232
7. Dogruer T, Nusret T (2018) Design of PI controller using optimization method in fractional order control systems. IFAC-Papers Line 51(4):841–846
8. Singh S, Singh B (2015) Solar PV water pumping system with DC-link voltage regulation. Int J Power Electron 7(1–2):72–85
9. Mani P, Mahadevan SK, Johnson AR, Kullan M (2022) An optimized design modelling of PV integrated SEPIC-based four-switch inverter for sensorless PMBLDC motor control. Automatika 63(1):90–101
10. Rajan K, Singh B (2017) Single stage solar PV fed brushless DC motor driven water pump. IEEE J Emerg Select Top Power Electron 5(3):1377–1385
11. Najet R, Hmidet A, Gammoudi R, Hasnaoui O (2015) Implementation of photovoltaic water pumping system with MPPT controls. Front Energy 9:187–198
12. Chauhan U, Rani A, Kumar B, Singh V (2019) A multi verse optimization based MPPT controller for drift avoidance in solar system. J Intell Fuzzy Syst 36(3):2175–2184

An Intelligent MPPT Technique for a Three-Stage Battery Charge Controller for Standalone System

Alina Hussain and Manisha

Abstract This paper presents a comparative analysis between two different MPPT algorithms for a 3-stage battery charge controller (BCC) using a standalone system. A DC–DC buck converter is used as a step-down converter. Two different algorithms are used to extract the maximum power from solar PV panel. The maximum power from the solar photovoltaic panel is extracted using a conventional approach of Perturbation and Observation, as well as an intelligent MPPT technique called Fuzzy logic control. A battery charge controller (BCC) is used to charge the battery by using three different stages of the charging strategy. The different stages of charging incorporate Stage1—Bulk charging, Stage2—Absorption charging and Stage3—Float charging stage. The overall performance of the model is measured in terms of MPP tracking, lead-acid battery charging and controller efficiency. The output shows that the MPPT charge controller can track the MPP within 0.5 s regardless of solar irradiation variation. The concept of charging the battery in stages is implemented. The efficiency of the battery charge controller is attained up to 98.86% with a Fuzzy logic controller.

Keywords Fuzzy logic controller (FLC) · Perturb and observe (P&O) · Battery charge controller (BCC)

1 Introduction

Over the last few years, researchers from all over the world have been working furiously in the renewable energy area to provide clean and eco-friendly energy. Because PV systems depend on sunlight to create electricity, they can only be used during the day when sunlight is present. Therefore, one of the alternatives to store energy is batteries. As a result, the solar photovoltaic charge controller plays a very important role in allowing this solution to be possible. The MPP charge controller for solar photovoltaics is made up of a BCC and an MPP tracker. MPPT controller extracts the maximum amount of power from the solar photovoltaic panel and transfers it to the

A. Hussain (✉) · Manisha
Department of Instrumentation and Control Engineering, Netaji Subhas University of Technology, New Delhi 110078, India
e-mail: alinah.ie20@nsut.ac.in

A. Rani et al. (eds.), *Signals, Machines and Automation*, Lecture Notes in Electrical Engineering 1023, https://doi.org/10.1007/978-981-99-0969-8_34

BCC. The battery is charged using a multi-stage charging process. To avoid damage from excessive charge gassing and overheating issues, this procedure is utilized to charge the battery. In this research paper, MPPT performance is evaluated in terms of tracking time and tracking efficiency with two different MPPT algorithms.

2 Methodology

This model is comprised of a solar photovoltaic panel, a buck converter, a battery and an MPPT charge regulator system. Figure 1 gives an outline of the solar PV MPPT battery charge control system configuration. The block of the MPPT charge control system contains a P&O MPPT algorithm as well as a 3-stage charge regulator for lead-acid batteries. For the implementation of an intelligent technique in solar PV battery charge control system Fuzzy logic is also implemented with 3-stage charge regulators with lead-acid battery. This system configuration is fit to charge a battery of 48 V from the 2-kW solar photovoltaic power source.

2.1 *Solar PV System*

The SPV system's efficiency and output power are entirely dependent on PV array configurations and different weather conditions such as sun irradiation and temperature changes. The work in this study is based on a constant temperature of 25 °C and variable solar irradiations of 600–1000 W/m^2 (Table 1).

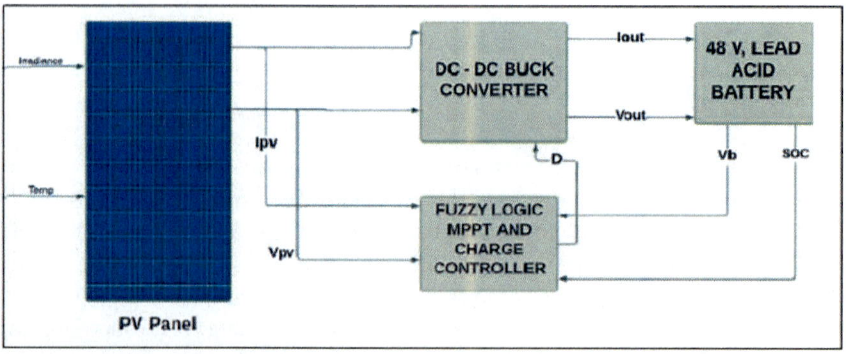

Fig. 1 MPPT system configuration block diagram

Table 1 Solar PV panel parameter specifications

Measuring parameter	Values
Maximum power (P_m)	250 W
Maximum current (I_m)	8.1 A
Maximum voltage (V_m)	30.9 V
Short circuit current (I_{sc})	60 A
Open circuit voltage (V_{oc})	36.6 V
Number of cells in parallel (N_p)	2
Number of cells in series (N_s)	4

2.2 *Buck Converter*

A buck converter is used in this model because the voltage of the solar PV panel is higher than the voltage of the battery. Therefore, a buck converter is used to reduce the PV panel's voltage while sustaining power transfer to the battery. The buck converter's design equation can be obtained from Eqs. (1) to (4).

$$D = \frac{V_{out}}{V_{in}} \tag{1}$$

$$R_{in} = \frac{R_{load}}{D} \tag{2}$$

$$I_L = \frac{V_{in} D(1 - D)}{fswL} \tag{3}$$

$$V_C = \frac{V_{in} D(1 - D)}{8Lf2C} \tag{4}$$

where D represents the buck converter's duty cycle, R_{in} represents the input resistance, I_L represents the inductor ripple current and V_C represents the buck converter's output capacitor voltage (Table 2).

Table 2 Parameter specification of the buck converter

Parameters	Values
Value of the capacitor, C	1000 Uf
Value of the Inductor, l	10 Mh
Input voltage of the converter, V_{in}	120 V
Switching frequency of converter, F_s	1000 Hz
Duty cycle of the converter, D	0.4
Capacitor ripple voltage, dV_c	2.3 Nv
Inductor ripple current, di_l	0.288 Ma

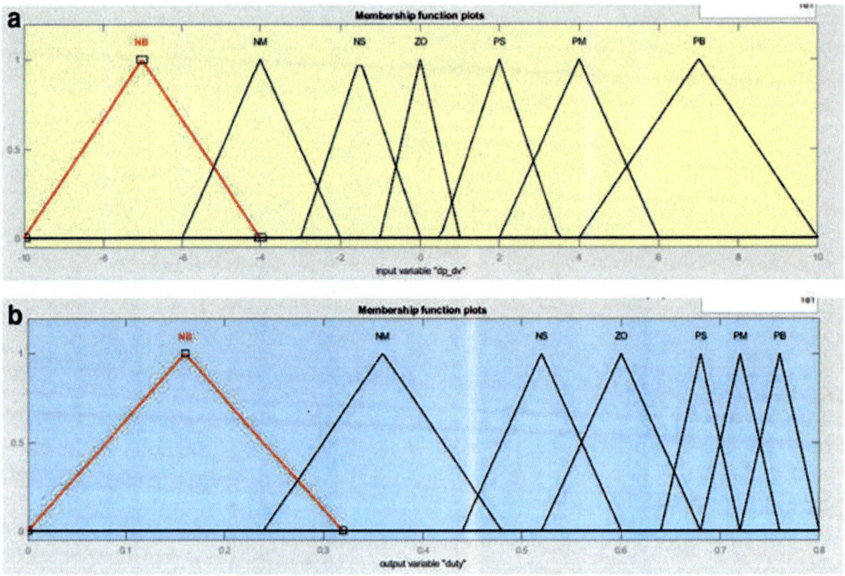

Fig. 2 **a** Input MF dp_dv. **b** Ouput MF 'duty cycle'

2.3 MPPT Algorithm–Fuzzy Logic Controller

To deal with non-linearity, imprecise input and exact output in an inaccurate system model, an MPPT–FLC is used. A fuzzy logic controller can be designed in three steps. Fuzzification is the initial stage, and it is used to convert crisp input into a linguistic variable. The inference system, which consists of a rule base, is the second step, and defuzzification, which converts the fuzzy output into crisp output, is the third step. Membership functions include Neg. Big (NB), Neg. Medium (NM), Neg. Small (NS), Zero (ZO), Pos. Small (PS), Pos. Medium (PM), and Pos. Big (PB) are used for both input error and change in error (Figs. 2 and 3).

2.4 MPPT Algorithm–Perturb and Observe

Many industrial solar Photovoltaic charge controllers use the P&O MPPT because of the ease with which it can be tracked and implemented. This MPPT algorithm measures the PV array's maximum power and delivers a duty cycle proportional to that power to the battery charge controller.

ΔP \ ΔV	NB	NM	NS	ZE	PS	PM	PB
NB	NB	NB	NB	NB	NM	NS	ZE
NM	NB	NB	NB	NM	NS	ZE	PS
NS	NB	NB	NM	NS	ZE	PS	PM
ZE	NB	NM	NS	ZE	PS	PM	PB
PS	NM	NS	ZE	PS	PM	PB	PB
PM	NS	ZE	PS	PM	PB	PB	PB
PB	ZE	PS	PM	PB	PB	PB	PB

Fig. 3 Fuzzy logic controller rule base

3 Results and Discussion

For performance analysis, the maximum power point battery charge regulator for the standalone PV system model was successfully implemented. A user-defined method with variable steps is used to configure the Simulink model. This system configuration is fit to charge a battery of 48 V from a 2-kW solar photovoltaic power source also controlling the charging by utilizing a 3-stage battery charging technique. This model gives the overall efficiency up to 98.86% which is similar to some top-of-the-line business solar photovoltaic maximum power point trackers for charge controllers (Fig. 4).

3.1 *Performance of the Solar PV Panel*

Figure 5a Shows the comparative analysis between the output current of the PV panel with P&O and FLC-based MPPT algorithm. With FLC smooth DC output current is obtained whereas there is some disruption in output current in P&O.

Figure 5b Shows the comparative analysis between output voltage obtained from the PV panel with P&O and FLC. In each step, FLC has less oscillations, faster response time and precise tracking as compared with the P&O MPPT algorithm.

Figure 5c Shows the comparison b/w maximum output power obtained from the PV panel. Both P&O and FLC MPPT effectively track the maximum power as shown in Fig. 5c. A fuzzy logic controller must be chosen over the P&O controller for practical application because of its superior performance. As a result, the FLC has higher performance and is closer to the P&O.

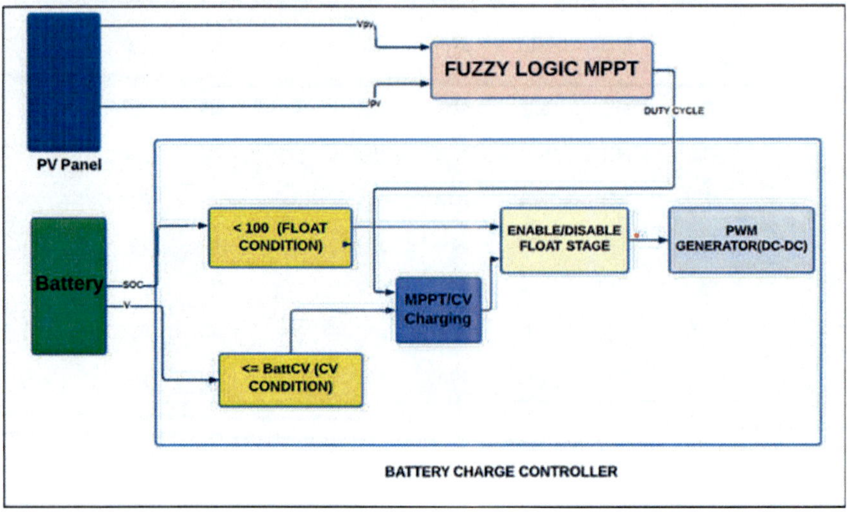

Fig. 4 Block diagram of a battery charge controller with FLC

3.2 Performance of the Lead/Acid Battery

Figure 6a shows the battery's output current with P&O and FLC MPPT algorithm. With FLC smooth DC output current is obtained whereas with P&O there is a lot of oscillations in output current is obtained.

Figure 6b shows the comparative analysis between output voltage of the battery with P&O and FLC MPPT algorithm.

Figure 6c shows the comparison between SOC of the battery. SOC obtained with FLC is higher than the SOC obtained from the P&O MPPT algorithm (Table 3).

4 Conclusion

In this paper, a complete simulation for a solar photovoltaic MPPT charge regulator model is carried out effectively in MATLAB. This accomplished the overall efficiency up to 98.86% which is similar to some top-of-the-line business solar photovoltaic Simulink. A fuzzy logic technique-based MPPT controller, P&O-based MPPT tracking algorithm, DC–DC step-down converter and 3-stage battery charge regulator are demonstrated. This system configuration is fit to charge a battery of 48 V from a 2-kW solar photovoltaic power source also controlling the charging by utilizing a 3-stage battery charging technique with maximum power point trackers for charge controllers. This MATLAB-Simulink system model introduced can be easily modified to match the needs of any modern Maximum power point charge controller having a similar configuration. According to the simulation, the PV system can give

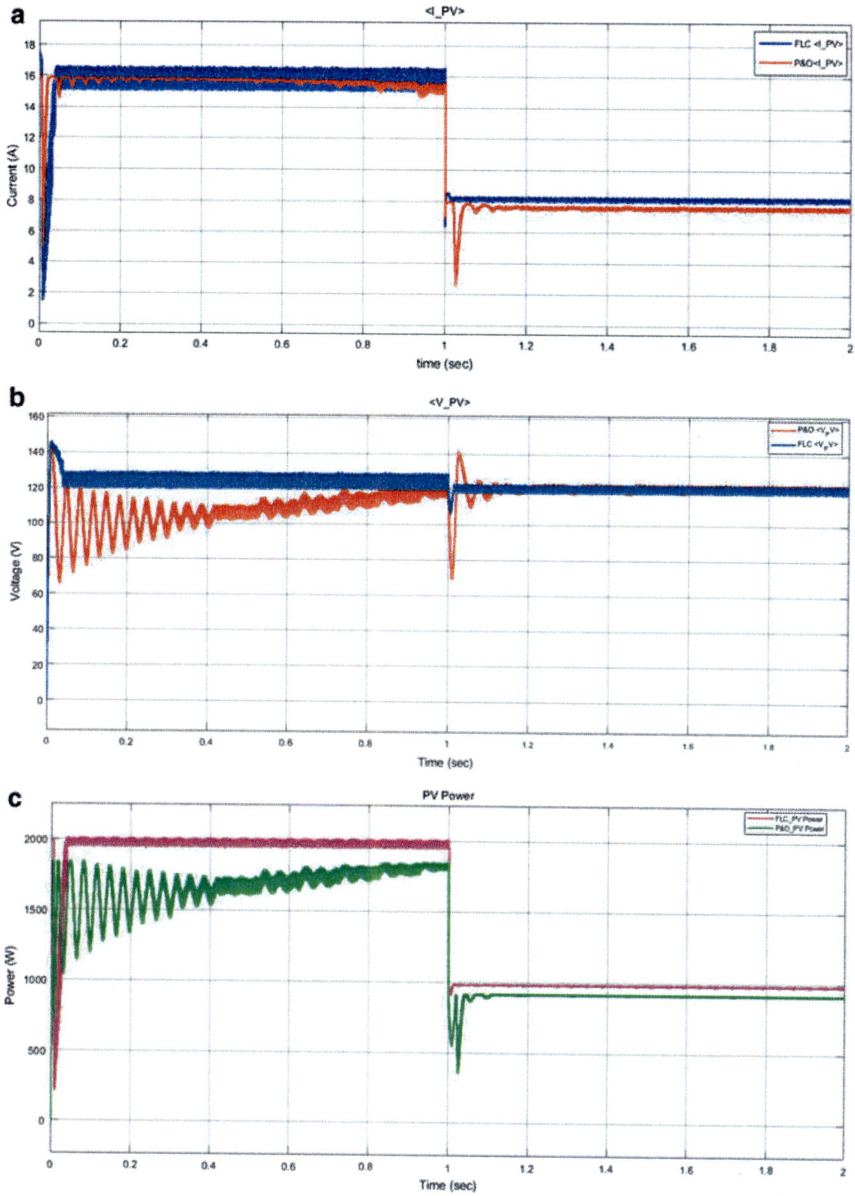

Fig. 5 **a** The output current of the PV panel with P&O and FLC. **b** The output voltage of the PV panel with P&O and FLC. **c** Maximum power of the PV panel with P&O and FLC

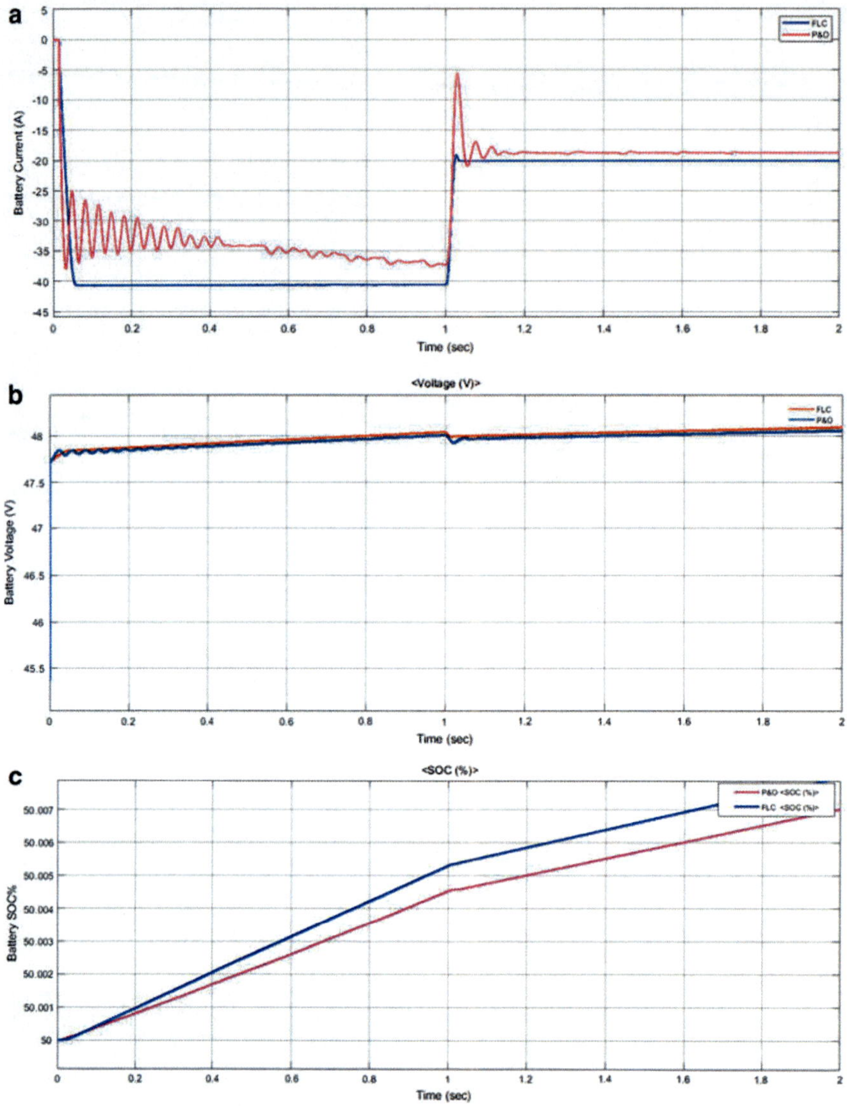

Fig. 6 **a** Battery's output current with P&O and FLC. **b** Battery's output voltage with P&O and FLC. **c** SOC of the battery with P&O and FLC

the maximum power with both MPPT controllers. Fuzzy MPPT, on the other hand, outperforms standard controllers when it comes to nonlinear systems. When MPP is detected, it has the capacity to reduce perturbed voltage. In contrast to the traditional MPPT, where the output power fluctuates about MPP, this activity saves a more consistent outcome power.

Table 3 Comparative analysis using FLC and P&O MPPT techniques

Parameter	P&O	FLC
The efficiency of the controller	98.85%	98.86%
Ripple in output voltage (SPV)	High	Low
Ripple in output current (SPV)	High	Low
Ripple in output power (SPV)	High	Low
State of charge (battery)	50.06	50.07
Output current (battery)	Fluctuating	Smooth DC
Ripple in output voltage (battery)	Negligible	Negligible
Control strategy	Voltage control	Intelligent control

References

1. Tripathi S, Shrivastav A (2020) Fuzzy logic controller based solar MPPT and battery charging for hybrid vehicle application. AIP Conf Proc
2. Rodney HGT, Chee Kang E, Solanki1 SG (2020) Modeling of photovoltaic MPPT lead acid battery charge controller for standalone system applications. E3S Web Conf 182:03005
3. Manisha, Gaur P (2021) The survey of MPPT under non-uniform atmospheric conditions for the photovoltaic generation systems. Int J Inf Technol 13:767–776
4. Salman S, Al X, Zhouyang W (2018) Design of a P&O algorithm based MPPT charge controller for a stand-alone 200W PV system. Protect Contr Modern Power Syst 3:1–8
5. Gaur P, Pathak D, Kumar B, Chauhan YK (2016) PI and fuzzy logic controller based tip speed ratio control for smoothening of output power fluctuation in a wind energy conversion system. In: 7th India international conference on power electronics (IICPE)
6. Tripathi SK, Shrivastava KC (2019) Jana Robust throttle control of hybrid electric vehicle. IOP Conf Ser Mater Sci Eng
7. Manisha, Gaur P (2018) A review of MPPT techniques for photovoltaic system. IEEE Conf INDIACom
8. Pathak PK, Yadav AK (2019) Design of battery charging circuit through intelligent MPPT using SPV system
9. Patil R, Anantwar H (2017) Comparative Analysis of fuzzy-based MPPT for buck and boost converter topologies for PV application
10. Enany MA, Farahat MA, Nasr A (2016) Modelign and evaluation of main maximum power point tracking algorithms for photovoltaics systems. Renew Sustain Energy Rev 58:1578–1586
11. Karami N, Moubayed N, Outbib R (2017) General review and classification of different MPPT techniques. Renew Sustain Energy Rev 68:1–18
12. Mohapatra BN, Dash A, Jarika BP (2017) Power saving solar street lights. Int J Emerg Technol Eng Res 5:105–109
13. Lopez J, Seleme SI Jr, Donoso PF, Cortizo PC, Severo MA (2016) Digital control strategy for a buck converter operating as a battery charger for stand-alone photovoltaic systems. Sol Energy 140:171–187

14. Pathak D, Gaur P (2019) A fractional order fuzzy-proportional-integral-derivative based pitch angle controller for a direct-drive wind energy system. Comp Electr Eng 78

Performance Analysis of BLDC Motor Drive Using Various Speed Controller

Akash Deep Karnwal, Surya Kant, and Nidhi Singh

Abstract DC motors with brushless technology are permanent magnet synchronous motors with trapezoidal back electromotive forces. High torque, minimal losses, and high performance make BLDC motors a popular choice in industries, but BLDC motor has some limitations such as current ripples, low control accuracy, and torque ripples. By using conventional PI, PID, and fuzzy logic controller, these problems can be overcome. This paper presents the mathematical modeling and implementation of different types of speed controllers for speed control of BLDC motor drive. The controllers used in this paper are traditional controllers along with fuzzy logic controller. Normal PID controllers are difficult to tune and achieve satisfied control characteristics. Fuzzy logic controllers have the ability to achieve and maintain satisfactory control characteristics. The speed of the BLDC motor is analyzed and compared at variable speed under various controlling strategies using MATLAB/Simulink. Two different controllers are used to analyze and compare the dynamic performance of brushless DC motor drive regarding constant torque operation at different speed.

Keywords Brushless DC motor (BLDCM) · PI controller · PID controller · Fuzzy logic controller (FLC) · MATLAB/Simulink

1 Introduction

Brushless DC motor is broadly employed in lots of systems that call for robust torque with fast response speed and is famous because of its low price, durability, particularly flexible response, and low maintenance. We cannot simply dismiss electrical motors and their operation on the basis that they convert electrical energy to mechanical energy. We need to take into account the importance of technologies

A. D. Karnwal · S. Kant (✉) · N. Singh
Department of Electrical Engineering, Gautam Buddha University, Greater Noida 201312, India
e-mail: suryakantshukla8@gmail.com

S. Kant
Electrical Engineering Department, National Institute of Technology, Raipur, India

A. Rani et al. (eds.), *Signals, Machines and Automation*, Lecture Notes in Electrical Engineering 1023, https://doi.org/10.1007/978-981-99-0969-8_35

that depend on this mechanism [1, 2]. Because of their high overall performance, excessive power element, and low noise operation, BLDC motors are rapidly gaining traction in industries such as electronics, HV AC, hard disc drives, electronics, vehicles, and instruments. Since few years, continuing technological improvements in power semiconductors, microprocessors, adjustable speed driver control schemes, and permanent magnet brushless electric motor production were combined to provide a reliable, cost-effective solution for a huge range of adjustable speed applications [3, 4]. A BLDCM is a conventional DC motor without brushes. It is far much like a permanent magnet synchronous motor, with the exception that a PMSM produces sinusoidal back electromotive force, however, a brushless DC motor produces trapezoidal back electromotive force. BLDC motors have permanent magnets on the rotor and coils on the stator, to cause the rotor to rotate. The position of rotor is detected using Hall Effect sensors as it moves [5]. Motor speed is the parameter that needs to be regulated in the proposed work. Thus, the different controllers are provided with the correct error signal depending on the reference speed and motor speed. Upon receiving the error, the controller generates the necessary corrective signal by tuning the parameters of the controller in use [6]. Adjustment or tuning of PID controllers is typically carried out using proportional, integral, or derivative terms. The Ziegler–Nichols approach depends on trial and error to tune the parameters of proportional, integral, and deferential (PID) controller design (K_p, K_i, and K_d). Unexpected movement of the set point and variation of the parameters is the main obstacles facing PID technique. This causes poor performance. [7, 8]. Fuzzy logic controllers (FLC) are able to alleviate this problem because unlike conventional control systems, FLCs do not require any in-depth knowledge of the transmission characteristic of the system. Unlike a fuzzy logic device-based entirely on expert databases, fuzzy logic control devices require less calculations, but they lack the capabilities for the new rules [9, 10].

As a result of this paper's detailed comparison of BLDC motors with conventional PID, PI, and fuzzy logic controllers, we gain a deeper insight into their capabilities and limitations.

This paper is organized in four sections, first, there is a brief introduction about this paper, then second section is all about mathematical modeling after that third section consists of various controllers used in this model, and last, there is a detailed performance analysis of various speed controlling strategies.

2 Mathematical Modeling of BLDC Motor

$$\begin{bmatrix} v_a \\ v_b \\ v_c \end{bmatrix} = \begin{bmatrix} R & 0 & 0 \\ 0 & R & 0 \\ 0 & 0 & R \end{bmatrix} \begin{bmatrix} i_a \\ i_b \\ i_c \end{bmatrix} + \begin{bmatrix} L-M & 0 & 0 \\ 0 & L-M & 0 \\ 0 & 0 & L-M \end{bmatrix} \frac{d}{dt} \begin{bmatrix} i_a \\ i_b \\ i_c \end{bmatrix} + \begin{bmatrix} e_a \\ e_b \\ e_c \end{bmatrix} \quad (1)$$

where the phase voltages are v_a, v_b, and v_c, the phase currents are i_a, i_b, and i_c, and the phase back-EMF waveforms are e_a, e_b, and e_c. M represents the mutual inductance between two phases, whereas R represents the phase resistance, L represents the phase self-inductance, the electromagnetic torque may be determined and it is given as:

$$T_{e=\frac{1}{w_r}}(e_a i_a + e_b i_b + e_c i_c) \tag{2}$$

where the rotor's mechanical speed is represented by W_r.

An equation of motion can be stated as follows:

$$\frac{d}{dt}w_r = (T_e - T_L - Bw_r)/J \tag{3}$$

According to the following equation, the electrical speed W_e of a motor with P poles is related to its mechanical speed.

$$w_{e=(\frac{P}{2})w_r} \tag{4}$$

3 Speed Controlling Methods

The speed of BLDC can be controlled with a variety of controllers, such as PI, PID, and FLC.

3.1 PI Controller

PI controllers use an error signal resulting from the difference between input and output values. Input speed and motor speed are used to provide the PI controller with the desired error signal, since motor speed is the parameter that needs to be regulated. Once the controller receives this error signal, it gives the necessary corrective signal on the basis of tuning of K_p and K_i.

$$\text{Output} = K_p.e(t) + K_i \int_0^t e(t)dt \tag{5}$$

e(t) is the instantaneous error in the signal, while K_p represents the proportional gain, and K_i represents the integral gain.

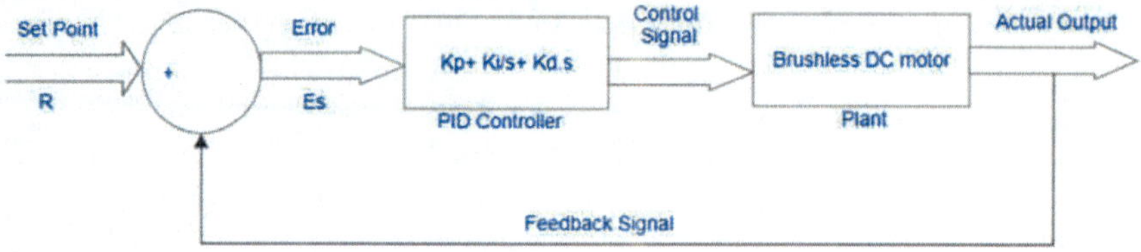

Fig. 1 BLDC motor with PID controller system block diagram

3.2 PID Controller

In the Ziegler–Nichols (ZN) technique, the parameters governing the proportional, integral, and deferential) controller design (K_p, K_i, and K_d) are tuned by trial and error. The objective of this technique is achieved by first determining the motor parameter values of an open-loop system for computing the PID factor gains using the BLDC motor model with a purpose to achieve the optimum response.

The command characteristic is written as:

$$\text{Output} = K_p.e(t) + K_i \int_0^t e(t)\text{d}t + K_d \frac{\text{d}e(t)}{\text{d}t} \tag{6}$$

Here the effects of control parameters are if K_p is increased independently, then rise time will decrease, the steady-state error will decrease, and there is a slight change in settling time and overshoot will increase. If we increase the integral constant, then rise time will increase, settling time will increase, and overshoot will also increase, but it will eliminate steady-state error. Last, if the derivational constant is increased, then rise time will slightly increase, settling time will decrease, overshoot will decrease, and there will be no effect on steady-state error. An example of a closed loop controller is shown in Fig. 1.

3.3 Fuzzy Logic Controller

FL toolbox in MATLAB gives an understandable processing controller which has been developed. In Fig. 2, there are four main components of the FLC: fuzzification, knowledge base, inference engine, and defuzzification interface. Fuzzification is a procedure of adding fuzziness to information in fuzzy logic. Fuzzy linguistic descriptions are a kind of formal representation of a system this is created using fuzzy IF–THEN rules. They used sentences like IF (a set of conditions) is satisfied, then (a set of consequents) may be deduced to encode information about a system. Defuzzification is the procedure of converting a fuzzy set right into a single crisp value. Defuzzification produces a crisp result in Mamdani type FIS, which may be utilized for both a couple of inputs and single output and more than one input and multiple outputs structures.

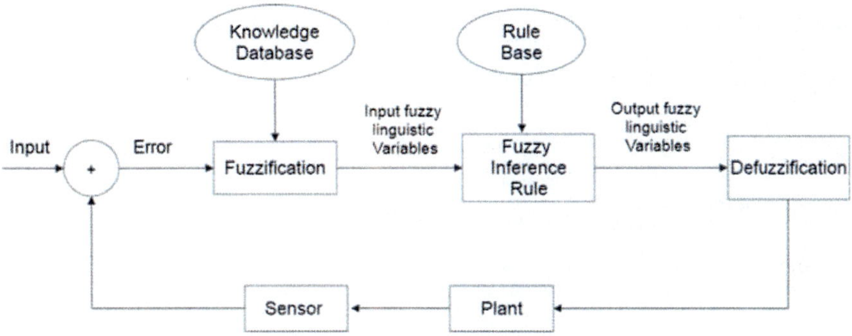

Fig. 2 Block diagram of fuzzy logic controller

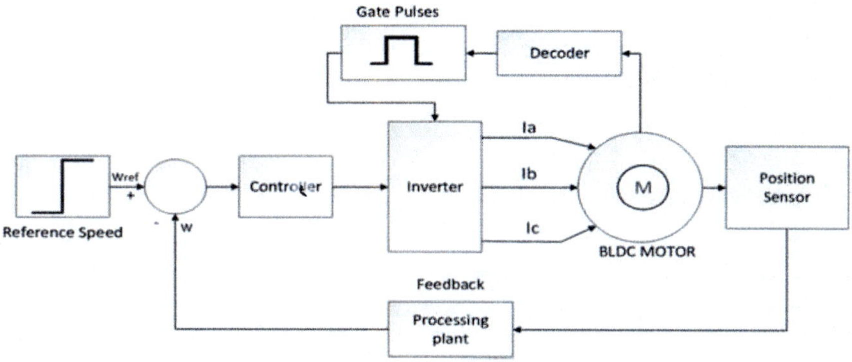

Fig. 3 Block diagram of BLDC motor drive

4 Control of BLDC Motor Drive

Figure 3 shows a block diagram of BLDCM drive, in which first reference speed is given as step function which is then compared with motor speed measured from position sensor via processing plant feedback then if any error is detected, it gets minimized by using different controller types (PI, PID, and FLC) as well as providing a reference signal. The inverter gate signals are generated by comparing controller output and decoded signal of Hall Effect sensor then BLDC motor is fed through PWM inverter.

5 Results and Discussions

MATLAB/Simulink is required to simulate brushless DC motor (BLDCM) drive speed control with constant torque and variable speed using different controllers. The

Table 1 BLDC motor parameters used in simulation model

Parameters	Ratings
Stator phase resistance (*R*)	2.8750 Ω
Stator phase inductance (*L*)	8.5 mH
Number of slots (*N*)	18
Voltage constant (*V*)	40 V/Krpm
Number of poles (*P*)	16
Number of phases (Ph)	3
Voltage input (V_{in})	24 V
Power input (P_{in})	40 W
Proportional constant (K_{p})	0.015
Integral constant (K_i)	0.25

simulation study uses a three-phase 40 W, 1800 rpm BLDCM with recommended ratings given in Table1.

5.1 *Performance Analysis of BLDCM Drive at Rated Speed*

Figure 4a–c shows the speed, torque, and current characteristics of BLDC motor at rated speed using proportional-integral (PI), PID controller, and fuzzy logic controller. At starting the motor is running at no load with rated speed. After 0.15 rated load is applied to the motor. It is evident from Fig. 4, that speed and motor torque follow the reference smoothly. Fuzzy logic controller provides improved dynamic characteristic of speed torque and current compared to other two controller that is PI and PID.

5.2 *Performance Analysis of BLDCM Drive at Step Changes in the Speed*

Figure 5 shows speed, torque, and current characteristics of BLDC motor at step changes in speed using proportional-integral (PI) controller, PID controller, and fuzzy logic. At starting motor is running at no load with speed of 1500 rpm after 0.15 load is applied to motor. There is some dip in speed due to application of load. The speed and torque track the reference speed and torque smoothly. At 0.25 speed is changed from 1500 to 1800 rpm. The torque of motor increases in transient condition after some time it will settle at reference as the speed reached at 1800 rpm here the fuzzy logic controller gives improved dynamic performance than PI and PID.

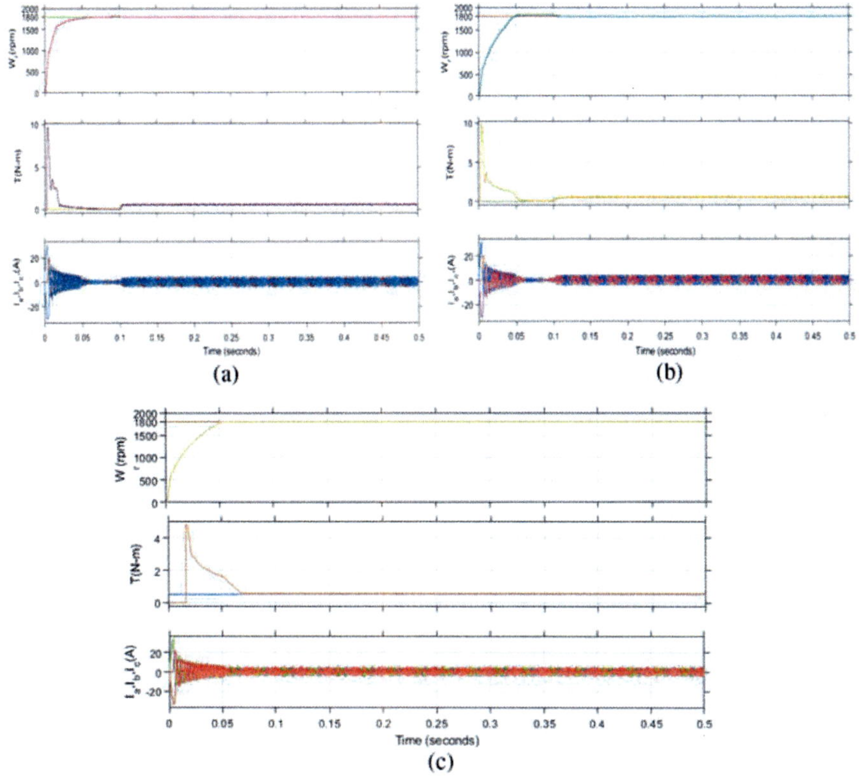

Fig. 4 Speed, torque, and stator current of BLDC motor using **a** PI controller **b** PID controller **c** fuzzy logic controller

6 Conclusion

This research paper compares the various speed controlling techniques and presents detailed and easy to understand results. First, we use a PI controller in this paper to control the speed by changing K_{p} and K_i values then, a PID controller is implemented which provides less settling time and more stability by adjusting K_{p}, K_i, and K_{d} values, and at last, FLC was applied to achieve the desired speed with negligible steady-state error and a more stable output; we have performed Simulation at two speeds one at a rated speed which is 1800 rpm and other is random speed 1500 rpm, and the outcomes are very stable and accurate. In contrast to PI controllers, PID controllers rise more rapidly. When the BLDC motor is controlled with a PID controller, its steady-state error is zero, but when the motor is controlled with a PI controller, it is considerably higher. When the load torque is implemented as part of the PID response, the motor speed does not decrease after the fuzzy-based BLDC rotor response is used. Thus, fuzzy-based controllers are highly efficient. In terms of

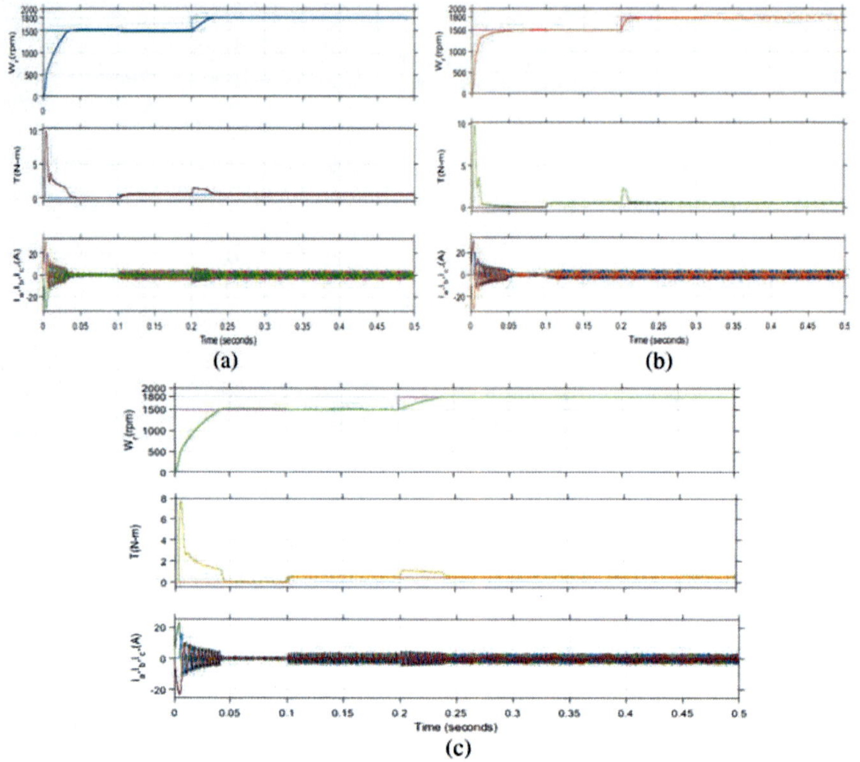

Fig. 5 Speed, torque, and stator current of BLDC motor with a step change in speed using **a** PI controller **b** PID controller **c** fuzzy logic controller

the motor's dynamic performance, the results reveal the motor is reasonably accurate irrespective of the loading condition.

References

1. Agrawal L, Chauhan B, Banerjee G (2018) Speed control of brushless DC motor using fuzzy controller. Int J Pure Appl Mathe 119:2689–2696
2. Potnuru D, Saibabu C (2018) Design and implementation methodology for rapid control prototyping of closed loop speed control for BLDC motor. J Electr Syst Inf Technol 5(1):99–111
3. Suryakant S, Sreejeth M, Singh M (2018) IEEE 2018 International Students' Conference on Electrical, Electronics and Computer Science (SCEECS)—Bhopal (2018.2.24–2018.2.25)]—performance analysis of PMSM drive using hysteresis current controller and PWM current controller
4. Suganthi P, Nagapavithra S, Umamaheswari S (2017) Modeling and simulation of closed-loop speed control for BLDC motor. In: 2017 conference on emerging devices and smart systems (ICEDSS). IEEE

5. Kumar B, SK Swain, Neogi N (2017) Controller design for closed-loop speed control of BLDC motor. Int J Electr Eng Inf 9(1):146
6. Sarala P, Kodad SF, Sarvesh B (2016) Analysis of closed loop current controlled BLDC motor drive. In: 2016 International conference on electrical, electronics, and optimization techniques (ICEEOT). IEEE
7. Shamseldin, Abdelbar M, EL-Samahy AA (2014) Speed control of BLDC motor by using PID control and self-tuning fuzzy PID controller. In: 15th international workshop on research and education in mechatronics (REM). IEEE
8. Kandiban R, Arulmozhiyal R (2012) Speed control of BLDC motor using adaptive fuzzy PID controller. Proc Eng 38:306–313
9. Singh H, Gupta M, Meitzler T, Hou, ZG, Garg K, Solo A, Zadeh L (2013) Real-life applications of fuzzy logic. Adv Fuzzy Syst. https://doi.org/10.1155/2013/581879
10. Zadeh L (2013) Fuzzy logic. In: Computational complexity: theory, techniques, and applications. 1177–1200. https://doi.org/10.1007/978-1-4614-1800-9_73
11. Dwivedi A, Tiwari A (2013) A review of speed control of BLDC motor with different controllers. In: National conference on recent advances in database systems and applications
12. Rao A, Chandra P, Obulesh YP, Sai Babu C (2012) Mathematical modeling of BLDC motor with closed loop speed control using PID controller under various loading conditions. ARPN J Eng Appl Sci 7(10):1321–1328
13. Neethu U, Jisha VR (2012)Speed control of Brushless DC motor: a comparative study. In: 2012 IEEE international conference on power electronics, drives and energy systems (PEDES)
14. Moraga C (2005) Introduction to fuzzy logic. Facta universitatis—series. Electron Energ 18:319–328. https://doi.org/10.2298/FUEE0502319M

Liquid Level Control of Coupled Tank System Using FOPID Controller

Sharmistha Mandal and Amrin Afza

Abstract In this paper, a fractional order proportional-integral-derivative (FOPID) controller has been designed for liquid level control of coupled tank system. The desired performance specifications have been expressed by a fitness function and has been optimized using particle swarm optimization (PSO) technique. The outputs show that the designed FOPID controller controls liquid level of coupled tank system very efficiently in terms of maximum overshoot, settling time and steady state error. The results of FOPID controller have been compared with other designed controllers. The robustness of FOPID controller is verified by varying the system parameters and that result also gives satisfactory result.

Keywords Liquid level control · Fractional order proportional-integral-derivative controller · Particle swarm optimization

1 Introduction

Liquid level control system has numerous applications in industry like food processing, biochemical, petro-chemical, etc. The liquid level control of coupled tank system has been attacked by many researchers and designed different controllers to control liquid level of the system [1–3].

In this study, a fractional order proportional-integral-derivative (FOPID) controller has been designed for liquid level control of coupled tank system. The gain of FOPID controller has been determined using different techniques [4–10] by many researchers. In present case, the gain of the FOPID controller have been calculated using bio-inspired efficient optimization technique, particle swarm optimization (PSO). The results show that FOPID controller controls the liquid level of coupled tank system and fulfills the required performance specifications. The performance of FOPID controller has been compared with proportional-integral (PI), proportional-integral-derivative (PID) and fractional order proportional-integral

S. Mandal (✉) · A. Afza
Electrical Engineering Department, Aliah University, Salt Lake, Kolkata, India
e-mail: itssharmistha@gmail.com

A. Rani et al. (eds.), *Signals, Machines and Automation*, Lecture Notes in Electrical Engineering 1023, https://doi.org/10.1007/978-981-99-0969-8_36

(FOPI) controllers and responses show that FOPID gives better result than the other controllers. The robustness of the controller has been checked by varying the parameters by ± 50% from its nominal value.

2 Coupled Tank System

Tank 1, tank 2, a pump and a water basin are the different parts of coupled tank system as shown in Fig. 1. Both tank 1 and tank 2 have uniform cross section. Liquid from basin to tank 1 goes with the help of pump, then flows to tank 2 and finally goes to the water basin. The values of parameters of coupled tank system is given in Table1. The non-linear dynamics [11] of the tank 1 is given by

$$\frac{\partial L_1}{\partial t} = f(L_{1,} V_P) \tag{1}$$

where V_P is pump voltage (input) and L_1 is water level of tank 1(output).

After linearization of Eq. (1), the open-loop transfer function of tank 1 is given by

$$G_1(s) = \frac{L_1(s)}{V_P(s)} = \frac{k_{dc1}}{T_1 s + 1} \text{ where } k_{dc1} = \frac{K_P\sqrt{2gL_{10}}}{A_{01}g};\ T_1 = A_{t1}\sqrt{2gL_{10}}$$

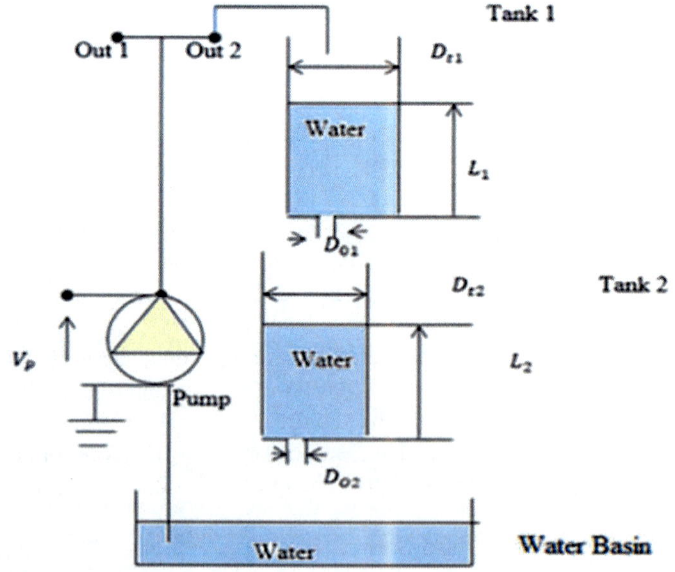

Fig. 1 Coupled tank system

Table 1 Values of the parameters of coupled tank system [11]

Description of the parameters	Value
Out 1 orifice diameter, D_{Out1}	0.635 cm
Out 2 orifice diameter, D_{Out2}	0.47625 cm
Flow constant of pump, K_p	3.3 cm^3/s/V
Inside diameter of tank 1, D_{t1}	4.445 cm
Inside diameter of tank 2, D_{t2}	4.445 cm
Tank 1 height (i.e., water level range), L_{1max}	30 cm
Tank 2 height (i.e., water level range), L_{2max}	30 cm
Gravitational constant on earth, g	981 cm/s^2

Similarly, open-loop transfer function of tank 2 is given by $G_2(s) = \frac{L_2(s)}{L_1(s)} = \frac{K_{dc2}}{T_2s+1}$ where L_2 is water height of tank 2.

The transfer function of coupled tank system is $G(s) = \frac{0.01397}{s^2+0.1313s+0.0043}$.

3 Fractional Order PID (FOPID) Controller

The transfer function of FOPID controller [6] is given by

$$G_{FOPID}(s) = K_p + \frac{K_i}{S^\lambda} + K_d S^\mu, (\gamma, \mu > 0)$$

where K_p, K_i, K_d are proportional, integral, derivational controller gain, respectively, μ and λ are real number and order of differentiator and integrator.

FOPID controllers are more flexible than PID controllers and less sensitive to parameter variations of the controlled system. The problem with the FOPID controller is that here five parameters have to be tuned whereas in case classical PID controller, three parameters are tuned. FOPID controller is used in different applications like temperature control, process control spacecraft control, etc.

4 Particle Swarm Optimization (PSO)

Particle swarm optimization (PSO) is simple, efficient and popular bio-inspired optimization technique. The concept of PSO was given in 1995 by Kennedy and Eberhart [12]. Consider a swarm has n no. of particles and moving around n-dimensional search space. Initialization of each particle is done with a random position and a random velocity. Based on own and companions experience, each particle alters its flying in every iteration.

Individual particles position is updated [12] as

$$x_{k+1}^i = x_k^i + v_{k+1}^i \tag{2}$$

The velocity of the particle is given by

$$v_{k+1}^i = v_k^i + c_1 r_1 \left(p_k^i - x_k^i\right) + c_2 r_2 \left(p_k^g - x_k^i\right) \tag{3}$$

where x_k^i & v_k^i are particle position and velocity, respectively; p_k^i and p_k^g are the best individual particle position and best swarm position, respectively; $c_1$and c_2 are acceleration of particles and r_1, r_2 are random numbers between 0 and 1.

5 FOPID Using PSO

Block diagram of coupled tank system with FOPID controller is shown in Fig. 2. $G_{\text{FOPID}}(s)$ is the controller transfer function and $G_{\text{Pr}}(s)$ is the transfer function of coupled tank system. In this method, closed-loop dominant poles have been determined from the desired closed-loop performance specifications. The desired transient response specifications for the present problem are considered as.

- Maximum overshoot M_{p} less than 10%.
- Rise time t_{r} is equal to 0.25 s.

The dominant poles for the closed-loop system are $(-4.56 \pm j7.69)$. The characteristics equation of closed-loop system is given by following equation.

$$1 + \left[K_{\text{p}} + K_i S^{-\lambda} + K_{\text{d}} S^{\mu}\right] \frac{0.01397}{S^2 + 0.1313S + 0.01828} = 0 \tag{4}$$

The fitness function is taken [4] as $f_{\text{Fit}} = |R_{\text{e}}| + |I_{\text{m}}| + |P|$ where R_{e} and I_{m} are real part and imaginary part of Eq. (4) and P $\left(\tan^{-1} \frac{I_{\text{m}}}{R_{\text{e}}}\right)$ is phase. For dominant pole $(-4.56 + j7.69)$, the fitness function is determined and minimized using PSO algorithm. This algorithm gives the optimal value of the five parameters

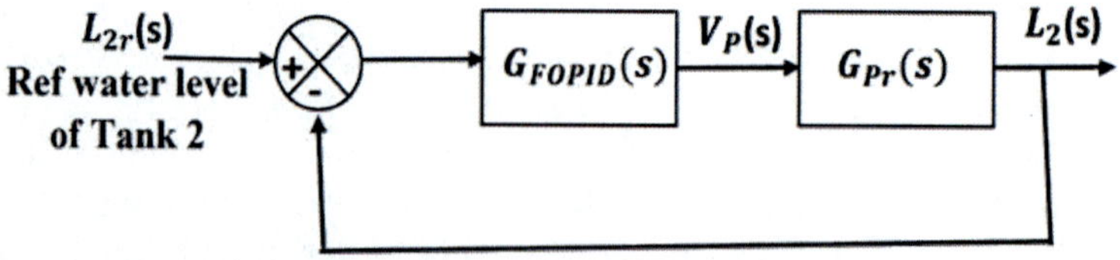

Fig. 2 Coupled tank system with FOPID controller

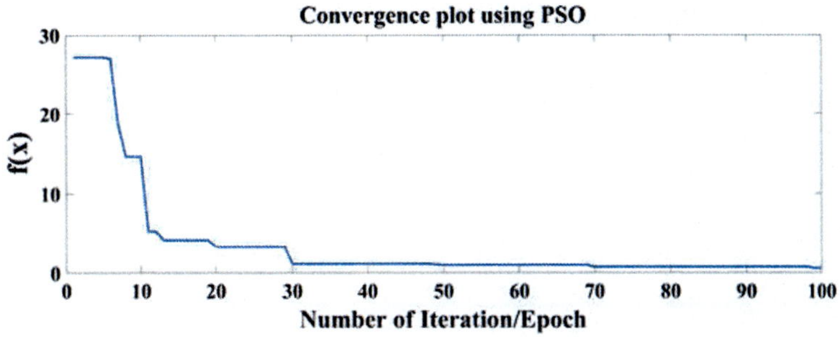

Fig. 3 Convergence characteristic of the PSO-FOPID controller

$\{K_p, K_i, K_d, \lambda, \mu\}$ that minimized the fitness function. For PSO algorithm acceleration are taken as $c_1 = c_2 = 1.479$, maximum iteration = 100, population size = 50, maximum velocity = 4 and dimension = 5. Figure 3 shows the convergence plot of FOPID.

The different parameters of the controller that are obtained using PSO algorithm are given in Table 2. Figures 4 and 5 shows the tank 2 water level and pump voltage respectively with FOPID controller when desired water level of tank 2 is taken as 10 cm. From the response of water level of tank 2, it is clear that the settling time is within 25 s and there is no undershoot and overshoot.

Table 2 Parameters of the different controllers

Controller	K_P	K_i	K_d	λ	μ
PI	0.5909	0.020472	–	1	–
FOPI	500	80	–	0.1442	–
PID	2	0.1	10	1	1
FOPID	100.4486	51.3139	740.8807	0.01	0.9324

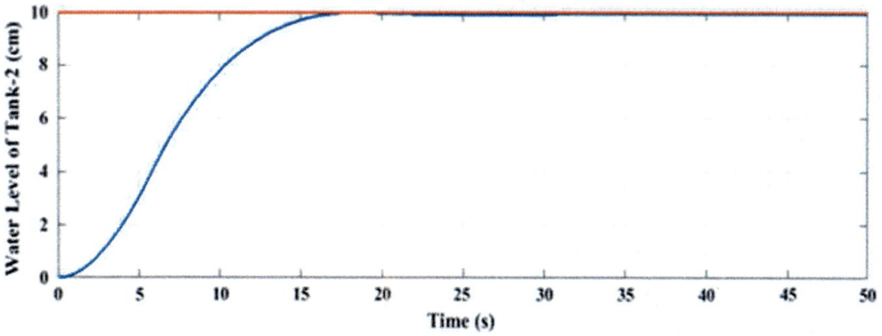

Fig. 4 Water level of tank 2 with FOPID controller

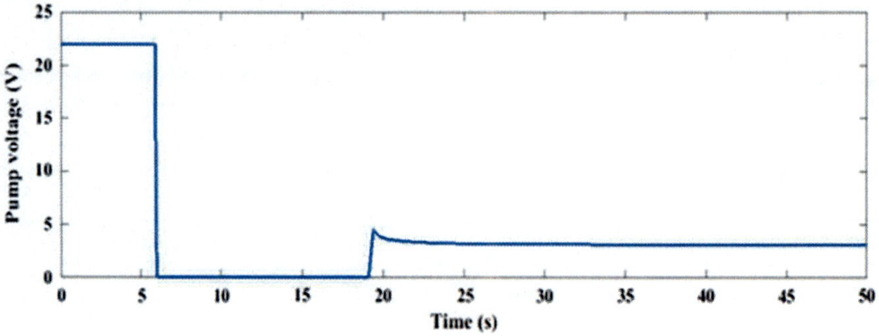

Fig. 5 Pump voltage with FOPID controller

Table 3 Time response specifications of different controllers

Controller	Overshoot (%)	Settling time (s)	Steady state error	Integral time absolute error (ITAE)
PI	15	50	0.09848	0.9848
FOPI	75	45	5.69	569
PID	25	50	0.0043	0.4309
FOPID	0	18	0.0207	2.099

The performance of the FOPID controller has been compared with PI, PID and FOPI controllers designed using PSO algorithm. The time response specifications for different controllers are given in Table 3 and from the data, it is clear that FOPID controller gives better result than PI, FOPI and PID controllers in terms of steady state error, settling time and maximum overshoot.

Figure 6 shows the performance of FOPID controller with variation in parameter. The value of the system parameters may not be perfect. The controller has been tested by varying equivalent DC gain $\pm$ 50% from its nominal value. From response, it can be observed that the controller also gives the robust performance.

6 Conclusion

The FOPID controller has been designed for coupled tank system using PSO. The responses show that the controller tracks the liquid level and gives satisfactory transient response specifications. Robust performance is also obtained from the controller in presence of parameters variations. In future the designed FOPID controller may be applied to real time liquid level control system.

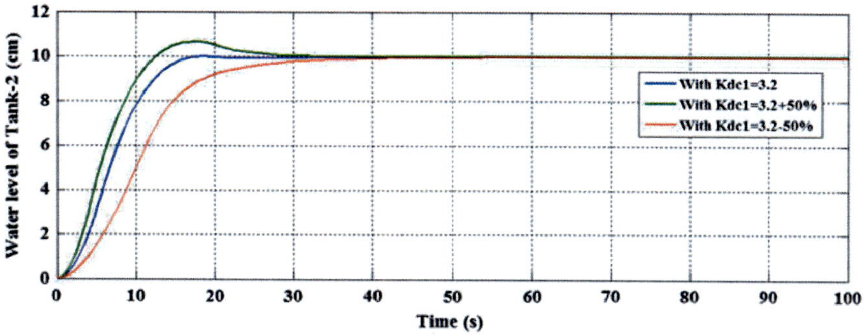

Fig. 6 Tank 2 water level in presence of parameter variation

References

1. Sharma S, Arora M, Kaushik K (2015) Analysis of liquid level control of coupled tank system by PI, PD, PID controller. Int J Sci Technol Res 4:11
2. Mursyitah D, Faizal A, Ismaredah E (2018) Level control in coupled tank system using PID-fuzzy tuner controller. In: 2018 electrical power, electronics, communications, controls and informatics seminar (EECCIS), Batu, East Java, Indonesia, pp 293–298
3. Barat P, Mandal S (2021) Robust control of coupled-tank system using uncertainty and disturbance estimator. In: 2021 IEEE second international conference on control, measurement and instrumentation (CMI), pp 128–132
4. Maiti D, Biswas S, Konar A (2008) Design of a fractional order PID controller using particle swarm optimization technique. In: 2nd National conference on recent trends in information systems (ReTIS-08)
5. Tepljako E, Petlenkov JB, Finajev J (2013) Fractional-order controller design and digital implementation using FOMCON toolbox for MATLAB. In: IEEE conference on computer aided control system design (CACSD), Hyderabad, India, August 28–30
6. Edet E, Katebi R (2018) On fractional-order pid controllers. IFAC-Papers Online 51(4):739–744
7. Rajesh R (2019) Optimal tuning of FOPID controller based on PSO algorithm with reference model for a single conical tank system. SN Appl Sci 1:758
8. Aseem K, Kumar SS (2022) Hybrid k-means grasshopper optimization algorithm based FOPID controller with feed forward DC–DC converter for solar-wind generating system. J Ambient Intell Humaniz Comput 13:2439–2462
9. Zhuo-Yun N, Yi-Min Z, Qing-Guo W, Rui-Juan L, Lei-Jun X (2020) Fractional-order PID controller design for time-delay systems based on modified bode's ideal transfer function. IEEE Access 8:103500–103510
10. Micev M, Calasan M, Oliva D (2020) Fractional order PID controller design for an AVR system using chaotic yellow saddle goatfish algorithm. Mathematics 8(1182):1–21
11. Apkarian J, Lacheray H, Abdossalami A, Instruction manual of couple tank. Quanser Version (v1.0)
12. Kennedy J, Eberhart R (1995) Particle swarm optimization. In: Proceedings of ICNN'95—international conference on neural networks, Vol 4, pp 1942–1948

Investigation of an Intelligent Control Technique Applied in NPK Model of the AHWR System

Devbrat Gupta, Vishal Goyal, and Jitendra Kumar

Abstract The advanced heavy water reactor (AHWR) is the advanced type of nuclear reactor which has nonlinear coupled multi-input multi-output (MIMO) dynamical characteristics, making it difficult for researchers to manage the world-wide demand power. In order to control the demand power efficiently, an intelligent controller, i.e. a fractional-order fuzzy proportional plus derivative (FOFPD) controller is suggested along with the control rod dynamics in this work. The suggested FOFPD controller is utilized in the normalized point kinetics (NPKs) model of the AHWR system. To make the FOFPD controller more stable and robust, the gains of the controller are tuned with the help of a meta-heuristic genetic algorithm (GA)-based optimization approach that uses the integral of absolute error (IAE) as the performance criteria. To validate the simulation findings, the recommended controller's performance is compared to that of fractional-order PD (FOPD) and conventional PD for the trajectory tracking of demand power. It can be stated that the FOFPD controller's response outperforms among all the investigated controllers.

Keywords AHWR · FOFPD · GA · IAE · NPK

1 Introduction and Literature Studies

The population is growing at a much faster rate resulting inadequate supply of energy for consumption. In order to meet the long-term energy needs of such a huge country as India, an indigenous resource-based approach is required [1]. As a result, nuclear energy has a significant role to play as a long-term solution [2]. India's nuclear power strategy was devised about 60 years ago, taking into account the country's relatively limited uranium and substantial thorium supplies. Geophysical surveys revealed that the country's uranium supply was fairly restricted, while the country's thorium stock

D. Gupta · V. Goyal · J. Kumar (✉)
Department of ECE, GLA University, Mathura 281406, India
e-mail: jitendra.kumar@gla.ac.in

V. Goyal
e-mail: vishal.goyal@gla.ac.in

A. Rani et al. (eds.), *Signals, Machines and Automation*, Lecture Notes in Electrical Engineering 1023, https://doi.org/10.1007/978-981-99-0969-8_37

was vast [3]. So, for commercial power generation, the use of an advanced heavy water reactor (AHWR) is the best bet among various nuclear reactors due to the vast reserve of thorium fuel. The AHWR is a vertical, pressure tube type, heavy water moderated, and boiling light water-cooled natural circulation reactor designed to produce 300 MWe of electricity. The reactor power which is highly nonlinear in nature is controlled by an autonomous reactor control system [4].

Researchers in the nuclear sector have long been preoccupied with load following and control challenges. A variety of techniques was offered to accomplish this goal [5–7]. Among these, the proportional plus integral plus derivative (PID) controller is best known which is also being extensively utilized in the nuclear power plant sector because of its simple construction and strong performance under a wide variety of operating situations. FOPID controllers with particular gain-phase margins have been built utilizing the Balanced Truncation approach to lower the order of PHWR models to a more manageable level. It was showcased that the FOPID controller performs better at 30 per cent control loss and varying beginning power levels when compared to a conventional controller [8, 9].

In reactor core modelling, power control, as well as trajectory tracking for load following control, decades of study, have been committed. This work has the potential to lead to real-world applications. In this study, the incorporation of a FOFPD controller is proposed into the NPK model of the AHWR system where the gains are tuned with GA based on the IAE performance index. The results are simulated MATLAB environment and further compared as well as validated with various controllers.

The rest of the research paper is arranged as follows. The problem formulation along with mathematical modelling of AHWR system is presented in Sect. 2. The controller design with GA tuning is shown in Sect. 3. Section 4 provides an investigation of trajectory tracking for demand power control. At the end of Sects. 2–4, we offer a series of relative discussions. Section 5 includes the conclusion and references, respectively, for this work.

2 Problem Formulation and Plant Modelling

From the past analysis of nuclear reactors by researchers, it is observed that neutrons are generated at various energies, and they are absorbed and ejected at various energies as well [10]. Here, the reactor's mathematical model will be based on a continuous function of neutron energy, which will make analysis more difficult. All spatial and angle dependencies in equations are ignored, and only the time dependency is left behind when using point kinetics as a term for the method. So for this work, the normalized point kinetics (NPKs) model of the AHWR system is utilized as expressed in equations [11, 12].

$$\frac{dn_r}{dt} = \frac{\rho - \beta}{l} n_r + \sum_{i=1}^{6} \frac{\beta_i}{l} c_{r_i} \tag{1}$$

$$\frac{dc_{r_i}}{dt} =_i n_r -_i c_{r_i} \tag{2}$$

where n_r and c_{r_i} is neutron and precursor densities for $i = 1, 2, 3, \ldots, 6$ groups. ρ is reactivity. β_i is delayed-neutron fraction for $i = 1, 2, 3, \ldots, 6$ group; β is an effective delayed-neutron fraction [11].

The preceding Eqs. (1) and (2) are expanded and conveniently structured as a standard state-space model $\dot{X} = AX + BU$, $y = CX$ as illustrated in Eq. (3),

$$A = \begin{bmatrix} \frac{-\beta}{l} & \frac{\beta_1}{l} & \frac{\beta_2}{l} & \frac{\beta_3}{l} & \frac{\beta_4}{l} & \frac{\beta_5}{l} & \frac{\beta_6}{l} \\ \lambda_1 & -\lambda_1 & 0 & 0 & 0 & 0 & 0 \\ \lambda_2 & 0 & -\lambda_2 & 0 & 0 & 0 & 0 \\ \lambda_3 & 0 & 0 & -\lambda_3 & 0 & 0 & 0 \\ \lambda_4 & 0 & 0 & 0 & -\lambda_4 & 0 & 0 \\ \lambda_5 & 0 & 0 & 0 & 0 & -\lambda_5 & 0 \\ \lambda_6 & 0 & 0 & 0 & 0 & 0 & -\lambda_6 \end{bmatrix}$$

$$B = \left[\frac{n_r}{l} \; 0 \; 0\,0\,0\,0\,0 \right]^T; \; U = [\rho] \tag{3}$$

The control rod's (CR) position and reactivity are expressed in Eq. (4) as

$$\dot{\rho}_{\text{rod}} = k\rho_{\text{rr}}; \;\; \rho_{\text{rr}} = \rho \tag{4}$$

where $\dot{\rho}_{\text{rod}}$ is defined as % out movement of CR position, k is the constant term taken from the paper [13]. ρ is reactivity. The neutron lifetime (l), effective delayed-neutron fraction (β), delayed-neutron fractions ($\beta_1, \beta_2, \beta_3, \beta_4, \beta_5, \beta_6$), and decay constants ($\beta_1, \beta_2, \beta_3, \beta_4, \beta_5, \beta_6$) of 1st to 6th group precursors are considered from [10], respectively.

3 Intelligent Controller Design and Tuning Using GA

The use of conventional PID controllers in any kind of control system has been going on for decades. And it is extended further by applying fractional-order derivative and integral operators, or by using various fuzzy logic control (FLC) layers by using other intelligent techniques [14]. In this section, an intelligent controller, i.e. fractional-order fuzzy proportional plus derivative (FOFPD) controller is suggested along with the GA tuning process [15]. The integral controller is omitted in this work because the AHWR is a highly nonlinear coupled complex system and integrates with itself [16].

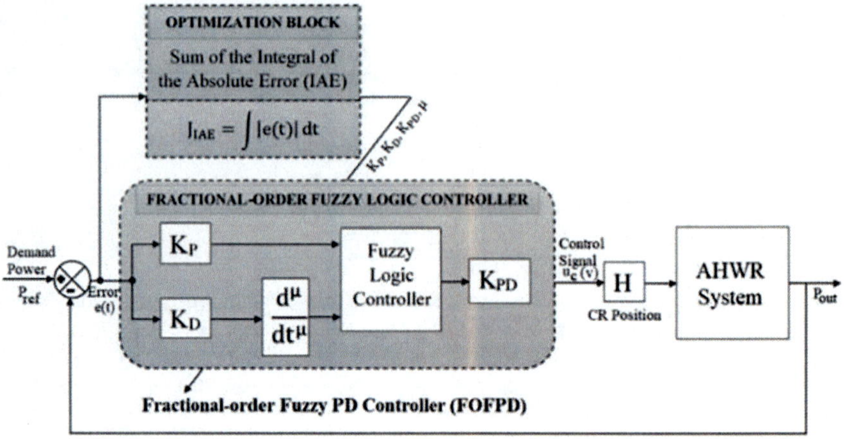

Fig. 1 Block diagram representation of a tuned FOFPD controller incorporated into the AHWR system

The basic structure of the fuzzy controller has two inputs error and fractional-order derivative (FOD) of the error and one output as shown in Fig. 1.

The mathematical expression of the FOFPD controller can be illustrated as,

$$K_{P}e(t) + K_{D}\frac{d^{\mu}e(t)}{dt^{\mu}} = K_{PD}u(t);\ u_{FOFPD}(t) = K_{PD}u(t);\ u_{c} = u_{FOFPD}(t) \qquad (5)$$

where the two inputs of the FOFPD controller are error $K_{P}e(t)$ and rate of change of error $K_{D}\frac{d^{\mu}e(t)}{dt^{\mu}}$ while the output of the FOFPD controller is $K_{PD}u(t)$ in absolute form. $u_{c}(t)$ is the controller output measured in volts within the range of [−1 1].

For FOD gain (μ) implementation, the Oustaloup approximation (OSA) method is utilized in the FOFPD controller [17]. An FLC layer is showcased in Fig. 1 that includes fuzzification block, fuzzy inference block, knowledge base block, and defuzzification block. To construct the basic FLC, Gaussian membership functions are employed. A two-dimensional rule base is formulated according to the problem. Centre of gravity defuzzification, a Mamdani-type inference process, and the minimal and maximum operator types are utilized in the FLC layer [14]. To get the desired system performance, the gains must be fine-tuned. This research work uses the GA optimization approach [18] to obtain the tuned gains for all three controllers. The GA [12], which is based on natural selection, the mechanism that propels biological evolution, is a technique for resolving both limited and unconstrained optimization issues. A population of unique solutions is repeatedly modified by the genetic algorithm. The tuning problem's objective function (J_{IAE}) was considered to be the sum of the integral of an absolute error (IAE) index with GA. It was used to attain a global optimal solution to minimize this J_{IAE}. The expression for the J_{IAE} is shown in Eq. (6) as follows:

Table 1 J_{IAE} values of PD, FOPD, and FOFPD controllers based on IAE metric

Controller	Objective function values (J_{IAE})
PD	0.9157
FOPD	0.7967
FOFPD	0.2139

Table 2 Gains of PD, FOPD, and FOFP controllers based on IAE metric

Controller	Controller gains			
	K_P	K_D	μ	K_{PD}
PD	−49	−9	–	–
FOPD	−49	−9	0.1	–
FOFPD	−49	−9	0.1	143

$$J_{IAE} = \int_0^{350} |e(t)|dt \tag{6}$$

4 Investigation of Trajectory Tracking for Demand Power

The MATLAB/Simulink environment is used to simulate all of the outcomes. The Euler explicit technique is used to simulate the plant model. The J_{IAE} and gain values for PD, FOPD, and FOFPD controllers are shown in Table 1 and Table 2, respectively, along with the convergence curve in Fig. 2. The trajectory tracking curve is shown in Fig. 3. It can be observed that the blue curve which is applied FOFPD controller into the system, offered a stable trajectory tracking for demand power along with less overshoot and fewer oscillations when compared to PD and FOPD controllers.

5 Conclusion

For proper demand power control of the Advanced Heavy Water Reactor (AHWR), an intelligent control scheme was suggested in the normalized point kinetics (NPKs) model of the system along with control rod dynamics. In this research article, the utilized intelligent control scheme was a fractional-order fuzzy proportional plus derivative (FOFPD) controller. A meta-heuristic Genetic Algorithm (GA)-based optimization approach using Integral of absolute error (IAE) as the performance criteria were used to tune the controller gains to make the suggested controller more robust and susceptible to disturbances. The simulation results were validated by comparing the performance of the FOFPD controller with the fractional-order proportional plus

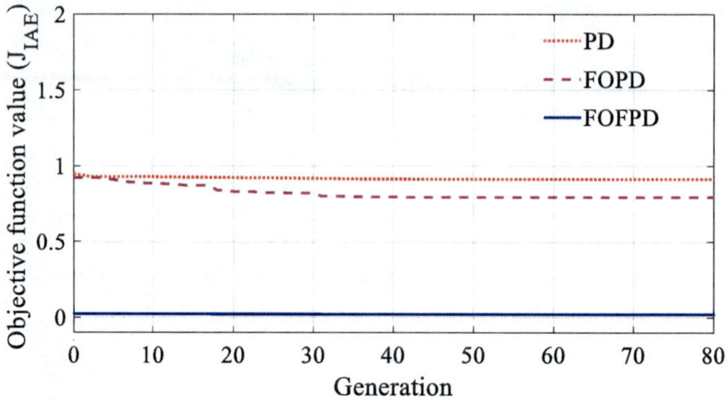

Fig. 2 J_{IAE} values versus generation curve for various controllers

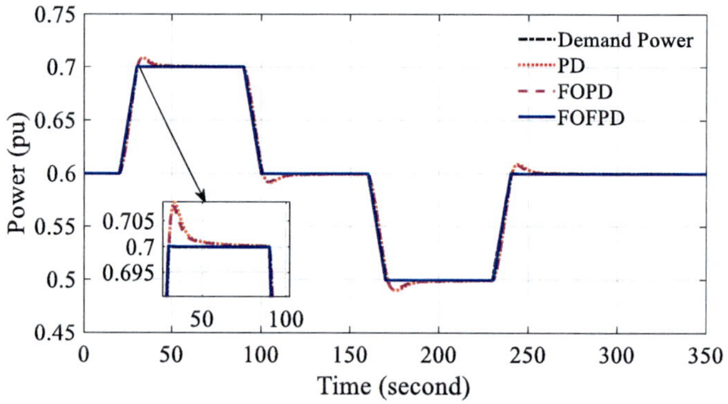

Fig. 3 Trajectory tracking curve for various controllers

derivative (FOPD), and conventional PD controllers. It could be observed that the FOFPD controller provided satisfactory results over the other two controllers.

References

1. Lavagne d'Ortigue O, Whiteman A, Elsayed S (2015) Renewable energy capacity statistics 2015. Int Renew Energy Agency (IRENA) 44
2. Vijayan PK, Dulera IV, Krishnani PD, Vaze KK, Basu S, Sinha RK (2016) Overview of the thorium programme in India BT—thorium energy for the world. Presented at the, 2016
3. Banerjee S, Gupta HP (2017) The evolution of the Indian nuclear power programme. Prog Nucl Energy 101:4–18
4. Sinha RK, Kakodkar A (2006) Design and development of the AHWR—the Indian thorium fuelled innovative nuclear reactor. Nucl Eng Des 236:683–700
5. Vadivel M, Karthigaivel R, Selvakumaran S (2014) Control of PWR nuclear reactor power using soft computing techniques based proportional-integral-derivative design. Int J Appl Eng

Res 9:30605–30619
6. Ghazali AK, Hassanb MK, Soh AC (2018) PID controller for nuclear reactor power control system. Int J Pure Appl Math 118
7. Mousakazemi SMH, Ayoobian N (2019) Robust tuned PID controller with PSO based on two-point kinetic model and adaptive disturbance rejection for a PWR-type reactor. Prog Nucl Energy 111:183–194
8. Lamba R, Sondhi S, Singla SK (2020) Reduced order model based FOPID controller design for power control in pressurized heavy water reactor with specific gain—phase margin. Prog Nucl Energy 125:103363
9. Lamba R, Singla SK, Sondhi S (2017) Fractional order PID controller for power control in perturbed pressurized heavy water reactor. Nucl Eng Des 323:84–94
10. Hetrick DL (1971) Dynamics of nuclear reactors. Country unknown/Code not available 1971
11. Mousakazemi SMH, Ayoobian N, Ansarifar GR (2018) Control of the reactor core power in PWR using optimized PID controller with the real-coded GA. Ann Nucl Energy 118:107–121
12. Gupta D, Goyal V, Kumar J (2022) Design of fractional-order NPID controller for the NPK model of advanced nuclear reactor. Prog Nucl Energy 150:104319
13. Mousakazemi SMH (2021) Comparison of the error-integral performance indexes in a GA-tuned PID controlling system of a PWR-type nuclear reactor point-kinetics model. Prog Nucl Energy 132:103604
14. Nithilasaravanan K, Thakwani N, Mishra P, Kumar V, Rana KPS (2018) Efficient control of integrated power system using self-tuned fractional-order fuzzy PID controller. Neural Comput Appl 9:1–19
15. Mishra P, Kumar V, Rana KPS (2015) A fractional order fuzzy PID controller for binary distillation column control. Expert Syst Appl 42:8533–8549
16. Munje RK, Patre BM, Londhe PS, Tiwari AP, Shimjith SR (2016) Investigation of spatial control strategies for AHWR: a comparative study. IEEE Trans Nucl Sci 63:1236–1246
17. Tang Y, Cui M, Hua C, Li L, Yang Y (2012) Optimum design of fractional order PI λD μ controller for AVR system using chaotic ant swarm. Expert Syst Appl 39:6887–6896
18. Deb K (1999) An introduction to genetic algorithms. Sadhana 24:293–315

Healthcare Technologies

Brain Tumor Recognition and Classification of MRI Images Using a Convolutional Neural Network Model

Anand Upadhyay, Niharika Tiwari, and Rishi Singh

Abstract A brain tumor is one of the conditions in which a patient's brain develops abnormal cells. They are called tumors and there are several varieties of them. In this paper, brain tumor segmentation and detection methods are being developed that use images from an MRI series as input images to identify the type of tumor. The use of a convolutional neural network (CNN) model to recognize brain tumors from X-ray images pictures is described in this paper. This investigation was carried out with the help of Google Collab. The MRI images of brain tumors are categorized into glioma, meningioma, and pituitary, and no tumor is one such difficult problem. In this work, the datasets comprise 388 images for testing purposes. With an accuracy of 70%, the projected arrangement achieves a commendable performance. The successful conclusion demonstrates the suggested algorithm's ability to classify brain tumors.

Keywords Brain tumor · Multi-class classification · Convolutional neural network · MRI images · Tumor identification

1 Introduction

The brain is a crucial organ that controls our body's ideas, memory, feelings, touch, vision, inhaling, temperature, appetite, and several other functions. A brain tumor places tremendous pressure on the brain, resulting in brain damage. It pushes on the skull, covering much of your brain and destroying all of its nerves. A patient with a brain tumor may experience symptoms such as headaches, morning sickness, laziness, speech problems, visual loss, and ascending pressure, among others [5]. The type of tumor and where it is located in the brain determine the sort of treatment necessary. There are two levels of education: primary and secondary. Other forms of

A. Upadhyay · N. Tiwari (✉) · R. Singh
Department of IT, Thakur College of Science and Commerce Thakur Village, Kandivali(E), Mumbai, Maharashtra 400101, India
e-mail: niharikatiwari6307@gmail.com

A. Upadhyay
e-mail: anandhari6@gmail.com

A. Rani et al. (eds.), *Signals, Machines and Automation*, Lecture Notes in Electrical Engineering 1023, https://doi.org/10.1007/978-981-99-0969-8_38

brain tumors include glioma, meningioma, and pituitary tumors. Glioma tumors, which are the most frequent type of primary brain tumor, are divided into four categories, with the higher the grade, the more malignant the tumor, and they begin in the brain's glial cells. Meningiomas are benign tumors that arise from the meninges, a layer of tissue that surrounds the brain and spinal cord. The expansion of this species is slow and limited. The pituitary tumor continues to grow on the pituitary gland. These tumors are also benign and have a smaller distribution [4]. Researchers used brain magnetic resonance imaging (MRI) as the most effective imaging technique for detecting brain tumors and modeling tumor growth in both the detection and therapy stages [1]. Magnetic resonance imaging (MRI) paired with CNN gives a strong combination for accurately detecting the type of brain tumors. MRI is a preferable approach for emerging over CT scan for detecting these tumors in the early stages. The latter can expose the patient to excessive radiation and has a slow data collecting rate. The CNN model is trained with the images to reach a specific level of accuracy. Majorly the data generated by the MRI scan is large and it is considered okay for a small number of images. CNN is gauged for higher accuracy and uses a vast set of images for its training purposes which ensures that the tumor can be localized in identify due to detail-rich data from MRI scans, the CNN can identify the type of tumor and can specify the region at which it is present.

2 Literature Review

In image processing approaches, intelligent systems, and learning techniques are commonly used for segmenting, identifying, and classifying MRI images, as well as for classifying and detecting brain tumors. There have been numerous studies on the categorization of brain MRI images. Some of the worldwide articles we looked at for deep learning-based brain tumor detection and classification are Shaik Basheera et al. [2] suggested a method for diagnosing brain tumors in which the tumor is first segmented from an MRI image, and the segmented area is then retrieved using stochastic gradient descent by a pre-trained convolutional neural network. Hemantha Kumar et al. [6] Combination models that could accurately classify brain cancers were implemented. Resnet50, Vgg16, and InceptionV3 were the models used in their research. With Resnet50, they were able to obtain 95% accuracy in their research. Seetha et al. [8] suggested a model that uses Fuzzy C Means (FCM)-based segmentation, texture and form feature extraction, and SVM and DNN-based classification to classify brain tumors.

3 Algorithm

The convolutional neural network architecture is highly useful for binary and multi-class image classification. CNN naughty coins are artificial neural networks that use convolutional layers in at least one of their layers [7]. In a variety of fields, CNN has been recognized as a competitive technique for picture classification, segmentation, and identification. A multilayer perceptron is one in which all of the CNN layers networks are connected. In this CNN model control, we are using a sparse categorical method for the multi-class dataset. Data can be handled faster and more accurately utilizing the sparse categorical technique, resulting in superior results (see Fig. 1).

The first layer uses 16 fairly large filters (3 × 3) due to the small size of the input images. The input shape is set to [200, 200, 3] because the images are 200 × 200 pixels. Next, in the max layer, the pool size is (2, 2) which means the spatial dimension is segmented by (2, 2). In structure, the convolutional layer is repeated twice preceded by the max-pooling layer. We can repeat this structure many times for a large dataset of images. As we progress up the CNN toward the output layer, the number of layers increases (from 64 to 128, then 256). Next, the fully associated network nodes are made up of two hidden dense layers and a dense output layer. Because a dense layer requires a 1D array for each iteration, it must flatten its input. Finally, we apply the SoftMax activation algorithm to create a dense output layer with four neurons, because the classes are exclusive. This is called multi-class classification. The reason CNN is so popular is their architecture—the best thing is that there is no need for feature extraction. The next layer features is convoluted with different filters to create more abstract and invariant features, and the process continues until the final. Also, deep convolutional networks work well on image data since they're flexible. With the use of CNNs the number of parameters can be reduced while maintaining the quality of the models.

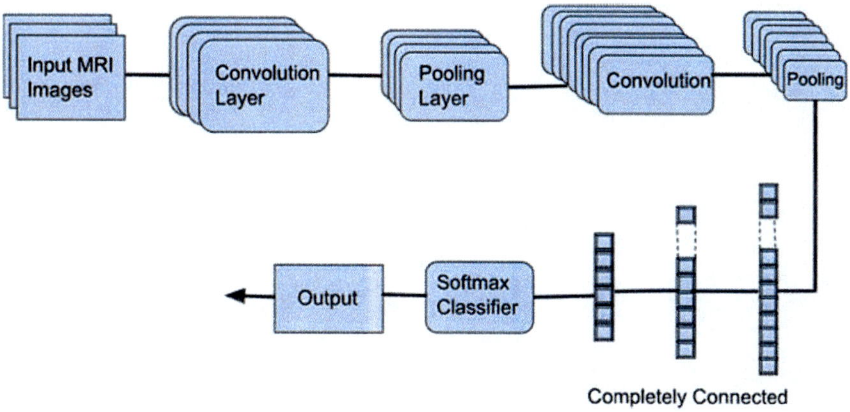

Fig. 1 Classic convolutional neural network architecture

4 Methodology

The purpose of this investigation is to enhance the exactness of brain MRI images identification by using deep learning algorithms and convolutional neural network approach. The Pseudo code for the proposed method for detecting and identifying the types of tumor are as follows:

Step 1: Start the process.

Step 2: Take the MRI Images as input and resize.

Step 3: The image is classified into two, i.e., training dataset and testing dataset.

Step 4: These two datasets undergo different convolutional layers such as Conv2D, MaxPool2D, and Flattens.

Step 5: The model is compiled and fit and gives the accuracy and loss.

Step 6: Finally, the model identifies the types of brain tumor.

To understand the methodology in a better way, here is the conceptual diagram for identifying the tumors (see Fig. 2).

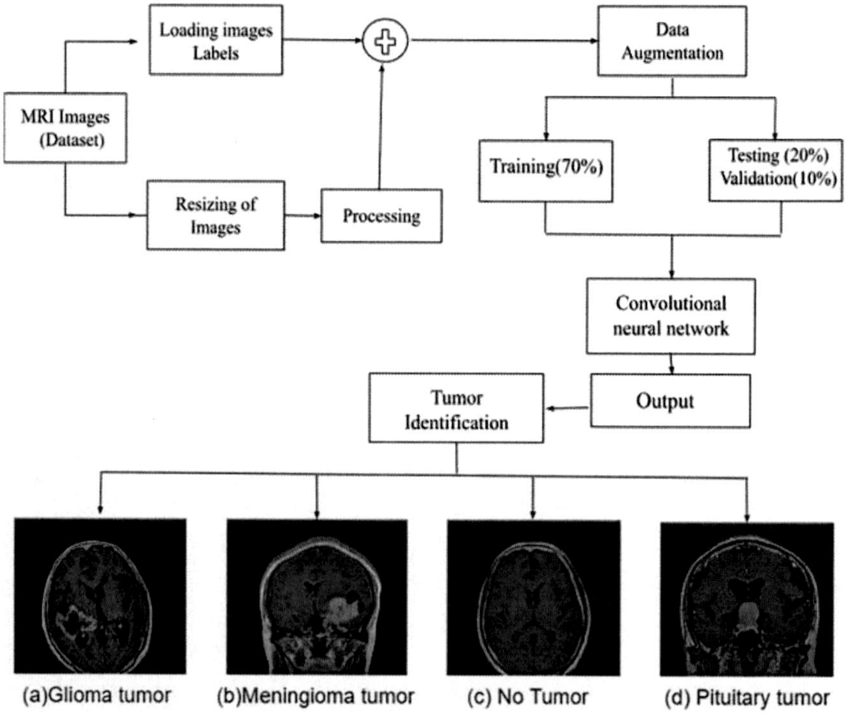

Fig. 2 Semantic diagram of brain tumor classification

```
Epoch 95/100
10/10 [==============================] - 30s 3s/step - loss: 0.3884 - accuracy: 0.8587 - val_loss: 3.0109 - val_accuracy: 0.5593
Epoch 96/100
10/10 [==============================] - 30s 3s/step - loss: 0.1977 - accuracy: 0.9484 - val_loss: 2.9717 - val_accuracy: 0.6392
Epoch 97/100
10/10 [==============================] - 30s 3s/step - loss: 0.2079 - accuracy: 0.9333 - val_loss: 3.2161 - val_accuracy: 0.6443
Epoch 98/100
10/10 [==============================] - 31s 3s/step - loss: 0.1786 - accuracy: 0.9516 - val_loss: 3.6547 - val_accuracy: 0.6340
Epoch 99/100
10/10 [==============================] - 31s 3s/step - loss: 0.1913 - accuracy: 0.9453 - val_loss: 3.4608 - val_accuracy: 0.6340
Epoch 100/100
10/10 [==============================] - 30s 3s/step - loss: 0.1661 - accuracy: 0.9422 - val_loss: 3.0021 - val_accuracy: 0.6907
```

Fig. 3 Model accuracy and loss

4.1 Data Augmentation

Data augmentation enhances the amount of training dataset, by creating multiple replicas of every training sample. As a result, oversampling is reduced, and it becomes a normalization approach. The examples that are constructed should be as real as practical. Ideally, a person should not be able to identify whether the image from the training dataset was enhanced or not. Adding background noise alone will not assist; the changes must be learnable [9].

Training and Testing set. In this paper a dataset of 3352 MRI pictures, of which 2870 images were used for training, 388 images were used for testing, and 94 images were used to validate the program's results. Kaggle has provided us with the following dataset. To improve accuracy and performance, we scaled the image from its original size to 200 × 200 pixels.

Performance Evaluation. The evaluation of performance is an important part of the machine learning process. It is however a difficult task as a result here there are 5 sets of epochs from 95 to 100 with different accuracy and loss (see Fig. 3).

5 Result

The dataset contains four classes glioma tumor, meningioma tumor, pituitary tumor, and no tumor MRI images. The collection of datasets contains 3352 real brain pictures from genuinely affected patients. The dataset used in this paper is accessible on Kaggle. The dataset is separated into three categories: training data, validation data, and testing data. The data consist of 2870 images to train, 94 images to validate, and 388 to test and evaluate the model accuracy. Firstly, we use data augmentation to improve our dataset by making minor changes to our MRI pictures and extracting these images and using the suggested sparse categorical cross-entropy. The model is trained by 100 epochs with a batch size of 64. The model was 94% accurate in training data and 70% accurate in validation data. After fitting the model and evaluating the data the result (see Fig. 4).

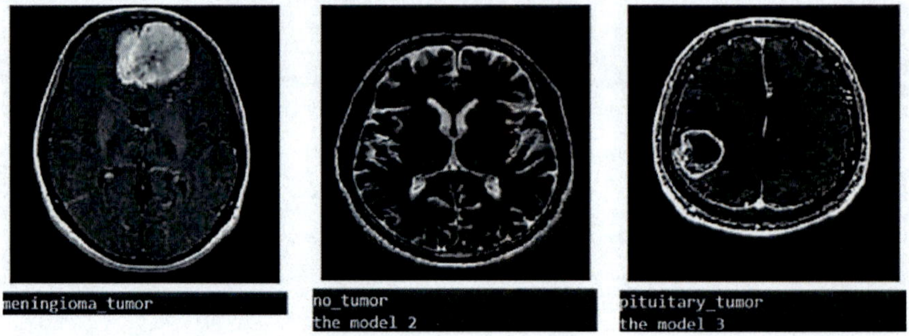

Fig. 4 Final output of classified tumor

6 Conclusion

A new approach to classifying brain tumors was presented in this research. Initially, to expand the size of our training data, we used the data augmentation technique. Further, we propose a sparse categorical cross-entropy in this paper to fit the model for multi-class purposes. A vast quantity of data is required to train a neural network for sophisticated and accurate outputs however, our results indicate that even with a small dataset, our accuracy rate is 70%. Khan HA, Jue W research provides a method for classifying brain tumors into cancerous and non-cancerous using data augmentation techniques and convolutional neural network models, including VGG-16, ResNet-50, and Inception-v3, with an overall accuracy of 96% on training data and 89% accuracy on validation data [10]. But in this paper, the different types of tumors are classified and identified. Due to the minimum number of datasets, the accuracy is overall fine. The proposed approach is designed for multi-class classification issues; however, further research can be done in the future to utilize different classifiers with optimization methodology to increase performance by combining more accurate identification and extraction algorithms with actual images.

References

1. Mohsen H, El-Dahshan ES, El-Horbaty ES, Salem AB (2018) Classification using deep learning neural networks for brain tumors. Future Comput Inf J 3(1):68–71
2. Sharif MI, Li JP, Khan MA, Saleem MA (2020) Active deep neural network features selection for segmentation and recognition of brain tumors using MRI images. Patt Recogn Lett 1(129):181–189
3. Manjunath S, Sanjay Pande MB, Raveesh BN, Madhusudhan GK (2019) Brain tumor detection and classification using convolution neural network. Int J Rec Technol Eng (IJRTE) 2277–3878
4. Sapra P, Singh R, Khurana S (2013) Brain tumor detection using neural network. Int J Sci Mod Eng (IJISME) 2319–6386
5. Sajja VR, Kalluri HK (2020) Classification of brain tumors using convolutional neural network over various SVM methods. Ingénierie des Systèmes d Inf. 25(4):489–495

6. Mehrotra R, Ansari MA, Agrawal R, Anand RS (2020) A transfer learning approach for AI-based classification of brain tumors. Mach Learn Appl 15(2):100003
7. Seetha J, Selvakumar Raja S (2018) Brain tumor classification using convolutional neural networks. Biomed Pharmacol J
8. Rehman A, Naz S, Razzak MI, Akram F, Imran M (2020) A deep learning-based framework for automatic brain tumors classification using transfer learning. Circ Syst Sign Process 39(2):757–775
9. Géron A (2019) Hands-on machine learning with Scikit-Learn, Keras, and TensorFlow: concepts, tools, and techniques to build intelligent systems. O'Reilly Media, Inc.
10. Khan HA, Jue W, Mushtaq M, Mushtaq MU (2020) Brain tumor classification in MRI image using convolutional neural network. Math Biosci Eng 17(5):6203–6216

Effect of Biohazard Waste on Human Health

Shraddha Sharma, Geeta Singh, and Gauhar Mehmood

Abstract Due to increasing population the solid waste is increasing day by day and it is posing a problem in management before Municipal Corporation of India. A number of methods are being used to collect and dispose of the solid municipal waste. Most of the solid waste is either reusable or can be converted to either energy or fertilizer but biohazard waste is indeed causing a major problem toward human health and environment. Hence, it is essential to develop effective methods to properly deal with biohazard waste. The ineffective disposal methods target primarily pre-school children, workers at waste site, hospital staff, workers at chemical factory, etc. Biohazard waste can pose problems like: low birth weight, vomiting, neurological disease, cancer, etc. This article deals types of biohazard waste, its various methods of disposal and future prospects of biohazard treatment.

Keywords Biohazard waste · Incineration · Radiation · Shredding

1 Introduction

Bio-hazardous waste contains biological materials which pose risk of transmission of infectious disease. Radioactive substances, antibiotics, corrosive chemicals, etc. contribute largely toward biomedical waste which are generated at hospitals, clinics and health care centers [1]. Out of total waste generated in India 1–1.5% is biomedical waste in which 10–15% is infectious. According to Radha et al. [2] in India out of total waste generated 85% waste is non-infectious while 5% is highly hazardous. Current situation in India is dangerous because most of the bio-waste from hospitals is disposed along with normal municipal solid waste. Sharp biological waste like needle, etc. is being collected non-specified plastic bags. Although a number of hospitals have developed their own system of separating biohazard from the normal municipal solid

S. Sharma (✉) · G. Singh
Department of Environmental Engineering, Delhi Technological Uni Versity, Delhi, India
e-mail: shraddha.sharma@iilm.edu

G. Mehmood
Department of Civil Engineering, Jamia Millia Islamia, New Delhi, India

A. Rani et al. (eds.), *Signals, Machines and Automation*, Lecture Notes in Electrical Engineering 1023, https://doi.org/10.1007/978-981-99-0969-8_39

waste but government needs to be highly focused on the effective technologies for proper disposal of biomedical waste apart from normal waste. Improper management causes grave environmental problems like water, land, and air pollution. Biomedical waste can be classified as biological, radioactive, and chemical waste [3]. All over the world the biomedical waste management is in its nascent stage and the policy makers and executors are puzzled over the handling of biomedical waste. Waste generates pathogens and they can remain in air for a long time which will cause in-door as well as outdoor air pollution. Biomedical waste in liquid form when put into sewer can cause water pollution and it may change the pH value of water and biomedical waste in the form of discarded medicines and chemical causes soil pollution which can affect the fertility of the soil. Heavy metals like cadmium and Mercury may be absorbed by the plants and can enter the food chain to pose health problems for humans and animals. Medical wastes are affecting urban areas in much larger intensity as compared to the rural areas, hence scientific management of medical waste is necessary to maintain cleanliness in hospitals and other health care centers [4].

The case study of Mysore health care center shows that medical waste consists of 11% human anatomical waste, 12.1% laboratory waste, 4.4% medicine shop waste, 1.8% corporation waste, 15.4% paper waste, 25.1% plastic waste, 23.3% metal waste, 3.3% glass waste, and other waste is 3.6%. It is highly required to segregate the hospital waste at the source and color coded bins should be placed at a specified place in the hospital so that proper disposal can be done to avoid the spreading of various diseases due to infectious bio-hazardous waste. This paper is a review article in which studies performed by various authors have been incorporated. In Sect. 3 types of biohazard solid waste has been discussed while Sect. 4 discusses about the collection and disposal methods of biohazard solid waste. A brief discussion on how developed nations treat biohazard waste has been presented in Sect. 5. Section 6 discusses about the future possibilities of biohazard treatment and conclusion has been given in Sect. 7.

2 Literature Review

Depending upon the quantity biomedical waste is categorized as primary or secondary [5]. Major sources of biomedical waste are Hospitals, Labs, Research centers, Animal research, blood banks, and nursing homes while minor sources of biomedical waste include ambulance services, cosmetic piercing, funeral services, vaccination centers, and dental clinics. Biomedical waste consist of 10–15% of hazardous waste while non-hazardous waste is 75–90%. Hazardous infectious waste includes sharp, non-sharp, plastic disposables, and liquid wastes while other hazardous waste consist of radioactive, chemical, and cytotoxic waste. In India biomedical waste rule 1998 has divided biomedical waste into 10 categories namely: Human anatomical waste, microbiology waste, animal waste, discarded medicine, sharp waste, normal solid waste, infectious waste, chemical waste, liquid waste, and incineration waste.

Before disposal proper disinfection is necessary for bio-hazardous and biomedical wastes [6]. Thermal disinfection is done for dry or wet waste by autoclaving, chemical disinfection is performed using Formaldehyde, Sodium Hypochlorite, Ethyl Oxide, and Bleaching powder. Irradiation process by exposure to ultraviolet rays is also used for disinfection and for small quantities of laboratory waste requires small microwaves while for large quantities large units are used along with grinding and steam spraying. In India the waste is normally collected in open containers in mixed form [7]. A number of communicable diseases such as Hepatitis-A and B, skin problems and respiratory infections are caused due to medical wastes sometimes directly or sometimes sharp waste injuries. Responsible microorganism hosts which cause infection include non-hemolytic streptococci, HIV, HBV, enterococci, and clostridium tetani. One of the major risk to health care workers is due to handling of sharps because 60% of them get injury knowingly or unknowingly by sharps. In some of the cases of improper waste management working at sub-optimal conditions add to the health hazard.

3 Types of Biohazard Solid Waste

Biohazard has been categorized as Physical, Biological, Chemical, Radiation, and Violence [8].

Physical Hazard

Needle stick injuries amount to 40% of the total injurics. Various types of injuries caused at the health care centers and their percentage involvement is as follows (Table 1 and Fig. 1).

Injuries are also caused in the process of patient handling, ramps, and even due to slip on the wet floor. Health care personals (HCP) are at high risk of getting exposed to bio-hazards that can cause stress among HCP. A study carried out at Obafemi Aowlowo Hospital, Nigeria at the Department of Obstetrics and Gynecology [9] shows that health care personals 83.3% of getting risk of stress, 75.6% of needle-stick injury, health problems may also be caused blood strain in the skin with a percentage of 73.1, drugs cause 47.4% health risk, sleep disturbance problem is

Table 1 Injury percentage by physical processes

Process	Injury percentage
Manipulating a needle	26
Sharp disposal	21
Collision with a worker	10
Clean up	9
Recapping needles	5

Fig. 1 Distribution of various types of injuries

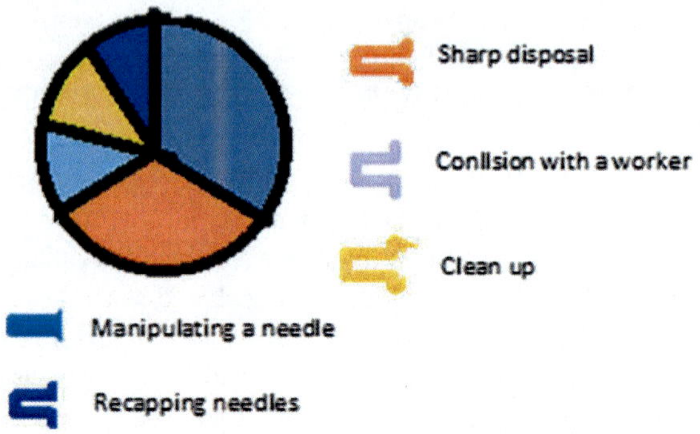

42.3%, skin reaction is 37.2%, health hazard caused due to patient assault is 24.3% while hepatitis problem is 8.9%.

Biological Hazard

The basic ways of getting affected by the biological hazards is either by direct contact or droplets or it may be air-born. Infectious agents through droplets come into play via sneezing or cough or certain medical processes. According to Baussano et al. [10], Due to the effect of biohazard the latent tuberculosis among health are persons is 20% while in developing nations it varies from 40 to 70%. According to report of WHO the annual needle stick injuries in 2003 were 83,000 out of which 1000 were found to be infected with HIV, 16,000 by HCV and 66,000 by HBV. In china every year nearly one million HCP get needle stick injuries the rate at which one HCP gets infected is every 30 s.

Chemical Hazards

In hospitals a number of cleaning agents are used to clean floor, washrooms, windows, carpets, etc. These chemicals in the form of droplets in vapors or aerosols are exposed to skin or eyes causing irritation. Some of the important chemicals used at hospitals and places of health care are: Detergents, complexing agents like water softener, alkaline agents, and acids. Such chemicals cause irritation to skin, eyes, and mucous membrane. Certain solvents used to dissolve fatty acids are neurotoxic while some chemicals like ethanol amine and benzalkonium chloride produce sensitization. Mucosal membrane irritation leads to a number of respiratory symptoms such as cough, shortness of breath and rhinitis. Some chemical are neuorotoxic or carcinogenic and may act as endocrine disruptors.

Radiation Hazard

Patients involved in diagnostic imaging have to confront low or moderate radiation doses. The intensity of radiation is measured in terms of micro- or millisieverts (μSv or mSv). Dental or chest radiographs have doses of only few μSv, however, doses for organ can be tens of mSv for CT scan and fluoroscopic examinations. National and international organizations time to time conduct study on low dose radiation effect. According to report of UNSCEAR 2000 about epidemiological studies in two designs, viz., cohort and case control. In cohort investigation a correlation study of disease level with different radiation level is done while in case control process individuals who have already developed the disease are studied. It has been explained that there is enhanced cancer risk for human who have been exposed to X-rays of doses 10–50 mSv. Some studies show that there is cancer risk to children who have fetal exposure to X-rays.

Metal Hazard

Earlier the focus was only on fetal exposure but later it was realized that exposure to biomedical can affect reproductive health like ovulation, menstruation, fertility. Fertility is affected not only in female but males also. Metals like mercury, lead, nickle, and manganese are toxic to the reproduction health. The risk involves repeated abortions, delay in conceiving, still birth, and toxicity development in offspring. In metal industries most of the employees are males, hence spermatogenesis studies were performed on males. Low doses of heavy metals are found in urban air which causes negative effect on the reproductive health. Initially, the exposure of PbB (blood lead concentration) of the range 40–60 μg/dl was considered to be safe but studies have related these concentrations with reduction in sperm quality.

4 Collection and Disposal of Biohazard Solid Waste

The process of collection and disposal must of high qualitative grade because one error and it can cause human health hazard. In India a methodological system has been developed to discard medical waste generate from various hospitals so as to maintain the hygiene.

Segregation

Segregation provides separation of non-hazardous waste from hazardous waste and it can help the waste transportation to the suitable sites. This method is also helpful in reducing the toxicity of the waste because most of the hazardous waste is already separated from the hazardous waste and non-hazardous waste is either sent for recycling or it can be dumped safely as compared to the other wastes.

Separation of waste according to the characteristics of waste

Sharp, infectious, and pathological wastes are placed in separate containers. For wet waste the containers are water tight and for various categories of wastes are kept in different colors of bins. For example infectious waste which are disposed by the method of incineration or landfill burial are kept in plastic bags of yellow color. Wastes which are treated in microwave or autoclave are kept in plastic bags of red color. In autoclave the saturated stream decontaminates the waste at high temperature and pressure. Waste kept in white or blue plastic bags is treated by chemicals or shredding.

Disinfection

Disinfection is necessary to reduce the toxicity of some of the medical wastes. This is done by using disinfectants like chlorine-di-oxide, per acetic acid, or sodium hypochlorite. Low temperature disinfection is done by vapors between 100 and 180 °C or by using infrared radiation.

Incineration

The process of disposing waste by burning is known as incineration. The process is accomplished at high temperature furnaces. This process reduces the mass of the waste and turns it into ash which is not hazardous. Incineration process reduces the mass from 50 to 400 times for disposal.

Plasma Disinfection

In this process medical waste is mixed with low temperature plasma produced in generators using air as working fluid. Thermal plasmas can be generated either by electric arc or with the help of radio frequency induction. There are many advantages of plasma treatment process. For example due to high energy densities and high temperatures helps in rapid heating and melting of materials with high melting points. Electricity is used as energy source which results in the treatment of processing environment.

Technological developments

With the help of new technology hazardous medical waste is converted to solid municipal waste with the help of shredding and grinding. The main steps included in the process are loading, shredding, heating, sterilization, draining, vacuum, and unloading.

5 Treatment by Developed Nation

In USA environmental issues and public health have been in great public attention. Basically four types of methods are used for medical waste disposal. These methods include incineration, steam sterilization, sanitary sewer disposal, and landfill disposal.

In general three types of incinerators are used to incinerate medical waste, viz., controlled air, multiple chamber air and rotary kiln model. This process reduces volume of medical waste and very small processing is required before treatment but this process is quite expensive. Incineration is a dry oxidation process at high temperature. This process converts combustible organic waste into incombustible inorganic material. The process is used to treat wastes which are not recyclable.

Steam sterilization is a process in which medical wastes are sterilized by autoclaving before landfill disposal. In this process infectious waste bags are placed in a chamber at temperature 250–270 °F for approximately 30 min. After this process the waste can be disposed of in landfill. Steam sterilization process is not as complex as incineration. Autoclave is less expensive and takes less space and emits very less toxins. Most of the landfill sites don't accept post sterilized waste because of incomplete sterilization, hence steam sterilization should be applied only when the medical waste does not contain materials like sharps, tissues, chemotherapy waste, etc.

Disposal of liquid waste is done by sanitary sewer. At first it seems to be very careless but a number of reasons suggest that disposal of liquid waste into sanitary sewer is acceptable. This method is good to avoid any infection even because of wastes containing blood. This method reduces microbial content from 90 to 99% when conventional sewage treatment method is applied. Blood is heavily diluted by sewage liquid and reduced dangers of infection.

6 Future Suggestions

Hospitals are shelter of all kind of patients and hence it may also become generator and distributor of all kind of diseases, if the wastes are not treated properly. Conventional treatment medical waste water needs a lot of cost, space, and time. Ozonization is an effective way of re-moving hazardous chemicals from water. Ozone produces hydroxyl radical (–OH) by dissolving in air and oxidizes organic compounds like pesticides, atrazine, phenol, etc. Ozonization combines small solid particle into larger one by chemical reaction through oxidation process. The objective of ozone-based technologies is to remove pollutants in water and improve the disinfection.

7 Conclusion

From above discussion it is obvious that a number of technologies are available to treat medical waste and stringent regulations have been passed in this regard. But the main reason is that the regulations are not being followed. New technologies are being developed in developed countries, hence India needs to invest in the research of treatment of medical waste. Ozone technology has emerged as one of the best technologies, but due to development in other fields new types of wastes are produced and hence continuous research is required to cope up with these problems.

References

1. Tiwari AV, Kadu PA (2013) Biomedical waste management practices in India—a review. Int J Curr Eng Technol 3(5):2030–2033
2. Radha KV, Kalaivani K, Lavanya R (2009) A case study of biomedical waste management in hospitals. Glob J Health Sci 1(1): 82–88
3. Babu BR, Parande AK, Rajlakshmi R, Suriyakala P, Volga M (2009) Management of biomedical waste in India and other countries: a review. J Int Environ Appl Sci 4(1):65–78
4. Shashikala K, Basavrajappa S (2018) Hospital waste disposal practices at few health care centers in Mysore city, Karnataka, India: a case study. World J Adv Health Care Res 2(5):191–197
5. Kalpana VN, Prabhu DS, Vinodhini S, Devirajeshwari V (2016) Biomedical waste and its management. J Chem Pharma Res 8(4):670–676
6. Raj MRR (2009) Biomedical waste management: an overview. J Indian Acad Oral Med Radiol 21(3):139–142
7. Devi KS, Krishna SR (2019) Biomedical waste management practices in India—a review. J Emerg Technol Innov Res 6(2):364–372
8. Chhabra SA (2016) Health hazard among health care personal. J Mahatma Gandhi In stit Med Sci 21:19–24
9. Orji EO, Fassubaa OB, Onwudiegwu U, Dare FO, Ogunniyi SO (2002) Occupational health hazard among health care workers in an obstetrics and gynaecology unit of a Niderian Teaching hospital. J Obstet Gynaecol 22(1):75–78
10. Baussano I, Bugiani M, Carosso A, Mairano D, Barocelli AP, Tagna M, Cascio V, Piccioni P, Arossa W (2007) Risk of tuberculin conversion among the healthcare workers and the adoption of preventive measures. Occup Environ Med 64:161–166
11. Odumosu BT (2015) Biomedical waste: its effects and safe disposal. Appl Environ Microbiol 37(3):550–553
12. Djalante R, Shaw R, DeWit A (2020) Building resilience againstbiological hazards and pandemics: COVID-19 and its implications for the Sendai Framework. Progr Disast Sci 6:100080

Game-Stress-Tracker: EEG-Based Smart Advisor Bot for Stress Detection During Playing BGMI Game

Sima Das, Sayantan Malick, Sovan Sundar Dey, Amin Sarkar, Fardin Hossain, and Abdus Samad

Abstract In today's fast-moving world, different ages peoples are suffering from mental stress, among them video games are the most popular activity to reduce psychological stress; but some games reduce stress and some games induce stress. In this paper our proposed system will detect the stress level of a game during playing. Chronic stress consequences in cancer, cardiovascular disease, depression, and diabetes consequently are unfavorable to physiological fitness and mental health. Battle Ground Mobile India is a popular game and it is an edition of PUBG. Kraft released the BGMI game on July 2, 2021, which supports android devices. This game is based on the multiplayer battle royal game which is available online. This game is very popular among teenagers but the game has some limitations also. Limitations of BGMI games are as follows: not hearing anyone while playing the game, eyes being affected, stress on the brain, disbursing money, addiction to the game, and bad effects on education. So, in this paper, we are discussing how the BGMI game affects the human brain. We used electroencephalography to collect brain signals, and alpha (α) range brain signals filtered by the band pass filtering method. After that feature extraction is done by the principal component analysis method. Classification is done by a support vector machine and classified by the following categories: normal or stressed, stressed further divided into medium and high-level stress. The proposed system was validated using accuracy, f1 score, and precision with values of 89.53, 89.51, and 89.52%. The proposed system will be notifying the guardian using Telegram Bot about the current stress of their child, based on that result, the guardian can avoid or allow their child to play games. So, the system will help to reduce the stress of children as well as their guardians.

Keywords Electroencephalography · Principal component analysis · Support vector machine · Physio care · Stress detection · Brain computer interaction · Bot

S. Das (✉) · S. Malick · S. S. Dey · A. Sarkar · F. Hossain · A. Samad
Camellia Institute of Technology and Management, Hooghly, West Bengal, India
e-mail: simadas@ieee.org

A. Rani et al. (eds.), *Signals, Machines and Automation*, Lecture Notes in Electrical Engineering 1023, https://doi.org/10.1007/978-981-99-0969-8_40

1 Introduction

In this era, video games are the most popular activity among more than 2 billion people [1]. Sometimes playing violent video games causes aggression and stress [2]. Chronic stress is potentially health-threatening [3, 4]. An EEG is one of the major diagnostic assessments for epilepsy. An EEG also can play an important role in diagnosing different brain disorders [5]. In the present world stress is a common factor for a child to old-aged people. Like all habituated game BGMI additionally have a few barriers: it's addictive, so could probably waste a variety of a while just by using gambling it, the competitiveness of the sport would possibly have an impact on your highbrow fitness as you could lose someday, wastage of money, it is the mode of amusement and you need to no longer play an excessive amount of as it has a bad effect on thoughts and eyes. So, these are disadvantages of gambling BGMI but all of these are based upon your ways to play the game and the manner intensely, approximately triumphing in it.

The current work focused on analyzing the stress by comparison of training and testing of EEG-based game datasets. The dataset is collected from electroencephalography which is connected to the subject's scalp while they are playing a BGMI game. The raw EEG dataset needs to be cleaned, for that purpose alpha (α) range filtering is used. Features are selected by principal component analysis; classification is done by using a Support vector machine and classified as stress or normal load. Telegram bot used to send messages to the user and their parent's mobile about the mental condition of the user. The rest of the chapter is as follows, Sect. 2 for the literature survey, Sect. 3 for the proposed work, Sect. 4 for experimental results of the current system, and the last conclusion and future works are discussed.

2 Literature Survey

The article was designed by Lee et al. [6], to study intellectual and physical strong sports in virtual reality via brainwave detection. To mitigate strain, it's critical to relax the mind by way of enough sleep. Professionals furthermore perceive that relaxation is vital to enhance the mind's reminiscence, cheering the concentration to accomplish time to arrange the length over an intellectual composer. It is a digital authenticity content that may construct a sensible atmosphere, and it recommends satisfied material which could verify the method of resting over mild actions and elements that support relaxation as a technique of brain wellbeing.

Shirke et al. [7] are introduced to the stress of varied depth while performing each day's duties. Excessive strain may be injurious to humanoid fitness. Hence, stress analysis is crucial in avoiding harmful lengthy-period consequences. In this examination, they examine the probability of EEG intended for the size of severe intellectual pressure. As well as up-strain evaluation by 4 samples on strain-caused themes is performed. The tests have been performed with an EEG sensor, Thing

Speak, and an application program on mobile. Challenge is needed to perform a cell sport that encourages strain for the game developments. As soon as the severe stage of pressure is perceived, the character becomes informed through the cell utility to produce a peaceful song and remaining eyes. Inclusive, the test was resolved by decreased strain ranges of the problem subsequently remaining eyes accompanied by track and unaccompanied by track.

In our previous work, Das et al. [8] are designing this paper to measure cognitive load and the analytical burden is considered for young people whilst sporting a one android game involving easy complicated brain action. The recording of human brain action given analytical burden during the data achievement method; EEG is using. EEG is human brain signals acquired through this electric field generate invariants owing to neuronal firings also acquired through the sensor placed by the skull in one manner. Then acquired human brain signals are processed via different steps with the maximum target of assorting through analytical burden; one subject separated the three groups: lower, medium, and high.

Dharmapuri et al. [9], are developed AI-based chatbot to detect mental stress for preventing suicidal tendencies and help to psychiatric treatment.

3 Proposed Work

In this era, most teenagers or young people are interested in playing mobile or computer games which affect their brain and affect stress. In this section, we will propose a method that will be helpful to detect stress during playing BGMI games. Figure 1 shows that EEG is used to collect data, an α-range filter is used for cleaning data, the principal component analysis is used to select features, and the support vector machine is used for classification as normal or stressed. As the result is stressed then it will be further categorized as medium or high-level stress using the color bar, brain map, and SVM, the collected result is sent to the user as well as parents' mobile to alert them.

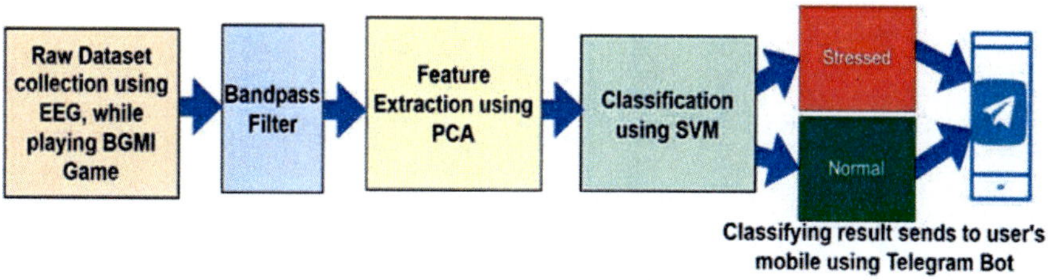

Fig. 1 Proposed work for "game-stress-tracker"

3.1 Dataset Collection

The current work has been applied to 20 numbers of subjects with their consent whose age is above 18. For the EEG test, a small metal disc is used, which consists of electrodes with thin wire pasted onto the brain scalp. The electrodes perceive small electrical signals controlled by the movement of brain cells. O1, O2, T3, T4, FP1, FP2, A1, A2, FZ, CZ, and PZ electrodes are positioned on the scalp of the participant which measures voltage and sends a numeric value via the Brain Tech Traveler system.

3.2 Filtering

In this paper, we used an 8–12 Hz bandpass filter which means the lowest frequency of the filter is 8 and the highest is 12 Hz, which is the range of the alpha bandpass filter. In Eq. (1), F is the band pass filter, $F(H)$ means the highest frequency, and $F(L)$ is the lowest frequency band pass allowed for the proposed system.

$$(F)^2 = F(H) \times F(L) \tag{1}$$

3.3 Feature Extraction

The data collected while testing is usually in a large data set. Reduce the large data sets into smaller ones that still contain large data, using this method feature extraction, and selection are done. Here features are O1, O2, T3, T4, FP1, FP2, A1, A2, FZ, CZ, and PZ. In Eq. (2–4), X is the after-filtering dataset, C is used for covariance matrix, Λ for eigenvalue, and an eigenvector is E.

$$X = [\text{After Filtering\_EEG Dataset}] \tag{2}$$

$$C = X^{\mathrm{T}} X \tag{3}$$

$$\Lambda E = CE \tag{4}$$

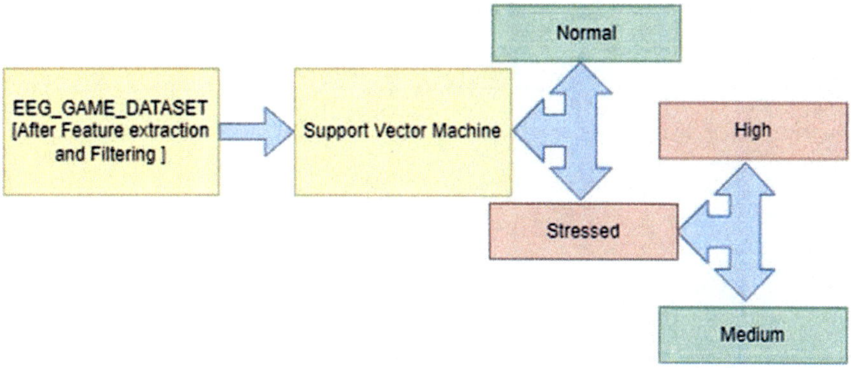

Fig. 2 Classification using support vector machine

3.4 *Classification*

Classification is done by using a support vector machine and is primarily classified as normal or stressed. After primary classification, stress is further classified as medium stress or high stress by comparing a color map with a brain map. The red color indicates high stress and the blue color indicates medium stress. In Eqs. 5 and 6, ΛE is the after-feature extraction dataset, W is the weightage value, and P is the threshold value. If the ΛE. W is greater than the threshold value than positive means normal, otherwise stressed, stressed datasets are further categorized as high and medium as shown in Fig. 2.

$$\Lambda E.W > P \tag{5}$$

$$\Lambda E.W < P \tag{6}$$

3.5 *Game Rating Calculation*

In this paper, the rating of the game is calculated by classifying stress levels during playing games. The total number of data collected during playing games of each level and each person within several days. After collecting data, stress is detected as medium and high stress by using color maps. In this paper, we assume that 1 is the lowest and 5 is the highest rating of the game. In this system, 2 were used for medium weightage value, and 5 were used for highest weightage value for stress detection which is shown in Eq. 7.

FP1-A1	FP2-A2	O1-A1	O2-A2	T3-A1	T4-A2	FZ-A1	FZ-A2	CZ-A1	CZ-A2	PZ-A1	PZ-A2
0	0	0	0	0	0	0	0	0	0	0	0
0	0	0	0	0	0	0	0	0	0	0	0
13.5	15.6	7.8	8.1	2.6	3.5	4	5.1	6	7.3	0.3	1.3
23.5	32.6	18.1	17.6	14.3	10	15.1	17	15.6	13.8	15.3	13.9
32.3	41.3	36.1	25.6	18.9	14	17.3	20	23	23.2	18.9	16

Fig. 3 Dataset collection using EEG

$$\text{Game Stress Rating} = \left(\frac{2 \times \text{medium\_ stress}}{\text{medium\_ stress} + \text{high\_ stress}} + \frac{5 \times \text{high\_ stress}}{\text{medium\_ stress} + \text{high\_ stress}} \right) \tag{7}$$

4 Result

The current work has been applied to 20 numbers of subjects with their consent whose age is above 18. The proposed system was validated using accuracy, f1 score, and precision with values 89.53, 89.51, and 89.52% which are shown in Fig. 5, and was initiated and improved upon existing ones. The system analyses stress levels to be useful for psychologists and mental healthcare and sends the report using the Telegram bot.

4.1 EEG Placement and Dataset Collection

The standard 10–20 systems are used to collect data. In the current work, we used O1 and O2 in the Occipital lobe used to see the game, T3 and T4 in the temporal lobe used for try to temporal solution of the game, FP1, and FP2 in the prefrontal lobe used for trying to fetch the solution, and A1 and A2 for ground and reference, FZ, CZ, PZ are common reference electrodes as shown in Figs. 3 and 4.

4.2 Brain Map and Telegram Bot

The brain activity is captured by EEG and generated brain map and also detect stress level by using a color map. In this proposed system we used the Occipital, Prefrontal, and Temporal lobe for stress detection. As shown in Table 1, the Brain map was generated with blue and red colors by using Loreta. Blue color indicates that they have low to medium stress and red indicates that high-level stress occurs

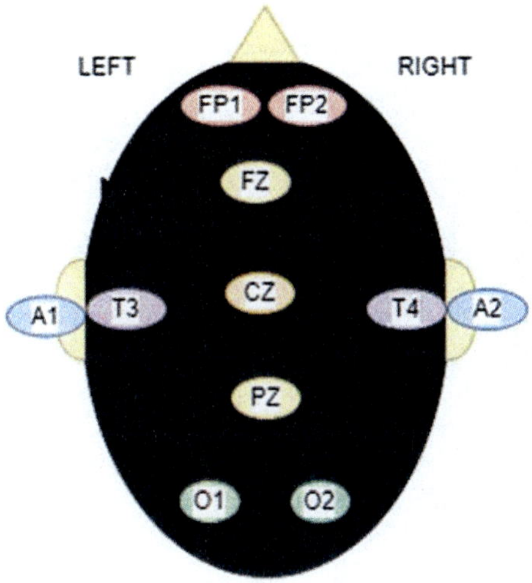

Fig. 4 10–20 system of electrode

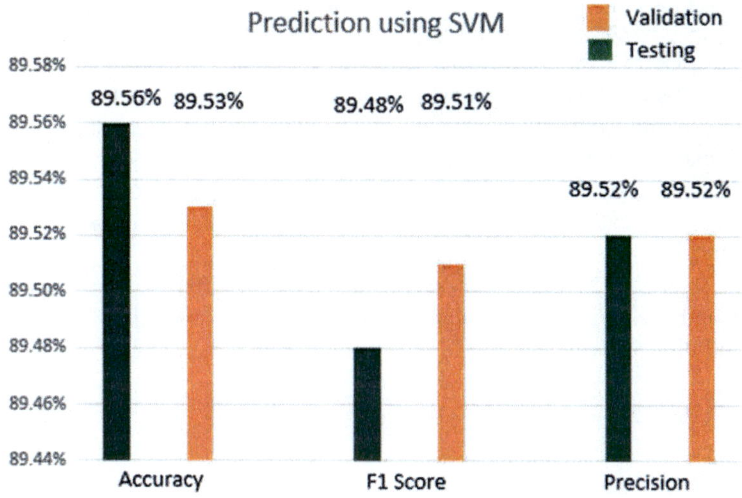

Fig. 5 Stress detection using a support vector machine placement using EEG

during playing the BGMI game. After calculating the stress level results are sent to the user's mobile by using Telegram Bot.

Table 1 Identification of Brain Activation while Playing BGMI Game

Brain map	Screen shot of telegram bot
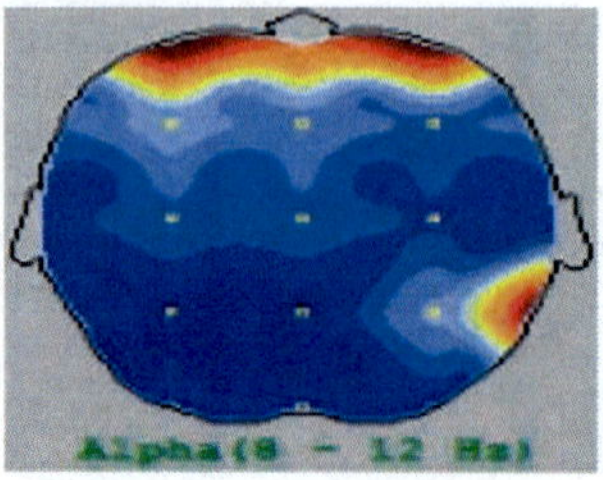	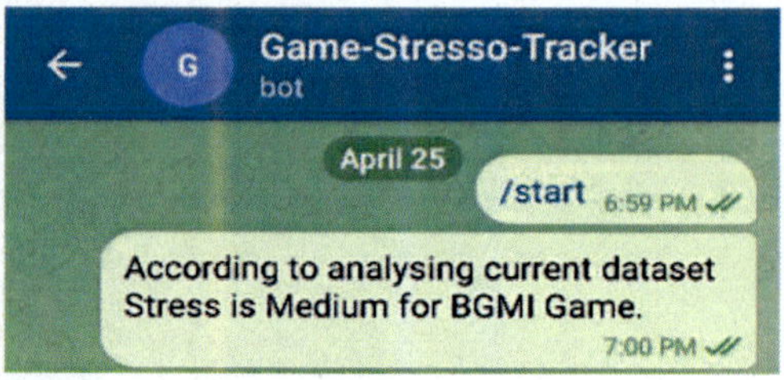

5 Conclusion and Future Work

BGMI, is an imitation of the world's most popular game, and is now ubiquitous. Some games are very aggressive, the human brain is under a lot of pressure while playing the game, addicted to the game has a bad impact on the human mind. Electroencephalography is used to collect the brain signal, α range bandpass filtering is used to filter, and the principal control analysis method has been used to extract the features. Support vector machines have been used to classify brain signals as normal or stressed. If stresses are found then they will be further categorized as medium and high stress by comparing color bars with generated brain maps. In this paper, we used different classification techniques like back propagation neural network, k-nearest neighbor, linear discriminant analysis, and quadratic discriminant techniques, but the support vector machine gives 89% and above accuracy which indicates that it is the best technique for the proposed system.

In the future, we will use this current method with different types of machine learning and deep learning techniques for better result. The future method will be used at different ages and applied to different kinds of games.

References

1. Von der Heiden Juliane M, Beate B, Müller WK, Egloff B (2019) The association between video gaming and psychological functioning. Front Psychol 10:1664–1078. https://doi.org/10.3389/fpsyg.2019.01731
2. Rong S, Yunqiang W (2019) The relation of violent video games to adolescent aggression: an examination of moderated mediation effect. Front Psychol 10. https://doi.org/10.3389/fpsyg.2019.00384
3. Dai S, Mo Y, Wang Y, Xiang B, Liao Q, Zhou M, Li X, Li Y, Xiong W, Li G, Guo C, Zeng Z (2020) Chronic stress promotes cancer development. Front Oncol 10:1492. https://doi.org/10.3389/fonc.2020.01492
4. Cao L, Zhang H, Li N, Wang X, Ri W, Feng L (2022) Category-aware chronic stress detection on microblogs. IEEE J Biomed Health Inform 26(2):852–864. https://doi.org/10.1109/JBHI.2021.3090467

5. Ghosh A, Das S, Saha S (2022) Stress detection for cognitive rehabilitation in COVID-19 scenario. https://doi.org/10.1049/PBHE042E_ch12
6. Lee J, Lee D, Jeong I, Cho J (2021) A study on the content of mental and physical stability game in virtual reality through EEG detection. In: 2021 International conference on information and communication technology convergence (ICTC), pp 693–696. https://doi.org/10.1109/ICTC52510.2021.9620932
7. Shirke B, Wong J, George K (2019) Acute mental stress measurement using brain-IoT system. IEEE First Int Conf Cogn Mach Intell (CogMI) 2019:81–86. https://doi.org/10.1109/CogMI48466.2019.00019
8. Das S, Ghosh L, Saha S (2020) Analyzing gaming effects on cognitive load using artificial intelligent tools. https://doi.org/10.1109/CONECCT50063.2020.9198662
9. Dharmapuri CM, Agarwal A, Anwer F, Mahor J (2022) AI Chatbot: application in psychiatric treatment and suicide prevention. In: 2022 International mobile and embedded technology conference (MECON), pp 41–44. https://doi.org/10.1109/MECON53876.2022.9752126

An Efficient EMG Denoising Technique Based on the W-NLM Method

Rinku Bhobhriya, Ritu Boora, Manisha Jangra, and Priyanka Dalal

Abstract Denoising of the electromyography (EMG) signal is critical in the diagnosis of muscle illnesses and several EMG-based mechatronics applications. This paper presents an improved EMG denoising method based on the discrete wavelet transform (DWT) and the nonlocal means (NLM) estimates. The DWT-based denoising method is quite effective in reducing the noise present in high-frequency regions. Unfortunately, this method demands large decomposition levels to mitigate the noise in low-frequency regions. The NLM method is efficient in mitigating noise from low-frequency regions, but its performance is limited by the rare-patch effect that results in signal distortion. Henceforth, these techniques are unable to meet the growing demand of the new generation applications. Subsequently, an improved denoising technique that effectively integrates the capabilities of both NLM and DWT is proposed in this paper. The performance of the proposed method is evaluated and compared to the proposed work on the signal taken from the EMGLAB database.

Keywords Electromyography (EMG) · Discrete wavelet transform (DWT) · Denoising · Wavelet thresholding · NeighShrinkSURE (NSS) · Nonlocal means (NLM)

1 Introduction

Electromyography (EMG) is a myoelectric signal that specifies the information on the electrical activities of the human neuromuscular system. These signals have a vital role in several applications including psychomotor and neuromuscular research, neurological diagnostics, and robot limb control [1]. However, these signals often get contaminated with artifacts during their acquisition, recording, and transmission process. The noises that affect these biopotential signals generally come from electrodes, cables, data collection equipment, and amplifiers. Moreover, even the body itself influences the signal acquisition from the body surface such as motion artifacts.

R. Bhobhriya (✉) · R. Boora · M. Jangra · P. Dalal
Department of ECE, GJUS&T, Hisar, India
e-mail: rinku13152923@gmail.com

A. Rani et al. (eds.), *Signals, Machines and Automation*, Lecture Notes in Electrical Engineering 1023, https://doi.org/10.1007/978-981-99-0969-8_41

Several mathematical frameworks for EMG denoising in the literature are based on empirical mode decomposition (EMD) [2], DWT [3], NLM [4], variational mode decomposition (VMD) [5], and generalized variational mode decomposition (GVMD) [6]. The EMD-based denoising [2] approaches tend to enhance the signal-to-noise ratio (SNR) but are less efficient in preserving the morphological structure of EMG signals [7]. Furthermore, these methods are computationally expensive when it comes to extracting intrinsic mode functions (IMFs) [8]. The VMD is used to decompose the signal into a set of modes generated using a non-recursive process such that the spectrum of each mode is concentrated around center frequencies. However, its center frequencies cannot be flexibly adjusted [6]. The DWT-based denoising approach is effective in removing only the high-frequency noise which results in signal distortion and information loss. Moreover, accessing the very low-frequency components requires substantial decomposition levels that lead to computational overhead [3]. Inversely, the NLM-based methods effectively remove the low-frequency noise but suffer from the rare-patch effect due to its incompetence in high-frequency regions [4]. This paper introduces a competitive method utilizing the efficacy of both DWT- and NLM-based techniques for EMG signal denoising.

This paper is structured as: Sect. 2 explains the materials and methods, while Sect. 3 describes the proposed wavelet-nonlocal means (W-NLM) EMG denoising approach and performance metrics. Next, Sect. 4 discusses the qualitative analysis of the results. Finally, the conclusion and future scope are presented in Sect. 5.

2 Materials and Methods

The performance of the proposed work is evaluated on synthetic EMG signals available on the EMGLAB database [18]. Three signals S00101, S00111, and S00121 were taken from the database, and each signal consists of 50,000 samples. The purpose of this research is to remove noise from a corrupted EMG signal, $S(i)$ without impacting the EMG's morphological components. This section briefly explains the DWT- and NLM-based EMG denoising techniques.

2.1 Discrete Wavelet Transform (DWT)-Based Denoising

The wavelet-based denoising techniques are widely preferred for denoising the EMG signals due to their inherent time–frequency resolution [9]. It has become a powerful tool for nonstationary signal analysis due to the availability of multiple wavelet functions. The DWT-based denoising methods are implemented in three steps where the signal is decomposed into detail and approximation coefficients in stage 1. In stage 2, the signal is denoised by implementing thresholding on the obtained detailed coefficient. Finally, the signal is reconstructed from the modified coefficients.

In stage 2, after decomposition, the detailed coefficients are exposed to the thresholding techniques for denoising the signal. These techniques are broadly classified into two categories namely hard and soft thresholds. The hard threshold is a keep or kills rule, and it is more suited only when the detail coefficient is either a signal or a noise coefficient which is generally not the case. On the contrary, the soft threshold (cf. Eq. (1)) is preferred when the detail coefficient contains both signal and noise [10].

$$\hat{D}_{j,k} = \begin{cases} \text{sign}(D_{j,k})(|D_{j,k}| - \lambda), & |D_{j,k}| \geq \lambda \\ 0, & |D_{j,k}| < \lambda \end{cases}, \tag{1}$$

where λ is the threshold value, $D_{j,k}$ and $\hat{D}_{j,k}$ denote the detail and estimated coefficients, respectively. This method denoises the high-frequency noise well but demands large decomposition levels to diminish the noise from low-frequency regions. A higher decomposition level means more filter banks, which increases computing time and complexity.

2.2 *Nonlocal Means (NLM)-Based Denoising*

The NLM algorithms were originally developed for image denoising. However, the algorithms have evolved and have been applied to EMG signal denoising due to their repetitive characteristics similar to that of the image [11]. The NLM method calculates an estimate for each sample in the noisy EMG signal. The estimated mth sample value $\hat{S}(m)$ can be expressed as the weighted sum of n samples in the search neighborhood. The centers of the local patches that occur inside a search neighborhood of $R(m)$ correspond to m and n for a given signal. Each patch will have B_δ samples ranging from $-P$: P. Subsequently, $B_\delta = (2P + 1)$ samples where δ represents the patch number. The following is a representation of the estimated signal:

$$\hat{S}(m) = \frac{1}{s(m)} \sum_{n \in R(m)} w(m, n) S(n). \tag{2}$$

Here, $s(m) = \sum_n w(m, n)$ is the total of the weight values over a search neighborhood ($n \in [-Q : Q]$) and $w(m, n)$ is the weight value specified as

$$w(m, n) = \exp\left(-\frac{\sum_{\Delta \in \delta} (S(m + \Delta) - S(n + \Delta))^2}{2 B_\delta \tau^2} \right), \tag{3}$$

where δ is the patch width ($-P$: P) in Eq. (3) and Δ is a variable that ranges over B_δ. The bandwidth parameter, τ determines how much smoothing will be applied to the signal. The discrepancy between the data points of the patches centered at m

and n, accordingly, is represented by d. The weight value is calculated by summing the difference value over δ and normalizing it. There have been other weighting approaches presented, but the most accepted is squared patches with a central reference point. A patch correction technique is used to achieve better results in the instance of image denoising as follows:

$$w(m, m) = \max_{n \in S(m), n \neq m} w(m, n). \tag{4}$$

The under averaging of high-frequency areas due to the infrequent patch effects [12] results in signal distortion. It a computationally demanding as each sample is estimated over the entire search neighborhood.

3 Proposed W-NLM EMG Denoising Method

The DWT and NLM algorithms are two powerful denoising approaches with a complementary set of advantages and limitations. Subsequently, combining these approaches can result in an effective EMG denoising technique. However, a direct cascading of these approaches will result in an ineffective and computationally expensive denoising system. Henceforth, this research work presents an effective method of combining these approaches to get the desired results.

The proposed W-NLM method is implemented in the following three steps: (i) signal decomposition using DWT, (ii) denoising using NLM and thresholding, and (iii) signal reconstruction. A block diagram representation of the proposed work is illustrated in Fig. 1.

Wavelet Decomposition: Initially, the acquired signal (given in Eq. (1)) is decomposed using DWT to get detailed (D_1)-and approximation (A_1)-coefficients. The low-pass representation of the signal is used for the approximation coefficients, while the wavelet coefficients are used for the details. The signal is decomposed to a level such that it maintains the signal's morphological structure and keeps computations low. A single-level decomposition is selected after several experimental analysis,

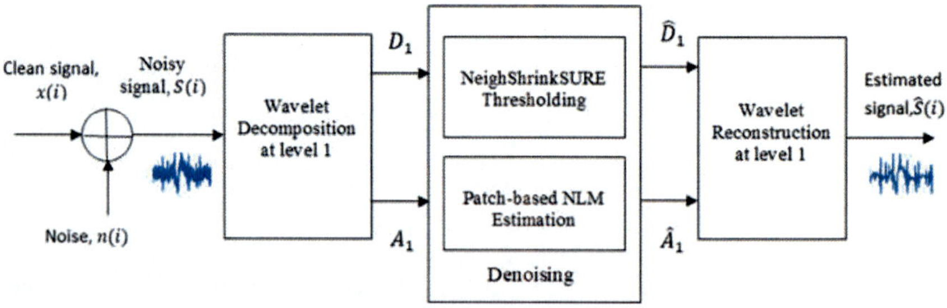

Fig. 1 Block diagram of the W-NLM EMG denoising approach

in this paper. Furthermore, a wide range of mother wavelet function's performance was compared in terms of SNR output (SNR_{out}), root mean square error (RMSE), and percent root distortion (PRD) as performance measures. On basis of preliminary experimental analysis, Symlet wavelet function of order 9 ('sym9') was chosen for decomposition of the signal [8].

Denoising the Signal: In this stage, the decomposed signal is denoised using NLM estimation and thresholding. The DWT's detailed coefficients are exposed to soft thresholding techniques for denoising the signal. The various strategies for thresholding are available in the literature such as VisuShrink [10], SURE-Shrink [12, 13], NeighShrink [14], and NeighShrinkSURE [15]. The NeighShrinkSURE is an upgraded version of NeighShrink thresholding approach [16]. A performance comparison of various threshold selection techniques was conducted at different SNR_{in}. On basis of preliminary results, NeighShrinkSURE is selected for soft thresholding of detail coefficients.

The approximation coefficients (A_1) being low-frequency coefficients are denoised by applying the NLM method. The selection of the following NLM parameters is crucial for the proposed work: patch half-width (P), the search neighborhood half-width (Q), and the bandwidth parameter (τ). The patch half-width P chooses the scale to compare the patches. Moreover, raising the neighborhood half-width Q (which results in a "less local" search) should improve performance, in principle. Although a wider search neighborhood ($2Q + 1$) provides better estimation, it increases the computational load. The smoothness degree provided to the given signal is determined by the parameter τ. Over-smoothing and patch similarity problems will occur if τ is too large or too small. The value of τ is chosen in the majority of the previous research [4] so that it is proportionate to noise standard deviation. SNR_{out} for various combinations of P and Q values averaged across the developmental set at a given SNR level of 5 dB. The ideal P and Q values were determined to be 6 and 900 samples, respectively. Similarly, on comparing the SNR_{out} for various values of τ, its value is taken as 0.65σ.

Wavelet Reconstruction: Following noise removal, the enhanced EMG signal is subjected to a reverse decomposition (inverse DWT). The signal is reconstructed using the modified detail ($\widehat{D}_1$) and approximation coefficients ($\widehat{A}_1$).

Performance Measures: We have used the *SNR_{out}*, *RMSE*, and *PRD* as the performance measures [17]. SNR_{out} measures the signal-to-noise ratio after denoising. In contrast, *RMSE* is a measure of error between the original and denoised signal, and the *PRD* identifies the distortion present in the denoised output. Equations (5–7) represents the performance measures used.

$$\text{SNR}_{\text{out}}(dB) = 10\log_{10}\left[\frac{\sum_{i=1}^{N}\hat{S}^2(i)}{\sum_{i=1}^{N}(\hat{S}(i) - x^2(i))^2}\right], \tag{5}$$

$$\text{RMSE} = \sqrt{\frac{1}{N}\sum_{i=1}^{N}(\hat{S}(i) - S(i))^2}, \tag{6}$$

$$\text{PRD}(\%) = 100\sqrt{\frac{\sum_{i=1}^{N}(\hat{S}(i) - S(i))^2}{\sum_{i=1}^{N} S^2(i)}}. \tag{7}$$

4 Experimental Results

This section presents the performance validation of the proposed work while exploring a comparative study of reference methods [4, 18] in different scenarios. The AWGN is added to synthetic EMG signals to generate different *SNR* levels. The comparison of denoised signal with W-NLM, DWT, and NLM is presented in Fig. 2. Here, the noisy EMG signal (Fig. 2b) is created by adding AWGN to the clean EMG signal (Fig. 2a) to attain an input *SNR* of 0 dB for this experiment. As distinctly shown in Fig. 2(e), the W-NLM has most effectively denoised the signal when compared to the other two in Fig. 2 (c, d). The reference methods have also distorted the morphological structure of the signal.

In Fig. 3, a quantitative study based on SNR_{out} for a given range of SNR input is carried out for the three methods discussed above. Here, the performance is evaluated at SNR_{in} of -5 dB, 0 dB, 5 dB, and 10 dB. The results show that as SNR_{in} drops, the SNR_{out} degrades gradually for the W-NLM, whereas the reference methods show a sharp decay in SNR_{out} for the drop-in input *SNR*.

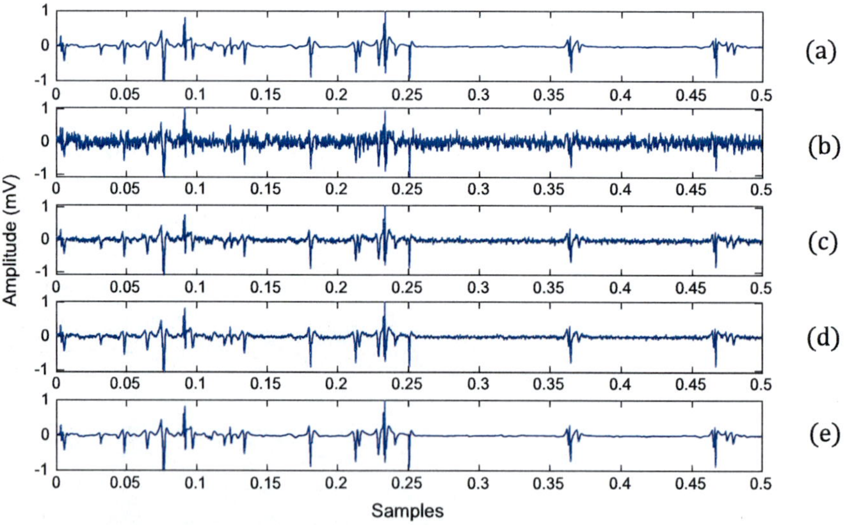

Fig. 2 **a-e**. Denoising of the EMG signal with W-NLM and reference methods **a** clean EMG signal, **b** noisy signal at 0 dB input SNR, **c** DWT, **d** NLM, and **e** W-NLM

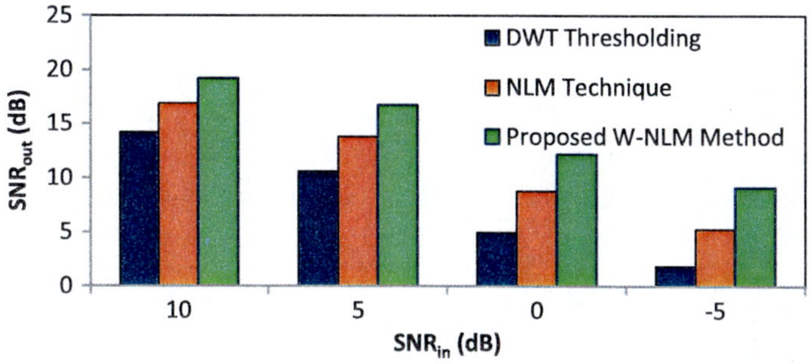

Fig. 3 Performance comparison in terms of SNR_{out} of the proposed with existing work for various input SNR

Table 1 shows the quantitative analysis in terms of SNR_{out}, RMSE, and PRD values concerning three EMG signals at different SNR_{in}. The PRD metric is employed to determine the amount of distortion present, which should be bare minimum. The results show that, for all input SNR levels, the W-NLM method has the highest SNR_{out}, and least-RMSE and *PRD* values. The performance of the other two methods degrades in noisy condition as they observe high distortions with drop-in SNR.

Table 1 Performance comparison of the proposed work with existing state-of-the-art of technologies at different SNR_{in}

EMG records	$(\frac{S}{N})_{in}$	DWT-based denoising			NLM-based denoising			Proposed W-NLM method		
		$(\frac{S}{N})_{out}$	RMSE	PRD	$(\frac{S}{N})_{out}$	RMSE	PRD	$(\frac{S}{N})_{out}$.	RMSE	PRD
S00101	−5	1.92	0.073	27.6	5.32	0.063	20.5	9.12	0.053	14.4
	0	5.01	0.061	22.9	8.81	0.051	16.5	12.2	0.042	11.5
	5	10.6	0.045	14.8	13.8	0.036	10.7	16.7	0.028	7.53
	10	14.2	0.027	9.21	16.2	0.022	7.13	19.2	0.016	5.23
S00111	−5	2.01	0.070	27.4	5.51	0.061	20.3	9.32	0.051	14.2
	0	5.23	0.058	22.6	9.01	0.049	16.3	12.4	0.039	11.3
	5	10.8	0.043	14.6	14.1	0.034	10.5	16.9	0.025	7.34
	10	14.4	0.025	8.99	17.2	0.021	6.88	19.4	0.015	4.98
S00121	−5	1.73	0.075	27.7	5.12	0.065	20.7	8.89	0.055	14.6
	0	4.82	0.063	22.9	8.61	0.053	16.8	11.9	0.044	11.7
	5	10.4	0.048	14.9	13.6	0.038	10.9	16.5	0.030	7.72
	10	13.9	0.029	9.42	16.7	0.024	7.37	18.9	0.018	5.45

5 Conclusion and Future Scope

The efficiency of DWT- and NLM-based denoising techniques is exploited to propose an EMG denoising method in this paper. In other words, the proposed method retains the efficacy of NLM in low-frequency zone and DWT methodology in reducing high-frequency noises. The DWT decomposes into the detailed and approximation coefficients. Subsequently, the detail coefficient is soft thresholded to diminish the high-frequency noise. Further, the signal is exposed to the NLM technique to remove the low-frequency noise. The W-NLM significantly reduces the computing time to that of NLM and DWT methods. The performance of the proposed method has been validated on various EMG signals in different noisy conditions. In the future, this proposed work can be extended to denoise other noise sources such as baseline wander (BW), power-line–interference (PLI), electrocardiogram (ECG), and motion artifacts.

References

1. Fukuda O, Kim J, Nakai I, Ichikawa Y (2011) EMG control of a pneumatic 5-fingered hand using a Petri net. Artif Life Robot 16(1):90–93
2. Chatlani N, Soraghan JJ (2012) EMD-based filtering (EMDF) of low-frequency noise for speech enhancement. IEEE Trans Audio Speech Lang Process 20(4):1158–1166
3. Internationale C (2007) A comparative study of wavelet denoising of surface electromyographic signals. Annu Int Conf IEEE Eng Med Biol Soc 1868–1871
4. Tracey BH, Miller EL (2012) Nonlocal means denoising of ECG signals. IEEE Trans Biomed Eng 59(9):2383–2386. https://doi.org/10.1109/TBME.2012.2208964
5. Dragomiretskiy K, Zosso D (2014) Variational mode decomposition. IEEE Trans Signal Process 62(3):531–544. https://doi.org/10.1109/TSP.2013.2288675
6. Guo Y, Zhang Z (2021) Generalized variational mode decomposition: a multiscale and fixed-frequency decomposition algorithm. IEEE Trans Instrum Meas 70
7. Kabir A, Shahnaz C (2012) Denoising of ECG signals based on noise reduction algorithms in EMD and wavelet domains. Biomed Signal Process Control 7:481–489
8. Rakshit M, Das S (2018) An efficient ECG denoising methodology using empirical mode decomposition and adaptive switching mean filter. Biomed Signal Process Control 40:140–148. https://doi.org/10.1016/j.bspc.2017.09.020
9. Pandit D, Zhang L, Liu C, Aslam N, Chattopadhyay S, Lim CP (2017) Noise reduction in ECG signals using wavelet transform and dynamic thresholding. Emerg Trends Neuro Eng Neural Comput 193–206
10. Peng YH (1995) De-noising by soft-thresholding. In: IEEE Asia-Pacific Conf Circuits and systems—proceedings 41(3):760–762
11. Fan L, Li X, Guo Q, Zhang C (2018) Nonlocal image denoising using edge-based similarity metric and adaptive parameter selection. Sci. China Inf. Sci. 61(4):5–7
12. Buades A, Coll B, Morel J, A JM (2005) A review of image denoising algorithms, with a new one to cite this version : a SIAM Interdiscip. J Soc Ind Appl Math 4(2):490–530
13. Taylor P, Donoho DL, Johnstone IM, Donoho DL (2012) Adapting to unknown smoothness via wavelet Shrinkage. J Am Stat Assoc 90(432):1200–1223
14. West DM, Hg C (2004) Image denoising using neighboring wavelet coefficients. ICASSP, IEEE. II:910–920

15. Panchaxri P, Jagadale BN, Priya BS, Nargund MN (2021) Image denoising using adaptive NL means filtering with method noise thresholding. Indian J Sci Technol 14(39):2961–2970. https://doi.org/10.17485/ijst/v14i39.1532
16. Heo YC, Kim K, Lee Y (2020) Image denoising using non-local means (NLM) approach in magnetic resonance (MR) imaging: a systematic review. Appl Sci 10(20):1–16
17. Gualsaquí M, Vizcaíno I, Proaño V, Flores M (2018) ECG signal denoising using discrete wavelet transform: a comparative analysis of threshold values and functions. MASKANA. 9(1):105–114. https://doi.org/10.18537/mskn.09.01.10
18. Marateb HR (2011) EMGLAB signals. Available at http://www.emglab.net/emglab/Signals/signals.php

2DOF PID-Based Controller for Chemotherapeutic System

Disha Mondal, Asha Rani, and Vijander Singh

Abstract This paper aims to analyze the capability of 2DOF PID controller while dealing with multiple and conflicting objectives problem. A basic PID control system fails to track the reference input and reject the disturbance simultaneously. This problem may be better dealt with the use of a two-degree of freedom controller which combines the effect of two controllers in separate loops to meet the two criteria. The problem considered in this study is the drug dose control in chemotherapy. In this study, multi-objective GA and multi-objective swarm optimization algorithms are implemented to obtain the optimum amount of the drug delivery for chemotherapeutic treatment. The designed controller is compared with the conventional PID controller. Various analyses like step response analysis, bode analysis, parameter perturbation analysis and disturbance analysis are carried out to justify the performance of the designed controller.

Keywords 2DOF · NSGA-II · Step response · Disturbance · Parameter perturbation

1 Introduction

Proportion integral and derivative (PID) controller is a common control algorithm used in almost every engineering process for the effectiveness and simplicity of the controllers. PID controller is the basic controller that has been in use since decades. With time many advancements are made in the PID control scheme for better performance. These controllers are tuned using several methods and algorithms. A

D. Mondal (✉) · A. Rani · V. Singh
Department of Instrumentation and Control Engineering, Netaji Subhas University of Technology, New Delhi, India
e-mail: disham.ie20@nsut.ac.in

A. Rani
e-mail: asha.rani@nsut.ac.in

V. Singh
e-mail: vijaydee@nsut.ac.in

A. Rani et al. (eds.), *Signals, Machines and Automation*, Lecture Notes in Electrical Engineering 1023, https://doi.org/10.1007/978-981-99-0969-8_42

conventional PID controller fails to effectively reject the system disturbance which causes sudden change in the system dynamics. Such changes affect the output of the system. Such disturbing effects may be countered by a two-degree of freedom PID controller (2DOF PID).

A 2DOF PID controller uses two control loops which lead to a robust control system. Taguchi and Araki [1] provided several configurations of a 2DOF PID controller. Another study [2] shows the supremacy of a 2DOF PID over conventional PID controller in the field of power electronics. In several other studies, 2DOF controllers are preferred over the conventional ones [3]. The controller performance for a particular application depends upon the tuning parameters. The most classical techniques used are the Zeigler-Nichols and Cohen-Coon method of tuning. With the advancement of intelligent techniques, the controllers may be efficiently tuned using the intelligent optimization algorithms [4, 5]. The problem considered in this study is the control of drug dose in chemotherapeutic treatment. This problem deals with two conflicting objective functions. The multi-objective evolutionary algorithms, i.e., multi-objective GA and multi-objective swarm optimization, are incorporated to control the drug delivery in the crucial chemotherapy problem [6–8]. The supremacy of the 2DOF PID controllers are established using several validation techniques.

The study shows cancer model and its description in Sect. 2. The controller design is shown in Sect. 3 followed by the validations in Sect. 4. The conclusion is mentioned in Sect. 5.

2 Cancer Model

Chemotherapeutic drug control deals with two objectives which are in opposition to each other. The drug injection in the body increases the drug content of the body which successfully kills the cancer cells, however, with the cancer cells the normal cell are also killed due to the harmful effect of the drug. Thus, regulating the drug content to control the harmful effect of drug in the body is equally important. A master slave cascade control system is implemented to achieve both the targets where two individual controllers are used to control the two parameters [9].

The model reported by Martin is used quite often to understand the proliferation of cancer cells. The model is represented in terms of ODE [9] (Eq. 1–3).

$$\frac{d}{dt}Z(t) = -\lambda Z(t) + k(D(t) - \alpha)H(D(t) - \alpha), \tag{1}$$

$$\frac{d}{dt}D(t) = u(t) - \gamma D(t), \tag{2}$$

$$\frac{d}{dt}T(t) = D(t) - \eta T(t),$$

λ, γ, α, η and k are tumor growth rate ($1.5*10^4$ cells/day), drug decay rate (0.27 per day), drug threshold level (10 drug days), elimination rate constant (0.4 per day) and cells killed/time/drug concentration ($9.9*10^{-3}$ per day per drug unit). $D(t)$ and $T(t)$ are drug concentration and toxicity level. Z is the transformed variable. Two conflicting objectives are achieved by introducing the following constraints in the modeling:

- The drug concentration should be regulated between 10 and 50 drug unit.
- The harmfulness or toxicity of the body should be below 100 drug unit.

3 2DOF Proportional Integral Derivative Controller

The degree of freedom of a controller gives the number of closed loops that can be adjusted independently. A two-degree of freedom controller attempts to produce the desired output and reject any disturbance in the system simultaneously. Thus, it is more robust as compared to a single degree of freedom PID controller. A 2DOF PID controller is implemented in the system that deals with multi-objective problem. Literature [1] reports the design of several configurations of 2DOF PID controller. The configuration considered in this study is a conventional PID controller and a feedback compensator. The conventional PID controller tracks the reference point, and the feedback compensator handles the disturbances. The feedback compensator consists of proportional and derivative control action along with the derivative filter. The derivative filter is used to avoid the derivative kick. The block diagram of the 2DOF PID-based control system is given in Fig. 1.

where K_{por} = proportional control action, K_{int} = integral control action, K_{der} = derivative control action and s_1 and s_2 are the set point weights. The control action of the controller is obtained by multi-objective algorithm tuned controller. The multi-objective genetic algorithm, non-dominated sorting genetic algorithm-II and multi-objective PSO are used to tune the controller. The output obtained is analyzed using various methods to study the effectiveness of the 2DOF PID controller and chooses the best tuned 2DOF PID controller for the desired drug control. The objective functions are

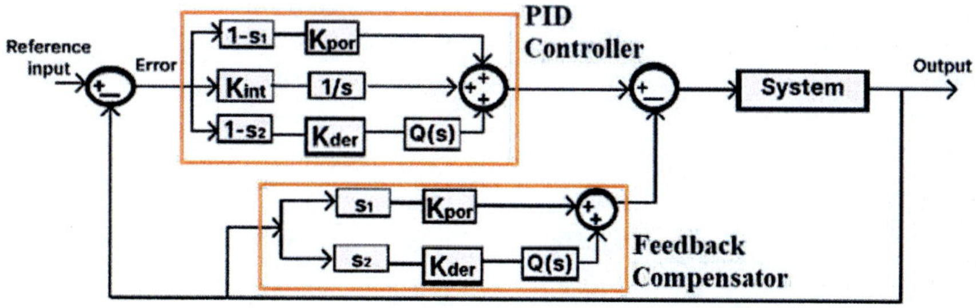

Fig. 1 2DOF PID controller

1. Sum of absolute error in toxicity level of the body, $\sum |E1(nT)|$,
2. Sum of absolute error between actual and desired drug concentration in the body, $\sum |E_2(nT)|$.

4 Validation

Four types of validation analysis are done in this paper using MATLAB Simulink. The methods include step response analysis for the toxicity of the body, bode analysis to find the stability of the system, disturbance analysis and parameter perturbation to study the effect of parameter changes on the number of cancer cells killed.

4.1 Step Response Analysis

Initially, time response analysis is carried out by considering the constraint on maximum allowable toxic level of the body as a step function with a step size of 100. The quality of the step response is measured in terms of various parametric values listed in the Table 1. A step signal is treated as a sudden input and the response of the controller to this sudden change is recorded. Step response gives the information about the stability of the controller and its ability to switch from one state to another state. The step response obtained using different 2DOF PID controllers and the conventional PID controller is shown in Fig. 2, and the quantitative analysis is given in Table 1.

It is observed from the results that the 2DOF PID controller provides a significantly better response than PID in terms of rise time, settling time and peak value. Further, the output for 2DOF PID controller settles faster than the PID controller. Further, the toxicity level of the body is best limited by NSGA-II tuned 2DOF PID controller. As the increase in toxicity in the body above 100 can be lethal for cancer patients,

Table 1 Step response analysis of the controllers

2DOF PID	Rise time (days)	Settling time (days)	Peak time (days)	Peak value	Steady state value
NSGA-II	4.723	18.420	13.954	99.96	99.96
MOGA	5.594	20.068	14.102	100	100
MOPSO	5.482	19.094	14.033	99.97	99.97
PID					
NSGA-II	6.410	21.961	14.200	100	100
MOGA	5.355	21.100	15.000	100	100
MOPSO	5.918	20.365	15.979	100.6	100

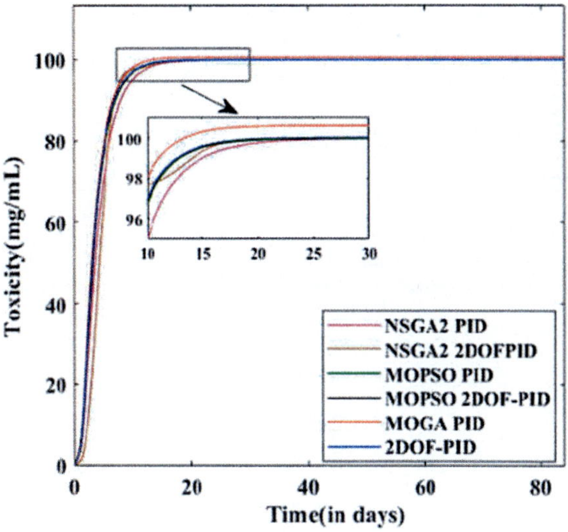

Fig. 2 Step response for toxicity level

hence the controller with minimum overshoot must be considered to control the drug injection.

4.2 *Bode Analysis*

Bode analysis is done to study the dynamics of the system. Bode plot measures the magnitude and phase of the output as a function of frequency. The gain and phase margin of the 2DOF PID controllers is evaluated. The values for gain margin are 70.7 dB, 54.2 dB and 53.3 dB for controllers tuned by NSGA-II, MOGA and MOPSO, respectively. Similarly, the phase margin for the controllers are 179 degrees for NSGA-II and 180 degrees for the other designed controllers. It is observed from the analysis that all the designed controllers lead to a stable system. However, NSGA-II tuned controller is more stable as compared to other controllers (Fig. 3).

4.3 *Parameter Perturbation*

While the treatment is carried out the model parameters are considered as fixed and constant. However, the parameter values can change owing to any physiological change or model approximation. Thus, a small change in the parameters can cause a prominent effect on the final output of the system. This issue is analyzed by observing the variation of maximum cancer cells reduced with the parametric change. The model parameters are perturbed from their nominal values (Table 2).

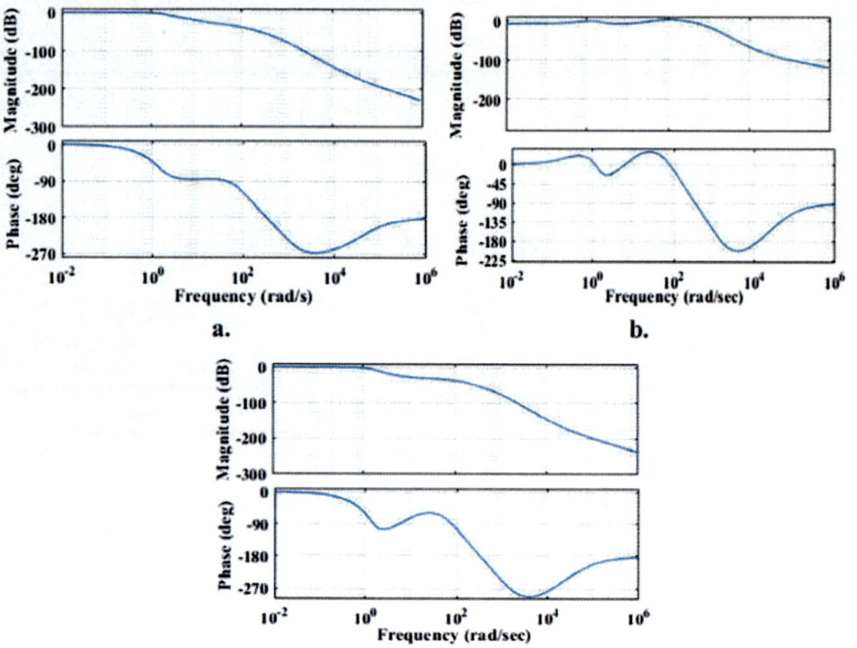

Fig. 3 Bode plot of 2DOF PID using **a**. NSGA 2, **b**. MOGA and **c**. MOPSO

Table 2 Variation of cancer cells with disturbance signal

2DOF PID (10^9)	5% decrease	10% decrease	5% increase	10% increase
NSGA -II	7.663	87.62	0.0518	0.0027
MOGA	9.029	93.04	0.0577	0.0033
MOPSO	8.914	89.64	0.0565	0.00731
PID (10^9)				
NSGA -II	10.44	133.38	0.0941	0.0067
MOGA	11.66	134.38	0.0986	0.00731
MOPSO	11	135.55	0.0972	0.00692

The toxicity obtained using different controllers is given in Table 3. It is revealed from the results that the change in body toxicity does not vary much with the changes in parameters for the 2DOF controller. Further, NSGA-II tuned 2DOF PID controller proves to be the most robust as compared to the other controllers.

Table 3 Peak value of toxicity level with disturbance in drug dose

2DOF PID controller	Sin (3,3,1)	Sin (3,15,1)	Pulse of amplitude 0.5	Pulse of amplitude 1
NSGA-II	100	99.98	99.98	100.01
MOGA	100.35	100.01	100.01	100.01
MOPSO	100.20	100.15	100.14	100
PID controller				
NSGA-II	100.42	100.10	100.17	100.26
MOGA	100.54	100.15	100.14	100.14
MOPSO	101	100.75	100.23	100.80

4.4 Disturbance Analysis

Disturbance is a common phenomenon in a practical control system. A well-designed controller must be able to handle the disturbance without causing any deviation in the desired output. Thus, disturbance analysis provides information about controller response to an unknown disturbance. In a chemotherapeutic treatment, the disturbance can arise while the drug is injected in the body using dc motors. Four different cases of disturbance are considered in this study, and the final toxic level is noted. For a better control, the controller should be able to restrict the peak and final toxic level to 100 and lesser output depicts better disturbance rejection. The disturbance signals are introduced halfway in the treatment period, i.e., on 42nd day of the treatment. Sinusoidal and pulse waves are used as disturbance in the drug dose (Fig. 4).

The variation in the output because of these disturbance signals are shown in Table 3. The waveforms of corresponding output responses are shown in Fig. 3.

Results reveal that the toxic content of the body rises above the safe threshold level in the presence of different disturbance signals. However, the rise in toxicity for 2DOF PID controller-based system is less as compared to the PID controller (Table 4).

5 Conclusion

This article presents the 2DOF PID control system for a multi-objective problem. The advantage of 2DOF PID controller is that it gives more robust performance by utilizing two feedback loops. During step change in PID controller, there is a rapid change caused due to proportional and derivative actions, which can be better handled by 2DOF PID controller. In this paper, the 2DOF PID controller is used in cascade configuration to control two objections which are conflicting with each other. The controller parameters are tuned using optimization algorithms. The stability and robustness of 2DOF PID controllers are validated using four validation techniques, i.e., step response analysis for body toxicity measurement, bode plot analysis for

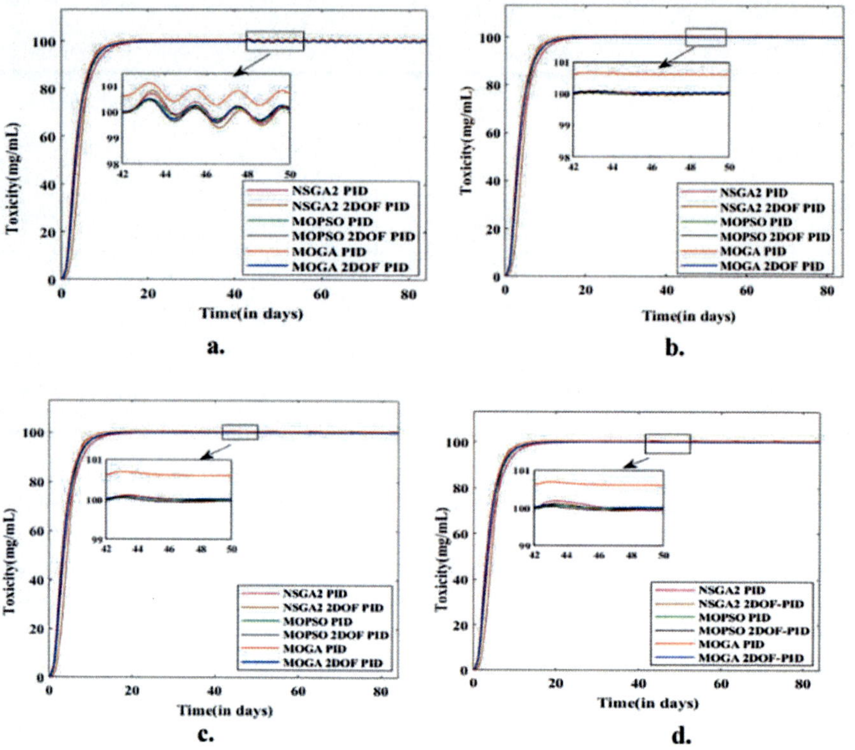

Fig. 4 Toxicity regulation by the controllers in response to **a**. pulse disturbance of 0.5 width, **b**. pulse disturbance of 1 width, **c**. 3Sin3t disturbance signal and **d**. 3Sin15t disturbance signal

Table 4 Controller parameters for the designed controllers

	Controller parameter	NSGA 2 2DOF PID	NSGA 2 PID	MOPSO 2DOF PID	MOPSO PID	MOGA 2DOF PID	MOGA PID
	Kpor	0.1381	0.3598	1.6122	1.8605	1.8612	1.9996
	Kint	1.2024	0.7228	0.8443	0.6287	1.0206	0.5402
C1	Kder	0.1758	0.2294	0.6630	0.7021	0.6381	0.2294
	s1	0.9411	–	0.3223	–	0.2083	–
	s2	0.2084	–	0.4376	–	0.1127	–
	Kpor	0.5673	1.1279	1.6850	1.8736	1.9910	1.9766
	Kint	0.6234	0.2615	1.5467	0.4977	1.3271	0.6944
C2	Kder	0.0172	0.0002	0.1051	0.2832	0.0393	0.0003
	s1	0.8354	–	0.3880	–	0.2017	–
	s2	0.0172	–	0.4139	–	0.3712	–

stability, disturbance analysis to check the robustness of the controllers and parameter perturbation to study the count of tumor cells with parameter variation. It is revealed from the complete analysis that 2DOF PID controller performs better than the conventional PID controller. Further, the 2DOF PID controller is best tuned by NSGA-II algorithm. Hence, it is concluded that the introduction of fractional order to conventional PID controller provides the robust and efficient control of drug delivery for chemotherapeutic treatment.

References

1. Taguchi H, Araki M (2000) Two-degree-of-freedom PID controllers—their functions and optimal tuning. IFAC Proc Vol 33:91–96
2. Azaharahmed M, Raja K, Patan MK et al (2020) Invasive weed optimized area centralized 2 degree of freedom combined PID controller scheme for automatic generation control. J Electric Eng Technol 16:31–42
3. Yuan Y, Lv H, Zhang Q (2021) DNA strand displacement reactions to accomplish a two-degree-of-freedom PID controller and its application in subtraction gate. IEEE Trans Nanobiosci 20:554–564
4. Mishra S, Prusty RC, Panda S (2020) Design and analysis of 2DOF-PID controller for frequency regulation of multi-microgrid using hybrid dragonfly and pattern search algorithm. J Control Autom Electric Syst 31:813–827
5. Eke I, Saka M, Gozde H et al (2021) Heuristic optimization based dynamic weighted state feedback approach for 2DOF Pi-controller in automatic voltage regulator. Eng Sci Technol Int J 24:899–910
6. Soussi Y, Rokbani N, Wali A, Alimi (2021) A multi-objective PSO with pareto neighborhood topology for clustering
7. Rahman CM, Rashid TA, Ahmed AM, Mirjalili S (2022) Multi-objective learner performance-based behavior algorithm with five multi-objective real-world engineering problems. Neural Comput Appl 34:6307–6329
8. Al-Tashi Q, Abdulkadir SJ, Rais HM et al (2020) Binary multi-objective grey wolf optimizer for feature selection in classification. IEEE Access 8:106247–106263
9. Panjwani B, Mohan V, Rani A, Singh V (2019) Optimal drug scheduling for cancer chemotherapy using two degree of freedom fractional order PID scheme. J Intell Fuzzy Syst 36:2273–2284

A Comparative Analysis of CNN Architectures for Classifying Pneumonia Variants and COVID-19 Using CXR Images

Aayush Garg, Aakash Kumar Singh, Aniket Patel, and Manjeet Kumar

Abstract The first coronavirus disease (SARS-CoV-2 COVID-19) originated in Wuhan, China and spread globally, resulting in millions of deaths. The detection of COVID-19 is very important especially at an early stage in order to provide proper care to the infected person as well as to prevent further spread of this disease. This paper attempts to provide a comparative analysis of different convolutional neural networks (CNN)-based architectures to identify the COVID-19 and pneumonia using chest X-ray (CXR) images. Further work has been done in order to identify the chest X-ray images which are infected with two variants of pneumonia. A large dataset using "COVID-19 important dataset radiography" and "chest X-ray images (pneumonia)" has been compiled from Kaggle repository in order to provide adequate learning to the CNN-based network. The model has been trained using pretrained models (Resnet50V2, Inception V3 and Densenet201), and as a result, Densenet201 model has achieved the highest accuracy (96.39%) among all the models.

Keywords Chest X-ray (CXR) · Convolutional neural network (CNN) · Deep learning · COVID-19 · Pneumonia · DenseNet · Inception V3 · Resnet50V2

1 Introduction

The coronavirus disease originated in Wuhan, China and spread globally, resulting in millions of deaths, and some of the common indications of COVID-19 includes fatigue, difficulty in breathing, fever and cough. The extensive outburst of this virus has caused it to be declared as a pandemic by the world health organization (WHO). The symptoms are similar to that of influenza, and hence, medical practitioners deliberately reliant on other methods of detection like medical images. Computer aided diagnosis (CAD) has played a critical role in the detection and diagnosis of

A. Garg · A. K. Singh · A. Patel · M. Kumar (✉)
Delhi Technological University, Delhi, India
e-mail: manjeetchhillar@gmail.com; akshath12345678912@gmail.com

A. Garg
e-mail: aayushgarg08012000@gmail.com

A. Rani et al. (eds.), *Signals, Machines and Automation*, Lecture Notes in Electrical Engineering 1023, https://doi.org/10.1007/978-981-99-0969-8_43

this disease. Chest X-ray images (CXR) are being used for the accurate recognition and diagnosis of the virus infection by healthcare practitioners all over the world.

Apart from COVID-19, pneumonia is also very dangerous and deadly. It fills the lungs with a fluid known as pus and make it harder for a person to breathe (especially in case of older adults). It has two main variants known as bacterial and viral pneumonia in which bacterial pneumonia is treated with antibiotics, whereas viral gets better by itself.

Although it is accepted that the work carried out earlier [1–4] is effective but none of them have assessed the classification of pneumonia variants from COVID-19. This paper majorly focusses on the comparison of pretrained models upon fine-tuning, after which they can be considered reliable in classification of COVID-19, bacterial and viral pneumonia. To explore in this project, we worked with 5269 CXR images of normal, COVID-19 and pneumonia diagnosed patients.

2 Literature Review

In Ref. [1], local binary patterns (LBP) were taken as input for extracting the features and classifying the disease. It uses these patterns to extract features and engages the model deeply on the texture features of an input image, and then, a histogram is constructed using the binary strings obtained from comparison of every pixel in the image, which in turn is used to generate a feature set. Different algorithms such as decision tree (DT), random forests (RFs), Naive Bayes (NB), support vector machine (SVM), K-nearest neighbors (KNNs) and their ensemble models have been compared on various parameters in order to choose the best. Ensemble methods found to perform better than the individual classifiers on various performance parameters.

Ouyang et al. [2] suggested that computerized tomography imaging (also known as CT scan) procedure found out to be a more reliable method to test for COVID-19, and in order to automate the process of detection and distinction of COVID-19 from pneumonia, models based on deep learning have been developed to detect COVID-19.

Ahmed et al. [3] mainly focused on the deep learning approaches using convolutional neural networks (CNN). Here, generalization gap on COVID-19 X-rays-based classification was done using CNN. In the model training, the pretrained ResNet50 was fine-tuned and the accuracy of ResNet50 after fine-tuning was 98.1%.

Nahiduzzaman et al. [4] mainly focused on a new method for multivariate classification using principal component analysis (PCA) based on hybrid CNN-PCA-based feature extraction using extreme learning machine (ELM) with the CXR images. Here, CXR images were used because of the swelling of the lungs due to the virus, bacteria or fungi attack. As a result, the classification model using extreme learning machine (ELM) model without contrast limited adaptive histogram equalization (CLAHE) and PCA technique gave 97.09% accuracy and after that the model gave 99.83% accuracy when both CLAHE and PCA were used.

This paper is organized into six sections such that Sect. 1 consists of basic introduction about COVID-19 and pneumonia and followed by the literature survey in Sect. 2. In Sect. 3, the dataset description is provided with sample images of each class. Section 4 consists of the research methodology in which the data preprocessing techniques and implemented CNN architectures are described briefly. In Sect. 5, the results and analysis is presented regarding the model training and testing. Section VII concludes the paper.

3 Experimentation Environment

Several datasets are available online containing CXR images, and this section brief about the dataset used in this paper. A large dataset using "COVID-19 important dataset radiography" and "chest X-ray images (pneumonia)" databases has been compiled from Kaggle repository in order to provide adequate learning to the CNN-based network. The dataset has been taken from [12, 13] and then combined together. This dataset contains four classes namely normal, COVID-19, bacterial pneumonia (BP) and viral pneumonia (VP) as shown in Fig. 1. A brief dataset information is given in Table 1.

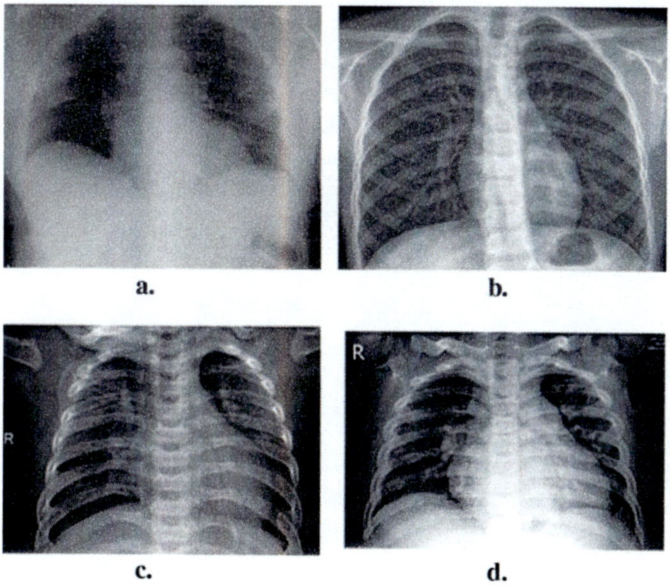

Fig. 1 Dataset information

Table 1 Dataset Information

Dataset	Data analytics			
	Normal	COVID	Bacterial pneumonia	Viral pneumonia
Training	1170	900	2055	1134
Testing	344	145	742	421

4 Research Methodology

Several methodologies have been used in the past to classify either COVID-19 or pneumonia but not both simultaneously as there is a very less difference between the COVID-19 and pneumonia-infected lungs. In this paper, we have fine-tuned the pretrained CNN architectures to classify between normal, COVID-19 and pneumonia-infected lungs. The research methodology consists of following techniques.

4.1 Data Preprocessing

Originally, the images are of varying sizes in the dataset. As the images given as input to the model should be identical in size to avoid the irregularity, hence the CXR images are resized to a new and common size of 224 × 224 (standard size of input image used while training). Now, to avoid any overfitting of the model and to make the CNN architecture model more generic with the predictions, techniques such as data augmentation and data preprocessing are used. By using these types of techniques, we ensure that our CNN model gets new variations of data at every iteration during the training phase. The data augmentation and preprocessing techniques implemented in this paper are as follows in Table 2.

Table 2 Data preparation techniques

Rescaling	1.0/255
ZCA whitening	True
Range of rotation	50
Zooming range	0.1
Width shifting range	0.2
Height shifting range	0.2
Shearing range	0.25
Horizontal flipping	True

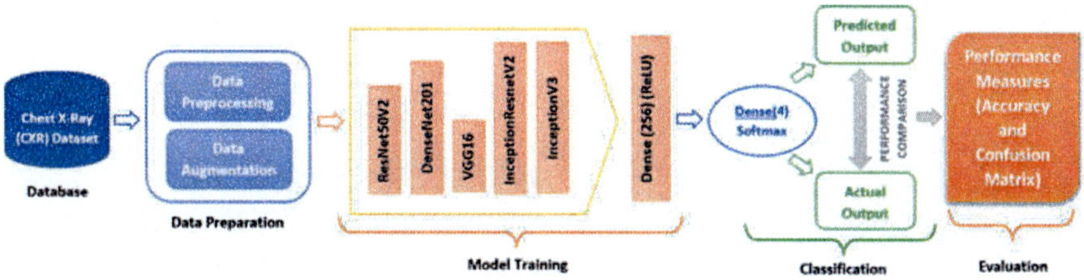

Fig. 2 Experimental design architecture (flow process)

4.2 *CNN Architectures*

In this section, we will briefly discuss about the CNN architectures which were used in the classification of CXR images to detect COVID-19, bacterial pneumonia, viral pneumonia and normal lungs. It is a multi-class classification problem. The experimental design architecture (i.e., flow process) of CNN is shown in Fig. 2. The CXR image is taken as input and being preprocessed and forwarded to model training using pretrained models. After creating the models, a dense layer is used for classification using softmax activation function and the accuracy is calculated finally. The brief information about the used pretrained models and their architectures are as follows:

Resnet50V2—Resnet50V2 is a 50 layers deep CNN architecture. This network comprises a novel approach pathway called skip connection. This version is a modified form of ResNet50 that performs much better than ResNet50 and ResNet101 on the ImageNet dataset. In this version, the most effective modification was done with the promulgation formulation of the connections between blocks. This network is able to achieve good accuracy by using multiple features which were extracted by the two robust networks in [9].

Inception V3—This CNN architecture is a widely used model for image recognition and object detection [5]. The Inception V3 is an improvement on inception-V1 as it further decreases the number of parameters by factoring the convolution layers into asymmetric convolutions. All these changes resulted in a model which requires less computation power, is faster and is more accurate, while also allowing deeper networks. The model consists of different sizes of convolution filters, max pooling, concatenations and average pooling [6]. The basic inception module is shown in Fig. 3.

Densenet201—This network was proposed by Huang et al. [10]. It is a deep CNN architecture, known for its excellent performance in competitive image recognition. One of the important features of this model is its ability to reuse the features in order to provide better parameter and training efficiency. In this model, the input in each layer is the combination of input from all the previous layers as well as its own feature maps [7]. Due to this, the feature maps generated by previous layers can be easily

Fig. 3 Inception module

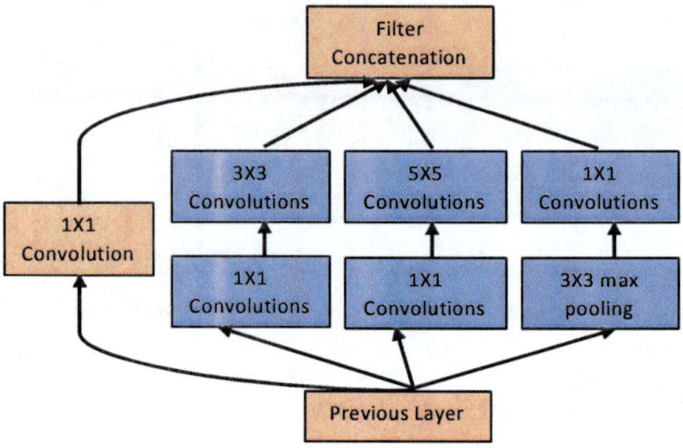

accessed by the deep network layers and thus the features are reused. This model has also found use in various healthcare applications [8].

5 Results and Discussion

In this section, we present the final results of the classification. The publicly available CXR datasets [12, 13] are used. The final combined dataset consists of 6911 CXR images from a large unevenly distributed population among the classes "COVID", "VP" (i.e., viral pneumonia), "BP" (i.e., bacterial pneumonia) and "normal". The images are randomly distributed into training and testing datasets in which 80% is used for training and 20% is for testing from overall dataset. Further, 20% of the training dataset is used as validation dataset. A comparative performance evaluation is conducted to obtain the superior of the three models. Further, to evaluate and analyze the performance of all CNN architectures, the accuracy has been computed and confusion matrix plots are represented as heatmaps as shown in Figs. 4, 5, 6.

According to the confusion matrices obtained, the models performed very well in differentiating normal, COVID-infected and pneumonia-infected lungs but struggles a somewhat in distinguishing the viral pneumonia and bacterial pneumonia-infected lungs. We have achieved good accuracies with all the above architectures. Table 3 shows the accuracy comparison between three different CNN models.

The following Table 4 shows the comparison between existing and proposed work. The work reported in the literature is only based on COVID-19 and pneumonia detection using data augmentation and pretrained CNN models, whereas the proposed work classifies the COVID-19 and pneumonia variants with higher accuracy using pretrained CNN models.

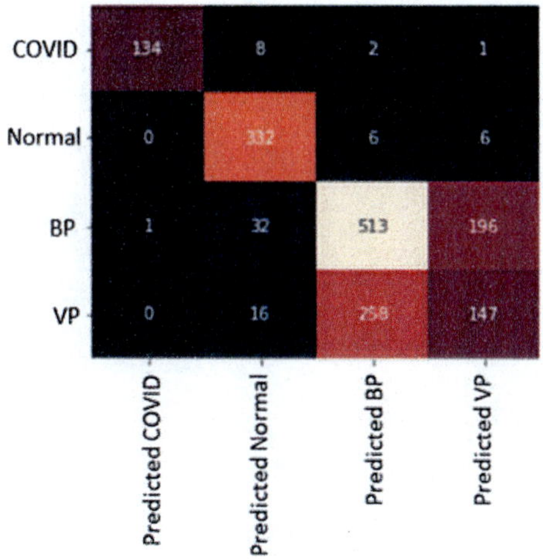

Fig. 4 Confusion matrix for DenseNet 201

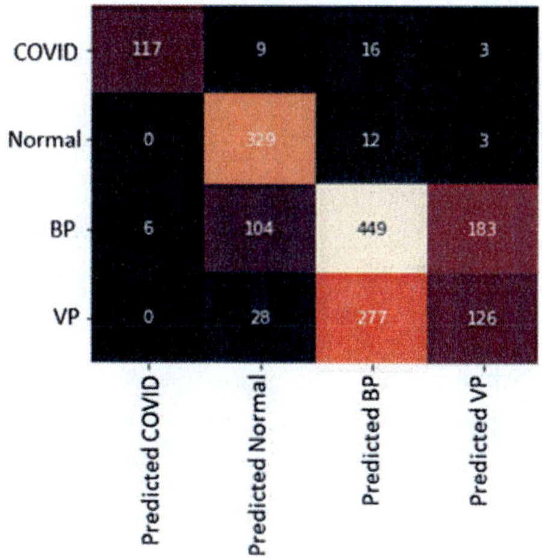

Fig. 5 Confusion matrix for Inception V3

6 Conclusion

In this paper, three CNN architectures are successfully compared by fine-tuning their pretrained base models that are designed in order to detect pneumonia from CXR images. Techniques such as data augmentation and preprocessing (rescaling, zca-whitening, etc.) are used before training the model. As a result, the models are able to predict COVID-19 with very high precision. The experimental results shows that the CNN architecture namely DenseNet201 performs better than all the other models with 96.39% accuracy followed by ResNet50V2 (93.4%), Inception V3 (85.6%) and InceptionResNetV2 (82.03%). Limitation of the work is that the pretrained models are able to perform well with four or less classes only. In the future, ensemble of

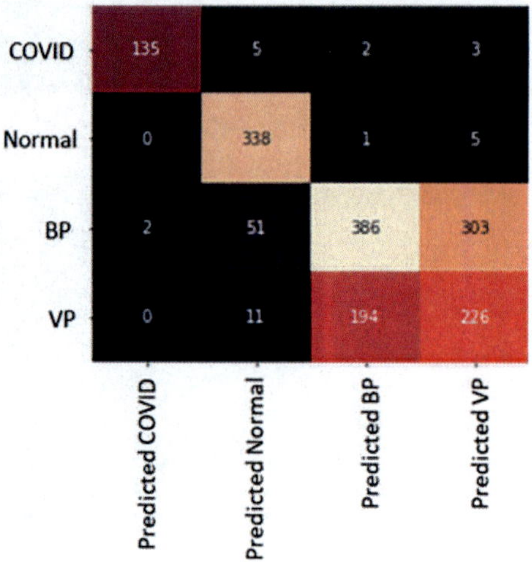

Fig. 6 Confusion matrix for ResNet50V2

Table 3 Accuracy comparison

CNN models	Proposed model accuracy (%)
DenseNet201	96.39
Inception V3	85.60
Resnet50V2	93.40

Table 4 Existing versus proposed work comparison

Property	Ref. [11]	Ref. [3]	Proposed work
Accuracy	90%	94%	96.39%
Approach	Using augmentation and dropout with CNN	Fine-tuning ResNet50	Fine-tuning DenseNet201
Type	Binary classification	3-class classification	4-class classification

pretrained models can be used for classification of higher number of variants of COVID-19 (omicron) and pneumonia.

References

1. Thepade, Sudeep D, Jadhav K (2020) Covid19 identification from chest X-ray images using local binary patterns with assorted machine learning classifiers. In: 2020 IEEE Bombay section signature conference (IBSSC), pp 46–51, IEEE
2. Ouyang X, Huo J, Xia L, Shan F, Liu J, Mo Z, Yan F et al (2020) Dual-sampling attention network for diagnosis of COVID-19 from community acquired pneumonia. IEEE Trans Med

Imag 39(8):2595–2605
3. Ahmed KB, Goldgof GM, Paul R, Goldgof DB, Hall LO (2021) Discovery of a generalization gap of convolutional neural networks on COVID-19 X-rays classification. IEEE Access 9:72970–72979
4. Nahiduzzaman M, Goni MOF, Anower MS, Islam MR, Ahsan M, Haider J, Gurusamy S, Hassan R, Islam MR (2021) A novel method for multivariant pneumonia classification based on hybrid CNN-PCA based feature extraction using extreme learning machine with CXR images. IEEE Access 9:147512–147526
5. Google Cloud, Advanced guide to inception V3", [Online]. Available: https://cloud.google.com/tpu/docs/inception-V3-advanced#introduction [Last Accessed 10 Apr 2022]
6. Sundar, Sai KV, Bonta LR, Baruah PK, Sankara SS (2018) Evaluating training time of Inception-v3 and Resnet-50,101 models using TensorFlow across CPU and GPU. In: 2018 second international conference on electronics, communication and aerospace technology (ICECA), pp 1964–1968, IEEE
7. Jasil, Godlin SP, Ulagamuthalvi V (2021) Skin lesion classification using pre-trained DenseNet201 deep neural network. In: 2021 3rd international conference on signal processing and communication (ICPSC), pp 393–396, IEEE
8. Adam K, Mohamed II, Ibrahim Y (2021) A selective mitigation technique of soft errors for dnn models used in healthcare applications: Densenet201 case study. IEEE Access 9:65803–65823
9. Karar ME, Hemdan EED, Shouman MA (2021) Cascaded deep learning classifiers for computer-aided diagnosis of COVID-19 and pneumonia diseases in X-ray scans. Complex Intell Syst 7(1):235–247
10. Huang G, Liu Z, Pleiss G, Van Der Maaten L, Weinberger K (2019) Convolutional networks with dense connectivity. IEEE Trans Patt Anal Mach Intell
11. Sharma H, Jain JS, Bansal P, Gupta S (2020) Feature extraction and classification of chest x-ray images using cnn to detect pneumonia. In: 2020 10th international conference on cloud computing, data science and engineering (Confluence), pp 227–231, IEEE
12. Liu S, Deng W (2015) Very deep convolutional neural network based image classification using small training sample size. In: 2015 3rd IAPR Asian conference on pattern recognition (ACPR), pp 730–734, IEEE
13. Szegedy C, Ioffe S, Vanhoucke V, Alemi AA (2017) Inception-v4, inception-resnet and the impact of residual connections on learning. In: Thirty-first AAAI conference on artificial intelligence

Multi-objective Control-Based Artificial Pancreas for Type-1 Diabetic Patients

Sharmistha Mandal and Ashoke Sutradhar

Abstract Artificial pancreas (AP) is an artificial organ which is used to regulate blood glucose (BG) in Type-1 diabetic mellitus (T1DM) patient. In this study, a multi-objective controller has been suggested for AP system. The controller is designed based on Bergman's intravenous minimal model using linear matrix inequality (LMI) technique. The simulation outcomes show that the controller regulates BG level very proficiently, avoids hypoglycemia and post-prandial hyperglycemia effects in presence of unannounced meal disturbance, glucose sensor noise and insulin pump error. The performance of the suggested controller has been compared with attractive ellipsoid method (AEM) observer-based controller designed earlier, but suggested controller gives better performance than the AEM-based controller.

Keywords Artificial pancreas · Multi-objective control · Linear matrix inequality

1 Introduction

Diabetic patients suffer from COVID-19 may grow more severe complications. Type-1 diabetes mellitus (T1DM) is also named as juvenile diabetes, is occurred due to the short of insulin production by the pancreas of the human body. Uninterrupted and regulated delivery of insulin is compulsory for the T1DM patient to sustain blood glucose (BG) level between 70 mg/dl to 130 mg/dl on fasting and not exceeding 140 mg/dl after two hours of taking meal in presence of normal activities [1].

Researches have designed efficient and accurate controllers for artificial pancreas (AP) to provide calculated amount of insulin to T1DM patient body. Using intravenous (IV) or subcutaneous (SC) route, insulin can be infused into the patient body. Researchers developed different IV [2, 3] and SC mathematical models of physiological process of T1DM patients [4]. To design the efficient controller for BG

S. Mandal (✉)
Electrical Engineering Department, Aliah University, Salt Lake, Kolkata, India
e-mail: itssharmistha@gmail.com

A. Sutradhar
Electrical Engineering Department, IIEST, Howrah, India

A. Rani et al. (eds.), *Signals, Machines and Automation*, Lecture Notes in Electrical Engineering 1023, https://doi.org/10.1007/978-981-99-0969-8_44

regulation in T1DM patient, different control algorithms like H-infinity (H_∞) [5, 10], multi-objective [6, 8, 11], model predictive [12], intelligent online feedback [7], adaptive [13] and robust observer-based controller [9, 14] have been used based on IV and SC model.

The main aim of this study is to design an efficient, simple, multi-objective and robust controller for AP system for BG regulation in TIDM patient using linear matrix inequality (LMI) technique. Performance of the designed controller has been verified in presence of unannounced meal. The simulated outputs show that the designed controller regulates BG level tightly. The outputs in presence of glucose sensor noise, and insulin pump error show that the controller gives also robust performance.

2 Dynamics of Type-1 Diabetics

The artificial pancreas (AP) system with diabetic patient is shown in Fig. 1. A glucose sensor, a controller and an insulin pump are the three components of AP. Blood sugar level is continuously measured by glucose sensor and directs the data to a controller. The controller computes the optimum insulin delivery rate and sends the command to the insulin pump. Bergman's third order nonlinear intravenous model [2] has two parts where the dynamics of glucose uptake is described by one part and the insulin kinetics is described by another part. The modified Bergman's minimal model is given by the following state equations [8].

$$\dot{x}_G(t) = -p_1[x_G(t) - G_b] - x_G(t)x_{RI}(t) + d(t), \tag{1}$$

$$\dot{x}_{RI}(t) = -p_2 x_{RI}(t) + p_3[x_I(t) - I_b], \tag{2}$$

$$\dot{x}_I(t) = -p_4[x_I(t) - I_b] + u(t), \tag{3}$$

where x_G (mg/dl) is plasma glucose concentration; x_{RI} (min-1) and x_I (mU/l) represent remote insulin concentration and plasma insulin concentration, respectively. The basal levels of glucose and insulin are G_b and I_b, respectively. The external intravenous insulin infusion rate which is the control input, is denoted by $u(t)$. Here, $d(t)$ denotes the rate at which glucose is absorbed to the blood after food intake and is given by the dynamics $\dot{d}(t) = -p_5 d(t)$. The rate of appearance of meal disturbance in the plasma glucose is $p_5(\text{min}^{-1})$. For the design of controller of AP system for T1DM patient, the values of the parameters of the minimal model are taken from article [9] and are given in Table 1.

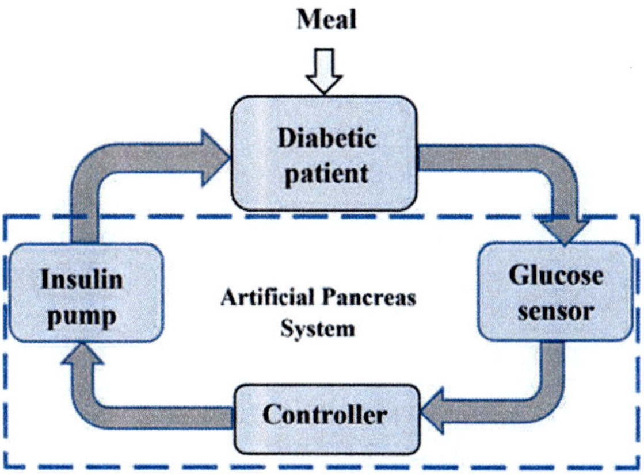

Fig. 1 Artificial pancreas system with diabetic patient

Table 1 Description and values of the parameters for T1DM patient [11]

Description of parameters	Value
The insulin-independent rate constant of glucose uptake in muscles and liver (p_1)	0
The rate for decrease in tissue glucose uptake ability (p_2)	0.015 (min^{-1})
The insulin-dependent increase in glucose uptake ability in tissue per unit of insulin concentration above the basal level (p_3)	2×10^{-6} ($mU/l/min^2$)
The insulin degradation rate (p_4)	0.2 (min^{-1})

3 Multi-objective Control of AP System Using LMI Technique

The dynamics of nonlinear intravenous Bergman's model of T1DM patient (Eqs. (1)-(3)) is implemented using MATLAB SIMULINK® toolboxes. The nonlinear model is linearized around the basal value of plasma glucose concentration (80 mg/dl). In this work, the multi-objective constraints H_∞ performance, pole-placement and H_2 performance [16] have been taken for considering both robustness and performances objectives and have been solved using linear matrix inequality (LMI) technique. The control arrangement of open-loop plant G and the controller K is shown in Fig. 2. The output vectors are $z_\infty = \begin{bmatrix} z_P & z_T & z_u \end{bmatrix}^T$ and $z_2 = \begin{bmatrix} z_P & z_u \end{bmatrix}^T$. The objective of H_∞ control problem is to minimize the sensitivity function, the control effort and complementary sensitivity function. In this case, the performance weights are W_P, W_T, W_u and are selected as $W_P = \frac{0.6667s+0.00001}{s+10^{-8}}$; $W_T = \frac{s+53.33}{0.001s+80}$; $W_u(s) = \frac{s+0.04667}{0.001s+0.07}$. Here, objective is to design a multi-objective output-feedback controller gain K that.

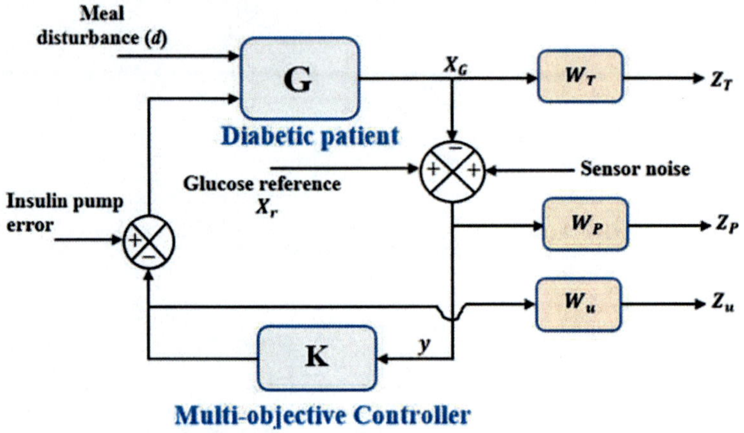

Fig. 2 T1DM patient under multi-objective output-feedback control

- Guarantees the H_∞ norm of $T_{z_\infty\omega}(s)$ less than the given $\gamma \succ 0$ where $T_{z_\infty\omega}(s)$ is closed-loop transfer function of the system from w to z_∞.
- Confines the closed-loop poles in some prescribed LMI region.
- Minimizes the H_2 performance $J_2 = T_{z_2w_2}$ where $T_{z_2w}(s)$ is closed-loop transfer function of the system from w to z_2.

The LMI region in this case is selected as combination of half plane and conic sector a is shown in Fig. 3 (shaded region). The optimized values of H_∞. And H_2. performances are 70.72 and 4.17, respectively. For simulation, reduced order (fifth order) controller is used.

Fig. 3 Selected LMI region

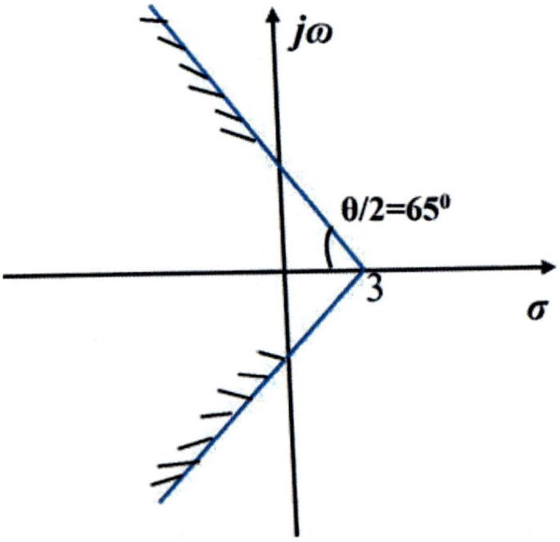

Table 2 Meal protocol 1

Meals	Breakfast	Lunch	Dinner
Time (Hrs.)	07:00	13:00	20:00
$d(0)$ (mg/dl/min)	5	7	5

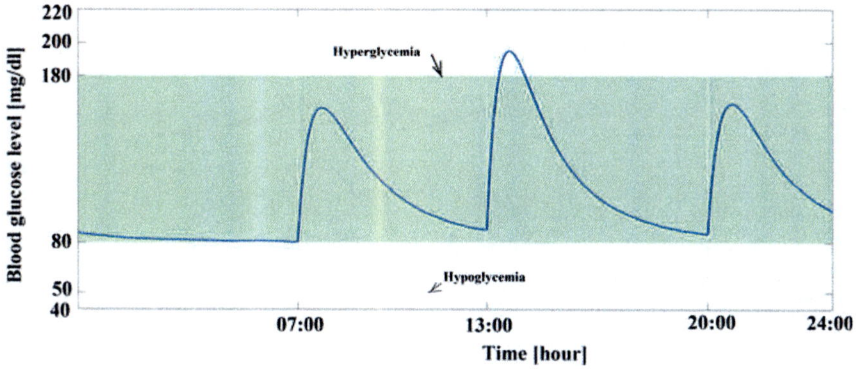

Fig. 4 BG level under meal protocol1

4 Simulation

The designed controller is tested on nonlinear Bergman model in presence of unannounced meal, sensor noise and insulin pump error. The meal size and time are given in Table 2. The glucose sensor noise and insulin pump error are taken as band-limited white noise and both noise power are taken as 0.001. The resulting responses are shown in Figs. 4 and 5, respectively. From the responses, it is clear that designed controller avoids post-prandial hyperglycemia effect and hypoglycemia effect is also absent. The BG level remains mostly within 80 mg/dl to 180 mg/dl. From insulin infusion rate, it can be seen that maximum insulin infusion required during lunch time is 16.5 mU/l/min.

The performance of the controller is compared with attractive ellipsoid method observer-based controller [9]. Here, patient is subjected to high initial meal disturbance and initial condition of the state variables [9] are taken as $\left[x_{Go}\ x_{RI0}\ x_{I0}\right]^T = \left[200\ 0.001\ 7\right]^T$. The BG level and plasma insulin are given in Figs. 6 and 7, respectively, and performances of both controllers are given in Table 3.

5 Conclusion

In this study, LMI-based multi-objective controller regulates BG level very tightly. The BG responses show that there is no chance of hypoglycemia. The designed controller also gives robust performance in presence of glucose sensor noise and

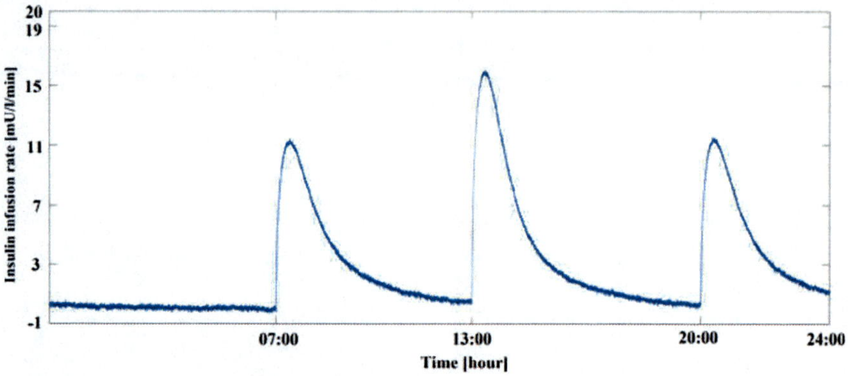

Fig. 5 Insulin infusion rate using IV route under meal protocol1

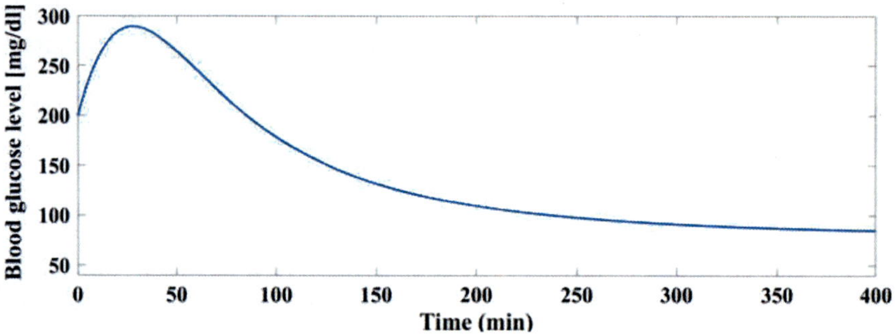

Fig. 6 BG level under high initial meal disturbance

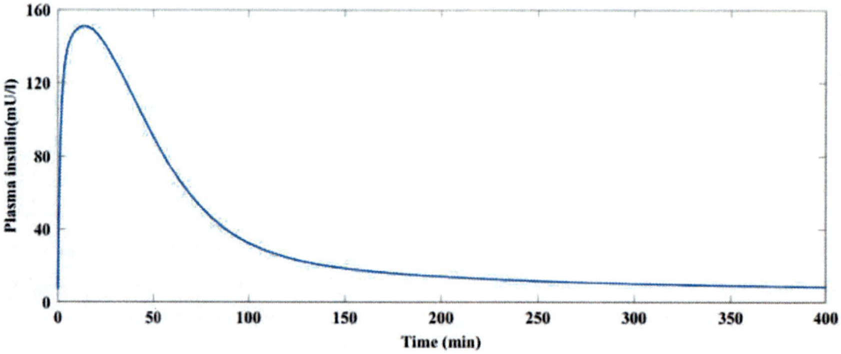

Fig. 7 Plasma insulin under high initial meal disturbance

Table 3 Comparison with attractive ellipsoid method observer-based controller [9]

Performance metrics	Observer-based controller [9]	Proposed controller
Maximum BG (mg/dl)	Greater than 300	Less than 300
Time taken to become less 180 mg/dl (min)	110	90

insulin pump error. The proposed controller gives better results than the attractive ellipsoid method observer-based controller. In future, the controller performance may be tested in presence of other physical activities like exercise.

References

1. Yang X, Nathan DM (2014) The diabetes control and complications trial/epidemiology of diabetes interventions and complications study at 30 years: overview. Diabetes Care 37:9–16
2. Bergman RN, Phillips LS, Cobelli C (1981) Physiologic evaluation of factors controlling glucose tolerance in man: measurement of insulin sensitivity and β-Cell glucose sensitivity from the response to intravenous glucose. J Clin Invest 68(6):1456–1467
3. Sorensen JT (1985) A physiologic model of glucose metabolism in man and its use to design and assess improved insulin therapies for diabetes. Ph.D. thesis, Department of Chemical Engineering, MIT
4. Man CD, Rizza RA, Cobelli C (2007) Meal simulation model of the glucose-insulin system. IEEE Trans Biomed Eng 54(10):1740–1749
5. Parker RS, Doyle FJ III, Ward JH, Peppas NA (2000) Robust H_∞ glucose control in diabetes using a physiological model. Bioeng Food Natl Prod 46(12):2537–2549
6. Mandal S, Sutradhar A (2017) Multi-objective control of blood glucose with H_∞ and pole-placement constraints. Int J Dyn Control 5(2):357–366
7. Ali SF, Padhi R (2009) Optimal blood glucose regulation of diabetic patients using single network adaptive critics. Optimal Control Appl Methods 32:196–214
8. Dua P, Doyle FJ III, Pistikopoulos EN (2009) Multi-objective blood glucose control for type 1 diabetes. Med Biol Eng Compu 47(3):343–352
9. Nath A, Dey R (2019) Robust observer based control for plasma glucose regulation in type 1 diabetes patient using attractive ellipsoid method. IET Syst Biol 13(2):84–91
10. Colmegna P, S´anchezPe˜na RS, Gondhalekar R, Dassau E, Doyle III FJ (2014) Reducing risks in type 1 diabetes using H_∞ control. IEEE Trans Biomed Eng 61(12):2939–2947
11. Mandal S, Sutradhar A (2019) Robust multi-objective blood glucose control in Type-1 diabetic patient. IET Syst Biol 13(3):136–146
12. Bhattacharjee A, Easwaran A, Leow MK-S, Cho N (2019) Design of an online-tuned model based compound controller for a fully automated artificial pancreas. Med Biol Eng Compu 57:1437–1449
13. Belmon AP, Auxillia J (2020) An adaptive technique based blood glucose control in type-1 diabetes mellitus patients. Int J Numer Meth Biomed Engg 36(8):e3371
14. Golestani F, Tavazoei MS (2022) Delay-Independent regulation of blood glucose for type-1 diabetes mellitus patients via an observer-based predictor feedback approach by considering quantization constraints. Eur J Control 63:240–252
15. UVa/Padova T1DMS simulator Type-1 Diabetes Metabolic Simulator. User Guide v3.2. TEG The Epsilon Group, April 14, 2014.
16. Chilali M, Gahinet P (1996) H_∞ design with pole placement constraints: An LMI approach. IEEE Trans Autom Control 41(3):358–367

A Hybrid Ensemble Deep Learning Model for COVID-19 Detection Using CT Scan

Vrinda Aggarwal, Devendra Kumar Jat, Vaibhav Maheshwari, and Manjeet Kumar

Abstract COVID-19 is a highly communicable disease with various variants after various mutations, having different effects from person-to-person. Hence, early detection and diagnosis of the infection can help in controlling the spreading rate. In the present paper, a weighted average ensemble deep learning using ResNet50 and InceptionV3 is proposed for a binary detection of COVID using CT scan images. The model is created with Google Colab using Keras module and trained using graphics processing unit (GPU). Here, SARS-CoV-2-CT-Scan-Dataset is accustomed to validate the performance of the proposed algorithm. The proposed model achieved an accuracy of 94.23% compared to individual models.

Keywords COVID-19 · Machine learning · CNN · ResNet50 · InceptionV3 · Image processing · Ensemble learning

1 Introduction

Coronavirus disease 2019 also known as COVID-19 has been announced high risk by the World Health Organization (WHO) and has started the decade with a new strain of a respiratory disease. People who are afflicted with this ailment might have cold, fever, and chest tightness as the symptoms [1]. The recent variant omicron and deltacron are expected to have mild effects with a great communicability factor among the population, resulting in a high spread rate.

The mild symptoms or the asymptomatic patients are the easiest one to cure during the pandemic [2]. The most important test for finding out the infection is RT-PCR, but the problem is the shortage of the testing kits [3]. People who are at high risk due to the infection can be helped out by detecting COVID-19 early and so they can get their treatment start without waiting for the result of RT-PCR which will help them to recover soon and will help in decreasing the mortality rate. It can help identify patients with high levels of COVID and test them without RT-PCR [4]. Based on

V. Aggarwal · D. K. Jat · V. Maheshwari · M. Kumar (✉)
Delhi Technological University, Delhi, India
e-mail: manjeetchhillar@gmail.com

A. Rani et al. (eds.), *Signals, Machines and Automation*, Lecture Notes in Electrical Engineering 1023, https://doi.org/10.1007/978-981-99-0969-8_45

extensive clinical criteria, the CT technician should make the careful decision to operate on a CT to verify an aberrant diagnosis from a chest X-ray [5].

The authors of [6] have suggested a unique detailed-oriented capsule network architecture capable of recognizing fine-grained and discriminative picture characteristics for classification of patients with COVID-19 by following data augmentation model. The accuracy they obtained by their model is 87.6% and F1-score is 0.871.

In [7], authors have suggested a model named as EffiecintCovidNet which has a method based on voting and a cross-dataset exploration. By using their model for identification of COVID-19 using CT reports, they are getting the accuracy of 87.60% on their dataset. In [8], the authors build a CNN design model for distinguishing COVID-19 CT scans from others and detecting COVID-19. Their model CNN-2 is much better then original SqueezeNet with accuracy of 85.03% having F1-score of 0.862. The authors of [9] performed ensemble learning on the combined dataset using ResNet50, Inception V4, and EfficientNetB0 and produced predictions based on majority vote, and by using their model they are getting the accuracy of 95.36%. The convolution neural network (CNN) technique in deep learning [10] has shown significant utility in image classification and thus is most widely used by researchers today.

For CT-scan analysis [11] of the chest, deep learning techniques are popular because everyone may use them with low-cost imaging techniques and have a large amount of relevant data to train models with. Manual evaluation of the images of CT scans reports and X-rays reports requires their specialized knowledge, and the process is also time consuming and might be inaccurate sometimes, that is why there comes the algorithms of deep learning and machine learning which can help to extract the relevant information and perform the same task of evaluation of COVID-19 in an easy way [12]. The goodness of the two most effective pre-trained deep CNN models, namely InceptionV3 and ResNet50, has been explored for ensemble learning before too in some research, and their performances have been analysed on the basis of accuracy and other parameters [13]. Their study's goal is to provide a transfer-learning strategy based on CNN for identifying COVID-19 utilizing multiple models which can be more accurate [14]. Ultrasounds, dermoscopy, X-rays, magnetic resonance imaging (MRI), cognitive analytic therapy (CAT), and positron emission tomography (PET) are dynamic and developing domains for research, especially in image-processing techniques and algorithms [15]. The model which is been worked upon in this research is created with Google Colab using Keras module and trained using graphics processing unit (GPU).

The following are the primary contributions of this planned research. In this research, there is a hybrid model which is using the weighted average ensemble method so that the models can be trained and there will be the qualities of both of the model, i.e. InceptionV3 and ResNet50, and this will increase the accuracy of the detection. In InceptionV3, the accuracy is 90.23% and in ResNet50 the accuracy is 89.65% and so the hybrid model accuracy is 94.23% and having F1-score of 91.56%.SARS-CoV-2-CT-Scan-Dataset has been used for training and testing purposes with 80:20 ratio.

The paper is subdivided into the subsequent sections. Section 2 provides the information about the methodology of the research and information about the models used. Section 3 discusses the result of the proposed model along with detailed discussion. Finally, Sect. 4 concludes the paper.

2 Methodology

In this section, the detailed methodology used for COVID detection is proposed using ensemble deep learning algorithms.

2.1 *ResNet50*

ResNet50 [16] uses the concept of skip connection that resolves the vanishing gradient problem. In skip connection connects the input of the model to the output of the convolution block by either training the layer or just skipping the layer. It mainly consist two types of block namely identity block as shown in Fig. 2 and convolution block as shown in Fig. 3. Identity block works when the output size of the layer is equal to the input size of the layer, whereas convolution block works when the size of input and output layer differs by the method of 1 × 1 convolution block in the skip connection part which provides the solution of making the size equal. ResNet50 as shown in Fig. 1 is more precise with a large dataset as compared to the small dataset [17]. The architecture of ResNet50 is shown in Fig. 1.

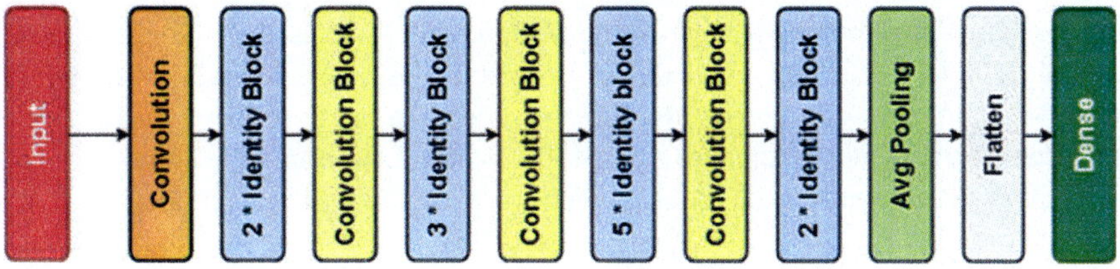

Fig. 1 Architecture of ResNet50 with identity and convolutional block

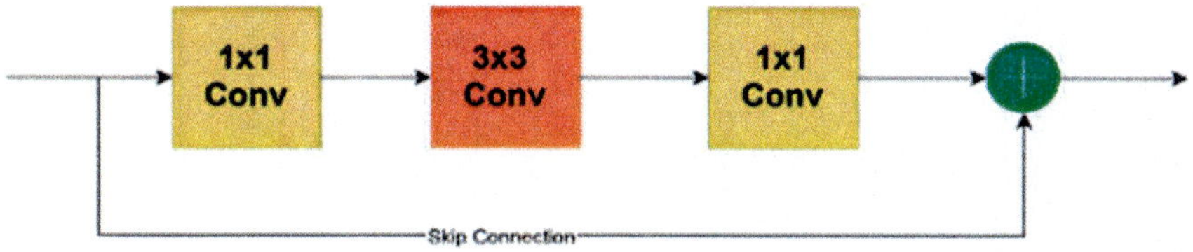

Fig. 2 Identity block

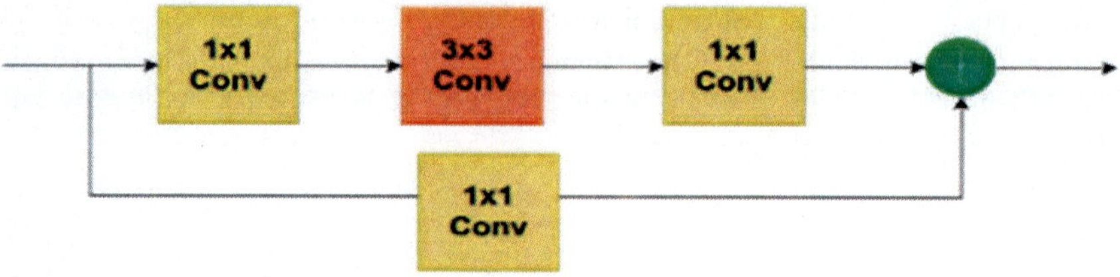

Fig. 3 Convolution block

In the following equation, $f(x)$ is minimized which symbolizes the difference between the input and output variables for the provided layers. Hence, providing the minimum data loss for different layers used in the model.

$$y = f(x) + x, \tag{1}$$

where in Eq. (1) x defines input for the convolution layer.

2.2 *InceptionV3*

InceptionV3 [18] model is represented in Fig. 4. Learns from different sized filters parallel at multiple stages. The model uses both small and big filters to assess all vital information from the images. This can happen due to variations in the location of the information. Inception covers a wider network by using parallel filters of different sizes rather than deeper networks. The number of parameters is reduced to increase the computational complexity by using factorization of convolution in smaller convolution and asymmetric convolution and makes the algorithm more memory efficient. A small CNN model is used in the middle of the layers, while training the model and the loss computed is added to the main network loss thus reducing the vanishing gradient problem [19]. Figure 2. describes the architecture of InceptionV3 and the different blocks which were used to build this model.

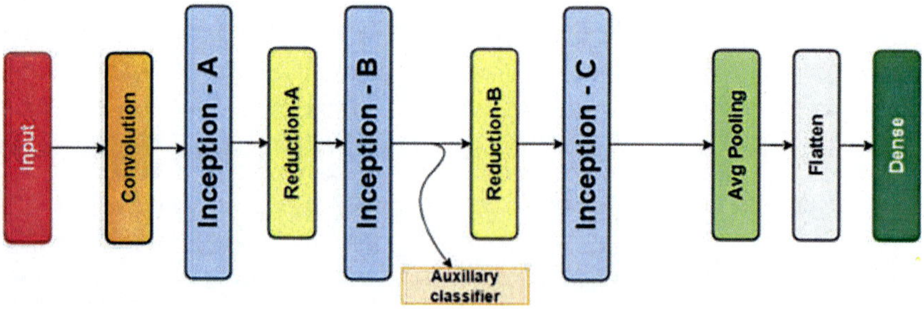

Fig. 4 Architecture of InceptionV3

2.3 Hybrid Ensemble Model

Ensemble learning [20] provides better performance and reduces the diffusion, i.e. decrease in variance, of the predictions with respect to any single model. Ensemble learning has three types: bagging, boosting, and stacking. This paper used the bagging type ensemble learning approach in which different models trained on the same dataset and then by averaging, validating the results, the ensemble model will predict the final class.

Deep learning procedure is used which bundle predictions from different models by taking individual model's features in proportion to the model's effectiveness or capabilities. This model is commonly known as weighted sum/average ensemble. To achieve greater accuracy, authors proposed this hybrid weighted average approach and predicted the result. The training set will be used to fit the hybrid model in Fig. 5, and the validation set will be used to evaluate it. The model weighting will be determined by the validation set's accuracy.

$$W_1 = \frac{\text{Accuracy}_1}{\text{Accuracy}_1 + \text{Accuracy}_2} \tag{2}$$

$$W_2 = \frac{\text{Accuracy}_2}{\text{Accuracy}_1 + \text{Accuracy}_2} \tag{3}$$

$$Y = \frac{\sum_{i=1}^{n=2} \text{Accuracy}_i * W_i}{/\sum_{i=1}^{n=2} W_i}. \tag{4}$$

Here, Accuracy_1 is the accuracy of ResNet50 and Accuracy_2 is the accuracy of InceptionV3. W_1 in Eq. (2) and W_2 in Eq. (3) are calculated weight for the ResNet50 and InceptionV3, respectively. Equation (4) describes the hybrid weighted approach to predict the results.

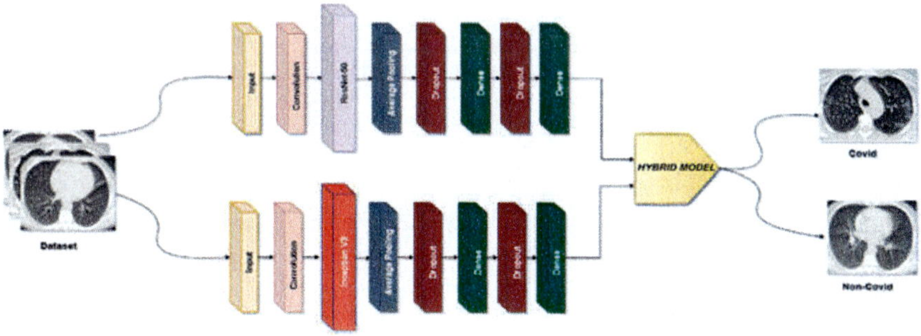

Fig. 5 Architecture of proposed hybrid model using weighted average ensemble learning

3 Results and Discussions

3.1 Dataset

In the present paper, the proposed model is used to improve the accuracy of COVID and non-COVID detection from CT scan image reports. SARS-CoV-2-CT-Scan-Dataset [17] from Kaggle is utilized for model training, validation, and testing. The experiment has been performed on Google Colab by using Keras library. The dataset comprises 2481 CT scan images fetched from 120 patients including 1252 COVID images and 1229 non-COVID images. The training and testing dataset is divided into 80:20 ratio, whereas the training dataset is further bisect into training and validation by the ratio of 80:20.

3.2 Accuracy

The model has been parameterized according to accuracy, specificity, sensitivity, precision, and F1-score, defined below. The proposed hybrid model is compared with individual model is summarized in Table 1. Following evaluation parameters are used to compare the proposed model with the individual model.

$$\text{Accuracy} = \frac{\text{TP} + \text{TN}}{(\text{TP} + \text{TN} + \text{FP} + \text{FN})}, \tag{5}$$

$$\text{Sensitivity} = \frac{\text{TP}}{\text{TP} + \text{FN}}, \tag{6}$$

$$\text{Specificity} = \frac{\text{TN}}{\text{TN} + \text{FP}}, \tag{7}$$

$$\text{Precision} = \frac{\text{TP}}{\text{TP} + \text{FP}}, \tag{8}$$

$$F1\ \text{Score} = \frac{(2 * \text{TP})}{2 * \text{TP} + \text{FN} + \text{FP}}, \tag{9}$$

where TP, TN, FP, and FN act as true positive, true negative, false positive, and false negative values, sequentially.

Table 1 Comparison between InceptionV3, ResNet50, and the proposed hybrid model

Model	Accuracy (%)	Specificity (%)	Sensitivity (%)	Precision (%)	F1-score (%)
InceptionV3	90.23	95.12	85.87	84.78	90.17
ResNet50	89.65	91.54	83.21	80.45	89.65
Proposed model	94.23	90.10	84.97	89.95	91.56

Table 2 Comparison between the accuracy of models from different research works

Research paper	Accuracy (%)	Specificity (%)	Sensitivity (%)	Precision (%)	F1-score (%)
Ref. [6]	87.6	85.2	NP	84.3	87.1
Ref. [21]	93	93	93	93	93
Proposed model	94.23	90.10	84.97	89.95	91.56

3.3 Discussion

Table 2 shows the comparisons between previous work done in this field and the proposed model solution to resolve the problem with better and more accurate results. InceptionV3 and ResNET50 are deep learning algorithms which have great impact on the vanishing gradient problem of dataset and helps us to produce better accuracy than other CNN algorithms. Hybrid model uses the qualities of both the algorithms which are combined using the weighted average ensemble method that also helps to achieve a step ahead accuracy and prediction then the single model. Decaps and Peekaboo [6] model proposed an accuracy of 87.6% on the dataset prepared by Zhaoetal consisting of 746 CT scan images which is lesser by 7.035% that the proposed methodology. Another method namely DRE-Net [21] executed over a small database of only 88 CT scan reports which have reached an accuracy of 93%. It showed an unexpected result with all the parameters having the same value. The proposed model uses the best techniques ResNet50, InceptionV3, and ensemble learning models.

The confusion matrix represented in Fig. 6. helps us to get a visual idea of the accuracy represented by the proposed model.

4 Conclusion

In this paper, a hybrid ensemble model is proposed using InceptionV3 and ResNet50 to detect COVID and non-COVID patients CT scan images. The experiment has been executed on Google Colab by the help of Keras library implemented over the dataset extracted from Kaggle named as SARS-CoV-2-CT-Scan-Dataset. The proposed model uses bagging ensemble learning which predicts the final accuracy

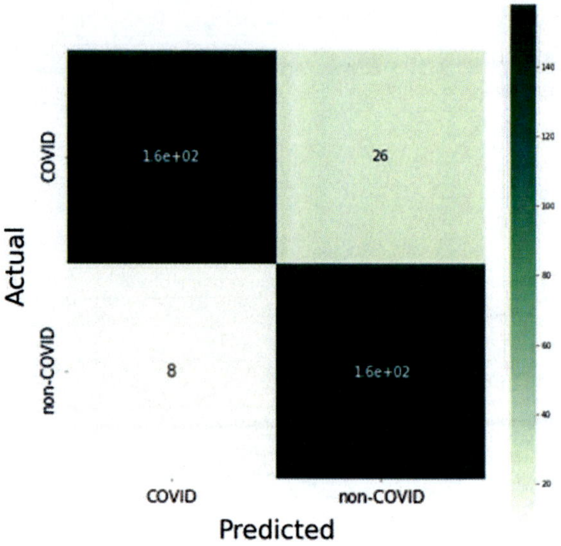

Fig. 6 Confusion matrix for the proposed model

by the weighted average of the individual model accuracy. It provides an accuracy of 94.23% which is higher than the InceptionV3 and ResNet50 model which have an accuracy of 90.23% and 89.65%, respectively. This research can be further improved to segregate different lung diseases.

References

1. Hussain A, Kaler J, Tabrez E, Tabrez S, Tabrez SSM (2020) Novel COVID-19: a comprehensive review of transmission, manifestation, and pathogenesis. Cureus. https://doi.org/10.7759/CUREUS.8184
2. Hashmi HAS, Asif HM (2020) Early detection and assessment of covid-19. Front Med 7:311. https://doi.org/10.3389/FMED.2020.00311/BIBTEX
3. Wang W, Xu Y, Gao R, Lu R, Han K, Wu G, Tan W (2020) Detection of SARS-CoV-2 in different types of clinical specimens. JAMA 323(18):1843–1844. https://doi.org/10.1001/JAMA.2020.3786
4. Mangal A, Kalia S, Rajgopal H, Rangarajan K, Namboodiri V, Banerjee S, Arora C (2020) CovidAID: COVID-19 detection using chest X-ray. http://arxiv.org/abs/2004.09803
5. Tenda ED, Yulianti M, Asaf MM, Yunus RE, Septiyanti W, Wulani V, Pitoyo CW, Rumende CM, Setiati S, Tenda ED (2020) The importance of chest CT scan in COVID-19: a case series. Acta Med Indones-Indones J Intern Med 52
6. Mobiny A, Cicalese PA, Zare S, Yuan P, Abavisan MS, Wu CC, Ahuja J, de Groot PM, Nguyen HV (2020) Radiologist-level COVID-19 detection using CT scans with detail-oriented capsule networks
7. Silva P, Luz E, Silva G, Moreira G, Silva R, Lucio D, Menotti D (2020) COVID-19 detection in CT images with deep learning: a voting-based scheme and cross-datasets analysis. Inf Med Unlocked 20. https://doi.org/10.1016/j.imu.2020.100427
8. Polsinelli M, Cinque L, Placidi G (2020) A light CNN for detecting COVID-19 from CT scans of the chest. Pattern Recogn Lett 140:95–100. https://doi.org/10.1016/j.patrec.2020.10.001

9. Shrivastava P, Singh A, Agarwal S, Tekchandani H, Verma S (2021) Covid detection in CT and X-ray images using ensemble learning. In: Proceedings—5th international conference on computing methodologies and communication, ICCMC 2021, pp 1085–1090. https://doi.org/10.1109/ICCMC51019.2021.9418308
10. Modi S, Guhathakurta R, Praveen S, Tyagi S, Bansod SN (2021) Detail-oriented capsule network for classification of CT scan images performing the detection of COVID-19. Mater Today: Proc. https://doi.org/10.1016/j.matpr.2021.07.367
11. Nayak SR, Nayak DR, Sinha U, Arora V, Pachori RB (2021) Application of deep learning techniques for detection of COVID-19 cases using chest X-ray images: a comprehensive study. Biomed Signal Process Control 64. https://doi.org/10.1016/j.bspc.2020.102365
12. Hasan NI (2021) A hybrid method of covid-19 patient detection from modified CT-scan/chest-X-ray images combining deep convolutional neural network and two- dimensional empirical mode decomposition. Comput Methods Programs Biomed Update 1:100022. https://doi.org/10.1016/j.cmpbup.2021.100022
13. Tekchandani H, Verma S, Londhe ND (2018) Severity assessment of lymph nodes in CT images using deep learning paradigm. In: 2018 second international conference on computing methodologies and communication (ICCMC), pp 686–691
14. Thakur S, Kumar A (2021) X-ray and CT-scan-based automated detection and classification of covid-19 using convolutional neural networks (CNN). Biomed Signal Process Control69. https://doi.org/10.1016/j.bspc.2021.102920
15. Aslan MF, Unlersen MF, Sabanci K, Durdu A (2021) CNN-based transfer learning–BiLSTM network: a novel approach for COVID-19 infection detection. Appl Soft Comput 98. https://doi.org/10.1016/j.asoc.2020.106912
16. Mascarenhas S, Agarwal M (2021) A comparison between VGG16, VGG19 and ResNet50 architecture frameworks for image classification. In: 2021 international conference on disruptive technologies for multi-disciplinary research and applications (CENTCON), pp 96–99. https://doi.org/10.1109/CENTCON52345.2021.9687944
17. SARS-COV-2 Ct-Scan Dataset | Kaggle (n.d.) Retrieved March 12, 2022, from https://www.kaggle.com/plameneduardo/sarscov2-ctscan-dataset
18. Szegedy C, Vanhoucke V, Ioffe S, Shlens J, Wojna Z (2016) Rethinking the inception architecture for computer vision. In: Proceedings of the IEEE computer society conference on computer vision and pattern recognition, pp 2818–2826. https://doi.org/10.1109/CVPR.2016.308
19. Paperspace Blog (n.d.) from https://blog.paperspace.com/popular-deep-learning-architectures-resnet-inceptionv3-squeezenet. Last accessed 2022/04/24
20. Ensemble learning Simplified. by Sunila Gollapudi | Retrieved March 27, 2022, from https://medium.datadriveninvestor.com/ensemble-learning-simplified-868a99b6d945
21. Song Y, Zheng S, Li L, Zhang X, Zhang X, Huang Z, Chen J, Wang R, Zhao H, Chong Y, Shen J, Zha Y, Yang Y (2021) Deep learning enables accurate diagnosis of novel coronavirus (COVID-19) with CT images. https://doi.org/10.1109/TCBB.2021.3065361

Communication and Network Technologies

Integrated Frequency Reconfigurable Elliptical Slot Antenna and Inverted-L Slot MIMO Antenna for 4G and 5G Wireless Handheld Devices

Ranjana Kumari, V. K. Tomar, and Ankit Sharma

Abstract In this article, elliptical and inverted-L slot antennas are integrated at same substrate to achieve 4G and 5G applications. The proposed antenna consists of elliptical and inverted-L slot on ground plane. A varactor diode (SMV1233) is loaded on elliptical slot to achieve frequency tunability in dual band from 1.89 GHz to 2.65 GHz for lower band and 5.27–5.6 GHz for higher band (4G, WLAN) with maximum gain of 4.55dBi. A L-shape slot is excited by inverted-L-shape monopole antenna for 5G applications with bandwidth 3.26–3.75Ghz. To increase the functionality, an elliptical slot loaded with varactor diode is easily extended into four port and L-shape slot also easily extended in two port for multiple input and multiple output (MIMO) applications. The six elements are implemented on Rogers substrate with size of 120 * 60 * 0.76 mm^3. Due to integration of 4G and 5G operation on slots, the proposed antenna structure is compact, simple in structure. Proposed elliptical frequency reconfigurable MIMO antenna is designed for wireless handheld devices and mobile terminals in cognitive radio applications.

Keywords MIMO antenna · Varactor diode · Frequency reconfigurable · 5G

1 Introduction

To increase data rate and channel capacity, multiple input and multiple output (MIMO) technology is utilized in wireless communication system. In recent trends, various handheld devices are operating on multi-frequency standards. Due to increase the number of users, frequency congestion spectrum might be increased. To reduce the congestion problem on particular frequency spectrum, researchers have focused to design a frequency reconfigurable MIMO antenna to efficiently utilize the frequency spectrum with high throughput at cognitive radio (CR) platforms [1]. Due to low cost,

R. Kumari (✉) · V. K. Tomar
Department of Electronics and Communication, GLA University, Mathura, India
e-mail: ranjana.biet@gmail.com

R. Kumari · A. Sharma
Galgotias College of Engineering and Technology, Greater Noida, Uttar Pradesh, India

A. Rani et al. (eds.), *Signals, Machines and Automation*, Lecture Notes in Electrical Engineering 1023, https://doi.org/10.1007/978-981-99-0969-8_46

fabrication simplicity and simple integration, patch radiators have been employed to design frequency reconfigurable antenna in [2–7]. In [8–10], reconfigurability is achieved by PIN diode, varactor diode and microelectromechanical (MEMS) system. Xu et al. [11] proposed a frequency reconfigurable two port MIMO antenna with four resonant mode, and reconfigurability is achieved by three PIN diode. Soltani et al. [12] has been proposed a port reconfigurable MIMO antenna to cover three frequency spectrum (2.5, 5, 5.5 GHz) with substrate dimensions 40 * 20 * 1.6 mm^3. However, size is compact but reconfigurability is not in continuous range. In [13], frequency reconfigurable MIMO antenna is designed for metal frame smart phone applications. Four pin diodes have been connected on feed slots to achieve the frequency reconfigurability. Abovementioned antennas are frequency reconfigurable but most of them are not in continuous range. Therefore, to achieve continuous range frequency reconfigurability, varactor diode is good one choice. In [14], four port frequency reconfigurable slot antenna has been proposed and continuous range frequency agility is done by implanting varactor diodes on ground of slots. [15] has been proposed integrated frequency reconfigurable slot antenna with connected slot array for 4G and 5G mobile applications. In [16], stacked Yagi-like MIMO antenna has been presented with frequency reconfigurability. Frequency reconfigurability is continuous in frequency range (1.5–2.1 GHz) which is achieved by varactor diodes.

This article integrated four port frequency reconfigurable elliptical slot MIMO antenna with two ports Inverted-L slot MIMO antenna for 4G and 5G wireless handheld devices and mobile terminals in cognitive radio applications. The six-element MIMO antenna has been proposed to increase channel capacity and data rate of MIMO system. The isolation among elements is achieved by spatial as well as pattern diversity. The four port of proposed structure covers the wide tuning range from 1.89 to 2.65 GHz with other band 5.5 GHz is almost fixed and other two port are integrated for resonant frequency 3.5 GHz. The proposed MIMO antenna design procedure and results are discussed in next section.

2 Proposed Antenna Design

Initially, a single element is designed on 60 * 30 mm^2 substrate and an elliptical slot is etched out from ground having dimensions Ex and Ey as major and minor axis as illustrated in Fig. 1. The fundamental frequency f_0 is 3.6 GHz without reactive loading when optimized values of Ex and Ey are 20.2 mm and 8.5 mm. To acquire frequency reconfigurability in proposed antenna, a varactor diode (SMV1233) is loaded elliptical ring slot on ground of antenna. When reverse bias voltage is applied across the varactor diode then values of R, L are constant and Cj (junction capacitance) will be varied. The simple RLC model of varactor diode (SMV1233) is shown in Fig. 1. After optimizing the single element with reconfigurability, single element is easily extended in to four elements on 120 * 60 mm^2 substrate.

To miniaturization of proposed antenna, two more antennas (Ant5–Ant6) are placed on vertical edges to achieve good isolation among elements in frequency

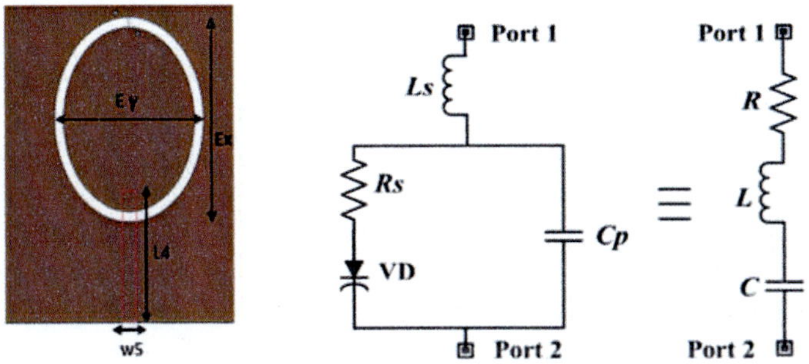

Fig. 1 Proposed single element and equivalent diagram of varactor diode (SMV1233)

spectrum ranging from 3.4 GHz to 3.6 GHz. Resonant lengths of Ant5 and Ant6 at 3.5 GHz are 22.5 mm. The front and back view of proposed antenna with detailed parameters are shown in Fig. 2a, b, and physical dimensions of proposed antenna are also described in Table 1.

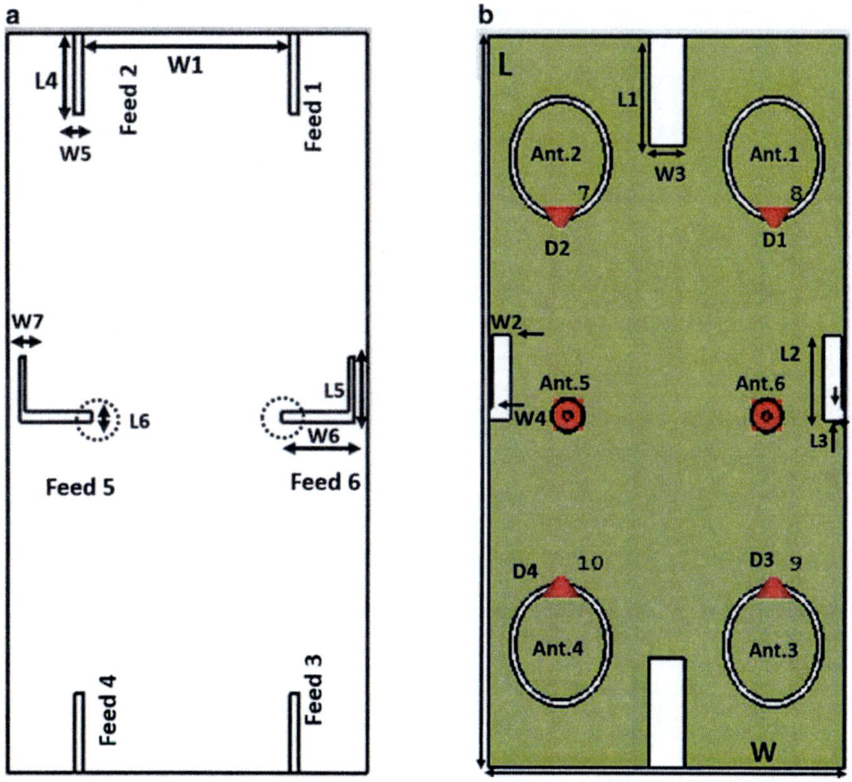

Fig. 2 **a**, **b** Front view and back view of proposed antenna

Table 1 Parameters and dimensions

Parameters	W	L	W1	W2	W3	W4	W5	W6
Units (mm)	60	120	34.5	3	6	0.5	1.55	12
Parameters	L1	L2	L3	L4	L5	L6	W7	
Units (mm)	18	14	2	13	10.5	1.55	1	

3 Result and Discussion

The schematic design of proposed MIMO antenna at 5.08 pf is illustrated in Fig. 3a, and simulation is done by CST2019 software. Simulated S-parameters are illustrated in Fig. 3b. At 5.08pf, proposed reconfigurable antenna is producing dual band response at 1.89 GHz and 5.27 GHz, respectively. When varactor diode reverse bias voltage is varied from 0 to 15 V, then voltages correspond to different capacitance values of varactor diode will be produced. At different values of capacitances, Simulated S11 (dB) parameters are described in Fig. 4. The capacitance values largely impact in the lower band as compared to higher band. Moreover, two Ant5 and Ant6 elements are also simulated, and according to simulated results, antenna elements are producing single band with resonant frequency 3.5 GHz as illustrated in Fig. 5.

The different applications of proposed antenna can be utilized at different capacitance value of varactor diode SMV1233. Isolation among the radiating elements is desirable for good MIMO performance. The closely spaced antenna elements isolation has been improved by DGS structure as described in Figs. 6 and 7. Without using decoupling structure, isolation was 12db in lower band but by introducing DGS technique (defected ground system) less surface current flow between adjacent elements. This improves isolation greater than 15 dB in lower band and 22 dB in higher band.

The simulated gain (dBi) of six-element MIMO antenna has been illustrated in Figs. 8 and 9. As can be seen from the illustration, positive gain value more than 3 dBi observed in lower band and more than 4 dBi in higher band. Maximum gain values have achieved 4.55dBi in lower band and 4.27 dBi in higher band. Antenna radiation

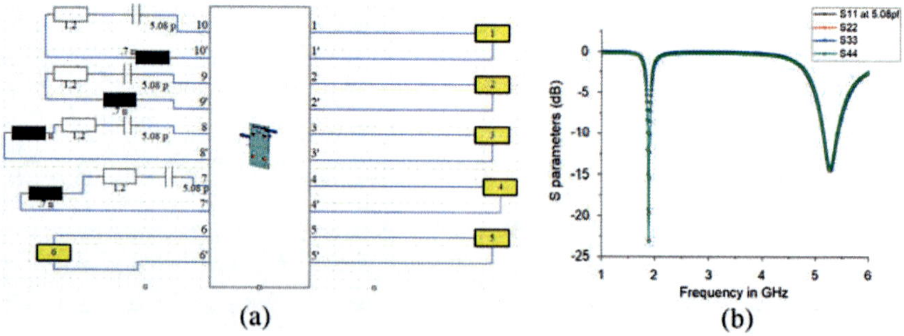

Fig. 3 **a**, **b**. Schematic design and simulated result (S11) of proposed antenna at 5.08 pF

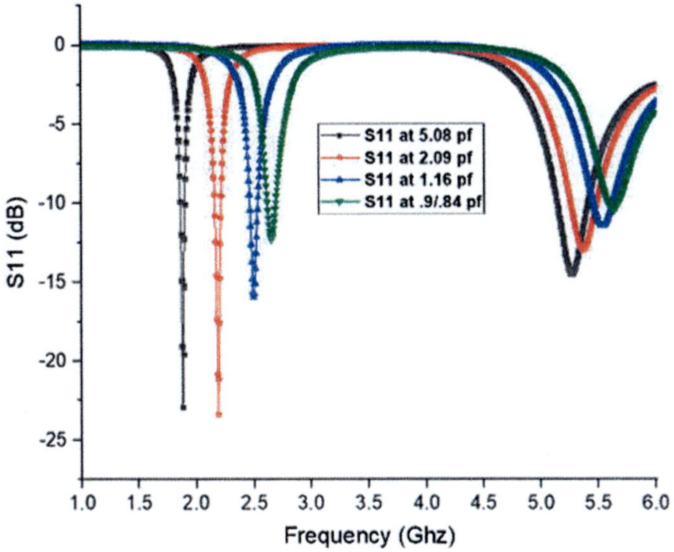

Fig. 4 Simulated results of S11 (dB)

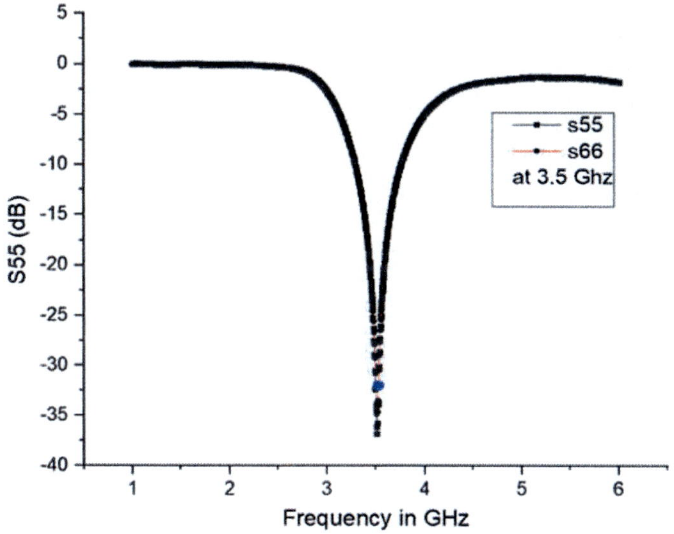

Fig. 5 Simulated result of S55/S66

efficiency is an important parameter for MIMO antenna system. Radiation efficiency of proposed antenna is more than 90% in entire frequency range. The simulated 2D gain patterns of antenna elements, Ant1 and Ant2 at 1.8 GHz, are shown in Figs. 10a, and antenna elements (3 and 4) at 1.88 GHz are shown in Fig. 10b. Due to effect of large ground, radiation pattern is tilted and this improves the correlation between antenna elements.

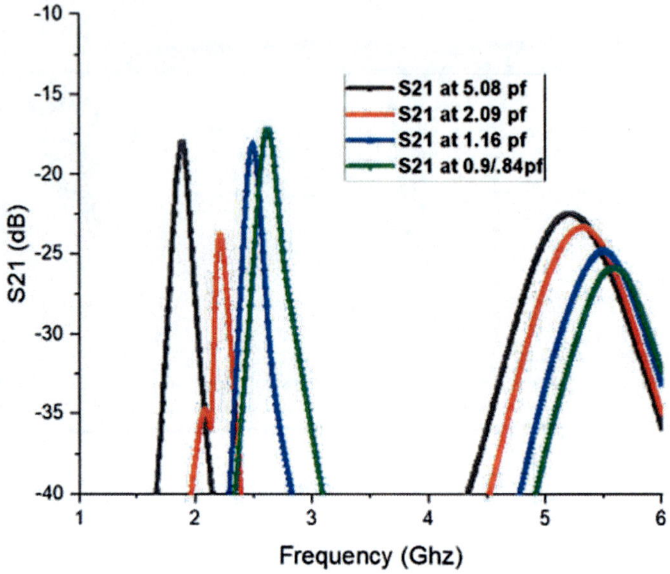

Fig. 6 Simulated results of isolation S21 (dB)

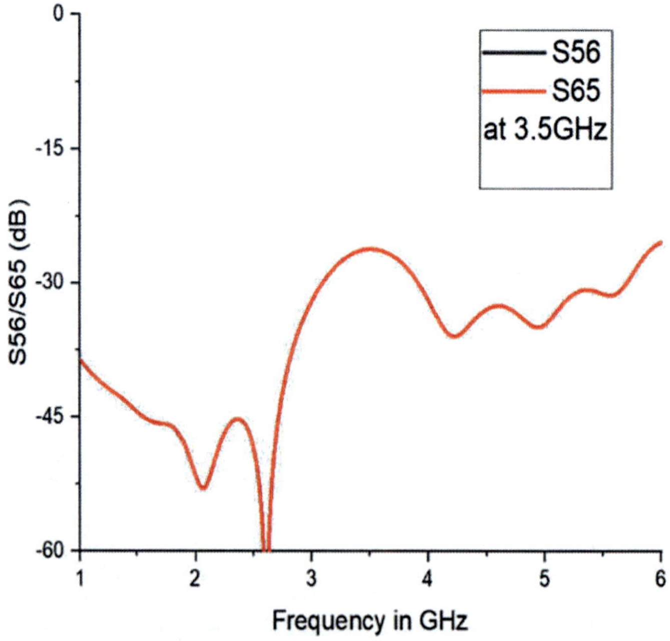

Fig. 7 Simulated isolation (dB)

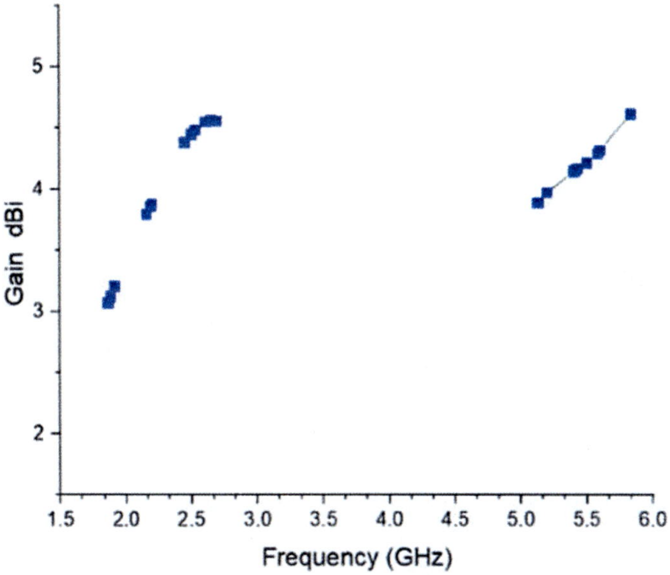

Fig. 8 Simulated gain (dBi) at S11 port

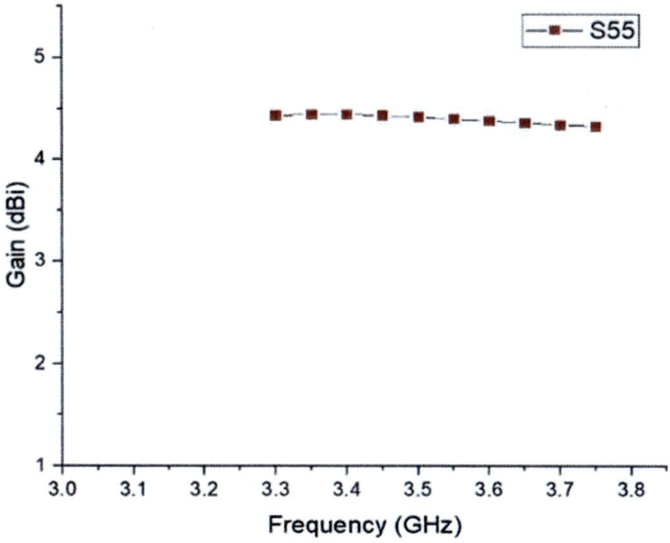

Fig. 9 Simulated gain (dBi) at S55 port

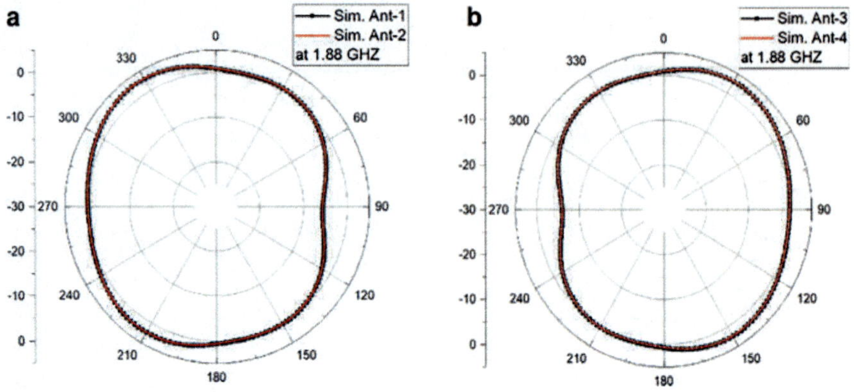

Fig. 10 2D pattern at 1.88 Ghz **a** Ant1 and Ant2 for phi = 0^0 **b** Ant3 and Ant4 for phi = 0^0

3.1 *Diversity Performance of Proposed Antenna*

The diversity performance analysis of proposed antenna has been discussed by envelope correlation coefficient (ECC), diversity gain (DG) and mean effective gain (MEG). Envelope correlation coefficient (ECC) measures the correlation between antennas in a MIMO system. The overall antenna module ensures correlation values which should be less than 0.5 as illustrated in Fig. 11a. As per observation from Fig.11a, the value of ECC is less than 0.5 in intended spectrum. One more diversity paramter is diversity gain which is directly depended on correlation among the radiating elements. The results of DG should be approximately 10 dB throughout the band, which ensures good diversity performance of the antenna. The proposed antenna simulated diversity gain (DG) is illustrated in Fig. 11b. As per observation from Fig.11b, the value of DG(dB) is similar to ideal value in intended spectrum which shows that proposed antenna is good contender for 4G and 5G application.

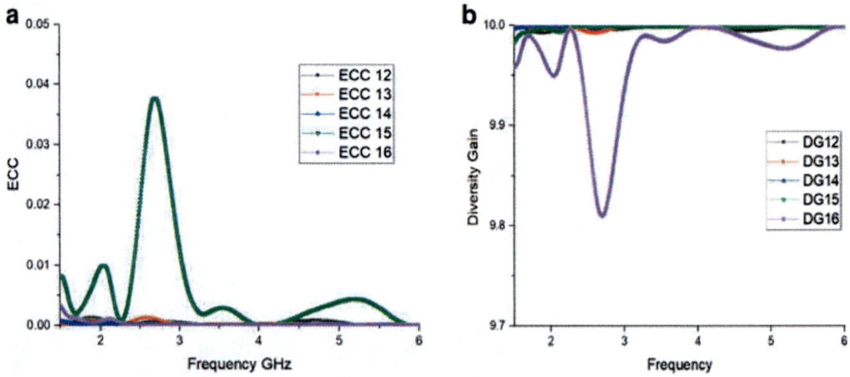

Fig. 11 Simulated results of ECCs and DG (diversity gain)

Table 2 Mean effective gain (MEG) for Ant1 to Ant6

Frequency (GHz) 1.88	MEG Ant1 (dB)	MEG Ant2 (dB)	MEG Ant3 (dB)	MEG Ant4 (dB)	MEG Ant5 (dB)	MEG Ant6 (dB)
Indoor XPR = 1	−3.01	−3.03	−3.01	−3.13	−3.21	−3.24
Indoor XPR = 5	−4.02	−4.10	−4.20	−4.22	−4.02	−4.19

Mean effective gain (MEG) is the ability of antenna to collect electromagnetic power in multipath environment. It is ratio of mean received power to incident power of antenna. The simulated values of (MEG i/MEG j) are less than 3 dB which exhibit the satisfactory diversity performance in indoor and outdoor environments. The simulated values of MEGs at 1.88 GHz are discussed in Table 2. As per simulated values of MEGs in Table 2, MEGs values of proposed antenna exhibit the performance of MIMO antenna for 4G and 5G applications. Furthermore, proposed antenna is also compared with existing MIMO antennas in Table 3. The size of proposed antenna is compact as compared to existing antenna except one. Gain and efficiency of proposed antenna is higher than existing antennas.

Table 3 Comparison of proposed antenna with existing antennas

Year	Ant. size (mm^2)	Number of elements	Frequency bands (GHz) covered	Switches per element	Gain (dBi)	Efficiency %	Integration with 5G ant
2017	120 × 60	4	1.8–2.5 continuous range	1	−2.6, 0.5, 2.1,2.36, 2.43	51, 67, 71.5, 72, 73	No
2018	160 × 85	4	824 to 960 MHz 1710 to 2690	1	–	55–73	No
2020	100 × 100	4	1.5 to 2.1 GHz	1	4 dBi	76	No
[P]	120 × 60	6	1.88–2.65 GHz and 5.27–5.6 GHz (continuous range) 3.5 GHz	1	3.12 to 4.55 and 4.03 to 4.27 and 4.41 at 3.5 GHz	79–96 and 95–96 and 96	Yes

[P] proposed antenna

4 Conclusion

The six-element MIMO antenna is proposed for LTE-7, LTE-23 and LTE-35 and UNII bands. Integration of frequency reconfigurable elliptical slot antenna and inverted-L slot antenna which is designed for 4G and 5G wireless handheld devices. The four varactor diodes (SMV1233) are loaded on elliptical slots to achieve frequency tunability in dual band from 1.89 GHz to 2.65 GHz for lower band and 5.27–5.6 GHz for higher band (4G, WLAN) with maximum gain of 4.55 dBi and stable radiation patterns. The two elements inverted-L slots are etching out from ground to achieve 3.26–3.75Ghz spectrum. Due to integration of 4G and 5G operation on slots, the proposed antenna structure is compact, simple in structure. The simulated results of ECCs and DGs exhibit the good diversity performance of proposed MIMO antenna which shows that proposed antenna is good contender for 4G and 5G wireless applications.

References

1. Christodoulou CG, Tawk Y, Lane S, Erwin S (2012) Reconfigurable antennas for wireless and space applications. Proc IEEE 100(7):2250–2261
2. Hung C, Chiu T (2015) Dual-band reconfigurable antenna design using slot-line with branch edge. IEEE Trans Antennas Propag 63(2):508–516
3. Cetiner BA, Crusats GR, Jofre L, Biyikli N (2010) RF MEMS integrated frequenc reconfigurable annular slot antenna. IEEE Trans Antennas Propag 58(3):626–632
4. Majid HA, Rahim MKA, Hamid MR, Murad NA, Ismail MF (2013) Frequency- reconfigurable microstrip patch-slot antenna. IEEE Antennas Wireless Propag Lett 12:218–220
5. Majid HA, Rahim MKA, Hamid MR, Ismail MF (2014) Frequency reconfigurable microstrip patch-slot antenna with directional radiation pattern. Prog Electromagn Res 144:319–328
6. Majid HA, Rahim MKA, Hamid MR, Ismail MF (2014) Frequency and pattern reconfigurable slot antenna. IEEE Trans Antennas Propag 62(10):5339–5343
7. Sharma S, Tripathi CC (2015) Frequency reconfigurable U-slot antenna for SDR application. Prog Electromagn Res Lett 55:129–136
8. Khan MS, Iftikhar A, Capobianco AD, Shubair RM, Ijaz B (2017) Pattern and frequency reconfiguration of patch antenna using pin diodes. Microw Opt Technol Lett 59(9):2180–2185
9. Khidre A, Yang F, Elsherbeni AZ (2015) A patch antenna with a varactor-loaded slot for reconfigurable dual-band operation. IEEE Trans Antennas Propag 63(2):755–760
10. Rajagopalan H, Kovitz JM, Samii YR (2014) MEMS reconfigurable optimized E-shaped patch antenna design for cognitive radio. IEEE Trans Antennas Propag 62:1056–1064
11. Xu ZQ, Sun YT, Zhou QQ, Ban YL, Li YX, Ang SS (2017) Reconfigurable MIMO antenna for integrated-metal-rimmed smartphone applications. IEEE Access 5:21223–21228
12. Soltani S, Lotfi P, Murch RD (2016) A port and frequency reconfigurable MIMO slot antenna for WLAN applications. IEEE Trans Antennas Propag 64(4):1209–1217
13. Zhang YH, Yang SR, Ban YL (2018) Four-feed reconfigurable MIMO antenna for metal-frame smartphone applications. IET Microwaves Antennas Propag 12:1477–1482
14. Hussain R, Ghalib A, Sharawi MS (2017) Annular slot-based miniaturized frequency-agile mimo antenna system. IEEE Antennas Wireless Propag Lett 16:2489–2492
15. Ikram M, Abbas EA, Nguyen-Trong N (2019) Integrated frequency-reconfigurable slot antenna and connected slot antenna array for 4G and 5G mobile handsets. IEEE Access 67
16. Hussain R, Jehangir SS, Khan MU, Sharawi MS (2020) Stacked frequency reconfigurable Yagi-like MIMO antenna system. IEEE Access 5

Design and Investigation of an Improved FTTH Network for Superior Quality Triple-Play Services

Deepak Garg and Abhimanyu Nain

Abstract The rising high demand for multimedia services with ultra-high-definition videos which require large bandwidth and data rate over long distances with seamless transmission capacity can be realized using various optical communication techniques like radio over fiber (RoF), fiber to the home (FTTH), etc. FTTH is the most common and cost-effective communication network system with effortless service delivery but does suffer from the fiber impairments like signal noise, unwanted frequencies and increased latency. In this paper, a compensated broadband passive optical network (BPON) FTTH is designed including an in-line combination of fiber Bragg grating (FBG) and pair of dispersion compensation fiber (DCF) for 32 subscribers to overcome the limitation of conventional FTTH network. The proposed network is simulated successfully for 35 km transmission at a 7 Gbps data rate for voice, video and data, i.e., triple-play services. The improved performance is measured in terms of bit error rate (BER), Q-factor and scattering diagrams.

Keywords FTTH · BPON · DCF · Fiber grating · Triple-play services

1 Introduction

FTTH access networks are rising to meet the growing demand and provide multimedia services like voice and high-definition videos with high-speed connectivity at the user sites. FTTH replaces coaxial cable with fiber optic to connect subscribers through a more secure and very low attenuation network [1]. FTTH is a PON from the central station (CS) to the end-user, delivering video on demand and multi-Gbps video conferencing data formats with low battery and power usage. For an FTTH network to possess characteristics like security, flexibility and reliability, it requires minimal cost and power with large bandwidth compared to conventional ultra-long-distance data transmission systems [2]. Since it utilizes bend-insensitive flexible fibers, thus may suffer from leaky propagation over long distances and can cause

D. Garg (✉) · A. Nain
Guru Jambheshwar University of Science and Technology, Hisar, Haryana, India
e-mail: garg.deepu2018@gmail.com

A. Rani et al. (eds.), *Signals, Machines and Automation*, Lecture Notes in Electrical Engineering 1023, https://doi.org/10.1007/978-981-99-0969-8_47

multi-path interference (MPI) noise which may give rise to disruption in fiber-based optical networks [3].

Desired characteristics of greater bandwidth with limited latency and seamless service delivery can be achieved using RoF [4, 5] and free space optics (FSO) [2, 6] on integration with wavelength division multiplexing (WDM) technology. While using FTTH, multiple services direct to the premises can be achieved in a single bundle. Thus, BPON meets the desired outputs with all the characteristics.

Various PON FTTH scheme types based on network features such as data values, usage topologies and signal formats are available [2]. Major requirements like large transmission capacity with high bit rate and user scalability can be resolved using dense wavelength division multiplexing (DWDM) PON but integrating it with the system requires one-time costing infrastructure for the replacement of single-mode components with multimode components.

The FTTH architecture used in FTTH networks is designed to meet the needs of multiple subscribers through flexible design and cost-effectiveness [7]. A good structure lies in the proper placement of the splitter when following two paths that can be centered or distributed.

A review of the requirements for the FTTH network components and their contribution to the network structure is discussed [2]. It also presented the design and implementation of a secure GPON FTTH Type B network for 1000 users and ensured design using optical power limits in an arbitrary area.

A hybrid access network without RF fading is shown in [8]. An ultra-wide band (UWB) pulse generator is used to suppress the effects of frequency fluctuations due to induced dispersion [9], and low-pulse RF switch and baseband error-free transmission using the multi-band flexibility method associated with the phase module [10] are proposed in the literature.

For improving the performance of FTTH networks, several methodologies and techniques are used in the literature for the large bandwidth channels using WDM PON with spectral overlap [11], multi-pump Raman amplifier and advanced modification schemes [12] to satisfy the need for large bandwidth, good spectral performance and reduce power consumption without affecting performance. The BER penalty measured can be reduced by the variance of the received strength as obtained by variable rates of various interference [3].

Various multiple access techniques are widely used in optical networks with various types of PON FTTH like Ethernet PON, GPON and vertical cavity surface emitting laser (VCSEL) to deploy PON FTTH [13]. The optical code division multiple access [14] technique generates a harmonic way to accommodate several multi-users over TDM PON and WDM PON FTTH architecture.

In this paper, Sect. 2 describes the methodology and simulation setup of the proposed model where a compensated BPON FTTH network is implemented for overcoming the limitation of conventional FTTH networks by using FBG and a pair of DCF expandable up to 32 subscribers. Section 3 discusses the results obtained through the simulation, and finally, the article is concluded in Sect. 4.

2 Methodology and System Design

The design and deployment of FTTH networks involve active opto-electronics at both ends of the transmitting network. Optical line terminal (OLT) is deployed in CS and optical network terminal (ONT) around each home or commercial center at the other end. Fiber distribution is accomplished using a hierarchical structure in which data is sent from CS to a feeder network via OLT, where the feeder network distributes the channels with a two-level splitter in a ratio of 1:32 and finally, it drops through the drop cables to ONT's home premises.

The proposed model is a BPON FTTH design expandable up to 32 subscribers and a range of 35 km. Figure 1 depicts the block diagram for a downstream link with a single ONT device connected to simplify the proposed model simulated on OptSim.

For 32 subscribers, CS is connected to ONT with a 1:4 splitter using single-mode fiber over a transmission length of 30 km. Each of the four outputs received is fed to another fiber of length 4.5 km before entering the ONT with a 1:8 splitter. Thus, final outputs received are connected via 300 m drop cables to 8 end users at ONTs.

2.1 Simulation Setup

Desired simulation framework for the proposed model shown in Fig. 1 is developed using OptSim. FTTH BPON network is deployed as an integrated voice, video and data signal service. For optimizing BPON bandwidth, transmission through fiber-optic paths uses a coarse wavelength division multiplexing (CWDM) with data and

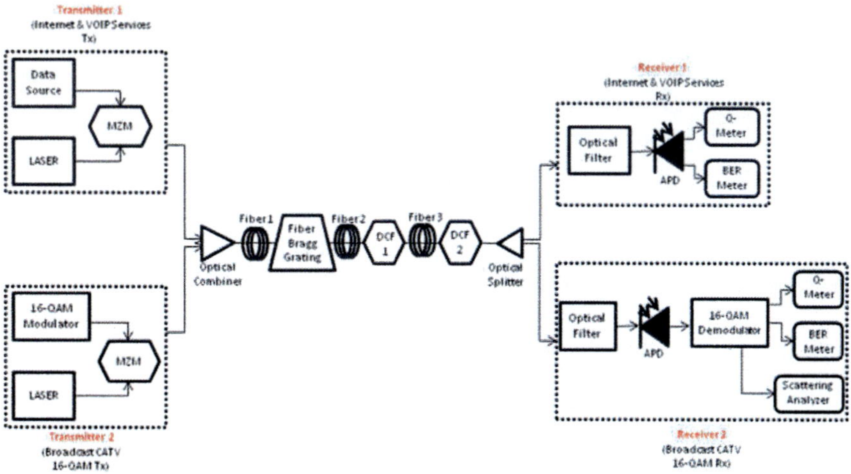

Fig. 1 Block diagram of proposed BPON FTTH network

Table 1 Simulation parameters

Parameter	Value
Internet and VOIP wavelength	200 THz/1499 nm
CATV wavelength	193.4 THz/1550 nm
MZM Vpi	5 V
MZM V_{offset} for video signal	2.5 V
Fiber 1 length	5–30 km
Fiber 2 length	4.5 km
Fiber 3 length	0.3 km
DCF 1 length	1 km
DCF 2 length	0.1 km
APD dark current	0.1 nA

voice components transmitted at 1499 nm and video at 1550 nm wavelength. The high-speed network is illustrated via a data link with various bit rates from 2 to 14 Gbps downlink bandwidth. Voice components can be expressed as VOIP service instead of the conventional public switched telephone network (PSTN) on the receiver end. The video component is represented through 16-QAM sub-carrier multiplexed (SCM) system. In general, 64-QAM is used to transmit digital video/data streams on RF channels, but 16-QAM is used in this paper for simplicity.

Broadcast cable television (CATV) transmitter generates a quality 16-QAM encoded sub-carrier signal at 2 Gbps. The electrical signal is sent to an external optical modulator and combined with the VOIP signal and fed to the optical fiber using FBG and DCF with a dispersion of −72 ps/nm/km and −48 ps/nm/km for a fiber length of 1 km and 0.1 km, respectively. The FBG used here is considered ideal in nature which is centered around 1550 nm. Two corresponding wavelengths are separated by respective filters, used before the APD. The sub-carrier filtered signal is subsequently directly detected by APD at each receiver end. For broadcast CATV, it is further demodulated by 16-QAM demodulation and analyzed through various parameters as discussed in the next section.

Various parameters used for the simulation of the proposed model are given in Table 1.

3 Results and Discussion

The simulation model shown in Fig. 1 has been run and explored with the OptSim simulation tool to measure the outputs.

Figure 2 shows Q-factor versus fiber trunk length of conventional and compensated link for Internet and VOIP data transmission at 1.25 Gbps. Q-value decreases from 14.5 dB to 14.1 dB and 22.9 dB to 20.7 dB for the conventional link and compensated link, respectively, for a varying trunk length from 5 to 30 km.

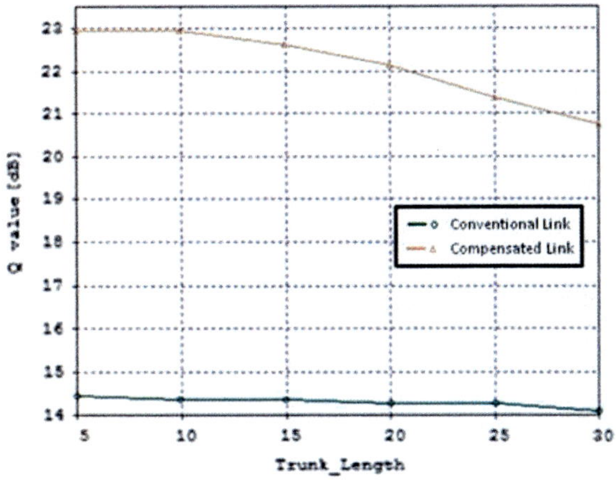

Fig. 2 Q-factor versus fiber trunk length of conventional and compensated link for Internet and VOIP data transmission

Figure 3 shows BER versus fiber trunk length of conventional and compensated link for Internet and VOIP data transmission at 1.25 Gbps. BER increases from 10^{-4} to 10^{-2} and 10^{-40} to 10^{-27} for the conventional link and compensated link, respectively, for a varying trunk length from 5 to 30 km.

Figure 4 shows Q-factor versus bit rate of conventional and compensated link for Internet and VOIP data transmission at 30 km distance. Q-value degrades from 26 dB to 7 dB and 40 dB to 7 dB for the conventional link and compensated link, respectively, when the bit rate is increased from 2 to 14 Gbps.

Figure 5 shows BER versus bit rate of conventional and compensated link for Internet and VOIP data transmission at 30 km distance. BER increases from 10^{-40} to 10^{-2} for the conventional link and compensated link for a varying bit rate of 2 to 14 Gbps. It can be observed from the figure that compensated links produce good quality transmission till 7 Gbps as compared to 4 Gbps of conventional links.

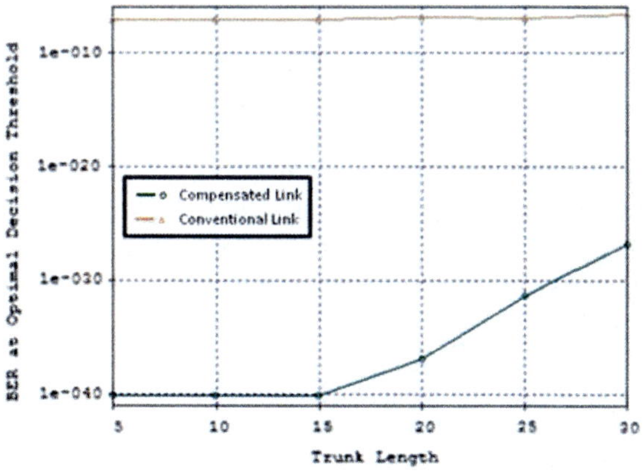

Fig. 3 BER versus fiber trunk length of conventional and compensated link for Internet and VOIP data transmission

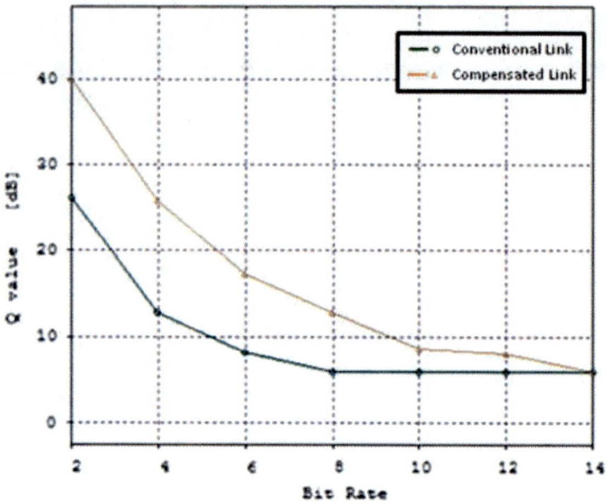

Fig. 4 Q-factor versus bit rate of conventional and compensated link for Internet and VOIP data transmission

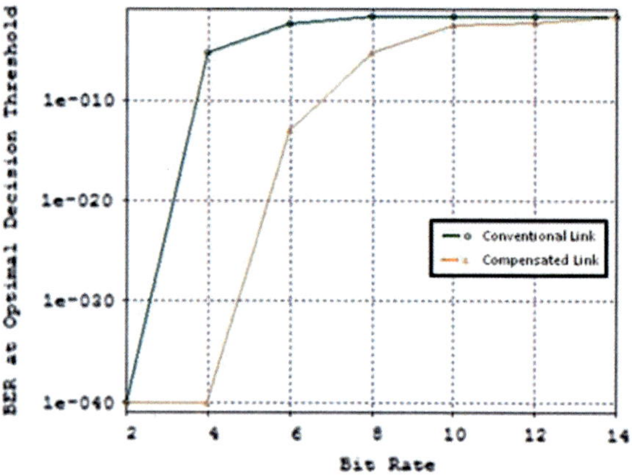

Fig. 5 BER versus bit rate of conventional and compensated link for Internet and VOIP data transmission

To evaluate the quality of the detected signal, an eye diagram is obtained against time for Internet and VOIP data transmission as shown in Fig. 6. In the conventional model, eye-openings of 0.008 units were observed, whereas in the compensated model with FBG, eye-openings of 0.021 units were observed over varying fiber transmission from 5 to 30 km with the data rate of 1.25 Gbps.

Figure 7 shows a scattering diagram of conventional and compensated links for 16-QAM-based CATV video broadcast at 2 Gbps with a 30 km distance. The constellation for the conventional link is rotated in the plane which shows that phase of the signal is distorted. While the constellation of proposed compensated link is almost linear in the plane which implies distortion-less transmission. Further, the less dispersed constellation of a compensated link implies a better transmission compared to conventional links.

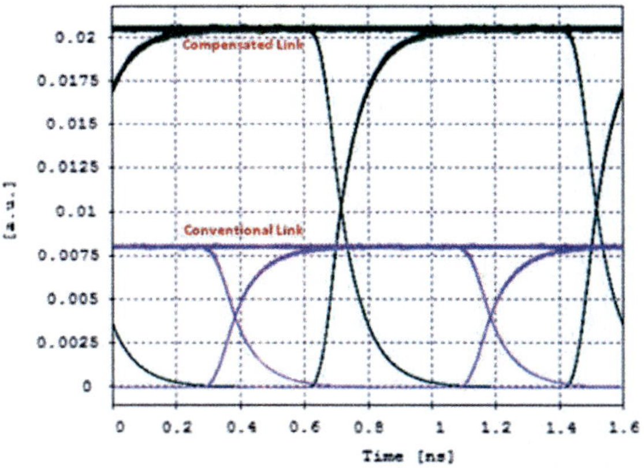

Fig. 6 Eye-opening comparison of conventional and compensated link for Internet and VOIP data transmission

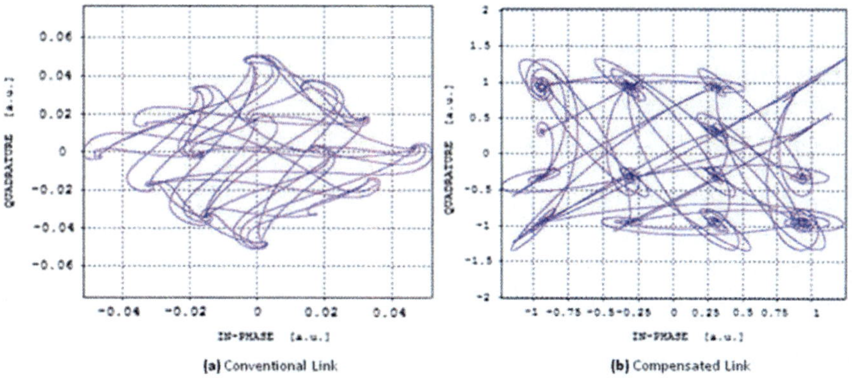

Fig. 7 Scattering diagram of conventional and compensated link for 16-QAM-based CATV video broadcast

Figure 8 shows the comparison of eye-openings for conventional and compensated links for 16-QAM-based CATV video broadcast at 2 Gbps with 30 km distance which marked an improvement in link performance due to compensation of dispersion by FBG and DCFs deployed.

The proposed compensated system design yields significant improvement compared to conventional design in terms of increased Q-values and decreased BER values which is also observed from the above graphs. This improved performance of the FTTH network can be credited to the prominent combination of FBG and a pair of DCF. The utilization of FBG and DCF has compensated fiber dispersion and suppressed fiber nonlinearities which makes the proposed system faster and reliable.

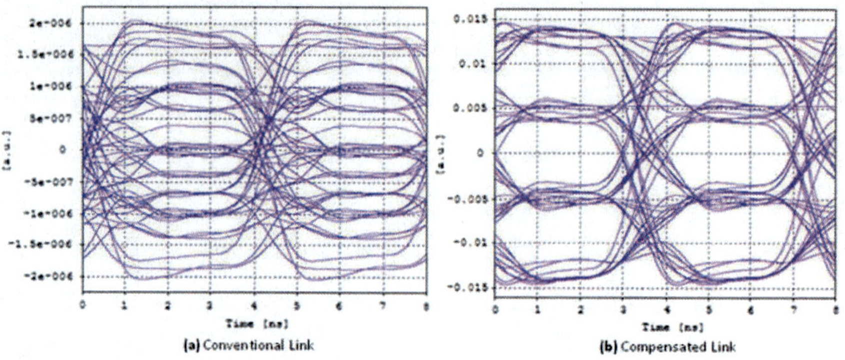

Fig. 8 Eye-opening of conventional and compensated link for 16-QAM-based CATV video broadcast

4 Conclusion

This paper demonstrated a detailed design and simulation of an improved BPON FTTH network serving 32 users for triple-play services from service providers to the home and business within 35 km reach. Q-factor measured from analysis reported an improvement of 6.6 dB and 13 dB, BER obtained improved by 10^{-25} and 10^{-35} at 30 km distance with the data rate of 1.25 Gbps and 4 Gbps, respectively, for Internet and VOIP data transmission. A significant eye-opening improvement was also obtained for Internet and VOIP data transmission, and a clear eye can be seen for 16-QAM-based CATV broadcast at 2 Gbps data rate with 30 km transmission length against conventional broadcast at 0.8 Gbps only. It is concluded that the proposed compensated FTTH network by incorporating FBG and multiple DCFs significantly improved the quality of the triple-play service experience.

References

1. Gopal P, Jain VK, Kar S (2014) Modulation techniques used in earth-to- satellite and inter-satellite free space optical links. Proc SPIE—Int Soc Opt Eng 9248:10
2. Garg D, Nain A (2021) Next generation optical wireless communication: a comprehensive review. J Opt Commun
3. Travagnin M (2013) BER penalty induced by coherent MPI noise in FTTH optical links. J Lightwave Technol 31(18):3021–3031
4. Kumar S, Sharma D, Nain A (2017) Evaluation of sub carrier multiplexing based rof system against non-linear distortions using different modulation techniques. Int J Adv Res Comput Sci Softw Eng 7:454–461
5. Nain A, Kumar S, Singla S (2017) Performance estimation of wdm radio-over-fiber links under the influence of srs induced crosstalk. In: Singh R, Choudhury S (eds) Proceeding of international conference on intelligent communication, control and devices. Singapore, Springer Singapore, 279–284

6. Garg D, Nain A (2022) Analysis and mitigation of photodiode non- linearity under the influence of optical modulator in multitone rof link. J Opt
7. Commscope (2017) FTTH architecture white paper series
8. Jiang W, Lin C, Shih P, WangHe L, Chen J, Chi S (2010) Simultaneous generation and transmission of 60-GHz wireless and baseband wireline signals with uplink transmission using an RSOA. IEEE Photonics Technol Lett 22(15):1099–1101
9. Abtahi M, Rusch LA (2011) RoF delivery over PONs of optically shaped UWB signals for gigabit/s wireless distribution in the home. IEEE J Sel Areas Commun 29(6):1304–1310
10. Shao T, Paresys F, Le Guennec Y, Maury G, Corrao N, Cabon B (2012) Convergence of 60 GHz radio over fiber and WDM-PON using parallel phase modulation with a single mach-zehnder modulator. J Lightwave Technol 30(17):2824–2831
11. Singh A, Tharani L (2018) Design and performance evaluation of bidirectionalWDM-PON using FTTH architecture. In: 2018 international conference on communication and signal processing (ICCSP), pp 0648–0652
12. Muciaccia T, Gargano F, Passaro VMN (2015) A TWDM-PON with advanced modulation techniques and a multi-pump raman amplifier for cost-effective migration to future UDWDM-PONs. J Lightwave Technol 33(14):2986–2996
13. Pfeiffer T, Dom P, Bidkar S, Fredricx F, Christodoulopoulos K, Bonk R (2022) Pon going beyond ftth. J Opt Commun Netw 14(1):A31–A40
14. Ashour I, Shaari S, Menon PS, Bakarman H (2012) Optical code-division multiple-access and wavelength division multiplexing: hybrid scheme review. J Comput Sci 10:1718–1729

An Electronically Tunable Digitally Controlled Current Mode Quadrature Oscillator Using DC-VDTA

Ruchin Sharma, Dinesh Prasad, and Ravendra Singh

Abstract An electronically tunable digitally controlled current mode quadrature oscillator is designed employing one new digitally controlled voltage differencing transconductance amplifier (DC-VDTA), two grounded capacitors, and one grounded resistor. The condition of oscillation and frequency of oscillation can be adjusted independently and electronically. Also, the FO can also be controlled through an n-bit digital input. By introducing the digital control technique, the proposed oscillator can produce 2^n different oscillating frequencies without altering the circuit topology. The functioning of the proposed design is validated through Pspice simulation using 180 nm technologies.

Keywords Quadrature oscillator · VDTA · Digital control unit (DCU)

1 Introduction

The oscillator is a closed loop system with positive feedback which follows the Barkhausen criteria to get a sustained oscillation. If the oscillator circuit can generate two outputs with 90° phase difference, then it is known as a quadrature oscillator (QO). Oscillators are very important and integral part of the various communications and control systems [1, 2]. An enormous variety of quadrature oscillators using different active building blocks (ABB) were already reported in [3–29]. Oscillators reported in [3–10] are able to generate current-mode outputs while voltage mode outputs are obtained in [11–23]. Also, among these oscillators, none is able to generate both voltage mode (VM) and current mode (CM) signals at the same time. This issue is compensated in the circuits reported in [24–27]. In [28], a digitally programmable VMQO was firstly introduced. But, these oscillators have one or more than one of the following drawbacks

R. Sharma · D. Prasad
Department of ECE, Jamia Millia Islamia, New Delhi, India

R. Singh (✉)
Galgotia College of Engineering and Technology, Uttar Pradesh, Greater Noida, India
e-mail: ravendransit@gmail.com

A. Rani et al. (eds.), *Signals, Machines and Automation*, Lecture Notes in Electrical Engineering 1023, https://doi.org/10.1007/978-981-99-0969-8_48

(i) Require two or more ABBs [25–27];
(ii) Require four or more passive components [24, 26, 27];
(iii) Require floating passive components for realization [26];
(iv) Lack of electronic tuning [24, 26]
(v) Except [28] no other circuits can provide digital control feature.

So in this manuscript, an electronically tunable, digitally controlled CM QO is designed to overcome all the drawbacks mentioned above. The proposed QO topology requires one DC-VDTA with three grounded passive elements. As this circuit utilizes only passive elements, so the presented design is acceptable for IC realization. The functioning of DC-VDTA is explained in the next section.

2 DC-VDTA

The DC-VDTA is a modified version of VDTA [29] as shown in Fig. 1. The aspect ratio of all the transistors utilized in realization of DC-VDTA is defined in Table 1. In VDTA, the digital controllability is achieved by adding an n-bit DCU between first and second stage

$$I_x = \mathrm{K}g_{m2}V_Z \quad \text{where } K = \frac{\beta N}{2^n} \tag{1}$$

where K is the n-bit digital control input, n is number of control lines in DCU, and N is the digital control word.

The other port characteristics of DC-VDTA is given as

$$I_N = I_P = 0, \quad I_Z = g_{m1}(V_N - V_p) \tag{2}$$

where g_{m1} and g_{m2} are the transconductances of DC-VDTA.

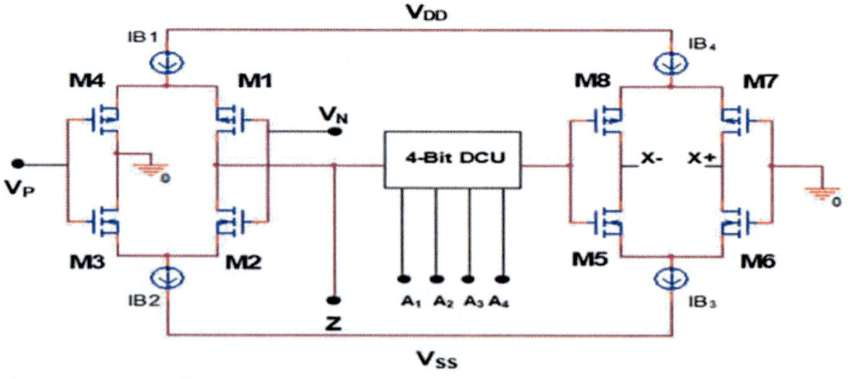

Fig. 1. CMOS structure of DC-VDTA

Table 1 Aspect ratio

Transistor	Width (μm)	Length (μm)
M2,M3	3.6	0.36
M1,M4	16.64	0.36
M5,M6	3.6	0.36
M7,M8	16.64	0.36

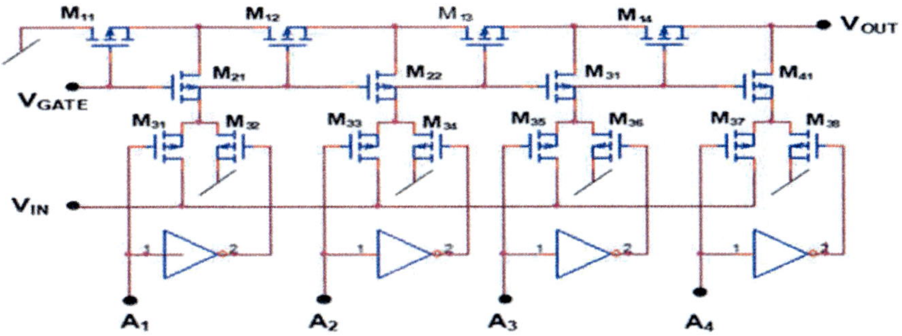

Fig.2. 4-bit DCU

3 Digital Control Unit (DCU)

In this manuscript, a 4-bit DCU is utilized. The CMOS structure of DCU is illustrated in Fig. 2. The output of DCU is determined as

$$V_{\text{out}} = \frac{V_{\text{in}}}{2^4}\left(A_1 + 2A_2 + 2^2 A_3 + 2^3 A_4\right) = K V_{\text{in}} \tag{3}$$

where A_1, A_2, A_3 and A_4 are the bit values of digital control word.

4 Proposed QO

The proposed structure of QO using DC-VDTA is illustrated in Fig. 3. Subsequent by applying KCL at different node, we get the characteristic equation,

$$s^2 + \frac{s}{C_1}\left(\frac{1}{R_1} - g_{m1}\right) + \frac{\beta N g_{m1} g_{m2}}{2^n C_1 C_2} = 0 \tag{4}$$

The CO and FO are determined from Eq. (4)

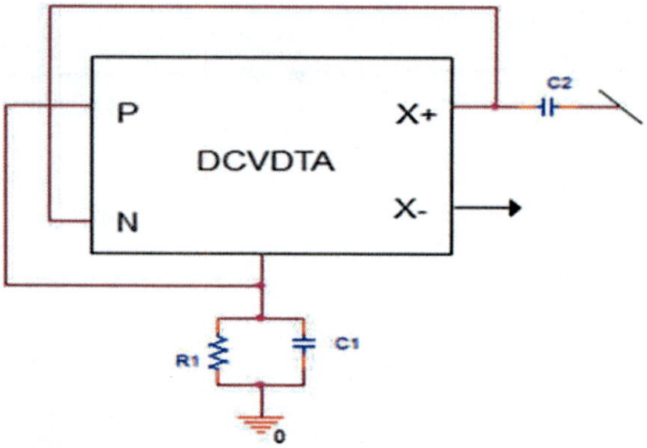

Fig. 3 Proposed QO

$$\text{CO}: \quad \frac{1}{R_1} - g_{m1} \leq 0 \quad \text{and} \quad \text{FO}: \quad \omega = \sqrt{\frac{\beta N g_{m1} g_{m2}}{2^n C_1 C_2}} \tag{5}$$

The proposed circuit is able to provide the quadrature current signal, and their amplitude ratio is mentioned in Eq. (6)

$$\frac{I_{C2}(j\omega)}{I_{C1}(j\omega)} = K \frac{g_{m2}}{\omega C_2} \angle -90^\circ, \quad \left| \frac{I_{C2}(j\omega)}{I_{C1}(j\omega)} \right| = \sqrt{\frac{2^n g_{m2}}{\beta N g_{m1}}} \tag{6}$$

5 Non-ideal and Sensitivity Analysis

A non-ideal analysis of the proposed DC-VDTA is also investigated with considering the parasitic port element. A DCVDTA with parasitic element is shown in the Fig. 4. By considering the parasitic elements, the characteristic equation will be given as

$$\begin{aligned} s^2(C_1 + C_{p1})(C_2 + C_{p2}) + s\Bigg\{(C_1 + C_{p1})\frac{1}{R_{p2}} \\ + (C_2 + C_{p2})\left(\frac{1}{R_1} + \frac{1}{R_2} - g_{m1}\right)\Bigg\} + \frac{1}{R_1 R_{p2}} + \frac{1}{R_{p1} R_{p2}} \\ - \frac{g_{m1}}{R_{p2}} + K g_{m1} g_{m2} = 0 \end{aligned} \tag{8}$$

As $C_1 \gg C_{p1}$ and $C_2 \gg C_{p2}$ then the capacitance C_1 and C_2 will eliminate the effect of C_{p1} and C_{p2} respectively. So, the parasitic capacitance does not affect the functioning of QO, but the parasitic resistance will affect its functioning.

The active and passive sensitivities of proposed oscillator are given as

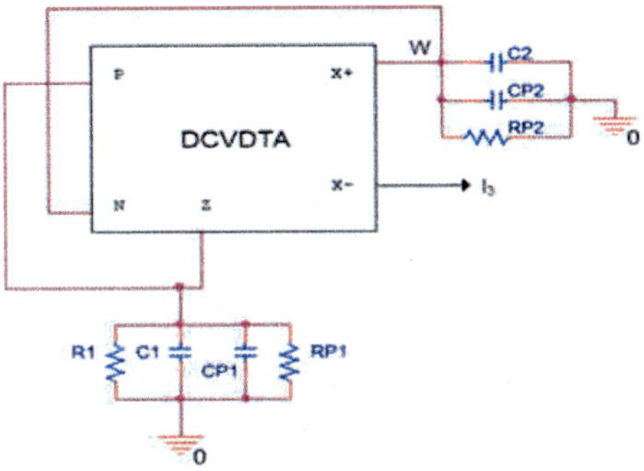

Fig. 4 Non-ideal model of DC-VDTA

$$S^{\omega}_{C_1,C_2} = -0.5, \quad S^{\omega}_{g_{m1},g_{m2}} = 0.5 \tag{9}$$

So, the sensitivity values are low and lie within the specified range.

6 Simulation Results

To observe the functionality of proposed QO, the Pspice simulation is carried out using 180nm technology. To design the QO 4-bit DCU is used with the components values are $C_1 = C_2 =$ 50pF, $R_1 =$ 1.7KΩ, $V_{DD} =$ 0.9V, $V_{SS} =$ -0.9V, and for $I_{B1} = I_{B2} = I_{B3} = I_{B4} =$ 150μA the value of $g_{m1} = g_{m2} =$ 611μA/V. The 4-bit DCU will help to generate sixteen different frequency signals. The transient response of designed QO for control words N =1 and 6 are demonstrated in Fig. 5. The steady state characteristic is also been shown in Fig. 6. The quadrature outputs I_{C1} and I_{C2} are having a phase shift of 89.3°. The variation in calculated and measured frequency is also shown in Fig. 7 that indicates that the both values are closely related.

7 Conclusion

A digitally controlled VDTA (DC-VDTA) and its application as digitally controllable QO have been reported for the first time. The proposed QO requires only one DC-VDTA and three grounded passive elements that make the circuit suitable for fabrication. The 4-bit DCU can generate sixteen different frequencies of quadrature output for different combinations of control word. The FO and CO can be controlled electronically and independently. The FO can also be modulated through the both grounded capacitors. The functioning of the circuit is validated through the Pspice simulations.

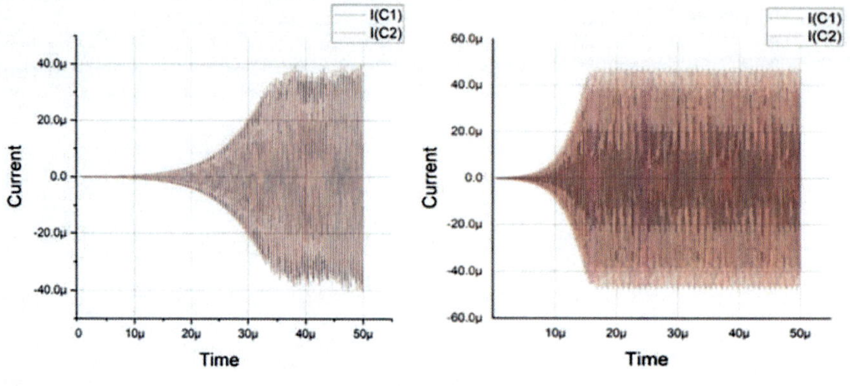

(a)N=0001 (b)N=0110

Fig. 5 Steady state response of the QO for **a** $N = 1$ **b** $N = 6$

Fig. 6 Steady state response

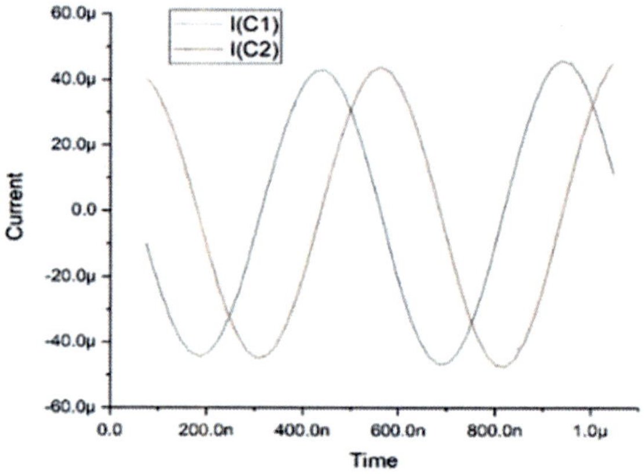

Fig. 7 Frequency Deviation

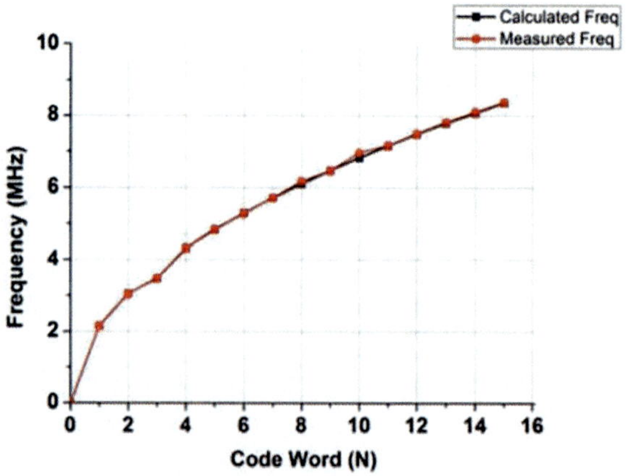

References

1. Horang JW (2005) Current conveyors based all pass filters and quadrature oscillators employing grounded capacitors and resistors. Comput Electric Eng 31:81–92
2. Prasad D, Singh R, Ranjan A, Kumar HT (2020) Grounded capacitors single resistance controlled oscillator using single FTFNTA. Indian J Pure Appl Phys 58(7):525–530
3. Keskin AU, Biolek D (2006) Current-mode quadrature oscillator using current differencing transconductance amplifiers (CDTA). IEE Proc Circuits Devices Syst 153(3):214–218
4. Jaikla W, Siripruchyanun M, Bajer J, Biolek D (2008) A simple current-mode quadrature oscillator using single CDTA. Radioengineering 17(4):33–40
5. Lahiri A (2009) New current-mode quadrature oscillator using CDTA. IEICE Electron Express 3(3):135–140
6. Tangsrirat W, Tanjaroen W (2010) Current-mode sinusoidal quadrature oscillator with independent control of oscillation frequency and condition using CDTAs. Indian J Pure Appl Phys 48(5):363–366
7. Maheshwari S, Chaturvedi B (2011) High output impedance CMQOs using DVCCs and grounded components. Int J Circuit Theory Appl 39(4):427–435
8. Jin J, Wang C (2012) Single CDTA-based current-mode quadrature oscillator. Int J Electron Commun (AEU) 66(11):933–936
9. Li YA (2014) Electronically tunable current-mode biquadratic filter and four-phase quadrature oscillator. Microelectron J 45(3):330–335
10. Jin J, Wang C (2014) Current-mode universal filter and quadrature oscillator using CDTAs. Turkish J Electric Eng Comput Sci 22(2):276–286
11. Minhaj N (2007) Current conveyor-based voltage-mode two-phase and four-phase quadrature oscillators. Int J Electron 94(7):663–669
12. Tangsrirat W, Pisitchalermpong S (2007) CDBA-based quadrature sinusoidal oscillator. Frequenz 61(3–4):102–104
13. Tangsrirat W, Prasertsom D, Piyatat T, Surakampontorn W (2008) Single-resistance-controlled quadrature oscillator using current differencing buffered amplifiers. Int J Electron 95(11):1119–1126
14. Tangsrirat W, Pukkalanun T, Surakampontorn W (2008) CDBA-based universal filter and quadrature oscillator. Active Passive Electron Components 6, Article ID 247171
15. Herencsar N, Koton J, Vrba K, Lahiri A (2009) New voltage-mode quadrature oscillator employing single DBTA and only grounded passive elements. IEICE Electron Express 6(24):1708–1714
16. Maheshwari S, Mohan J, Chauhan DS (2010) High input impedance voltage-mode universal filter and quadrature oscillator. J Circuits Syst Comput 19(7):1597–1607
17. Herencsar N, Vrba K, Koton J, Lahiri A (2010) Realisations of single-resistance-controlled quadrature oscillator using a generalised current follower transconductance amplifier and a unity-gain voltage-follower. Int J Electron 97(8):897–906
18. Horng JW (2011) Quadrature oscillators using operational amplifiers. Active Passive Electron Components, 4, Article ID 320367
19. Sotner R, Herencsar N, Jerabek J, Dvorak R, Kartci A, Dostal T, Vrba K (2013) New double current controlled CFA (DCC-CFA) based voltage-mode oscillator with independent electronic control of oscillation condition and frequency. J Electric Eng 64(2):65–75
20. Khateb F, Jaikla W, Kubanek D, Khatib N (2013) Electronically tunable voltage-mode quadrature oscillator based on high performance CCCDBA. Analog Integr Circuits Signal Process 74(3):499–505
21. Yesil A, Kacar F, Gurkan K (2016) Design and experimental evaluation of quadrature oscillator employing single FB-VDBA. J Electric Eng 67(2):137–142
22. Arora TS, Gupta S (2018) A new voltage-mode quadrature oscillator using grounded capacitors: an application of CDBA. Eng Sci Technol Int J 21(1):43–49
23. Banerjee K, Singh D, Paul SK (2019) Single VDTA based resistorless quadrature oscillator. Analog Integr Circuits Signal Process 100(2):495–500

24. Horng JW, Hou CL, Chang CM, Chou HP, Lin CT, Wen YH (2006) Quadrature oscillators with grounded capacitors and resistors using FDCCIIs. ETRI J 28(4):486–494
25. Lahiri A (2009) Novel voltage/current-mode quadrature oscillator using current differencing transconductance amplifier. Analog Integr Circuits Signal Process 61(2):199–203
26. Maheshwari S, Khan IA (2007) Novel single resistor controlled quadrature oscillator using two CDBAs. J Active Passive Electron Devices 2:137–142
27. Srivastava M, Prasad D (2016) VDCC based dual-mode quadrature sinusoidal oscillator with outputs at appropriate levels. Theor Appl Electric Eng 14(2):168–177
28. Prasad D, Singh S, Kumari A, Srivastava M (2017) A novel digitally programmable quadrature oscillator using DCVDGA. In: 2017 international conference on applied system innovation (ICASI). IEEE, Sapporo, Japan, 1391–1394
29. Yeşil A, Kaçar F, Kuntman H (2011) New simple CMOS realization of voltage differencing transconductance amplifier and its RF filter application. Radioengineering 20(3):632–637

Investigation of 4 × 4 Spatial Multiplexed Transceiver FSO Link Under Strong Turbulence Condition

Shivaji Sinha and Chakresh Kumar

Abstract One of the major issues faced by free space optical link is the signal strength variation at the receiver end due to atmospheric conditions. The performance of the communication link is degraded with the increase in optical link in terms of bit error rate (BER). In this paper, the performance of 4 × 4 spatial multiplexed 10 Gbps outdoor optical link is investigated for the propagation range up to 45 km under strong atmospheric turbulence which is characterized by gamma-gamma distribution. The proposed spatial multiplexed FSO link along with erbium-doped fiber amplifier (EDFA) at the receiver side is found to be useful to mitigate the turbulence induced signal fading. Performance has been reported on the basis of quality factor (Q), total received power and the BER of the received optical signal for different propagation distance. The results for the proposed multiplexed FSO system with QAM modulation scheme can significantly lower the BER nearly 10^{-6} for 45 km link range.

Keywords Free space optics (FSO) · Spatial multiplexing · Atmospheric turbulence · Quality factor (Q) · Bit error rate (BER)

1 Introduction

Free space optical (FSO) communication has been proved as a cutting edge promising solution to meet growing demand of huge bandwidth, high speed data transmission and channel capacity for various applications especially in multimedia services [1]. Although it offers many advantages such as license free spectrum, minimum power consumption and excellent security, the atmospheric turbulence induce scintillation effect during beam propagation [2, 3]. This result in attenuation which varies from

S. Sinha (✉)
University School of Information, Communication and Technology, New Delhi, India
e-mail: chakreshk@gmail.com

C. Kumar
Guru Gobind Singh Indraprastha University, New Delhi, India
e-mail: chakreshk@ipu.ac.in

A. Rani et al. (eds.), *Signals, Machines and Automation*, Lecture Notes in Electrical Engineering 1023, https://doi.org/10.1007/978-981-99-0969-8_49

30 dB/Km to 300 dB/Km depending on the weather conditions. Air pockets formation due to temperature difference and the wind velocity results in refractive index variations and induce atmospheric turbulence known as scintillation. The received signal intensity changes randomly at the receiver and increases with change in link distance. This will ultimately affects the BER performance of the overall system [4]. The best suitable wavelength candidate which found to minimize the losses in this scenario is 1550 nm. The amount of scintillation is measured in terms of normalized intensity variance given by Eq. 1

$$\sigma_I^2 = \frac{\langle I^2 \rangle - \langle I \rangle^2}{\langle I \rangle^2} \tag{1}$$

where I is the received optical signal intensity. Strong scintillation characterized by $\sigma_I^2 > 1$ is considered in our FSO model [7]. The received optical power P_R for the transmitted power P_T is given by Eq. 2

$$\mathrm{P}_R = 10 - \left(\frac{A_a}{\pi \left[\frac{\theta}{2} \times L\right]^2} \right) \times \eta \times 10^{-\left(\frac{\alpha L}{10}\right)} P_T + P_{\text{back}}. \tag{2}$$

Equation 2 can be rearrange by

$$P_r = 10 - (\alpha_{\text{geo}} + \alpha_{\text{att}}) P_T + P_{\text{back}}, \tag{3}$$

where $\alpha_{\text{att}} = 4.34(\ln\langle I \rangle - 2.17\sigma_I^3)$.

In above equation, the receiver aperture area is A_a, the beam divergence angle is θ and the effective efficiency is η. The total channel attenuation for the propagation distance L is α dB/Km. Here, P_{back} is considered as the background radiation noise at the receiver.

Previous research has proposed various techniques to mitigate the effect of atmospheric turbulence, but the space multiplexing is reported to lead higher Q factor and lower BER in wireless optical link. MIMO optical wireless link using OOK and PPM modulations over K distributed Rayleigh and log normal fading channels has been reported to offer high diversity gain and power efficiency compared to single-input single-output (SISO) optical link. Although the PPM found to offers low bandwidth and the OOK is power inefficient scheme [5].

The proposed model described in next section is analyzed using spatial multiplexing. The QAM modulation offers high optical data rate and excellent spectral efficiency compared to OOK and PPM schemes [6]. At the detector side, we opt to choose the avalanche photo detector (APD) in place of PIN detector because of its better average BER performance under the gamma-gamma distribution [7].

Rest of this article is organized as follows. In Sect. 2, the proposed spatial multiplexed system is explained in detail while the Sect. 3 presents the simulation parameter taken in our system. The simulation results are compared and also reported in this section. Section 4 concludes the presented work in brief.

2 Proposed System Model Description

Figure 1 presents a 4 × 4 spatial multiplexed transceiver FSO system consist of an optical transmitter, FSO channel and an optical receiver. The pseudo random bit sequence block (PRBS) generates the input binary information which further converted into electrical format by the non-return to zero (NRZ) section [8]. This gray coded binary electrical stream is mapped into QAM-4 signal $x(t)$which composed of two independent in-phase $x_I(t)$ and quadrature phase $x_Q(t)$ components [6] and expressed by

$$x(t) = x_I(t)\cos(2\pi f_c t) - x_Q(t)\sin(2\pi f_c t). \tag{4}$$

In the above Eq. 4, $x_I(t) = \sum_{-\infty}^{\infty} a_i(t)p(t - iT_s)$ and $x_Q(t) = \sum_{-\infty}^{\infty} b_i(t)p(t - iT_s)$ are in-phase and Q-phase components of $x(t)$. Where $p(t)$ is the unit strength gate pulse of symbol duration T_s and $a_i(t)$, $b_i(t)$ are the ith symbol amplitude of I and Q components, respectively. The subcarrier frequency of QAM is f_c. The intensity modulation (IM), the QAM signal is represented by Eq. 5

$$s(t) = P_s[1 + \mu x(t)], \tag{5}$$

where $\mu = 1$ is the modulation index and the average transmitted optical power for each symbol is P_s. The narrow beam of independent modulated components is transmitted through the turbulent channel to minimize the link losses and is modeled by intensity fluctuated gamma-gamma distribution.

$$f_I(I) = \frac{2(ab)^{\left(\frac{a+b}{2}\right)}(I)^{\left(\frac{(a+b)}{2}\right)-1}K_{a-b}(2\sqrt{abI})}{\Gamma(a)\Gamma(b)}. \tag{6}$$

In Eq. 6, $K_m(.)$ indicates the second kind modified Bessel function of order m and $\Gamma(.)$ shows the gamma function. The size of turbulence cells is decided by the

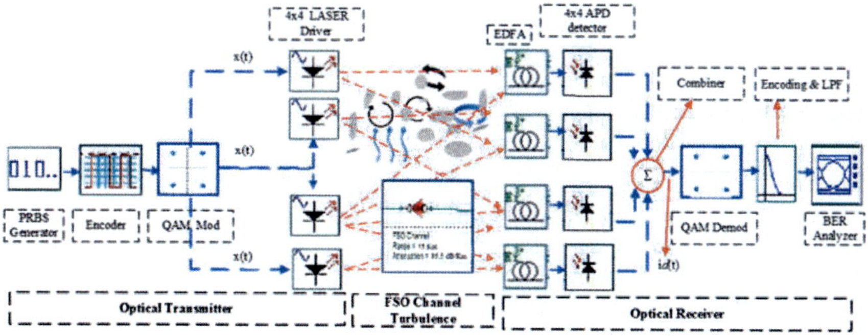

Fig. 1 Spatial multiplexed 4 × 4 optical link in strong atmospherics turbulence

shaping parameter ‘a’ and ‘b’. At the receiver side, these distorted components are amplified by the EDFA before detection and then converted into electrical signal by the APD array. The output of APD current components is summed up by combiner and given by

$$i_d(t) = R\alpha P_s \mu x(t) \sum_{n=1}^{N} \frac{I_N}{N} + n(t), \tag{7}$$

where α and R are the channel attenuation coefficient and APD detector responsivity, respectively. I_N is the optical intensity in individual branch, and the white Gaussian noise (AWGN) with variance $\sigma_n{}^2$ is the $n(t)$ [9].

The next section of this paper discusses about the design parameters, and the numerical results obtained from the simulation results in terms of Q factor and average BER parameters.

3 Simulation Parameters and Result Discussion

This section investigates the simulation results obtain for the simplex, duplex and the 4×4 spatial multiplexed transceivers FSO system on the basis of Q factor, received optical power and BER. The proposed system parameters are shown in Table 1. A 10 Gbps FSO link operating at 1550 nm wavelength with refractive index structure parameter 10^{-13} $\text{m}^{-2/3}$ and variable link length up to 45 km is simulated using Optisystem environment.

Table 1 System simulation parameters

Parameters	Symbol	Values
Transmission rate (Gbps)	B_T	10
Link distance (Km)	L	0- 45
Operative wavelength (nm)	λ	1550
Responsivity (A/W)	R	1
Aperture diameter (m)	D	0.04
Extinction coefficient (dB/Km)	β_v	10^{-1}
Divergence angle (rad)	φ	10^{-3}
Transmitted average power (dBm)	P_s	0
EDFA amplifier gain	a	3
Attenuation level (dB/Km)	α	85.5
Antenna aperture area (cm^2)	A	400
Refractive index structure parameter ($\text{m}^{-2/3}$)	$C_n{}^2$	10^{-13}

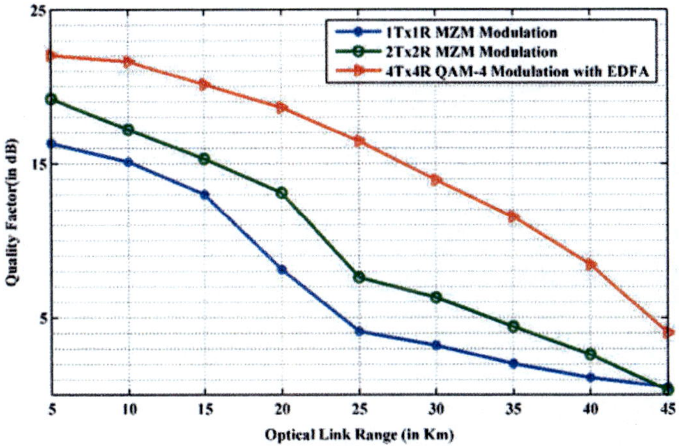

Fig. 2 Q factor versus optical link range for NxN FSO transceivers system

As shown in Fig. 2, the Q factor decreases continuously from [16.3–0.5]dB, [19.2–0.3]dB and [2.2–4]dB for 1 × 1 Tx/Rx, 2 × 2 Tx/Rx and 4 × 4 Tx/Rx pairs, respectively, with increase in link range from [5–45]Km. The results shows that the 4 × 4 spatial multiplexed system offer an improved performance of 8.8 dB compare to 2 × 2 system and 12.3 dB better than simplex system at 35 km link range. At 45 km 1 × 1 and 2 × 2, system performance becomes almost same but 4 × 4 still offers improved Q factor.

The received optical power decreases from [−15 to −40] dBm for 1 × 1 FSO system, from [−5.3 to 30] dBm for 2 × 2 and [−0.7 to −35] dBm for 4 × 4 transceivers FSO system, respectively, as shown in Fig. 3. The 4 × 4 FSO system performs outstanding compare to other transceivers system. Between [20 and 35] Km distances, the 4 × 4 FSO optical receiver is showing almost 3 dBm improvement in received power level compare to 2 × 2 system and 21 dBm improvement with respect to 1 × 1 system.

Figure 4 illustrates that with increase in propagation distance, the performance of all the three system degraded exponentially. Under strong turbulence conditions, 4 × 4 transceivers system offers better performance around 10^5 compare to 1 × 1 system while the 2 × 2 system offers approximately 10 times better performance than simplex FSO system between [25–35] Km ranges. The results also show that for short link range 4 × 4 system offers even better performance compare to large distance.

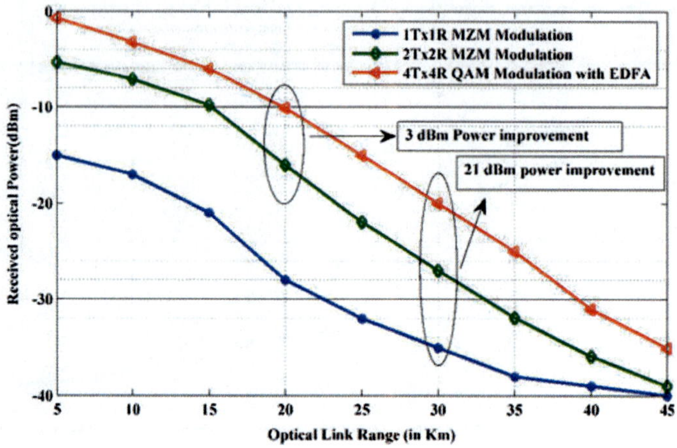

Fig. 3 Received power versus optical link range for NxN FSO transceivers system

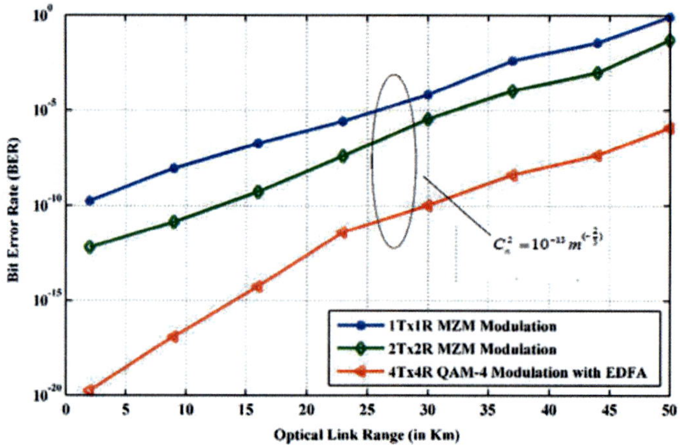

Fig. 4 BER versus optical link range for NxN FSO transceivers

4 Conclusion

The improved performance of a 10 Gbps multi-transceivers FSO system for MZM and QAM modulation schemes with EDFA as a preamplifier at the detector end is compared in this paper. The results explained in above section indicate that the link range can be extended up to 45 km using spatial multiplexed transceiver system. Therefore, increasing the number of transceivers system from 1 to 4, the Q factor increases from 0.4 to 4 dB, the received optical power increased by 5 dB and BER improved 10_6 times for 45 km.

References

1. Al-Gailani SA et al (2021) A survey of free space optics (FSO) communication systems, links, and networks. IEEE Access 9:7353–7373
2. Liu W, Ding J, Zheng J, Chen X (2020) Relay-assisted technology in optical wireless communications: a survey. IEEE Access 8:194384–194409
3. Vats A, Kaushal H (2014) Analysis of free space optical link in turbulent atmosphere. Optik-Int J Light Electron Opt 125(12):2776–2779
4. Kumar N, Rana AK (2013) Impact of various parameters on the performance of free space optics communication system. Optik 124(22):5774–5776
5. Tsiftsis TA, Sandalidis HG, Karagiannidis GK, Uysal M (2008) In FSO links with spatial diversity over strong atmospheric turbulence channels. IEEE international conference on communications, 5379–5384
6. Yang L, Gao X, Alouini M (2014) Performance analysis of relay-assisted all-optical FSO networks over strong atmospheric turbulence. J Light Technol 32(23):4011–4018
7. Dat PT, Kazaura ABK, Wakamori K, Suzuki T, Matsumoto M, Higashino T, Tsukamoto K, Komaki S (2009) Studies on characterizing the transmission of RF signals over a turbulent FSO link. Opt Express 17(10):7731–7743
8. Vu BT, Thang TC, Pham AT (2016) BER analysis of MIMO/FSO systems using rectangular QAM over gamma- gamma channels. ICUFN, 2013, pp 370–374
9. Miridakis NI, Vergados DD, Michalas A (2018) All-optical FSO relaying under mixture-gamma fading channels and pointing errors. Wireless Personal Commun 771–781

Current-Mode Full-Wave Rectifier Based on Transconductance-Boosted Bulk-Driven CDTA and Two Diodes

Shailendra Bisariya and Neelofer Afzal

Abstract A full-wave rectifier employing a transconductance-boosted bulk-driven current differencing transconductance amplifier (BDCDTA) is proposed. The BDCDTA is a widely acceptable design for low-voltage and low-power applications. This BDCDTA is first designed to achieve high transconductance by using a split transistor network for sufficient gain as per the requirement. The full-wave rectifier circuit which contains only one such transconductance-boosted BDCDTA and two diodes is proposed which can be easily tuned for its gain by properly selecting the number of transistors in the split network. The full-wave rectifier circuit performance is depicted through PSPICE simulations with 0.18 um CMOS process.

Keywords Current mode · Full-wave rectifier · Bulk driven · Current differencing transconductance amplifier

1 Introduction

Bulk-driven current differencing transconductance amplifier (BDCDTA) [1] is a circuit that operates with very low voltage and consumes much less power which is highly required particularly in the case of using portable devices. A number of such active elements have been reported in the technical literature as in [2–6]. Generally, for analog signal processing circuits, the techniques that are used for this purpose are the subthreshold technique [2], floating gate technique [3], quasi-floating gate technique [4], and bulk-driven technique [5, 6]. In terms of IC fabrication, the bulk-driven technique consumes a smaller area of chip in comparison to other mentioned techniques as it contains a comparatively less number of transistors and has less associated capacitance value. However, it also has a drawback of 3–4 times smaller transconductance and suffered from a higher amount of input referred noise in comparison to the basic gate driven approach.

S. Bisariya (✉) · N. Afzal
ECE Department, Faculty of Engineering and Technology, Jamia Millia Islamia University, New Delhi 110025, India
e-mail: shailendraindian@gmail.com

A. Rani et al. (eds.), *Signals, Machines and Automation*, Lecture Notes in Electrical Engineering 1023, https://doi.org/10.1007/978-981-99-0969-8_50

The transconductance improvements are essentially required for a number of applications so as to provide more gain to properly strengthen the weak signals for further processing. A full-wave rectifier circuit is required in a number of electronic devices for converting low-voltage alternating source into high-voltage direct current source with sufficiently high amplitude of the output signal. A BDCDTA is suitable for such cases, but it suffers from the drawback of having much less transconductance. By improving the transconductance of basic BDCDTA, it can be widely accepted for such application areas.

Full-wave rectifier design proposed to date [8–14] does not include the transconductance improvement in parallel to rectification. CDTA-based full-wave rectifier circuit proposed in [8] and [9] which uses two transistors and in [10] which utilizes bipolar technology instead of MOS, all consume comparatively higher power from the supply and no current gain is provided. Winner takes all circuit using bulk-driven concept proposed in [11] which shows good results, especially for half-wave rectification but again lacks in providing any transconductance improvement. A CCII-based full-wave rectifier as proposed in [12] includes a total of six more transistors and two resistors to provide the rectified output while the circuit proposed in [14] also includes a higher number of components for implementation. OTRA-based full-wave rectifier as proposed in [13] utilizes four diodes and works on a 2.5 V power supply.

Our design first proposes an improvement in the transconductance of basic BDCDTA [1] by employing a split transistor network approach in the OTA circuit, and then, we proposed a full-wave rectifier circuit for analog signal processing applications based on this modified design of BDCDTA. This full-wave rectifier circuit is thus able to rectify very small alternating currents and can provide higher output current in comparison to basic BDCDTA.

This paper is organized as follows: in Sect. 2, first of all, BDCDTA's brief description is provided. In Sect. 3, the transconductance-boosted BDCDTA implementation approach is illustrated. In Sect. 4, a full-wave rectifier circuit based on this transconductance-boosted BDCDTA is proposed, and in Sect. 5, simulated results of the proposed transconductance-boosted full-wave rectifier circuit are described. Finally, a conclusion is drawn for the same at the end.

2 Bulk-Driven Current Differencing Transconductance Amplifier (BDCDTA)

The BDCDTA was introduced in 2011 [1]. The schematic symbol of the BDCDTA is shown in Fig. 1. It can be built up from current followers and mirrors or by using three OTAs such that the first two OTAs form the current conveyor section to produce the difference of current followed by the third OTA to provide sufficient transconductance gain.

The transistor level implementation of the BDCDTA is shown in Fig. 2. It consists of two bulk-driven CCIIs and a bulk-driven OTA with dual output. The current

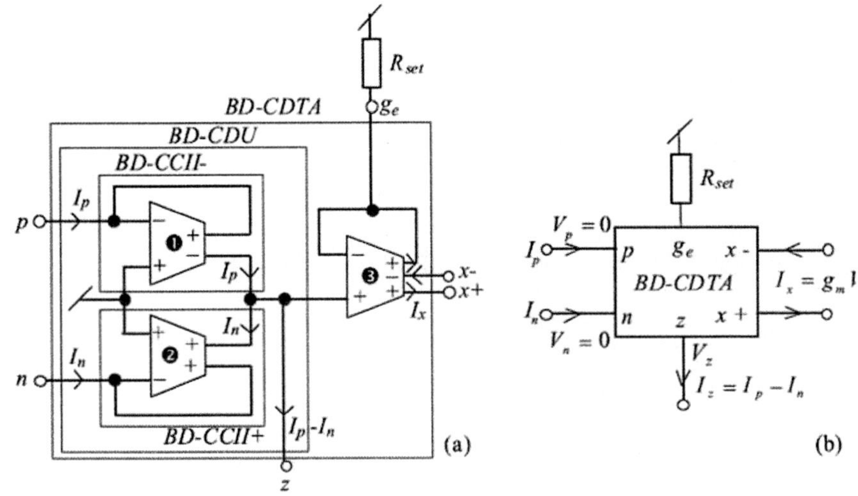

Fig. 1 **a** CDTA built from bulk-driven OTAs, **b** its schematic symbol [1]

differencing unit consists of the two current conveyors. For the first CCII, the applied current I_n moves toward its $z+$ terminal, and for the lower CCII, the current I_p moves toward its z-terminal, but both are in the opposite direction. Hence, the current, obtained at the Z terminal of this CDTA, comes out to be simply the difference between these two currents I_p and I_n.

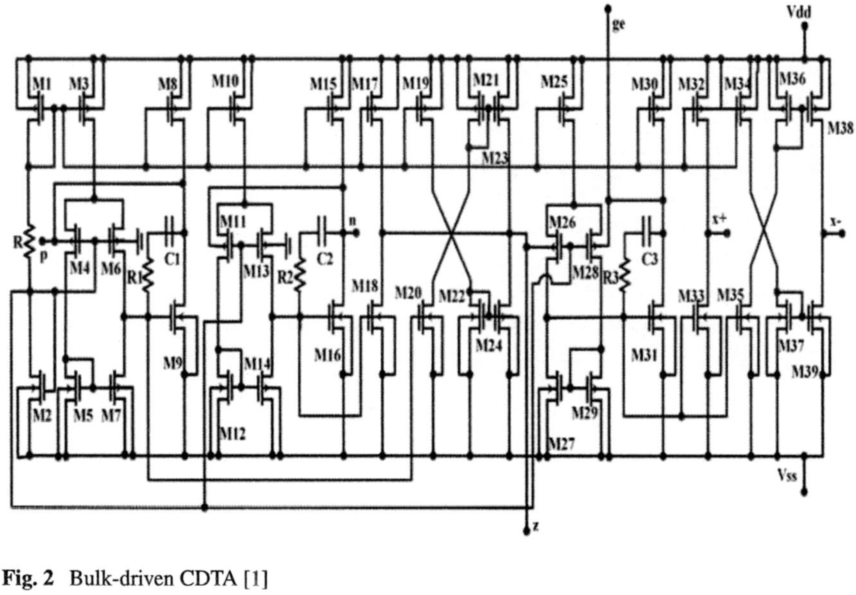

Fig. 2 Bulk-driven CDTA [1]

The transconductance of the output OTA stage for this basic BDCDTA was set via an external resistor R_{set}. Such that, $gm_{set} = 1/R_{set}$. This topology improves the linearity range but limits the value of transconductance. This BDCDTA works with a power supply of ± 0.6 V, and very low-power dissipation in the range of 143 μW is achieved in this case.

3 Transconductance-Boosted BDCDTA Approach

In order to achieve a higher transconductance value, the very first step is to obtain independency from R_{set}. This will reduce the linearity range, which can again be obtained by a suitable selection of bias currents. The elimination of R_{set} can abruptly increase the transconductance up to the gm value of the OTA. This will however reduce the frequency range of the device.

To further increase the value of transconductance, this circuit can be modified again by connecting the 'N' number of symmetrical MOS transistors in parallel as shown in Fig. 3 as 'Network 1' and 'Network 2' in place of M26 and M28 transistors of CDTA of Fig. 2 [7]. The use of 'Network 1' and 'Network 2' is useful because all the MOSFETs which are in parallel draw the same amount of current since their gates are connected, so V_{GS} is the same and all are kept working in the saturation region. Thus, the total current coming out of either 'Network1' or 'Network 2' will be the addition of individual current drawn by each MOSFET. Thus, even if the input is not changed, the output currents are boosted. Hence, we can achieve a very high value of transconductance (gm) without having much power dissipation.

This approach improves the transconductance value of bulk-driven CDTA which can be easily set as per the design requirement by suitably selecting the value of 'N', i.e., the total number of transistors in the split network. It is observed that the value of 'n' can give a good result for a range between 2 and 4.

The basic BDCDTA was designed using this transconductance-boosted scheme as shown in Fig. 4. Initially, the circuit was designed with $N = 1$ only to check the improvement in transconductance by eliminating R_{set} only. This circuit was further redesigned with a split transistor network in the OTA section.

The simulation result of this BDCDTA is shown in Fig. 5. The simulation was done to obtain the frequency response of the same first by eliminating the R_{set} only. The result shows a significant improvement in the value of the transconductance as expected. As shown in Fig. 5, the transconductance obtained in this case is 3.33 mS

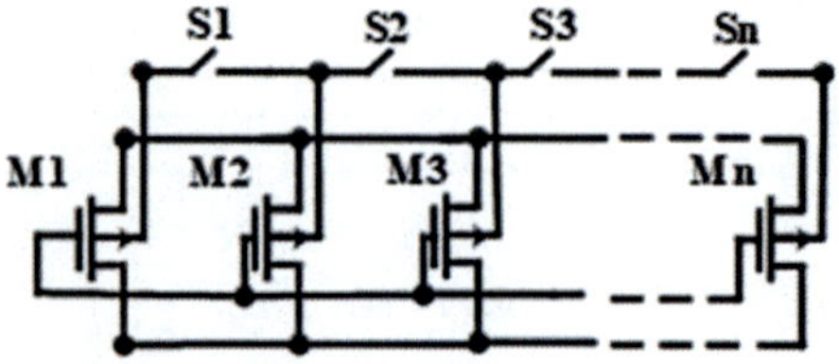

Fig. 3 Bulk-driven N PMOS transistors in parallel with select switches [7]

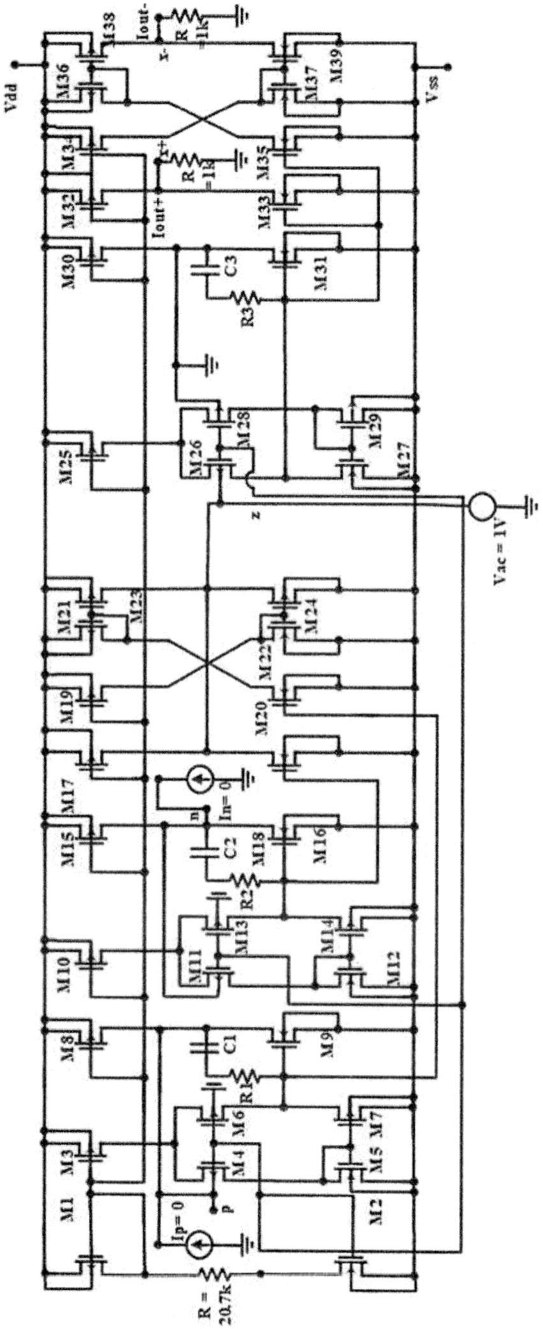

Fig. 4 Schematic of transconductance-boosted basic BDCDTA without split network

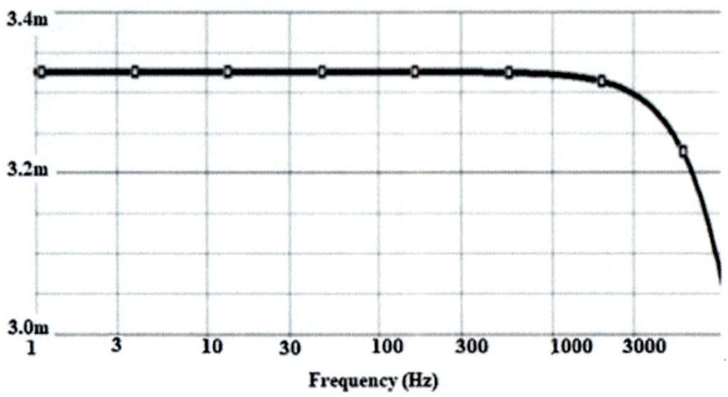

Fig. 5 Frequency response plot for transconductance-boosted basic BDCDTA without split network

which is much higher in comparison to the basic version with a transconductance value of 99 uA only with R_{set} as 10 kΩ.

4 Proposed Full-Wave Rectifier Design Based on Transconductance-Boosted BDCDTA

A full-wave rectifier circuit requires the use of two diodes such that only one diode conducts in the half cycle of an applied alternating input signal and the second diode should conduct for the remaining half cycle. So, two diodes should be connected in opposite manner such that the swing for both half cycles can be obtained in the same direction. This concept is explained in Fig. 6 for the designed transconductance-boosted BDCDTA circuit.

The equations for describing Fig. 6 are presented as follows:

$$I_{x+} = \begin{cases} K I_{\text{in}} & \text{if } I_{\text{in}} > 0 \\ -K I_{\text{in}} & \text{otherwise} \end{cases},$$

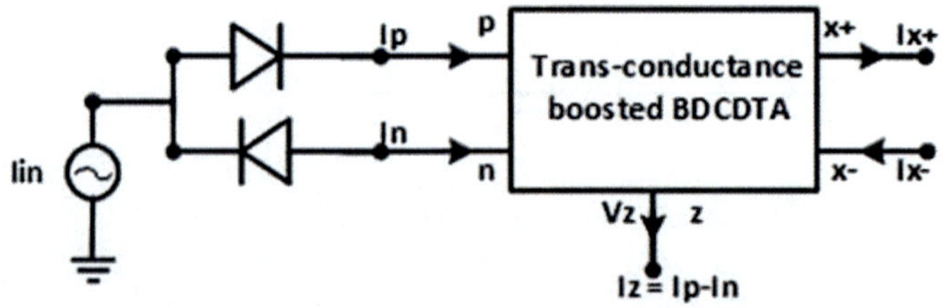

Fig. 6 Full-wave rectifier based on transconductance-boosted BDCDTA and two diodes

$$I_{x-} = \begin{cases} -K I_{\text{in}} & \text{if } I_{\text{in}} > 0 \\ K I_{\text{in}} & \text{otherwise} \end{cases},$$

where I_{in} is the applied value of input current and K is the factor by which its amplitude gets increasing because of TBBDCDTA which further depends upon the number of transistors used in the split network. This number can be set as per the requirement of the application, and hence, suitable improvement in the value with tunability in hands can be obtained easily just by reconfiguring the structure.

The complete schematic of the full-wave rectifier based on this transconductance-boosted BDCDTA is presented in Fig. 7. It has two diodes at the input section which is followed by bulk driven current differential transconductance amplifier which includes a transistor split network in its output OTA section, which can provide sufficiently large transconductance value, and hence, the much higher value of output current in comparison to the basic bulk driven transconductance amplifier. We have chosen from one to four transistors in the split network, and the corresponding improvement is mentioned in the result section for all cases.

Figure 8 shows the simulation results obtained from the designed full-wave rectifier circuit. Results clearly indicate that output current is much large in comparison to the case of basic full-wave rectifier design which utilizes BDCDTA design without transconductance improvement scheme. The results were further verified by varying the transistor count in the split network which proved further enhancement in the output current by increasing the number of transistors in parallel. This approach can be utilized in practical application requirements conveniently just by suitably selecting the number of transistors as per the field requirement.

5 Simulation Results

The PSPICE simulations were done using 180 nm CMOS technology parameters. First, the basic BDCDTA of [1] was simulated with known parameter values. In the second step, the R_{set} was eliminated and biasing conditions were checked for linearity. Then, the achievement in transconductance was measured as plotted in Fig. 5. In the next step, the split network was introduced with variation in the transistor count, and the corresponding increase in transconductance was measured and verified. This step showed a clear picture of transconductance improvement by increasing the transistor count. The measured values of improved transconductance are 3.33 mS for a single OTA without a split network, while it increases from 3.33 mS to 3.96 mS for two transistors in a split network and up to 4.46 mS for four transistors in the network which is much higher in comparison to the transconductance value of basic BDCDTA.

After verifying the design of transconductance-boosted BDCDTA, the schematic of full-wave rectifier was designed and the simulation was done with a supply voltage of 0.6 V only, and two diodes were supplied with an alternating sinusoidal current input of 7 uA only with a frequency of 500 Hz. Results show rectified output with

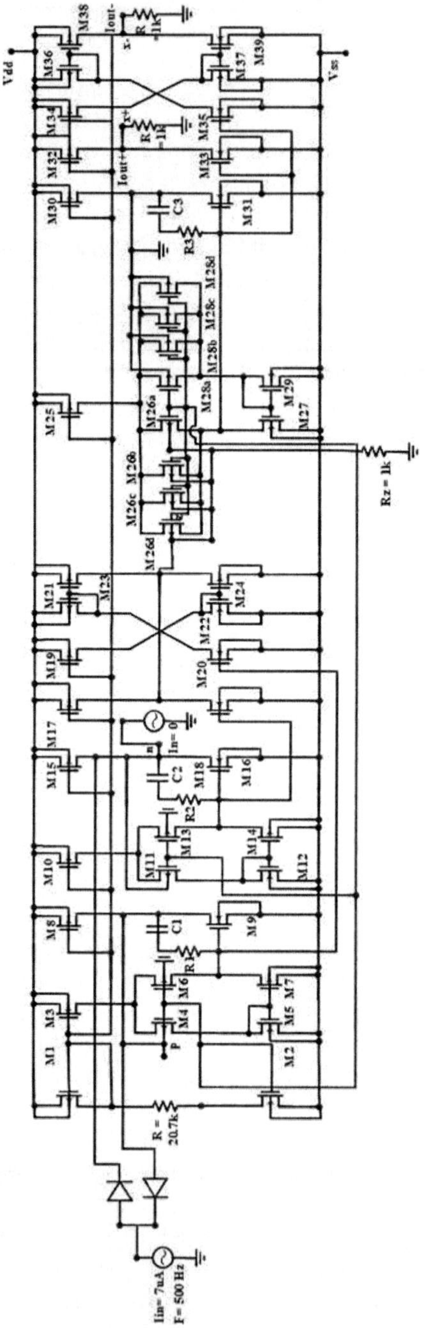

Fig. 7 Complete schematic of full-wave rectifier based on transconductance-boosted BDCDTA and two diodes

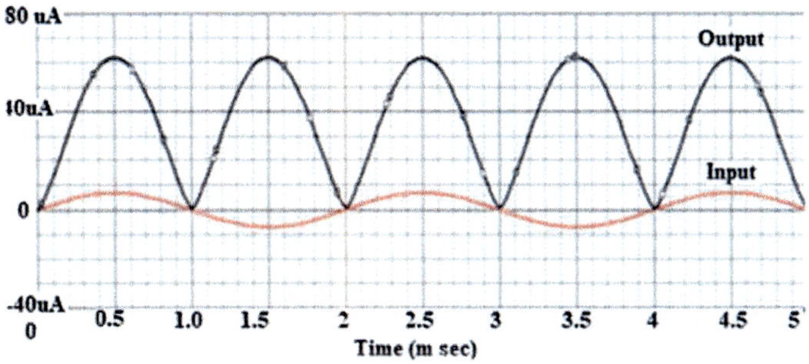

Fig. 8 Simulation results for full-wave rectifier based on transconductance-boosted BDCDTA and two diodes

Table 1 Comparison of proposed TBBDCDTA-based full-wave rectifier with other's work

	Proposed work	R[9]	R[10]	R[12]	R[13]	R[14]
Analog building block (ABB)	TBBDCDTA	CDTA	CDTA	CCII	OTRA	CCII
Number of ABB	1	1	1	2	1	1
Number of diodes	2	–	2	–	4	–
Number of transistors	–	2	–	6	–	4
Number of resistors	1	–	1	2	1	1
DC power supply (V)	0.6	1.8	5	12	2.5	1.5
Technology (μm)	0.18	0.35	Bipolar	Bipolar	0.25	0.35
Power consumption (mW)	0.144	14	–	–	2.31	1.18
Output transconductance improvement	Yes	No	No	No	No	No

an amplified current of value approximately 62 uA which is much larger in comparison to the results obtained from basic BDCDTA design. Table 1 further shows the comparison of this proposed TBBDCDTA-based full-wave rectifier design with other's already designed circuits.

6 Conclusion

The simulation results for full-wave rectifier design clearly indicate a huge improvement in the value of output current and hence justify its viability. This circuit can easily be used for rectification as well as amplification of such alternating sources which have much less amplitude with a lower frequency range and need a larger improvement in its current value in parallel to the rectified output.

Acknowledgements The authors would like to thank the ECE department of Jamia Millia Islamia for providing wonderful facilities to pursue this work.

References

1. Khateb F, Biolek D (2011) Bulk-driven current differencing trans-conductance amplifier. Circuits Syst Signal Process 30:1071–1089
2. Bernal MRV, Celma S, Medrano N, Calvo B (2012) An ultra-low power low-voltage class-AB fully differential op amp for long-life autonomous portable equipment. IEEE Trans Circuits Syst II 59:643–647
3. Khateb F, Khatib N, Kubanek D (2011) Novel low-voltage low-power high precision CCII± based on bulk-driven folded cascode OTA. Microelectron J 42:622–631
4. Moradzadeh H, Azhari SJ (2011) Low-voltage low-power rail-to-rail low- Rx wideband second generation current conveyor and a single resistance controlled oscillator based on it. IET Circuits Devices Syst 5:66–72
5. Kulej J (2013) 0.5-V bulk-driven CMOS operational amplifier. IET Circuits, Devices Syst 7:352–360
6. Raikos G, Vlassis S, Psychalinos C (2012) 0.5V bulk-driven analog building blocks. Int J Electron Commun 66:920–927
7. Rai SK, Gupta M (2015) A trans-conductance boosted CMOS current differencing trans-conductance amplifier (TBCDTA) and its application. Analog Integr Circ Sig Process 84:75–88
8. Kacar F, Basak ME (2014) A new mixed mode full wave rectifier realization with CDTA. J Circuits Syst Comput 23:1450101–10
9. Pandey N, Pandey R (2013) Current mode full wave rectifier based on a single MZC-CDTA. Act Passive Electron Compon 2013:1–5
10. Khateb F, Vavra J, Biolek D (2010) A novel current mode full-wave rectifier based on one CDTA and two diodes. Radioengineering. 19:437–445
11. Khateb F, Spyridon VS (2013) Low-voltage bulk-driven rectifier for biomedical applications. Microelectron J 44:642–648
12. Monpapassorn A (2018) A new current conveyor full-wave rectifier for low frequency/small signal medical applications. Circuits Syst 9:58–65
13. Chien HC (2019) Full-phase operation transresistance-mode precision full-wave rectifier designs using single operational transresistance amplifier. Act Passive Electron Compon 2019:1–18
14. Petrovic PB (2020) Current/voltage mode full-wave rectifier based on a single CCCII. Int J Circuit Theory Appl 48(2):1–14

Effective Coverage Analysis for Wireless Multihop Sensor Network Incorporate Overlapping

Bhawna Kankane, Rajesh Mishra, and Sandeep Sharma

Abstract Wireless multihop sensor network (WMSN) has a wide range of applications in various fields, especially with Internet of Things (IoT) and other smart devices that may include weather forecasting, health monitoring, military surveillance, etc. This paper proposes an analytical model for effective coverage estimation that incorporate a redundant area for a defined region of interest with uniform random distribution for sensor node. Proper coverage is one of the major concerns for WMSN. The model is proposed in two steps. First, the effective coverage area is evaluated with redundant possibility using Euclidean distance between two neighboring nodes and excluding the overlap area to reduce redundancy. Finally, the coverage is estimated with detection probability for a known effective coverage area. The simulation result shows that the proposed model gives better coverage than previous work. It extends the detection probability significantly for a wireless multihop sensor network.

Keywords Effective coverage · Overlap area · Wireless multihop sensor network · Euclidean distance · Detection probability · Boolean sensing model · Sensor nodes

1 Introduction

With emerging technology, miniaturization of the devices has developed an interest in the design of tiny sensors that have a wide range of applications in various field such as military surveillance, industrial application and health monitoring system [1]. Sensors are the basic building block for any wireless network and primarily comprised a sensing unit, transceiver, memory and power unit. In most applications, coverage efficiency reflects the system's performance, especially in IoT-based applications or in other smart devices; thus, it becomes necessary to enhance coverage of such wireless networks. In literature, different approaches with different sensing

B. Kankane (✉) · R. Mishra
Gautam Buddha University, Greater Noida, India
e-mail: bhawnakankane@gmail.com

S. Sharma
Madhav Institute of Technology and Science, Gwalior, India

A. Rani et al. (eds.), *Signals, Machines and Automation*, Lecture Notes in Electrical Engineering 1023, https://doi.org/10.1007/978-981-99-0969-8_51

models have been used to analyze coverage. Sensor nodes' deployment is the key step for a network design, and various deployment strategies have been used to achieve maximum coverage. The wireless sensor network is highly applicable for the hostile region, so random deployment is widely used. Nowadays, coverage has become a major concern of research for researchers. The area of coverage sensed by a sensor is determined by the fraction of sensed area to the total area of the region. In 2016, Katti and Lobiyal [2] studied and discussed different ways to enhance the coverage efficiency using a deterministic model. Jin et al. [3] developed a model for coverage and connectivity analysis under border effect in a circular region. The assumption has taken that the sensor is equipped with an omni-directional antenna with uniform sensing range. Cai et al. [4] have proposed an approach for energy conservation for wireless sensor networks where the sensor nodes are defined in three different modes, i.e., low power, sleep and active modes for a disk sensing model. Sensors in different modes prolong the network lifetime. Similarly, in 2020, Suparna et al. [5] have presented a Monte Carlo simulation approach for area coverage with sensors in three different states, viz, sleep, active and relay. The proposed model establishes connectivity for mobile sink based on the state of the sensors and defined coverage for every established connection. In this article, the node position is predetermined and is not applicable for random distribution. Sensor nodes collect and process sensed data which are further forwarded to the sink node. In literature, various coverage and connectivity-based reliability models have been proposed. In 2013, Liu et al. [6] have done coverage analysis using percolation theory. In 2020, S. Chakraborty et al. [7] proposed an area coverage-based model for multi-state sensor nodes where Voronoi cells are used for sensor nodes' representation. The connectivity between the sensor nodes is shown using Euclidean distance and energy matrix. Author has also applied a sum of disjoint approach to determine area coverage. In 2019, Amutha et al. [8] reviewed different strategies of wireless sensor network to improve coverage and energy efficiency. In 2017, Al-Karaki and Gawanmeh [9] have done a performance evaluation and maximize network endurance with the mentioned parameters like nodes' availability, connectivity and coverage. In 2021 [10], Nagar et al. have formulated an approach to define connectivity in a multihop scenario using node degree distribution and node isolation probability. The author has also considered the border effect. Most of the work has been done for coverage analysis for wireless sensor networks considering different sensing models and different performance parameters. However, none of the researchers has considered a redundant area for coverage. In the proposed model, the effective coverage is estimated by considering the overlap area. The formulating result shows the redundancy is reduced to the greater extent and required coverage estimation is highly improved.

2 System Model

This paper presents a mathematical model for the analysis of effective coverage area without redundancy. The proposed model considers a random square region of area $(A = a^2)$ with dimension 100 x 100 m^2. We first assume that two random nodes 'A' and 'B' as shown in Fig. 1 with equal and uniform sensing range 'R_S' are randomly deployed. Due to random distribution of sensor nodes, some of the regions of ROI are covered by both sensors and considered as redundant area. The redundancy for coverage area depends on the distance between the neighboring nodes which is given by Euclidean distance between the nodes. The redundancy occurs if the distance between the nodes is less than or equal to twice the sensing range. The distance 'D' in Eq. (1) between the two neighboring sensor nodes with coordinates (A_i, A_j) and (B_i, B_j) is determined by the Euclidean distance [11]. The expression for 'D' is given by:

$$D = \sqrt{(B_i - A_i)^2 + (B_j - A_j)^2}. \tag{1}$$

The coverage area by the sensor nodes is estimated by the ratio of the effective area sensed by the sensor to the total area of RoI. In the presence of overlap area, the effective coverage area A_{eff} is evaluated as in Eq. (2).

$$A_{eff} = A_{2S} - A_O, \tag{2}$$

where A_{2S} is sensing area covered by two sensor nodes for uniform sensing range as sensors are representing with a disk sensing model. If the sensing range of a sensor is R_S, then the area covered by each sensor is πRs^2; similarly, for n number of sensors, it will be $(n \times \pi R^2)$. The sensing area for two sensor nodes will be $(2 \times \pi R^2)$, and A_O is defining the overlapping area or common area between the sensors.

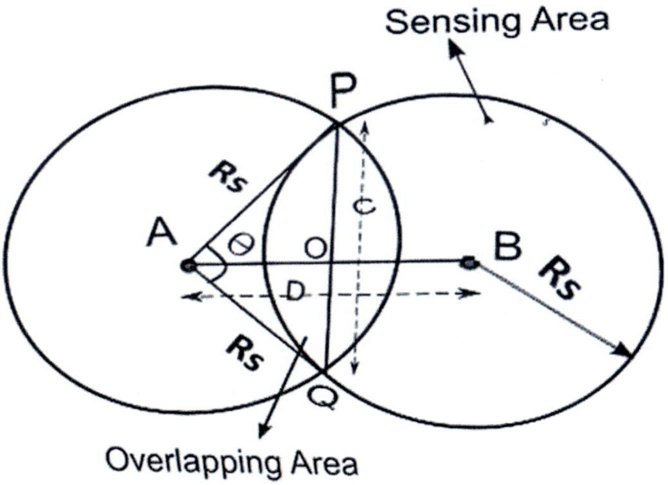

Fig. 1 Sensor node coverage with overlapping area

2.1 The Mathematical Expression for Overlap Area

The overlapping area is decided by the Euclidean distance 'D' and can be mathematically expressed by considering three conditions as discussed below:

$$A_O = \begin{cases} \pi R_S^2 & \text{for } D = 0 \\ 0 & \text{for } D \geq 2R_S \\ A_O & \text{for } D \leq 2R_S \end{cases} \tag{3}$$

Overlapping area can be evaluated if the value of D lies between $0 < D < 2Rs$ and is given by the equation.

$$A_O = \{\text{Area of sector} - \text{Area of triangle}\}, \tag{4}$$

$$A_O = \left\{\widehat{APQ} - \Delta(APQ)\right\} \tag{5}$$

The area of sector is given by $\frac{1}{2}\theta R_S^2$, where the value of θ is evaluated by using cosine rule. Similarly, the area of triangle in the overlap region is evaluated by Heron's formula [12].

$$\Delta APQ = \sqrt{s(s - R_S)^2(s - C)}, \tag{6}$$

where 'C' is the chord length which is estimated as $C = \sqrt{2R_S^2(1 - \cos\theta)}$ and 's' is given by $\frac{2R_S + C}{2}$. By using Eq. (2), the effective coverage area is estimated as

$$A_{eff} = 2 \times \pi R^2 - \frac{1}{2}\left\{\cos^{-1}\left(\frac{2R_S^2 - C^2}{2R_S^2}\right)R_S^2 - \sqrt{s(s - R_S)^2\left(s - \sqrt{2R_S^2(1 - \cos\theta)}\right)}\right\}. \tag{7}$$

2.2 Coverage Estimation Using Boolean Sensing Model

In this model, the sensing area for each sensor is defined by $A_S = \pi R_S^2$. A sensor in the given area can sensed the target if the target exists under the sensing range of a sensor node. The detection probability of a sensor node for a particular target is given as $p_d = \frac{A_S}{A}$, where 'A' is the total area of square region of interest [13]. If incorporating the overlap area between two sensor nodes [14], the detection probability for the system model will be expressed as $p_d = \frac{A_{eff}}{A}$, and hence, the equation will become

$$P_d = \frac{2 \times \pi R^2 - \frac{1}{2}\left\{\cos^{-1}\left(\frac{2R_S^2 - C^2}{2R_S^2}\right)R_S^2 - \left\{\sqrt{\left\{\frac{2R_S + C}{2}\right\}\left\{\frac{2R_S + C}{2} - R_S\right\}^2\left\{\frac{2R_S + C}{2} - \sqrt{2R_S^2(1 - COS\theta)}\right\}}\right\}\right\}}{A}. \quad (8)$$

The coverage probability p_c by at least one sensor node among the 'k' set of sensor nodes can be expressed as

$$P_C = 1 - (1 - P_d)^K. \quad (9)$$

As the value of 'N' is very large, so taking approximation of Eq. (9) and the equation is approximated as shown below in (10):

$$P_C = 1 - \exp\left[\frac{-K\left(2\pi R_S^2 - A_O\right)}{A}\right]. \quad (10)$$

The coverage probability is estimated using the above Eq. (10), where it shows that as the overlap area increases, the redundancy increases and reduces the effective coverage area. Thus, the detection probability is reduced significantly.

3 Simulation Result and Discussion

This section describes performance evaluation for coverage analysis of a wireless multihop sensor network in a square region with $100 \times 100\,\mathrm{m}^2$ dimensions as shown in Fig. 2. Here, we have randomly deployed 20, 30 and 40 active sensor nodes with a maximum sensing range of 10 m in a square region. The performance is estimated in an effective coverage area over several sensor nodes and sensing ranges. It has been observed that the effective coverage area is being reduced with an increase in the number of nodes as shown in Fig. 3. Another important parameter is the detection probability. The detection probability increases with the increase in number of nodes and sensing range. The analytical result is formulated using Eqs. (2), (3) and (10). Finally, the performance analysis for coverage is done in MATLAB R2015a simulation environment.

4 Conclusion

In this paper, we have estimated the effective coverage area under different performance parameters. The proposed model eliminates the overlap area and thus reduces the redundancy in the coverage analysis. The analytical formulation is done using a

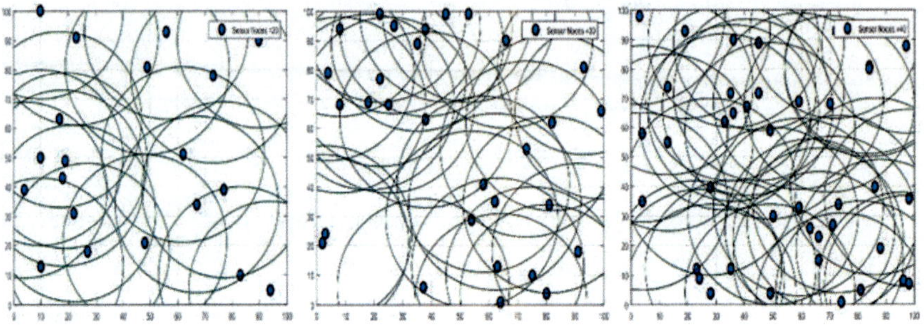

Fig. 2 Random node deployment in $100 \times 100\ m^2$ with $K = 20, 30, 40$

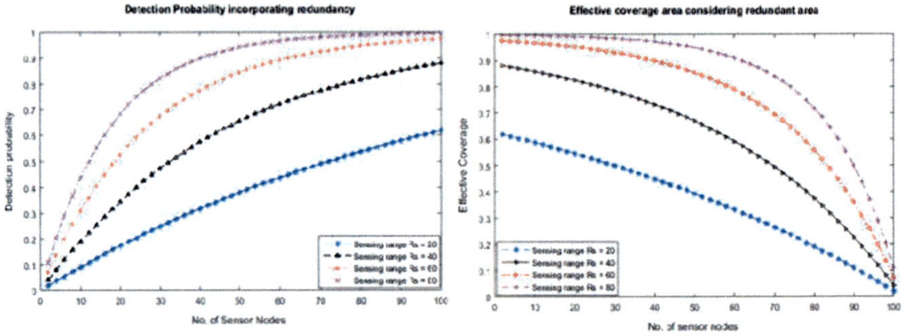

Fig. 3 Detection probability and effective coverage area incorporating redundancy with respect to number of sensor node for different sensing ranges

mathematical model and simulated under MATLAB environment. In future, the work can be extended using different sensing models together with considering boundary effects.

References

1. Zennaro M, ICTP Trieste-Italy (2012) Introduction to wireless sensor networks
2. Katti A, Lobiyal DK (2016) Sensor node deployment and coverage prediction for underwater sensor networks. In: 2016 3rd international conference on computing for sustainable global development (INDIACom). IEEE, pp 3018–3022
3. Jin Y, Jo J-Y, Wang L, Kim Y, Yang X (2008) Eccra: an energy-efficient coverage and connectivity preserving routing algorithm under border effects in wireless sensor networks. Comput Commun 31(10):2398–2407
4. Cai Y, Li M, Shu W, Min-You W (2007) ACOS: an area-based collaborative sleeping protocol for wireless sensor networks. Ad Hoc Sens Wirel Netw 3(1):77–97
5. Chakraborty S, Goyal NK, Mahapatra S, Soh S (2020) A Monte-Carlo Markov chain approach for coverage-area reliability of mobile wireless sensor networks with multistate nodes. Reliab Eng Syst Saf 193:106662

6. Liu L, Zhang X, Ma H (2013) Percolation theory-based exposure-path prevention for wireless sensor networks coverage in internet of things. IEEE Sens J 13(10):3625–3636
7. Chakraborty S, Goyal NK, Soh S (2020) On area coverage reliability of mobile wireless sensor networks with multistate nodes. IEEE Sens J 20(9):4992–5003
8. Amutha J, Sharma S, Nagar J (2020) WSN strategies based on sensors, deployment, sensing models, coverage and energy efficiency: review, approaches and open issues. Wirel Pers Commun 111(2):1089–1115
9. Al-Karaki JN, Gawanmeh A (2017) The optimal deployment, coverage, and connectivity problems in wireless sensor networks: revisited. IEEE Access 5:18051–18065
10. Nagar J, Chaturvedi SK, Soh S (2021) An analytical framework with border effects to estimate the connectivity performance of finite multihop networks in shadowing environments. Cluster Comput 1–16
11. Priyadarshi R, Gupta B, Anurag A (2020) Deployment techniques in wireless sensor networks: a survey, classification, challenges, and future research issues. J Supercomput 76(9):7333–7373
12. Weisstein EW (2003) Heron's formula. https://mathworld.wolfram.com/
13. Tsai Y-R (2008) Sensing coverage for randomly distributed wireless sensor networks in shadowed environments. IEEE Trans Veh Technol 57(1):556–564
14. Singh A, Sharma S, Singh J, Kumar R (2019) Mathematical modelling for reducing the sensing of redundant information in WSNs based on biologically inspired techniques. J Intell Fuzzy Syst 37:1–11, 10

Performance Investigation of Photonic Radar for Autonomous Vehicles' Application Under Various Degrading Conditions

Abhishek Sharma and Jyoteesh Malhota

Abstract Autonomous vehicles are projected to be the future of the automotive industry. Photonic radars have emerged as one of the most reliable sensors in the development of self-driving automobiles. The effect of different degrading conditions such as atmospheric turbulence and impact material surface in the target recognition by photonic radar is analyzed in this work. For simplicity, we have considered four scenarios in which the target has turbulence in terms of atmospheric attenuation and reflectivity of the surface, while two scenarios of the scintillation effect are studied. With an operating range of 250 m and attenuation up to 75 dB/km, the reported results show successful detection of the target.

Keywords Autonomous vehicle · Photonic radar · Reflectivity · Atmospheric turbulences · Scintillations

1 Introduction

Sensor based upon the optical signals has proved to be diverse and found their applications in various sectors be it autonomous vehicles or geography sensing, land mass observation, aerial navigation and subaquatic fluxes [1, 2]. It is expected in the automotive sector which will grow with 31% of CAGR in 2021–2028 period, particularly in autonomous vehicle's (AV's) sector [3]. Key demand of autonomous sector is higher resolution-based system to detect and range stationary or moving objects with high degree of accuracy in all weather conditions [4, 5]. Furthermore, the AVs need high-security provisions for redundant and shared measurements. However, under varying weather conditions, exact measurements of road conditions become tedious

A. Sharma (✉)
Department of Electronics Technology, Guru Nanak Dev University Amritsar, Amritsar 143005, India
e-mail: er.abhisheksharma07@gmail.com

J. Malhota
Department of Electronics and Communication Engineering, Guru Nanak Dev University, Regional Campus, Jalandhar, India
e-mail: jyoteesh.ecejal@gndu.ac.in

A. Rani et al. (eds.), *Signals, Machines and Automation*, Lecture Notes in Electrical Engineering 1023, https://doi.org/10.1007/978-981-99-0969-8_52

job. Primary requirement for AVs is dependent upon a sensory system that is proficient of providing accurate measurement particularly up to 500 m with high resolution. Likewise, power requirement should also be minimized particularly below 20W as limited power is available from car batteries. Many sensors have been deployed in auto motives to gather surrounding conditions such as RADARs, Ultrasonic sensors, LiDAR, Cameras and Photonic radars. Out of these currently used sensors, photonic radar stands to be more reliable and efficient as it offers higher bandwidth that enables better range resolution, low beam divergence, smaller size, low input power requirement and immunity from electromagnetic interference (EMI) [6, 7].

Photonic radar typically utilizes low input power continuous wave (CW) laser source with comparatively extended surveillance duration. The information signal is encoded in triangular format to decrease the sweep duration which is further frequency modulated using RF signal. This FM-modulated signal is exploited to figure the object distance and speed. The FM-modulated RF signal is combined with continuous wave light source using Mach Zander interferometer, and output signal is obtained in form of light. There are two types of configuration used in photonic radars, viz non-coherent or direct detection and coherent detection. Along with transceiver design issues, operating frequency selection is also important as RF poses transmission limit as well as inability in penetrating through dense targets. Also, higher bandwidth enables to attain better angular resolution. To solve this issue, use of millimeter wave band (mm-band) is recommended. On the contrary, higher bandwidth of mm band tends to be affected with higher attenuation with various influences of atmosphere [8–24].

Though AVs are still a growing field of study, many researchers have published considerable reports in this area. Reference [25] reported the impact of suspended particles in air such as smoke and dust on the working of AVs. Another study [26] discussed target simulation discussing irregular reflectivity inconsistencies in atmosphere. Thus, reported work presents impact of atmospheric influences. The photonic radar operates in free space, and hence, various attenuation factors such as rain haze snow and fog must be considered thoroughly while designing the channel modeling. Authors have studied in detail impact of atmospheric effects of free-space optics [11, 16, 17, 27] and proposed various mitigation schemes to lower their impact [28]. Thus, researchers need to consider the atmospheric conditions while designing the system. Frequency band is another important parameter as resolution improves with higher bandwidth [29]. Various configurations of photonic radar using LFMCW techniques are discussed with their advantages [30].

Along with different atmospheric factors such as turbulences and scintillations, effectiveness of photonic radar also depends upon the material reflectivity of different targets. There is limited work reported in finding impacts of attenuation, scintillation and material reflectivity. In this work, we have considered linear FMCW-based photonic radar to analyze the influence of attenuation, scintillation and material reflection altogether on detection of target. Present work is organized as: Sect. 1 designates the introduction of photonic radars and previous work; Sect. 2 explains the system description of the recognized photonic radar; Sect. 3 presents the material impact upon the developed system, and concluding part of this study is given in Sect. 4.

2 System Description

As presented in Fig. 1, basic photonic radar scheme comprises a transmitter where linear frequency modulated (LFM) saw tooth signal and continuous wave source are modulated together using dual-port Mach Zander modulator into the free space toward the target via optical lens.

The operating frequency of LFM generator is 77 GHz and bandwidth for this system is 2 GHz. The range frequency that is frequency of replicated echo from the object is calculated as per Eq. (1):

$$f_R = \frac{2 \times R \times B}{T_m \times C}, \tag{1}$$

where distance between object and radar equipped vehicles is given by R, the operational bandwidth of the system is given by B and time taken by the signal to travel from transmitter to object and back to the receiver is given as T_m known as sweep time. Numerous aspects, precisely scattering, propagation under climatic conditions and object material have deteriorating impact on the echo signal. The intensity of the echo signal P_r is calculated as [31]:

$$P_r = \begin{cases} P_t \frac{\rho_t D^2 \tau_{opt} \tau_{atm}^2}{4R^2} & \text{for extended target} \\ P_t \frac{\rho_t A_t D^2 \tau_{opt} \tau_{atm}^2}{4R^2 A_{ill}} & \text{for any target} \end{cases}, \tag{2}$$

where D is the width of the transmitter lens, reflectivity of target is given by ρ_t, effective area of the target is given by A_t, A_{ill} is the lit part at target, and τ_{opt} and τ_{atm} designate transmission and atmospheric loss factors. At the receiver, photodiode

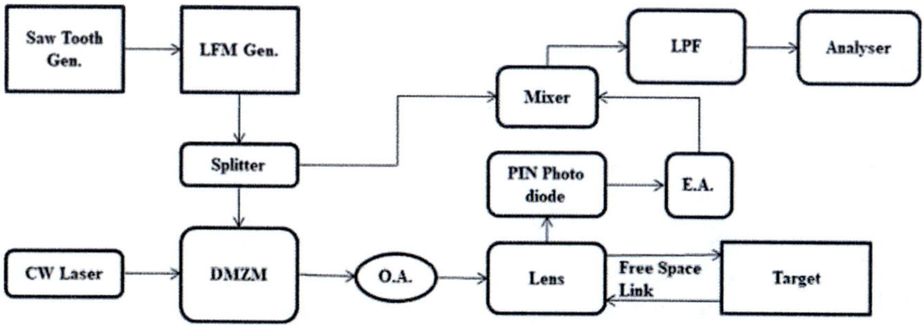

Fig. 1 Block diagram of proposed photonic radar

has been used to capture the echo signal, and based upon echo signal intensity, corresponding electrical current is generated. Photodiode current with responsivity $\Re$ is [32]:

$$i_{ph}(t) = \Re.P_r\left(1 + \frac{\beta}{2}\cos(2\pi f_c(t-\tau) + \frac{\pi B}{T_m}(t-\tau)^2\right)^2. \quad (3)$$

Equation (3) is also presented in filtered form as [33]:

$$i_{ph}(t) = I_{dc} + i_{sig}(t) \approx \Re.P_r\left(1 + \frac{\beta}{2}\cos(2\pi f_c(t-\tau) + \frac{\pi B}{T_m}(t-\tau)^2\right)^2, \quad (4)$$

where I_{dc} is direct current generated from the photodiode, while i_{sig} is alternating current. We have utilized PIN-type photodiode to detect the echo signal having responsitivity of 1 AW^{-1}. The output of PIN diode is analyzer using electrical analyze for evaluation SNRvalues. The SNR is measured as detector as [31]:

$$SNR_{dir} = \frac{\beta^2\Re^2P_r^2/2}{2q\Re P_r B_{rx} + 4k_b T_r B_{rx}/R_L}, \quad (5)$$

where B_{rx} is the receiver bandwidth, q is the electrical charge $\approx 1.6 \times 10^{-19}$ c, and k_b is the Boltzmann constant $\approx 1.38 \times 10^{-23}$ J/K.

The signal is further subjected to electrical amplifier. We have used amplifier with 40 dB gain, and this amplified signal is mixed with FM-modulated input signal using a mixer. The output of mixer is subjected to the low-pass filter where beat signal is obtained as shown in Eq. (6) [31]:

$$S_b(t) = A_c\Re P_r\beta\cos\left(2\pi f_c\tau - \frac{\pi B}{T_m}\tau^2 + 2\pi f_r t\right) \quad (6)$$

3 Results and Discussion

The photonic radar system is designed using OptiSystem™ software. Firstly, the system range frequency is calculated mathematically using Eq. (1). Then, the proposed system is tested firstly for various atmospheric attenuations caused by rain fog snow and haze. For simplicity of the operation, we have considered four values of the attenuation that is 0.2, 25, 50 and 75 dB/km [34, 35]. The reported results as shown in Fig. 2 depicts successful detection of target under varying attenuation with decreasing received power.

It can be observed that range frequency is at 333.33 MHz which is same as calculated using Eq. (1). This testifies that the system detection is theoretically correct. Further, it is clear from Fig. 2 that system withstands the varying attenuation and

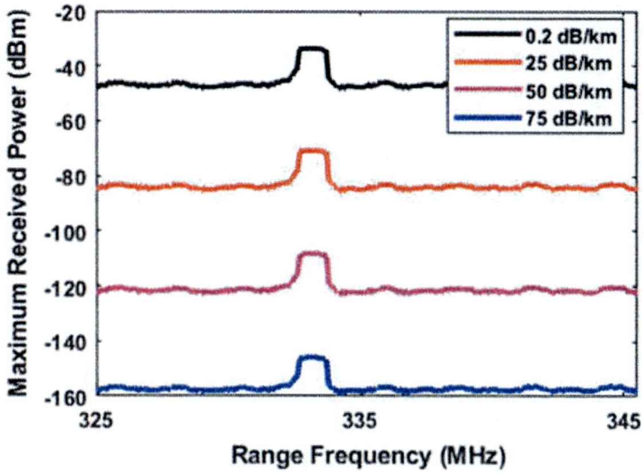

Fig. 2 Max received power vs. frequency under varying attenuation

successfully detected the target. Further, the proposed system is tested for material surface effects in detection of the target. For this, we have considered four scenarios in which material reflectivity is predefined as 15%, 50%, 80% and 99%. The results are shown in Fig. 3.

As shown in Fig. 3, with increasing reflectivity of target surface, the maximum received power also increases. Again range frequency is 333.33 MHz which matches theoretical value. Lastly, the system is detected for impact of scintillation effects. Two scenarios are considered with high scintillation of 10^{-6} and low scintillation of 10^{-22}.

As shown in Fig. 4, the scintillation has minimal impact of the signal intensity and results obtained in low as well as high scintillations are comparable. Thus, it is reported that from the simulation results, attenuation and material reflectivity have high impact on the performance of photonic radar as compared to the scintillation impact.

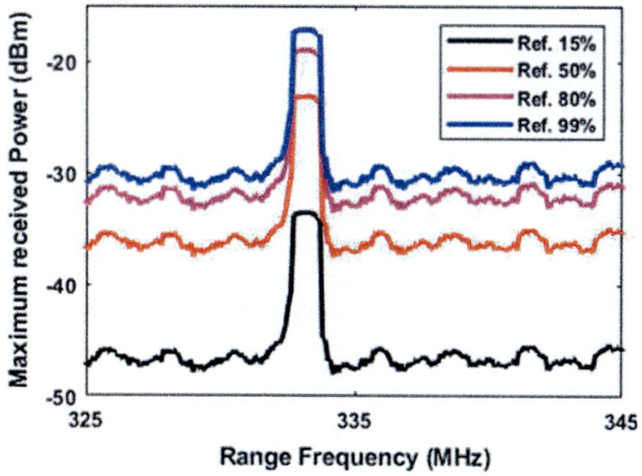

Fig. 3 Max received power versus frequency under varying reflectivity of target material. Ref: Reflectivity of the surface material of target

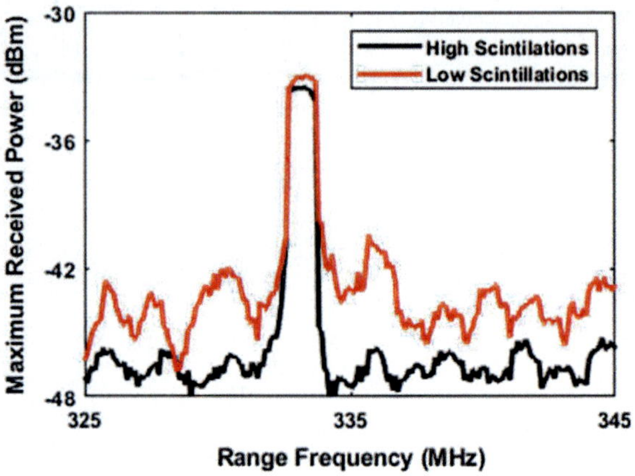

Fig. 4 Max received power versus frequency under varying scintillation

4 Conclusion

In this work, linear frequency-modulated photonic radar has been modeled in direct detection configuration for range detection of stationary target by analyzing the echo signal. Proposed system has successfully detected the stationary target placed at 250 m. Further impact of attenuation is observed with attenuation of 0.2–75 dB/km, and surface material impact is observed by modeling target having reflectivity of 15%, 50%, 80% and 99%. The reported results show that material reflectivity and attenuation have degrading impact of performance of the photonic radars, while scintillation effects are not severe in autonomous vehicle applications. The performance of the system under varying degrading conditions is tested and verified that system withstands all these conditions with successful detection of the target.

References

1. Poggio L, Gimona A (2017) Assimilation of optical and radar remote sensing data in 3D mapping of soil properties over large areas. Sci Total Environ 579:1094–1110
2. Chaudhary S, Wuttisittikulkij L, Saadi M, Sharma A, Al Otaibi S, Nebhen J et al (2021) Coherent detection-based photonic radar for autonomous vehicles under diverse weather conditions. PLoS ONE 16:e0259438
3. Insights FB. Autonomous car market [Online]. Available: https://www.fortunebusinessinsights.com/industry-reports/autonomous-cars-market-100141
4. Sharma A, Malhotra J (6 June 2022) Performance enhancement of photonic radar sensor for detecting multiple targets by incorporating mode division multiplexing. Opt Quantum Electron 54:410
5. Sharma A, Chaudhary S, Malhotra J, Parnianifard A, Wuttisittikulkij L (1 July 2022) Measurement of target range and Doppler shift by incorporating PDM-enabled FMCW-based photonic radar. Optik 262:169191

6. Sharma A, Chaudhary S, Malhotra J, Saadi M, Otaibi SA, Nebhen J et al (2021) A cost-effective photonic radar under adverse weather conditions for autonomous vehicles by incorporating frequency modulated direct detection scheme. Front Phys 467
7. Sharma A, Chaudhary S, Malhotra J, Parnianifard A, Kumar S, Wuttisittikulkij L (2022) Impact of bandwidth on range resolution of multiple targets using photonic radar. IEEE Access 10:47618–47627
8. Chaudhary S, Wuttisittikulkij L, Nebhen J, Tang X, Saadi M, Al Otaibi S et al (7 Sept 2021) Hybrid MDM-PDM based Ro-FSO system for broadband services by incorporating donut modes under diverse weather conditions. Front Phys 9
9. Chaudhary S, Sharma A, Tang X, Wei X, Sood P (1 Feb 2021) A cost effective 100 Gbps FSO system under the impact of fog by incorporating OCDMA-PDM scheme. Wirel Pers Commun 116:2159–2168
10. Sharma A, Chaudhary S, Thakur D, Dhasratan V (2020) A cost-effective high-speed radio over fibre system for millimeter wave applications. J Opt Commun 41:177–180
11. Shakthi Murugan KH, Sharma A, Malhotra J (7 Nov 2020) Performance analysis of 80 Gbps Ro-FSO system by incorporating hybrid WDM-MDM scheme. Opt Quantum Electron 52:505
12. Chaudhary S, Tang X, Sharma A, Lin B, Wei X, Parmar A (30 April 2019) A cost-effective 100 Gbps SAC-OCDMA–PDM based inter-satellite communication link. Opt Quantum Electron 51:148
13. Chaudhary S, Sharma A, Singh V (1 Jan 2019) Optimization of high speed and long haul inter-satellite communication link by incorporating differential phase shift key and orthogonal frequency division multiplexing scheme. Optik 176:185–190
14. Chaudhary S, Kapoor R, Sharma A (1 April 2019) Empirical evaluation of 4 QAM and 4 PSK in OFDM-based inter-satellite communication system. J Opt Commun 40:143–147
15. Chaudhary S, Chauhan P, Sharma A (2019) High speed 4×2.5 Gbps–5 GHz AMI-WDM-RoF transmission system for WLANs. J Opt Commun 40:285–288
16. Zhou Z, Zhang H, Lin C, Sharma A (2018) Performance analysis of duobinary and CSRZ modulation based polarization interleaving for high-speed WDM-FSO transmission system. J Opt Commun 1
17. Sood P, Sharma A, Chandni (2018) Analysis of FSO system and its challenges–a review. Int J Comput Appl 179:42–45
18. Sharma SRA, Rana S (2017) Comprehensive study of radio over fiber with different modulation techniques–a review. Int J Comput Appl 170:22–25
19. Sharma A, Thakur D (2017) A review on WLANs with radio-over-fiber technology. Int J Electron Commun Eng (IJECE) 6:1–6
20. Sharma A, Kapoor R (2017) Study of various challenges in is OWC: a review. Int J Res Appl Sci Eng Technol (IJRASET) 5:802–807
21. Chaudhary S, Sharma A, Chaudhary N (1 Dec 2016) 6×20 Gbps hybrid WDM–PI inter-satellite system under the influence of transmitting pointing errors. J Opt Commun 37:375–379
22. Sharma A, Kumar V, Gupta V. A review on inter-satellite optical wireless communication. Int J Comput Appl 180:13–17
23. Sharma A, Malhotra J, Chaudhary S, Thappa V (1 Feb 2021) Analysis of 2×10 Gbps MDM enabled inter satellite optical wireless communication under the impact of pointing errors. Optik 227:165250
24. Sharma A, Chauhan P. A study of radio over fiber technology in WLAN applications
25. Peynot T, Underwood J, Scheding S (2009)Towards reliable perception for unmanned ground vehicles in challenging conditions. In: 2009 IEEE/RSJ international conference on intelligent robots and systems, pp 1170–1176
26. Rasshofer RH, Spies M, Spies H (2011) Influences of weather phenomena on automotive laser radar systems. Advan Radio Sci 9:49–60
27. Study of radio over fiber with different coding channel–a review. Int J Comput Appl
28. Zhang H, Tang X, Lin B, Zhou Z, Lin C, Chaudhary S et al (2019) Performance analysis of FSO system with different modulation schemes over gamma-gamma turbulence channel, vol 11048. SPIE

29. Zhang F, Gao B, Pan S (2018) Photonics-based MIMO radar with high-resolution and fast detection capability. Opt Express 26:17529–17540
30. Sharma A, Malhotra J (21 March 2022) Simulative investigation of FMCW based optical photonic radar and its different configurations. Opt Quantum Electron 54:233
31. Elghandour AH, Ren CD (2013) Modeling and comparative study of various detection techniques for FMCW LIDAR using optisystem. In: International symposium on photoelectronic detection and imaging 2013: laser sensing and imaging and applications, p 890529
32. Keiser G (2003) Optical communications essentials: McGraw-Hill education
33. Hui R, O'Sullivan M (2009) Fiber optic measurement techniques. Academic Press
34. Awan MS, Csurgai-Horváth L, Muhammad SS, Leitgeb E, Nadeem F, Khan MS (2009) Characterization of fog and snow attenuations for free-space optical propagation. JCM 4:533–545
35. Chaudhary S, Wuttisittikulkij L, Nebhen J, Sharma A, Rodriguez DZ, Kumar S (2022) Terabyte capacity-enabled (10 x 400 Gbps) Is-OWC system for long-haul communication by incorporating dual polarization quadrature phase shift key and mode division multiplexing scheme. PLoS ONE 17:e0265044

An Enhanced IoT and LoRa-Based Communication System for Underground Mines

Sandi Kumar Reddy and Anil S. Naik

Abstract The mining industry is well known for valuable minerals all over the world and the way to get minerals is by mining below the surface. Real-time monitoring of environmental parameters from underground mines is a very complex task to ensure safety. Accidents in the underground area are due to hazardous toxic gas spread beyond the threshold limit value at excavations. These mine gases are a threat to both mine workers' health and the mine environment and are also very hard for miners to detect. Automation in the mining industry reduces hazardous accidents while improving efficiency and productivity. Adopting the Internet of Things (IoT) with Long Range (LoRa) wireless communication technology to build a robust long-range wireless communication system to monitor environmental parameters and hazardous events. To ensure a safe mine environment for mineworkers at the mine site, Artificial Intelligence techniques are an add-on to automation systems to build predictive modeling for an intelligent early warning alert system. However, current solutions are limited to monitoring, low computing power and processing capability, battery lifetime, suitable at the laboratory level, or reports events in a few cases. The proposed system is comprised of the IoT-based sensor modules, communication protocols, Long-Range (LoRa), and machine learning technologies to establish safe, reliable communication from the underground mine to the surface area. A reliable, robust, and cost-effective communication system should be required that monitors environmental parameters and predicts reports of unusual events by adopting new technology.

Keywords Internet of Things · LoRa · Underground mine · Environmental monitoring · Machine learning

---

S. K. Reddy (✉) · A. S. Naik
National Institute of Technology Karnataka, Surathakal 575025, India
e-mail: skreddy@nitk.edu.in

A. S. Naik
e-mail: anilsnaik.217mn001@nitk.edu.in

A. Rani et al. (eds.), *Signals, Machines and Automation*, Lecture Notes in Electrical Engineering 1023, https://doi.org/10.1007/978-981-99-0969-8_53

1 Introduction

The primary goal of any mining industry is to ensure the safety of mineworkers at the mine site. The safety issue comes first in the underground mining industry because of the complex process to extract minerals from underground mine areas. Excavation areas are exposed to toxic gases such as Carbon Monoxide, Methane, Hydrogen Sulfide, Nitrogen Dioxide, Carbon Dioxide, Sulfur Dioxide, and Hydrogen; also, other factors such as temperature, humidity, and fire accidents make mine areas unsafe for mine workers, so continuous monitoring of environmental parameters is required.

Most of the mining industry in India has adopted wired or hybrid-based communication systems at underground mines. It operates perfectly in normal conditions, but due to its inappropriate mine structure, any uncertainty, wires get damaged due to mining operations and can cause accidents. A reliable communication system inside the mines is required to exchange information on environmental conditions, track/detect the position of mineworkers, to alert for the safe and unsafe areas in mines. Implementation of automation in mining industry gives confidence of safe environment to mine worker. It is a challenging task to implement a reliable and robust communication system due to the adverse dynamic nature of the mining environment [1].

To alleviate this problem, Internet-based real-time monitoring systems are used in advanced countries to keep a track of the environmental condition of the underground mines. Wireless radio communication technologies Wireless Fidelity (Wi-Fi) and ZigBee cause radio frequency signal path loss in the underground mine area. Unlicensed frequency band characteristics of ZigBee or LoRa technology for long-range and reliable data transmission are appropriate to use with IoT for data transmission from underground to the surface area.

The comparative analysis of communication technologies such as Bluetooth, Wi-Fi, ZigBee, and Low Power Wide Area Network (LP-WAN) technology LoRa is widely used for monitoring underground mining as shown in Table 1 [2, 3].

A reliable, low-cost, portable online monitoring system for the analysis of underground mining environmental parameters is required. This paper includes related work section which describes current solutions of IoT-based methodology to send

Table 1 Comparative analysis of communication technologies

Parameter	ZigBee	LoRa
Frequency band	2.4 GHz	Unlicensed
Max_data transmission rate	250 kb/s	50 kbps
Transmission distance	10–100 m	5 km (urban) 20 km (rural)
Power	(−25) ~ 0 dBm	14 dBm
Network extension	Tree/mesh topology	Chirp spread spectrum (CSS)

early warnings in underground mines and also described the architecture of the proposed system by integrating IoT and LoRa-based systems for monitoring the environmental parameters and other aspects of underground mining. The detailed review and analysis, challenges, and further research and limitations are discussed. Finally, the paper is summarized with conclusions and some useful remarks.

2 Related Work

The Conventional Smart Digital underground mine system is not fully automated due to its limited communication and involves the contribution of mine supervisors. Underground mine automation is broadly classified as a wireless communication system, database system, mine workers' location and equipment tracking, and intelligent system for early warning and rescue at the mine site. This section describes the current solutions of IoT in underground mining with their methodology, advantages, and limitations proposed by researchers.

IoT-based safety system using a microcontroller ESP8266 and the ThingSpeak platform to monitor environmental data is implemented. Wireless systems are economically feasible and efficient to use but lack validation and data analysis of the proposed system [4]. ZigBee-based energy-efficient system for safety monitoring of the environment is implemented, and simulation results are discussed [5]. ZigBee-based wireless and cost-effective mine monitoring system with an early warning system is proposed [6]. IoT-based system to monitor environmental parameters safety system is proposed and the system is integrated with PLX-DAQ (Parallax microcontroller data acquisition add-on tool for Microsoft Excel) to acquire and read environmental data in real time with high security [7]. ZigBee-based smart helmet system is designed to monitor hazardous gases and Head Injury Criteria (HIC) value is calculated by accelerometer software and Contiki operating system to control sensors [8]. An Integration of Sensors, IoT and Cloud computing system to monitor real-time mining environmental parameters, ventilation control, early warning, and communication effectively. Data analysis is performed to predict measured data by adopting a theory of probability [9]. A ZigBee-based continuous monitoring and alert system to measure environmental parameters at mine site such as heartbeat, temperature, humidity, toxic gas detection is used in the safety jacket of mine workers [10]. A cost-effective intelligent IoT-based helmet to detect hazards in the mines and an early warning system for the safety of the mine environment is implemented at laboratory level [11]. WSN and ZigBee system to monitor environmental parameters in the mining area to reduce potential safety problems in coal production is implemented at the laboratory level [12]. A smart helmet-based proximity warning system (PWS) to provide visual proximity warning alerts to both the equipment operator and the pedestrian. Authors performed various kinds of experiments to compare the signal strength in the underground mine site visual proximity alerts to mineworkers and machine operators [13]. IoT with an IP-enabled gateway for real-time environmental monitoring of the underground environment is implemented by establishing maximum of 800 m distance to transfer to surface level [14]. WSN and LORA-based systems are used to monitor toxic gases, and measurements can be communicated from underground to the surface Area [15]. A LoRa-based cyber-physical system for monitoring the concentration of gases and controlling the ventilation system for the safety of mine workers to establish maximum distance of 100 m depending on curve and bend in mine site is implemented [16]. IoT and machine learning-based hybrid CNN-LSTM models are used for real-time monitoring and forecasting of flammable nontoxic gas methane CH4 gas concentration mine hazards for the safety and production efficiency at underground mine sites [17]. IoT-based environmental conditions are using different sensors and GIS tools to display information at the laboratory setup level [18]. Multilayer Perception- Artificial Neural Network(MLP-ANN) Modelling based mine air quality monitoring and forecasting system to predict air quality in underground mine safety through early warnings [19]. Portable gas detectors and IoT-based low-cost gas sensors and microcontrollers are used to measure environmental parameters in underground mines. Microcontroller-based system with sensors is used to acquire the data and send data to the smartphone using Bluetooth. Smartphone plays the role of visualization to monitor the data. Acquired data from smartphone to cloud are transferred once the Wi-Fi or Internet is connected [20]. A novel conceptual model is proposed to monitor underground mine parameters from the surface area. Single-wire energy transfer and wireless energy transfer along the monorail track are used to monitor the energy transmission network and access points for environmental parameters in underground mines along with an innovative power supply system [21].

Improvement in communication technologies at the mine site is increasing by adopting IoT with ZigBee modules, LoRa technology, and Artificial Intelligence to support industry 4.0 smart digital mine. Observations from the current solutions that researchers focused on the design and development of full-fledged, accurate, reliable real-time monitoring systems are required. Enhancement in sensor technology in terms of efficiency, computing power, and lifetime is needed for sensing and effective communication at the mine site because of the harsh environment, curving, and bend structure of the underground mine cause signal path attenuation. A real-time monitoring system generates a huge amount of data leading to the use of data analysis techniques to predict uncertainty from historic data and also to send notifications or alerts of early warning to the concerned authority. The adoption of multiple gateways can reduce the number of sensor nodes to build wireless communication at the mine site, and loss is reduced in case of damage to sensor nodes due to accidents. The existing real-time monitoring system is designed for detection of toxic gases, temperature, and humidity. The system can be extended to monitor the excavator's health conditions in real time and also data analysis by comparing with historic data to generate a notification if there are any unusual changes in physiological data of mine workers. Validation of real-time monitoring data with portable devices will be an add-on to enhance the safety system.

3 Proposed Model System Architecture

The architecture of the proposed system is described as shown in Fig. 1. The system is composed of a transmitter unit and a monitor unit. Data processing is carried out at the transmitter unit and the microcontroller is integrated with sensors MQ7—Carbon Monoxide (CO) Gas Sensor, MQ4—Methane Natural Gas Sensor Module, DHT22, AM2302—Digital Temperature and Humidity Sensor, MQ136—Hydrogen Sulfide Gas Sensor, 110–508—NO_2 Nitrogen Dioxide Sensor 5 ppm, 110–602—Sulfur Dioxide (SO_2) Sensor 20 ppm, MQ8—Hydrogen H2 Gas Sensor Module, MG811—Carbon Dioxide CO_2 Sensor Module. The data are visualized at the PC monitor unit with the necessary details. To establish a safe, comfortable, and efficient environment for mine workers in underground mining, IoT-based Smart Safety Jackets can be designed for the safety of mineworkers and the system module can fix on their safety jackets as shown in Fig. 2. The safety jacket is integrated with the main processing device and sensor modules to acquire environment parameters in real time to process. Once the sensed parameters exceed the threshold value, then alert mine workers through a buzzer sound when a hazard occurs at the mine site, and mine workers can take immediate action to shift to a safer place at the mine site.

Arduino UNO/Nano microcontroller board with integration of LoRa module and DHT-22 temperature and humidity sensor to measure temperature up to 50° C and humidity up to 90% is shown in Fig. 3. Air quality index ranges of MQ135 sensor are Good: 0–50 ppm, Moderate: 51–100 ppm, Unhealthy to sensitive group: 100–150 ppm, Unhealthy: 151–200 ppm, Very unhealthy: 201–300 ppm, Hazardous: 301–500 ppm.

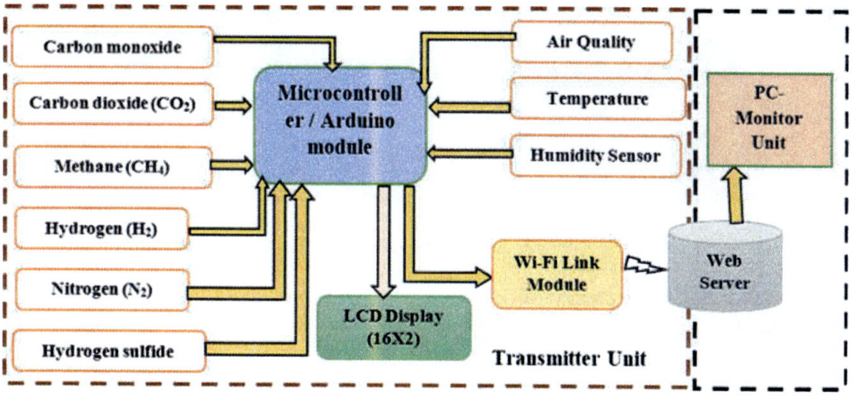

Fig. 1 Block diagram of the transmission and monitor systems

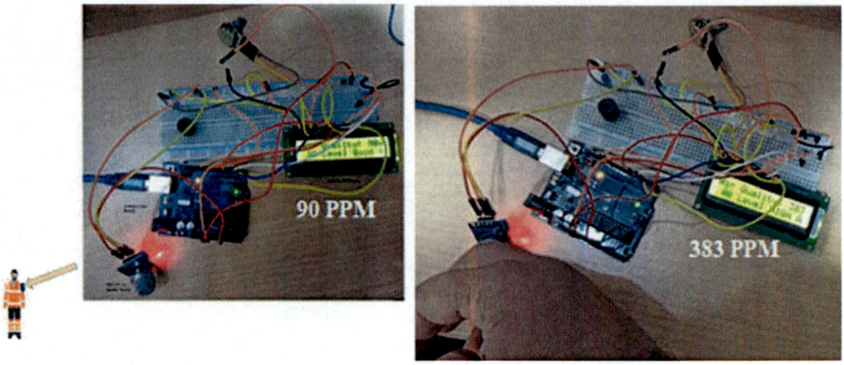

Fig. 2 Case 1: IoT-based smart safety jacket for underground mine workers (AQI: 90 PPM in the indoor environment) and rise of AQI to 383 PPM in the indoor environment when a liquid spray hand sanitizer is placed near the sensor

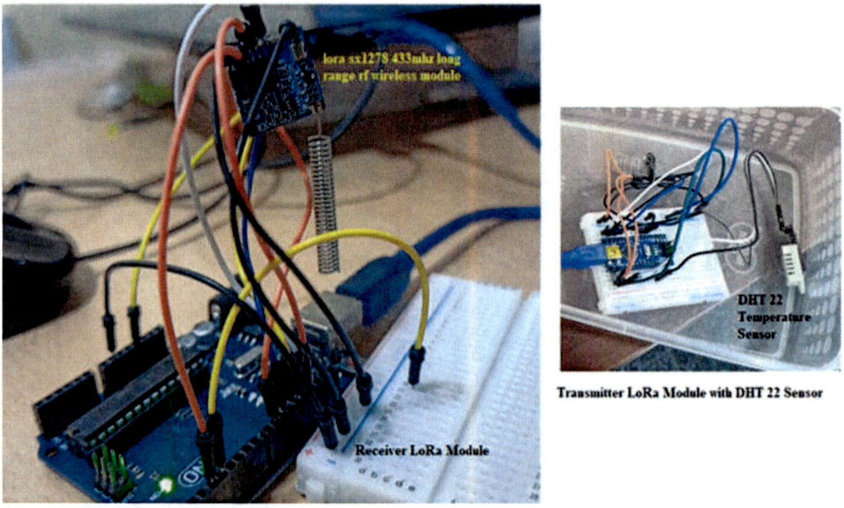

Fig. 3 IoT-based LoRa module with DHT11/22 temperature and humidity sensors

4 Results and Discussion

The circuit is assembled on a breadboard and code should be uploaded using Arduino IDE and open the serial monitor tool and monitor the sensed data displayed on the 16X2 LCD crystal display. Figure 2 shows the AQI value as 90 PPM in an indoor environment and the rise of sensed value to 383 PPM when a liquid spray of hand sanitizer contents is spread on hands. The result analysis is shown in Fig. 4. The value is raised to 620 PPM at the highest level, and the value represented in a graph

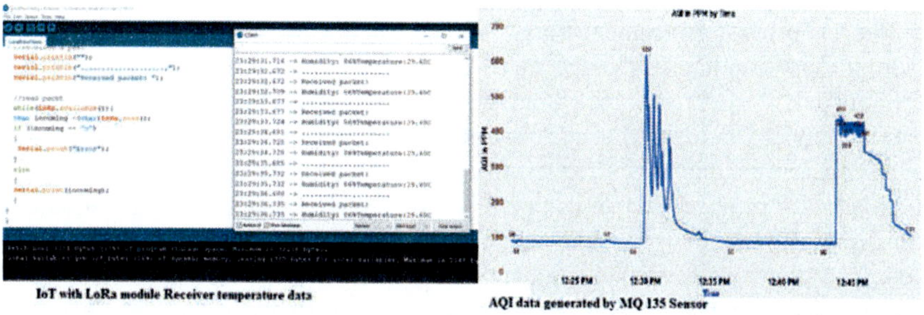

Fig. 4 Result analysis of sensed data of MQ135 sensor represented in a graph and IoT with LoRa module-based DHT22 temperature and humidity data display

is considered from the serial monitor tool. The value of temperature and humidity sensed by DHT22 embedded with LoRa module is also shown.

The datasets used in this study were obtained from MQ135 Air Quality Index Sensor, LoRa module with DHT22 temperature sensors. The dataset consists of 1735 observations, and data are partitioned to 70% and 30% for training data and testing data, respectively. Machine learning algorithms Support Vector Machine (SVM) and K-Nearest Neighbor (KNN) models are developed using R-language (R-Studio) to predict the Air Quality Index and accuracy of algorithms and described below. SVM and KNN algorithms' rate of correct predictions is accurate compared to other algorithms. Validation of the trained model based on test data by SVM and KNN algorithms: (i) SVM-Type: eps-regression and SVM-Kernel type: radial and linear, Accuracy of model = 99.80%. (ii) SVM-Type: eps-regression and SVM-Kernel: sigmoid, Accuracy of model = 98.21%. (iii) KNN: Accuracy rate is 99.62% for K = 5.

5 Conclusion

In the underground mining industry, the main focus is on the safety of mine workers with the prevention of accidents. Enhancement of automation and efficient communication in the mining industry by adopting IoT technology will benefit mine workers and organizations. The system model mainly contributes to the design and development of low-power and reliable solutions to ensure a safe environment at the mine site. As a part of the implementation, a system model is designed with the integration of the MQ135 sensor to Arduino UNO/Nano board to monitor the result in a serial monitor tool and 16X2 LCD. LoRa (433 MHz)-based wireless module with DHT22 temperature sensor is implemented to monitor the value in a real time in mining environment. The system can also configure as an integration of ESP8266 Wi-Fi node microcontroller unit with air quality sensor MQ135, and ThingSpeak IoT analytics platform to monitor air quality of the mine environment. The sensor nodes acquire the environmental parameters and send the acquired data to the cloud to store and process. Machine learning-based prediction models are developed using R-Studio platform and KNN algorithms give a best accuracy rate. Both SVM and KNN models provide good rate accuracy in prediction compared to other machine learning models. In future, the system can configure to send an alert or notification to the concerned authority for requesting emergency assistance and enhancement of automation to provide reliable communication in the mining industry by adopting IoT with LoRa technology to address safety issues that occur in the mining environment. The adoption of artificial intelligence is a benefit to IoT to automate the various processes in the mining industry. However, it takes time to make all mines smart, but the future needs a robust infrastructure to overcome the difficulties associated with the integration of technologies in mining.

References

1. Jha A, Tukkaraja P (2020) Monitoring and assessment of underground climatic conditions using sensors and GIS tools. Int J Min Sci Technol 30(4):495–499
2. Wang L, Xu S, Qiu J, Wang K, Ma E, Li C, Guo C (2020) Automatic monitoring system in underground engineering construction: review and prospect. Advan Civil Eng 1–16
3. Miles B, Bourennane EB, Boucherkha S, Chikhi S (2020) A study of LoRaWAN protocol performance for IoT applications in smart agriculture. Comput Commun 164:148–157
4. Bhuvaneswari R, Roy R (2021) Design of coal mining safety monitoring and alerting system based on IoT. Paideuma J 14(5):371–378
5. Joshi HC, Das S (2017) Design and simulation of smart helmet for coal miners using ZigBee technology. Int J Emerg Technol (Special Issue NCETST-2017) 196–200
6. Ansari AH et al (2021) IOT based coal mine safety monitoring and alerting system. Int J Sci Res Sci Eng Technol 8(3):404–410
7. Savitha G, Deepak NA, Deepak DJ (2021) Data acquisition using IoT to monitor coal mining environment. In: 3rd international conference on integrated intelligent computing communication and security (ICIIC 2021). Atlantis Press, pp 158–166
8. Behr CJ, Kumar A, Hancke GP (2016) A smart helmet for air quality and hazardous event detection for the mining industry. IEEE, pp 2026–2031
9. Jo BW, Khan RMA, Javaid O (2019) Arduino-based intelligent gases monitoring and information sharing internet-of-things system for underground coal mines. J Ambient Intell Smart Environ 183–194
10. Kavitha D, Chinnasamy A, Devi AS, Shali A (2021) Safety monitoring system in mining environment using IoT. J Phys: Conf Ser 1724(1):012022. IOP Publishing
11. Rajalakshmi R, Vidhya J (2019) Toxic environment monitoring using sensors based on Arduino. IEEE, pp 1–6
12. Anitha K, Seshagiri T (2019) Implementation of wireless sensor in coal mine safety system using ZigBee. Int Res J Eng Technol 1467–1472
13. Kim Y, Baek J, Choi Y (2021) Smart helmet-based personnel proximity warning system for improving underground mine safety. Appl Sci 11(10):4342
14. Mishra PK, Kumar S, Kumar M, Kumar J (2019) IoT based multimode sensing platform for underground coal mines. Wireless Pers Commun 108(2):1227–1242
15. Suganthi SU, Valarmathi G, Subashini V, Janaki R, Prabha R (2021) Coal mine safety system for mining workers using LoRa and WUSN. Mater Today: Proc 46:3803–3808
16. Nikolakis N, Kantaris G, Bourmpouchakis K, Alexopoulos K (2021) A cyber-physical system approach for enabling ventilation on-demand in an underground mining site. Procedia CIRP 97:487–490

17. Dey P, Chaulya SK, Kumar S (2021) Hybrid CNN-LSTM and IoT-based coal mine hazards monitoring and prediction system. Process Saf Environ Prot 152:249–263
18. Jha A, Tukkaraja P (2020) Monitoring and assessment of underground climatic conditions using sensors and GIS tools. Int J Min Sci Technol 30:495–499
19. Jo B, Khan RMA (2018) An internet of things system for underground mine air quality pollutant prediction based on azure machine learning. Sensors 18(4):930
20. Ziętek B, Banasiewicz A, Zimroz R, Szrek J, Gola S (2020) A portable environmental data-monitoring system for air hazard evaluation in deep underground mines. Energies 13(23):6331
21. Kianfar AE, Sherikar M, Gilerson A, Skora M, Stankiewicz K, Mitra R, Clausen E (2022) Designing a monitoring system to observe the innovative single-wire and wireless energy transmitting systems in explosive areas of underground mines. Energies 15(2):576

Performance Comparison of Various Optical Amplifiers for DP-QPSK-Based Ultra-Dense WDM System at a Narrow Channel Spacing of 25 GHz

Rajeev and Chakresh Kumar

Abstract In this research paper, we have compared the performance of various conventional amplifiers such as erbium-doped fiber amplifier (EDFA), semiconductor optical amplifier (SOA) and RAMAN amplifier used in an ultra-dense wavelength division multiplexing (UD-WDM) system, for 400 channels at a narrow channel spacing of 25 GHz and a bit rate of 100 Gb/s per channel, based on the high spectral efficient dual-polarized quadrature-phase-shift keying (DP-QPSK) modulation technique. The performance is compared in terms of quality factor (Q-factor), bit error rate (BER) and received output power with respect to varying fiber lengths from 50 to 400 km and applied input powers. From the simulation results, it is reported that the RAMAN amplifier gives better performance in all the above-mentioned parameters. The maximum Q-factor of 13.5 dB, minimum BER of around 10^{-10} and received output power of -34 dBm is reported for a fiber length from 50 to 400 km for RAMAN amplifier which is better than all the other amplifiers. Furthermore, the maximum Q-factor of 11 dB and minimum BER of around 10^{-13} are reported when measured with the applied input power.

Keywords DP-QPSK · Ultra-dense WDM · BER · Quality factor · SOA · EDFA · RAMAN

1 Introduction

In today's world, optical fibers are playing a very important role for high data rate transmission. The demand for high-capacity bandwidth-efficient transmission systems is increasing day by day due to various data-hungry services [1]. The growing demand for high bandwidth can be fulfilled by using optical fiber-based ultra-dense wavelength division multiplexing (UD-DWDM) system along with some high spectral efficient modulation techniques [2]. The deployment of an ultra-dense DWDM system to expand the capacity of optical communication systems is an appealing

Rajeev · C. Kumar (✉)
University School of Information, Communication and Technology, Guru Gobind Singh Indraprastha University, New Delhi 110078, India
e-mail: chakreshk@gmail.com

A. Rani et al. (eds.), *Signals, Machines and Automation*, Lecture Notes in Electrical Engineering 1023, https://doi.org/10.1007/978-981-99-0969-8_54

option. The channels of the UD-WDM system are densely packed by a few GHz and transmitted over many kilometers with the help of optical fiber [3]. Although optical fiber is the best choice for high data rate transmission but it has a drawback in terms of attenuation. Attenuation means the signal strength is decreased while transmitted through a long distance [4]. To minimize the attenuation, some kind of amplifier is required between transmitter and receiver to amplify the attenuated signal. Some famous conventional amplifiers such as erbium-doped fiber amplifier (EDFA), RAMAN amplifier and semiconductor optical amplifier (SOA) are available for signal amplification in optical domain [5]. Kheraliya et al. [6] analyzed the performance of SOA, EDFA and RAMAN amplifiers for a WDM system of 32-channels at bit rate of 10 Gb/s for a distance of 40–280 km. Pradhan et al. [7] analyzed a 16 channel DP-QPSK-based DWDM system using EDFA amplifier with bit rate of 400 Gb/s per channel. Furthermore, Kassegne et al. [8] proposed a DP-QPSK-based 32-channel DWDM system with 50 GHz channel spacing. Kumar et al. [9] analyzed the performance of RAMAN, EDFA and SOA amplifiers for a WDM system of 96-channels at a bit rate of 10 Gb/s for a distance of 300 km. Mahesh et al. [10] analyzed the performance of SOA, EDFA and RAMAN amplifiers for a WDM system of 80-channels at a bit rate of 10 Gb/s for a distance of 280 km. Furthermore, Kheraliya et al. [11] analyzed the performance of EDFA, RAMAN and SOA amplifiers for a WDM system of 64-channels at a bit rate of 10 Gb/s for a distance of 40–280 km. In this paper, we have designed a 400 channel UD-WDM system based on DP-QPSK modulation technique with bit rate of 100 Gb/s per channel and a narrow channel spacing of 25 GHz. The performance is compared for EDFA, SOA and RAMAN amplifier in terms of Q-factor, BER and received output power with varying fiber lengths from 50–400 km and applied input powers. This paper is organized in four sections. Section 1 covered the introduction part. Section 2 detailed the simulation model and set up. Results and discussions are explained in Sect. 3 and finally the conclusion in Sect. 4.

2 Simulation Model and Set up

Figure 1 shows the block diagram for 400 channels DQPSK modulated UD-WDM system. It consists of mainly three parts as the transmitter side, the transmission medium and the receiver side. At the transmitter side, 400 channels having 400 optical DP-DQPSK signals are multiplexed using a dense multiplexer to form a single multiplexed signal which propagates through the transmission medium consisting of an arrangement of different amplifiers such as EDFA, SOA and RAMAN and optical fiber cables such as single-mode fiber (SMF) and dispersion compensated fiber (DCF). After passing through the transmission medium, the signals are demultiplexed at the receiver side with the help of a dense demultiplexer as shown in Fig. 1. The transmission link also includes a single-mode fiber (SMF) of length 40 km and a dispersion compensating fiber (DCF) of length 10 km, for a total of 50 km in a

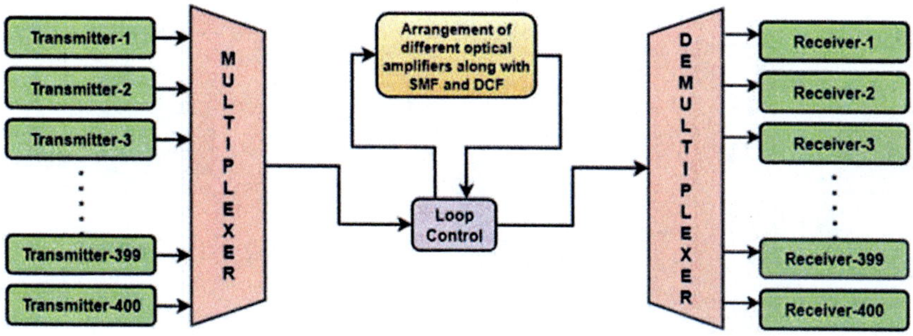

Fig. 1 Block diagram for DP-QPSK modulated UD-WDM system for 400 channels

single span. Total 8 spans are used for simulation in this paper to cover a distance of 400 km.

The block diagram for DP-QPSK transmitter is shown in Fig. 2. A laser signal is applied to the input of a polarization beam splitter which split the laser signal into two signals which are transverse magnetic (TM) and transverse electric (TE) signals as shown in Fig. 2. These two signals are further modulated in in-phase (I) and quadrature-phase (Q) components to provide X-polarized and Y-polarized signals [12]. Finally, these X-polarized and Y-polarized signals are combined to generate the optical DP-QPSK signal.

Figure 3 represents the DP-QPSK receiver. Here, the optical DP-QPSK signal is passed through a polarization splitter to extract the TM and TE signals. Further, in-phase and quadrature-phase signals in electrical form are detected using the arrangement of four PIN photodetectors as shown in Fig. 3. Low pass Bessel filters are used to minimize the noise. Further, the analog signals are converted into digital signals by using the arrangement of analog to digital converters (ADCs). Finally, a parallel to serial converter is used to extract the original data [12]. Simulation parameters of 400 channel UD-WDM system are shown in Table 1.

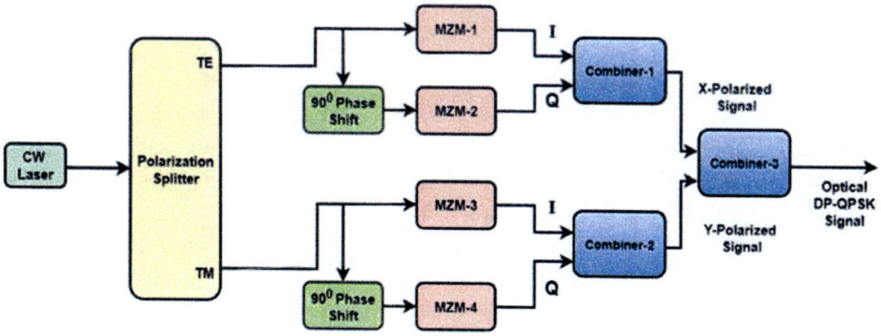

Fig. 2 DP-QPSK transmitter

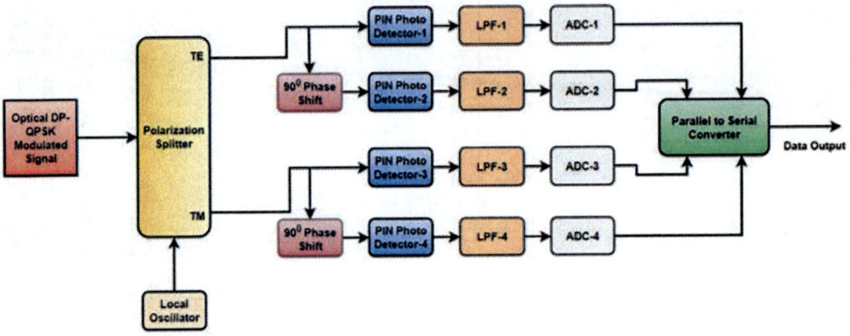

Fig. 3 DP-QPSK receiver

Table 1 Simulation parameters for the proposed UD-WDM system

Parameters	Values
Wavelength	1550 nm
Bit rates	100 Gb/s
Modulation technique	DP-QPSK
Number of channels	400
Channel spacing	25 GHz
Amplifiers used	EDFA, SOA, RAMAN
Distance	50–400 km
Sine generator frequency	Bit rate/2
CW laser noise threshold	−100 dB
CW laser noise dynamics	3 dB

3 Results and Discussions

In this paper, Optisystem software is used to simulate the proposed DP-QPSK modulated UD-WDM system with 400 channels at a narrow channel spacing of 25 GHz with bit rate of 100 Gb/s per channel. The performance is compared for EDFA, SOA and RAMAN amplifiers in terms of BER, Q-factor and received output power with varying fiber lengths from 50 to 400 km and applied input powers. Figure 4 represents the variation in bit error rate with fiber length. It is clearly observed from the Fig. 4 that for a fiber length of 50 to 400 km, the performance of RAMAN amplifier is better than EDFA and SOA amplifier. The reported value of minimum BER for RAMAN amplifier is around 10^{-10} which is quite better than the minimum BER values of SOA and EDFA amplifier which are 10^{-3} and 10^{-1}, respectively.

Figure 5 represents the variation in quality factor with fiber length. Here also, the performance of RAMAN amplifier is better than EDFA and SOA amplifier. The reported value of maximum Q-factor for RAMAN amplifier is 13.8 dB which is

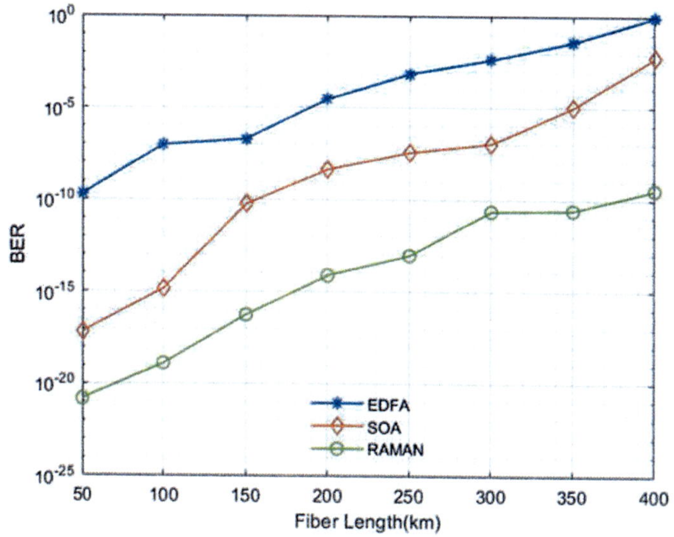

Fig. 4 Bit error rate (BER) vs fiber length

better than the maximum Q-factor values of SOA and EDFA amplifier which are 4.8 dB and 2.9 dB, respectively, for the same fiber length of 50–400 km.

Figure 6 represents the variation in received output power with fiber length. As we can see from Fig. 6, the RAMAN amplifier outperformed the EDFA and SOA

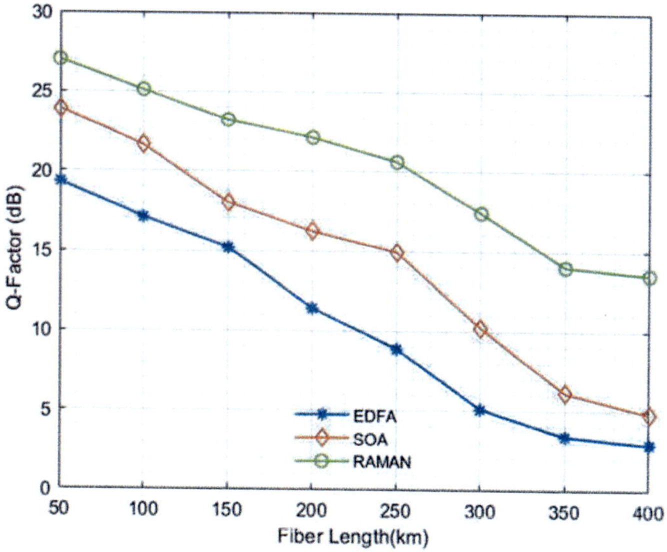

Fig. 5 Quality factor (Q-factor) vs fiber length

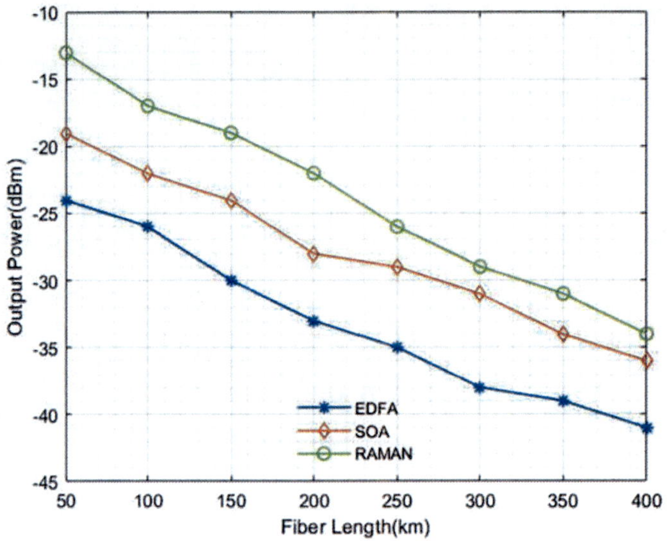

Fig. 6 Output power vs fiber length

amplifiers. The reported value of received output power for RAMAN amplifier is -34 dBm which is better than the received output power of SOA and EDFA amplifier which is -36 dBm and -41 dBm, respectively, for the same fiber length of 50–400 km.

Figures 7 and 8 represent the variation in BER and Q-factor with applied input powers. Here also, the performance of RAMAN amplifier is better than EDFA and SOA amplifier. The reported value of minimum BER is around 10^{-13} and that of the maximum Q-factor is 11 dB for RAMAN amplifier which is quite better than the minimum BER and maximum Q-factor values of SOA and EDFA amplifier. The minimum value of BER and maximum value of Q-factor for SOA amplifier are 10^{-7} and 3.3 dB. The minimum value of BER and maximum value of Q-factor for EDFA amplifier are 10^{-9} and 1.6 dB. Table 2 shows the comparison of the present research work with the existing one.

4 Conclusion

In this paper, a simulation set up of a DP-QPSK modulated UD-WDM system of 400 channels with channel spacing of 25 GHz at bit rates of 100 Gb/s per channel using EDFA, RAMAN and SOA amplifiers is presented. The performance of all the three amplifiers is compared in terms of Q-factor, BER and received output power with varying fiber lengths from 50 to 400 km and applied input powers. The simulation results reported that the RAMAN amplifier outperformed the EDFA and SOA amplifiers in all the performance parameters mentioned above. On the basis of

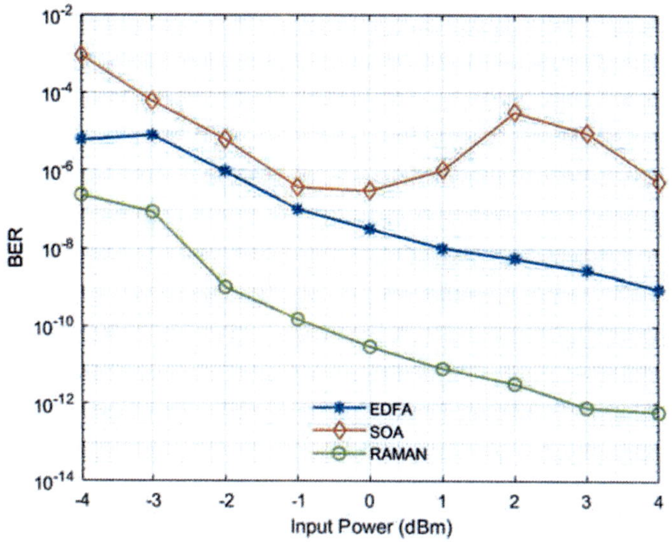

Fig. 7 Bit error rate (BER) versus input power

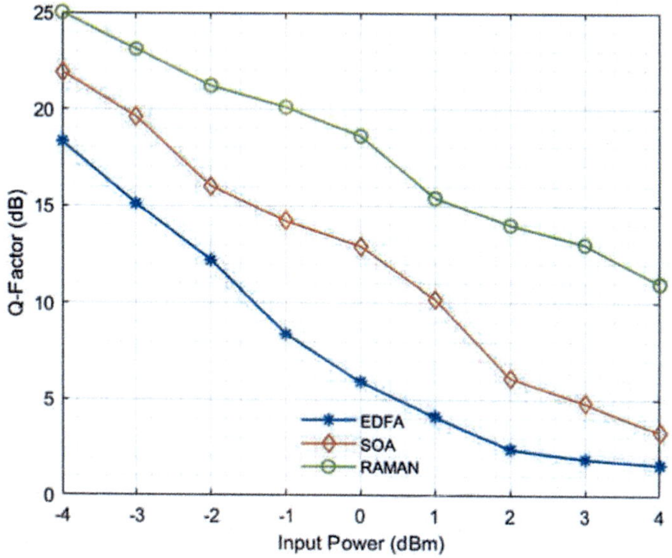

Fig. 8 Quality factor (Q-factor) versus input power

Table 2 Comparison of present research work with existing research works

Parameters	Kheraliya et al. [6]	Pradhan et al. [7]	Kassegne et al. [8]	Kumar et al. [9]	Present work
Modulation technique	QPSK	DP-QPSK	DP-QPSK	DPSK	DP-QPSK
Number of transmitters	32	16	32	96	400
Fiber length	280 km	900 km	110 km	300 km	400 km
Channel spacing	100 GHz	1 THz	50 GHz	100 GHz	25 GHz
Bit rate/channel	10 Gb/s	400 Gb/s	200 Gb/s	10 Gb/s	100 Gb/s
Amplifier	EDFA, SOA, RAMAN	EDFA	EDFA	EDFA, SOA, RAMAN	EDFA, SOA, RAMAN

the reported results, we can recommend the RAMAN amplifier as a good approach for minimizing the attenuation problem in optical communication.

References

1. JM Senior. Optical fiber communications: principles and practice, 3rd edn
2. Okamura K, Mori Y, Hasegawa H (April, 2021) Pre-filtering techniques for spectrum narrowing caused by optical node traversal in ultra-dense WDM networks. IEEE Photonics J 13(2):1–13. Art no. 7200513
3. Kumar G, Kumar S (2021) Effect of different channel spacings for DWDM system using optical amplifiers. Natl Acad Sci Lett 44:415–418
4. Kumar C, Kumar G (2019) Mitigate the dominating signals for super dense optical communication using HOA. SN Appl Sci 1:260
5. Kumar C, Kumar G (2019) Impact of adaptive modulated OOFM signals for SD-WDM system using HOA. Appl Phys A 125:216
6. Kheraliya S, Kumar C (2021) Comparative study of various optical amplifiers for 32-channel WDM system. J Opt Commun 42(2):201–209
7. Pradhan S, Patnaik B, Panigrahy R (2018) DP-QPSK based 400 Gbps/channel fiber optic DWDM system. In: 2018 IEEE 5th international conference on engineering technologies and applied sciences (ICETAS), pp 1–5
8. Kassegne D, Singh S, Ouro-Djobo SS, Mao B-M (2019) Influence of nonlinear effects on 6.4Tb/s dual-polarization quadrature phase shift keying modulated dense wavelength division multiplexed system. Int J Commun Syst 32:e4021
9. Kumar C, Kumar G (2022) Performance investigate and analysis of 96 × 10 Gbps DWDM system using suitable rating from optical amplifiers. J Opt Commun 43(2):171–179
10. Mahesh, Kumar C (2021) Study the performance of various optical amplifiers for 80 channels WDM system using attenuator. J Opt Commun 42(2):189–199
11. Kheraliya S, Kumar C, Kumar G, Meena SR (2021) Influence of conventional optical amplifiers for 64 × 10 Gbps WDM system. J Opt Commun 42(4):565–574
12. Pradhan S, Patnaik B, Panigrahy RK (2019) DP-QPSK technique for ultra-high bit-rate DWDM FSO system. J Opt Commun

High-Computing Technologies

Attaining Real-Time for Microscopic Images Using GAN

Vibhu Bhatia and Yatender Kumar

Abstract In the last years, several deep learning models, especially generative adversarial networks (GAN) have received a lot of attention to perform the task of single image super-resolution (SISR). These methods build up an end-to-end framework, which shows super resolution (SR) for a given low resolution (LR) image in a single step and achieves state-of-the-art performance. This paper develops an application which can perform super-resolution microscopy in real-time in a standard *GPU*. In this work, we propose a speed up in training process of the network. Furthermore, we suggest changes to an existing super resolution network to improve the results. We then explore the possibility of the trained network to produce high-resolution HR outputs for different domains. Finally, we evaluate our model on various images and videos to show that it performs better than the baseline network and other approaches and can perform inference in real time, thus opening its applications in different areas like low-end benchtop and even mobile microscopy.

Keywords Confocal microscopy · Deep learning · Generative adversarial networks · Super resolution

1 Introduction

Single image super-resolution (SISR), a widely studied problem in computer vision aims to generate a high-resolution (HR) image given a low-resolution (LR) counterpart [1, 2]. It is a highly challenging task, but it has been receiving increasing attention from the community over the past few years [3–6]. More recent works employ the use of generative adversarial networks (GAN), which have outperformed all other methods. Imaging on conventional benchtop microscopes produces images, which are typically limited in the resolution captured by the imaging software used. Recent advances in super-resolution microscopy [7] allow us to achieve better magnification. However, these methods rely on sophisticated setups. In this paper, we propose

V. Bhatia · Y. Kumar (✉)
Department of Biological Sciences and Engineering, Netaji Subhas University of Technology Dwarka, New Delhi 110078, India
e-mail: yatender.kumar@nsut.ac.in

A. Rani et al. (eds.), *Signals, Machines and Automation*, Lecture Notes in Electrical Engineering 1023, https://doi.org/10.1007/978-981-99-0969-8_55

a method which generates a machine learning model for super resolution on microscopic images. Furthermore, we demonstrate that model can achieve results in real time.

2 Related Work

Prediction-based methods were among the first methods to tackle the single image super-resolution (SISR). While filtering approaches like Linear, Bicubic, or Lanczos [8, 9] filtering can be very fast, they oversimplify the SISR problem and usually yield solutions with overly smooth textures. A lot of statistical methods have been developed which focus on edge preservation [10] and gradient profiles [11]. Early work on the lines of using training data to establish a correlation between differing resolution images was carried out by Freeman et al. [12]. In Glasner et al. [2], the authors exploit the property of multiple patch occurrences within the image to drive the SR. Some methods also focus on building an internal database of patches to construct images [13].

Deep learning (DL)-based approaches are also being used to tackle the problem of SISR. Dong et al. [14] used bicubic interpolation to upscale an input image and trained a three-layer deep fully convolutional network end-to-end to achieve SR. Similar works by Johnson et al. [15] and Bruna et al. [16], rely on a loss function using perceptual similarities. Ledig et al. [17] propose the use of residual blocks and content loss to train a GAN based network to perform resolution known as SRGAN. Following this, Lim et al. [9] propose the use of deeper networks without the use of batch normalization (BN) layers. Wang et al. [18] propose a new network ESRGAN, which works on residual in residual dense blocks (RRDB), in addition to removing the BN layers and propose a relativistic discriminator. Oquab et al. [19] have shown using global max pooling layer to obtain a point on the boundary of the object. This concept is further taken by Zhou et al. [20] which use Global Average Pooling Layer to find out regions in an image containing the object. Wang et al. [20] present a deep learning-enabled super-resolution network which uses a U-net [21]-like architecture as the generator as opposed to the SRGAN-based generator and can convert 10X resolution images to 20X. Zhou et al. [22] proposed a convolutional neural net (CNN)-based approach for super-resolution localization microscopy. Zhang et al. [23] propose a registration-free GAN-based model for providing resolution enhancement in conventional microscopy.

3 Method

3.1 Data Generation and Tiling Strategy

The Human Protein Atlas Image Classification Challenge [24] is a competition hosted in Kaggle, which encourages people to develop models which can classify proteins in microscopic images. For our purpose, this dataset provided the most suitable images that we could obtain. These images were gathered using immunofluorescence confocal microscopy. The training dataset used contains a total of 124, 288 illustrations. The images are given in four channels: red, blue, green, and yellow which represent antibody-based staining of microtubules DAPI staining of the nucleus, protein localizations and endoplasmic reticulum. The images were generated with differing number of channels in an image to increase the diversity of samples available for the dataset. A total of 2,000 images of (512 $\times$ 512) were finally used which were down sampled 4X using bicubic sampling. The images were divided into tiles of size 256 and 64 pixels, respectively, exploit the tradeoff between an optimal field of view, localization of cells, and faster running time for our generator.

3.2 GAN Architecture and Improvements

Ledig et al. [17] proposed the architecture for SRGAN, which we used as the base model and over which we build up our improvements. The SRGAN training process starts producing results with checkerboard-like patterns (see Fig. 1 left). This is due to the fact that while deconvolution layers are used for upscaling the image, uneven overlaps may exist if kernel size is not divisible by stride [25]. When this operation is performed over two dimensions like in an image channel, these overlaps get multiplied along the length and width of the image (see Fig. 1 Right). To overcome this problem, a subpixel convolution layer [3] is proposed, which works on using a kernel size divisible by the strides. This method is used in SRGAN, but even then, artifacts can occur. To address these issues, we use an interpolation layer followed with a regular convolution layer as shown by Dong et al. [26].

Another change that we implemented was the removal of batch normalization (BN) layers from the residual blocks. As demonstrated by Lim et al. [4], removing the batch normalization layer seems to reduce the computational time, and increase the flexibility available to a generator. For the discriminator, we propose the addition of a global average pooling (GAP) layer to the output of the last convolutional block. GAP layers have been shown to have excellent localization abilities [20]. It has also been shown to act as an excellent regularizer and helps to prevent overfitting while reducing the total number of parameters [1]. These reasons made using a GAP layer as an option worth exploring in place of the flatten layer proposed in the original architecture.

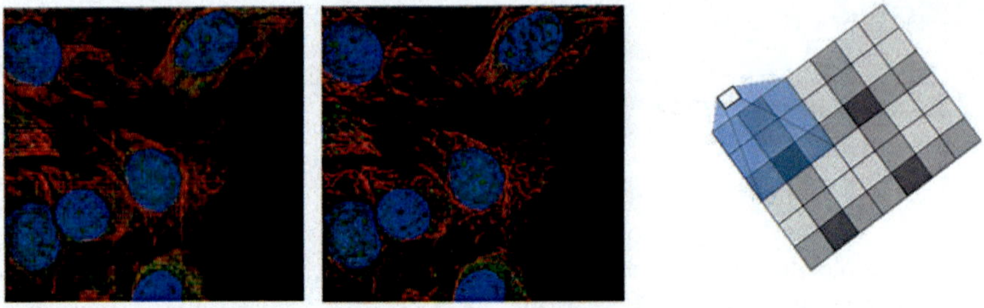

Fig. 1 (Left) Checkerboard pattern produced on training SRGAN model after the 200th vs. the ground truth image. (Right) Overlaps produced in standard deconvolution algorithm

3.3 *Loss Function and Label Smoothing*

The loss function for the discriminator is the perceptual loss function, which consists of the adversarial loss and content loss. The adversarial loss trains the GAN to produce more solutions which resemble that of the dataset. The content loss, on the other hand, compares high-level feature representations of the images using a separate pre-trained VGG network and enables the network to produce results which are perceptually similar [16].

For the discriminator, we propose using label smoothing proposed by Salimans et al. [27]. Target labels were a random number between 0.8 and 1.2 for real and 0 and 0.2 for fake labels. This helps in reducing confidence in real and fake classes for discriminator and improves the stability of discriminator during training process.

3.4 *Training Schedule and Inference Process*

The modified SRGAN network which consists of a discriminator and a generator is used for training on the dataset of confocal microscopy images from Sect. 3.1. The images are first divided into tiles, down sampled using bicubic filter to obtain HR and LR images, and then fed into the generator. The generated SR images and the HR images are then fed to the discriminator, which outputs the probability of the generated image as an actual HR image.

All the experiments were performed by downsampling the HR images $4 \times$ times via bicubic interpolation to obtain the noisy low-resolution images. Each HR image which is of 512×512 pixels, is divided into tiles of 256 pixels. A mini batch of 12 was used to train models. In our experiments, we tried different patch sizes, such as that of 64, 128, 256, and 512 pixels per image. We found that an enormous patch size does not improve the finer details but increases the overall quality of the image, whereas a smaller patch size improves the finer details but increases the computational time per image. We settled on a patch size of 256. We trained our network on a Nvidia Tesla K80 GPU on a sample of 10,000 images obtained from the Human Protein

Atlas Image Classification Challenge [24]. All models were trained for a total of 200 epochs of 1000 iterations. We used learning rate scheduler which used value of 1e-4 for the first half of the optimization process and then reduced to 1e-5 for the next half, while using Adam optimizer with beta1 = 0.9 and beta2 = 0.99. The trained model was thereafter integrated into a web-based portal was developed for the inference process.[1]

3.5 Image Similarity Measures

Different metrics for testing the quality of the image obtained have been used. For our experiments, we have used three of them: PSNR, SSIM, and MOS. PSNR is used as a quality ratio for measuring the reconstruction quality from a loss compression. SSIM also is known as the structural similarity measure, is also a metric of measuring the perceived quality of digital images and videos. In this method, the measurement of prediction of image quality is based upon the original image as the reference. It is an improvement over the PSNR discussed above. MOS is known as Mean Opinion Score, is a metric which is used to measure the quality of experience of an Image. The quantity to be measured, rated on a scale of 1–5 with one being poor and five being excellent. It is a subjective quality evaluation metric; i.e., the ratings calculated from different unbiased sources.

4 Experimental Results

4.1 Qualitative Results

Figure 2 compares the results obtained on images from the validation set with different modifications we applied to the SRGAN network and used to train our model. The changes to the baseline were stacked on top of one another. The final architecture, which is SRGAN with nearest neighbor Interpolation in the up-sampling block and no batch normalization, produces perceptually better results than the other methods used.

4.2 Quantitative Results

For calculating the MOS, we conducted a survey of around 98 anonymous people who were provided with a set of 36 images, which were generated from the outputs

[1] Source Code for training can be found at: https://github.com/vibss2397/Real-Time-Super-Resolution-For-Microscopic-Images-Using-GAN.

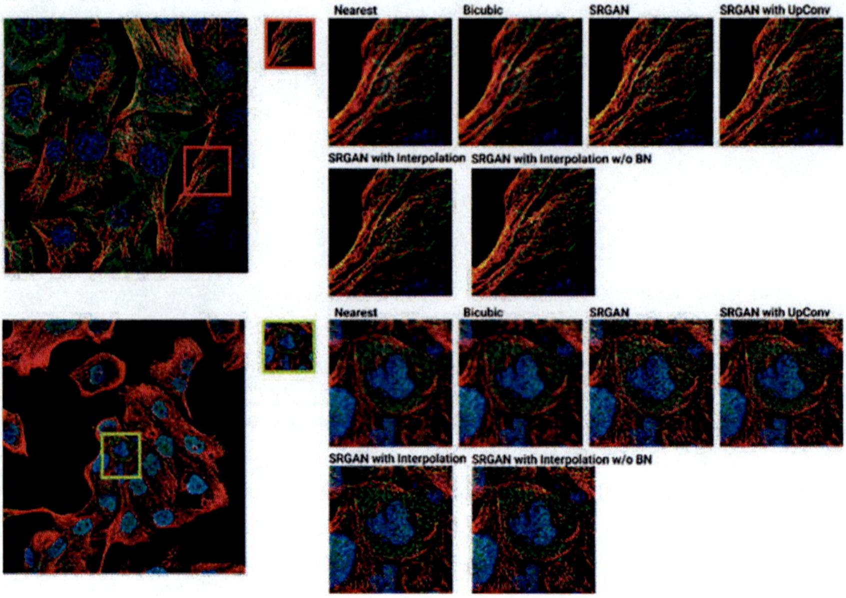

Fig. 2 (Top–Bottom): image created by upscaling LR image using nearest neighbor interpolation, bicubic interpolation, SRGAN, and different modifications proposed for the generator

of the generator, original SRGAN, bicubic approach, nearest neighbor approach, and ground truth image. The images were created and presented in random order, and the raters were asked to rate the images on a scale of 1–5. These results are aggregated in Table 1. Furthermore, PSNR and SSIM were also calculated on the test set and their results collected. As we can see that for the test set, SRGAN performed better than all the other methods in tests such as PSNR, SSIM, and MOS, and was very close to even the ground truth image in the MOS.

Table 1 Benchmarking and comparison of image similarity measures for upscaling methods

Method	PSNR (dB)	SSIM	MOS
Ground truth	∞	1	4.8
Nearest neighbor	27.556	0.771	2.7
Bicubic interpolation	28.498	0.810	2.9
SRGAN	32.585	0.809	3.9
Our method	**33.873**	**0.832**	**4.2**

Table 2 Evaluation of runtime for obtaining SR images with different modifications to the network for single patch, whole image, and for a single 64 × 64 patch in the video

Generated Data	Single Patch	Whole Image	Video Patch
SRGAN	0.067	0.30	19.9
SRGAN w/upconv	0.069	0.33	20.1
SRGAN w/interpolation	0.089	0.34	19.1
SRGAN w/o BN	0.041	0.124	25.7
Our method	**0.051**	**0.14**	**24.2**
Nearest neighbor	0.0009	0.00042	99.9
Bicubic interpolation	0.00016	0.0035	134.5

4.3 Evaluation of Running Time for Modified SRGAN

We evaluated the running time of our model for different images in the test set. The running time recorded for two tasks—upscaling a single patch and upscaling an entire image (4 patches). All the experiments were performed on a Nvidia Tesla K80 GPU. We further test the capabilities of our network to perform the inference in real-time for a given image. We tested the ability of our network to perform super-resolution for a given ROI in the frame of the video. The video that we used is publicly available as Light Sheet Fluorescence Microscope of a Zebrafish Heart [28]. Our model took 0.05 s to convert a single frame. The results are presented in Table 2, as we can see that our model can produce results in a frame rate of 24.2 fps. One thing to notice is that removal of the batch normalization layer increases the fps of the video generated. These results are consistent with our initial assumptions that batch normalization results in the addition of extra parameters to our model and with its removal, our model produces effect faster.

5 Conclusion

This paper demonstrates some of the problems that modern deep learning-based architectures for SISR, specifically SRGAN face. We propose changes in the upsampling and the residual blocks of the generator and the addition of GAP layer to the discriminator, which showed improved results for super resolution. We then train this modified SRGAN architecture on the dataset provided by Atlas Human Protein Identification Challenge for its use in enhancing microscopic images. Further, we test the viability of the trained generator in other cases for images in different domains, such as natural images, biomedical histology, and pathology images. We also validated our results by performing MOS tests that the network does produce outputs which lie closer to the ground truth image in terms of the overall structure and perceptual quality.

References

1. Lin M, Chen Q, Yan S. Network in network
2. Glasner D, Bagon S, Irani M (2009) Super-resolution from a single image. In: Proceedings of the IEEE international conference on computer vision, pp 349–356
3. Shi W et al (Sept 2016) Real-time single image and video super-resolution using an efficient sub-pixel convolutional neural network
4. Lim B, Son S, Kim H, Nah S, Lee KM (July 2017) Enhanced deep residual networks for single image super-resolution
5. Kim J, Lee JK, Lee KM (Nov 2015) Accurate image super-resolution using very deep convolutional networks
6. Kim J, Lee JK, Lee KM (Nov 2015) Deeply-recursive convolutional network for image super-resolution
7. Browning ND et al (1999) Scanning transmission electron microscopy: an experimental tool for atomic scale interface science
8. Keys RG (1981) Cubic convolution interpolation for digital image processing
9. Duchon CE (1979) Lanczos filtering in one and two dimensions. J Appl Meteorol 18(8):1016–1022
10. Li X, Orchard MT. New edge directed interpolation
11. Sun J, Sun J, Xu Z, Shum HY (2008) Image super-resolution using gradient profile prior. In: 26th IEEE conference on computer vision and pattern recognition, CVPR
12. Freeman WT, Jones TR, Pasztor EC. Example-based super-resolution
13. Freedman G, Fattal R, Fattal R, Selim T. Image and video upscaling from local self-examples
14. Dong C, Loy CC, He K, Tang X (2014) Learning a deep convolutional network for image super-resolution. In: Fleet D, Pajdla T, Schiele B, Tuytelaars T (eds) Computer vision–ECCV 2014. ECCV 2014. Lecture notes in computer science, vol 8692. Springer, Cham. https://doi.org/10.1007/978-3-319-10593-2_13
15. Johnson J, Alahi A, Fei-Fei L (2016) Perceptual losses for style transfer and SR. Arxiv, pp 1–5
16. Bruna J, Sprechmann P, LeCun Y (Nov 2015) Super-resolution with deep convolutional sufficient statistics
17. Ledig C et al (Sept 2016) Photo-realistic single image super-resolution using a generative adversarial network
18. Wang X et al (Sept 2018) ESRGAN: enhanced super-resolution generative adversarial networks
19. Oquab M, Bottou L, Laptev I, Sivic J. Learning and transferring mid-level image representations using convolutional neural networks
20. Zhou B, Khosla A, Lapedriza A, Oliva A, Torralba A. Learning deep features for discriminative localization
21. Wang H et al (2019) Deep learning enables cross-modality super-resolution in fluorescence microscopy. Nat Methods 16(1):103–110
22. Ronneberger O, Fischer P, Brox T. 2015-U-net, pp 1–8
23. Zhang H et al (2019) High-throughput, high-resolution deep learning microscopy based on registration-free generative adversarial network. Biomed Opt Express 10(3):1044
24. Human protein atlas image classification|Kaggle. [Online]. Available: https://www.kaggle.com/c/human-protein-atlas-image-classification. Accessed 30 Sept 2019
25. Odena A, Dumoulin V, Olah C (Jan 2017) Deconvolution and checkerboard artifacts. Distill 1(10)
26. Dong C, Loy C, He K, Tang X (2016) Image super-resolution using deep convolutional networks. IEEE Trans Pattern Anal Mach Intell 38(2):295–307
27. Salimans T, Goodfellow I, Zaremba W, Cheung V, Radford A, Chen X, Chen X (2016) Improved techniques for training gans. In: Lee DD, Sugiyama M, Luxburg UV, Guyon I, Garnett R (eds) Advances in neural information processing systems 29. Curran Associates, Inc., pp 2234–2242
28. Lightsheet Z.1 by Carl Zeiss microscopy–zebrafish heart. [Online]. Available: https://zeiss-microscopy.uberflip.com/youtube/lightsheet-z-1-by-carl-zeiss-microscopy-zebrafish-heart. Accessed 30 Sept 2019

Grey Wolf Optimization-Based PV Array Reconfiguration to Enhance Global Maximum Power Point Under PSCs

Rupendra Kumar Pachauri, Raktangi Swain, and Piyush Kuchhal

Abstract In this paper, conventional photovoltaic (PV) array configurations, e.g., series–parallel (SP) and total-cross-tied (TCT), are considered and compared with the Grey Wolf optimization (GWO) algorithm during shading scenarios. Various performance measures are used to assess the efficiency of the suggested algorithms, including fill factor (FF), mismatch losses (ML), and percentage of power enhancement (PE). Following successful shade dispersion, the overall energy produced by the GWO algorithm is computed and compared to the TCT reconfigured method. Per day energy generation from conventional and GWO-based PV array configurations is analyzed using a shady scenario for day (sunny hours).The energy estimations and income outcomes described here confirm that the GWO generated higher power by 13% compared to conventional configurations. Furthermore, the power-voltage (P–V) graph presented demonstrates a reduction in the quantity of numerous peaks in the scheme. As a result, the suggested GWO method can be approved as better than other current approaches.

Keywords Fill factor · Power loss · Power gain · Renewable energy · Solar photovoltaic system

1 Introduction

In today's world, renewable energies have received much interest for a variety of purposes, such as the usage of non-renewable power sources, environmental issues, and the desire to live in a green and pleasant environment [1]. Photovoltaic (PV) systems are a substitute method of generating power that does not require external storage of fuel. Because of this, photovoltaic systems are a viable option not only for mobile applications but also for stand-alone energy production in remote areas [2].

R. K. Pachauri (✉) · R. Swain · P. Kuchhal
School of Engineering, University of Petroleum and Energy Studies, Dehradun 248007, India
e-mail: rupendrapachauri@gmail.com

P. Kuchhal
e-mail: pkuchhal@ddn.upes.ac.in

A. Rani et al. (eds.), *Signals, Machines and Automation*, Lecture Notes in Electrical Engineering 1023, https://doi.org/10.1007/978-981-99-0969-8_56

Photovoltaic systems experience low power generation during shading environmental conditions. PV module interconnection methods are beneficial to reduce power loss due to shadow conditions [3]. As solar panels become a substitute for non-renewable power sources, the tremendous intrusion of solar PV into electric supply systems poses multiple challenges in the current era. Partial shading, on the other hand, is a crucial concern that impacts the performance and life of PV plants. PV restructuring is well-known as a proven method for dealing with this effect. It is accomplished by reconfiguring the PV arrays based on their temperature and shade stages [4, 5].

1.1 Literature Review

A modern physical skyscraper puzzle is appropriate for every photovoltaic module structure in [6], and this technique is attributed to improved shade distribution across the whole photovoltaic system. The authors of [7] compared their Su-Do-Ku puzzle-based photovoltaic module configurations with hybrid BL-TCT, bridge link-total cross-tied (BL-TCT), and honeycomb (HC) and acquired the output throughout shading instances. The 4 × 4 size Su-Do-Ku arrangement is compared with total cross-tied and hybrid SP-TCT setups in [8] during gradual shadowing situations of 350 W/m^2 and 1000 W/m^2. When compared to conventional approaches, our Su-Do-Ku configuration obtained the finest GMPP indices of 2279 W, 2139 W, 1806 W, and 1680 W, 2279 W, 2139 W, 1806 W, 1680 W, 2139 W, and 1680 W throughout all ten shading conditions (three patterns). In [9, 10], a hardware arrangement for a 7 × 5-size Futoshiki puzzle-dependent realigned photovoltaic structure was designed to confirm the MATLAB/Simulink outcomes for the six shading conditions. The comparison of the efficiency of the total cross-tied arrangement is done, and the percentage progress of the GMPP position is observed for experimenting in percentage (30.89, 10.98, and 38.19) respectively.

2 PV Technology and Modeling

Development of 6 × 6 size PV array is shown with equivalent circuit in Fig. 1 [11] as follows,

$$V_C = \frac{AkT_C}{e}\ln\left(\frac{I_{ph} + I_d - I_c}{I_d}\right) - I_c\left(\frac{R_{se}R_{sh}}{R_{se} + R_{sh}}\right) \tag{1}$$

For the development of MATLAB/Simulink models, PV module specifications are considered as: $P_m = 5$W, $I_m = 0.52$A, $V_{OC} = 11.25$ V and $I_{SC} = 0.55$A.

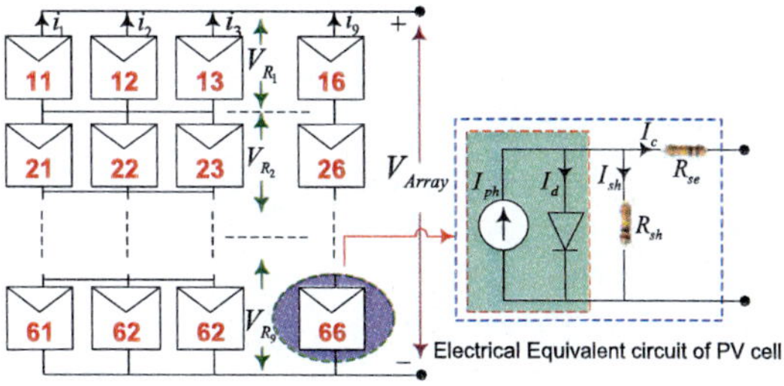

Fig. 1 Development of 6 × 6 size PV array

2.1 Conventional SP and TCT Configurations

The SP array is depicted in Fig. 2a using a straightforward procedure. The cross-tied methodology between the parallel strings is adapted to change SP connection into TCT (6 × 6 size) configuration. TCT configuration and PV module placement are shown in Fig. 2b. The generated voltage and current by the PV array system is expressed in Eq. (2) as,

$$V_{\text{array}} = \sum_{k=1}^{6} V_{mk},\ m = 1, 2 \ldots 9\ \&\ I_{\text{array}} = \sum_{k=1}^{6} I_k = 0 \tag{2}$$

The sun irradiance (S_x) is mainly responsible to generate PV module current. The shading factor $\left(S_f\right)$ and module current (I_m) are given in Eq. (3) as,

$$S_f = \frac{S_x}{S_{\text{STC}}} \quad \& \quad I_m = S_f I_{\max} \tag{3}$$

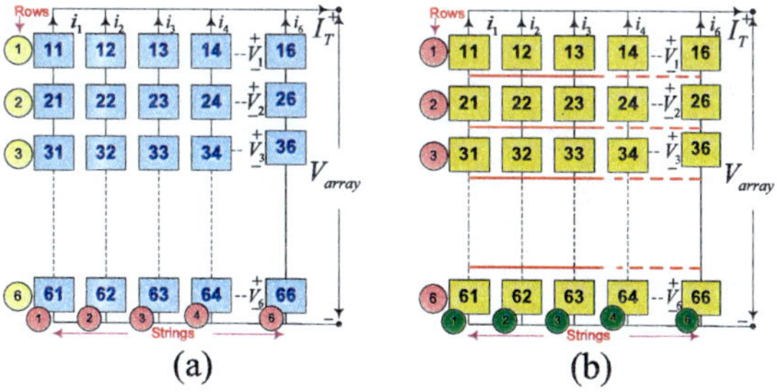

Fig. 2 **a** SP connections **b** TCT connections

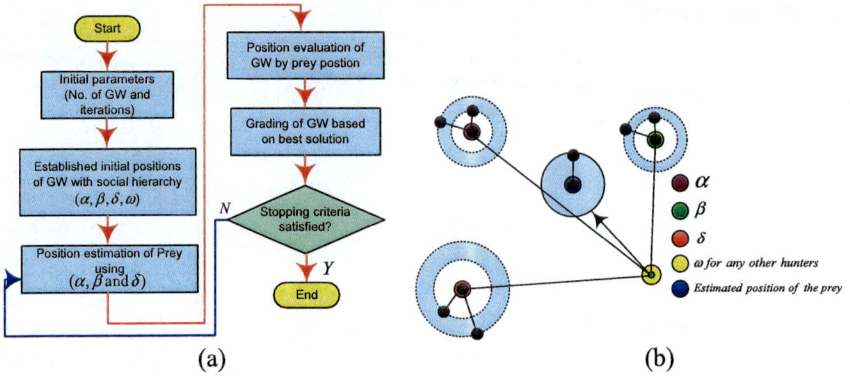

Fig. 3 Methodology of GWO

2.2 *Grey Wolf Optimization-Based PV Array*

In this perspective, these tasks might be treated as a complex optimization problem with two opposing objectives [12]. There are two primary goals: (i) total generated energy and (ii) absolute difference between highest and lowest current ruling values. These two goals can be shown in Eq. (4) as follows:

$$\text{Max(obj)} = \text{sum}(P_x) = \sum_{x=1}^{6} I_x V_x \quad \& \quad \text{min(obj)} = |I_{\max} - I_{\min}| \tag{4}$$

where PV array current and voltage for the xth row are represented by I_x and V_x respectively. Moreover, highest and lowest current values as $I_{\max}$ and $I_{\min}$ are shown above [12, 13]. The methodology of GWO is given in Fig. 3 as.

2.3 *Shading Scenarios*

Distinguishing between varying levels of solar irradiance is made easier in the first shading pattern. In Fig. 4a, b the shading patterns are depicted, and in Fig. 4a, the GWO technique for current normalization is shown in this way.

3 Performance Assessment: Power Loss and Fill Factor

The differentiation between the maximum powers generated at standard test conditions (STC) and under shading scenarios. The power loss and FF can be expressed through Eq. (5) as,

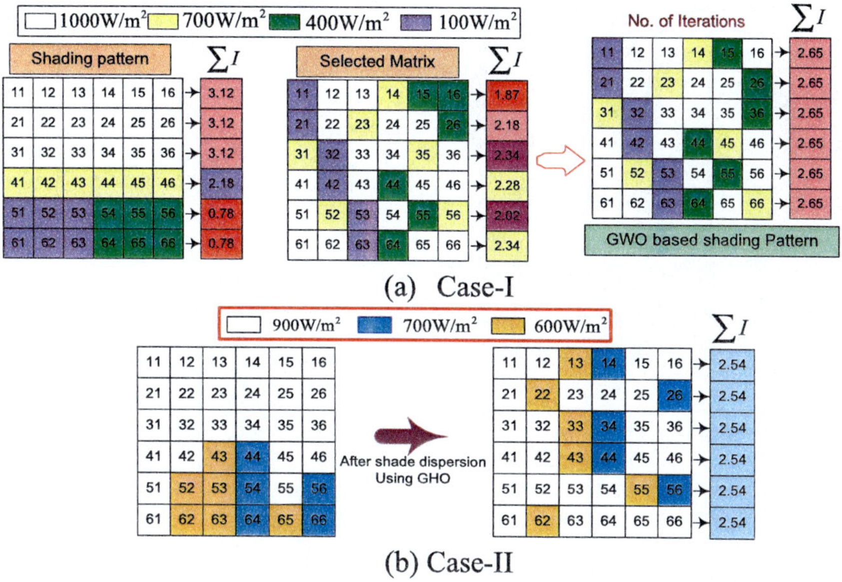

Fig. 4 Shading cases

$$\text{PL} = P_m \text{ at STC} - \text{GMPP at PSCs} \quad \& \quad \text{FF} = \frac{V_m I_m}{V_{OC} I_{SC}}. \tag{5}$$

4 Results and Discussion

Under ideal conditions, the maximum power and voltage are obtained from a 6 × 6-sized PV array at 180W and 57.65 V, respectively. In addition, the amount of maximum current is generated at 3.12A during a similar ideal environment.

4.1 I–V and P–V Curves During Shading Cases: I

Figure 5 shows the non-linear behavior of P–V and I–V curves to reflect the MPPs. Moreover, GWO-based configuration has smooth nature also.

Quantitative observations are made under both shading scenarios, and depicted in Table 1. All of the key performance indicators are as follows:

The GWO-based PV arrangement has a PL as low as 28.7 W and 39.2 W, respectively, when compared to existing SP and TCT array systems. During both shading cases I–II, the FF is evaluated and found to be higher for GWO-based PV systems at

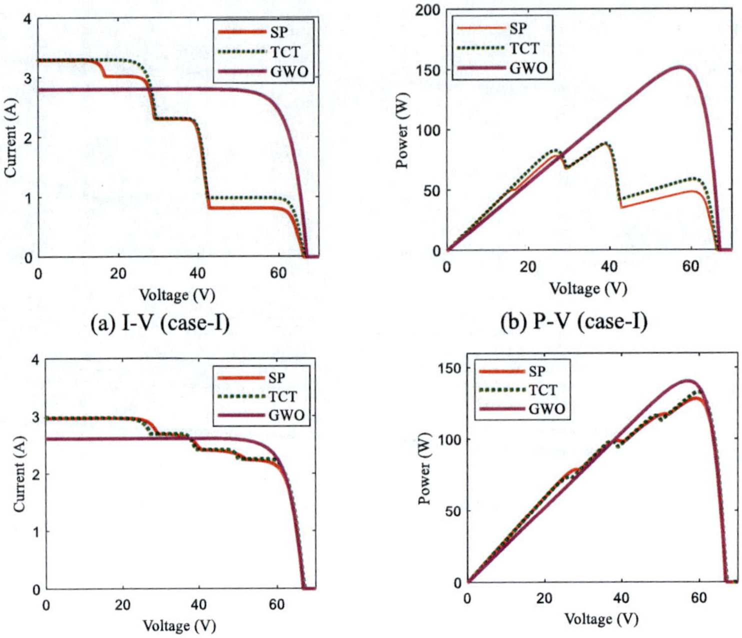

(a) I-V (case-I) (b) P-V (case-I)

(c) I-V (case-II) (d) P-V (Case-II)

Fig. 5 Performance characteristics for shading cases (I–II)

Table 1 Quantitative analysis of PV array systems during shading cases I-II

Shading case-I				Shading case-II		
	SP	TCT	GWO	SP	TCT	GWO
$P_{GMPP}(W)$	87.87	88.47	151.5	128.4	134.9	141
$V_{GMPP}(V)$	39.09	39.09	57.65	59.22	60.25	57.29
$I_m(A)$	2.247	2.263	2.62	2.16	2.239	2.461
$V_{OC}(V)$	66.80	66.80	66.93	66.48	66.45	66.31
$I_{SC}(A)$	3.29	3.29	2.801	2.96	2.96	2.61
FF	0.399	0.402	0.80	0.652	0.685	0.814
PL (W)	92.33	91.73	28.7	51.8	45.3	39.2
PG(%) w.r.t. SP	–	0.68	72.4	–	5.06	9.81

0.80 and 0.814 compared to SP (0.399 and 0.65) and TCT (0.402 and 0.685) configurations. The power and voltage at GMPP, PL, FF, and PG comparisons are depicted in Fig. 6a–e using a bar chart representation.

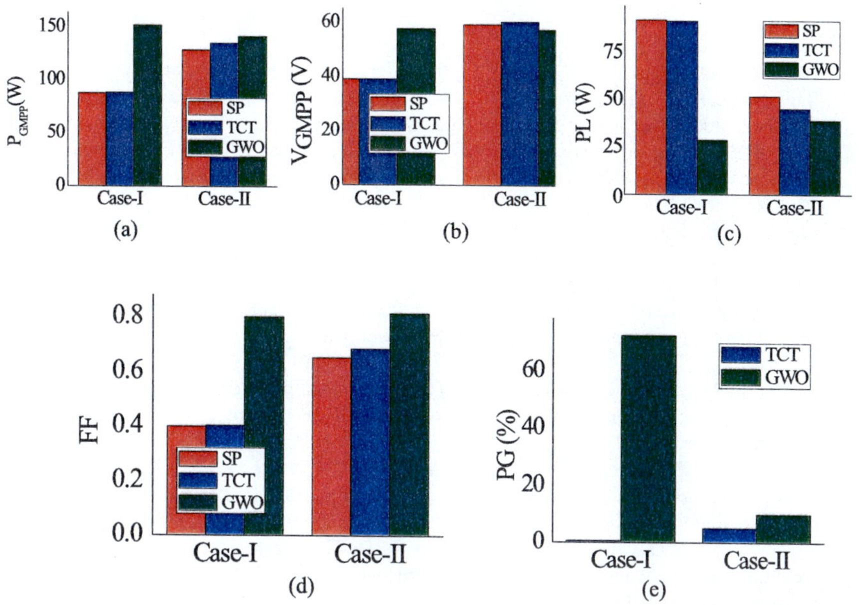

Fig. 6 **a** Power at GMPP **b** voltage at GMPP **c** PL **d** FF **e** PG

5 Conclusion

Conventional SP, TCT, and GWO based PV configurations were examined in depth during two distinct shadowing conditions. Important performance key points of study are as follows:

- Higher GMPP locations, such as 151.5 W and 141 W, were observed for GWO under both shading scenarios when compared to SP (87.87 W, 128) and TCT (88.47 W, 134.9 W) configurations, respectively.
- GWO-based configurations show reduced PL of 28.7 W and 39.2 W when compared to SP (92.33 W, 51.8 W) and TCT (91.73 W, 45.3 W) configurations shading conditions I–II.
- During the study, the FF values for the GWO configuration are also 0.80 and 0.814 when compared to the standard SP and TCT arrangements.

All array systems were shown to be more capable of dispersing shade using GWO optimization.

References

1. Bai J, Sun L, Pachauri RK, Wang G (2021) Investigate on photovoltaic array modeling and the MPPT control method under partial shading conditions. Int J Photo Energy 2021:1–16
2. Yadav P, Kumar A, Gupta A, Pachauri RK, Chauhan YK, Yadav VK (2016) Investigations on the effects of partial shading and dust accumulation on PV module performance. Adv Intell Syst Comput 479:1005–1012
3. Bleicher A, Pachauri RK, Kuchhal P, Bansal K (2020) Experimental study on electrical connections of PV system for improved performance under shadow test cases. Energy Web Inf Technol J 7(30):1–11
4. Gupta V, Sharma M, Pachauri RK, Babu KND (2020) A Low-cost real-time IOT enabled data acquisition system for monitoring of PV system. Energy Sources, Part A: Recovery, Utilization, Environ Eff 23:1–24
5. Kumari A, Pachauri RK, Chauhan YK (2016) Passive islanding detection approach for inverter based DG using harmonics analysis. In: Proceedings IEEE conference on power electronics, instrumentation control and energy system at DTU, Delhi, pp 1562–1568
6. Nihanth MSS, Ram JP, Pillai DS, Ghias AMYM, Garg A, Rajasekar N (2019) Enhanced power production in PV arrays using a new skyscraper puzzle based one-time reconfiguration procedure under partial shade conditions (PSCs). Sol Energy 194:209–224
7. Sagar G, Pathak D, Gaur P, Jain V (2020) A Su-Du-Ko puzzle based shade dispersion for maximum power enhancement of partially shade hybrid bridge-link-total-cross-tied PV array. Sol Energy 204:161–180
8. Yadav AS, Pachauri RK, Chauhan YK. Comprehensive investigation of PV arrays under different shading patterns by shade dispersion using puzzled pattern based Su-Du-Ko puzzle configuration. In: Proceedings IEEE conference on next generation computing technologies (NGCT) at India on 4–5 September 2015, pp 1–6
9. Sahu HS, Nayak SK (2016) Extraction of maximum power from a PV array under nonuniform irradiation conditions. IEEE Trans Electron Dev 63(12):4825–4831
10. Sahu HS, Nayak SK, Mishra S (2016) Maximizing the power generation of a partially shaded PV array. IEEE J Emerg Select Top Power Electron 4(2):626–637
11. Pachauri RK, Thanikanti SB, Bai J, Yadav VK, Aljafari B, Ghosh S, Alhelou HH (2022) Ancient Chinese magic square based PV array reconfiguration methodology to reduce power loss under partial shading conditions. Energy Convers Manage 253:115–148
12. Yousri D, Babu TS, Balasubramanian K, Osama A, Fathy A (2020) Multi-objective grey wolf optimizer for optimal design of switching matrix for shaded PV array dynamic reconfiguration. IEEE Access 8:159931–159946
13. Pachauri RK (2022) Imperative role of optimization techniques to reconfigure solar photovoltaic array systems to diminish shading effects. Int J Circuit Theory Appl 50(1):317–353

A Modified Ion-Motion Optimization Algorithm for Optimal Generation Scheduling in Multi-area Power System

Mohit Kumar, Rupendra Kumar Pachauri, Piyush Kuchhal, Ahmad Faiz Minai, and Shashikant

Abstract In this paper, a novel modified ion-motion optimization technique has been proposed to cope with the real-world multi-area economic load dispatch issue. The original ion-motion optimization techniques work on the principle of attraction–repulsion between the ions. However, in the liquid phase, the effect of repulsion has been ignored in its mathematical model. To overcome this deficiency of ion-motion optimization, a more accurate liquid phase has been developed in the proposed method. The derived numerical outcomes show the dominancy of the proposed algorithm.

Keywords Power system · Optimization techniques · Load dispatch · Performance analysis

1 Introduction

A multi-area (MA) economic load dispatch (ELD) issue divides the thermal generating units of the power system into various areas interconnected by tie-lines. The

M. Kumar
EIE, Sant Longowal Institute of Engineering and Technology Longowal, Sangrur, Punjab 148106, India
e-mail: mohit.dhiman60@gmail.com

R. K. Pachauri (✉) · P. Kuchhal
EED, School of Engineering, University of Petroleum and Energy Studies, Dehradun 248007, India
e-mail: rpachauri@ddn.upes.ac.in

P. Kuchhal
e-mail: pkuchhal@ddn.upes.ac.in

A. F. Minai (✉)
Electrical Engineering Department, Integral University, Lucknow 226026, India
e-mail: fzminai@gmail.com

Shashikant
EED, School of Engineering, Babu Banarasi Das University, Lucknow 226028, India
e-mail: shashikant52@gmail.com

A. Rani et al. (eds.), *Signals, Machines and Automation*, Lecture Notes in Electrical Engineering 1023, https://doi.org/10.1007/978-981-99-0969-8_57

graphical representation of a MA power system (PS) is shown in Fig. 1. In this problem, the PS network is divided into several interconnected areas. The interconnection of areas is done with the help of unique transmission lines. Each part has self-thermal generating units and predefined power demand. However, generators may generate greater or lesser power than the power demand in that area. The surplus or deficiency of power from the power demand in interconnected areas is met with tie-lines [1]. When there is a surplus of generated power in any area, tie-lines transfer surplus power to those areas with a deficiency of power. Each tie-line is also bounded by its minimum and maximum power transfer limits. The presence of the tie-lines dramatically improves the reliability of the power system operation. However, finding an optimal generation schedule for the MA-ELD issue with transmission lines becomes a challenging optimization task [2].

In [1], a covariance matrix-adapted evolution strategy (CMAES) was proposed for the MA-ELD issue, and the optimality was verified by Karush–Kuhn–Tucker (KKT) conditions. Sharma et al. introduced a differential evolution (DE) with time-varying mutation (DE-TVM) for three types of MA-ELD issues [2]. In [3], the authors have applied an artificial bee colony (ABC) with the variable elimination method to handle each area's equality constraint. A teaching-learning-based optimization (TLBO) [4] was applied on multi-area ELD problems. A dynamically controlled particle swarm optimization (DCPSO) [5] was also suggested for MA-ELD issues. The DCPSO algorithm varies the inertia weight using a decaying exponential function. In [6], the authors have suggested a chaotic global best ABC (CGBABC) with an iterative equality constraint handling scheme for the multi-area ELD problem. The CGBABC algorithm analyzed different chaotic maps. A hybrid of DE and PSO (DE-PSO) for the MA-ELD problem was proposed in [7]. The equality constraint of each area was handled by a quadratic penalty function method in DE-PSO. In [8], a new symbiotic organisms search algorithm (NSOS) was proposed to solve the multi-area ELD problem. This scheme solves the MA-ELD problem in dual step. In the first

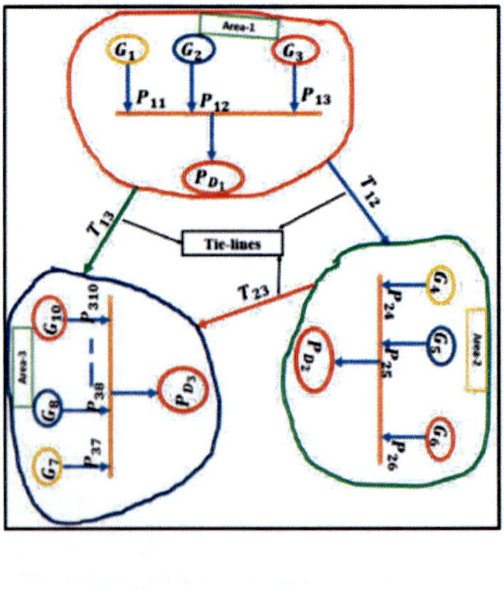

Fig. 1 Graphical representation of a Multi-area Power system

step, the MA-ELD problem is considered a single area (SA) and MA in the later stage. In [9], implemented an evolutionary particle swarm optimization (E-PSO) algorithm for the MA-ELD issue. The E-PSO algorithm introduces crossover and mutation operators in the basic PSO algorithm. The controlling factors of PSO were tuned dynamically. In the tuning process, the authors introduced additional parameters known as shape-shift coefficients of inertia weight.

Lin and Wang proposed an improved stochastic fractal search (ISFS) algorithm for the MA-ELD issue in [10]. The authors have presented a hybrid optimization algorithm, thereby hybridizing the JAYA algorithm with the TLBO algorithm and named JAYA-TLBO [11]. In the JAYA-TLBO scheme, firstly, JAYA was improved and then hybridized with TLBO.

2 MA-ELD Issue

The MA-ELD issue is composed of taking a power system having total N areas. Each area in the MA-ELD issue consists of a total, M_n; $(n = 1, 2, \ldots N)$ generating units. So, total production units in the MA-ELD issue is the summation of all the production units in each area, i.e., $\sum_n^N M_n$. A vector presents the design variables of the multi-area ELD problem to be optimized, $[PT]$ consisting of two sub vectors $[P]$ and $[T]$ such that $[PT] = [[P], [T]]$, having the following in Eq. (1)–(2) as,

$$[P] = [[P_1], [P_2], \ldots, [P_n], \ldots, [P_N]] \tag{1}$$

$$[T] = \left[[T_1], [T_2], \ldots, [T_n], \ldots, \left[T_{N-1}\right]\right] \tag{2}$$

where $[P_n] = \left[P_{n1}, P_{n2}, \ldots, P_{ni}, \ldots, P_{nM_n}\right]$ is the power response vector of production units belonging to the area, n; $[T_n] = \left[T_{n(n+1)}, T_{n(n+2)}, \ldots, T_{np}, \ldots, T_{nN}\right]$ is the powers interchange or tie-line vector between n (system's area) and p (the other areas).

2.1 Cost Function (CF)

The CF, $(Cost([P], [T]))$, which is the objective function in this case of the MA-ELD issue consists of two parts: the fuel cost (FC) for all the system's areas $(FuelCost([P]))$ and the tie-line power transfer cost $(LineCost([T]))$. The mathematical expression is given in Eq. (3)–(6) as follows [12]:

$$Cost([P], [T]) = FuelCost([P]) + LineCost([T]) \tag{3}$$

$$FuelCost([P]) = \sum_{n=1}^{N}\sum_{i=1}^{M_n} F_n(P_{ni}) \tag{4}$$

$$F_n(P_{ni}) = a_{ni}P_{ni}^2 + b_{ni}P_{ni} + c_{ni} + \left|d_{ni}\sin\left\{e_{ni}\left(P_{ni}^{\min} - P_{ni}\right)\right\}\right| \tag{5}$$

where $a_{ni}, b_{ni}, c_{ni}, d_{ni}$ and e_{ni} are the fuel cost coefficients of ith generating unit from the area, n.

$$LineCost([T]) = \sum_{n=1}^{N-1}\sum_{p=n+1}^{N} C_{np}T_{np} \tag{6}$$

Further, in all previous studies, the $LineCost([T])$ is not considered. Hence, for the sake of comparison, the $LineCost([T])$ is also neglected in the present research work.

2.2 Constraints

For a given area, n the MA-ELD issue has various inequality constraints and one equality constraint. A detailed discussion of the constraints of the multi-area ELD problem is given as follows [11, 12].

The limits of the power output of the generators: The mathematical expression of the limits for power generation in the MA-ELD issue is given in Eq. (7) as,

$$P_{ni}^{\min} \le P_{ni} \le P_{ni}^{\max} \quad (i = 1, 2, \ldots, M_n; n = 1, 2, \ldots, N) \tag{7}$$

where $P_{ni}^{\min}$ and $P_{ni}^{\max}$ are the lower and higher power generation limits of ith generator of nth area, respectively.

The limits of the capacity of the Tie-lines: The power flow of tie-line,T_{np} from n to p must lie between the lower and higher transfer limits as shown in Eq. (8) as,

$$T_{np}^{\min} \le T_{np} \le T_{np}^{\max} \quad (n = 1, 2, \ldots, N; p = 1, 2, \ldots, N\forall n \ne p) \tag{8}$$

where $T_{np}^{\min}$ and $T_{np}^{\max}$ are the lower and higher power transfer capacities of the tie-line for the power flow from n to p.

Active power balance equation: In the MA-ELD issue, the active power balance equation for n^{th} area can be written in Eq. (9) as,

$$\sum_{i=1}^{M_n} P_{ni} = P_{Dn} + \sum_{p=1;p\ne n}^{N} T_{np} \quad (n = 1, 2, \ldots, N) \tag{9}$$

Equation (9) is alternatively represented by Eq. (10). The term, Tol_n is the acceptable tolerance between power generated and power demand in the nth area. Ideally, $Tol_n = 0$.

$$Tol_n = P_{Dn} + \sum_{p=1;p\neq n}^{N} T_{np} - \sum_{i=1}^{M_n} P_{ni} \quad (n = 1, 2, \ldots, N) \tag{10}$$

The FC function,F_n of the nth area is expanded with the equality constraint of the nth area using the following expression in Eq. (11) as,

$$F_n = \begin{cases} F_n, & IF\ P_{ni} \in FR \\ F_n + rTol_n^2, & \text{otherwise} \end{cases} \tag{11}$$

where FR is the feasible power production region, F_n is calculated by using Eq. (11), and r is a positive integer. In this research, = 500.

3 Ion-Motion Optimization (IMO)

Ion-Motion optimization (IMO) [13] was suggested by Javidy et al. The mathematical model of IMO comprises two states: liquid and solid. The initial ions, P_{ji} randomly produced as given below in Eq. (12)–(14) as,

$$P_{ji} = P_i^{\min} + \left(P_i^{\max} - P_i^{\min}\right) * rand \tag{12}$$

The population of ions is sub-divided into anions, PA_{ji} and cations, PC_{ji} equally as:

$$PA_{ji} = P_{ji} \quad (i = 1, 2, ..., N_G; j = 1, 2, ..., N_P) \tag{13}$$

$$PC_{ji} = P_{ji} \quad (i = 1, 2, ..., N_G; j = N_P + 1, N_P + 2, ..., 2N_P) \tag{14}$$

where N_P is the number of population of ions anions and cations, N_G is the no. of decision variables.

The liquid state of IMO is given in Eq. (15)–(16) as follows:

$$A_{ji}^F = \frac{1}{1 + \exp^{\left(-\frac{0.1}{|PA_{ji} - Cbest_i|}\right)}} \tag{15}$$

$$C_{ji}^F = \frac{1}{1 + \exp^{\left(-\frac{0.1}{|PC_{ji} - Abest_i|}\right)}} \tag{16}$$

where A_{ji}^{F} and, C_{ji}^{F} are the overall attraction force of anions and cations.

New locations of anions, PA_{ji}^{new} and cations, PC_{ji}^{new} are obtained via the expression as in Eq. (17)–(18) as, [13].

$$PA_{ji}^{\text{new}} = PA_{ji} + A_{ji}^{F}\left[Cbest_i - PA_{ji}\right] \tag{17}$$

$$PC_{ji}^{\text{new}} = PC_{ji} + C_{ji}^{F}\left[Abest_i - PC_{ji}\right] \tag{18}$$

where $Abest_{iiesti}$ is the best anions and $Cbest_{iiesti}$ represents the best cations.

In the crystal phase, the updating of A_{ji}^{new} and C_{ji}^{new} is done conditionally. The position is only updated if the best anion fitness of both of them is greater than the half of worst finesses in Eq. (19)–(20) as,

$$PA_{ji}^{\text{new}} = \begin{cases} PA_{ji} + \phi_2(Cbest_i - 1); & z_{ji} > 0.5 \\ PA_{ji} + \phi_2 Cbest_i; & z_{ji}\langle 0.5 \text{ and } z_{ji}\rangle 0.05 \\ P_i^{\min} + z_{ji}\left(P_i^{\max} - P_i^{\min}\right); & z_{ji} < 0.05 \end{cases} \tag{19}$$

$$PC_{ji}^{\text{new}} = \begin{cases} PC_{ji} + \phi_1(Abest_i - 1); & z_{ji} > 0.5 \\ PC_{ji} + \phi_1 Abest_i; & z_{ji}\langle 0.5 \text{ and } z_{ji}\rangle 0.05 \\ P_i^{\min} + z_{ji}\left(P_i^{\max} - P_i^{\min}\right); & z_{ji} < 0.05 \end{cases} \tag{20}$$

where ϕ_1 and ϕ_2 are random no. in the range $[-1, 1]$, and z_{ji} is a function that returns a random number in [0, 1] and $(i = 1, 2, \ldots, N_G; j = 1, 2, \ldots, N_P)$.

4 Modified IMO (MIMO)

The IMO completely ignores repulsion forces among the identical charges in its modeling, which is a crucial issue related to the liquid phase. Hence, this problem needs to be solved to procure more approximate mathematical model of IMO in nature. The mathematical model of the liquid phase of the proposed modified IMO optimizer is given by Eqs. (21) and (22). In these equations, repulsion force factor, R_{FF} among like ions is represented as mathematically as: $(R_{RF} = \alpha \times rand)$, where α is a constant which can have a value between 1.5 and 2 and *rand* lies between 0 and 1.

The developed mathematical model of the liquid state of MIMO is represented in Eq. (21)–(22) as,

$$PA_{ji}^{\text{new}} = PA_{ji} + R_{FF}A_{ji}^{F}\left[\left(Cbest_i - PA_{ji}\right) + \left(PA_{ji} - PC_{ji}\right)z_{ji}\right] \tag{21}$$

$$PC_{ji}^{\text{new}} = PC_{ji} + R_{FF}C_{ji}^{F}\left[\left(Abest_i - PC_{ji}\right) + \left(PC_{ji} - PA_{ji}\right)z_{ji}\right] \tag{22}$$

The proposed MIMO technique only modifies the liquid state of the IMO. All all other steps remain the same as in IMO.

5 Results and Discussions

In the MIMO, a solution or ion population P_j; $(j = 1, 2, \ldots, 2N_P)$ is represented as the possible generation schedule of the multi-area ELD problem. The mathematical expression of jth ion population is given as follows:

$$P_j = [[P_n], [T_n]] = \left[\left[P_{n1}, P_{n2}, \ldots, P_{ni} \ldots, P_{nM_n}\right], \left[T_{n1}, T_{n2}, \ldots, T_{np} \ldots, T_{nN}\right]\right]$$

The number of ions or dimensions,N_G of jth ion population, P_j in MA-ELD problem is given as, $N_G = \sum_{n=1}^{N} M_n + N(N-1)/2$. The multi-area test system has four areas, i.e., $N = 4$. Hence, the total number of tie lines becomes six. This leads to the dimension, N_G equals to 46 as $N_G = \sum_{n=1}^{N} M_n + N(N-1)/2$. The best power generation schedules, along with tie-line power transferred between the areas obtained by IMO and MIMO are given in Table 1. It is shown in Table 1 that the suggested MIMO algorithm provides a smaller FC value as compared to the IMO algorithm for the multi-area test system.

It is also noticed from Table 1 that the obtained best generation schedules obtained by IMO and MIMO give a satisfactory performance while handling the active power balance condition in all the areas.

The convergence curves of IMO and MIMO for the multi-area test system are depicted in Fig. 2. This figure confirms that the proposed MIMO converges earlier than the IMO algorithm and becomes almost stable after 50,000 function evaluations. Boxplots visualize the distribution of the cost of IMO and MIMO over 30 independent runs for the multi-area test system in Fig. 2. The better fuel cost distribution is represented by the proposed MIMO algorithm throughout 30 runs compared to the IMO algorithm. The boxplot of this algorithm is more compressed than the IMO algorithm (Fig. 3).

Table 2 is about the comparison for the multi-area test system. In Table 2, the output are matched with algorithms like; ABC [3], TLBO [4], DCPSO [5], CGBABC [6], E-PSO [9], and JAYA-TLBO [11]. It can be observed from Table 2 that the different FC was reported by E-PSO as finalized by the researchers on the given production schedule.

Thus, considering the feasible solution's acceptable feasibility, the proposed MIMO generates the best average and maximum fuel among all the methods, shown in Table 2. The CGBABC is the best algorithm with respect to minimum FC. Further, the MIMO gives the value of SD (standard deviation) less than ten, proving the robustness of the suggested algorithm over other algorithms in solving the multi-area test system.

Table 1 Best production schedules achieved by IMO and MIMO on the multi-area test system

$[PT]$ MW	IMO	MIMO	$[PT]$ MW	IMO	MIMO	$[PT]$ MW	IMO	MIMO
$P_{1,1}$	102.78	110.80	$P_{3,3}$	518.57	523.27	$T_{2,4}$	−91.43	−97.62
$P_{1,2}$	106.46	110.80	$P_{3,4}$	522.49	523.27	$T_{3,4}$	−3.89	−42.75
$P_{1,3}$	92.34	97.40	$P_{3,5}$	510.84	523.30	$\sum_{i=1}^{M_1} P_{1,i}$	1724.40	1690.80
$P_{1,4}$	176.87	179.73	$P_{3,6}$	436.76	523.27	$\sum_{i=1}^{M_2} P_{2,i}$	3952.05	3736.31
$P_{1,5}$	94.13	93.27	$P_{3,7}$	25.79	10.00	$\sum_{i=1}^{M_3} P_{3,i}$	3164.00	3257.52
$P_{1,6}$	137.96	140.00	$P_{3,8}$	35.81	10.00	$\sum_{i=1}^{M_4} P_{4,i}$	1659.54	1815.37
$P_{1,7}$	277.44	259.59	$P_{3,9}$	18.43	10.00	$\sum_{n=1}^{N} \sum_{i=1}^{M_n} P_{ni}$	10,499.99	10,500.02
$P_{1,8}$	284.64	284.59	$P_{3,10}$	83.78	87.82	P_{D1}	1575	1575
$P_{1,9}$	230.23	284.60	$P_{4,1}$	166.68	162.32	P_{D2}	4200	4200
$P_{1,10}$	221.50	130.00	$P_{4,2}$	158.70	190.00	P_{D3}	3150	3150
$P_{2,1}$	245.60	168.79	$P_{4,3}$	148.86	190	P_{D4}	1575	1575
$P_{2,2}$	318.69	168.79	$P_{4,4}$	158.96	164.79	$\sum_{p=1;p\neq 1}^{N} T_{1,p}$	149.40	115.81
$P_{2,3}$	219.28	214.75	$P_{4,5}$	161.94	164.80	$\sum_{p=1;p\neq 2}^{N} T_{2,p}$	−247.94	−463.69
$P_{2,4}$	390.15	394.27	$P_{4,6}$	152.30	164.80	$\sum_{p=1;p\neq 3}^{N} T_{3,p}$	14.00	107.50
$P_{2,5}$	306.17	394.27	$P_{4,7}$	91.52	89.11	$\sum_{p=1;p\neq 4}^{N} T_{4,p}$	84.54	240.37
$P_{2,6}$	484.08	394.27	$P_{4,8}$	83.43	89.12	Tol_1	0.00	0.0032
$P_{2,7}$	481.37	489.28	$P_{4,9}$	92.98	89.12	Tol_2	−2E-05	−0.0087
$P_{2,8}$	490.03	489.28	$P_{4,10}$	444.12	511.27	Tol_3	0.00	−0.016
$P_{2,9}$	513.89	511.27	$T_{1,2}$	99.04	198.63	Tol_4	−0.00	0.00089
$P_{2,10}$	502.75	511.28	$T_{1,3}$	39.57	17.18	Cost ($/h)	126,445.5	121,596.6
$P_{3,1}$	522.65	523.27	$T_{1,4}$	10.78	−99.99			
$P_{3,2}$	488.81	523.27	$T_{2,3}$	−57.46	−167.44			

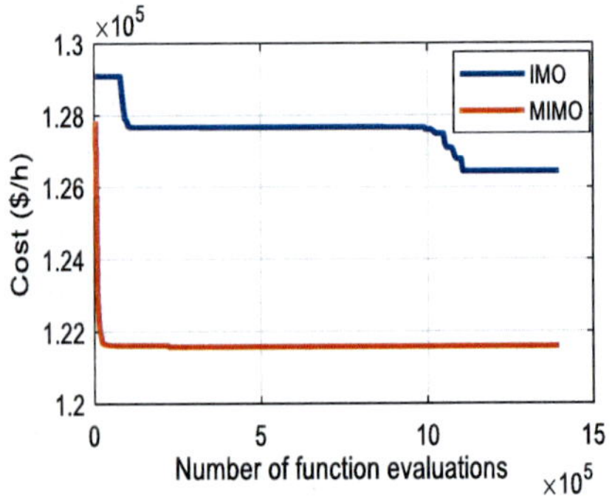

Fig. 2 Convergence curves of proposed MIMO and IMO

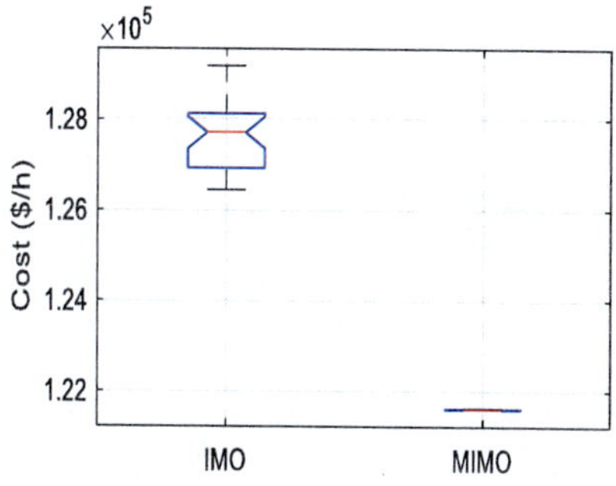

Fig. 3 Boxplots of MIMO and IMO

Table 2 Cost comparison with other methods suggested by researchers

Name of algorithm	Cost ($/h)			
	Minimum	Average	Maximum	SD
ABC [3]	124,009.4	NA	NA	NA
DCPSO [5]	121,948.8	123,450.8	125,772.4	NA
CGBABC [6]	121,595.83	121,814.7	122,169.1	116.21
E-PSO [9]	117,080.62 (123,274.5)[a]	119,568.73	122,655.1	1464.788
TLBO [4]	121,760.5	NA	NA	NA
JAYA-TLBO [11]	121,694.4	NA	NA	NA
IMO	**126,445.5**	**127,643.4**	**129,182.1**	**797.558**
MIMO	**121,596.6**	**121,604.7**	**121,623.1**	**9.49063**

NA stands for Not Available

6 Conclusions

The proposed algorithms IMO and MIMO have been successfully implemented to solve multi-area test systems with tie lines. It has been observed that the IMO fails to obtain an optimal solution for a considered multi-area test system. When this optimization technique is modified then it gives robust results. A detailed comparison of results produced by proposed algorithms with other recently published and established algorithms proves that MIMO provides better or comparable results.

References

1. Manoharan PS, Kannan PS, Baskar S, Iruthayarajan MW (2009) Evolutionary algorithm solution and KKT based optimality verification to multi-area economic dispatch. Int J Electr Power Energy Syst 31(7):365–373
2. Sharma M, Pandit M, Srivastava L (2011) Reserve constrained multi-area economic dispatch

employing differential evolution with time-varying mutation. Int J Electr Power Energy Syst 33(3):753–766
3. Basu M (2013) Artificial bee colony optimization for multi-area economic dispatch. Int J Electr Power Energy Syst 49:181–187
4. Basu M (2014) Teaching–learning-based optimization algorithm for multi-area economic dispatch. Energy 68:21–28
5. Jadoun VK, Gupta N, Niazi KR, Swarnkar A (2015) Multi-area economic dispatch with reserve sharing using dynamically controlled particle swarm optimization. Int J Electr Power Energy Syst 73:743–756
6. Secui DC (2015) The chaotic global best artificial bee colony algorithm for the multi-area economic/emission dispatch. Energy 93:2518–2545
7. Ghasemi M, Aghaei J, Akbari E, Ghavidel S, Li L (2016) A differential evolution particle swarm optimizer for various types of multi-area economic dispatch problems. Energy 107:182–195
8. Secui DC (2017) Large-scale multi-area economic/emission dispatch based on a new symbiotic organisms search algorithm. Energy Convers Manage 154:203–223
9. Mohammadian M, Lorestani A, Ardehali MM (2018) Optimization of single and multi-areas economic dispatch problems based on evolutionary particle swarm optimization algorithm. Energy 161:710–724
10. Lin J, Wang ZJ (2019) Multi-area economic dispatch using an improved stochastic fractal search algorithm. Energy 166:47–58
11. Mokarram MJ, Niknam T, Aghaei J, Shafie-khah M, Catalao JP (2019) Hybrid optimization algorithm to solve the nonconvex multiarea economic dispatch problem. IEEE Syst J 13(3):3400–3409
12. Kumar S, Kumar V, Katal N, Singh SK, Sharma S, Singh P (2021) Multiarea economic dispatch using evolutionary algorithms. Math Probl Eng 2021:1–14
13. Javidy B, Hatamlou A, Mirjalili S (2015) Ions motion algorithm for solving optimization problems. Appl Soft Comput 32:72–79

Using GPU and TPU Hardware Accelerators to Develop a Cloud-Based Genetic Algorithm System

Qabas Abdal Zahraa Jabbar, Narjis Mezaal Shati, Nada Thanoon Ahmed, and Yasmin Makki Mohialden

Abstract Genetic algorithms (GAs) are widely used in the scientific and industrial sectors due to their speed and flexibility in solving real problems. This article discusses the creation of a cloud-based genetic algorithm system. Utilizing Google Colab to discover optimum solutions and test the system's overall performance across a variety of fundamental cloud run-time types that contain Tensor Processing Unit (TPU) and Graphics Processing Unit (GPU) resources. The suggested technique demonstrated that when implementing mathematical equations, TPU outperforms GPU by achieving a (460.35) average of generation with a (13.95) average of run time in TPU versus a (484) average of generation with a (14.32) average of run time in GPU.

Keywords Genetic algorithms · Google Colab · Google Colaboratory · Run-time evaluation · GPU · TPU · PyGAD · Fully-connected neural networks (FCNNs)

1 Introduction

An evaluation technique for computer models that is based on genetic principles is known as the genetic algorithm. To take advantage of this technique, one must first create "genes" that include possible examples of actions. In each generation, current models are assessed and based on their fitness level, are given permission to mate and reproduce in the following generation. Crossings and mutations may arise as a result of the genes' being transmitted during mating. The current population has been destroyed, and its descendants have been chosen to form the next generation. Additionally, the term "genetic algorithm" refers to a collection of modeling or optimization methods that, in their quest for the optimal solution, attempt to imitate certain aspects of biological modeling in their search for the optimal answer. The modeled object is generally represented in such a way that it is simple to update it on an ongoing basis by computer software. Following that, a huge number of possible

Q. A. Z. Jabbar · N. M. Shati · N. T. Ahmed · Y. M. Mohialden (✉)
Computer Science Department, Collage of Science, Mustansiriyah University, Baghdad, Iraq
e-mail: ymmiraq2009@uomustansiriyah.edu.iq

A. Rani et al. (eds.), *Signals, Machines and Automation*, Lecture Notes in Electrical Engineering 1023, https://doi.org/10.1007/978-981-99-0969-8_58

models are created and evaluated in light of the available information. Only the "best" models are preserved for use in the following generation of models once each model is tested. Preservation can be either deterministic (select the best *k* models) or a random process (choose the best *k* models with a probability proportional to the score). The procedure is repeated endlessly until convergence is attained when these models are randomly disrupted (as in asexual reproduction). If the model is built in such a way that the winners have "genes," it is possible that they will "mate," resulting in the creation of the next generation. In scientific and commercial fields, genetic algorithms (GAs) are very popular search techniques for tackling tough optimization issues. This is due to the efficiency and flexibility with which GAs can find advantageous solutions to challenging optimization problems [1–5]. Colab is a cloud-based notepad. Google Docs allows you to collaborate on documents. Colab gives you access to popular research libraries. Google's Colaboratory promotes machine learning and research. Jupyter notebooks work well in the cloud or without any further setup. This project aims to integrate Google Colab for cloud-based machine learning and deep learning applications [6].

Google Colaboratory (colab) simplifies data analysis by not requiring installation or a web browser. The Oogle Colaboratory web application allows you to write code in the browser, keep results and share them, and use advanced computing resources for free. Free GPU support for machine learning models in Google Colab is quite useful.

Python libraries can develop genetic algorithms. Among these are the Python libraries. This section summarizes these libraries' goals and limitations [7]. PyGAD, an open-source Python library for genetic algorithm optimization, was used to create the genetic algorithm system. PyGAD premiered in April 2020. This essay is divided into sections. Section 2 reviews the available literature. Section 3 gives a technique; Sect. 4 explains it. In Sect. 5 Conclusion and future research ideas.

2 Literature Review

Mehanović et al. [2] data preparation is becoming increasingly important as data production, storage, and processing rise. Preprocessing functions like feature selection take time. In this study, a wide range of commodity hardware components are combined to eliminate this issue. The research largely used G-algorithms. Hadoop was used as an open-source library for genetic algorithms. Hadoop The machine learning characteristics support the vector machine, ANN, RT, regression logistic, and nave bays. Using feature selection methods on four NSL-KDD data sets reduced feature usage by 90.45%. It has broad ramifications. However, using MapReduce to parallel the evolutionary technique is tricky. A random selection of characteristics is also possible using evolutionary algorithms, and parallelization reduces data preparation time. The choice of functions goes beyond current methods [2].

Quiza [8] this chapter addresses the use of cloud-based optimization tools to address digital manufacturing concerns. The first section summarizes key numerical optimization techniques. The second portion explores swarm intelligence evolutionary algorithms. It also covers contemporary manufacturing digital sectors like cloud computing and Industry 4.0. A case study on cloud-based turning optimization concludes the final section [8].

Kamel et al. [9] the article develops a framework for assessing and improving time-based transit prices in large networks, as these charges affect mode, departure time, and route, as well as vehicle-to-vehicle and vehicle-to-traffic interactions. The Greater Toronto Area, more precisely the city of Toronto, is used as a case study. According to the research, Time-based fares are effective for spreading out transportation demand during peak periods. However, when compared to the huge increase in peak-hour fares, the decreases in weighted average multimodal door-to-door travel lengths are modest [9].

Sharma et al. [10] this article compares CNN's performance on GPU and TPU (TPU). You can use Google Colab's GPU and TPU. On MNIST, TPUs need fewer epochs than GPUs for various batch sizes. This work may be studied by HPC experts (HPC) [10].

Li et al. [11] the authors propose to speed up genetic algorithm evolution by improving parallel genetic algorithms. With a rising population, the evolutionary technique uses more computer resources. It uses the GPU's numerous cores to provide a parallel genetic technique. Using GPUs for parallel genetics allows for rapid population growth. The huge volume of the next generation's chromosomal population can be partitioned into blocks for faster selection and crossover. TSP computes GPU and CPU without using algebraic optimization [11].

3 The Proposed Methodology

We present a way of creating a genetic algorithm system for solving arithmetic problems using Google Colab. Intuitive setup of a deep learning or machine learning environment. Google calls it "collab." Colaboratory is a no-setup cloud-based notebook environment. The browser-based Colaboratory allows you to achieve this.

PyGAD is the fastest-growing library.

PyGAD is made up of three parts:

1. Py denotes a Python library. A standard Python library name convention.
2. The genetic algorithm stands for "GA".
3. Since the library initially only supported decimal genetics, D., this includes binary genetic algorithms. PyGAD stands for PyGAD. The software was intended for two purposes: simplicity and novice-friendliness. User-centric design the library provides single-objective optimization with flexible parameters. It follows a life-cycle from population generation to the optimal solution. Any operator can be added to the lifetime, and the order can be altered. Both decimal and binary

formats are supported by PyGAD. PyGAD is a user-friendly toolkit for applying the genetic algorithm for optimization. It was created with two key goals in mind:

(1) Attempting to make it all as simple as possible for inexperienced users.
(2) Giving the user full control this simplifies the naming of classes, methods, properties, and parameters. The evolutionary algorithm may be developed quickly and easily, with clear and understandable configuration options. Compared to other libraries, it has fewer classes, methods, and functions to call.

PyGAD has an elastic lifecycle that allows users to change the evolution loop in any situation. This involves enabling or disabling crossover and mutation operators, while also overriding them to create new research operators.

PyGAD contains the following seven modules:

1. (Pygad.pygad): The core module for the GA. It is also imported when the library is imported.
2. (Pygad.nn): creates FCNNs entirely in NumPy.
3. (Train.gann): The NN module is used to design trains.
4. (Pygad.cnn): similar to nn, but with the exception of CNNs.
5. (Pygad.gacnn): In the same way as the GAN module, but designed specifically for CNN training.
6. (Pygad) is used to train Keras models.
7. PyTorch models are trained using the pygad module.

Figure 1 the proposed system, life cycle. Each problem requires the specification of five mandatory parameters:

(1) Iterations and generations are both considered.
(2) Sol per population: the size of the population (number of solution/chromosomes/individuals).
(3) No. of parents mating: The number of solutions selected as parents for mating and offspring generation in the population.
(4) No. of genes: A solution's total number of genes.
(5) Fitness function: the ability to exercise.

These are the PyGAD minimal parameters. The parameter names are self-explanatory. All PyGAD modules' classes, parameters, attributes, methods, and functions are described in the documentation.

After the run() method, other approaches that may be used to learn more about solution of PyGAD's. Two of them are:

(1) best_solution (): It gives you the finest option. PyGAD discovered the information regarding the ideal solution:

    (a) The parameters of the best solution (as an example, the three weights for the problem solved in Listing)
    (b) The best values of fitness

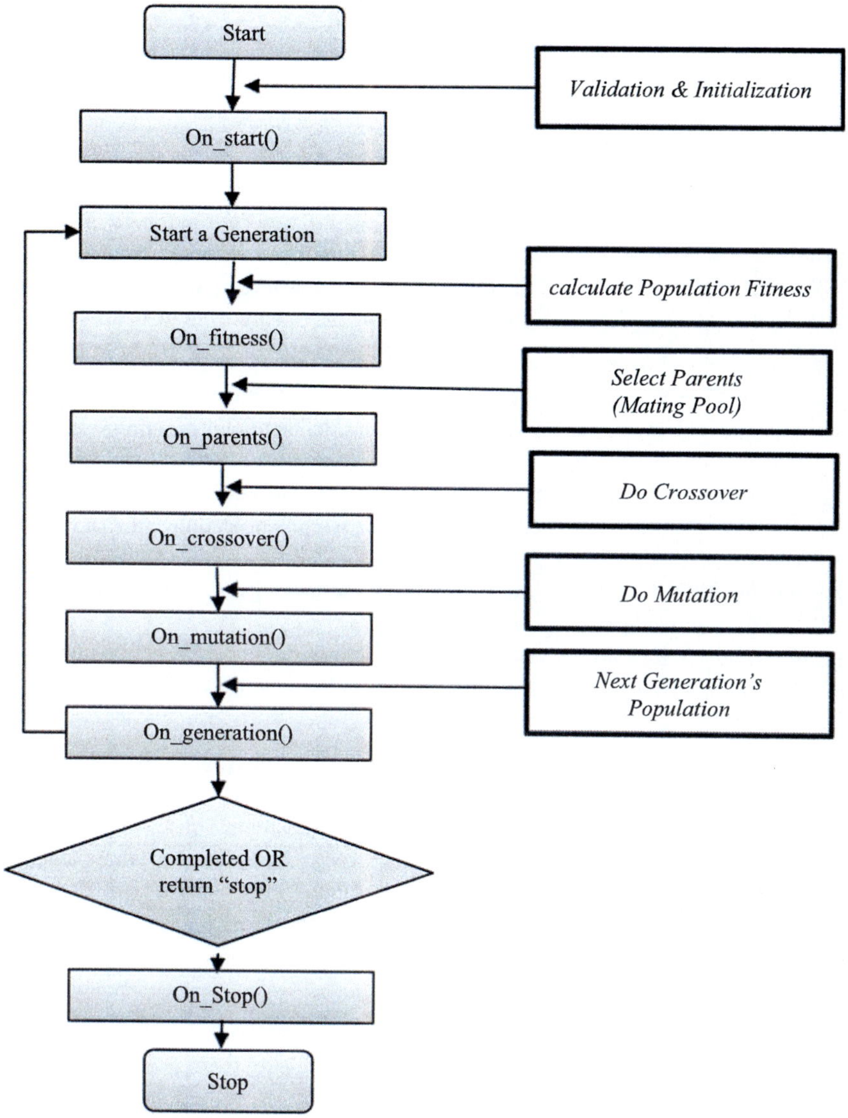

Fig. 1 Proposed system flowchart

(c) The index of solution's in its.

(2) plot_result(): creates a graph that shows how the fitness value changes with each generation If the user wants to save the image, this method returns it. We can use PyGAD, see the following equation with five inputs, to determine the values of w_1, w_2, w_3, w_4 and w_5 that fulfill this equation.

$$Y = w_1x_1 + w_2x_2 + w_3x_3 + w_4x_4 + w_5x_5 \tag{1}$$

4 Results and Discussions

Many variables affect the performance of the proposed evolutionary algorithm system. One of the most critical variables is the availability of Google Colab GPU and TPU. Your application can use the GPU or TPU at runtime. Change Cloab Runtime type. Table 1 shows when to use CPU, GPU, and TPU. Lifecycle of PyGAD.

There are seven callback functions available:

(1) on_start(): after the run () function has been executed once.
(2) on_fitness (): the population fitness has been calculated, this function is executed after each generation's.
(3) on_parents(): In each generation, the parents are chosen.
(4) on_crossover (): After each generation's crossover procedure, this function is called.
(5) on_mutation(): This function is executed after each generation has passed the mutation procedure.
(6) on_generation (): At the end of each generation, this function is executed.
(7) on_stop(): After the run () method has completed its execution, it is executed once more.

After extensive preliminary experiments to solve the above mathematical equation (Eq. 1) by using the GPU and the TPU cloud base methods. These methods were implemented for four different outputs with five times repetition for each output's execution using a random input function, as shown in Tables 2 and 3, where it shows the details of the implementation of the two methods.

Table 1 Utilization of CPU, GPU, and TPU

Processor type	Utilization
CPU	• Rapid prototyping that is flexible • Simple models that need little training • Tiny models, short lifetimes • More advanced TensorFlow models • Restricted input and output and network bandwidth
GPU	• TensorFlow models that need advanced processing require an external tool • Models that are resource-free or unmanageable • TensorFlow models that need the execution of a significant number of specific TensorFlow operations must be run partially on CPUs (where possible)
TPU	• Matrix-intensive models • Models with TensorFlow operations are not part of the main training cycle • Training for weeks or months • Models with very large effective cluster sizes

Table 2 Result of Google Colab TPU

Required output	Run's	Generation	Fitness	Output	Error	Time/Sec
23.33	Min	48	6353.62	23.3298	0.00004	10
	Max	681	36,662.35	23.3301	0.0002	25
	Average	368.2	16,849.586	23.32996	0.00004	13
−55	Min	245	8638.93	−55.00	0	9
	Max	773	132,895.01	−55.00	0	22
	Average	518	57,000.2054	−55.00	0	13
55	Min	219	9150.55	54.9999	0	9
	Max	822	49,089.87	55.0001	0.0001	37
	Average	528.2	32,270.66	55	0	14.8
−23.33	Min	33	4811.8	−23.33	−0.01	9
	Max	798	30,120.02	−23.32	0	35
	Average	427	12,418.066	−23.33	0	15

Table 3 Result of Google Colab GPU

Required output	Run's	Generation	Fitness	Output	Error	Time/Sec
23.33	Min	124	4560.5	23.329	−0.0001	11
	Max	975	83,185.5	23.3301	0.001	12
	Average	450	30,474.6254	23.32942	0.00058	11.8
−55	Min	42	2872.75	−55	0	12
	Max	972	47,143.22	−55	0.0003	30
	Average	473.8	19,609.554	−55	0.000074	15.8
55	Min	543	9973.52	54.9998	0.0001	12
	Max	710	44,843.29	54.9998	0.0002	30
	Average	615.6	25,656.498	54.99988	0.00012	16.2
−23.33	Min	212	4338.37	−23.33	−0.00003	12
	Max	612	425,170.62	425,170.6	0.0002	12
	Average	396.6	109,987.802	−23.33	0.000054	13.5

According to the result listed in Tables 2 and 3 the Google Colab TPU method is slightly better than the Google Colab GPU method in terms of generation number, error, and run time, as illustrated and summarized in Table 4.

From the previously listed results in Table 4, it could be noticed that the two methods are good in implementing mathematical equations because the difference in the results is very small that shows average of generation in TPU equivalent to (460.35) as for the GPU equal to (484), and in average of run time, the result obtained in TPU = (13.95), GPU = (14.32) also notes from the obtained result that the TPU method is always superior to the GPU method, as shown in Figs. 2 and 3 that illustrate

Table 4 Result summarization

Required output	TPU generation	GPU generation	TPU fitness	GPU fitness	TPU Error	GPU Error	TPU Time/Sec	GPU Time/Sec
23.33	368.2	450	16,849.586	30,474.6254	0.00004	0.00058	13	11.8
−55	518	473.8	57,000.2054	19,609.554	0	0.000074	13	15.8
55	528.2	615.6	32,270.66	25,656.498	0	0.00012	14.8	16.2
−23.33	427	396.6	12,418.066	109,987.802	0	0.000054	15	13.5
Average	460.35	484	29,634.62935	46,432.11985	0.00001	0.000207	13.95	14.325

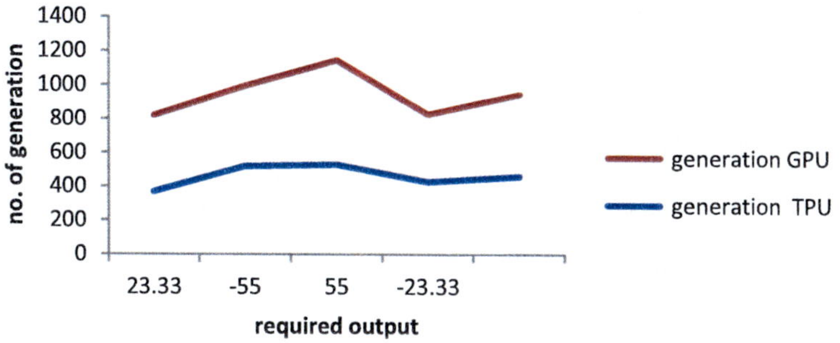

Fig. 2 Comparison in terms of the number of generations

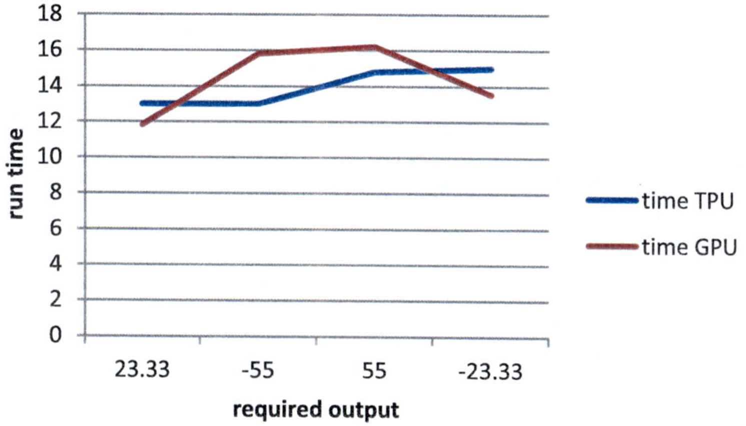

Fig. 3 Comparison in terms of run time

the comparison in terms of the number of generations, and the comparison in terms of run time for the two methods, respectively.

5 Conclusion

This work proposes utilizing PyGAD, which is an open-source Python library for single-objective genetic optimization. A formalized Create a fitness function, instantiate the pygad.GA class, and use PyGAD to run the "run" method. A wide range of parameters and characteristics can be used to tailor the genetic algorithm. PyGAD has a lifetime that allows you to follow the evolution from beginning to end. This helps solve problems faster and with fewer lines of code than with other libraries. As a programming environment, we are using Google Colab, which is cloud-based. A free-of-charge Python programming environment contains two types of hardware accelerators, which are called GPU and TPU, that are used in the running of the genetic programming optimizer program for mathematical equations. The experimental work in many cases proved that the success of the genetic algorithm program depends primary on the type of the processor. From the obtained result, infer that using TPU gives superior result rather than GPU indicating that in TPU the average of generation with average of run time equivalent to (460.35), (13.95), respectively, and in GPU obtained (484) as average of generation with average of run time equal to (14.32).

Acknowledgements The authors would like to thank Mustansiriyah University (https://uomustansiriyah.edu.iq/) Baghdad–Iraq for its support in the present work.

References

1. Otubamowo K, Egunjobi TO, Adewole AP (Oct 2012) A comparative study of simulated annealing and genetic algorithm for solving the travelling salesman problem. [cited 29 April 2022]; Available from: https://ir.unilag.edu.ng/handle/123456789/5486
2. Mehanović D et al (2021) Feature selection using cloud-based parallel genetic algorithm for intrusion detection data classification. Neural Comput Appl 33:11861–11873
3. Ahmed N, Mohialden Y, Alshibani D (2018) A new method for self-adaptation of genetic algorithms operators. Int J Civil Eng Technol 9:1279–1285
4. Ahmed NT, Mezaal Shati N (2019) A new method for solving deadlock using genetic algorithms. https://papers.ssrn.com/abstract=3533677
5. Adnan FA, Ab Saad S, Yahya ZR, Wan Muhamad WZA. Genetic algorithm method in examination timetabling problem: a survey. In: Regional conference on science, technology and social sciences (RCSTSS 2016)
6. Prashanth B, Mendu M, Thallapalli R (2021) Cloud based Machine learning with advanced predictive analytics using Google Colaboratory. Mater Today: Proc https://doi.org/10.1016/j.matpr.2021.01.800
7. Gad AF (2021) PyGAD: an intuitive genetic algorithm python library. arXiv:2106.06158 [cs, math]
8. Quiza R (2020) Cloud-based optimization in manufacturing digital industries. Book chapter/Manufacturing in Digital Industries, vol 6
9. Kamel I, Shalaby A, Abdulhai B (2020) A modelling platform for optimizing time-dependent transit fares in large-scale multimodal networks. Transp Policy 92:38–54
10. Sharma V, Gupta GK, Gupta M (11 May 2021) Performance benchmarking of GPU and TPU on Google Colaboratory for convolutional neural network. Applications of artificial intelligence in engineering proceedings of first global conference on artificial intelligence and applications (GCAIA 2020). Springer, pp 639–646
11. Li C-C, Liu J-C. On the accelerated convergence of genetic algorithm using GPU parallel operations: computer science & IT book chapter|IGI global. https://www.igi-global.com/chapter/on-the-accelerated-convergence-of-genetic-algorithm-using-gpu-parallel-operations/161064
12. Yacob N, Mohd Noor N. Regional conference on science, technology and social sciences (RCSTSS 2016)

Pattern Recognition of Human Fingerprint Utilizing an Efficient Artificial Intelligence Algorithm

Nisreen Abd Alhadi Jabr

Abstract The fingerprint identification has great effectiveness in forensic science and helps in the criminal investigations. Fingerprints are distinctive and remain enduring throughout a person's life. The automatic fingerprint recognition systems are dependent upon hills and its characteristics known as minutiae. Hence, it is highly essential to score these minutiae accurately then refuse the improper parts. In this work a ridge ending and ridge ramify have been utilized as minutiae for fingerprint recognition system. At the time of analysis of algorithms, the approaches of attributes impart better results. The recognition rate is increased and the error rate is diminishing with the aid of this technique. The ultimate crucial stride here in matching of automatic fingerprint is to securely extractor specifics from the binary images of captured fingerprints. There are already a variety of techniques available to extract fingerprint details. The rate of recognition for such intended approach of fingerprint recognition system using artificial neural networks is 93%. From the extricate outcome, we may infer about a very affirmative impact of artificial neural networks on the comprehensive recognition rate, specifically in low excellence images.

Keywords Fingerprint recognition · Human fingerprint · Artificial neural networks · Back propagation

1 Introduction

This study providing specifics against the achievement of a pattern matching procedure to coordinate and validate fingerprints. Neural Network insight algorithm has been utilized to perform compelling pattern matching to recognize fingerprints. The word biometric come up to from the Greek words profiles-life also measurements examine [1]. People utilize somebody uniqueness, for example, iris, face, finger, eyes, hand, or voice to distinguish one another. Biometric can't be acquired, taken, or neglected. Iris recognition, fingerprint identification, speaker recognition,

N. A. A. Jabr (✉)
Branch of Theoretical Sciences, College of Physical Education and Sports Science, Mustansiriyah University, Baghdad, Iraq
e-mail: nisreen1987@uomustansiriyah.edu.iq

A. Rani et al. (eds.), *Signals, Machines and Automation*, Lecture Notes in Electrical Engineering 1023, https://doi.org/10.1007/978-981-99-0969-8_59

and face recognition are different biometric frameworks that exist. With modern advances, various biometric frameworks are developed. Fingerprint ID is part of the normal biometric frameworks. Biometric is the method of recognizing a singular in light of their particular qualities. The manual fingerprint check is exceptionally muddled (drawn-out), tedious, and costly which is unequipped for meeting the present expanding execution prerequisites. In biometrics, there are two unmistakable fingerprint matching techniques those are fingerprint recognizable proof as well as fingerprint check [2–5].

We will illustrate the methodology in Sect. 2, the proposed model's general structure in Sect. 3, simulation results, and discussion in Sect. 4. Lastly, the conclusions are given in Sect. 5.

2 Methodology

The architectural layout of suggested framework, for example, earlier fingerprint recognition framework involving neural network upon dual distinct information bases as FVC 2002 also for investigation of the fingerprint recognition framework displayed in Fig. 1 that incorporates picture dataset to the concentrate include set, and finally highlight characterization to fingerprint recognition framework picture as match or nonmatch [6–10].

Information base Specification for Fingerprint as under [11–13] Datasets for fingerprint recognition framework as FVC 2002 fingerprint. FVC 2002: Four particular datasets are utilized as DB1, DB2, DB3, and DB4. Every information base is of 150 fingers and 12 examples each (1800 fingerprint pictures). Some example pictures from the FVC2002 Fingerprint information base are displayed in the accompanying Table 1.

2.1 Fingerprint Recognition System Utilizing Neural Network

The investigated work of fingerprint recognition in light of minutiae-based. The interaction comprises of the accompanying advances.

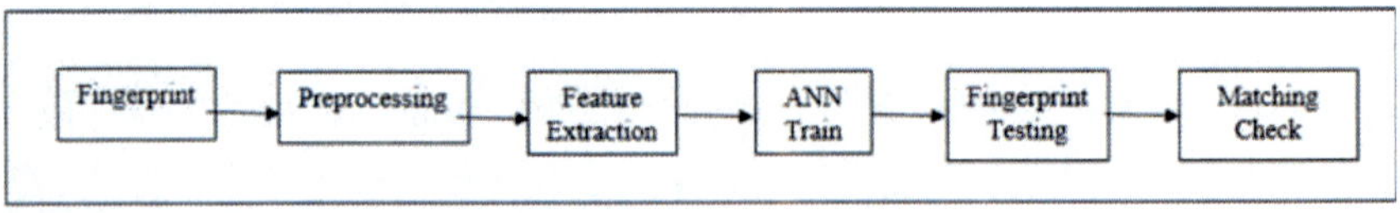

Fig. 1 Block diagram showing the procedure steps for fingerprint recognition system utilizing neural network

Table 1 Element pictures of FVC2002 DB1, DB2, DB3, DB4 database of fingerprint

Sample Images of FVC2002 DB1 Fingerprint Database				
Image Name	1DB1.tif	2DB1.tif	3DB1.tif	4DB1.tif
Image				
Sample Images of FVC2002 DB2 Fingerprint Database				
Image Name	1DB2.tif	2DB2.tif	3DB2.tif	4DB2.tif
Image	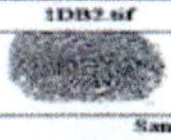	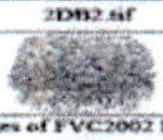		
Sample Images of FVC2002 DB3 fingerprint Database				
Image Name	1DB3.tif	2DB3.tif	3DB3.tif	4DB3.tif
Image			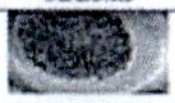	
Sample Images of FVC2002 DB4 Fingerprint Database				
Image Name	1DB4.tif	2DB4.tif	3DB4.tif	4DB4.tif
Image				

To catch an advanced image of the fingerprint patterns, a fingerprint sensor is an electronic gadget is utilized. The standard data sets are utilized, for example, FVC 2002.

Pre-processing images are useful for improving the image quality by separating and eliminating undesirable commotions. The underlying advance in the proposed work is by to apply limit an incentive for the calculation of the parallel image from the information grayscale fingerprint image.

The fingerprint binarization algorithm creates the 1-bit class of image with 1 as valleys and 0 as edges. In any case, the versatile binarization strategy is support on a limit with gray-level pixels, lower than relegated to one and zero. In MATLAB, a worth 1 represents as white pixel, and worth 0 represents a dark pixel. By utilizing the edge interaction, we can get a grayscale image to double image, as shown in Fig. 2. By applying a limit cycle to n picture, every pixel esteems are breaking down to the information edge. Those pixel esteems are more modest over the limit esteem is set to 0 and those pixel esteem which is more prominent than the edge esteem is put to 1 [14].

The categorization of every perspective is done by tracked down the worth of every pixel, and every one edge is diminished to its midpoint pixels. The reason at the back thinning is to diminish the fingerprint dependent upon one pixel wide, and thinning is a morphological activity accomplished upon double pictures. Such thing is accomplished by progressive pixels' deletions along various edges of every picture [15, 16].

Fingerprint Image upgrade is utilized to cause picture more clear for additional utilization. Here, we can apply Histogram balance to upgrade the difference of the image by changing the power upsides of the image. Histogram leveling utilizes all the reach from 0 to 255, and the representation impact is improved [17–20]. This segment investigation the aftereffects of the examination. For fingerprint improvement, we might analyzed different channel procedures as displayed in underneath Table 2.

The better fingerprint image is binarized for thinning up to 1 pixel broad. The parallel image is used to remove minutiae focuses, and that are the marks of edge finish and bifurcations. The areas of minutiae are focuses next to with the variation is

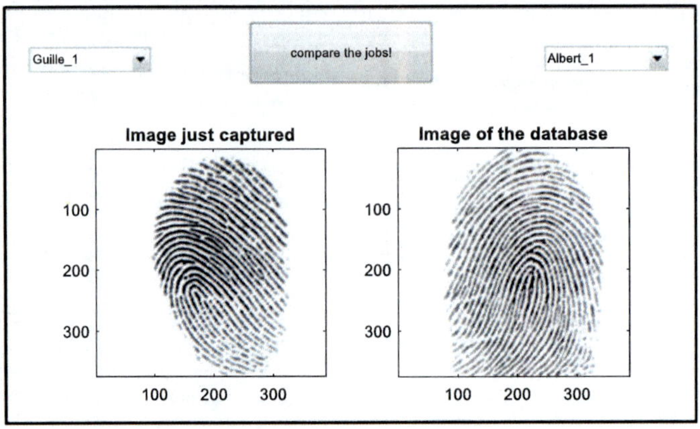

Fig. 2 Binarization implementing

Table 2 Various histogram filter approaches utilized for fingerprint recognition [21]

Methods	Mean square error (MSE)	Peak signal-to-noise ratio (PSNR)
Gabor filter	7.3	12.11
Median	356.48	18.22
Gaussian	124.92	22.79

extricated and gather to frame highlight sets. For extraction of minutiae point's eight associated pixels are utilized. The Crossing Number (CN) technique is utilized for performing minutiae extraction, which extricates the edge endings and bifurcations from the paired image by analyzing the nearby neighborhood edge pixel utilizing a 3×3 window. The (CN) Crossing Number for an edge P pixel is determined by minutiae relying upon the sensor goal and finger situation on the sensor. The fingerprint image captured through helpless scanners might contain a less number of minutiae focuses [22].

$$CN = 0.5 \sum_{i=1}^{8} |P_i - P_{i+1}| \tag{1}$$

The accompanying data is recorded in shown Fig. 3 for every minutia extricated.

1. The direction X with Y
2. Direction of the related edge section, as well as
3. Gathering of minutiae (edge finishing or complication).

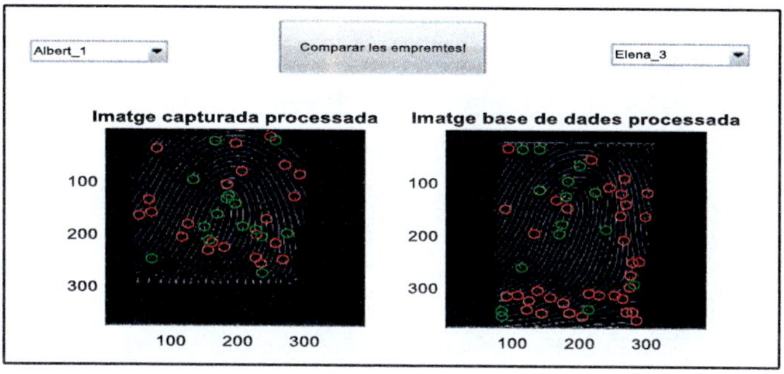

Fig. 3 Fingerprint sample minutiae extraction from [23]

3 The Proposed Model

In this section, it is explained the proposed model of research using artificial neural networks (ANN) including research design and research procedure. Equations should be placed at the center of the line and provided consecutively with equation numbers in parentheses flushed to the right margin, as in (1). The use of Microsoft Equation Editor or Math Type is preferred. There are many fingerprint matching methods are accessible. Here, we accomplish artificial neural networks (ANN) to confirm whether or not they have a place with the indistinguishable anybody. Utilizing neural network, minutiae platform fingerprint matching gives real matching result. There are various different sorts of the neural organizations, for example, perceptron, back spread organization, counter engendering network, Hopfield organizations, and so on. In this work, feed-forward back proliferation network technique is utilized. Ensuing plots delineate the result of fingerprint recognition framework at include extraction level at unmistakable fingerprint information bases utilizing artificial neural networks [24, 25]. The construction of the multilayer perceptron (MLP) comprises three layers, for example, input layer, stowed away layer, and result layer as shown in Fig. 4. The highlights extricated by fingerprint quality are melded and given as information (preparing). The information layer disperses inputted information to the processing components in the following layer. The following layer is covered up layer, which maps the nonlinearity conduct and change the loads and predisposition. The result layer involves the initiation work for testing the presentation of test include vectors. The back propagation algorithm is utilized to limit the mean absolute error (MAE). The exhibition accomplished with MLP it shows the huge improvement in the presentation.

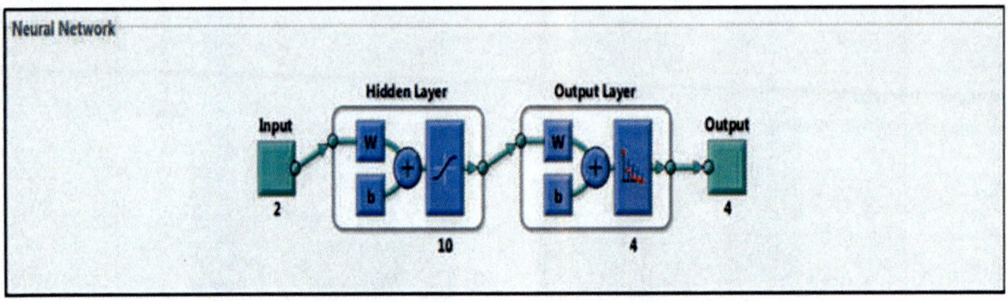

Fig. 4 Structure of the ANN architecture utilized in fingerprint pattern recognition process

4 Results and Discussion

In this section, we have implemented the finger print pattern recognition process utilizing feed-forward artificial neural network (ANN). The testing data sets have been applied for 1000 different finger prints shapes samples. Each one of the 1000 entered finger data shapes has been analyzed and binarized into 1000*2 input matrix and passed through the procedure steps for fingerprint recognition system shown previously in Fig. 2, starting from the fingerprint acquisition and finished with the matching step to produce resulting set of 1000*4 output matrix. The input data set vector will include information of contrast and mean values for each one of the 1000 fingerprint image sample. Similarly, the output data set vector will contain information of the tested fingerprint image sample such as variance, mean square error (MSE), peak signal to noise ratio (PSNR) and CN for each one of the 1000 fingerprint image sample. In this research, we have examined the scaled conjugate gradient (SCD) algorithm in applying the ANN technique. Figure 4 illustrates a block diagram of ANN train network with two input layers, ten hidden layers, and four output layers. Also, the implementation of the testing program utilizing the suggested ANN approach. The resulting performance of the ANN train network for the fingerprint model has been shown in Fig. 5.

It is obvious from the results illustrated in Figs. 5 and 6 that the ANN training network produce a very accurate matching to the fingerprint pattern recognition. The performance error and gradient of the training process are very small which indicate the high accuracy and perfect matching of the tested fingerprint data with the required constraints. It is obvious from the results illustrated in Fig. 5 that the ANN training network produce a very accurate matching to the fingerprint pattern recognition. The performance error and gradient of the training process are very small which indicate the high accuracy and perfect matching of the tested fingerprint data with the required constraints. By observing the ROC graph that might be considered as a subset of a true positive rate (sensitivity) against the false positive rate (specificity) as the limit is changed. The receiver operating characteristic (ROC) graph illustrates FAR vs. FRR for a gathering of edge esteems T. For ideal ROC graph, just have amounts which are located either upon the x-axis (FAR) or the y-axis (FRR). For FRRs other than 0, the FAR is 1, as well as the other way around. After the presentation of the

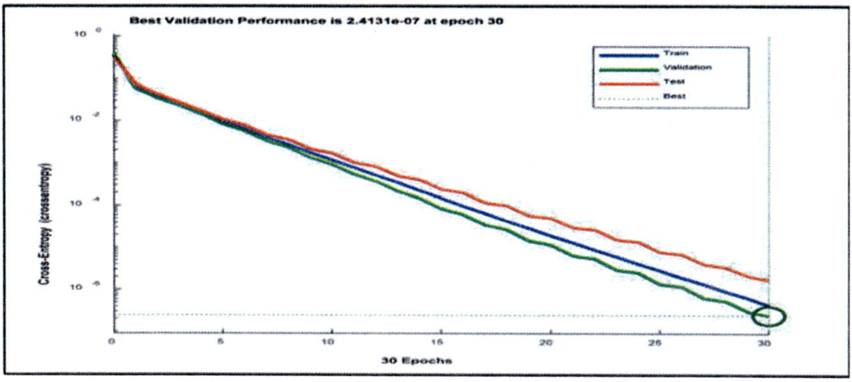

Fig. 5 Performance results of the ANN train network program

ideal test, at the upper-left corner, it will show 100 percent responsiveness and 100 percent explicitness in Fig. 6. The execution of the proposed strategy is contrasted and another method; the coordinating and non-matching depend on Euclidean distance which establishes limit esteem (i.e., on the off chance that Euclidean distance is not exactly the edge esteem, it is matched else not coordinated).

Some of the time, the biometric security framework might permit unapproved admittance to the unauthentic client, and to gage this kind of episode, FAR is utilized. FAR is determined like:

$$\text{FAR} = (\text{FA}/N) * 100 \tag{2}$$

FA Number of occasions of bogus acknowledgment.
N Total number of tests.

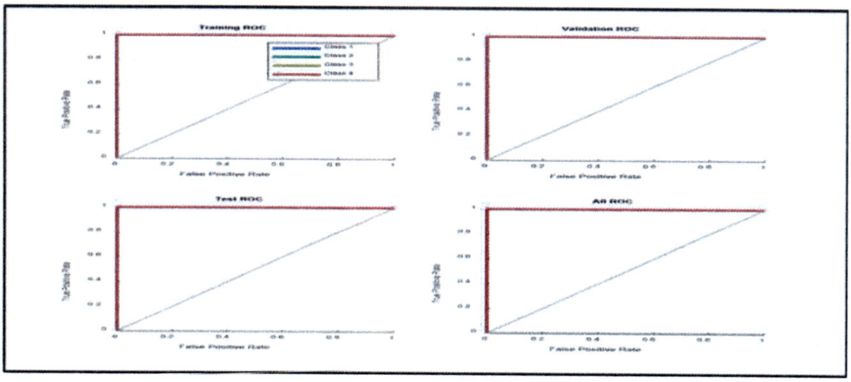

Fig. 6 ROC curves of the ANN train network program

In some cases, biometric security framework may reject the unapproved access of approved people and to quantify these kinds of episodes FRR is utilized. FRR is determined as:

$$\text{FRR} = (\text{FR}/N) * 100 \quad (3)$$

FR Number of occasions of bogus dismissals,
N Number of tests.

Three various types of tests are applied to the organization to perform various exercises:

- Preparing: Here the organization is changed by its blunder.
- Validation: It is utilized to gage network speculation and to end preparing when speculation quits moving along.
- Testing: It is utilized uniquely for testing the last answer for affirm the real prescient force of the organization. In our work, we have led a few instructional meetings. The measurements training the performance in view of Mean Squared Error. It is the mean squared contrast among result and aim as showed in Table 3. Lower rates of mean square errors are considered as better 1 while 0 indicates no blunder [26].

The proposed system accomplishes the recital for fingerprint recognition system rate using artificial neural networks is 93%, FAR 1.154%, and FRR 8.154% as shown in below Table 4.

Table 3 Result of the proposed system fingerprint recognition system for match and non-match persons

Name of classifier	Train samples	Test samples	Features moment invariant eccentricity, perimeter	Accuracy for match	Accuracy for non-match	Specificity
N. N	640	640	Ridge ramify, kurtosis	0.8625	0.8469	0.912727273
	219	219	Ridge ending, kurtosis, skewness	0.7431	0.6789	0.833333333
	640	640	Ridge ending	0.85313	0.871875	0.996363636

Table 4 Performance evaluation for fingerprint recognition system

Trait	Train samples	Test samples	Algorithm	Accuracy	FAR (%)	FRR (%)
Fingerprint	650	650	Minutiae extraction and artificial neural network	93	1.154	8.154

5 Conclusions

In this paper, a precise details identification for unique fingerprint matching has been applied for image pre-handling is the most evaluative pattern recognition issues. The execution of a unique mark includes extraction as well as matching algorithms relies on the nature of the information images fingerprint. Using the crossing number (CN) for the images framework after the decreasing phase minutiae extraction algorithm might recognize all the particulars, containing each valid and bogus details. The main advance in programmed fingerprint matching is to dependably separate the details from the caught parallel fingerprint pictures. There is a wide assortment of procedures accessible for extricating unique mark particulars. The proposed method accomplished fingerprint recognition system rate definitively by utilizing artificial neural networks was 93% from the achieved results, we deduce with respect to an extremely sure effect of artificial neural networks upon the global recognition rate clear in images inferior quality.

References

1. Zhou B, Han C, Liu Y et al (2020) Fast minutiae extractor using neural network. Pattern Recogn 103(107273):1–12. https://doi.org/10.1016/j.patcog.2020.107273
2. Illa S, Rohini S (2017) A survey on fingerprint minutiae extraction. Int J Adv Res, Ideas Innovations Technol 3(3):264–267
3. Szymkowski M, Saeed K (2019) Fingerprint feature extraction with artificial neural network and image processing methods. In: Pejaś J, El Fray I, Hyla T, Kacprzyk J (eds) Advances in soft and hard computing. ACS 2018. Advances in intelligent systems and computing, vol 889. Springer, Cham. https://doi.org/10.1007/978-3-030-03314-9_8
4. Choi et al (June 2011) Fingerprint matching incorporating ridge features with minutiae. IEEE Trans Inf Forensics Secur 6(2)
5. Kaur M et al (2008) Fingerprint verification system using minutiae extraction technique. World Acad Sci, Eng Technol 2(10):497–502
6. Abdullah HA (2012) Fingerprint identification system using neural networks. Nahrain Univ, Coll Eng J (NUCEJ) 15(2):234–244
7. Alonso-Fernandez F, Fierrez J, Ortega-Garcia J (2012) Quality measures in biometric systems. IEEE Secur Priv 10(06):52–62. https://doi.org/10.1109/MSP.2011.178
8. Manza RR, Gaikwad BP, Manza GR (2012) Use of edge detection operators for agriculture video scene feature extraction from mango fruits. Adv Comput Res 4(1):50–53. http://www.bioinfo.in/contents.php?id=33
9. Lippmann RP (1989) Pattern classification using neural networks. IEEE Commun Mag 27(11):47–50. https://doi.org/10.1109/35.41401

10. Mohamed MH (2021) Fingerprint classification using deep convolutional neural network. J Electr Electron Eng 9(5):147–152. https://doi.org/10.11648/j.jeee.20210905.11
11. Abiodun OI et al (2019) Comprehensive review of artificial neural network applications to pattern recognition. IEEE Access 7:158820–158846. https://doi.org/10.1109/ACCESS.2019.2945545
12. Fingerprint verification competition (2002) FVC2002. From http://bias.csr.unibo.it/fvc2002
13. Peter O et al (2016) Fingerprint authentication system using back-propagation with downsampling technique. In: 2nd international conference on science and technology-computer (ICST). Yogyakarta, Indonesia, pp 182–187. https://doi.org/10.1109/ICSTC.2016.7877371
14. Jayaraman B et al (2008) Fingerprint authentication using back propagation algorithm. Int J Soft Comput 3(4):282–287
15. Chatterjee A et al (2010) Fingerprint identification and verification system by minutiae extraction using artificial neural network. JCIT 1(1):12–16
16. Askarunisa A et al (2009) Fingerprint authentication using neural networks. MASAUM J Comput 1(2)
17. Nur-A-Alam et al (2021) An intelligent system for automatic fingerprint identification using feature fusion by Gabor filter and deep learning. Comput Electr Eng 95(107387). https://doi.org/10.1016/j.compeleceng.2021.107387
18. Marák P, Hambalík A (2017) Fingerprint recognition system using artificial neural network as feature extractor: design and performance evaluation. Tatra Mt Math Publ 67(1):117–134. https://doi.org/10.1515/tmmp-2016-0035
19. Viswanathan M et al (2020) IKP based biometric authentication using artificial neural network. In: AIP conference proceedings, vol 2271, No 1. https://doi.org/10.1063/5.0025229
20. Park SH, Lee SP (Dec 1998) EMG pattern recognition based on artificial intelligence techniques. IEEE Trans Rehabil Eng 6(4):400–405
21. http://www.rami-khushaba.com/electromyogram-emg-repository.html Accessed July 2015
22. Boostani R, Moradi MH (2003) Evaluation of the forearm EMG signal features for the control of a prosthetic hand. Physiol Meas 24(2):309–319. https://doi.org/10.1088/0967-3334/24/2/307
23. Kim KS et al (2011) Comparison of k-nearest neighbor, quadratic discriminant and linear discriminant analysis in classification of electromyogram signals based on the wrist-motion directions. Curr Appl Phys 11:740–745. https://doi.org/10.1016/j.cap.2010.11.051
24. Khushaba RN et al (2012) Toward improved control of prosthetic fingers using surface electromyogram (EMG) signals. Expert Syst Appl 39:10731–10738. https://doi.org/10.1016/j.eswa.2012.02.192
25. Lucas MF et al (2008) Multi-channel surface EMG classification using support vector machines and signal-based wavelet optimization. Biomed Signal Proc Control 3(2):169–174. https://doi.org/10.1016/j.bspc.2007.09.002
26. Awasthi G et al (30 May 2020) Analysis of fingerprint recognition system using neural network. In: 2nd international conference on communication and information processing (ICCIP) 2020. https://doi.org/10.2139/ssrn.3648835

A Systematic Review of Human Fall Detection Algorithms Using Machine Learning

Divya Singh, Meenu Gupta, Rakesh Kumar, and Ahmed J. Obaid

Abstract In the past century, due to healthy lifestyle of people, they suffered from very less disease. As era changes, the life style of people also changes. People are taking unhealthy food in today's era owing to a busy schedule, and they are suffering from various diseases such as human falls. The most common causes of human falls among the elderly include high (low) blood pressure, arthritis, osteoarthritis, coronary heart disease, dementia, depression, and stroke. As a result, intervening early in the event of a fall could cut morbidity and death dramatically. In recent years, the machine learning-based human fall model has made significant progress. A human fall detection model assists the elderly by recognizing the fall and saving their lives. Machine learning (ML)-based models have now become an important research subject because of its capacity to proactively assist the elderly. In this proposed study, the most recent research trends in fall detection system using the machine learning techniques have been analyzed. Moreover, the analysis focuses on the shortcomings associated with ML-based tactics in order to conduct deeper research. The goal of the study is to find out how various type of machine learning algorithms are detecting fall effectively. It is a comprehensive review of existing fall detection models and trends.

Keywords Human fall detection model (HFDM) · False alarm rate · Threshold-based approaches · Machine learning (ML) approaches · Deep learning (DL)

D. Singh · M. Gupta (✉) · R. Kumar
Department of Computer Science and Engineering, Chandigarh University, Punjab, India
e-mail: meenu.e9406@cumail.in

R. Kumar
e-mail: rakesh.e8623@cumail.in

A. J. Obaid
Faculty of Computer Science and Mathematics, University of Kufa, Kufa, Iraq
e-mail: ahmedj.aljanaby@uokufa.edu.iq

A. Rani et al. (eds.), *Signals, Machines and Automation*, Lecture Notes in Electrical Engineering 1023, https://doi.org/10.1007/978-981-99-0969-8_60

1 Introduction

The term "sudden collapse" or "fall" refers to an unintentional or unnatural alteration in the body's posture as it descends to the ground. Falls occur in the process of 3–5 per 1000 bed/days, according to epidemiological reports, as well as the Organization for Healthcare Research and Quality (HR&Q) appraise that 7 Lakh to 1 M hospitalized people collapse or fall per year [1].

Every year, approximately 6,84,000 people die due to falls around the globe. Out of this, more than 80% of these deaths were reported in middle and low-income nations. The rate of death falls among those over the age of 60 is the highest. Every year, 37.3 million people have catastrophic falls that necessitate medical assistance [2]. Falls are costly, and they add to rising healthcare expenditures. The Centre for Disease Control and Prevention (CDC) estimates that fall injury to persons 65 years or older will cost $19 billion, including fatal falls contributing to $0.2 billion. The falls frequently result in catastrophic injuries, including traumatic brain injuries that can result in death [3] (Fig. 1).

This article provides an overview of investigations that have utilized machine learning to detect falls. The rest of the study consists of the following outlined. Section 2 introduces the HFDM and its methodologies, while Sect. 3 presents a review of limitations and hurdles in non-ML-based systems. In Sect. 4, elaborate findings of existing research along with current and significant trends in the area of fall detection. Finally, conclude the research paper in Sect. 5.

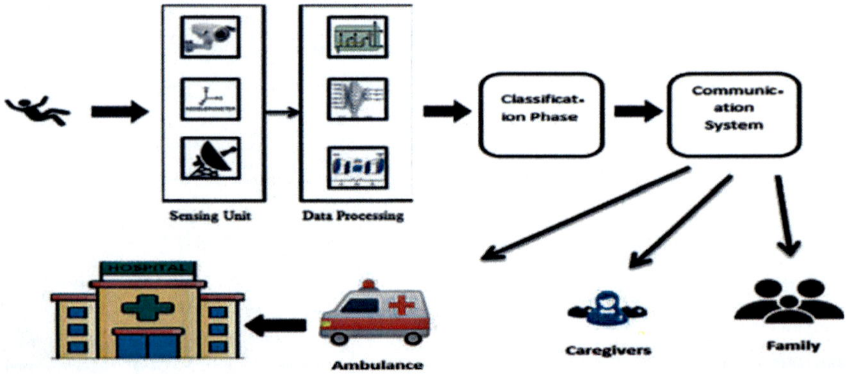

Fig. 1 Basic flow of human fall detection model [4]

2 Classification of Approaches and Method in Human Fall Detection

The development of HFDM has become a popular research area in recent years. Hausnotruf designed the first fall model (FM) in early 1970, which was formerly designated as the Personal Emergency Response System (PERS) [5] (Fig. 2).

2.1 *Threshold-Based Human Fall Detection Approaches*

A typical strategy frequently employed by the fall detection method is threshold-based categorization. The threshold-based technique is extremely effective, uses less energy, and is simple to incorporate into an embedded system (Table 1).

2.2 *Machine Learning-Based Human Fall Detection Approaches*

Machine learning (ML)-based HFDM is more accurate than the threshold model. Numerous investigations on fall detection approaches included technologies such as artificial neural networks (ANN), *k*-nearest neighbor (*k*-NN), and support vector machine (SVM) to characterize falls from ADL due to several advantages of ML. The preceding are the sub-divisions of the Machine Learning Based Model.

2.2.1 Wearable-Based Model

The wearable-based fall detection approach uses cutoff point classification, and this method has also encountered several issues, including sensor sensitivity to unwanted

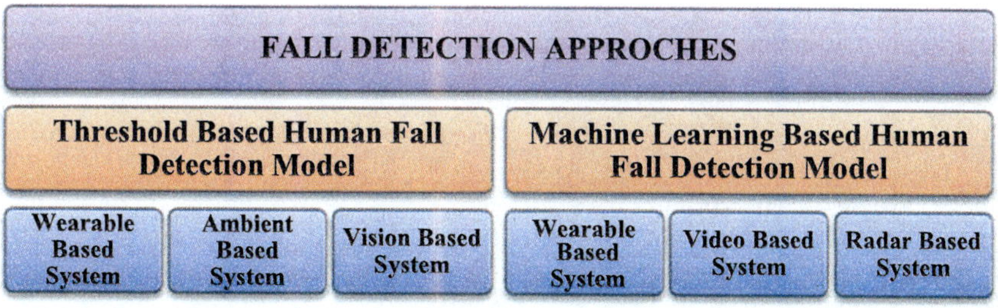

Fig. 2 Fall detection system classification using methodologies [4]

Table 1 Summary of threshold-based human fall detection model

Author	Methodology	ALG.	Performances
Kumar et al. [6]	Silhouette segmentation using histogram of oriented gradients (HOG)/global characterization and edge detection	Long short term memory (LSTM)	System has achieved a tracker precision of 94.67% and a fall accuracy rate of 98.01%
Kecskes et al. [7]	Using spatial–temporal graph convolutional network (ST-GCN), offer a comprehensive vision-based system that may be used immediately to new data	GCN	The nephelometric turbidity unit (NTU) red green blue-depth (RGB-D), the TST fall detection v2, and the fall free. We got 100% accuracy
Lo et al. [8]	Its creation of a field programmable gate array (FPGA-based) fall detection program employing a threshold-based analytical model is presented throughout this study	Verilog	The sensitivity of 97.45% and a specificity of 97.38%. The program's performance can be increased even more by adding a second gyroscope
Iguchi et al. [9]	Convolutional autoencoder, an unsupervised learning approach, and an Inertial Measurement Unit (IMU)	CA	As a result, we get sensitivity and specificity results of 100% and 99.1%

signals and many others. Machine learning algorithms extract the required characteristic from undesirable information and classify whether or not a fall has occurred (Table 2).

The authors would address a few of the wearable-based fall detection model limitations. In [9], multi ML-based model outperformed in all three models in this technique, and with the caveat that the odds of ambiguity are major causes to conflicting objects in the service area. The absence of real-time information. In [14], Data of experiments involve people of various ages was used to develop the method. Fall prevention should be more empirical evidence and warn about the risks of falling.

2.2.2 Video-Based Model

In the last several two decades, vision-based fall detection systems developed advanced quickly. Researchers are convinced that now the emergence of the Smarter Communities and IoT trends, combined with the growing amount of video equipment in everyday lives, creates a perfect environment for vision-based solutions (Table 3).

Table 2 Summary of wearable based model

Author	Sensor	Method	Performance
Whaeed et al. [10]	Accelerometer	RNN With BiLSTM	Accuracy = 97.21% and 97.41%, sensitivity = 96.97% and 99.77%, and specificity = 93.18% and 91.45% on SisFall and UP-fall dataset
Rodrigues et al. [11]	Accelerometer, gyroscope,	CNN	Only with 15 false positives, our data confirmed the efficacy of applying our detector in healthcare
Torti et al. [12]	Accelerometer	Recurrent neural network (RNN)	The highest reported fall detection accuracy was 83.68% before manual labeling) and 98.33% (after manual labeling)
Ramachandran et al. [13]	Accelerometer, gyroscope, magnetometer	KNN, Naïve-Bayes, SVM, ANN, decision tree	Without risk categorization, the decision tree scored 81%. About 85% for decision tree with risk classification

Table 3 Summary of vision-based model

Author	Methodology	ALG.	Performance
Chen et al. [15]	CNN and Bi-LSTM ANN/local characterization for foreground extraction	RNN-Bi-LSTM	The proposed model is with precision = 0.897, recall = 0.813, F1 = 0.852
Cai et al. [16]	ANN/local characterization feature maps were created using the (HCAE)	Softmax and HCAE	With a shallow-layer system with sensitivity = 100%, specificity = 93%, and accuracy = 96.2%
Ali et al. [17]	GMM/global characterization/foreground extraction through background subtraction	Boosted J48	Accuracies multi-camera 99% of the time multi-camera 99% of the time 99%
Ma et al. [18]	For safeguard confidentiality, face masking was utilized, and local features were generated produced utilizing CNN	Autoencoder SVM	Sensitivity = 93.3%, specificity = 92.8%, SVM sensitivity = 90.8%, specificity = 89.6%
Hasan et al. [19]	Open pose and RNN-LSTM ANN/local characterization found human connected components	Softmax and RNN-LSTM	URFD (Sensitivity-99%, specificity-96%), FDD (sensitivity-99%, specificity-97%), multicam (sensitivity 98%, specificity-96%)

The designing of vision-based fall detectors models encounters many significant issues and challenges, which are going to be discussed in this study. In [16], the mechanism for extracting features used in the approach is inadequate. A single individual is in charge of the training. In [20], the strategy focuses on numerous goals, the computation complexity is larger.

2.2.3 Radar-Based Model

Radar is cogitated as important procreation for fitness monitoring and fall detection in older people because a variety of characteristics are not more held by other sensing modalities. Radar's enticing characteristics, such as unanimous illumination, discreet sensing, lack of sensitivity to lighting circumstances, and confidentiality, have propelled electromagnetic radiation to the frontline of indoor positioning methodologies, displacing cameras and wearable gadgets. Hamid et al. [21] proposed a time-frequency and convolution neural radar-based fall detection system. Despite its inadequate classification performance, it differs between falls and non-falls with 98.37% accuracy. Saeed et al. [22], this system showed initial findings for a technique that uses the FMCW RADAR to distinguish a variety of human behaviors, such as falling, sitting, and standing, walking, and drinking. Using ResNet, ResNet decreased activity with 100% accuracy both in studies, acc to the results. Other accuracy rates % in experiments 1 and 2 were 96% and 85%. Hanif et al. [23], This paper, researchers employed a 24 GHz continuous-wave Doppler radar to build a minimal cost, greater accuracy Fall Detection Model (FDM) intended at tracking indoor human behavior and detecting fall. The presented program gives a 90% recall rate for fall detection and 97.7% and 95.3% correctness rate for heart rate and breathing monitoring.

The designing of radar-based fall detectors models encounters many significant issues and challenges, which are going to be discussed in this study. In [24], the radar must be placed with care, taking into account the environment and the patient's daily activities. In [25], the use of raw spectrograms is not recommended because it adds to the complexity. Rather than, the model is trained using a few carefully chosen features.

This review study explored both the threshold and ML-based classifications strategies to create an efficient HFDM and find that ML outperforms the threshold-based detection system.

3 Obstacles Approaching in Non-machine Learning System's

Inside this section, we've emphasized the various challenges that non-ML-based approaches face.

- In fall detection systems, camcorders such as RGB cameras, 3D-based equipment, and depth cameras that could operate throughout all light levels have replaced older camera systems. Conventional threshold techniques fail to identify detailed information from RGB cameras, resulting in a higher frequency of error rates [26].
- Due to several factors, such as appearance distortions, warped, and minimal picture information collected from CCTV, the existing system that depends on threshold-based categorization struggled to increase accuracy. Because of the sensitivity of sensors, ambient devices have significant false alarms due to undesired signals being recorded.
- Accelerometer sensor nodes in smartphone devices have a high false-positive percentage because of an assortment of constrictions. It increases efficacy only when the detectors are located at the midsection and heads. Threshold-based approaches struggled to identify the characteristics by resolving limitations such as accurate sensor placement in the body to improve accuracy, sensor sensitivity to undesired signals, and so on. The downside of gyroscopic sensors is that they have a time-varying zero shift, which causes mistakes. Conventional algorithms ended in failure to eliminate the dimensionality of data, resulting in a reduction in its ability to distinguish between fall and non-fall.

4 New Emerging Trends and the Research Findings of Fall Detection System's

Trends and research findings in fall detection systems were identified through a literature search. The most relevant points are described in this section.

4.1 Trends

The existing and future trends in fall detection systems are described first. As discussed in an earlier segment, amalgamating different kinds of algorithms and sensor devices can significantly increase fall dctection performance because these sensors and systems complement one another. Sensor fusion, which is a mix of numerous sensors, systems, and algorithms, is one of the most prevalent developments in the evolution of the fall models. Fortunately, an authentic fall repository is in the works [27], and enough actual fall data will soon be available to train and test all or most of the algorithms.

4.2 Research Findings

Throughout the paper, we discussed the benefits of ML-based fall detection versus threshold-based sensing technologies were explored, and also the disadvantages and shortcomings of the ML-based methodology. In earlier fall detection model improvement efforts primarily focused upon the improved technology that failed to reduce the false alarm rate. Due to the system's inability to discriminate between fall and non-fall, the traditional system has failed to be effective. Although when cameras create low-resolution images, ML-based categorization has been shown to extract the feature. Along with the assistance of an ML-based fall detection model, the challenge of sensor sensitivity to undesired signals has been resolved. In ML-based HFDM, it has been observed that time, expense, and complexities are all high. And another significant issue is the unavailability of real-time datasets. The majority of the datasets are created in a controlled setting. And in a real-life setting, self-adaptive-based HFDM has proven to be an effective technique. It has a memory-efficient characteristic that makes it ideal for wearable devices.

5 Conclusions and Future Work

In the end, this review paper concluded that robust fall detection should accurately extract the characteristics and limit the probability of false alarms. Previous techniques had various drawbacks when it came to enhancing the accuracy of a classification and lowering the false alert rate. As previously said, prior methods have improved accuracy despite taking into account the limitations of real-time applications. False alarms were drastically decreased when ML was used in fall detection, according to the study. The difficulty and cost of adopting ML-based solutions, on the other hand, are significant. Using data pre-processing, wavelet transforms techniques, or a thorough analysis of fall models that can modify themselves based on verifiable situations, the research will focus on lowering the complicated character of ML-based HFDM in the future.

References

1. PST patient safety network [cited 14 Jan 2022]. https://psnet.ahrq.gov/primer/falls
2. World Health Organization (2021) Falls. Geneva: WHO. [cited 26 April 2021]. http://www.who.int/mediacentre/factsheets/fs344/en/
3. Van Wey, Willems [cited 14 Sept 2021]. Whose fault is it when a patient falls in the hospital? vwpwlaw.com
4. Rastogi S, Singh J (2021) A systematic review on machine learning for fall detection system. Comput Intell
5. Greene S, Thapliyal H, Carpenter D (2016) IoT-based fall detection for smart home environments. In: 2016 IEEE international symposium on nanoelectronic and information systems

(iNIS). IEEE, pp 23–28; Mubashir M, Shao L, Seed L (2013) A survey on fall detection: principles and approaches. Neurocomputing 100:144–152
6. Kumar D, Ravikumar AK, Dharmalingam V, Kafle VP (2019) Elderly health monitoring system with fall detection using multi-feature based person tracking. In: 2019 ITU kaleidoscope: ICT for health: networks, standards and innovation (ITU K). IEEE, pp 1–9
7. Keskes O, Noumeir R (2021) Vision-based fall detection using ST-GCN. IEEE Access 9:28224–28236
8. Iguchi Y, Lee JH, Okamoto S (2021) Enhancement of fall detection algorithm using convolutional autoencoder and personalized threshold. In: 2021 IEEE international conference on consumer electronics (ICCE). IEEE, pp 1–5
9. Taramasco C, Rodenas T, Martinez F, Fuentes P, Munoz R, Olivares R, De Albuquerque VHC, Demongeot J (2018) A novel monitoring system for fall detection in older people. IEEE Access 6:43563–43574
10. Waheed M, Afzal H, Mehmood K (2021) NT-FDS—a noise tolerant fall detection system using deep learning on wearable devices. Sensors 21(6):2006
11. Rodrigues TB, Salgado DP, Cordeiro MC, Osterwald KM, Teodiano Filho FB, de Lucena Jr VF, Naves ELM, Murray N (2018) Fall detection system by machine learning framework for public health. Procedia Comput Sci 141:358–365
12. Torti E, Fontanella A, Musci M, Blago N, Pau D, Leporati F, Piastra M (2018) Embedded real-time fall detection with deep learning on wearable devices. In: 2018 21st Euromicro conference on digital system design (DSD). IEEE, pp 405–412
13. Ramachandran A, Adarsh R, Pahwa P, Anupama KR (2018) Machine learning-based fall detection in geriatric healthcare systems. In: 2018 IEEE international conference on advanced networks and telecommunications systems (ANTS). IEEE, pp 1–6
14. Saadeh W, Butt SA, Altaf MAB (2019) A patient-specific single sensor IoT-based wearable fall prediction and detection system. IEEE Trans Neural Syst Rehabil Eng 27(5):995–1003
15. Chen Y, Li W, Wang L, Hu J, Ye M (2020) Vision-based fall event detection in complex background using attention guided bi-directional LSTM. IEEE Access 8:161337–161348
16. Cai X, Li S, Liu X, Han G (2020) Vision-based fall detection with multi-task hourglass convolutional auto-encoder. IEEE Access 8:44493–44502
17. Ali SF, Khan R, Mahmood A, Hassan MT, Jeon M (2018) Using temporal covariance of motion and geometric features via boosting for human fall detection. Sensors 18(6):1918
18. Ma C, Shimada A, Uchiyama H, Nagahara H, Taniguchi R-I (2019) Fall detection using optical level anonymous image sensing system. Opt Laser Technol 110:44–61
19. Hasan MM, Islam MS, Abdullah S (2019) Robust pose-based human fall detection using recurrent neural network. In: 2019 IEEE international conference on robotics, automation, artificial-intelligence and internet-of-things (RAAICON). IEEE, pp 48–51
20. Khraief C, Benzarti F, Amiri H (2020) Elderly fall detection based on multi-stream deep convolutional networks. Multimedia Tools Appl 79
21. Sadreazami H, Bolic M, Rajan S (2021) Contactless fall detection using time-frequency analysis and convolutional neural networks. IEEE Trans Ind Inf
22. Saeed U, Shah SY, Shah SA, Ahmad J, Alotaibi AA, Althobaiti T, Ramzan N, Alomainy A, Abbasi QH (2021) Discrete human activity recognition and fall detection by combining FMCW RADAR data of heterogeneous environments for independent assistive living. Electronics 10(18):2237
23. Hanifi K, Karsligil ME (2021) Elderly fall detection with vital signs monitoring using CW Doppler radar. IEEE Sens J
24. Wang B, Guo L, Zhang H, Guo Y-X (2020) A millimetre-wave radar-based fall detection method using line kernel convolutional neural network. IEEE Sens J 20(22):13364–13370
25. He M, Nian Y, Zhang Z, Liu X, Hu H (2019) Human fall detection based on machine learning using a THz radar system. In: 2019 IEEE radar conference (RadarConf). IEEE, pp 1–5

26. Bosch-Jorge M, Sánchez-Salmerón A-J, Valera Á, Ricolfe-Viala C (2014) Fall detection based on the gravity vector using a wide-angle camera. Expert Syst Appl 41(17):7980–7986
27. Farseeing (FAll repository for the design of smart and self-adaptive environments prolonging independent living) project. Available online: http://farseeingresearch.eu/. Accessed 14 Nov 2021

Convolutional Neural Network to Predict Soil Texture Classes

Pallavi Srivastava, Aasheesh Shukla, and Atul Bansal

Abstract In the field of agriculture, image processing has been proved to generate significant results which is prediction and classification of various properties. Prediction of soil texture classes is one such application. This paper proposes a convolutional neural network to predict soil texture. It takes into account soil images collected from multiple fields of different crops. Hydrometer test and USDA triangle method are used to determine the texture classes of the samples. These samples were classified into nine textural classes. The network uses Leaky ReLU activation function to acknowledge the nonlinear property of the problem statement and Softmax function to as this a multiclass problem.

Keywords Soil texture · CNN · USDA triangle · Dropout layer

1 Introduction

Nowadays, there are many applications of image processing in agricultural field like soil texture classification [1], pH level prediction [2], fruit quality grading [3–5]. Agriculture and smart farming require fidelity in the acquisition of information on soil and crops' properties. In particular, soil properties significantly affect soil crop selection and growth management. Soil properties play important role in the agricultural sphere, for example, in land consolidation, soil erosion, soil irrigation, drainage management and soil improvement, and therefore influence agriculture engineering. Education about the various alterations in the properties of the soil could lead to significant information needed to formulate and deploy a reasonable execution plan, particularly in cultivated areas.

P. Srivastava · A. Shukla (✉)
GLA University, Mathura, Uttar Pradesh 281406, India
e-mail: aasheesh.shukla@gla.ac.in

A. Bansal
Chandigarh University, Mohali, Punjab 140413, India
e-mail: atul.bansal@cumail.in

A. Rani et al. (eds.), *Signals, Machines and Automation*, Lecture Notes in Electrical Engineering 1023, https://doi.org/10.1007/978-981-99-0969-8_61

The chemical and physical properties of soil are affected by various factors like biota, climate and geological history at larger scales. But, topography and human activities remain the major controlling factors of soil at smaller scales. It has been observed that there is a hike in the research domain of soil texture classification. In the past few years, there is escalation in research interest for soil-related properties like texture classification [1, 6] organic matter prediction, moisture content prediction [7] and pH level prediction [2, 8, 9].

Soil texture can be predicted on various parameters like void ratio liquid limit moisture content [10], etc. One of the methods to predict soil texture is collecting soil images to acquire soil images and apply different image processing and computer vision techniques. Azizi et al. [11] proposed a research to aggregates of soil samples. In this work, VggNet16, ResNet50 and InceptionV4 architectures were trained, out of which ResNet50 performed best with 98.72% accuracy. Swetha et al. [12] presented a novel setup made up of a smartphone camera, a dark chamber and a smartphone application for soil texture prediction. Image local, color, texture features are used along with random forest and CNN to predict the final soil texture. Srivastava et al. [13] presented a vast review of soil texture classification which included methods with conventional image processing techniques as well as deep learning algorithms. de Oliveira Morais et al. [14] proposed a method to predict and classify soil texture using soil images acquired from soil samples. Here, 63 soil samples are collected. The ground truth which is texture of the soil samples is determined by pipette method. This method attained 100% accuracy.

Since these techniques have been over explored with time, there is need to explore other beneficial methods like deep neural networks [7, 13–15]. These nets were employed to classify volunteer potato and sugar beet images taken under ambient varying light settings in agricultural environments. Next section discusses about the materials required for the proposed work. This paper discusses soil texture classification method which considers soil samples of five different crops collected from multiple fields [16, 17].

2 Experimental Setup and Image Acquisition

The suggested work considers five different crops, namely mustard, potato, sugarcane, wheat and rice. For each crop, ten fields of size 350 $\times$ 300 m are considered to collect soil samples. From each of the ten fields, five soil samples are collected. So, there is a collection of total 50 soil samples of each crop taken from ten different fields of a crop. Samples are collected from district Mau, Uttar Pradesh, India. The sample collecting site is located at 25°56′30″ N latitude and 83°33′40″ E longitude. Each field yields five samples with a 150-m separation. Samples are taken from a depth of 2 inches below the field's surface level.

After collecting soil images, soil image dataset is created. Motorola One Power smartphone with Android 10 version is used to capture soil images. Figure 1 shows some of the examples from the soil image dataset. This smartphone camera has a

Fig. 1 Some examples from soil image dataset

16-megapixel CMOS device. At the time of capturing an image, camera settings are kept in default condition such as exposure time = 1/30 s, F-stop = f/1.8, focal length = 1.12 μm. The soil images were clicked from 20 inches above.

Soil samples were taken out on white paper and arranged in the middle of it to capture the images. Soil samples are taken from fields of five different crops. Here, for every crop, ten fields are considered, and from every field, five soil samples are taken with a separation of 150 m. So, there are 50 samples per crop. At the time of creating soil image, dataset eight pictures are captured per soil sample. Dataset is created using an Android smartphone because farmers and other beneficiaries may not have good-quality cameras.

The objective of this proposed work is to classify soil texture for soil images of multiple crops. To fulfill this purpose, collected soil samples should be first tested out in the laboratory for their respective texture [18]. Along with the soil images, their respective soil texture is also needed as input to CNN architecture. Soil is composed of fraction of clay, sand and silt. This fraction or percentage of sand, silt and clay decides the soil texture like silty clay, loam, clay loam, etc. To do so, first hydrometer test is carried out, and then, USDA triangle is used to generate the final result. Radial graph in Fig. 2 shows the texture classes of the soil samples.

3 CNN for Soil Texture Classification and Result Discussion

A specific kind of deep neural network is the convolutional network, which is commonly referred to as CNN or ConvNet. It is a deep, feedforward artificial neural network. Feedforward neural networks are also called multi-layer perceptrons (MLPs). The models are called feedforward because information flows right through the model. There are no feedback connections in which outputs of the model

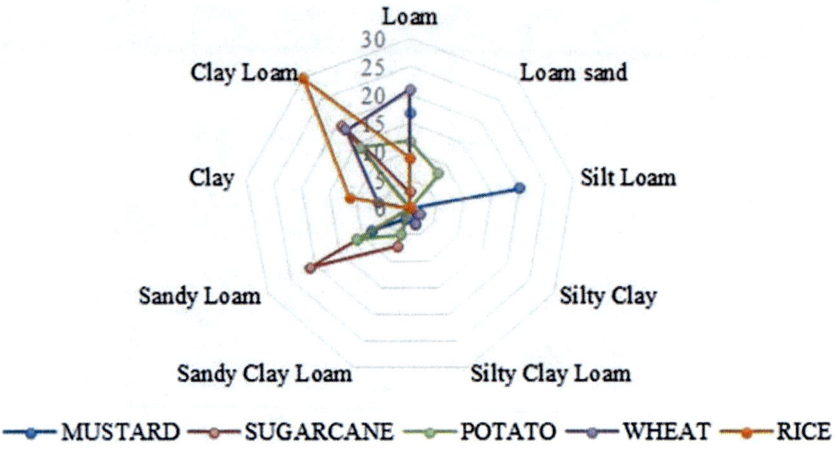

Fig. 2 Radial graph showing soil texture classes of samples from different crops

are fed back into itself. Figure 3 shows that an image is fed as an input to the network, which goes through multiple convolutions, subsampling, a fully connected layer and finally outputs something.

In this work, a CNN architecture is proposed which classifies the soil samples according to their textural classes. This CNN architecture takes in soil images and outputs the texture of the samples. As the dataset suggests, there are 2000 soil images acquired from five different crops. After performing the laboratory experiments, all the samples can be classified into nine textural classes. The entire dataset is divided in 80:20 ratio for training and validation purposes. For this work, first the image matrix is converted into an array and rescale is between 0 and 1, reshape it, and feed this as an input to the network. CNN uses three convolution layers: first layer has 32–3 $\times$ 3 filters, second layer has 64–3 $\times$ 3 filters and third layer has 128–3 $\times$ 3 filters. In addition to this, there are three max pooling layers of size 2 $\times$ 2.

The model is trained for 20 epochs. Leaky ReLU activation function is added which will help the network in learning nonlinear decision boundaries. Since there

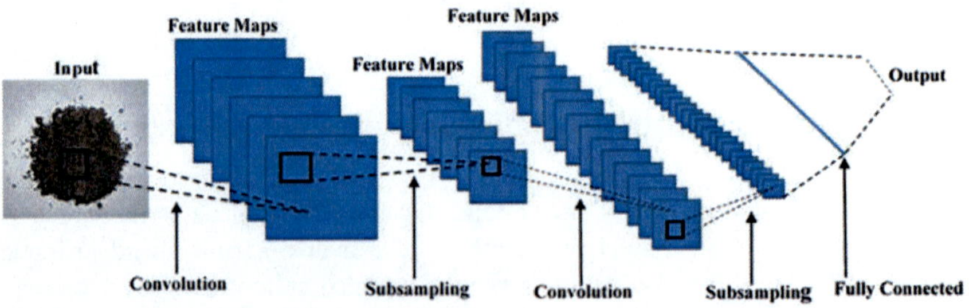

Fig. 3 Architecture of a convolutional neural network

are nine soil textural classes, Leaky ReLU activation function can separate these nine classes which are not linearly separable.

At last, a dense layer with a Softmax activation function with nine units is needed for this multiclass classification problem. It can be observed from Fig. 4 that the training accuracy is 99% and training loss is quite low. But, it seems like the model which is overfitting, as validation loss is 43.96% and validation accuracy is 92%. This overfitting problem can be handled by using dropout layer into the network and keeping all other layers unchanged. Figure 5 shows the results after adding dropout layer. The validation accuracy is 92.30% and the validation loss is 21.46%. Even though the validation accuracy did not improve significantly, validation loss decreased in comparison to the previous results.

The model is underperforming for class sandy loam regarding both precision and recall as can be seen in Table 1. For class loam and silt loam, model lacks in precision. In addition, for class silty clay loam, model lacks both precision and recall. This work proposes a CNN model which classifies the soil textural classes using soil sample images. For further research, this work can be extended with more number of samples. Number of layers can also be increased to obtain better validation accuracy.

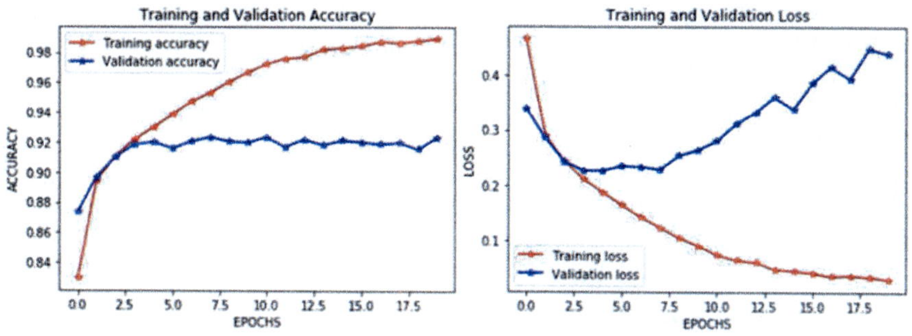

Fig. 4 Accuracy and loss generated from the model without dropout layer for 20 epochs

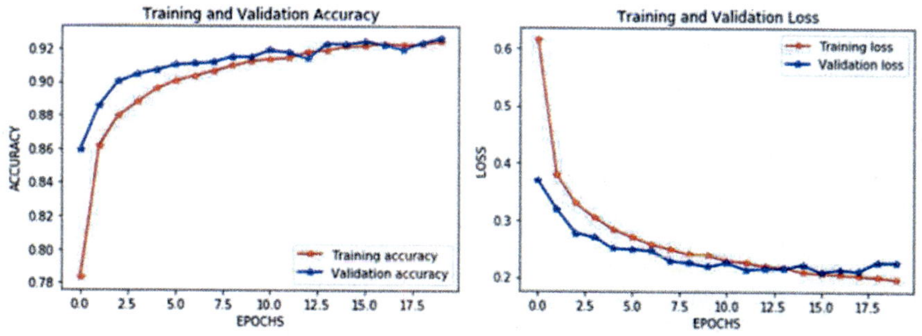

Fig. 5 Accuracy and loss generated from the model with dropout layer for 20 epochs

Table 1 Classification report class-wise

Class name	Precision	Recall	F1-Score
Loam	0.77	0.90	0.83
Loam sand	0.99	0.98	0.99
Silt loam	0.89	0.89	0.89
Silty clay	0.94	0.92	0.93
Silty clay loam	0.88	0.87	0.88
Sandy clay loam	0.99	0.98	0.98
Sandy loam	0.82	0.72	0.77
Clay	0.94	0.99	0.97
Clay loam	0.99	0.98	0.99
Average	0.92	0.92	0.92

References

1. Zhai Y, Thomasson JA, Boggess JE, Sui R (2006) Soil texture classification with artificial neural networks operating on remote sensing data. Comput Electron Agric 54(2):53–68. https://doi.org/10.1016/j.compag.2006.08.001
2. Barman U, Choudhury RD (2020) Soil texture classification using multi class support vector machine. Inf Process Agric 7(2):318–332. https://doi.org/10.1016/j.inpa.2019.08.001
3. Suh HK, IJsselmuiden J, Hofstee JW, van Henten EJ (2018) Transfer learning for the classification of sugar beet and volunteer potato under field conditions. Biosyst. Eng. 174:50–65. https://doi.org/10.1016/j.biosystemseng.2018.06.017
4. Dane JH, Topp GC (2020) Methods of soil analysis, part 4: physical methods, vol 20. Wiley
5. Bhargava A, Bansal A (2020) Automatic detection and grading of multiple fruits by machine learning. Food Anal Methods 13(3):751–761. https://doi.org/10.1007/s12161-019-01690-6
6. Zhao Z, Chow TL, Rees HW, Yang Q, Xing Z, Meng F-R (2009) Predict soil texture distributions using an artificial neural network model. Comput Electron Agric 65(1):36–48. https://doi.org/10.1016/j.compag.2008.07.008
7. Xie F, Gao Q, Jin C, Zhao F (2021) Hyperspectral image classification based on superpixel pooling convolutional neural network with transfer learning. Remote Sens 13(5):1–17. https://doi.org/10.3390/rs13050930
8. Taneja P, Vasava HK, Daggupati P, Biswas A (2021) Multi-algorithm comparison to predict soil organic matter and soil moisture content from cell phone images. Geoderma 385:114863. https://doi.org/10.1016/j.geoderma.2020.114863
9. Barman U, Choudhury RD, Uddin I (2019) Predication of soil pH using K mean segmentation and HSV color image processing. In: 2019 6th International conference on computing for sustainable global development (INDIACom), pp 31–36. IEEE
10. Ghazali MF, Wikantika K, Harto AB, Kondoh A (2020) Generating soil salinity, soil moisture, soil pH from satellite imagery and its analysis. Inf Process Agric 7(2):294–306. https://doi.org/10.1016/j.inpa.2019.08.003
11. Azizi A, Gilandeh YA, Mesri-Gundoshmian T, Saleh-Bigdeli AA, Moghaddam HA (2020) Classification of soil aggregates: a novel approach based on deep learning. Soil Tillage Res 199:104586. https://doi.org/10.1016/j.still.2020.104586
12. Swetha RK, Bende P, Singh K, Gorthi S, Biswas A, Li B, Weindorf DC, Chakraborty S (2020) Predicting soil texture from smartphone-captured digital images and an application. Geoderma 376:114562. https://doi.org/10.1016/j.geoderma.2020.114562
13. Srivastava P, Shukla A, Bansal A (2021) A comprehensive review on soil classification using deep learning and computer vision techniques. Multimedia Tools Appl 80:14887–14914. https://doi.org/10.1007/s11042-021-10544-5

14. de Oliveira Morais PA, de Souza DM, de Melo Carvalho MT, Madari BE, de Oliveira AE (2019) Predicting soil texture using image analysis. Microchem J 146:455–463. https://doi.org/10.1016/j.microc.2019.01.009
15. Jiang Y, Li Y, Zhang H (2019) Hyperspectral image classification based on 3-D separable ResNet and transfer learning. IEEE Geosci Remote Sens Lett 16(12):1949–1953. https://doi.org/10.1109/LGRS.2019.2913011
16. Bhargava A, Bansal A (2020) Quality evaluation of mono & bi-colored Apples with computer vision and multispectral imaging. Multimedia Tools Appl 79(11–12):7857–7874. https://doi.org/10.1007/s11042-019-08564-3
17. Bhargava A, Bansal A (2020) Machine learning based quality evaluation of mono-colored apples. Multimedia Tools Appl 79(31–32):22989–23006. https://doi.org/10.1007/s11042-020-09036-9
18. Bosilj P, Aptoula E, Duckett T, Cielniak G (2020) Transfer learning between crop types for semantic segmentation of crops versus weeds in precision agriculture. J Field Robot 37(1):7–19. https://doi.org/10.1002/rob.21869
19. Barman U, Choudhury RD, Talukdar N, Deka P (2018) Predication of soil pH using HSI colour image processing and regression over Guwahati,Assam, India. J Appl Nat Sci 10(2):805–809. https://doi.org/10.31018/jans.v10i2.1701
20. Pham BT, Nguyen MD, Nguyen-Thoi T, Ho LS, Koopialipoor M, Quoc NK, Armaghani DJ, van Le H (2021) A novel approach for classification of soils based on laboratory tests using Adaboost, Tree and ANN modeling. Transp Geotech 27:100508. https://doi.org/10.1016/j.trgeo.2020.100508

Customer Churn Analysis of a Financial Firm

Debani Prasad Mishra, Bibhuprasad Nanda, and Surender Reddy Salkuti

Abstract In the middle of a pandemic where most of the world has come to a halt, acquiring new customers is not an easy task. So, retaining the current customers has become the utmost priority for many organizations. Many methods have been applied in the past to predict the customer loyalty rate. However, no method has been discovered till now, which can predict 100% accurate results. The term churn is identified with predictions on which a client deserts his relationship with an organization; in this manner, churn analysis has gotten compulsory for most associations looking for manageable development and benefit. It has been discovered that obtaining new clients costs more than keeping the current ones. This paper is focused on a European Bank where churn analysis will be performed using geodemographic segmentation and further visualization will help to prevent possible customer churn.

Keywords Churn analysis · Geodemographic segmentation · Logistic regression · Supervised learning

1 Introduction

Customer churn refers to the situation when a customer stops using the products or services of an organization. Many authors have tried to tackle customer churn analysis in various sectors such as telecommunication, banking, subscription, Fintech, Ed-tech, gambling, and many also in retail. But, none of them were able to achieve 100% accuracy. The present paper will introduce a client churn analysis in a European Bank which has seen abnormal churn rates. Churn is when clients relinquish the organization and the bank has seen clients leaving at curiously high rates. Presently the bank needs to comprehend what the issue is and they need to

D. P. Mishra · B. Nanda
Department of Electrical Engineering, IIIT Bhubaneswar, Bhubaneswar, Odisha, India

S. R. Salkuti (✉)
Department of Railroad and Electrical Engineering, Woosong University, Daejeon, Republic of Korea
e-mail: surender@wsu.ac.kr

A. Rani et al. (eds.), *Signals, Machines and Automation*, Lecture Notes in Electrical Engineering 1023, https://doi.org/10.1007/978-981-99-0969-8_62

evaluate and address that issue. This bank works in Europe, in three nations, i.e., France, Spain, and Germany. They have just taken out a sample of 10,000 customers on which churn analysis will be performed. They have also taken out a sample of 1000 customers for testing purposes. The dataset contains various data of the customers such as their Customer ID, name, geography, gender, age, tenure, balance, estimated salary, and many more. The customers are randomly selected to avoid any influence of biasedness. First, the model will be built based on training datasets and then, the efficiency of the model will be tested based on testing datasets [1]. One more advantage of the churn analysis is that the organization will also get to know the loyalty rate of each customer. Through this, they can realize which individuals are bound to be reliable. What's more, this could administer the bank's choice on whether to give advances. Churn analysis can likewise be utilized to discover which exchanges are bound to be deceitful and which are more trustworthy.

Customer churn analysis, also known as customer attrition, has been developed under the pressure of retaining the most profitable customers. In the past, many studies have been conducted to predict the likelihood of customer churn. This paper aims to propose a model using geodemographic segmentation, which focuses on dynamic attributes and behavioral preferences [2]. Segmentation is one of the most widely used terms in the field of data analytics. It generally refers to the grouping of customers based on certain attributes. Also, segmentation leads the company to understand the customer requirements and to allocate their marketing campaigns. This paper intends to propose a model for customer churn analysis using geodemographic segmentation [3]. This paper aims to accomplish 100% accuracy for the model, yet most importantly, the focus will be on reducing the time complexity of the model. Thus, we can wind up getting an exceptionally effective model in moderately less time, hence making the model proficient, efficient, practical, and energy-saving.

2 Data Preprocessing

The data are gathered from the Kaggle site. Data preprocessing is an integral part of data mining. It occupies 70% of the whole data analysis process. Here, the extracted data need to be cleaned, and then, the data can be transformed into some other forms to get better results. Finally, the data are loaded into the database [4]. Then, SQL queries are usually used to extract the desired records from the database. Here, two datasets have been collected—one for training the model and the other for assessing our model. The training dataset contains 10,000 records, and the testing dataset contains only 1000 records. Both the datasets contain 14 columns [5]. The datasets have been collected from Kaggle. The data preprocessing was done using SQL Server Integration Services (SSIS). SQL stands for Structured Query Language. It is a powerful tool used for data extraction and data preprocessing. A few problems are occurred during data preprocessing. Some of the rows were shifted toward the right which resulted in the addition of an extra column. Some of the rows were shifted toward the left. Here, all the numeric columns were present in string format. In the age column, some of the rows had values of more than 200 which was a mistake. Some of these errors are human errors that can be avoided if one is alert during data input. Error handling is an integral part of data preprocessing. Having good domain knowledge has proved to be beneficial for data preprocessing. For example, if an organization is responsible for handling students' data, then they would not be usually dealing with records having an age of more than 30 [6, 7]. Sometimes, rare cases also occur and one needs to be alert at all times to prevent errors. Data preprocessing consists of three processes—Extract, Transform, and Load. Data need to be assessed carefully before and after the ETL process to remove any possible errors.

3 Data Modeling

The next section after data preprocessing is data modeling. This section is again divided into two sections—Modeling Building and Assessing the model where the accuracy of our model is found.

3.1 Model Building

The model is built using geodemographic segmentation. The term "Geodemographic" is formed from two words—"Geo" which stands for Geographic and "Demographics" which relates to statistical attributes. So, geodemographic segmentation means grouping the customers based on geographical and statistical attributes [8]. Some of the tools like Gretl and Microsoft Excel were used to build this model. First, a dummy is created for categorical variables such as Geography and Gender. Then, a logistic regression model is created, where the "Exited" column is passed as a dependent variable and other columns that could affect the dependent variables (such as "CreditScore", "Gender", "Tenure", "Balance", "NumOfProducts", "HasCrCard", "IsActiveMember", and "EstimatedSalary") are passed as Regressors [9, 10]. Then, the backward elimination method is performed by removing those regressors whose p-value is more than that of the level of significance. And, some of the independent variables are transformed to get a better result. The multicollinearity effect is also taken care of. A confusion matrix is a proportion of the precision of models in machine learning. The confusion matrix has four boundaries. True Positive implies that the forecast is positive and it is true. True Negative implies that the forecast is negative and it is true. False Positive implies that the expectation is positive; however, it is false [11]. This is otherwise called Type-1 Error. False Negative implies that the expectation is negative; however, it is false. This is otherwise called Type-2 Error (Table 1).

Table 1 Geodemographic segmentation model

Logit model considering records 1–10,000; responding variable: Exited					
	Coefficient	Standard error	z	p-value	
Const.	− 3.9125798	0.2371647	− 16.598	< 0.0001	***
Credit score	− 0.000674866	0.000280272	− 2.408	0.0160	**
Age	0.0726550	0.00257451	28.22	< 0.0001	***
Tenure	− 0.0158791	0.00934627	− 1.699	0.0893	*
Number of products	− 0.0950198	0.0475374	− 1.999	0.0456	**
Is active member	− 1.07578	0.0576458	− 18.66	< 0.0001	***
Germany	0.74739531	0.0651498	11.501	< 0.0001	***
Female	0.526721	0.0544591	9.672	< 0.0001	***
Log_Balance	0.0690263	0.0139592	4.945	< 0.0001	***
Mean-dependent var.		0.203700	S.D. dependent var.		0.402769
McFadden R-squared		0.152787	Adjusted R-squared		0.151006
Log-likelihood		− 4282.57	Akaike criterion		8583.141
Predicted					
Actual			0		1
		0	7687		276
		1	1597		440

Asterisk represents the significance level

3.2 *Assessing the Model*

The accuracy and precision of our model can be calculated with the help of a confusion matrix. The cumulative accuracy profile (CAP) curve is also taken into consideration to assess the model. CAP curve is more reliable and provides more accurate results. The CAP curve of a model addresses the aggregate number of positive results along the *y*-axis vs. relating combined number of a characterizing parameter along the *x*-axis [12]. The yield is known as the CAP curve. A model is built for training datasets. Now, the same model is run for test datasets. After running this model for the training and testing datasets, the CAP curves are obtained and they are depicted in Fig. 1.

4 Data Visualization

The presentation of numerical or categorical data into a pictorial or graphical format to make it easier for the human brain to process the data and come up with valuable insights is called data visualization. There are many tools available for data visualization such as Tableau, Power BI, Google Charts [13]. Tableau is used for

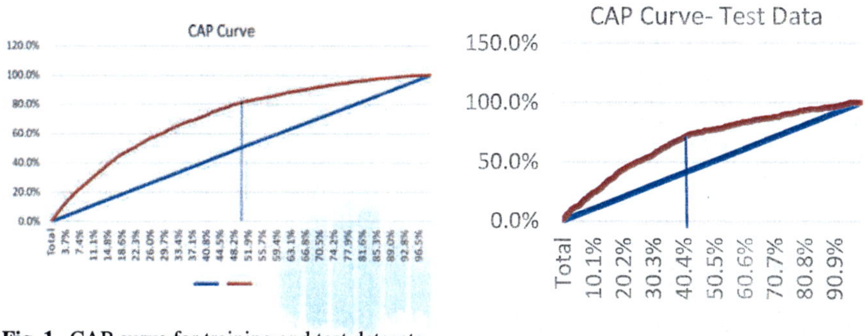

Fig. 1 CAP curve for training and test datasets

visualizing the data from the above dataset. Figure 2 presents the churn based on balance distribution. From Fig. 2, it can be noted that people having higher or lower balances tend to exit the bank. People having extreme balances in their accounts are highly inclined toward better investment plans.

Figure 3 presents the churn based on active members and the number of products. From Fig. 3, it can be seen that inactive customers tend to exit the bank frequently. Customers who have not been active for a few months might not be aware of the recent developments in the bank. From Fig. 3, it can also be safely assumed that customers having fewer products are more inclined to leave the bank. If a customer has bought multiple products, then it can be safely assumed that the customer has faith in the bank. They tend to be more loyal than regular customers [14]. To reduce the churning rate of regular customers, the bank needs to restore faith in them. The bank needs to make its customers feel that they are reliable and can be trusted.

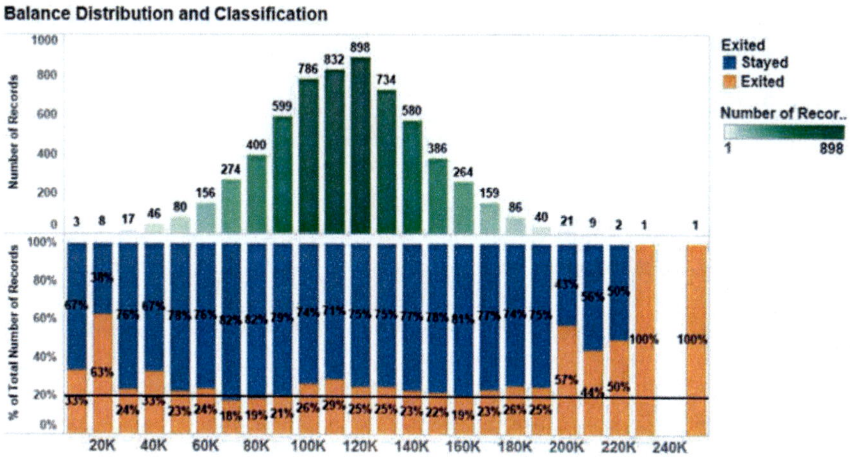

Fig. 2 Churn based on balance distribution

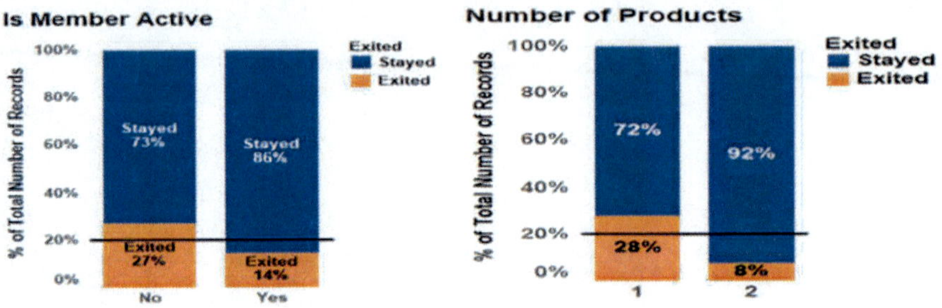

Fig. 3 Churn based on active and inactive members and the number of products

Therefore, it can be concluded that logistic regression has been performed using geodemographic segmentation. This work aimed to achieve 100% accuracy for the model. However, the accuracy of the above model is about 80%, which is not bad. It had been mentioned earlier in the introductory section that no authors have achieved 100% accuracy till now. Some of the authors managed to get 80% accuracy, but none of them exceeded 80. So, now, what differentiates our model from other models is that our model is relatively less expensive, less time taking, and relatively more accurate than other models [15]. So, it can be concluded that a model has been designed successfully that is highly accurate, cost-effective, and time and energy-saving.

5 Conclusion

Customer churn analysis has become a significant worry in pretty much every industry. Here, a model is built using geodemographic segmentation. An accuracy test was also performed to check the effectiveness of the model using the CAP curve. And, this model predicts almost 80% correct results by covering only 50% of the dataset. Customers having a very high balance or very low balance in their bank accounts are attracted to good investment plans which have a high return on investment. To prevent them from churning, the bank needs to come up with good advertisement campaigns ensuring maximum reach. The bank has also seen an increase in the churn rate of customers having only one account. Customers with multiple accounts tend to be more loyal. To prevent the churning of these customers, the bank needs to make them feel that they are reliable and can be trusted. Apart from all these, the bank also needs to think from the customer's perspective on how they can be benefitted from the bank.

Acknowledgements This research work was funded by "Woosong University's Academic Research Funding—2023".

References

1. He B, Shi Y, Wan Q, Zhao X (2014) Prediction of customer attrition of commercial banks based on SVM model. Procedia Comput Sci 31:423–430
2. Abdi F, Abolmakarem S (2019) Customer behavior mining framework (CBMF) using clustering and classification techniques. J Ind Eng Int 15:1–18
3. Yang S, Jiajia Y (2014) Using data mining technique for bank customer churn with decision tree prediction algorithm. Comput Knowl Technol 11:2533–2536
4. Huang Y, Kechadi T (2013) An effective hybrid learning system for telecommunication churn prediction. Expert Syst Appl 40(14):5635–5647
5. Carbo-Valverde S, Cuadros-Solas P, F, Fernandez (2020) A machine learning approach to the digitalization of bank customers: evidence from random and causal forests. PLoS ONE 15(10):1–39
6. Sayed H, Abdel-Fattah MA, Kholief S (2018) Predicting potential banking customer churn using apache spark ML and MLlib packages: a comparative study. Int J Adv Comput Sci Appl 9(11):674–677
7. Zoric AB (2016) Predicting customer churn in banking industry using neural network. Interdisc Descr Complex Syst 14(2):116–124
8. Singhal S, Sharma P, Passricha V (2018) K-means based SVM for prediction analysis. Int J Res Electron Comput Eng 6(3):255–259
9. Migueis VL, van den Poel D, Camanho AS, e Cunha JF (2012) Modeling partial customer churn: on the value of first product-category purchase sequence. Expert Syst Appl 39(12):11250–11256
10. Farooqi MR, Raza K (2011) A comprehensive study of CRM through data mining techniques. In: Proceedings of the national conference (NCCIST-2011), pp 1–5. arXiv:1205.1126 (arXiv:1205.1126v1). https://doi.org/10.48550/arXiv.1205.1126
11. Das SK (2012) Customer relationship management in banking sector: a comparative study of SBI and other nationalized commercial banks in India. Arth Prabhand: J Econ Manage 1(6):68–82
12. Tsiptsis K, Chorianopoulos A (2009) Data mining techniques in CRM: inside customer segmentation. Wiley, pp 1–13
13. Vivek B (2012) Data mining as a tool to predict churn behavior of customers. Proc Int J Comput Organ Trends 2(3):85–89
14. Prasad UD, Madhavi S (2012) Prediction of churn behavior of bank customers using data mining tools. Bus Intell J 5(1):96–101
15. Jhaveri S, Khedkar I, Kantharia Y, Jaswal S (2019) Success prediction using random forest, catboost, xgboost and adaboost for kickstarter campaigns. In: 2019 3rd International conference on computing methodologies and communication (ICCMC). IEEE, pp 1170–1173. https://doi.org/10.1109/ICCMC.2019.8819828

Consistency Assessment of OpenStreetMap Road Dataset of Haryana and Punjab Using K-means and Elbow Method

Saravjeet Singh, Jatin Arora, and Rishu Chhabra

Abstract Quality and reliability are the major concern of crowd-scoured based projects. OpenStreetMap being crowd-sourced project is likely to be prone to quality error and missing information. Many real-time applications use OpenStreetMap data as a source of spatial information. Incorrect information and missing data highly impact the performance of OpenStreetMap-based applications. In this study, an effort is made to assess the quality of OpenStreetMap data of the Indian region. This research paper presents an approach to identify the error in the OpenStreetMap road network using the K-means algorithm and Elbow method. Intrinsic parameters of the road network are identified using the tag/value combination, and further, these parameters were used to categorize the roads. For this experiment, 34,555 roads of length 51,671,905.83 km has been considered to identify the errors in road network using different quality indicators.

Keywords Intrinsic parameters · Topological error · Nodes · Links · Crowd sourcing

1 Introduction

With the advancement of technology, spatial information on real-world characteristics on the Earth's surface is easily available in the form of digital spatial data or digital maps. These digital spatial datasets are available for general-purpose mapping/development purposes and are been created, captured and maintained by the governmental officials and private firms. Crowd sourcing and Volunteered Geographic Information (VGI) are frequently used technology for geographical information creation and processing [1]. Among the different VGI-based projects, OpenStreetMap (OSM) is also crowd-sourced, and VGI-based extensive project

S. Singh (✉) · J. Arora · R. Chhabra
Chitkara University Institute of Engineering and Technology, Chitkara University, Punjab 140401, India
e-mail: saravjeet.2009@gmail.com

A. Rani et al. (eds.), *Signals, Machines and Automation*, Lecture Notes in Electrical Engineering 1023, https://doi.org/10.1007/978-981-99-0969-8_63

provides digital copy of the earth map. Completeness, logical consistency, positional and thematic accuracy, temporal quality are normally done by those who are not part of data collection and creation [2]. Quality of OSM dataset was evaluated by using both intrinsic and extrinsic parameters [3, 4]. Initially, extrinsic parameters were frequently used to assess the OSM with respect to other available datasets [5, 6]. But, with the advancement of technology and popularity of OSM, many researchers proposed different OSM quality assessment techniques using the intrinsic parameters [7]. Much work using intrinsic parameters was done for the quality assessment of different OSM objects in the US and European countries [8, 9]. Recently, many researchers provided different methods to assess the quality of OSM datasets. These assessments cover building dataset, basic tags, roads, linear features, contributors and their behaviors, incomplete data, etc., and all these studies were based on Indian dataset [10–12]. Based on the recent state of art, remaining sections of this research paper are as follows: objective and contribution are provided in Sect. 2, Sect. 3 provides complete methodology, Sects. 4 and 5 provide result analysis, conclusion and future scope.

2 Objective and Contribution

OSM is a crowd-sourced based project, and due to this feature, a cartographer with knowledge of OSM schema can contribute to OSM project. This collaborative and open-access feature causes many quality issues to OSM dataset. OSM is very popular in developed countries, but in developing countries like India, OSM has not got much popularity and very less research conducted related to OSM quality assessment. These quality issues mainly contain topological errors that include positional and characteristics' mismatch [4, 13–16]. Considering above-mentioned issues, this research provides a case study to assess the quality of OSM dataset. The main object of this research was to investigate and use unsupervised learning method to analyze the quality of OSM road dataset using intrinsic features only. To achieve this objective dataset of Haryana and Punjab, two states of India were considered. For this research study, road dataset was extracted using the overpass application interface of OSM. GeoJSON and GPX file of road dataset was first exacted, then parsed as per the requirement. These parsed files were further used to identify the quality issues related to road network. This work provides the quality of OSM road dataset, which would further help to correct the errors. In this work, K-means algorithm with Elbow method was used to identify the error in OSM road network.

3 Methodology

To conduct this case study, the considered methodology consists of two basic steps. The first step is data extraction and parsing, second step is data quality assessment,

Table 1 Specification of considered road network

	Haryana		Punjab	
Type	Count	Length (m)	Count	Length (m)
Path	275	93,710.855	400	101,069.121
Road	67	24,098.492	39	11,138.494
Motorway	1151	411,640.565	1229	531,258.67
Trunk	2727	4,151,086.836	2543	4,709,042.578
Primary	1229	2,141,366.489	1018	1,609,639.578
Secondary	1565	2,376,119.302	1262	2,537,928.228
Tertiary	9480	14,448,590.28	10,624	18,151,408.87
Motorway link	159	57,823.99	99	9835.9
Trunk link	309	122,755.871	379	183,391.712

after this result were analysis. Detailed description of these steps is provided in below subsections.

3.1 Data Selection and Collection

To perform the experiment, OSM road network of Haryana and Punjab (state of India) was selected. OSM road dataset was collected using overpass application interface of OSM, which provides current state of OSM road network. During this process, data were extracted using GeoJSON and GPX file format. These files include road information like road type, road id, nodes and tags. Road tags consist of different parameters like road alternate name, bridge, type of highway, number of lanes, layers, maximum speed, one-way or not, name, internal reference, etc. For this experiment, 34,555 roads of cumulative length 51,671,905.83 km were considered. Detailed descriptions of type of road and length of roads are provided in Table 1. This collected road information was parse to CSV files for the data analysis and processing. No addition reference dataset was considered for the quality assessment, and only the intrinsic parameters were used for the analysis. Intrinsic parameters/indicators are the characteristics of each element of the dataset.

3.2 Quality Assessment

For quality assessment, the K-means algorithm with Elbow method was used to identify the errors in the dataset. To analyze the quality, features were extracted from the dataset and converted to 0 and 1 format. For example, if a road has a bridge tag, then it must have a value, so corresponding each string value feature will contain 1, and for missing value, feature will contain 0. Similarly, for the roads, intersecting a building has type = "road" and contains building tag with value yes. One considered scenario of correct road dataset is as below:

"tags":

{"highway": "primary", "lanes": "2", "ref": "SH17", "surface": "asphalt"}.

Another scenario of dataset of road intersecting building will be as below:

"tags":

{"highway": "primary", "lanes": "2", "ref": "SH17", "surface": "asphalt", "building": "yes"}.

So, the feature value of road with intersecting building will be <value, 1,1,1,1>, whereas road with no intersecting building has feature like <value, 1,1,1,0>. Similarly, features were extracted for other parameters. K-means algorithm with Elbow method is used to group the dataset based on the feature similarity. To implement the K-means algorithm, normalized data (features were provided in the form 0 and 1) was provided to K-means algorithm. Some variables like node id, location and name were not converted to binary value. This method will provide the clusters based on the road characteristic, tags and basic parameters. Steps of this approach are provided in Algorithm 1.

$$C = \frac{\sqrt{\sum_{j=1}^{f} X_j}}{f}, \tag{1}$$

$$d(p, q) = \sqrt{\sum_{i=1}^{n} (q_i - p_i)^2}. \tag{2}$$

Algorithm 1: (*D*, *T*, *A*)

Input: |*D*| where D is complete data set having T number of classes and A is number of attributes in D. This data set contains feature element of OSM road network.

Result: Identified groups based on data set and features.

1. Identify count of clusters (*C*) using Elbow method and iterations (*N*).
2. Set initial centroid value using mean value of available data using Eq. (1) and (2).
3. While ($I != N$)
   - a. For each data entry, find similarly between data and each cluster centroid using Eq. 1.
   - b. Assignment data points to clusters based on maximum similarity and minimum distance.

To optimize and stable the value of number of groups/clusters, Elbow method was used. This method can be expressed using Eq. 3, d_j and C_i are the jth data item and centroid of ith cluster, respectively.

$$\text{Sum of square error} = \sum_{i=1}^{K}\sum_{j=1}^{n} d_j - C_i^2 \tag{3}$$

4 Result and Discussion

Based on the performed experiment, Table 2 shows the error identified in the considered road network. In this experiment, we observed road intersecting with building, missing roads, disconnect roads, dead ends, overshoot, undershoot and unclosed areas in the OSM data (also shown in Fig. 1). Apart from the above-mentioned error, the considered road networks do not include the updated information about the flyover and rural milestones and road data correctly as shown in Table 2.

Table 2 Errors identified in considered road network

Error type	Count	Error type	Count
Missing roads	144	Wrong flyovers	89
Overshoot	197	Old street name	203
Undershoot	211	Misspelled road name	129
Dead ends	322	Missing turnabout	23
Missing details	987	Missing link roads	24
Road with intersecting buildings	65	Incorrect milestone information	29

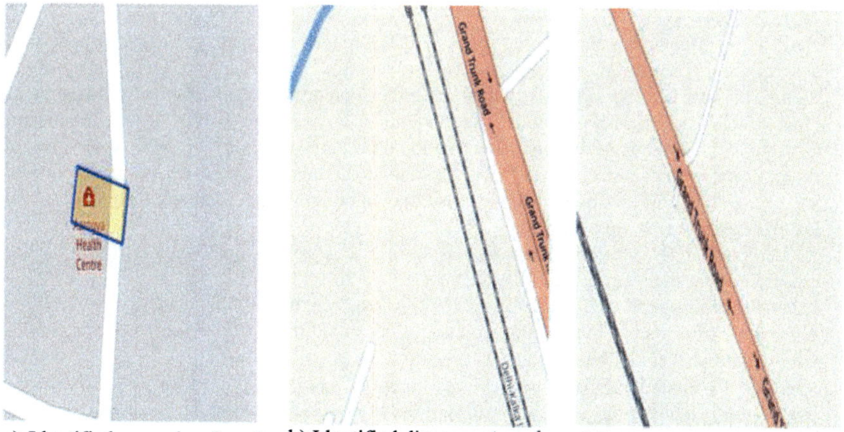

a) Identified scenario of road with intersecting building b) Identified disconnect road c) Error present in the form of overshoot

Fig. 1 Identified scenario of quality error in OSM road dataset

5 Conclusion and Future Scope

In this study, K-means algorithm with Elbow method was used to identify the error in OSM road network. These errors primarily come under the category of topological inconsistencies of OSM data. 34,555 roads having highway, sub-highway and local street of Haryana and Punjab OSM datasets were considered in this experiment. As OSM is crowd-sourced project, so data contain many errors. In this experiment, we identified building intersecting with roads, missing tags with road nodes, missing roads, overshoot, undershoot and dead ends present in the considered road network. Apart from these topological features, considered road network contains many errors in attributes and miscellaneous information. Any development and research project use these data without making the correction, which will suffer from many inaccuracies, so before using this dataset, error correct must be done. In future, we will try to incorporate automated mechanism to rectify the identified errors.

References

1. Neis P, Zielstra D, Zipf A (2011) The street network evolution of crowdsourced maps: OpenStreetMap in Germany 2007–2011. Future Internet 4(1):1–21
2. Colpaert P, Abelshausen B, Melendez JAR, Delva H, Verborgh R (2019) Republishing OpenStreetMap's roads as linked routable tiles. In: European semantic web conference, pp 13–17
3. Mondzech J, Sester M (2011) Quality analysis of OpenStreetMap data based on application needs. Cartographica Int J Geog Inf Geovisualization 46(2):115–125
4. Barron C, Neis P, Zipf A (2014) A comprehensive framework for intrinsic OpenStreetMap quality analysis. Trans GIS 18(6):877–895
5. Haklay M, Weber P (2008) Openstreetmap: user-generated street maps. IEEE Pervasive Comput 7(4):12–18
6. Helbich M, Amelunxen C, Neis P, Zipf A (2012) Comparative spatial analysis of positional accuracy of OpenStreetMap and Proprietary Geodata. In: Proceedings of GI_Forum, pp 24–33
7. Sehra SS, Singh J, Rai HS (2014) Assessing the topological consistency of crowdsourced OpenStreetMap data. Hum Comput 1(2):265–280
8. Brovelli MA, Minghini M, Molinari ME, Zamboni G (2016) Positional accuracy assessment of the Openstreetmap buildings layer through automatic homologous pairs detection: the method and a case study. In: International archives of the photogrammetry, remote sensing and spatial information sciences, pp 615–620
9. Zacharopoulou D, Skopeliti A, Nakos B (2021) Assessment and visualization of OSM consistency for european cities. ISPRS Int J Geo-Inf 10(6):361
10. Adabi R, Abbaspour RA, Chehreghan A (2021) Assessment of the completeness of volunteered geographic information focusing on building blocks data (case study: Tehran Metropolis). J Geospatial Inf Technol 9(1):41–64
11. Alghanim A, Jilani M, Bertolotto M, McArdle G (2021) Leveraging road characteristics and contributor behaviour for assessing road type quality in OSM. ISPRS Int J Geo-Inf 10(7):436
12. Madubedube A, Coetzee S, Rautenbach V (2021) A contributor-focused intrinsic quality assessment of OpenStreetMap in Mozambique using unsupervised machine learning. ISPRS Int J Geo-Inf 10(3):156
13. Sehra SS, Singh J, Rai HS (2017) Assessing OpenStreetMap data using intrinsic quality indicators: an extension to the QGIS processing toolbox. Fut Int 9(2):15

14. Shamas S, Panda SN, Sharma I (2022) K-Means clustering using fuzzy C-Means based image segmentation for Lung Cancer. In: 2022 3rd International conference on computation, automation and knowledge management (ICCAKM), pp 1–5. IEEE
15. Singh S, Singh J (2022) Map matching algorithm: empirical review based on Indian OpenStreetMap road network data. Int Arab J Inf Technol 19(2):143–149
16. Singh J, Singh S, Singh S, Singh H (2019) Evaluating the performance of map matching algorithms for navigation systems: an empirical study. Spat Inf Res 27(1):63–74

Performance Evaluation of Time Series Analysis on the Dataset of Airline Passengers: An Exploratory Data Analysis

Madhu Sree Sankaran, Alka Manvayalar Suresh, and Surender Reddy Salkuti

Abstract A surge in the amount of data makes it requisite to clean, interpret, and obtain useful results from large amount of datasets built over the years. Alongside, time series analysis plays the opportune role to record the points of data and concise its relevant trends periodically. This paper presents time series analysis involving statistical procedures applied to datasets to execute models based on prediction and also to produce expected results. It plays a propitious role to record the points of data and concise its relevant trends periodically. It also highlights the work of time series analysis, comprising the collation and analyzation of the series of data points. With learnings and understanding from different projects, we have presented the ideas on concepts of time series analysis along with exploratory data analysis (data visualization).

Keywords Time series analysis · Exploratory data analysis · Data visualization

1 Introduction

Exploratory data analysis (EDA) plays a primary role for ensuring what data can execute exempting the conventional modeling and provide a bigger picture for understanding the underlying pattern of data set variables and the existing relationship and to determine statistical analysis. Once the task of EDA is completed and insights are determined, its features are implemented for data analysis or even data modeling including machine learning and time series analysis. Eghtesady et al. [1] interprets and compares the patterns of syndromes occurring in the heart at birth with defects of the heart in the US. A detailed description of the innovation of temporary small water bodies (TSWB) maps depending on the time series of subjective well-being

M. S. Sankaran · A. M. Suresh
Department of Electrical and Electronics Engineering, Loyola-ICAM College of Engineering and Technology, Chennai, Tamil Nadu 600034, India

S. R. Salkuti (✉)
Department of Railroad and Electrical Engineering, Woosong University, Daejeon 34606, Republic of Korea
e-mail: surender@wsu.ac.kr

A. Rani et al. (eds.), *Signals, Machines and Automation*, Lecture Notes in Electrical Engineering 1023, https://doi.org/10.1007/978-981-99-0969-8_64

(SWB) for various periods of time in Haas et al. [2]. A random sampling approach has been employed along with one-stage cluster analysis to detect water resources. The rate of preciseness is increased, and error is reduced. In time series analysis, the analyst records the data points in a frequent time interval rather than observing the data points intermittently [3]. In economics, it tracks the gross domestic product (GDP), consumer price index, unemployment rates, and so on. In the field of social sciences, it can keep a track of the birth rates and population. Other fields include epidemiology, medicine, and physical sciences. The two divisions of time series analysis are univariate time series models, which are employed when a single dependent variable is characterized, whereas multivariate time series models are employed when multiple dependent variables are characterized [4].

2 Execution Requirements

Certain libraries are required to be imported in order to apply time series analysis alongside executing data analysis and visualization. datetime—provides access to classes to work with date and time, statsmodels.tsa.stattools—set of statistical tools to work with time series analysis [5, 6]. From the imported libraries, few functions are to be accessed, *to_datetime* to convert its data type from object to date and time, *adfuller* is a statistical test to analyze the impact of time-series defined by trend, *rolling( )* is used in signal processing and time-series data and presents rolling window calculation feature, *ewm( )* presents exponentially weighted functions, *shift( )* produce shifting in indices with required periods with an optional time–frequency, *trend* displays the overall rise and fall value in mean, *seasonality* displays a reoccurrence in the cycle [7], *residual* displays remaining random residuals, and *forecast* allows to predict future values using values of past and a lot of other factors.

3 Application of Time Series Analysis

'adfuller' stands for the augmented dickey-fuller test which is a statistical test and it is also known as the unit root test [8, 9]; parameter *'autolag'* is used to determine the lag length among the values present shown in Fig. 1. It is employed in order to identify if the provided time series is stationary or non-stationary. *'def'* is a keyword to define a function. *SYNTAX: def function_name(parameters):* In 'Series' function 'index' parameter is used to allot indices.

'rolling( )', works by taking a *'k'* size of the window at a time and executing the desired mathematical operation described in Fig. 2. The first 15 rows are displayed for which the mean is calculated using 'mean( )'. A single row is executed for which standard deviation is performed using 'std( )'. In the same graph, three values such as 'rollingmean', 'rollingstd', and 'actualdata' are plotted using 'plot( )' with different

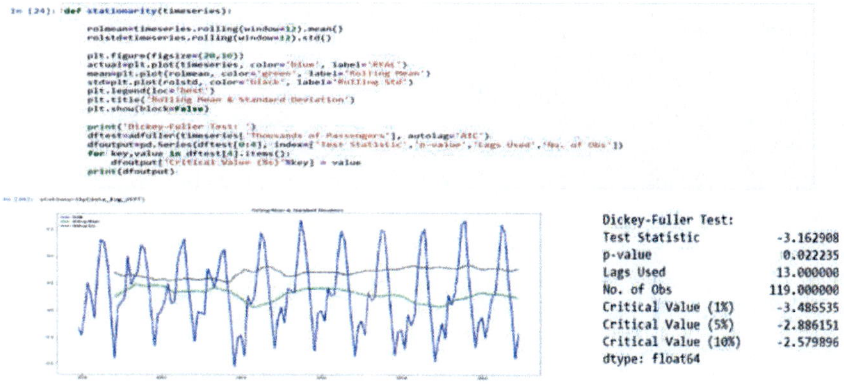

Fig. 1 Augmented dickey fuller test

colors and labels, enabling the parameters of the plot [10, 11], i.e., 'colors' and 'label'. 'show( )' is employed to display all figures.

In 'ewm( )' the parameter *'halflife'* specifies decay in means of half-life *'min_number'* specifies the least number of observations in the window essential to possess a value; otherwise, the outcome is null. *'adjust = True'*, the exponentially weighted function is calculated using weights in Fig. 3. *'Mavg'* stands for moving averages, which analyzes the time-series data by calculating averages of different subsets of the completed dataset provided. It is usually calculated by the product of

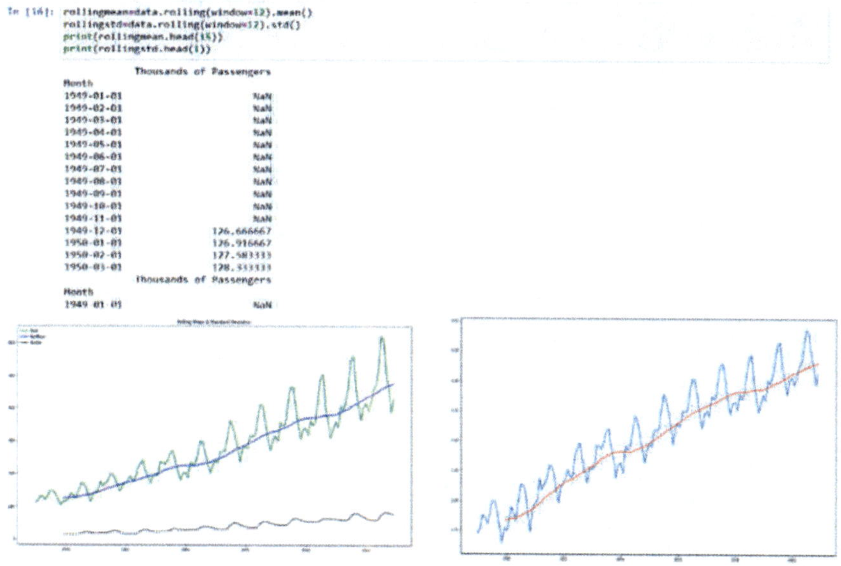

Fig. 2 rolling( ) plot

data weighted and points of data. The moving average is determined in a fashion where lower weights are provided to older observations. As the given data points get older, the weights are fallen exponentially.

Time series decomposition includes the practice of splitting time series data into components of the core. These include modules like potential in trend, seasonality, and residual. Almost every time series data are not stationary, referring that mean, variance, or covariance value is time-dependent. Hence, it is essential to separate the above-mentioned functions before applying them to a model. For time series decomposition, *'statsmodel'* library has a method *'seasonal_decompose( )'*. *'tight_layout'* is used to adjust padding between and around subplots in Fig. 4.

Autocorrelation (ACF) plot determines the correlation of time series with its own lags. Each vertical line in the plot illustrates the correlation between lags and series

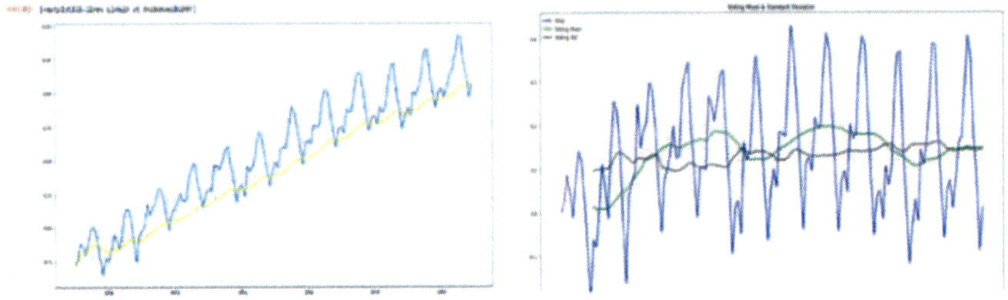

Fig. 3 ewm() plot

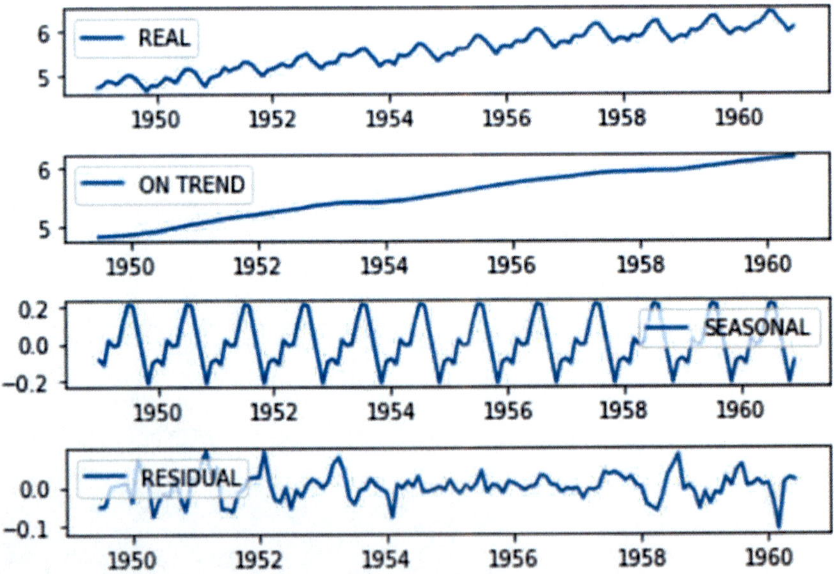

Fig. 4 Trend, seasonality, and residual plot

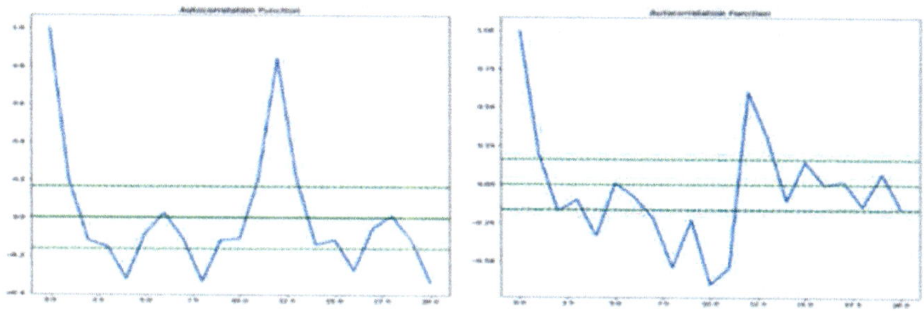

Fig. 5 Auto and partial correlation plot

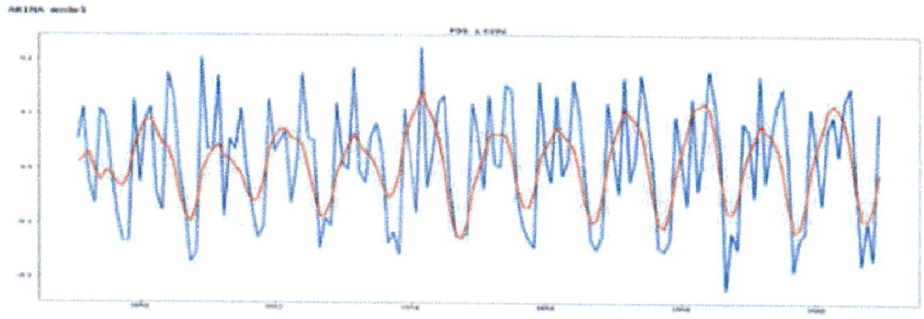

Fig. 6 Arima model

starting from lag 0. The green (could be any color mentioned in the parameter) shaded line is the 'significance level' and the lags that lie above and below are 'significant lags'. Partial Autocorrelation ('PACF'), represents the autocorrelation of any provided lag against the same/current series, but with the removal of lags-in between in Fig. 5 [4].

'Arima model' analyzes and forecasts time series data. Auto-Regressive Integrated Moving Average is the acronym of ARIMA. It also provides a prediction of future values based on past values. It also utilizes lagged moving averages to smooth time series data in Fig. 6 [5].

'forecast' allows predicting future values using values of past and other factors; here it is done in step 120. It involves model building through analysis and conducting observations, and supports future decision-making strategies. One forecasting model shown here is the time series model in Fig. 7 [6].

4 Discussion

In time series analysis, adfuller test is performed, and using time series decomposition, seasonality trend and residual data are determined along with moving averages. Arima model is executed for time series plot, and calculations are performed on

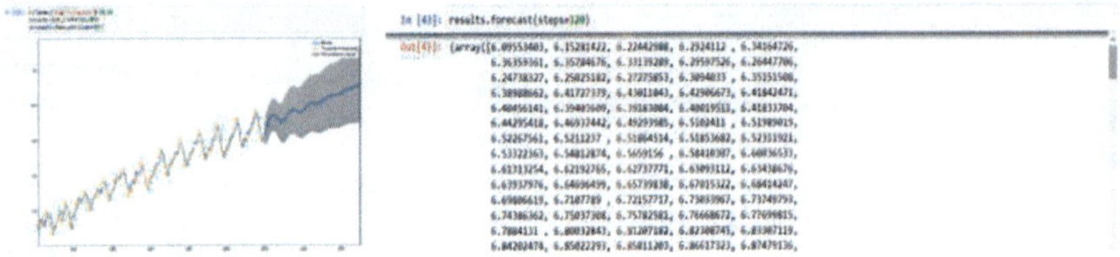

Fig. 7 forecast() plot

selected windows using rolling functions. Finally, correlation is performed to emphasize time series, and various plots are plotted using NumPy for a better understanding of insights into data. In this way, the pattern of the number of passengers is well studied for proper maintenance of airway transport, and the fashion followed in sales is determined by obtaining the central values using clustering for easy management.

5 Conclusion

To conclude, the time series prediction model using machine learning was useful to obtain the curve of the airline passenger. This way, the required context is served to build a suitable model to find solutions for the problems and to accurately interpret the results, and also forecast or predict values in order to save time and also for reference purposes.

Acknowledgements This research work was funded by "Woosong University's Academic Research Funding—2022".

References

1. Eghtesady P, Brar A, Hall M (2011) Seasonality of hypoplastic left heart syndrome in the United States: a 10-year time–series analysis. J Thorac Cardiovasc Surg 141(2):432–438
2. Haas EM, Bartholomé E, Combal B (2009) Time series analysis of optical remote sensing data for the mapping of temporary surface water bodies in sub-Saharan western Africa. J Hydrol 370(1–4):52–63
3. Brockwell PJ, Davis RA (1991) Time series: theory and methods, 2nd edn. Springer, New York, p 577
4. Flores JHF, Engel PM, Pinto RC (2012) Autocorrelation and partial autocorrelation functions to improve neural networks models on univariate time series forecasting. In: The 2012 International joint conference on neural networks (IJCNN), pp 1–8. IEEE. https://doi.org/10.1109/IJCNN.2012.6252470
5. Junaidi, Buliali JL, Saikhu A (2020) Improving ARIMA forecasting accuracy using decomposed signal on PH and turbidity at SCADA based water treatment. In: 2020 3rd International conference on information and communications technology (ICOIACT), pp 131–136. IEEE. https://doi.org/10.1109/ICOIACT50329.2020.9332080

6. Jenkins GM, Box GEP, Reinsel, Ljung (2008) Time series analysis: forecasting and control. Wiley, Hoboken
7. Ezugwu AE, Ikotun AM, Oyelade OO, Abualigah L, Agushaka JO, Eke CI, Akinyelu AA (2022) A comprehensive survey of clustering algorithms: state-of-the-art machine learning applications, taxonomy, challenges, and future research prospects. Eng Appl Artif Intell 110:104743
8. Ghosh A, Nashaat M, Miller J, Quader S, Marston C (2018) A comprehensive review of tools for exploratory analysis of tabular industrial datasets. Visual Inform 2(4):235–253
9. Ramsdale E, Snyder E, Culakova E, Xu H, Dziorny A, Yang S, Zand M, Anand A (2021) An introduction to machine learning for clinicians: how can machine learning augment knowledge in geriatric oncology? J Geriatr Oncol 12(8):1159–1163
10. Rohaan D, Topan E, Groothuis-Oudshoorn CGM (2022) Using supervised machine learning for B2B sales forecasting: a case study of spare parts sales forecasting at an after-sales service provider. Expert Syst Appl 188:115925
11. Fagroud FZ, Ajallouda L, Lahmar EHB, Toumi H, Achtaich K, Filali SE (2020) IOT search engines: exploratory data analysis. Procedia Comput Sci 175:572–577

Vehicle Classification for Autonomous Vehicles Using Transfer Deep Learning

Rishu Chhabra, Vikas Khullar, and Saravjeet Singh

Abstract Intelligent Transportation System (ITS) aims at improving the safety on roads, and with the advancement in vehicular technology, autonomous vehicles will soon be equally present on roads along with manual vehicles. Vehicle classification is used by autonomous vehicles to decide driving behavior. In this paper, we use transfer learning in deep learning for vehicle classification. Different algorithms are implemented and compared using Kaggle dataset. It is evident from the experimental results that the hybrid InceptionResNet algorithm exhibits the highest training accuracy of 92 percent and validation accuracy of 62 percent. Similarly, the validation of the InceptionResNet model results in 85 percent AUC. The classification results can be used by autonomous vehicles to improve driving behavior like lane changing, overtaking, etc., and improve the safety of the commute.

Keywords Autonomous vehicle · Deep learning · ITS · Transfer learning · Vehicle classification

1 Introduction

Autonomous vehicle technology has the prospective to enhance efficiency, safety, and cost of the current transportation system. They aim to provide greater safety benefits as compared to driver assistance systems that are capable to generate warnings in case of inappropriate driving behavior [1]. As shown in Fig. 1, the presence of autonomous vehicles is increasing on roads and the global market is growing rapidly. In developed nations, the highways are predicted to be driver-free by 2025 [2, 3]. Autonomous vehicle technology works on the ecosystem of sensors and is very helpful for people with disabilities and improves their mobility. It helps in reducing traffic congestion, the number of road crashes and increases the productivity of the drivers by allowing them to multi-task [4, 5]. Vehicle classification is the process of categorizing the identified vehicles into specific classes [6]. It is used in autonomous vehicles to

R. Chhabra (✉) · V. Khullar · S. Singh
Chitkara University Institute of Engineering and Technology, Chitkara University, Punjab, India
e-mail: rishu.miglani@chitkara.edu.in; rishuchh@gmail.com

A. Rani et al. (eds.), *Signals, Machines and Automation*, Lecture Notes in Electrical Engineering 1023, https://doi.org/10.1007/978-981-99-0969-8_65

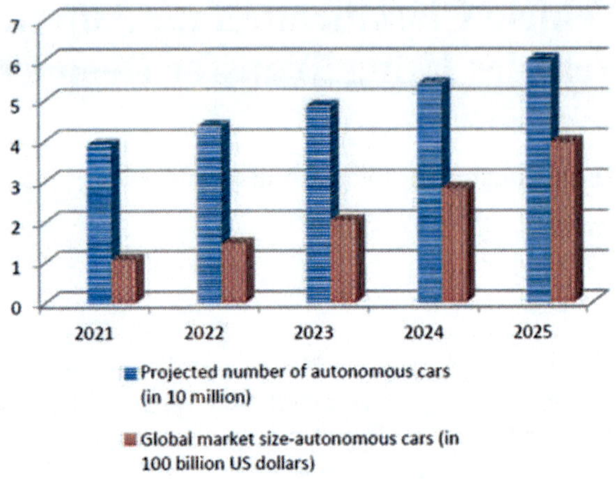

Fig. 1 Growth of autonomous vehicles and global market

decide and improve driving behavior like overtaking behavior, lane change behavior, acceleration, deceleration, etc.

Deep learning has been a major driving force behind the advancements in computer vision and has a major impact on autonomous vehicle technology [7]. Deep learning approaches have been instrumental in solving various real-time problems associated with autonomous driving [8]. Transfer learning is a technique to improve the prediction process by reusing a previously trained model on a new problem. In this paper, we propose to classify the vehicles using transfer learning for deep learning. The deep learning algorithms implemented are VGG16, DenseNet, and InceptionResNet. The model is trained using Kaggle dataset and classifies the vehicles into nine categories (car, bus, truck, taxi, ambulance, bicycle, bike, van) [9].

The remainder of the paper is organized as follows: Sect. 2 presents the related work. The methodology and experimental results are given in Sect. 3, and Sect. 4 concludes the work with future research directions.

2 Related Work

Autonomous driving technology is advancing from level-0 of no automation toward level-5 of full automation. Deep learning techniques have been instrumental in implementing several tasks related to autonomous driving like road detection, vehicle detection, lane detection, traffic signal detection, pedestrian detection, and driver's drowsiness detection [8].

Table 1 presents a holistic view of the state-of-the-art deep learning techniques implemented for autonomous driving tasks mentioned above. A detailed survey of different deep learning techniques applied in autonomous vehicle technology has been presented in [10]. Vehicle detection and classification information can be used by autonomous vehicles to improve driving behavior. The vehicle classification techniques implemented and compared in the paper can be applied to all five levels of automation (level-0 to level-4) [5].

3 Methodology and Experimental Results

The methodology in this paper includes the implementation of diverse transfer learning algorithms to classify the vehicles. The work is conducted using Kaggle dataset highlighting diverse categories of vehicles. As per the studied literature, transfer learning yields efficient results as compared to basic deep learning algorithms. In this paper, VGG16, DenseNet, and InceptionResNet models are implemented and analyzed for the vehicle category classification. As presented in Fig. 2, the InceptionResNet implementation results in the highest training accuracy of 92 percent and the validation accuracy of 62 percent. However, the AUC of the proposed implementation seems acceptable from 84 to 86 percent. The supporting detailed numeric results have been also presented Table 2.

4 Conclusions and Future Research Direction

Autonomous vehicle technology has been going on for a decade but is not widely implemented due to the cost involved in production. The research is going on to improve the development efforts and make autonomous vehicle technology benefit the majority of commute. In this paper, we implemented and compared several deep learning algorithms for vehicle classification that can be used by autonomous vehicles for improving driving behavior. It is concluded that the hybrid transfer learning approach InceptionResNet exhibits the highest training and validation accuracy between 92 and 62 percent, respectively. The classification results can be used by autonomous vehicles to make certain decisions related to lane changing, overtaking, etc., and improve safety on roads. One of the key challenges in the autonomous vehicle environment is the vehicle-to-vehicle connectivity between the vehicles from different manufacturers. Upcoming research studies should focus on inter-vehicle communication between manual vehicles with ADAS and automated vehicles (irrespective of the manufacturer).

Table 1 Review of deep learning approaches for autonomous driving

Ref. No. (Year)	Deep Learning Technique	System Input	Remarks
Limmer et al. [11] (2016)	CNN	Map database, camera images, and sensor data	The road detection system works efficiently during the nighttime
Cheng et al. [12] (2017)	CasNet	Remote sensing images	The system detects the road in complex background and presence of cars and trees
Chen and Chen [13] (2017)	RBNet	Road images	Road detection and road boundary detection
Wang et al. [14] (2018)	CNN	RGB images	The road boundaries are detected
Song et al. [15] (2018)	CNN	KITTI dataset	The system performs classification, lane detection and generated collision warning
Chu et al. [16] (2018)	CNN, RoI voting	KITTI and PASCAL2007	Vehicle detection is performed
Vijayan and Sherly [17] (2019)	InceptionV3, ResNet50, and VGG16	Video of driver	Driver drowsiness detection
Wang and Zhou [18] (2019)	CNN	HDR traffic light images	The system detects traffic lights from the image database irrespective of illumination constraints
Cai et al. [19] (2020)	CNN with object proposal technique	KITTI, Caltech datasets	Pedestrian detection
Ouyang et al. [20] (2020)	CNN	On-road tests	The system detects traffic lights and has been tested in real time
Li et al. [21] (2021)	CNN	Captured images	The system enhances the low-light images for connected autonomous vehicles
Lee and Liu [22] (2023)	CNN-PP	TORCS, LLAMAS, TuSimple	Lane and Path prediction using DSUNet

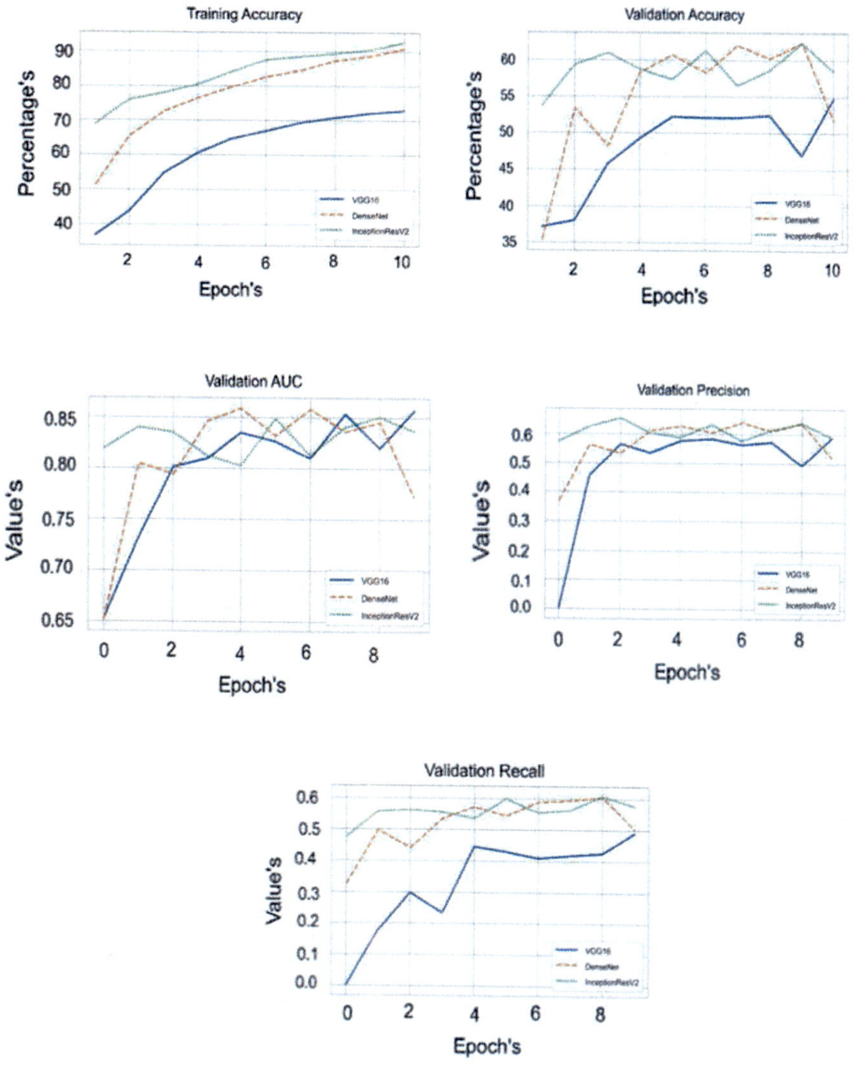

Fig. 2 Transfer learning-based result analysis for vehicle classification

Table 2 Comparative result analysis for vehicle classification implementation

Epoch's	Training Accuracy			Validation Accuracy			AUC			Precision			Recall		
	VG	DN	IRN	VG	DN	IRN	VG	DN	IRN	VG	DN	IRN	VG	DN	IRN
1	0.37	0.51	0.69	0.37	0.35	0.54	0.65	0.65	0.82	0.00	0.37	0.57	0.00	0.32	0.47
2	0.44	0.65	0.76	0.38	0.53	0.59	0.73	0.80	0.84	0.46	0.56	0.63	0.18	0.50	0.56
3	0.55	0.72	0.78	0.46	0.48	0.61	0.80	0.79	0.84	0.56	0.53	0.65	0.30	0.44	0.56
4	0.61	0.76	0.80	0.49	0.58	0.59	0.81	0.85	0.81	0.54	0.61	0.60	0.23	0.53	0.56
5	0.65	0.80	0.84	0.52	0.61	0.57	0.84	0.86	0.80	0.58	0.63	0.59	0.45	0.57	0.54
6	0.67	0.83	0.88	0.52	0.58	0.61	0.83	0.83	0.85	0.58	0.60	0.63	0.43	0.54	0.60
7	0.69	0.85	0.89	0.52	0.62	0.57	0.81	0.86	0.81	0.56	0.64	0.58	0.41	0.59	0.56
8	0.71	0.87	0.89	0.52	0.60	0.59	0.85	0.84	0.84	0.57	0.61	0.62	0.42	0.59	0.56
9	0.72	0.89	0.91	0.47	0.62	0.62	0.82	0.85	0.85	0.49	0.64	0.64	0.43	0.60	0.61
10	0.73	0.91	0.93	0.55	0.52	0.58	0.86	0.77	0.84	0.59	0.52	0.59	0.49	0.50	0.58

References

1. NHTSA (2019) Automated vehicles for safety
2. Statista (2021) Projected number of autonomous cars worldwide
3. Statista (2021) Worldwide - driverless car market size 2025
4. Kim M-K, Park J-H, Oh J, Lee W-S, Chung D (2019) Identifying and prioritizing the benefits and concerns of connected and autonomous vehicles: a comparison of individual and expert perceptions. Res Transp Bus Manag 32:100438
5. Bagloee SA, Tavana M, Asadi M, Oliver T (2016) Autonomous vehicles: challenges, opportunities, and future implications for transportation policies. J Mod Transp 24(4):284–303
6. Shokravi H, Shokravi H, Bakhary N, Heidarrezaei M, Koloor SSR, Petrů M (2020) A review on vehicle classification and potential use of smart vehicle-assisted techniques. Sensors 20(11):3274
7. Grigorescu S, Trasnea B, Cocias T, Macesanu G (2020) A survey of deep learning techniques for autonomous driving. J Field Robot 37(3):362–386
8. Muhammad K, Ullah A, Lloret J, Del Ser J, de Albuquerque VHC (2020) Deep learning for safe autonomous driving: current challenges and future directions. IEEE Trans Intell Transp Syst 22(7):4316–4336
9. Kaggle (2021) TAU vehicle type recognition competition
10. Kuutti S, Bowden R, Jin Y, Barber P, Fallah S (2021) A survey of deep learning applications to autonomous vehicle control. IEEE Trans Intell Transp Syst 22(2):712–733
11. Limmer M, Forster J, Baudach D, Schüle F, Schweiger R, Lensch HPA (2016) Robust deep-learning-based road-prediction for augmented reality navigation systems at night. In: 2016 IEEE 19th International conference on intelligent transportation systems (ITSC). IEEE, pp 1888–1895
12. Cheng G, Wang Y, Xu S, Wang H, Xiang S, Pan C (2017) Automatic road detection and centerline extraction via cascaded end-to-end convolutional neural network. IEEE Trans Geosci Remote Sens 55(6):3322–3337
13. Chen Z, Chen Z (2017) RBNet: a deep neural network for unified road and road boundary detection. In: International conference on neural information processing, pp 677–687
14. Wang Q, Gao J, Yuan Y (2017) Embedding structured contour and location prior in siamesed fully convolutional networks for road detection. IEEE Trans Intell Transp Syst 19(1):230–241
15. Song W, Yang Y, Fu M, Li Y, Wang M (2018) Lane detection and classification for forward collision warning system based on stereo vision. IEEE Sens J 18(12):5151–5163
16. Chu W, Liu Y, Shen C, Cai D, Hua X-S (2017) Multi-task vehicle detection with region-of-interest voting. IEEE Trans Image Process 27(1):432–441
17. Vijayan V, Sherly E (2019) Real time detection system of driver drowsiness based on representation learning using deep neural networks. J Intell Fuzzy Syst 36(3):1977–1985
18. Wang J-G, Zhou L-B (2018) Traffic light recognition with high dynamic range imaging and deep learning. IEEE Trans Intell Transp Syst 20(4):1341–1352
19. Cai Z, Saberian M, Vasconcelos N (2015) Learning complexity-aware cascades for deep pedestrian detection. In: Proceedings of the IEEE international conference on computer vision (ICCV), pp 3361–3369
20. Ouyang Z, Niu J, Liu Y, Guizani M (2019) Deep CNN-based real-time traffic light detector for self-driving vehicles. IEEE Trans Mob Comput 19(2):300–313
21. Li G, Yang Y, Qu X, Cao D, Li K (2021) A deep learning based image enhancement approach for autonomous driving at night. Knowl-Based Syst 213:106617
22. Lee D-H, Liu J-L (2023) End-to-end deep learning of lane detection and path prediction for real-time autonomous driving. SIViP 17:199–205

Self-attention-based Human Activity Detection Using Wearable Sensors

Shivin Diwakar, Dhirendra Dwivedi, Satya P. Singh, and Madan Sharma

Abstract Analyzing human activity through wearable sensors can assist applications connected to context—vigilance and health care. The proposed approach utilizes convolutional and recurrent modeling to express the space–time-dependent constraints from information provided by numerous sensors. We construct a neural network layout, a combination of convolutional neural network and recurrent neural network, which focuses on spatiotemporal characteristics provided by sensor time series data with short listing and learning crucial points by executing a self-attentive procedure. We exhibit validation of our designed strategy on the WISDM dataset and indicate that the self-attention technique achieves a remarkable amendment in presentation over deep neural networks through the union consisting of convolutional and recurrent network architectures. We also present that our designed algorithm offers a statistic performance improvement over other designed approaches for the information (dataset) under consideration. The presented technique allows precise interpretation of activity from numerous body part sensors through various time intervals. We evaluated our technique through various classification methods namely: MLP Classifier, gradient boosting, Random Forest Classifier, Conv2D–LSTM, and self-attention mechanism (proposed). The simulation results show that our proposed method achieved the highest accuracy of 89.58% out of all other methods. We compared the effects of the designed architecture scheme with a baseline constructed through ConvNN and RecNN models, and with earlier designed approaches. We figured out that our self-attention model outperformed the baseline model with high difference among all the considered metrics.

Keywords Self-attention · ConvNN · LSTM · Human activity detection

S. Diwakar · D. Dwivedi · S. P. Singh (✉)
Center of Excellence in Artificial Intelligence and Division of Electronics and Communication, Netaji Subhas Institute of Technology, University of Delhi, New Delhi, India
e-mail: satya.prakash@nsut.ac.in

M. Sharma
Department of Information and Communication Engineering, College of Engineering, University of Buraimi, Al Buraimi Governorate, Sultanate of Oman

A. Rani et al. (eds.), *Signals, Machines and Automation*, Lecture Notes in Electrical Engineering 1023, https://doi.org/10.1007/978-981-99-0969-8_66

1 Introduction

Analyzing human physical behavior has diverse ascending implementations for human well-being. The list includes health monitoring, lifestyle, and various other daily chores [1]. The most impressive techniques utilized for recording physical human activity are mainly based on artificial intelligence mechanisms, wearable sensors, deep learning (computer vision) techniques, and IoT-based strategies. As far as deep learning-related methods are concerned, usually, a camera is utilized to trace and monitor human activities [2]. These techniques demand regulating the lens at various angles executing deep learning algorithms on image data tracked through the camera, making it more complex to interpret the performed activity. However, wearable sensors consist of typically minute electronics or devices build on biochemistry which can be set on numerous parts as required or placed in pockets of the concerned [3]. Moreover, with recent upgrades in wireless technology, it is fetching possibilities of manufacturing energy efficient, mobile, and high precision wearable sensors. Typically, a wearable sensor comprises three aspects, namely (i) input data sensors like an accelerometer to fetch physical activity data, (ii) transfer signal medium (wireless or wired) to transfer the extracted information to immobile unit, and (iii) interpreting and analyzing information units for formulating appropriate constraints and information. Occasionally, the input signals fetched from the initial unit are erroneous and low in power, needing amplification filtering before performing fetching features and classifying techniques. For instance, the collection of information through accelerometers and magnetometers can drastically influence the required results through traces of electronic noises, causing varied abnormalities or erroneous values in the dataset [4].Thus, improvement can be performed through filtering processes.

Furthermore, the inclusion of extravagant GPUs authorizes the construction of deeper models for gaining insights into descriptive constraints from multiplexed data. Concerning this fact, convolution neural networks (ConvNN) acquired awareness over time. ConvNNs exhibit characteristics that result in performance improvement. They exhibit significant improvement in results but are computationally expensive and need a large amount of training data.

On the other hand, they cannot record complex exercises that needed to undergo time-dependent characteristics analysis of the time series wave. Time dependent characteristics are said to be constraints that are dynamic or switch through time. This reflects that the time series information is erroneous and that every sensor may not be taking part at a specific time. To overcome this limitation, we know that recurrent neural network (RecNN) considers temporal characteristics from the sensor data. Thus, recurrent neural networks (RecNNs) are executed in recognition activities from sensor data. RecNN is a unique subject of neural network that was specifically invented to overcome temporal sequences like machine translation, speech recognition, and sequential learning. RecNN struggled through the vanishing gradient problem. This got resolved through the arrival of LSTM networks. Considering the dominance of convolutional neural networks for tracking spatial domain data and the LSTM Network's advantage in capturing time-dependent information, many researchers suggest a combination of ConvNN and LSTMs to be used to improve in this field.

2 Related Work

In [5], the scheme focused on distributing 3D postures using both gyroscopic and accelerometric signals together from a portable device. The authors designed a specific convolutional neural network (CNN) structure having a mixture of max-pooling, averaging, and 1D convolutional mechanisms. It classified defined-length matrices head-on, which consisted of normalized sensor time series data, as one of the various postures to be recognized. There proposed mechanism examined different datasets and test/train combinations having 14 unique gestures and figured out the influence of diverse hyper-parameters like kernels and network size on the general performance. Their mechanism showed superiority in rates of recognition over other defined methods.

In [6], Yang discussed a structured feature learning technique for human activity detection. The scheme involved a deep convolutional neural-net (ConvNN) to automatize feature learning from input data in a structured manner. This architecture considers the learned features as high-level abstract presentations of unlabeled time series data. By grasping the labeled information through supervised learning techniques, these features are enhanced with more contrasting powers. It highly enhances the feature learning and classification techniques in a single model. Many models for human activity detection use various deep learning techniques to improve the performance of their proposed scheme. In [7], the author constructed a deep neural net model and executed transfer learning to distinguish between several human exercises from one or more lenses. The concept of using LSTM Network architecture to improvise the strategy for effective human activity recognition was also taken into account. While ConvNN and LSTMs drives focus on space–time-dependent data, we need to capture precise information from the implants gathered by the combined architecture of ConvNNs and LSTMs. This concept has not undergone consideration in this research field. In this paper, we pivot on the self-attention methodology. The self-attention assigns some weights to the implants of the input using a two-dimensional matrix in a manner in which each row of the 2D matrix serves a distinguished part of the information. Through the combination of ConvNNs and LSTM networks, we exhibit that the self-attention methodology results in a significant amendment in precision for Wireless Sensor Data Mining (WISDM) dataset [8].

Deep learning has drastically pushed the limits of human activity recognition through flexible wearable sensors. Comprehensive classification and encapsulation of present works that introduces deep learning techniques for wearable sensor-based human activity recognition, offering rigorous analysis of present amendments, development trends and major drawbacks, has been the discussed in major details in various researches [9]. Analyzing personalized deep and machine learning strategies, and comparison of performances through conventional deep learning techniques, has also been performed in the past [10]. 3D skeletal human activity recognition has also been practiced utilizing encoded space–time-dependent metrics and feature-engineering based on evolutionary algorithmic techniques [11] and BiLSTMs [12]. Alongside, radar-based human activity recognition has been performed in latest research works [13].

3 Proposed Scheme

The layout for the designed scheme is presented in Fig. 1. The input for the presented model is series time data in a time aperture of proportion S from M sensors. Let time series be in form of $f = (f(t))^T$ where $f(t) \in R^M$ is the initial input at a certain time stamp t.

3.1 *Embedding Layer*

The concept behind attaching a hidden layer is to know about the presentation of the sensors through B 1D space convolutional functioning. Let the weights assigned to filters for the bth filter be w_b and $g_c(t) = \left(g_{c,b}(t)\right)^B$ be outcome of the convolutional layout where $g_{c,b}(t)$ is

$$g_{c,b}(t) = f(t) * \mathrm{w}_b \tag{1}$$

The symbol $*$ represents convolution in the space domain, $\mathrm{w}_b \in R^{1*B}$.

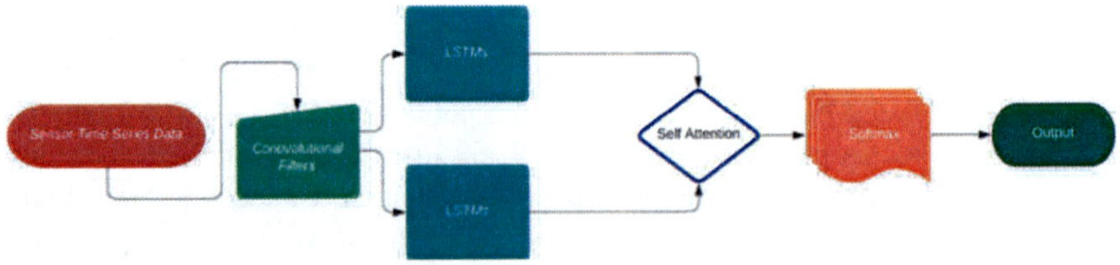

Fig. 1 Structure of presented technique exhibiting various steps based on the algorithm with a well-defined focus on outcomes linked to self-attention which is being explained by two sub-divided processes encapsulated in the LSTMs-based unit

3.2 Encoder (LSTM)

For a single LSTM network layer, the encoder outcome $\boldsymbol{u}_l(\boldsymbol{t}) \in \boldsymbol{R}^{1\times l}$, where L represents hidden layer units of the LSTM network encoder, and given through the equation:

$$u_l(t) = \mathrm{LSTM}(u_c(t), u_l(t-1)) \tag{2}$$

3.3 Attention Layer

After grasping the local context constraints and time-dependent dynamic information with combining ConvNN layout and LSTM network units from initial time data. We input the combined outcome of the encoder, $\boldsymbol{u}_l \in \boldsymbol{R}^{1\times L}$, for various timestamps to the attentive layout:

$$u_l = [u_l(t)]_{t=1}^{T} \tag{3}$$

The u_l matrix records the exhibition of the initial input sensor data over various times tamps T. The score, $S \in R^{T\times L}$, for example x is then transforms the equation to

$$\beta = \mathrm{Softmax}\left(Z_{\mathrm{att}}\tanh\left(Y_{\mathrm{att}}h_l^{'}\right)\right) \tag{4}$$

where $u_l^{'} = \mathrm{transpose}(u_l)$, $Y_{\mathrm{att}} \in R^{I\times L}$ and $Z_{\mathrm{att}} \in R^{J\times I}$ are matrices comprises weights, for the attentive set, I denotes the attentive length, J denotes the outcome length and $(.) =$ dotproduct taken in the space domain. $\beta \in R^{J\times T}$ denotes range of values between (0,1) and is utilized to distinguish weight constraints to each $u_l(t)$ from u_l, that is insight $h_l(t)$ discovered for every time stamp is evaluated in the attentive score $S(S = \alpha \cdot u_l)$.

3.4 Softmax (SM)

The leveling attentive-score vector $s \in R^{(J*L)*1}$ is employed as initial input to the outcoming SM layout. The SM layer outcome v^* is given as follows:

$$v^* = \mathrm{SM}(\mathrm{Q}s + a) \tag{5}$$

Here, $Q \in R^{B*(I*J)}a \in R^B$ denoting the weighing matrix and partial vector of the SM layer, and B represents the numerous possibilities of outcome states.

3.5 Learning Mechanism

The cross-entropy value, K, is reduced to train through the designed strategy, such that K represents an equation as

$$K = -\mathrm{Or}\big[\mathrm{dlog}\big(\nu\big(\gamma^{*}\big)\big)\big] \tag{6}$$

where Or represents the expected value over feeding inputs ν and γ, which denotes the state marker.

4 Simulation Results

See Tables 1 and 2.

All the experiments in this paper were executed in python using Keras [14] with TensorFlow. For coherence, we have bolded the max accuracy and second best with italics and underlines in both tables listed. Table 1 exhibits the performance of different classifiers that underwent the handcrafted feature extraction process on the WISDM dataset. We executed a tenfold cross-validation method for this experiment. The accuracy of MLP and gradient boosting classifier came out 76 and 63%, respectively. We calculated the mean accuracy for both the classifiers, which settled at 76.84% for MLP and 63.70% for gradient boosting classifier. Additionally, we performed our classification process with Random Forest Classification, which offered an improved mean accuracy of 85.10% with mean recall and F1-score of 80.37 and 71.83%, respectively. Instead of utilizing the handcrafted feature extraction

Table 1 Comparative analysis of different classification method on WISDM dataset

Classification methods	Mean accuracy (%)
MLP classifier	76.84
Gradient boosting	63.70
Random forest classifier	85.10
Conv2D–LSTM (baseline)	*85.44*
Self-attention model (proposed)	89.58

Significance of italics and underline are the result from the baseline network

Table 2 Comparative analysis of baseline approach versus proposed approach

	Baseline Conv2D–LSTM	Proposed self-attention model
Mean accuracy (%)	85.44	89.58
Mean recall (%)	85.40	88.56
Mean F-1 score (%)	81.80	87.02

process, we proposed a baseline model using Conv2D and LSTM networks for the feature extraction process. We recapitulate our results in terms of accuracy, recall and F1-score is tabulated in Table 2. According to Table 2 given, our scheme (proposed self-attention model) outperforms the baseline Conv2D–LSTM in all metrics, i.e., mean accuracy, mean recall, and mean F1-score. The amendments acquired by our presented scheme are analytically remarkable for the WISDM dataset. Ultimately, the proposed self-attentive mechanism along with the baseline model drastically improved the performance of our human activity detection model.

5 Conclusion and Future Work

In this paper, we designed a deep learning-based architecture that focuses on the space–time-dependent features of information extracted from numerous wearable sensors that utilize a self-attentive mechanism to discover hidden combinations for analyzing human activity. The ConvNN and RecNN architecture encodes the space and time aspects, respectively, along with the self-attentive layer that forms feature exhibition from the hidden information of sensor time data caused by the ConvNNs and RecNNs. We exhibit that the designed architecture gives a remarkable performance for the WIDSM dataset. After the comparative analysis of different classification method on the WISDM dataset, we conclude that our proposed self-attention model obtained the highest mean accuracy of 89.58%, second best was the baseline model Conv2D–LSTM with mean accuracy of 85.44%. We compared the effects of the designed architecture scheme with a baseline constructed through ConvNN and RecNN models, and with earlier designed approaches. We figured out that our self-attention model outperformed the baseline model with high difference among all the metrics (mean accuracy, mean recall, mean F1-score). Therefore, the designed layout has multiple scopes in human activity detection and can go over to a much larger number of times tamps and sensors. The future work for this scheme would be to include and analyze various attention methodologies along with the comparison of these self-attentive methods' performances with each other.

References

1. Heikenfeld J, Jajack A, Rogers J, Gutruf P, Tian L, Pan T, Li R, Khine M, Kim J, Wang J (2018) Wearable sensors: modalities, challenges, and prospects. Lab Chip 18(2):217–248
2. Fiore L, Fehr D, Bodor R, Drenner A, Somasundaram G, Papanikolopoulos N (2008) Multi-camera human activity monitoring. J Intell Rob Syst 52(1):5–43
3. Bonato P (2010) Wearable sensors and systems. IEEE Eng Med Biol Mag 29(3):25–36
4. Banos O, Garcia R, Holgado-Terriza JA, Damas M, Pomares H, Rojas I, Saez A, Villalonga C (2014) mHealthDroid: a novel framework for agile development of mobile health applications. In: International workshop on ambient assisted living. Springer, Cham, pp 91–98
5. Ordóñez FJ, Roggen D (2016) Deep convolutional and LSTM recurrent neural networks for multimodal wearable activity recognition. Sensors 16(1):115

6. Yang J, Nguyen MN, San PP, Li XL, Krishnaswamy S (2015) Deep convolutional neural networks on multichannel time series for human activity recognition. In: Twenty-fourth international joint conference on artificial intelligence
7. Putra PU, Shima K, Shimatani K (2022) A deep neural network model for multi-view human activity recognition. PlOS ONE 17(1):e0262181
8. Weiss GM, Yoneda K, Hayajneh T (2019) Smartphone and smartwatch-based biometrics using activities of daily living. IEEE Access 7:133190–133202
9. Zhang S, Li Y, Zhang S, Shahabi F, Xia S, Deng Y, Alshurafa N (2022) Deep learning in human activity recognition with wearable sensors: a review on advances. Sensors 22(4):1476
10. Ferrari A, Micucci D, Mobilio M, Napoletano P (2022) Deep learning and model personalization in sensor-based human activity recognition. J Reliable Intell Environ, pp 1–13
11. Basak H, Kundu R, Singh PK, Ijaz MF, Wozniak M, Sarkar R (2022) A union of deep learning and swarm-based optimization for 3D human action recognition. Sci Rep 12(1):1–17
12. Li Y, Wang L (2022) Human activity recognition based on residual network and BiLSTM. Sensors 22(2):635
13. Habib-Ur-Rehman K, Gorji A, Bourdoux A, Pollin S, Sahli H (2022) Multi-view CNNLSTM architecture for radar-based human activity recognition. IEEE Access 10:24509–24519
14. Chollet F (2018) Deep learning mit python und keras: das praxis-handbuch vom entwickler der keras-bibliothek. MITP-Verlags GmbH Co, KG
15. Singh SP, Sharma MK, Lay-Ekuakille A, Gangwar D, Gupta S (2020) Deep ConvLSTM with self-attention for human activity decoding using wearable sensors. IEEE Sens J 21(6):8575–8582
16. Singh SP, Urooj S (2015) Localized radon polar harmonic transform (LRPHT) based rotation invariant analysis of textured images. Int J Syst Dyn Appl (IJSDA) 4(2):21–41
17. Singh SP, Urooj S (2015) Wavelet packets based spectral estimation of textured images. In: 2015 IEEE International conference on computational intelligence communication technology. IEEE, pp 651–654
18. Singh SP, Urooj S (2019) A new computational framework for fast computation of a class of polar harmonic transforms. J Signal Proc Syst 91(8):915–922

Solar Irradiation Prediction Using Neural Networks

Vaibhav Baranwal, Smriti Srivastava, and Monika Gupta

Abstract The solar energy on the earth is abundant in nature; however, its full utilization is still a concern. Due to the high infrastructural cost of setting up a solar plant and uneven distribution of solar energy, it is difficult to switch to solar energy. But, the clean and abundant nature of solar energy is the most appropriate alternative for the conventional energy sources. The presented model predicts the Direct Normal Irradiance (DNI) and Diffused Horizontal Irradiance (DHI). The models which are employed are (i) forward backpropagation neural network, (ii) time series model, (iii) autoregression. The second and third models are time series model. The model validation has been done using correlation score and R-squared score.

Keywords Artificial neural network · Solar irradiance · Time series · Recurrent neural network · Regression

1 Introduction

In past few years, the improvement of renewable energy sources has emerged as a pressing difficulty because of the depletion of conventional fossil reserves, huge costs, and consequences of fossil gas intake at the environment. Solar radiation is the major and prime source of the Earth's energy, supplying more than 95% of the heat energy required for different processes and survival in different eco-systems. It plays an essential part as a renewable energy source. The output of photovoltaic

V. Baranwal
Department of EE, Netaji Subhas University of Technology, New Delhi, India
e-mail: vaibhavb.ee20@nsut.ac.in

S. Srivastava (✉)
Department of ICE, Netaji Subhas University of Technology, New Delhi, India
e-mail: smriti@nsut.ac.in

M. Gupta
Department of EEE, Maharaja Agrasen Institute of Technology, New Delhi, India
e-mail: monikagupta@mait.ac.in

A. Rani et al. (eds.), *Signals, Machines and Automation*, Lecture Notes in Electrical Engineering 1023, https://doi.org/10.1007/978-981-99-0969-8_67

cells highly depends on solar irradiance, temperature, and different weather parameters. Prediction of solar irradiance means that the output of PV cells is predicted one or more time steps ahead. Prediction helps us improve various applications of power systems [1]. An artificial neural network (ANN) is a sophisticated machine learning model which imitates the human nervous system. An ANN can be used for various applications such as predication, curve fitting, and regression. An ANN uses activation function or a transfer function to produce an output. The inputs are multiplied with weights, and their summation is given to the activation function to give the results [2].

Gupta et al. [3] listed various statistical and mathematical algorithms for predicting the solar irradiances along with different data preprocessing methods. Their study provides consolidated evaluation of different methods.

Reddy and Ranjan [4] built an ANN model for forecasting global radiation and compared it with different models by collecting various environmental parameters across the stations. The results obtained justified the application of ANN for predicting the solar irradiances. The ANN model consists of two hidden layers with eight and seven neurons, respectively.

Al-Alawi and Al-Hinai [5] created an ANN model to understand the relationship among various climatic factors and environmental factors. The model resulted in MAPE of 7.3% for training and 5.6% for testing datasets.

The various neural networks that are used are Feedforward ANN, recurrent neural network, and autoregression based on time series data.

2 Location and Parameter of Study

The location used to predict the irradiance is the capital city of India, Delhi. The exact coordinates are Latitude: 28.68 and Longitude: 77.22 with tilt angle of 26.27 degrees. The time stamp for time series model is 10 min. The data have been obtained from SolCast API for six days.

The parameters of concern for forward neural network are:

- Air temp: The temperature of air (2 m above ground level). Units in **Celsius**.
- Azimuth: The angle between the lines directing due north to sun's present position in the sky, for east is less than 0, for west is greater than 0, and 0 for due north. Units in **degrees**.
- Cloud opacity: It is the amount of how opaque the clouds are to the solar radiation in the specified location (0% = no clouds, 100% = full attenuation of incoming lights).
- Precipitable water: The total column precipitable water content. Units in **kg/m2**.
- Relative humidity: Percentage of moisture content relative to the amount that can held by atmosphere. Units in **percentage**.

- Surface pressure: The pressure at ground level. Units in **hPa**.
- Wind speed: The wind speed (10 m above ground level). Units in **m/s**.
- Zenith: The angle between lines perpendicular to the earth's surface and the sun (90 degrees for sunrise and sunset, 0 degree for overhead). Units in **degrees**.

Output Variables

DHI Diffused horizontal irradiance in watts/meter^2.
DNI Direct normal irradiance in watts/meter^2.

3 Methods

3.1 Feedforward Artificial Neural Network

The feedforward ANN consists of an input layer whose density depends on the input (features) features and an output layer whose density depends on the number of outputs (variables). Inside these two layers, there can be more than one layer that performs all the computation. The key component of a neural network layer is the activation function that takes the weighted summation of input. The initial weights are assigned randomly and converge to the solution with backpropagation [5]. In a model artificial neural network, the value of an output can be assessed with a given set of input and output data. The conjugate gradient (SCG) is used for training in a multi-layer forward network. The widely used activation functions are Rectified Linear Input (ReLU), Sigmoid, Tanh.

$$\mathbf{ReLu} = \mathbf{max}(\boldsymbol{x}, \mathbf{0}),$$

$$\mathbf{Sigmoid} = \frac{\mathbf{1}}{\mathbf{1}+\boldsymbol{e}^{-\boldsymbol{x}}},$$

where x is the weighted summation of inputs. The weights are converged with the help of backpropagation that uses a loss function. A loss function measures how far the predicted value is from true value. A loss function projects decisions to their associated costs. Loss functions are not constant, and they vary depending on the task in hand and the goal to be met [6]. Most commonly used loss function is Mean-Squared Error (MSE), and it is given by:

$$\mathbf{MSE} = \frac{1}{\boldsymbol{n}} \sum_{i=1}^{\boldsymbol{n}} \left(\boldsymbol{Y}_i - \boldsymbol{Y}_{\mathbf{predicted}}\right)^2.$$

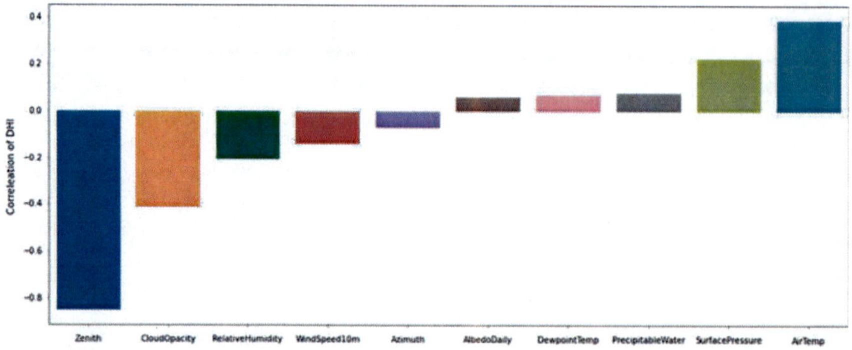

Fig. 1 Correlation of DHI with parameters

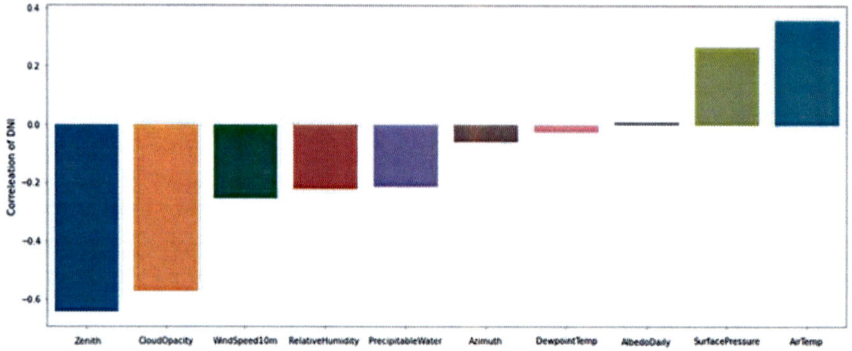

Fig. 2 Correlation of DNI with parameters

The parameters along the horizontal axis are zenith angle, cloud opacity, wind speed, relative humidity, precipitable water, azimuth angle, dew point temperature, albedo, surface pressure, and air temperature.

From the Figs. 1 and 2, DHI and DNI highly correlate with zenith angle and cloud opacity negatively and correlate positively with surface pressure and air temperature. During the day time, the azimuth angle is least because the sun is just above, having the least azimuth angle. For cloud opacity, lesser is the cloud density, the rays will get clear path for it to fall on the panel, and it will receive the maximum irradiance.

However, in case of precipitable water, the DHI correlates positively, whereas it correlates negatively for DNI.

3.1.1 Recurrent Neural Network (RNN)

The typical recurrent neural network language model generates a sequence of hidden states. Each hidden state ($h(t)$) depends on the previous state ($h(t-1)$) and the input/token at that state ($x(t)$), and this way it generates a sequential nature in the model [7]. This kind of structure is useful in language modeling because each token is dependent on its previous tokens for context. Example: "My dog is Labrador. He likes to play"; here, the token "he" is referring to the "dog", and the RNN model would be able to understand this information because the token "he" is dependent on the previous tokens for context.

This sequencing of input sometimes becomes an issue for the memory, especially for long sequences, because no parallelization can be done within a sequence training examples.

3.2 *Autoregressive*

An autoregressive (AR) model is employed when to describe certain time-varying process. It specifies that output variable depends linearly on its own previous value. The mathematical representation of AR model of order p can be written as:

$$y(t) = c + \emptyset 1 * y(t-1) + \emptyset 2 * y(t-2) + \ldots + \emptyset k * y(t-p).$$

The AR model uses autocorrelation of previous time stamps to make prediction. It calculates the autocorrelation of present time with past time stamp (Figs. 3 and 4).

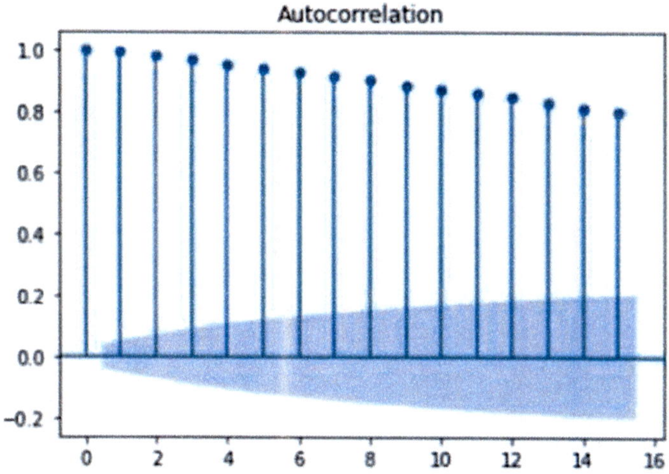

Fig. 3 Autocorrelation factor for DHI

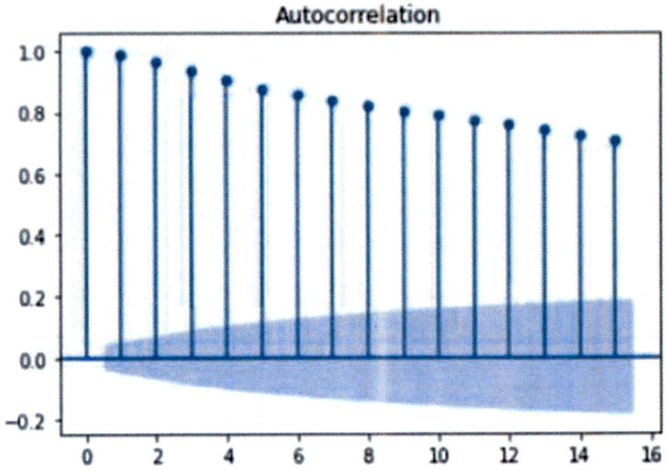

Fig. 4 Autocorrelation factor for DNI

4 Results

4.1 Using Feedforward ANN

See Fig. 5.

4.2 Using RNN

See Fig. 6.

4.3 Using Autoregressive

See Fig. 7.

4.4 Model Comparison

See Table no. 1.

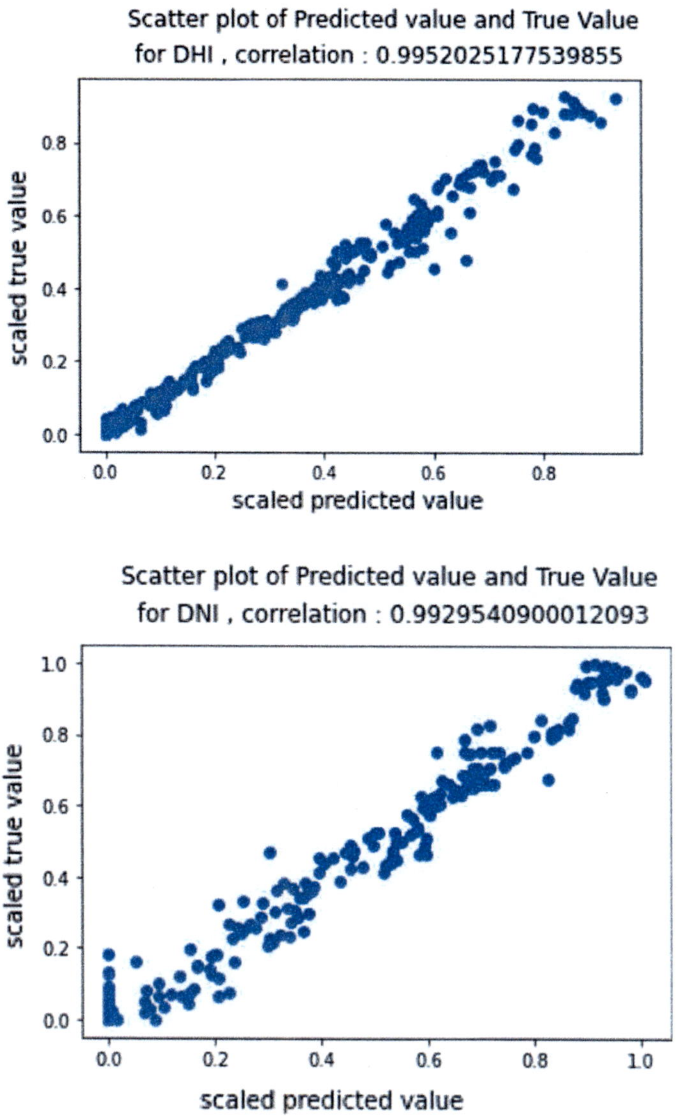

Fig. 5 Correlation score using ANN

5 Conclusion

Based on the result, graphically it is clear that the recurrent neural networks produce straight line for the correlation score compared to forward neural network and autoregressive model. The atmospheric parameters used for feedforward neural network are zenith angle, air temperature, wind speed, relative humidity, precipitable water,

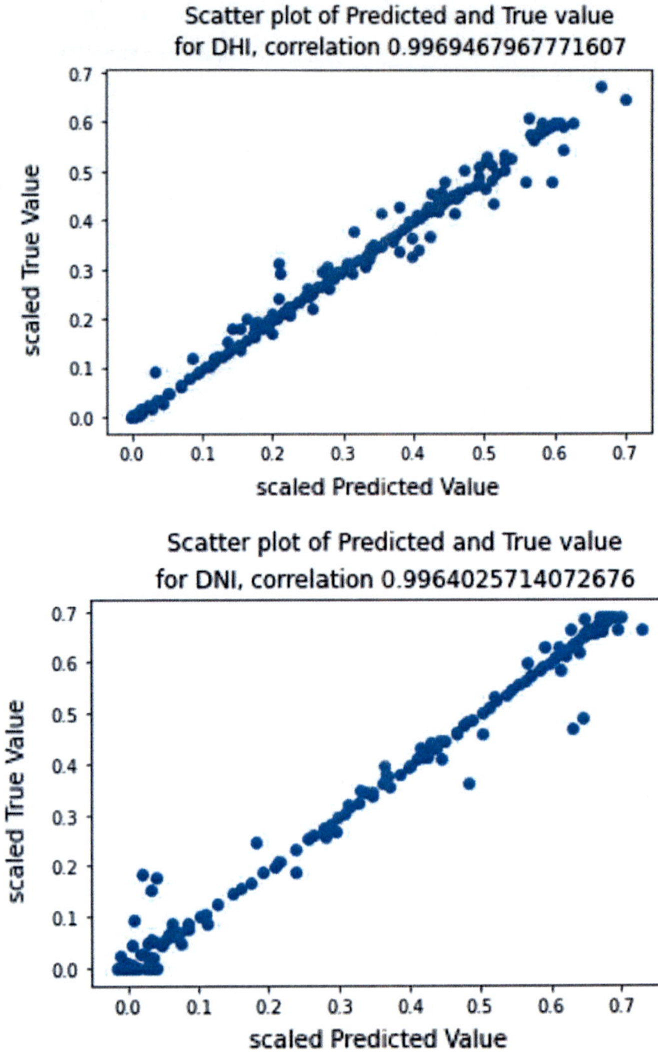

Fig. 6 Correlation score using RNN

cloud opacity, surface pressure, azimuth angle however zenith angle, cloud opacity, surface pressure; air temperature is more prominent in determining the irradiances.

However, the correlation scores for the three models are very close to each other. Therefore, employing the feedforward neural network is economical in terms of computing resource as compared to recurrent neural network given, and we have the environmental factors. If we have the historical record of the radiation, we can go with recurrent neural networks at the cost of computing resources.

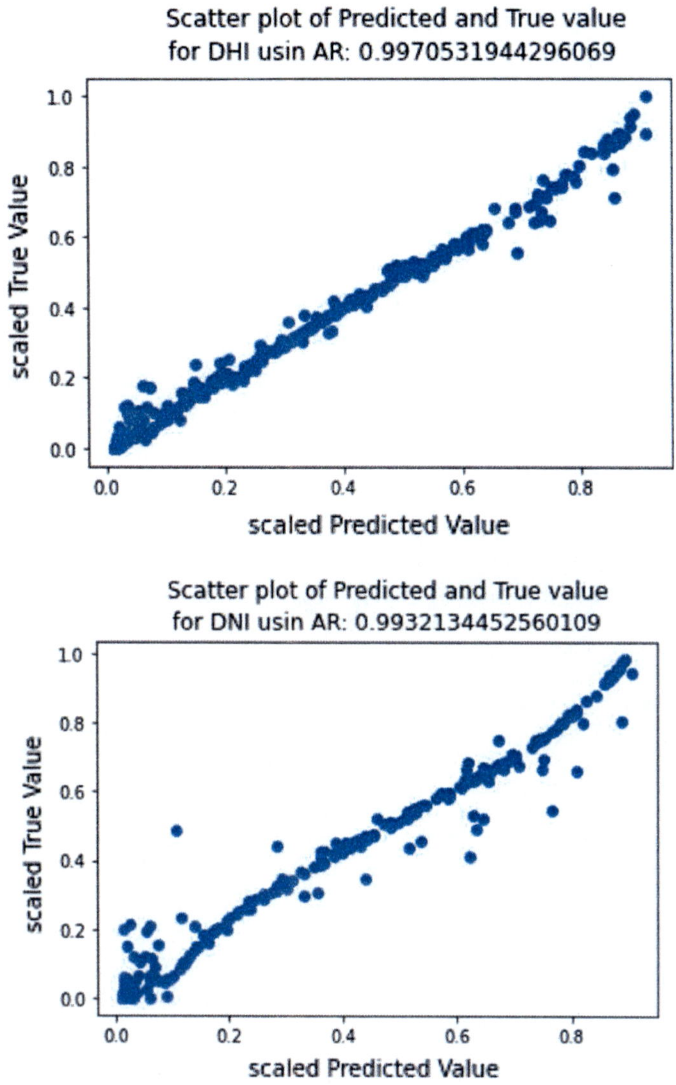

Fig. 7 Correlation score using AR

The complexity of ANN model can be reduced by employing dimensional reduction techniques such as principal component analysis and clustering. This will reduce the number of independent features while preserving the relevant information.

An RNN shares the parameters across different time steps popularly known as parameter sharing. This results in fewer parameters to train and decreases the computational costs. The states in RNN allow the model to capture the historical trend better than autoregressive model. Long Short-Term Memory (LSTM) class of RNN can be used if we are using large number of previous time stamps. Conventional RNN will

Table 1 Scores for different models

	DNI	DHI
Feed forward back propagation ANN	R-squared score: 0.9858905108	R-squared score: 0.989458810146
	Correlation score: 0.992	Correlation score: 0.995
Recurrent neural network	Correlation score: 0.9964	Correlation score: 0.9969
Auto-regressive model	Correlation score: 0.993	Correlation score: 0.997

have vanishing or exploding gradient problems. LSTM has input and forget gates which provides better gradient flow and preserve and long-range dependencies.

References

1. Alzahrani A, Kimball JW, Dagli C (2014) Predicting solar irradiance using time series neural networks. Procedia Comput Sci 36:623–628. ISSN 1877–0509. https://doi.org/10.1016/j.procs.2014.09.065
2. Premalatha N, Arasu AV (2016) Prediction of solar radiation for solar systems by using ANN models with different back propagation algorithms. J Appl Res Technol 14(3):206–2014
3. Gupta A, Gupta K, Saroha S (2021) Solar irradiation forecasting technologies: a review. Strateg Plann Energy Environ 39(3–4):319–354. https://doi.org/10.13052/spee1048-4236.391413
4. Reddy KS, Ranjan M (2003) Solar resource estimation using artificial neural networks and comparison with other correlation models. Energy Convers Manage 44(15):2519–2530
5. Al-Alawi SM, Al-Hinai HA (1998) An ANN-based approach for predicting global radiation in locations with no direct measurement instrumentation. Renew Energy 14(1–4):199–204
6. Pang Z, Niu F, O'Neill Z (2020) Solar radiation prediction using recurrent neural network and artificial neural network: a case study with comparisons. Renew Energy 156:279–289. https://doi.org/10.1016/j.renene.2020.04.042
7. Shboul B, Al-Alarfi I, Michailos S, Ingham D, Ma L, Hughes KJ, Pourkashanian M (2021) A new ANN model for hourly solar radiation and wind speed prediction: a case study over the north & south of the Arabian Peninsula. Sustainable Energy Technol Assess 46:101248. https://doi.org/10.1016/j.seta.2021.101248

Experimenting Transfer Learning and Machine Learning on Endoscopy Database with Augmentation

Mohit Angurala, Vikas Khullar, and Prabhdeep Singh

Abstract Endoscopy is being the most extensively used diagnostic tool for treatment of organs inside humans. However, many challenges exist that are faced by doctors (endoscopists) including: visual interpretation difficulty because of objects, problems in the identification of cancer abnormalities, and pre-cancerous precursors. In this research paper, an analysis has been performed for classification of diseases such as: non-dysplastic Barrett's esophagus (BE), subtle pre-cancerous lesion (suspicious), suspected dysplasia (HGD), adenocarcinoma (cancer), and polyp approach using ML and DL. Firstly, machine learning using AI is implemented on the dataset to train the model, but the model is not trained properly because the lesser number of images are used for training the model. Therefore, to ensure higher accuracy, image augmentation is applied on the given dataset of images so that number of images used to train the model can be increased. Here, a tremendous improvement can be vividly seen, but to further improve the accuracy, the augmentation and DL are amalgamated together and the comparison is then made with the existing ML-based augmentation approach. The outcome of DL with augmentation seems to dominate ML-based augmentation. As the epoch's value keeps on incrementing, the accuracy percentage also increases. Various transfer learning methods show higher accuracy in DL-based augmentation approach.

Keywords Deep learning · Endoscopy · Machine learning · Disease detection

M. Angurala
School of Computer Science and Engineering, Vellore Institute of Technology, University, Chennai, India

V. Khullar (✉)
Chitkara University Institute of Engineering and Technology, Chitkara University, Punjab, India
e-mail: vikas.khullar@gmail.com

P. Singh
Graphic Era (Deemed to be University), Uttrakhand, India

A. Rani et al. (eds.), *Signals, Machines and Automation*, Lecture Notes in Electrical Engineering 1023, https://doi.org/10.1007/978-981-99-0969-8_68

1 Introduction

One of the promising and reliable techniques to detect cancers or other diseases in organs of humankind is endoscopy which nowadays is an extensive clinical procedure. For detecting the cancers at an early stage, this technique provides higher accuracy imaging. During the past few years, deep learning (DL) techniques are being used in various tasks of endoscopic imaging which are in most cases inspired by artificial neural network-based solutions. These ANN-based solutions offer consistent localization, accuracy, and diseased segmentation of region of interests. These techniques can further be used to monitor and do surgical planning. Further, for a particular organ, endoscopic procedures are performed to look for into the rectum, esophagus, colon, small intestine, and stomach. Remarkably, few pre-cancerous abnormalities including dysplasia or inflammations and even cancer lesions could be detected with the endoscopic investigations. However, multiple types of abnormalities may exist, but automated systems are trained for few particular abnormalities, organ and imaging modalities only. Also, many image protocols are used in endoscopic procedures. In all endoscopic surveillances, artifacts are dominant which are generally confused with lesions leading to unreliable results. Further, by using high-definition endoscopic images (50 cancer, 50 Barrett), an automatic approach for detecting cinoma in esophagus for esophageal cancer was implemented [1]. The authors used convolutional neural networks (CNNs) using transfer learning technique. Further, [2] and [3] gave a complete review on few new literatures' survey for detecting and segmenting diseases using endoscopy method.

The most common and public datasets belong to class of polyp detections and segmentations' challenges which are specific to a single abnormality class [4]. The organ-specific challenges only assist in identifying one specific type of disease type. They are not similar to the clinical workflows in which the doctors (endoscopists) perform biopsies and treatment abnormalities incase threat. However, it is needed to recognize various stages of polyp for polyp class including dysplastic, benign or cancers. One problem with artifacts is that they are inevitable issue during endoscopic procedures as they often lead to confusion in detection of tissue abnormalities. Therefore, it is imperative to research in on these classes and restore frames where possible [5].

In this paper, a novel proposed analysis or comparison is made between ML-based augmentation and DL-based augmentation approaches. The remaining sections are organized in the following manner. Literature review is presented in Sect. 2 of this research work. Section 3 presents the problem formulation along with the proposed methodology. Finally, in Sect. 4, results are discussed, followed by the conclusion at the last Sect. 5.

2 Related Studies

In this section, the recent work as a state-of-the-art detection as well as segmentation or other relevant methods are discussed. However, out of all the available methods, we have evaluated two methods (ML-based augmentation and DL-based augmentation) and checked the efficacy of both these methods to detect and segment multiple classes. Also, additionally, a holistic analysis is performed to check their clinical applicability. But, before moving into this, a brief literature survey is presented in this section to see the work that has been done in this field.

The authors in [6] proposed some new features to classify the images of wireless capsule endoscopy in an automatic way. In this, the outcomes show 88.61% of overall accuracy of the proposed method. In [7], the author proposed a new colored feature extraction technique for discriminating the bleeding frames from the normal one. It is based on a two-fold system. They achieved the classification accuracy of 95.75%. Next, in [8], the authors achieved overall recognition accuracy of 90.7%, in which they proposed a joint-feature probabilistic latent semantic analysis (PLSA) model. In the proposed dual constraints: local image manifold and visual words importance are fixed into joint-featured PLSA model. The proposed idea guaranteed that each image is categorized by latent semantic topics instead of low-level characteristics. Thus, the proposed model showed 90.78% recognition accuracy. Alizadeh et al. [9] proposed a system that mainly consists of feature extraction and classification in which 32 features incorporating four statistical measures are calculated, and then, the mutual information is used to choose features with maximal dependence on the target class, but with reduced feature redundancy. Lastly, they used an adaptive neuro-fuzzy interface system (trained classifier system) which is implemented for classifying the endoscopic images as tumor, healthy and unhealthy classes. Further, for polyps' segmentation in wireless capsule endoscopy, the application of simple linear iterative clustering also known as "SLIC" super-pixel frame is assessed by the authors [10] in which various SLIC super-pixel numbers are evaluated for finding the highest sensitivity to detect polyps. Here, the classification outcomes showed 91% sensitivity.

Besides this, [11] exploited three vital understudied reasons to employ deep CNN to computer-aided recognition issues. They further explored and evaluated various architectures of CNN. Finally, they examined the reason as to why and when transfer learning can be useful. Moving ahead, in [12], the authors proposed an improved technique to train the proposed network by transferring knowledge from the analogous area of general texture classifications. They finally concluded that the proposed method caused 2% increase in the performance in comparison to its existing proposal. Zhang et al. [13] proposed deep CNNs to segment the isointense stage brain tissues using multi-modality MR images. They specifically used multi-modality information as inputs to generate segmentation maps as outputs. The outcomes finally illustrate significant improvement over prior methods. On the other hand, the authors in [14] tried to look for the answers of medical image analysis. The question of whether the use of pre-trained deep convolution neural network with enough fine tuning eradicates the requirement to train a deep CNN from scratch. For this, they took four separate medical image applications in radiology, cardiology, and gastroenterology that involves classification, detection, and segmentation from three different imaging modalities. Finally, in [15], the authors implemented a disease detection and semantic segmentation pipeline using a modified mask region-based convolution neural networks, and further, they used a selective and balanced augmentation stage for their architecture to extract more accuracy in detection and segmentation [16].

Therefore, the literature survey reveals that there is still requirement of robust and efficient mechanism to classify the disease class. For achieving this, we try to analyze ML-based augmentation approach and DL-based augmentation approach.

3 Problem Formulation and Methodology

From the literature survey, it is concluded that it is very difficult to accurately monitor the disease by just looking at the images. Sometimes, the doctors are unable to diagnose the issues of cancerous or pre-cancerous stages leading. Therefore, in order to overcome such issues, a robust automated methodology must be implemented to the datasets of images to diagnose the deadly problems. The endoscopy disease detection dataset with five classes (including BE, suspicious, HGD, cancer, and polyp) was collected from an open access online resource [15]. Therefore, we have proposed the mechanism as shown in Fig. 1.

In Fig. 1, it can be clearly seen that we are doing classification and detection of diseases using different approaches. At the very first step, we first take a dataset containing few images, but the model cannot be trained well using small set of images contained in the dataset if we apply ML using AI. Therefore, to increase the images of dataset, we apply augmentation approach and found that the results significantly improved. Finally, ML algorithms have several limitations, due to which we then applied DL-based transfer learning technique that automatically learns and trains. Further, again to improve the accuracy, the augmentation is again applied on DL method to further improve the efficiency of the proposed method. The outcome suggests further improvement over the previous method which is explained further in Sect. 4.

4 Results

Initially, machine learning algorithms were implemented on collected five classes' dataset and analyzed on basis of parameters, viz. Accuracy, Precision, Recall, F1-score, Zero One line, Matthew's Coefficient, and Kappa Statistics. The different machine learning algorithms (such as Decision Tree, Support Vector Machine, Quadratic Discriminant, Analysis, K-Nearest Neighbors, AdaBoost, Gaussian NB, Logistic Regression, Random Forest, Extra Trees, Bagging, Voting, and Histogram

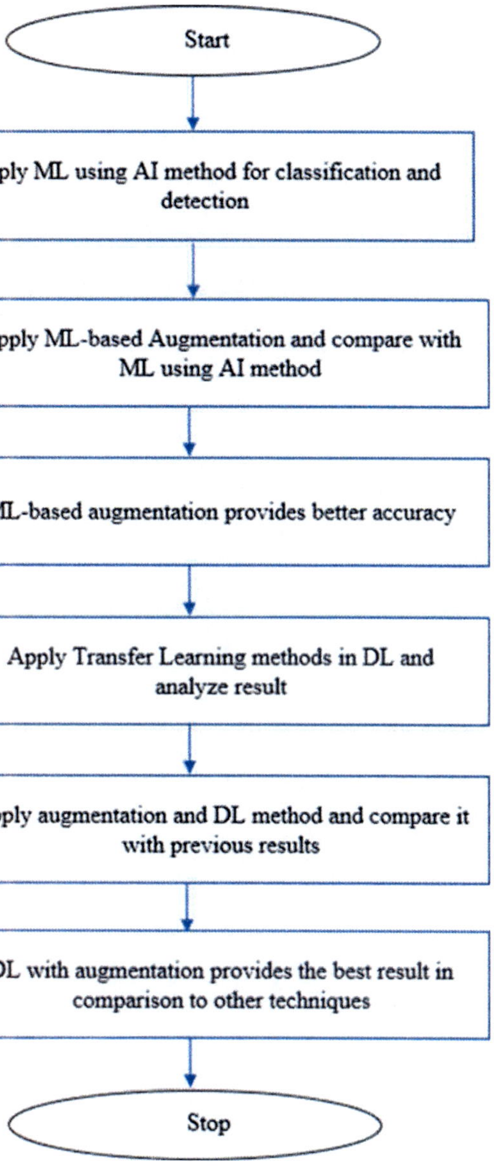

Fig. 1 Flowchart of the proposed methodology

Gradient Boosting) were implemented and analyzed. As shown in Table 1, the identified results revealed that the ensemble learning algorithms, i.e., Extra Tree and Random Forest classifier resulted highest accuracy and other parameters on both augmented and normal dataset.

Table 1 Comparative analysis of augmented and non-augmented data

Classification algorithm(s)	Augmented data							Non augmented data						
	Acc	Pre	Rec	F1S	Zol	Mc	Kp	Acc	Pre	Rec	F1S	Zol	Mc	Kp
Decision tree	0.72	0.49	0.50	0.49	0.28	0.59	0.59	0.60	0.43	0.43	0.43	0.40	0.43	0.43
Support vector machine	0.76	0.39	0.44	0.41	0.24	0.64	0.62	0.67	0.42	0.42	0.41	0.33	0.51	0.50
Quadratic discriminant analysis	0.60	0.41	0.34	0.35	0.40	0.40	0.39	0.33	0.28	0.22	0.22	0.67	0.08	0.07
K nearest neighbors	0.81	0.55	0.56	0.54	0.19	0.71	0.71	0.67	0.44	0.42	0.41	0.33	0.51	0.50
Ada boost	0.69	0.49	0.45	0.46	0.31	0.55	0.54	0.52	0.38	0.35	0.36	0.48	0.27	0.26
Gaussian NB	0.65	0.46	0.51	0.45	0.35	0.52	0.51	0.40	0.31	0.31	0.29	0.60	0.19	0.19
Logistic regression	0.72	0.38	0.41	0.39	0.28	0.57	0.57	0.45	0.29	0.28	0.28	0.55	0.17	0.17
Random forest	0.81	0.51	0.49	0.48	0.19	0.71	0.70	0.74	0.51	0.47	0.47	0.26	0.61	0.61
Extra trees	0.81	0.56	0.54	0.54	0.19	0.71	0.71	0.74	0.48	0.47	0.46	0.26	0.61	0.60
Bagging	0.71	0.36	0.38	0.37	0.29	0.56	0.56	0.45	0.29	0.28	0.27	0.55	0.16	0.16
Voting	0.81	0.63	0.52	0.50	0.19	0.71	0.71	0.65	0.41	0.41	0.41	0.35	0.48	0.48
Hist gradient boosting	0.79	0.41	0.44	0.42	0.21	0.68	0.67	0.74	0.48	0.46	0.46	0.26	0.62	0.61

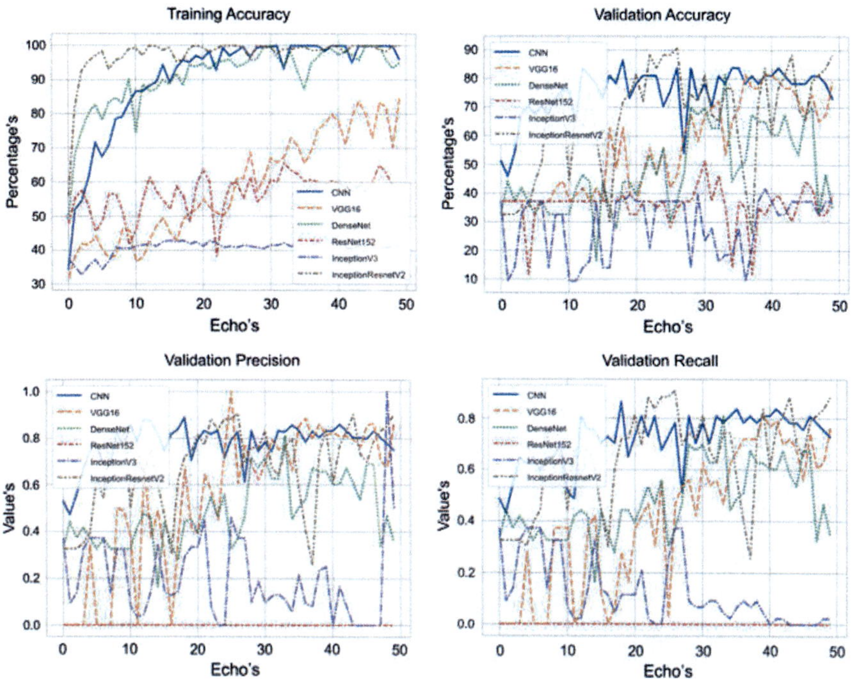

Fig. 2 Transfer learning results for non-augmented data classification

However, as earlier also seen the potential of deep transfer learning algorithms, similarly higher parameters were identified in comparison to machine learning algorithms. In transfer learning implementations, training and validation parameters resulted higher with augmented in comparison to non-augmented data. Figures 2 and 3 are representing transfer learning results with non-augmented and augmented data.

5 Conclusion

Correct identification of cancer kind of diseases is highly required. Manual human interventions may have chance for human error, whereas a trained deep learning or AI model has less chances of such unknown mistake. In this paper, machine learning and transfer learning-based models were utilized for identification of cancer images from endoscopy database. The collected database is an open-source data and readily available to use. Our implementations revealed that augmented data are resulting higher results on machine leaning as well as transfer learning algorithms. However, transfer learning implementations are resulting higher results in

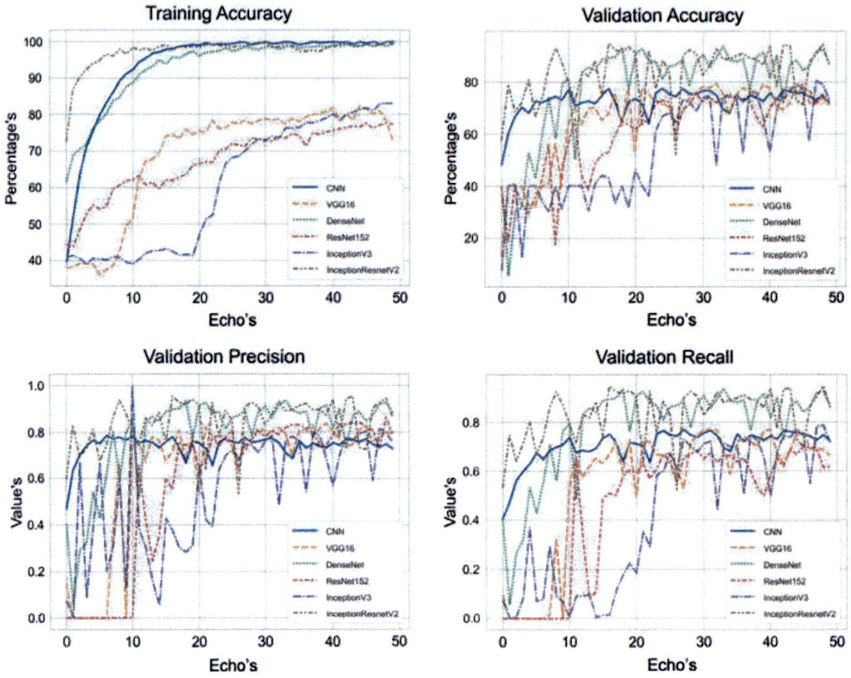

Fig. 3 Transfer learning results for augmented data classification

comparison to machine learning implementations. In future, such models could be implemented with real-time detection system.

References

1. Mendel R et al (2017) Barrett's esophagus analysis using convolutional neural networks. In: Image Processing for Medicine. Springer, pp 80–85
2. Mori Y, Berzin TM, Shin-ei K (2019) Artificial intelligence for early gastric cancer: early promise and the path ahead. Gastrointest Endosc 89(4):816–817
3. Min JK, Kwak MS, Cha JM (2019) Overview of deep learning in gastrointestinal endoscopy. Gut Liver 13(4):388–393
4. Bernal J et al (2017) Comparative validation of polyp detection methods in video colonoscopy: results from the MICCAI 2015 endoscopic vision challenge. IEEE Trans Med Imaging 36(6):1231–1249. https://doi.org/10.1109/TMI.2017.2664042
5. Ali S, Zhou F, Bailey A, Braden B, East JE, Lu X, Rittscher J (2021) A deep learning framework for quality assessment and restoration in video endoscopy. Med Image Anal 68:101900. https://doi.org/10.1016/j.media.2020.101900
6. Yuan Y, Li B, Meng MQ-H (2017) WCE abnormality detection based on saliency and adaptive locality-constrained linear coding. IEEE Trans Autom Sci Eng 14(1):149–159
7. Yuan Y, Li B, Meng MQ-H (2016) Bleeding frame and region detection in the wireless capsule endoscopy video. IEEE J Biomed Health Inform 20(2):624–630

8. Yuan Y, Yao X, Han J, Guo L, Meng MQ-H (2017) Discriminative joint-feature topic model with dual constraints for WCE classification. IEEE Trans Cybern 48(7):2074–2085
9. Alizadeh M, Maghsoudi OH, Sharzehi K, Hemati HR, Asl AK, Talebpour A (2017) Detection of small bowel tumor in wireless capsule endoscopy images using an adaptive neuro-fuzzy inference system. J Biomed Res 33(5):419–427
10. Maghsoudi OH (2017) Superpixel based segmentation and classification of polyps in wireless capsule endoscopy. In: 2017 IEEE Signal processing in medicine and biology symposium (SPMB). IEEE, pp 1–4. https://doi.org/10.1109/SPMB.2017.8257027
11. Shin H-C et al (2016) Deep convolutional neural networks for computer-aided detection: CNN architectures, dataset characteristics and transfer learning. IEEE Trans Med Imag 35(5):1285–1298
12. Christodoulidis S, Anthimopoulos M, Ebner L, Christe A, Mougiakakou S (2017) Multisource transfer learning with convolutional neural networks for lung pattern analysis. IEEE J Biomed Health Inform 21(1):76–84
13. Zhang W et al (2015) Deep convolutional neural networks for multi-modality isointense infant brain image segmentation. Neuroimage 108:214–224
14. Tajbakhsh N et al (2016) Convolutional neural networks for medical image analysis: full training or fine tuning?'. IEEE Trans Med Imag 35(5):1299–1312
15. Rezvy S et al (2020) Transfer learning for endoscopy disease detection and segmentation with mask-RCNN benchmark architecture. In: Proceedings of the 2nd International workshop and challenge on computer vision in endoscopy, EndoCV@ISBI 2020, Iowa City, Iowa, USA, vol 2595 of CEUR Workshop Proceedings CEUR-WS.org, pp 68–72
16. Ali S et al (2020) Endoscopy disease detection and segmentation (EDD2020). IEEE Data Port. https://dx.doi.org/10.21227/f8xg-wb80.s. Accessed on 14-May-2022

Optimal Planning of EV Charging Infrastructure in Distribution System

Rudraksh S. Gupta, Arjun Tyagi, and Sanjeev Anand

Abstract The ever-growing population and to limit CO_2 emissions have necessitated the development of alternate ways of transport, with electric cars being the most popular choice. The main stumbling block is the exacerbating influence on the power distribution network caused by indecorous charging station placement. This aims to discover the optimum planning of charging station infrastructure location. Because of limited battery capacity and extended charging times, it is critical to find the appropriate deployments of EV charging infrastructure in the distribution system. In this paper, IEEE 33 bus distribution system is used to demonstrate the methodology. The methodology uses Grey Wolf Optimization (GWO) technique for finding the optimal solution.

Keywords Charging infrastructure · Distribution system · Electric vehicles

Abbreviations

ARIMA	Autoregressive integrated moving average
EV	Electric vehicle
ICEV	Internal combustion engine vehicle
IEA	International energy agency
II	Inconvenience index
GWO	Grey wolf optimization
GHG	Green house gas
PHEV	Plugin hybrid electric vehicle
PV	Photovoltaic

R. S. Gupta (✉) · S. Anand
School of Energy Management, Shri Mata Vaishno Devi University, Katra, Jammu and Kashmir, India
e-mail: rudraksh.s.gupta@gmail.com

A. Tyagi
Department of Electrical Engineering, Netaji Subhas University of Technology, New Delhi, India

A. Rani et al. (eds.), *Signals, Machines and Automation*, Lecture Notes in Electrical Engineering 1023, https://doi.org/10.1007/978-981-99-0969-8_69

Nomenclature

P_G	Active power generated
P_{Loss}	System active power loss
P_D	Active Power demand
Q_G	Reactive power generated
Q_D	Reactive power demand
Q_{Loss}	System active power loss
$V_i^{\min}$	Minimum voltage
$V_i^{\max}$	Maximum voltage
$II_{\min}$	Minimum inconvenience value
$II_{\max}$	Maximum inconvenience value
$\sum_{j=1}^{N_b} P_{\text{loss},j,\text{WSPV}}$	Power loss with SPV
$\sum_{j=1}^{N_b} P_{\text{loss},j,\text{WOSPV}}$	Power loss without SPV
P^{SPV}	Solar plant PV output
SA_i	Surface area at desired location
GR	Standard radiation (1000 watts/m^2)

1 Introduction

The burning of conventional fuel for transportation and electricity produces undesirable greenhouse gas emissions (GHGs). Due to GHG emissions, climate change has grasped hazardous levels causing severe global warming and ocean rise due to the melting of icebergs. In order to limit the effect of climate change, immediate preventive measures need to be taken supported by climate change policy. In order to keep the expected average rise in temperature below 2 °C by 2050, International Energy Agency (IEA) proposed a future energy system setup for faster and better adoption of clean energy system [1]. In order to ensure trifling emissions from the transportation sector, manifold alternatives are being researched such as use of alternate fuels or incorporating clean technology such as electric vehicles keeping emphasis on maintaining or enhancing the vehicle performance. Electrifying transportation is a feasible stratagem with numerous advantages. Incorporation of electric vehicle can open a huge potential in terms of energy security, economic growth, and new research divisions for components and critically helps in reducing tailpipe emissions to conserve the environment. One of the major advantages known regarding EV is that it outperforms internal combustion engine (ICE) vehicles due to more efficient power trains and electric motors [2].

In many developed and developing countries, the demand and sales for battery electric vehicles (BEVs) and plug-in hybrid electric vehicles (PHEVs) have proliferated in recent years. Sustainable development against rising greenhouse gas emissions, financial savings in terms of fuel use, and enticements offered by the various government are all key factors in the public's changing perspective. In 2013, Government of India introduced National Electric Mobility Mission Plan (NEMMP) 2020 which included strategies, targets, and incentives to increase the sales of hybrid and battery electric vehicles with a target of 7 million on road vehicle by 2020. Even though battery charger technology is critical to the EV industry's progress, the integration of appropriately sized and situated charging stations is the first hurdle to overcome [3].

Many nations are encouraging electric vehicle adoption, despite the fact that the number of charging stations does not accommodate the rising number of EVs. Moreover, the incorrect positioning of charging facilities is one of the key impediments that contributes to EV inactivity. As a result, the best arrangement of EV charging stations/infrastructure must be given prominent consideration. There are three levels of charging for electric vehicle. Level-1 is slowest and mostly used in the home environment. Level-2 charging is frequently used at workplace or dedicated charging station, and Level-3 charging is used at dedicated fast-charging stations only. Level-1 is most cost friendly but takes time to charge as maximum output is 3.3 kW whereas Level-3 charges are most expensive but time of charging is very minimal in it as power output can reach up to 50–100 kW depending on the country it is installed in [4, 5].

In the literature, various algorithms are used to optimize the charging station placement such as genetic algorithm and particle swarm optimization. The authors in [6] focused on enhancing PV power generation by anticipating output and regulating power flow between the electric grid, photovoltaic, and electric vehicle. PV power production is predicted using an autoregressive integrated moving average (ARIMA), and electricity is allocated using MILP for cost reduction and grid stability. The authors in [7] utilized MATLAB to peak-shave and valley-fill the power consumption of a given building in order to achieve grid stability and steady building power consumption. During peak hours, it leverages the building's idle EV to assist the power input. Researchers and industry experts have been concerned about EV penetration levels. Higher penetration levels provide a multitude of challenges at both the local distribution and grid levels [8]. Integration of renewable energy system could be a standard to look out for grid stabilization. A stochastic method was designed in [9] to guarantee that optimum power supply was provided to existing loads while simultaneously charging EVs. This article recommends the best placement for a future public charging station based on renewable integration, current charging stations, distance between following charging stations, and the number of chargers placed at each charging station.

This paper has been bifurcated in 6 sections. Section 1 provides brief outline on the topic whereas Sect. 2 is based on the problem formulation. Sections 3 and 4 represent methodology and charging station designing data, respectively. Section 5 provides the results and outputs gathered from the optimization technique. Finally, in Sect. 6 conclusion has been provided with respect to this research work.

2 Problem Formulation

This section outlines the problem associated to charging station allocation. In this case, the EVs are added to the distribution network as a new source of load. The network's active power loss will rise as a result of the EV's increased power demand and penetration. The primary objective of the study is to reduce the overall increase in total active power loss as a result of the placement of an EV charging station in an acceptable location. This can be represented mathematically by Eq. (1).

$$F_1 = \text{minimization} \sum_{\forall N} P_{\text{Loss}} \tag{1}$$

where P_{Loss} and N represent the active power loss and total number of buses in the system, respectively. The system design needs to follow some constraints as expressed below by Eqs. (2), (3), (4), and (5).

- Voltage limits

$$V_i^{\min} \le V_i \le V_i^{\max} \qquad i \in \text{n} \tag{2}$$

- Power balance

$$P_G + P^{\text{SPV}} = P_D + P_{\text{Loss}} \tag{3}$$

$$Q_G = Q_D + Q_{\text{Loss}} \tag{4}$$

- Active power loss: System active power loss with SPV should be less than system loss without SPV

$$\sum_{j=1}^{N_b} P_{\text{loss},j,\text{WSPV}} < \sum_{j=1}^{N_b} P_{\text{loss},j,\text{WOSPV}} \tag{5}$$

- Power factor

$$pf_{\min} \le pf \le pf_{\max} \qquad i \in \text{n} \tag{6}$$

- Inconvenience Index (*II*): This constraint restricts placement of charging station nearby. If a customer has to travel across 5 buses (locations), then it includes in inconvenience value.

$$II_{\min} \le II \le II_{\max} \tag{7}$$

A few calculations are necessary to compute the power output of a solar power plant, which are stated in Eqs. (8) and (9).

- Surface area for installation

$$\mathrm{SA}_i = \left(\frac{\text{area}}{\text{port}}\right) * \text{number of ports} \tag{8}$$

- Solar PV plant output

$$\mathrm{PV}_o = \eta * \mathrm{SA}_i * \mathrm{GR} \tag{9}$$

3 Methodology

The Grey Wolf Optimization (GWO) approach is utilized to solve the objective function as mentioned in Eq. (1), subject under the provided restrictions. Because of its speedier convergence and variable algorithmic parameters, the Grey Wolf Optimization (GWO) method is selected in this work. The Grey Wolf Optimizer (GWO) uses our types of grey wolfs: alpha, beta, delta, and omega to replicate its natural leadership structure and hunting mechanism. In addition, three critical hunting strategies are implemented to optimize performance: seek for prey, encircling prey, and engaging prey. The best fitness value is the best optimized result found from the optimization [10]. A flow chart explaining the process and working of GWO is epitomized in Fig. 1.

4 Charging Infrastructure Designing

Currently, multiple AC/DC EV chargers are being established by various forerunner EV firms. This research looked at newly built and regularly used EV chargers by owners of electric vehicles. Table 1 shows the EV battery size and accompanying charger ratings. The total load is computed on the basis of their alternating current charging. Table 1 represents the data compiled from different sources regarding the charger specification of cars [11–14].

For this research purpose, it is considered that each charging station can accommodate 9 vehicles at a time. The charging station comprises 1-T Model S, 1-Nissan LEAF, 1-Hyundai KONA Electrics, 3-Tata NEXON EV, and 3-Tata TIGOR EV. This approximation is done based on the vehicle price and buying behaviors of people. The aggregate maximum possible load comes out to be 300 kW for a single charging station.

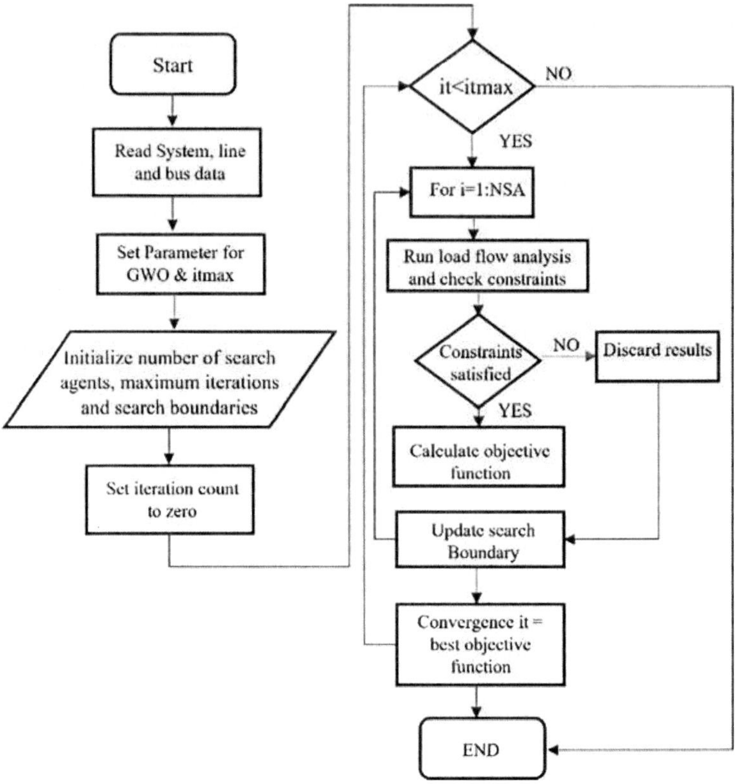

Fig. 1 Grey wolf optimization algorithm

Table 1 Electric vehicle battery and charging specification

S. No	Vehicle model	Battery rating (kWh)	Charger rating (kW) (Slow charging)	Charger rating (kW) (Fast charging)	Time for full charge from 0 to 100% (Slow charging) (Hrs.)	Time for full charge from 0 to 100% (Fast charging) (Hrs.)
1	Tata TIGOR EV	21.5	3	25	7.17	0.86
2	Tata NEXON EV	30.2	3.3	25	9.15	1.21
3	Hyundai-Kona Electrics	39.2	7.2	50	5.44	0.78
4	Nissan LEAF	40	6.6	50	6.06	0.8
5	Tesla Model S	100	22	50*	4.55	2

5 Results and Discussions

IEEE 33 bus distribution network is used to carry out the study using GWO algorithm. The system entails 33 buses and 32 lines as shown in Fig. 2. Bus-1 is classified as a slack bus, and no charging stations are permitted to be installed on that bus. The system voltage and base load is 12.66 kV and 3715 kW, respectively. All the optimization is done using MATLAB 2019b.

In this work, four 300 kW substations are considered to be installed at various locations across the distribution system to allow for the charging of nine EVs at a time at each charging station. As a result of the installation of these four charging stations, the instantaneous load will subsequently rise to 1200 kW. As the load expands, so will the system loss and the voltage in the load bus. The purpose of this research is to locate the optimal site for installation of four charging station in attempt to minimize the rising losses. The Grey Wolf Optimization (GWO) technique is employed in this work to address this minimization goal while addressing the constraints. The GWO algorithm is one of the most basic, with little algorithmic parameter adjustment. The following cases are explored in this study:

- ***Case 1***: No charging infrastructure is installed. This is considered as base case.
- ***Case 2***: Four charging stations are installed without taking into consideration the constraint of inconvenience index value.
- ***Case 3***: Four charging stations are installed without taking into consideration the constraint of inconvenience index value and with solar photovoltaic system installed.
- ***Case 4***: Four charging stations are installed taking into consideration all the constraints.
- ***Case 5***: Four charging stations are installed taking into consideration all the constraints with solar photovoltaic plant installed on the charging station.

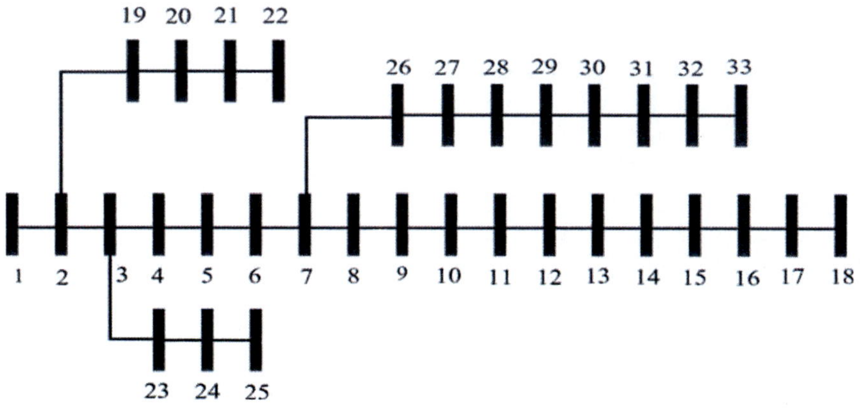

Fig. 2 IEEE 33 bus system

Table 2 Results for different case study

S. No	Case study	Optimal location	Power loss (kW)	Inconvenience index (%)	Processing time (sec)
1	Case 1	–	202.7	Not applicable	1.2
2	Case 2	2,19,20,21	218.04	59.37	35.79
3	Case 3	2,19,20,21	217.31	59.37	33.348
4	Case 4	2,5,12,19	266.5	18.75	40.44
5	Case 5	2,5,12,19	263.9	18.75	41.83

On solving the base case, i.e., without extra load from charging station, the total real power loss is 202.7 kW. In this study as discussed, four changing stations with total load of 300 kW each are considered. In order to reach the objective as explained in Eq. 8 taking into account all the constraints, optimal location in *Case 2* was found out to be at bus numbers 2, 19, 20, and 21 and total power loss of the system was determined as 218.04 kW. In *Case 3* when the solar photovoltaic plant in the charging station is accounted for energy generation in the system, the optimal location was found out to be at bus numbers 2, 19, 20, and 21 and total power loss of the system was determined as 217.31 kW. The total capacity of solar plant accounts to 252 kW. In *Case 4*, constraint on inconvenience index is included as the system is modeled for whole locality, therefore charging station cannot be installed at single section. Thus, inconvenience index is incorporated which ranges from $II_{\min}$, i.e., 0% to $II_{\max}$, i.e., 20%. Any customer who needs to travel more than 5 buses for charging of EV is taken as unhappy customer. A system where more than 20% of population in a locality have to face inconvenience for charging is discarded. Results gathered after calculations in *Case 4* represented optimal locations to be at bus numbers 2, 5, 12, and 19, and total loss of the system was determined to be 266.5 kW. Moreover, inconvenience index was 18.75% (zero the better). And finally in *Case 5*, it was seen that optimal locations were projected to be at bus numbers 2, 5, 12, and 19 with total loss of the system is 263.9 kW, and inconvenience index was 18.75%. In each case, it was found that power factor ranged between $pf_{\min}$, i.e., 0.9 and $pf_{\max}$, i.e., 1.1, thus satisfying the constraint value. The details of results are resented in Table 2. Figure 3 represents the voltage profile of each case study, and Fig. 4 provides graphical representation of results obtained in terms of power loss and inconvenience index value. Figures 5 and 6 represent the iteration count to reach minimum value for each case from 2 to 5.

6 Conclusion

A loss-based optimization problem is used to predict the optimal placement of electric vehicle charging stations in the distribution network. This work examines five scenarios: no charging station, charging stations location without any constraint,

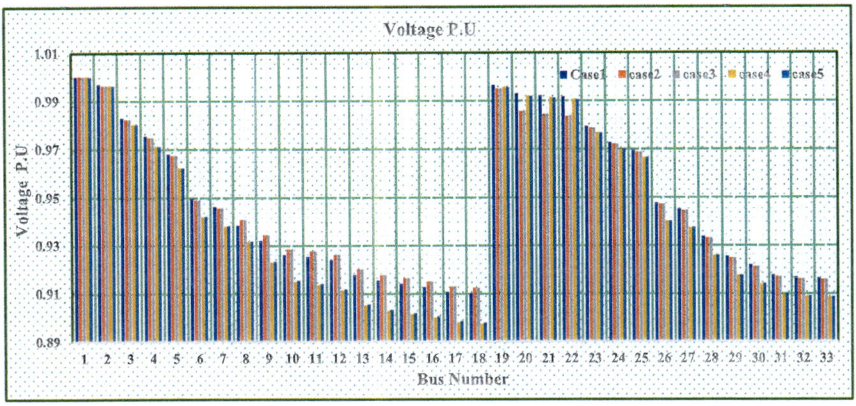

Fig. 3 Per unit voltage of each bus for each case study

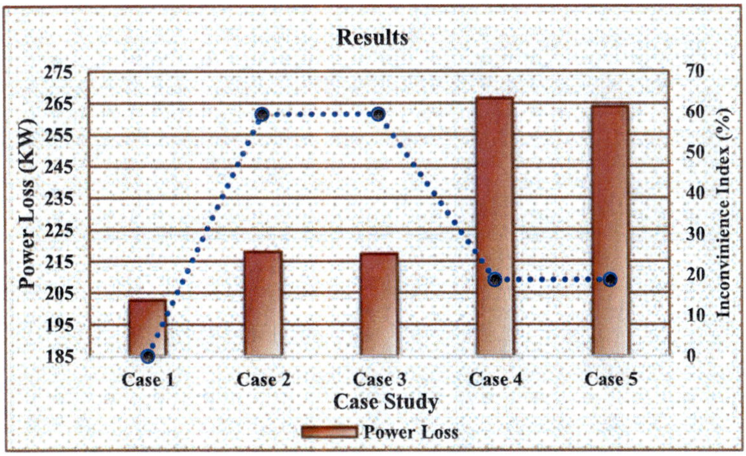

Fig. 4 Power loss and inconvenience index value for each case study

charging stations location without the constraint but with the solar photovoltaic system installed, charging stations location based on all the constraints, and charging stations location consideration all the constraints with a solar photovoltaic plant installed. With the aid of the Grey Wolf Optimization algorithm, the total system loss is reduced by arranging the four EV charging stations in the best feasible places while staying within the power system limits. Because the indicated sites are in four distinct feeders, it increased the charging station's usage accessibility. The findings suggest that there is roughly 30% increase in losses, with increase in accessibility for user convenience. Therefore, the final outcome keeping all of the constraints intact came out to be in case 5 with total loss of 263.9 kW.

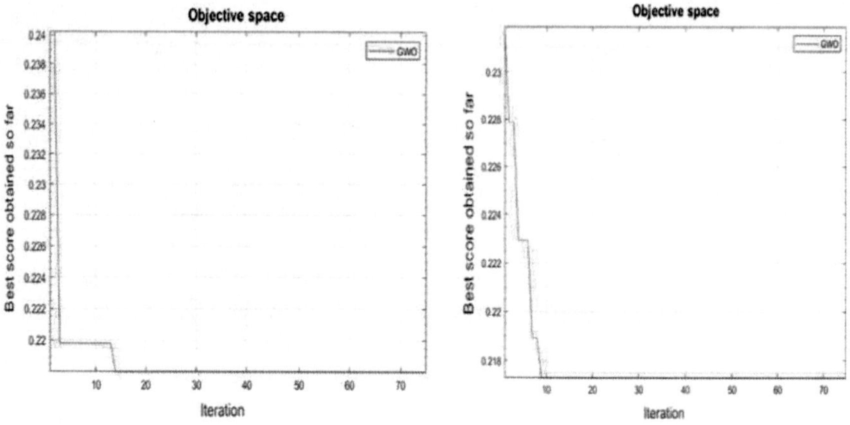

Fig. 5 Iteration count to reach final value for Case 2 and Case 3

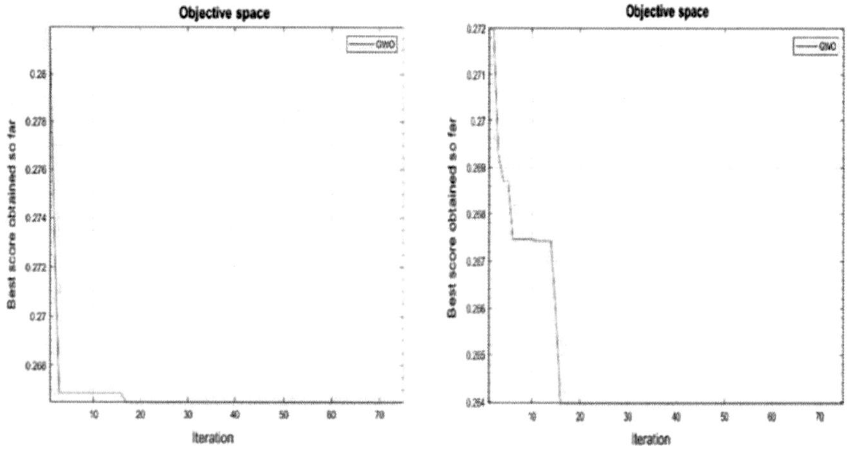

Fig. 6 Iteration count to reach final value for Case 4 and Case 5

Declarations Conflict of Interest

The authors declare that they have no conflict of interest.

Acknowledgements Rudraksh S. Gupta, one of the author(s), would like to acknowledge Ministry of New and Renewable Energy (MNRE) (Govt. of India), for financial assistance in the form of SRF to carry forward the research work.

References

1. Net Zero by 2050—Analysis—IEA. Retrieved from https://www.iea.org/reports/net-zero-by-2050. Accessed on 7 Apr 2022
2. Darabi Z, Ferdowsi M (2012) Impact of plug-in hybrid electric vehicles on electricity demand profile. Power Syst 53:319–349. https://doi.org/10.1007/978-3-642-21578-0_11
3. Status quo analysis of various sections of electric mobility and low carbon passenger road transport in India. Foreword by NITI Aayog
4. Awasthi A, Venkitusamy K, Padmanaban S, Selvamuthukumaran R, Blaabjerg F, Singh AK (2017) Optimal planning of electric vehicle charging station at the distribution system using hybrid optimization algorithm. Energy 133:70–78. https://doi.org/10.1016/J.ENERGY.2017.05.094
5. Gupta RS, Tyagi A, Anand S (2021) Optimal allocation of electric vehicles charging infrastructure, policies and future trends. J Energy Storage 43:103291. https://doi.org/10.1016/J.EST.2021.103291
6. Van Der Meer D, Mouli GRC, Mouli GM-E, Elizondo LR, Bauer P (2018) Energy management system with PV power forecast to optimally charge EVs at the workplace. IEEE Trans Industr Inf 14(1):311–320. https://doi.org/10.1109/TII.2016.2634624
7. Reddy MSK, Selvajyothi K (2020) Optimal placement of electric vehicle charging station for unbalanced radial distribution systems. In: Energy Sources, Part A Recovery, Utilization and Environmental Effects, pp 1–15. https://doi.org/10.1080/15567036.2020.1731017
8. Muñoz ER, Razeghi G, Zhang L, Jabbari F (2016) Electric vehicle charging algorithms for coordination of the grid and distribution transformer levels. Energy 113:930–942. https://doi.org/10.1016/J.ENERGY.2016.07.122
9. Tabatabaee S, Mortazavi SS, Niknam T (2017) Stochastic scheduling of local distribution systems considering high penetration of plug-in electric vehicles and renewable energy sources. Energy 121:480–490. https://doi.org/10.1016/J.ENERGY.2016.12.115
10. Mirjalili S, Mirjalili SM, Lewis A (2014) Grey wolf optimizer. Adv Eng Softw 69:46–61. https://doi.org/10.1016/J.ADVENGSOFT.2013.12.007
11. Tata Tigor EV features, specifications, interior & exterior design. Retrieved from https://tigorev.tatamotors.com/specs-features/. Accessed on 8 Apr 2022
12. Specifications & features of Tata Nexon EV & dark edition. Retrieved from https://nexonev.tatamotors.com/features/. Accessed on 8 Apr 2022
13. The Official Tesla Shop, Tesla. Retrieved from https://shop.tesla.com/. Accessed on 8 Apr 2022
14. 2022 Nissan LEAF specs & prices, Nissan USA. Retrieved from https://www.nissanusa.com/vehicles/electric-cars/leaf/specs/compare-specs.html#modelName=S%7C40kWh,SLPLUS%7C62kWh. Accessed on 8 Apr 2022

Perception-preserving Color Ocular-image Enhancement Through Multi-scale and Multiresolution Features

Gunjan Gautam and Susanta Mukhopadhyay

Abstract This paper presents a color ocular image enhancement method based on multi-scale and multiresolution features obtained using Mathematical Morphology (MM) and Wavelet Transform (WT), respectively. In addition to enhancing the image quality and highlighting the minute features, the proposed algorithm also preserves the perceptual appearance of the image. The algorithm first decomposes the luminance component of the input image into different sub-bands. All the coefficients (excluding approximation) of these sub-bands are remapped using local–global image statistics followed by performing morphological filtering on only coarsest-level coefficients. These modified coefficients are used to reconstruct the luminance component of the image using WT sub-band coding. Enhanced images are generated post reconstructing the luminance component. Experimental results on the UBIRIS and Berkeley datasets demonstrate that the proposed method outperforms various state-of-the-art techniques. Also, results of mobile-iris liveness detection demonstrate the potential and generalization ability of our method.

Keywords Color image enhancement · Mathematical morphology · Wavelets

1 Introduction

The visual appearance enhancement plays an important role in computer vision, pattern recognition, and digital image processing. Low quality of ocular images may result from various circumstances, including inadequacy of image capturing device/scanner, intensity of light during image acquisition (e.g., insufficient lighting, night-time, weather conditions), and non-uniform illumination. Consequently, such images lack in revealing crucial details such as iris texture, vein pattern, eyelid, and limbic contours, and the image fails to provide discriminatory information. Nevertheless, if some image properties are improved, these issues can be overcome. In this paper, we focus on improving the four image properties: contrast, sharpness,

G. Gautam (✉) · S. Mukhopadhyay
Indian Institute of Technology (ISM), Dhanbad, Jharkhand, India
e-mail: gautamgunjan29@gmail.com

A. Rani et al. (eds.), *Signals, Machines and Automation*, Lecture Notes in Electrical Engineering 1023, https://doi.org/10.1007/978-981-99-0969-8_70

perceptual quality, and colorfulness. Various global and local image enhancement techniques have been proposed in the literature. Global enhancement methods may induce over- or under-enhancement issues at some parts of the image [1]. Also, these methods may miss the fine details in an image [2, 3]. However, local techniques are more computationally expensive than their global counterpart and may produce some unwanted local artifacts [4]. Thus, we used a hybrid enhancement technique that takes local and global information into account which is suitable for a wide variety of images without fine-tuning. Contrast enhancement approaches have been utilized extensively in image processing, and among them, Histogram equalization (HE) and Adaptive histogram equalization (AHE) [5] are two often used techniques. Due to the variation of intensities and other characteristics across the entire image, HE being a global-natured method appears to be less effective. Various improved HE variants [6, 7] proposed in the literature are based on the division of histograms; therefore, they fail when this division is not accurate. On the contrary, AHE and its variants (e.g., clipped AHE, weighted AHE, etc.) are developed to overcome HE's issues through dealing with the local features of the image and have evidenced success. In base AHE, each pixel is mapped with an intensity proportional to its rank in the neighborhood pixels. Two major issues associated with AHE are over-enhancement of noise and slow computation time [8]. Thus, various improvements in base AHE have been presented so far, and CLAHE is one of its variants that limits the noise amplification, but the images start looking artificial. Gamma Correction (GC) is a non-linear adjustment technique to individual pixel values and has also been used to enhance the color images. Dependence on *gamma* value takes GC far from practicality, and it may lead to poor contrast if the *gamma* value is too small or too large. Also, it can result in the saturation of the image being processed. Huang et al. [9] presented a method Adaptive Gamma Correction with Weighting Distribution (AGCWD) to enhance a digital image by modifying its histogram. This technique improves the brightness of dimmed images using traditional gamma correction and the probability distribution of luminance pixels. However, while there is a lack of bright pixels, this approach may fail to enhance the image. The reason is as follows: the highest intensity of the output image is bounded by the maximum possible intensity of the input image, and therefore, the former will never cross the latter. Retinex (Ret) algorithm [10] is used to enhance the color images—consists of two key steps: (1) estimation and (2) normalization of illumination. The limitation of Ret is it can either achieve color rendition or dynamic range compression, but not both at the same time. However, Multi-scale Retinex (MulRet) [11, 12] can overcome these limitations by combining several Ret outputs to produce one enhanced image with good dynamic range compression, color consistency, and tonal rendition. MulRet tries to preserve most of the detail in the image; but still it suffers from color distortion [11]. Some of the latest image enhancement techniques can be found in [13–16].

2 Proposed Algorithm

First, we provide an overview of Undecimated Discrete WT (UDWT) and Mathematical Morphological Filtering (MMF) as a prerequisite for a better understanding of the proposed algorithm followed by the main steps in this (Fig. 1).

1. **Multi-resolution analysis (MRA) using UDWT:** UDWT is useful to '*fill in the gaps*' induced by the decimation stage in the standard DWT [17]. Unlike DWT, in UDWT the size of four different components after decomposition does not decrease; thus, more information can be preserved. The UDWT using filter banks (X,Y) leads to a set $H_1, V_1, D_1, \ldots A_j$ where H_j, V_j, and D_j are the detail coefficients and A_j is the approximation at scale j.

$$A_{j+1}[k+1] = \left(X^{(j)}X^{(j)} * A_j\right)[k,l] \tag{1}$$

$$H_{j+1}[k+1] = \left(Y^{(j)}X^{(j)} * A_j\right)[k,l] \tag{2}$$

$$V_{j+1}[k+1] = \left(X^{(j)}Y^{(j)} * A_j\right)[k,l] \tag{3}$$

$$D_{j+1}[k+1] = \left(Y^{(j)}Y^{(j)} * A_j\right)[k,l] \tag{4}$$

   where $XY * A$ is the convolution of A by the separable filter XY. At each scale, the three wavelet images, H, V, and D (i.e., **H**orizontal, **V**ertical, and **D**iagonal coefficients), are of the same size as the original image. Image reconstruction was achieved by a co-addition of all wavelet scales w such that $H_j, V_j, D_j \in$ w.
2. **Multi-scale analysis using MMF:** MMF is a popular and powerful tool to extract meaningful information of an image. Erosion, dilation, opening, closing, and top/bottom hat transformation are morphological operations useful in shape-based feature extraction and manipulation. *Dilation* fills the small holes present in objects and makes them more visible; however, *erosion* removes small objects to keep the essential ones intact. *Opening* removes small objects and preserves the size and shape of larger objects; however, *closing* fills the small holes and preserves the size/shape of different objects. Dilation (⊕) and erosion (⊖) of a 2D function $Z(r, c)$ by a 2D point set B can be represented using (5) [18]. Multi-scale *Opening* and *Closing* (with scale factor n) can be determined using (6).

$$\begin{gathered}\left(Z^{\oplus}B\right)(r,\ c) = \max\{Z(r-k,\ c-l)|(k,\ l) \in B\} \\ \text{and} \\ \left(Z^{\ominus}B\right)(r,\ c) = \min\{Z(r+k,\ c+l)|(k,\ l) \in B\}\end{gathered} \tag{5}$$

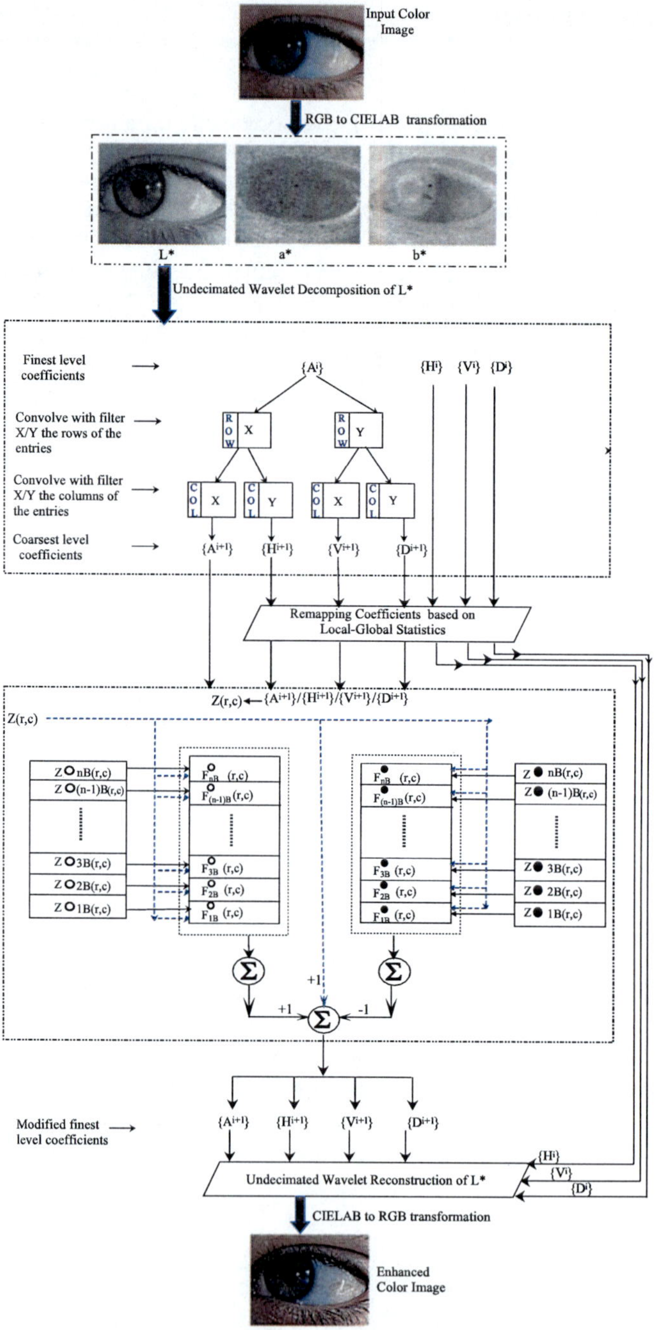

Fig. 1 Framework of the proposed approach

$$\begin{aligned}&\left(Z_{nB}^{\circ}\right)(r,\ c) = \left(\left(Z_{nB}^{\ominus}\right) \oplus\ nB\right)(r,c)\\&\text{and}\\&\left(Z_{nB}^{\bullet}\right)(r,\ c) = \left(\left(Z_{nB}^{\oplus}\right)\ominus nB\right)(r,c)\end{aligned} \tag{6}$$

Top-hat transform is an outstanding morphological tool to extract small details from an image. The two classes of top-hat transform, i.e., *white* and *black* top-hat alleviate feature extraction by obtaining brighter and darker features, respectively. Mathematically, white top-hat transform (T_w) is the difference between the image and its opening by an appropriate structuring element (SE); and its dual, i.e., black top-hat transform (T_b), is the difference between the closing and the image as shown in (7). Sequentially performing top-hat filtering using disk SE with varying radii captures scale-specific features, and these features can be exploited to accomplish local contrast stretching.

$$\begin{aligned}&T_{\mathrm{w}}(Z) = Z - (Z^{\circ}B)\\&\text{and}\\&T_b(Z) = (Z^{\bullet}B) - Z\end{aligned} \tag{7}$$

2.1 RGB to LAB Space Transformation of the Input Image

In Lab color space L^*, a^* and b^* represent the luminance, red/green, and yellow/blue values, respectively. The non-linear relations of L^*, a^*, and b^* are aimed to imitate the non-linear response of the eye. The conversion of *RGB* color space to *Lab* is obtained by converting the *RGB* to *XYZ* using (8) followed by *XYZ* to *Lab* using (9) considering a reference white as (X_r, Y_r, Z_r), ϵ as 0.008856 and κ as 903.3.

$$\begin{bmatrix} X \\ Y \\ Z \end{bmatrix} = [M]\begin{bmatrix} r \\ g \\ b \end{bmatrix} \tag{8}$$

$$\begin{aligned}L &= 116 f_y - 16\\a &= 500\left(f_x - f_y\right)\\b &= 200\left(f_y - f_z\right)\end{aligned} \tag{9}$$

where

$$f_x = \begin{cases} \sqrt[3]{x_r} & \text{if } x_r > \in \\ \frac{\kappa x_r + 16}{116} & \text{otherwise with } x_r = \frac{X}{X_r} \end{cases}$$

with

$$f_y = \begin{cases} \sqrt[3]{y_r} & \text{if } y_r > \in \\ \frac{\kappa y_r + 16}{116} & \text{otherwise with } y_r = \frac{Y}{Y_r} \end{cases}, \quad f_z = \begin{cases} \sqrt[3]{z_r} & \text{if } z_r > \in \\ \frac{\kappa z_r + 16}{116} & \text{otherwise with } y_r = \frac{Z}{Z_r} \end{cases}$$

2.2 *Sub-band Coding of Luminance Component L* Using UDWT*

Next, we performed two-level decomposition of L^* component by UDWT with *Haar* wavelet. Haar is conceptually memory-efficient, precisely reversible without affecting the edges, and computationally inexpensive. Employing UDWT divides the input image into four sub-bands decomposition, and various operations are performed on these sub-bands. Notably, these sub-bands hold the *fine* and *coarse*-scale information (in the form of wavelet coefficients) of the original image. The decomposed detail coefficients (i.e., H_j, V_j, and D_j) and approximation (i.e., A_j) of L^* are computed using (1–4).

2.3 *Re-map Detail Coefficients Using Local–global Statistics*

Global arithmetic mean $G_{AM}^{(\mathrm{w})}$ of all the detail coefficients w (i.e., $\{H_j, V_j, D_j\} \in \mathrm{w}$) are computed to obtain the global information. Then, all these coefficients undergo a division of 8×8 nonoverlapping blocks and the local arithmetic mean L_{AM} for each block is calculated. Afterward, the modified coefficient $\hat{\mathrm{w}}$ is computed using (10) where the local mean is defined over a fixed neighborhood. Afterward, these modified and remapped coarsest-level detail coefficients $\hat{\mathrm{w}}$ such that $\left\{\hat{H}_j, \hat{V}_j, \hat{D}_j\right\} \in \hat{\mathrm{w}}$ along with coarsest approximation (without modification) goes through the multi-scale MMF.

$$\hat{\mathrm{w}} = \sum\nolimits_{j=0}^{J} \sum\nolimits_{i=1}^{3} \mathrm{w}_j^{(i)} + \left| G_{\mathrm{AM}}^{\left(\mathrm{w}_j^{(i)}\right)} - L_{\mathrm{AM}} \right| \tag{10}$$

2.4 *MMF on Modified Coefficients Followed by Reconstruction*

In multi-scale MMF, the geometry and the size of the SE play a vital role in accomplishing desired results. Being rotation invariant, we used disk-shaped SE with increasing radii 3, 4, and 5 to perform the multi-scale top-hat transform. Multi-scale nT_w and nT_b images are computed for each SE using (7). After adding the outcome of multi-scale nT_w to the corresponding approximation or detail coefficients, the intermediate improved versions are obtained by subtracting the multi-scale nT_b from this result. Let for l: the coarsest-level S be the set such as $S = \left\{w_l^{(l)}, A^{(l)}\right\}$ $\forall\ i \in \{1, 2, 3\}$, then post performing MMF, the modified detail coefficients, and approximation at coarsest level can be computed using (11):

$$\hat{S}_k = \left(S_k + \sum_{n=1}^{N} T_{\mathrm{w}}^{(n)}(S_k)\right) - \sum_{n=1}^{N} T_{\mathrm{b}}^{(n)}(S_k) \tag{11}$$

where k denotes the index of every element of set S and N to represent the set of scales used in MMF such that $N = \{3,4,5\}$. In particular, this step produces $\hat{A}$ and $\hat{\mathrm{w}}$ which we will use to reconstruct the image. Finally, the image is reconstructed using approximation $\hat{A}$ and modified detail coefficients $\hat{\mathrm{w}}$. Conversion of *Lab* to *RGB* is performed using (8) and (9).

3 Experimental Results

Experiments are performed on UBIRIS2 dataset that contains 5799 images captured under visible wavelength with dynamic lighting conditions.

1. **Contrast:** We use co-occurrence matrix (CM) obtained from the HSI version of input image to quantify the contrast. Here, CM relies on the mean occurrence of a pixel pair at 2–5 pixel distances in four directions (0°, 45°, 90°, and 135°). Contrast is computed using $\sum_{i,j} |i - j|^2 p(i, j)$ where $p(i,j)$ is the $(i,j)th$ entry in CM.
2. **Sharpness:** Sharpness is the contrast between various colors. We computed sharpness S from image gradients using the following equation:

$$S = \frac{\text{Sum of all gradient norms}}{\text{Number of pixels gives sharpness}} \tag{12}$$

3. **PIQUE:** This can be computed using (13) [19]. where N_{SA} is the number of Spatially Active Blocks in the input image; C is the positive constant to deal with zero denominator; D_{sk} is the distortion for a for a distorted block. D_{sk} can be set (14); where ν_{blk} represents the variance feature, σ_{pq} is the standard deviation of the segment a_{pq} with p and q as edge and segment indices, T_{std} a threshold, σ_{blk} is the standard deviation of blocks, and β is a parameter to define noise criterion. Low PIQUE means a better perceptual image quality.

$$\text{PIQUE} = \frac{\left(\sum_{k=1}^{N_{\text{SA}}} D_{\text{sk}}\right)}{N_{\text{SA}} + C} \tag{13}$$

$$D_{\text{sk}} = \begin{cases} 1 & \text{if } \left(\sigma_{\text{pq}} < T_{\text{std}}\right) \text{ and } \left(\sigma_{\text{blk}} > 2 * \beta\right) \\ v_{\text{blk}} & \text{if } \left(\sigma_{\text{blk}} > 2 * \beta\right) \\ \left(1 - v_{\text{blk}}\right) & \text{if } \left(\sigma_{\text{pq}} < T_{\text{std}}\right) \end{cases} \tag{14}$$

4. **Colorfulness:** Colorfulness φ for a color image I_{rgb} is computed using:

$$\phi = \sqrt{\sigma_\alpha^2 + \sigma_\beta^2} + 0.3 * \sqrt{\mu_\alpha^2 + \mu_\beta^2} \tag{15}$$

where

$$\alpha = I_{\text{rgb}}(R) - I_{\text{rgb}}(G); \;\; \beta = 0.5 * \left(I_{\text{rgb}}(R) - I_{\text{rgb}}(G) - I_{\text{rgb}}(B)\right);$$

σ and μ represent standard deviation and mean for α and β.

Figure 2 shows a set of original images and the corresponding enhanced versions by different methods. It is observed that the proposed approach effectively enhances the minute details, makes edges more prominent and textures more well-defined, and avoids color distortion. HE and CLAHE are best known for the usage of the dynamic range of the pixel values for the highest contrast; however, it does not guarantee that the visual quality of enhanced image is better than the original image. CLAHE introduces some dark patchy effects in the image; hence, it has an unnatural look. HE decreases this patchy effect at the cost of troubling local contrast and makes dark regions darker and bright regions brighter; hence, the resulting image does not look pleasant. Though AGCWD preserves the natural look, it results in loss of details in the sclera region. Other techniques except proposed, produce undesired artifacts, degrade the visual quality, and increase the noise—this leads to an unnatural look. On the contrary, the proposed algorithm provides a controlled image enhancement with substantial improvement in the natural look. The visual appearance of the enhanced images through various methods demonstrates that the proposed method is highly effective in enhancing low-quality images and outperforms the state of the arts. We also examined the generalization ability of the proposed method by testing its performance on Berkeley dataset, a collection of natural scene images, and the result is shown in Fig. 3. Here, (a), (c), (e), and (g) represent the original images and (b), (d), (f), and (h) show the corresponding enhanced versions using the proposed algorithm. The image (a) looks smooth, and the text written on the canopies is not clear. However, this text in the enhanced image (b) is quite clear. Moreover, regions including floor texture, shadow, and edges are more prominent, informative, highly detailed, and visually defined. Likewise various unclear textures of image (c) such as rate-text and cauliflowers are highlighted in the enhanced image (d). Similarly, the enhanced images (f) and (h) evidence an improved quality of corresponding original images to a great extent.

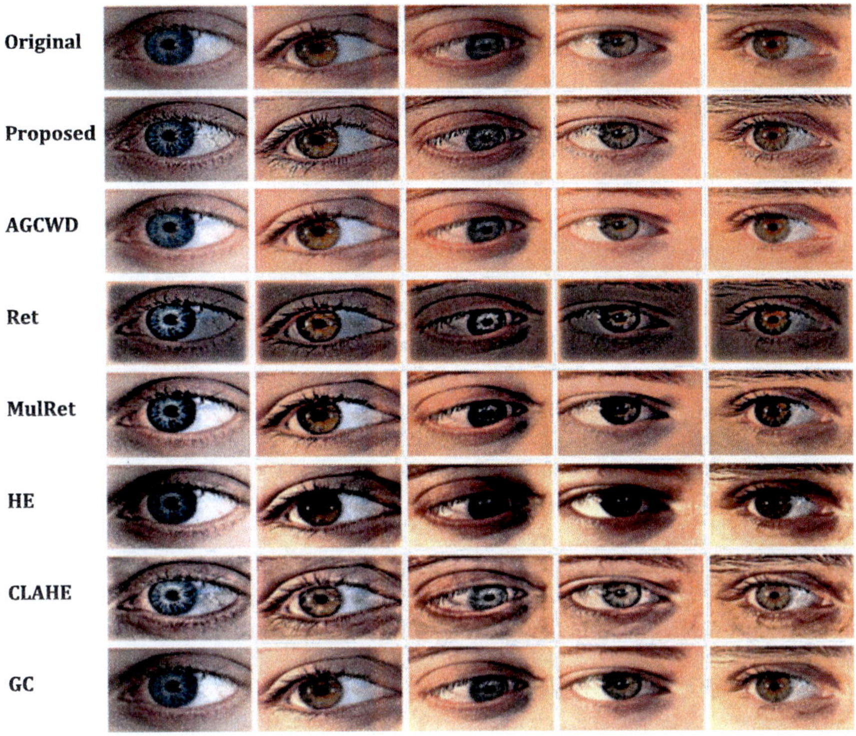

Fig. 2 Row-wise subjective results

Fig. 3 Subjective results on Berkeley dataset

Image quality results of estimated *contrast*, *sharpness*, *PIQUE*, and *colorfulness* are shown in Fig. 4. Figure 4a presents the *contrast* results indicating that the proposed method outperforms other methods in comparison. The maximum *contrast* for original images is 1 that increases and reaches up to 4.5 using proposed method. Figure 4b shows the *sharpness* scores. It is observed that the maximum and median *sharpness* scores of the proposed method are highest among all the methods in comparison. PIQUE results in Fig. 4c denote that our method obtains the lowest score—this indicates that our method has the potential of highlighting the small details and preserving the visual look. Figure 4d depicts the *colorfulness* results. We can see that AGCWD and MulRet outperform other methods; however, *contrast*, *sharpness,* and *PIQUE* values are substantially lower than that of the proposed approach. Also, our method successfully preserves the *colorfulness* by producing its values almost similar to the original images. Tables 1 and 2 present the effect of color image enhancement on mobile-iris liveness detection. We used MobBIOfake dataset having 800 real and 800 printed fake samples of iris biometrics. We fine-tuned the AlexNet on MobBIOfake to extract relevant features. Then, SVM classifier is used to classify the *real* and *fake* images. The Classification Accuracy (CR) and confusion matrix show that CR was 87.87% without enhancement which got rise of nearly 3% when the enhanced images using the proposed method were used. Also, true negatives and false positives with enhancement were 0.93 and 0.07, respectively. These values without enhancements were 0.82 and 0.18, respectively. This indicates that image enhancement using proposed method helps increase the CR.

4 Conclusion

In this paper, a framework for color image quality enhancement is presented that employs Undecimated Discrete Wavelet Transform and Multi-scale Morphological Filtering to highlight the minute details in the image. This framework combines the multiresolution and multi-scale information using local–global image statistics. The quality of the image is assessed using four measures: *Contrast*, *Sharpness*, *PIQUE*, and *Colorfulness*. Experimental results show that our method outperforms other state of the arts by producing artifact-free images. It also preserves the natural look and makes them visually pleasing. We have also demonstrated the generalization ability of our method via testing it on mobile-iris liveness detection.

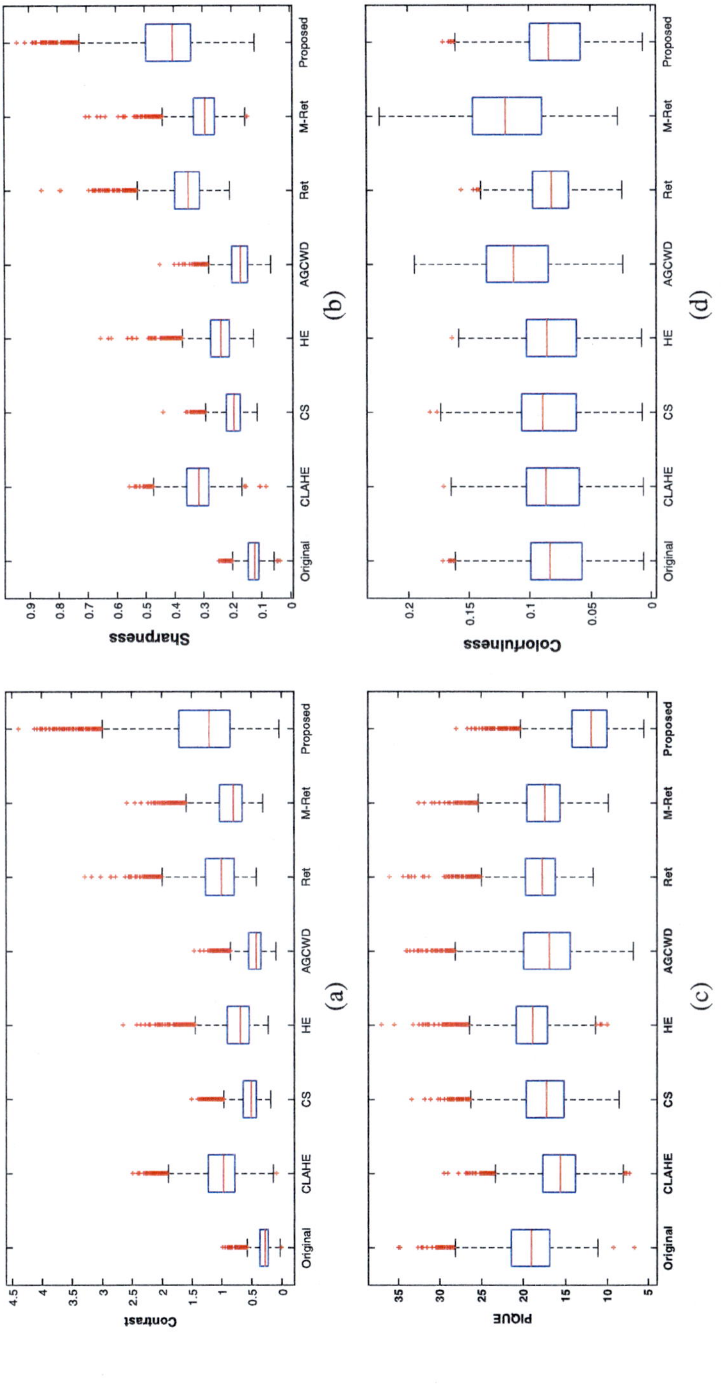

Fig. 4 Objective results for: **a** Contrast, **b** Sharpness, **c** PIQUE, and **d** Colorfulness

Table 1 Classification accuracy

Accuracy (%)		
Database	No enhancement	Enhancement
MobBIOfake	87.87	**90.13**

Best performance in bold

Table 2 Confusion matrix

Without/with enhancement		Predicted	
		Fake	Real
Known	Fake	0.82/0.93	0.18/0.07
	Real	0.06/0.13	0.94/0.87

References

1. Cheng HD, Shi XJ (2004) A simple and effective histogram equalization approach to image enhancement. Digital Signal Proc 14(2):158–170
2. Gautam G, Mukhopadhyay S (2018) Efficient contrast enhancement based on local–global image statistics and multiscale morphological filtering. In: Advanced computational and communication paradigms. Springer, pp 229–238
3. Mukhopadhyay S, Chanda B (2000) A multiscale morphological approach to local contrast enhancement. Signal Process 80(4):685–696
4. Celik T, Tjahjadi T (2011) Contextual and variational contrast enhancement. IEEE Trans Image Process 20(12):3431–3441
5. Zuiderveld K (1994) Contrast limited adaptive histogram equalization. In: Graphics gems IV. Academic Press Professional, Inc., pp 474–485
6. Kim M, Chung MG (2008) Recursively separated and weighted histogram equalization for brightness preservation and contrast enhancement. IEEE Trans Consum Electron 54(3):1389–1397
7. Wang Y, Chen Q, Zhang B (1999) Image enhancement based on equal area dualistic sub-image histogram equalization method. IEEE Trans Consum Electron 45(1):68–75
8. Pizer SM, Amburn EP, Austin JD, Cromartie R, Geselowitz A, Greer T, ter Haar Romeny B, Zimmerman JB, Zuiderveld K (1987) Adaptive histogram equalization and its variations. Comput Vis Graph Image Process 39(3):355–368
9. Huang S-C, Cheng F-C, Chiu Y-S (2012) Efficient contrast enhancement using adaptive gamma correction with weighting distribution. IEEE Trans Image Process 22(3):1032–1041
10. Jobson DJ, Rahman Z, Woodell GA (1997) Properties and performance of a center/surround retinex. IEEE Trans Image Process 6(3):451–462
11. Rahman Z, Jobson DJ, Woodell GA (1996) Multi-scale retinex for color image enhancement. In: Proceedings of 3rd IEEE International conference on image processing, vol 3. IEEE, pp 1003–1006
12. Wang W, Li B, Zheng J, Xian S, Wang J (2008) A fast multi-scale retinex algorithm for color image enhancement. In: 2008 International conference on wavelet analysis and pattern recognition. IEEE, pp 80–85
13. Ackar H, Almisreb AA, Saleh MA (2019) A review on image enhancement techniques. SE Eur J Soft Comput 8(1):42–48
14. Liu X, Pedersen M, Wang R (2022) Survey of natural image enhancement techniques: classification, evaluation, challenges, and perspectives. Digital Signal Proc 127:103547
15. Rahman T, Khandakar A, Qiblawey Y, Tahir A, Kiranyaz S, Kashem SBA, Islam MT, Al Maadeed S, Zughaier SM, Khan MS, Chowdhury MEH (2021) Exploring the effect of image

enhancement techniques on COVID-19 detection using chest X-ray images. Comput Biol Med 132:104319
16. Somvanshi SS, Kunwar P, Tomar S, Singh M (2018) Comparative statistical analysis of the quality of image enhancement techniques. Int J Image Data Fusion 9(2):131–151
17. Nason GP, Silverman BW (1995) The stationary wavelet transform and some statistical applications. In: Wavelets and statistics. Springer, pp 281–299
18. Maragos P (1989) Pattern spectrum and multiscale shape representation. IEEE Trans Pattern Anal Mach Intell 11(7):701–716
19. Venkatanath N, Praneeth D, Bh MC, Channappayya SS, Medasani SS (2015) Blind image quality evaluation using perception based features. In: 2015 Twenty first national conference on communications (NCC). IEEE, pp 1–6